Phosphate in Microorganisms

CELLULAR AND MOLECULAR BIOLOGY

Organizing Committee for the Symposium

A. Torriani-Gorini, Chairperson
A. Kornberg and S. Mizushima, Co-Chairs

J. Beckwith
W. Boos
P. L. Boquet
A. Demain
H. Halvorson
P. Maloney
N. Rao
H. Shinagawa
S. Silver
J. Tommassen
E. Yagil

Phosphate in Microorganisms

CELLULAR AND MOLECULAR BIOLOGY

Edited by

Annamaria Torriani-Gorini

DEPARTMENT OF BIOLOGY
MASSACHUSETTS INSTITUTE OF TECHNOLOGY, CAMBRIDGE, MASSACHUSETTS

Ezra Yagil

DEPARTMENT OF BIOCHEMISTRY
TEL AVIV UNIVERSITY, TEL AVIV, ISRAEL

Simon Silver

DEPARTMENT OF MICROBIOLOGY AND IMMUNOLOGY
UNIVERSITY OF ILLINOIS COLLEGE OF MEDICINE, CHICAGO, ILLINOIS

ASM PRESS • WASHINGTON, D.C.

1325 Massachusetts Ave., N.W.
Washington, DC 20005

Library of Congress Cataloging-in-Publication Data

Phosphate in microorganisms : cellular and molecular biology / edited by Annamaria Torriani-Gorini, Ezra Yagil, Simon Silver.
p. cm.
Includes index.
ISBN 1-55581-080-2
1. Phosphates—Metabolism—Congresses. 2. Microbial metabolism—Congresses. 3. Phospholipids—Congresses. 4. Phosphoproteins—Congresses. I. Torriani-Gorini, Annamaria. II. Yagil, Ezra. III. Silver, S. (Simon).
QR92.P45P46 1994
576′.1133—dc20 94-13574
CIP

Printed in the United States of America.

Cover photo: Structure of the GlpK-III^{Glc} complex, determined in the laboratory of Dr. S. J. Remington of the Institute of Molecular Biology, University of Oregon. The figure was generated by using the program RIBBONS 2.0, written by Mike Carson (see chapter 56).

CONTENTS

VI. POLYPHOSPHATES AND PHOSPHATE RESERVES

VII. PHOSPHOLIPIDS

VIII. PROTEIN EXPORT AND FOLDING

IX. SIGNAL TRANSDUCTION AND PHOSPHOPROTEINS

X. STRUCTURE/FUNCTION RELATIONSHIPS

PREFACE

In the preface to the monograph from the previous Phosphate Symposium (Torriani et al., 1987), I predicted that a future phosphate meeting would prove to be as fruitful as Pho-86. Now after seven years, we see that this is true. The progress of science has been extraordinary. The importance of this Symposium was as a wide window to new fundamental understanding of things such as the roles of signal transduction and cross-regulation in the global metabolism of the cell and the roles of the phosphoenolpyruvate phosphotransferase system (PTS) in regulation of metabolism, as well as in sugar transport. Observations made in microorganisms today resound in higher organisms tomorrow; that has been a lesson of molecular biology from its earliest days. That such cross-fertilization will continue is an easy prediction. In this book, Arthur Kornberg gives a new look at the origin of life, suggesting that polyphosphates, found now in prokaryotes, were basic to the initial support of the structure of DNA. Cells must build membranes to separate essential internal functions from the outside. Thus, substrate transport across membranes, a major role of the Pho regulon, continues to be studied in molecular elegance. Protein excretion and folding and the importance of phospholipid structures in controlling membrane function are presented in depth. The new excitement in the PTS, for which there had not been room in Pho-86, is fully developed in Pho-93. The essential (and historically important) crystallographic studies of alkaline phosphatase were added to the program as a workshop discussion of protein structural approaches and, together with several additional papers, comprise a section on protein structure.

The Pho-93 Symposium in Woods Hole (12 to 17 September 1993) was a success, thanks to the variety of topics discussed in depth and due to the novelty of new understanding. This monograph should therefore interest a wide range of scientists and students.

From the Marine Biological Laboratory in Concarneau (Bretagne, France), the site of Pho-86, we moved to the Marine Biological Laboratory on Cape Cod for Pho-93. We went from a terrific dinner of *huîtres* to one of lobsters, from a walled city of kings and pirates of the XIV century to the peaceful whale songs and the narrow sea hole between islands in the intricate harbor described by Mr. Woods . . . hence Woods Hole!

After almost 40 years of research on bacterial regulation of alkaline phosphatase synthesis (the Pho regulon), beginning in 1955 at the Institut Pasteur in Paris (Torriani, 1959), my financial support finally dried out. This negative fact of life had a positive effect: it allowed me time to organize Pho-93 and also to reflect on the development of our field of science. From 1955 to 1994, most of the founders of molecular biology have passed away or moved out. Their discoveries encouraged new people to bring about previously unimaginable progress, as presented in this monograph.

The idea for this meeting started in Konstanz in 1991, in the office of Winfried Boos, together with Jan Tommassen, Jon Beckwith, and Peter Maloney. The organizing committee evolved at a Gordon Conference a year later. The important step of making an attractive program and choosing the conveners was the real beginning of the work. Good choices of speakers were made, as the Table of Contents shows. The speakers were asked to provide overviews of their subjects at the Symposium and in writing for this monograph. Many speakers also individually funded their travel to New England, like the Pilgrims almost 375 years ago. The meeting was therefore cooperative, and each speaker provided brains, energy, and money. We also acknowledge the sponsorship of the International Union of Biochemistry and Molecular Biology (H. Kornberg, President), which provided travel funding for some speakers, together with the Foundation for Microbiology (B. Waksman, President) and the United States Department of Energy, along with the donors listed in the Acknowledgments. I would like to mention the special efforts of Arnie Demain, Arthur Kornberg, Chris Raetz, and Hideo Shinagawa as promoters of these financial sources.

At the end of the Symposium a request was made for a reunion of us all (if we have not perished) in the year 2000. Hopefully the science will continue its growth, and the asteroids will miss us.

ANNAMARIA TORRIANI-GORINI

REFERENCES

Torriani, A. 1959. Effect of inorganic phosphate (Pi) on the formation of phosphatases by *Escherichia coli*. *Fed. Proc.* **18:**339.

Torriani-Gorini, A., F. G. Rothman, S. Silver, A. Wright, and E. Yagil (ed.). 1987. *Phosphate Metabolism and Cellular Regulation in Microorganisms.* American Society for Microbiology, Washington, D.C.

ACKNOWLEDGMENTS

The International Symposium on Cellular and Molecular Biology of Phosphate and Phosphorylated Compounds in Microorganisms, held in Woods Hole, Mass., from 12 to 17 September 1933, was sponsored by:

Foundation for Microbiology, New York, N.Y.
International Union of Biochemistry and Molecular Biology, Clayton, Victoria, Australia (grant IG119)
United States Department of Energy, Washington, D.C.

The Organizing Committee gratefully acknowledges the support of the sponsors, as well as financial contributions from the following donors (in alphabetical order):

Bristol-Myers Squibb Pharmaceutical Research Institute, Princeton, N.J.
Fujisawa Pharmaceutical Company, Osaka, Japan
Genetics Institute, Inc., Cambridge, Mass.
Glaxo Research Institute, Research Triangle Park, N.C.
Merck Research Laboratories, Rahway, N.J.
Miles, Inc., West Haven, Conn.
Otsuka Pharmaceuticals Company, Tokyo, Japan
Parke-Davis Pharmaceutical Research, Ann Arbor, Mich.
Rhône-Poulenc Rorer, Antony, France
Sanwa-Riken Company, Osaka, Japan
Schering-Plough Research Institute, Kenilworth, N.J.
Takara Shuzo Company, Kyoto, Japan
Yuyama Medical & Scientific Company, Osaka, Japan

Special thanks go to Merck Research Laboratories for generous support.

The Gemma: a marine biological laboratory specimen-collecting boat and the Marine Resource Center of the Marine Biological Laboratories, site of the Phosphate Symposium. Original etching by Robert Golder.

Alphabetical listing of participants. Amemura, Mitsuko (92); Austin, Sara (not shown); Banta, Lois (47); Bayly, Ron (9); Beckwith, Jonathan (not shown); Begley, Gail (not shown); Boos, Winfried (75); Booth, James (20); Bostian, Keith (34); Bourne, Julie (42); Brennan, Cathy (45); Breukink, Eefjan (14); Casillas-Martinez, Lilliam (5); Chen, Huanfeng (29); Coleman, Joseph E. (18); Combie, Joan (30); Cox, Graeme B. (23); Demain, Arnold (61); Dowhan, William (49); Duffy, Richard (63); Dumora, Catherine (91); Dumsday, Geoff (22); Dunlop, Paul (not shown); Evans, I. Marta (60); Fang, Aiqi (90); Finan, Turlough M. (40); Goldstein, Alan (4); Gorkovenko, Alexander (46); Grossman, Alan (not shown); Grossman, Arthur R. (48); Halvorson, Harlyn (71); Hecht, Gregory (3); Hong, Jenshiang (17); Hulett, Marion (28); Izard, Jennifer (7); Jacobson, Gary R. (51); Jahreis, Knut (54); Justice, Michael (37); Kadner, Robert J. (11); Kantrowitz, Evan (70); Kasahara, Megumi (64); Kato, Junichi (79); Keller, R. C. A. (55); Kendall, Debra (not shown); Kim, Eunice (84); Kornberg, Arthur (87); Kornberg, Sir Hans (72); Kulaev, Igor S. (57); Lacoste, Anne-Marie (82); Lane, Todd (2); Lemos, Paulo C. (44); Lengeler, Josef W. (not shown); Makino, Kozo (93); Maloney, Peter C. (33); Martín, Juan (43); Martinez-Hackert, Erik (31); Matlin, Albert R. (19); Mekalanos, John (not shown); Merrick, Michael J. (not shown); Miller, Samuel I. (27); Mizushima, Shoji (85); Murphy, Jennifer (66); Nesmeyanova, Marina (86); Newton, Austin (not shown); Niere, Julie (39); Nikaido, Hiroshi (59); Ninfa, Alexander (62); Ohta, Noriko (not shown); Ohtake, Hisao (58); Oshima, Yasuji (83); Parent, Stephen (36); Pattus, Franc (not shown); Pereira, Helena (67); Persson, Bengt L. (38); Peterkofsky, Alan (74); Pettigrew, Donald W. (12); Phung, Le T. (89); Postma, Pieter (56); Randall, Linda L. (10); Rao, Narayana N. (1); Rasmussen, Beth (8); Reizer, Jonathan (77); Roberts, Mary F. (not shown); Robillard, George T. (16); Roseman, Saul (not shown); Rosen, Barry P. (69); Rosenbusch, Jurg P. (15); Rothman, Frank (not shown); Rusch, Sharyn L. (6); Saier, Milton H. (73); Santos, Helena (not shown); Shinagawa, Hideo (26); Silver, Simon (68); Smith, J. Donald (not shown); Soncini, Fernando (13); Spira, Beny (not shown); Stadtman, Thressa (81); Stock, Ann (not shown); Sukhan, Anand (21); Sun, Guofu (41); Surette, Michael (24); Titgemeyer, Friedrich (52); Toh-e, Akio (80); Tommassen, Jan (50); Torriani, Annamaria (88); Wanner, Barry (25); West, Ann (32); Whitton, Brian A. (53); Wickner, William (not shown); Wright, Andrew (not shown); Wyckoff, Harold W. (78); Xu, Xu (65); Yagil, Ezra (not shown); Yashphe, Jacob (not shown); de Cock, Hans (76); van Veen, H. W. (35); von Heijne, Gunnar (not shown).

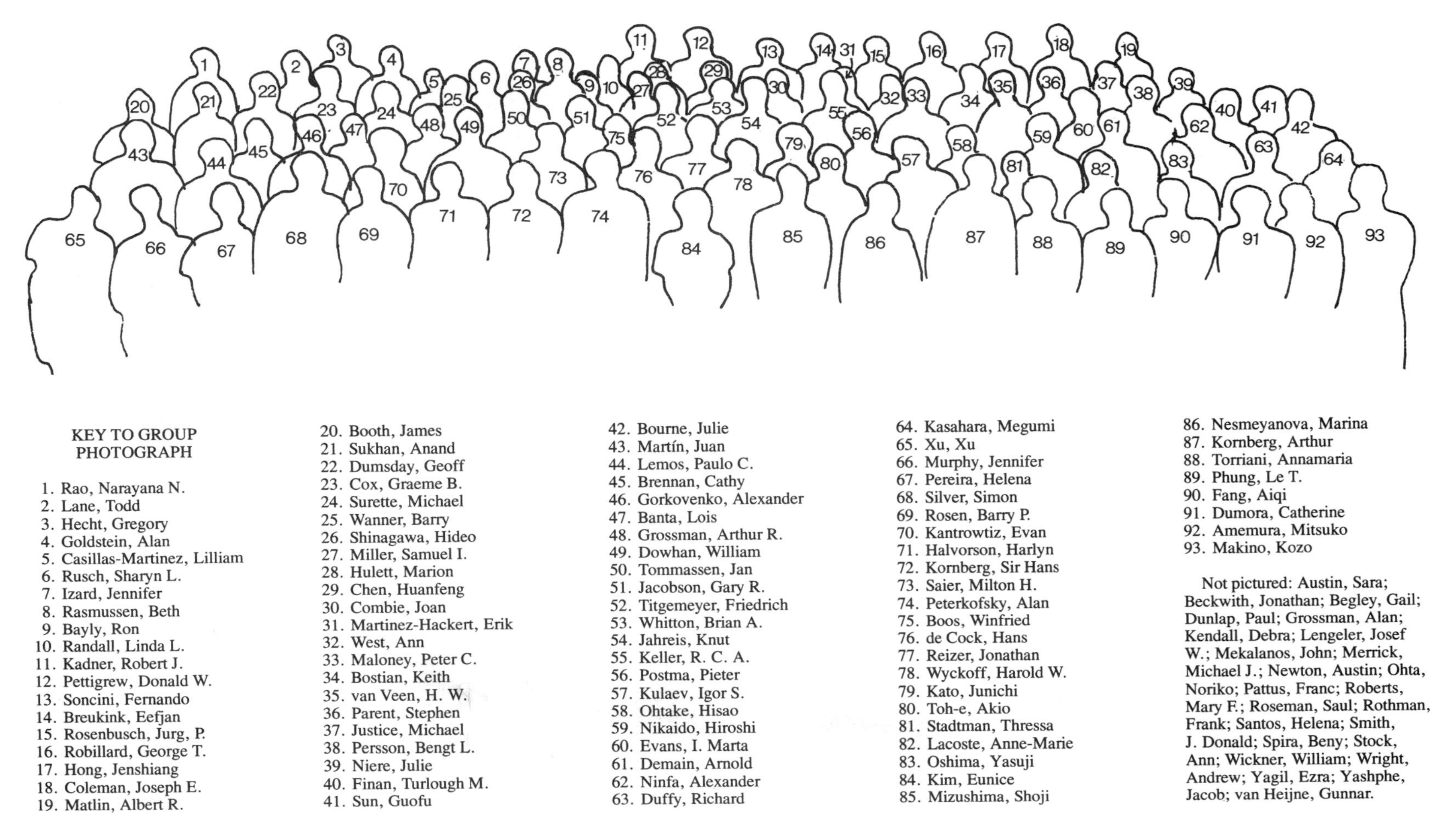

KEY TO GROUP PHOTOGRAPH

1. Rao, Narayana N.
2. Lane, Todd
3. Hecht, Gregory
4. Goldstein, Alan
5. Casillas-Martinez, Lilliam
6. Rusch, Sharyn L.
7. Izard, Jennifer
8. Rasmussen, Beth
9. Bayly, Ron
10. Randall, Linda L.
11. Kadner, Robert J.
12. Pettigrew, Donald W.
13. Soncini, Fernando
14. Breukink, Eefjan
15. Rosenbusch, Jurg, P.
16. Robillard, George T.
17. Hong, Jenshiang
18. Coleman, Joseph E.
19. Matlin, Albert R.
20. Booth, James
21. Sukhan, Anand
22. Dumsday, Geoff
23. Cox, Graeme B.
24. Surette, Michael
25. Wanner, Barry
26. Shinagawa, Hideo
27. Miller, Samuel I.
28. Hulett, Marion
29. Chen, Huanfeng
30. Combie, Joan
31. Martinez-Hackert, Erik
32. West, Ann
33. Maloney, Peter C.
34. Bostian, Keith
35. van Veen, H. W.
36. Parent, Stephen
37. Justice, Michael
38. Persson, Bengt L.
39. Niere, Julie
40. Finan, Turlough M.
41. Sun, Guofu
42. Bourne, Julie
43. Martín, Juan
44. Lemos, Paulo C.
45. Brennan, Cathy
46. Gorkovenko, Alexander
47. Banta, Lois
48. Grossman, Arthur R.
49. Dowhan, William
50. Tommassen, Jan
51. Jacobson, Gary R.
52. Titgemeyer, Friedrich
53. Whitton, Brian A.
54. Jahreis, Knut
55. Keller, R. C. A.
56. Postma, Pieter
57. Kulaev, Igor S.
58. Ohtake, Hisao
59. Nikaido, Hiroshi
60. Evans, I. Marta
61. Demain, Arnold
62. Ninfa, Alexander
63. Duffy, Richard
64. Kasahara, Megumi
65. Xu, Xu
66. Murphy, Jennifer
67. Pereira, Helena
68. Silver, Simon
69. Rosen, Barry P.
70. Kantrowtiz, Evan
71. Halvorson, Harlyn
72. Kornberg, Sir Hans
73. Saier, Milton H.
74. Peterkofsky, Alan
75. Boos, Winfried
76. de Cock, Hans
77. Reizer, Jonathan
78. Wyckoff, Harold W.
79. Kato, Junichi
80. Toh-e, Akio
81. Stadtman, Thressa
82. Lacoste, Anne-Marie
83. Oshima, Yasuji
84. Kim, Eunice
85. Mizushima, Shoji
86. Nesmeyanova, Marina
87. Kornberg, Arthur
88. Torriani, Annamaria
89. Phung, Le T.
90. Fang, Aiqi
91. Dumora, Catherine
92. Amemura, Mitsuko
93. Makino, Kozo

Not pictured: Austin, Sara; Beckwith, Jonathan; Begley, Gail; Dunlap, Paul; Grossman, Alan; Kendall, Debra; Lengeler, Josef W.; Mekalanos, John; Merrick, Michael J.; Newton, Austin; Ohta, Noriko; Pattus, Franc; Roberts, Mary F.; Roseman, Saul; Rothman, Frank; Santos, Helena; Smith, J. Donald; Spira, Beny; Stock, Ann; Wickner, William; Wright, Andrew; Yagil, Ezra; Yashphe, Jacob; van Heijne, Gunnar.

I. REGULATION OF PHOSPHATE METABOLISM AND TRANSPORT

Chapter 1

Introduction: the Pho Regulon of *Escherichia coli*

ANNAMARIA TORRIANI-GORINI

Department of Biology, Room 68-371, Massachusetts Institute of Technology, Cambridge, Massachusetts 02139

How Does *Escherichia coli* Survive Phosphate Starvation?

Koch stressed the precarious existence of a coliform clone in its natural setting: the intestinal gut or the soil (Koch, 1987). In the test tube, difficult conditions can be created for *Escherichia coli* by a synthetic medium rich in carbon source (e.g., glucose at 4 mg/ml) and with a limiting amount of P_i (100 μM). The cells live happily until the P_i concentration decreases to 0.16 μM. At this concentration the P_i transport system (Pit), a proton motive force-energized transmembrane protein with a K_m of 25 μM, becomes inefficient (Rosenberg, 1987; also see chapter 7). The cells, however, do not die of P_i starvation, but the alkaline phosphatase (AP of gene *phoA*) is suddenly, rapidly (1 h), and fully induced while the doubling time of the culture is steadily increasing from 80 min to 16 h at 37°C. The expression of *phoA* is easy to monitor, but it actually involves a chain of expression of other genes.

How Is AP Synthesis Regulated by P_i?

The study of the regulatory mechanisms of AP in *E. coli* started from the obvious question, how is AP synthesis regulated by P_i? Early results came from the selection of AP constitutive mutants isolated by Torriani and Rothman (1961) with Al and Sue Garen and Hatch Echols in the laboratory of Cy Levinthal at Massachusetts Institute of Technology. Two families of mutations, R1 and R2 (Echols et al., 1961; Garen and Echols, 1962), with a constitutive AP phenotype, were mapped at two remote genetic sites—9 and 83 min—on the *E. coli* circular chromosome. They clearly represented two independent levels of regulation. It was subsequently shown that R1 was actually a two-component operon of regulatory genes (*phoB* and *phoR*) (Morris et al., 1974; Stock et al., 1989) and R2 was an operon of five genes (now *pstS*, *pstC*, *pstA*, *pstB*, and *phoU*) of the phosphate-specific transport (Pst) system (Surin et al., 1985). Thus, phosphate regulation turned out to be complex and odd, since it involved a P_i transport operon as a regulatory element, besides the sensor-activator operon.

To start the chain of gene expression leading to AP, the genes *phoR* (the sensor) and *phoB* (the positive regulator, or activator) must be activated first (Fig. 1). PhoR is a transmembrane sensor (Scholten and Tommassen, 1993), synthesized and autophosphorylated when P_i becomes the growth-limiting factor. PhoR has a protein histidine kinase activity and *trans*-phosphorylates the transcriptional activator PhoB (also P_i starvation induced) (Makino et al., 1989). The phosphorylated form of PhoB specifically recognizes the "pho box" (a consensus sequence of 18 nucleotides), which is part of the promoter of a number of genes (more than 30). The genes controlled by P_i and activated by PhoB constitute the Pho regulon (Table 1).

The Many Proteins Induced Pertain to All Parts of the Cell

Phosphorylated compounds enter the periplasm through the pore proteins OmpC, OmpF, and the P_i starvation-induced porin PhoE of the outer membrane. Most of the phosphorylated compounds cannot cross the barrier of the cytoplasmic membrane and are dephosphorylated by proteins

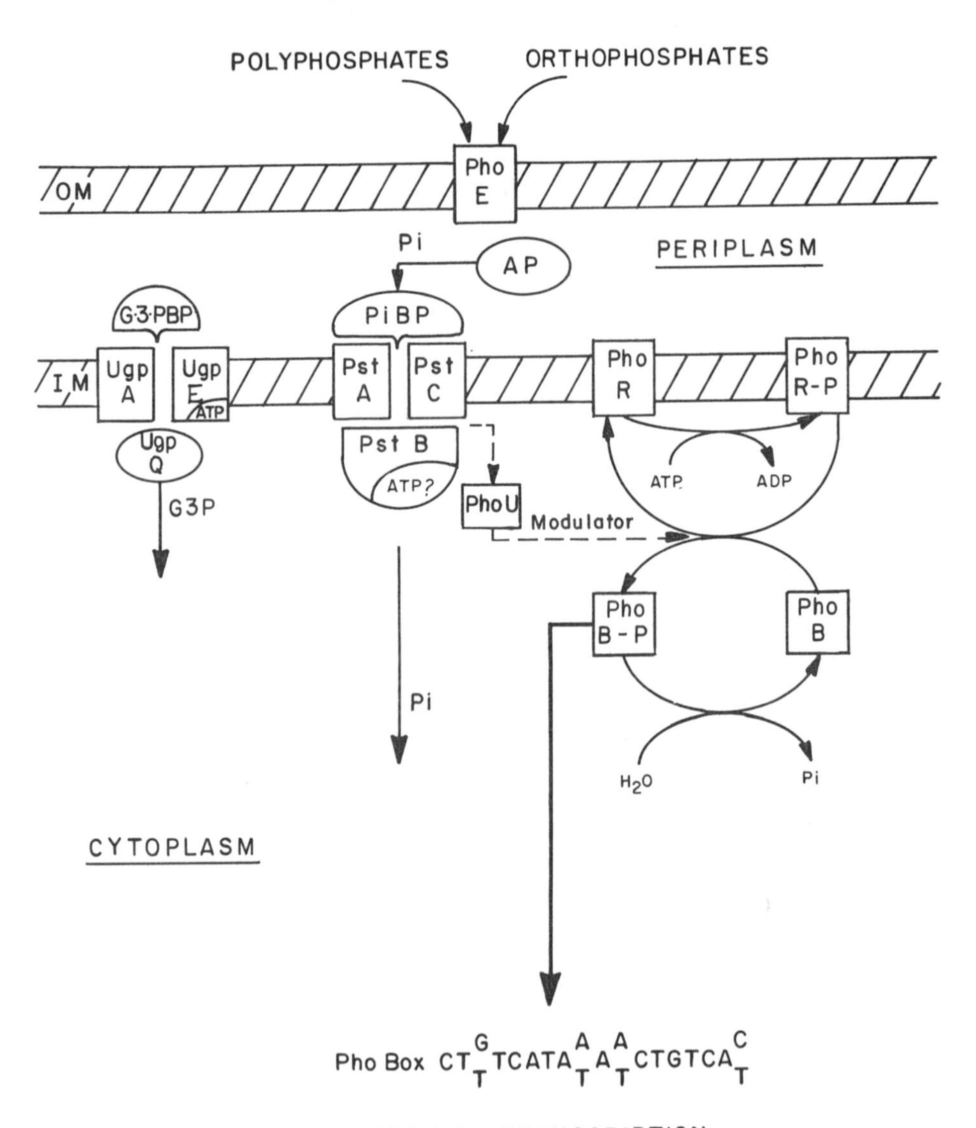

FIG. 1. The Pho regulon. Products of genes induced by PhoB during phosphate starvation. PhoE, porin E. G·3·P uptake: PBP periplasmic binding protein; inner membrane channel UgpA-UgpE; membrane-related UgpQ. Pst transport of P_i: P_iBP-binding protein; cytoplasmic membrane channel PstA-PstC; membrane-related ATP-bound supply of energy for transport, PstB; PhoU is the signal modulator of P_i starvation to the regulator operon *phoR-phoB*. PhoR protein kinase is autophosphorylated (PhoR.P) and phosphorylates the inducer PhoB (PhoB.P). OM, outer membrane; IM, inner membrane. This figure has been slightly modified from the one of Torriani (1990) to better fit the more recent results.

secreted to the periplasm (nucleotidases, phosphatases, and, most prominent, AP). Once P_i is liberated in the periplasm, it is captured by the high-affinity P_i-binding protein, PstS, which transfers it to the cytoplasm via the PstA-PstC channel in the cytoplasmic membrane. The P_i transport is probably energized by the membrane-associated protein PstB.

TABLE 1. Pho regulon proteins induced at P_i starvation

Cellular location	Function	Protein(s)
Outer membrane	Porin	PhoE
Periplasm	Phosphatases	Alkaline (AP) Acid (AppA)
	5′ Nucleotidase	5′ NU
	Binding proteins	
	P_i binding	PstS
	G3P binding (UGPB)	G3PBP
	Phosphonate transporter	PhnC, D, E
Cytoplasmic membrane	P_i channel	PstA, C
	G3P channel	UgpA, E
	P_i sensor protein kinase	PhoR
Membrane related	Component of active P_i transport	PstB
	Modulator of P_i transduction	PhoU
	G3P transport	UgpC
	Glycerophosphoryldiesterphosphodiesterase	UgpQ
	Polyphosphate synthesis	PolyP_i kinase (PpX)
	Polyphosphate utilization	PolyP_i phosphatase (PpK)
Cytoplasm	Pho regulon inducer	PhoB
	Phosphonate operon regulator	PhnO
	Phosphonate biodegradation	PhnF, G, H, I, J, L, M, N, P
	Unknown	PhoH

Is Cytoplasmic Free P_i the Repressor of the Regulon?

Rao et al. (1993; also see chapter 4) observed that the free P_i in the cytoplasm is present at a high concentration (10 mM) and has no effect on the expression of the *phoA* gene. Unlike most phosphorylated compounds, periplasmic glycerol 3-phosphate (G3P) is captured by the P_i starvation-induced G3P-binding protein B and transported intact to the cytoplasm via the G3P uptake UgpA-UgpE channel at the UgpQ membrane-associated protein (Brzoska and Boos, 1988; Kasahara et al., 1991) (Fig. 1). This system has no regulatory effect on the Pho regulon.

Brzoska and Boos (1988) and Kasahara et al. (1991) presented unexplained paradoxical results: G3P, transported by Ugp, can serve as the sole source of phosphate but not of carbon. In chapter 5, Brzoska et al. explain it by a negative effect of the cytoplasmic P_i on the activity of Ugp.

How Does the Signal of P_i Starvation Reach *phoR*?

If the cytoplasmic P_i level has no effect on the repression of the Pho regulon, how does the signal of P_i starvation reach *phoR*? A first hint came from the observation that mutants of PhoR (which we now know could not phosphorylate PhoB) produced AP in the presence of excess P_i; i.e., they were constitutive because PhoB was active (Echols et al., 1961). The answer came recently from the observation of cross-regulation with other protein kinases such as PhoM (now called CreC) (Wanner and Letterell, 1980; also see chapter 3) and acetyl kinase (Lee et al., 1990; also see chapter 48) which can phosphorylate PhoB but are not regulated by P_i.

Is PhoR the Protein Sensing P_i Starvation, or Is the Pst System the Sensor?

The more we know about PhoR, the less probable this direct relationship becomes. Most recently we learned that PhoR is essentially missing a segment spanning into the periplasm to sense the environmental stimulus (Scholten and Tommassen, 1993; also see chapter 46). We know that the Pho regulon is not regulated (repressed) if the Pst system is hampered. Experiments with mutations of the *pst* operon established that the two functions, P_i transport and Pho regulation, are independent (Cox et al., 1988). Thus, this chapter concentrates on the regulatory function which requires an integral structure of the five proteins of the operon: the P_i-binding protein of the periphasm (PstS); transmembrane and channel proteins (PstC and PstA); a membrane-associated protein, possibly ATP bound (PstB), tentatively supported by preliminary experiments of photo-cross-linking with 8-azido-[γ-^{32}P]ATP (Chang and Torriani, unpublished); and the fifth protein, whose function is still unclear (PhoU). Our recent results support the hypothesis that the complex structure of Pst is required to transfer the P_i signal. Possibly the cytoplasmic PhoU acts as a modulator of the PhoR phosphatase/kinase functions (Muda et al., 1992) (Fig. 1).

Many results indicate that some nucleotides may be involved directly or indirectly with the modulation of the Pho regulon (see chapter 4). This was unclear in 1987 (Torriani-Gorini, 1987), and although more information is now available (chapter 4), the basic mechanism remains to be solved in time for the symposium in the year 2000.

REFERENCES

Brzoska, P. and W. Boos. 1988. Characteristics of a *ugp*-encoded and *phoB*-dependent glycerophosphoryl diester phosphodiesterase which is physically dependent on the Ugp transport system of *Escherichia coli. J. Bacteriol.* **170:**4125–4135.

Chang, F.-Y., and A. Torriani. Unpublished data.

Cox, G., D. Webb, J. Godavan-Zimmerman, and H. Rosenberg. 1988. Arg-220 of the PstA protein is required for phosphate transport through the phosphate-specific transport system in *Escherichia coli* but not for alkaline phosphatase repression. *J. Bacteriol.* **170:**2283–2286.

Echols, H., A. Garen, S. Garen, and A. Torriani. 1961. Genetic control of repression of alkaline phosphatase in *Escherichia coli. J. Mol. Biol.* **3:**425–438.

Garen, A., and H. Echols. 1962. Properties of two regulating genes for alkaline phosphatase. *J. Bacteriol.* **83:**297–300.

Kasahara, M., K. Makino, M. Amemura, A. Nakata, and H. Shinagawa. 1991. Dual regulation of *ugp* operon by phosphate and carbon starvation at two interspaced promoters. *J. Bacteriol.* **173:**549–555.

Koch, A. 1987. Why *Escherichia coli* should be renamed *Escherichia ilei*, p. 300–305. *In* A. Torriani-Gorini, F. Rothman, S. Silver, A. Wright, and E. Yagil (ed.), *Phosphate Metabolism and Cellular Regulation in Microorganisms.* American Society for Microbiology, Washington, D.C.

Lee, T.-Y., K. Makino, H. Shinagawa, and A. Nakata. 1990. Overproduction of acetate kinase activates the phosphate regulon in the absence of the *phoR* and *phoM* functions in *Escherichia coli. J. Bacteriol.* **172:**2245–2249.

Makino, K., H. Shinagawa, M. Amemura, T. Kawamoto, M. Yamada, and A. Nakata. 1989. Signal transduction in the phosphate regulon of *Escherichia coli* involves phosphotransfer between PhoR and PhoB proteins. *J. Mol. Biol.* **210:**551–559.

Morris, H., M. J. Schlesinger, M. Bracha, and E. Yagil. 1974. Pleiotropic effects of mutations involved in the regulation of *Escherichia coli* K-12 alkaline phosphatase. *J. Bacteriol.* **119:**583–592.

Muda, M., N. Rao, and A. Torriani. 1992. Role of PhoU in phosphate transport and alkaline phosphatase regulation. *J. Bacteriol.* **174:**8057–8064.

Rao, N., M. F. Roberts, A. Torriani, and J. Yashphe. 1993. Effect of *glpT* and *glpD* mutations on expression of the *phoA* gene in *Escherichia coli. J. Bacteriol.* **175:**74–79.

Rosenberg, H. 1987. Phosphate transport in prokaryotes, p. 205–248. *In* B. Rosen and S. Silver (ed.), *Ion Transport in Prokaryotes.* Academic Press, Inc., San Diego, Calif.

Scholten, M., and J. Tommassen. 1993. Topology of the PhoR protein of *E. coli* and functional analysis of internal deletion mutants. *Mol. Microbiol.* **8:**269–275.

Stock, J., A. Ninfa, and A. Stock. 1989. Protein phosphorylation and regulation of adaptive responses in bacteria. *Microbiol. Rev.* **53:**450–490.

Surin, B. P., H. Rosenberg, and G. B. Cox. 1985. Phosphate-specific transport system of *Escherichia coli:* nucleotide sequence and gene-polypeptide relationship. *J. Bacteriol.* **161:**189–198.

Torriani, A. 1990. From cell membrane to nucleotides: the phosphate regulon in *Escherichia coli. Bioessays* **12:**371–376.

Torriani, A., and F. Rothman. 1961. Mutants of *Escherichia coli* constitutive for alkaline phosphatase. *J. Bacteriol.* **81:**835–836.

Torriani-Gorini, A. 1987. The birth and growth of the Pho Regulon, p. 3–11. *In* A. Torriani-Gorini, F. Rothman, S. Silver, A. Wright, and E. Yagil (ed.), *Phosphate Metabolism and Cellular Regulation in Microorganisms.* American Society for Microbiology, Washington, D.C.

Wanner, B., and P. Letterell. 1980. Mutants affected in alkaline phosphatase expression: evidence for multiple positive regulators of the phosphate regulon in *Escherichia coli. Genetics* **96:**353–366.

Chapter 2

Mechanism of Transcriptional Activation of the Phosphate Regulon in *Escherichia coli*

KOZO MAKINO, MITSUKO AMEMURA, SOO-KI KIM, ATSUO NAKATA, AND HIDEO SHINAGAWA

Department of Experimental Chemotherapy, Research Institute for Microbial Diseases, Osaka University, 3-1, Yamadaoka, Suita, Osaka 565, Japan

In *Escherichia coli* at least 31 genes, which are involved in the roles related to the transport and assimilation of phosphate and phosphorous compounds, are induced by phosphate starvation. They constitute a single phosphate (Pho) regulon and are under the same physiological and genetic control (Nakata et al., 1987; Shinagawa et al., 1987; Wanner, 1993). The proteins PhoB and PhoR, which are regulatory systems for the transcriptional regulation of the *pho* genes, belong to a family of two-component regulatory factors that respond to a variety of environmental stimuli in bacteria (Makino et al., 1986a, 1986b; Stock et al., 1989; Parkinson, 1993). PhoB is the transcriptional activator, which binds to the promoters of the *pho* genes (Makino et al., 1988, 1989). PhoR is a transmembrane protein that modulates the activity of PhoB by promoting specific phosphorylation and dephosphorylation of PhoB in response to the phosphate signal (Yamada et al., 1990; Makino et al., 1989, 1992; also see chapter 48). The phosphorylation of PhoB protein occurs concurrently with the acquisition of the ability to activate transcription from the *pho* promoters (Fig. 1). In the absence of the PhoR functions, PhoB is phosphorylated independently of the phosphate levels by PhoM, a PhoR-like protein (Makino et al., 1984; Amemura et al., 1986, 1990), which was renamed CreC by Wanner (1992).

In this article, we describe our recent studies on the mechanism of transcriptional regulation of the *pho* regulon.

Genes in the *pho* Regulon

Figure 2 shows the positions of the *pho* regulon genes on the genetic map of *E. coli*. The regulatory genes, *phoB* and *phoR*, are located at about 9 min and constitute a single operon (Makino et al., 1982, 1985, 1986a, 1986b). *phoA* encodes periplasmic alkaline phosphatase (Kikuchi et al., 1981), and *phoE* encodes an outer membrane protein, porin e (Overbeeke et al., 1983). The *phn* operon, which consists of 14 genes, encodes proteins related to the transport and utilization of phosphonates and is active in *E. coli* B but cryptic in *E. coli* K-12 because of a mutation (Wackett et al., 1987; Wanner and Boline, 1990; Chen et al., 1990; Makino et al., 1991; Metcalf and Wanner, 1993). *psiE* was identified by Wanner (1983) as one of the phosphate starvation-inducible genes, and the position of the gene was determined by Metcalf et al. (1990). The function of the gene is not known at present, but its expression is positively regulated by the product of *phoB* and negatively regulated by the product of *crp* (Wanner, 1983). Footprinting experiments with phospho-PhoB and cyclic AMP (cAMP)-cAMP receptor protein (CRP) indicated that the binding sites of these proteins on the promoter region partially overlap (unpublished data). The *pst-phoU* operon encodes proteins involved in the phosphate-specific transport system and signal transduction in the *pho* regulon (Magota et al., 1984; Surin et al., 1985; Amemura et al., 1985). The *ugp* operon encodes proteins involved in the transport and utilization of *sn*-glycerol-3-phosphate (Overduin et al., 1988). *phoH*, which had originally been named *psiH* by Wanner and McSharry (1982), encodes a protein that contains the ATP-binding motif and actually binds to ATP (Kim et al., 1993), but its function in vivo is not known.

The *pho* Box as a Consensus Sequence for the *pho* Promoters

The genes in the *pho* regulon are transcriptionally activated by PhoB. Therefore, they are likely to share a common regulatory element in the promoter regions which are recognized by PhoB. In agreement with this idea, the well-conserved 18-nucleotide sequence, 5′-CTGTCATA(A/T)A-(T/A)CTGTCA(C/T), which we named the *pho* box, is found 10 nucleotides upstream from the putative −10 regions in all cases, as shown in Fig. 3 (Overbeeke et al., 1983; Makino et al., 1986a, 1991; Kasahara et al., 1991; Kim et al., 1993).

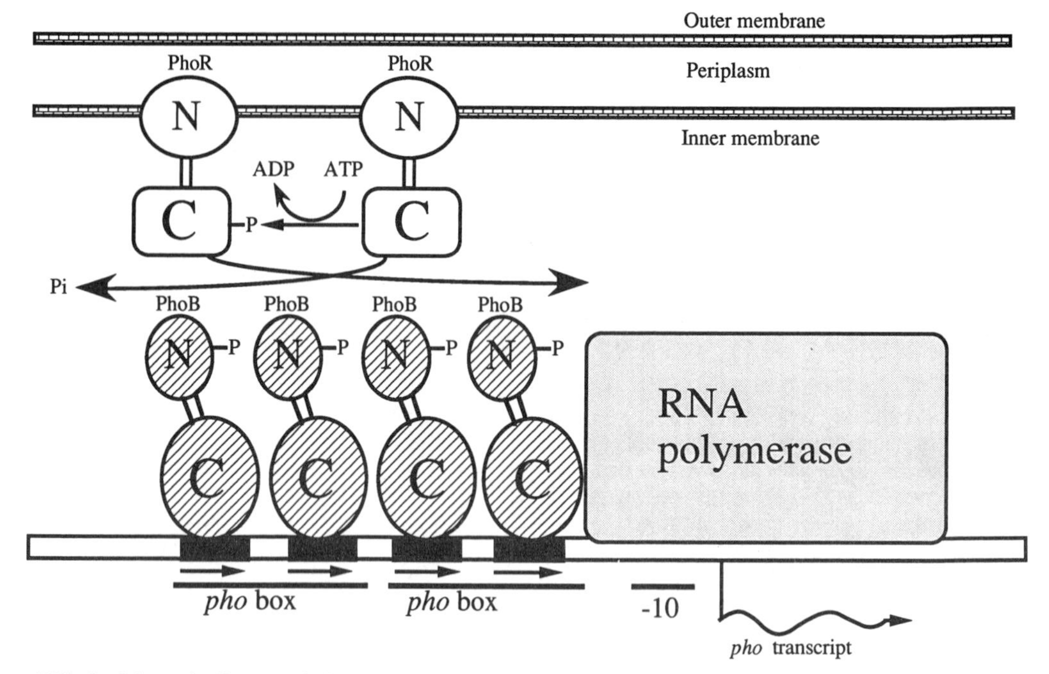

FIG. 1. Schematic diagram of PhoR-mediated phosphorylation and dephosphorylation of PhoB in relation to the transcriptional regulation of the *pstS* promoter. Solid boxes with arrows indicate repeated sequences (5′-CTGTC) in the *pho* boxes (for details, see the text). N and C indicate amino- and carboxy-terminal regions, respectively, of PhoB and PhoR. With limited phosphate, PhoR autophosphorylates by using ATP and promotes phosphorylation of PhoB. Phosphorylation of PhoB enhances its binding activity to the *pho* boxes and activates transcription from the *pstS* promoter in concert with RNA polymerase. The number of PhoB molecules bound to the *pho* boxes of the *pstS* promoter is putative. With excess phosphate, PhoR promotes dephosphorylation of the phospho-PhoB, and dephosphorylation of PhoB greatly decreases its affinity for the *pstS* promoter, which is inactive as the transcriptional activator. P_i indicates the product of dephosphorylation of phospho-PhoB by PhoR.

However, the sequences homologous to the −35 region are not found. DNase I footprinting with phospho-PhoB revealed that these *pho* boxes are the target sequences for the PhoB protein. The *pho* box is tandemly repeated at least twice in the regulatory regions of *pstS* and *ugpB*, respectively. Mutational analysis of the *pstS* promoter region in vivo and in vitro was performed (Kimura et al., 1989; unpublished data). Deletions extending into the upstream *pho* box but retaining the downstream *pho* box greatly reduced promoter activity, but the remaining activity was still regulated by phosphate levels in the medium and by the PhoB protein, indicating that each *pho* box is functional. No activity was observed in deletion mutants that lacked the remaining *pho* box or the −10 region. The binding region of PhoB in the *pstS* regulatory region was studied by a gel mobility retardation

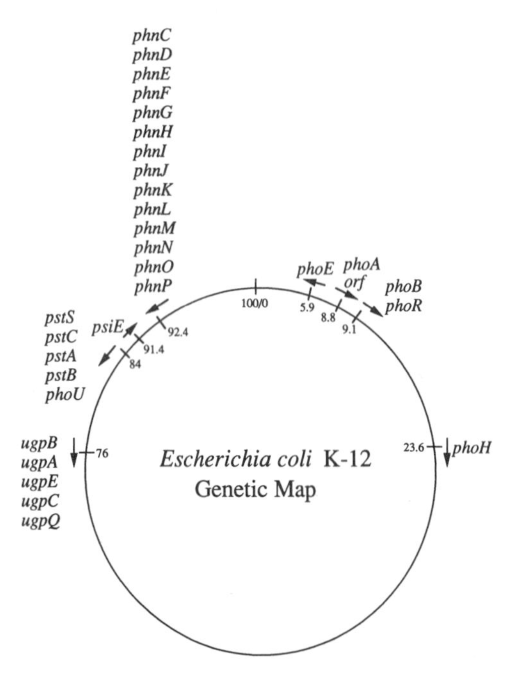

FIG. 2. Positions of the *pho* regulon genes on the genetic map of *E. coli* K-12. The circular map, nomenclature, and map positions are those according to Rudd (1992). The arrows show the direction of transcription.

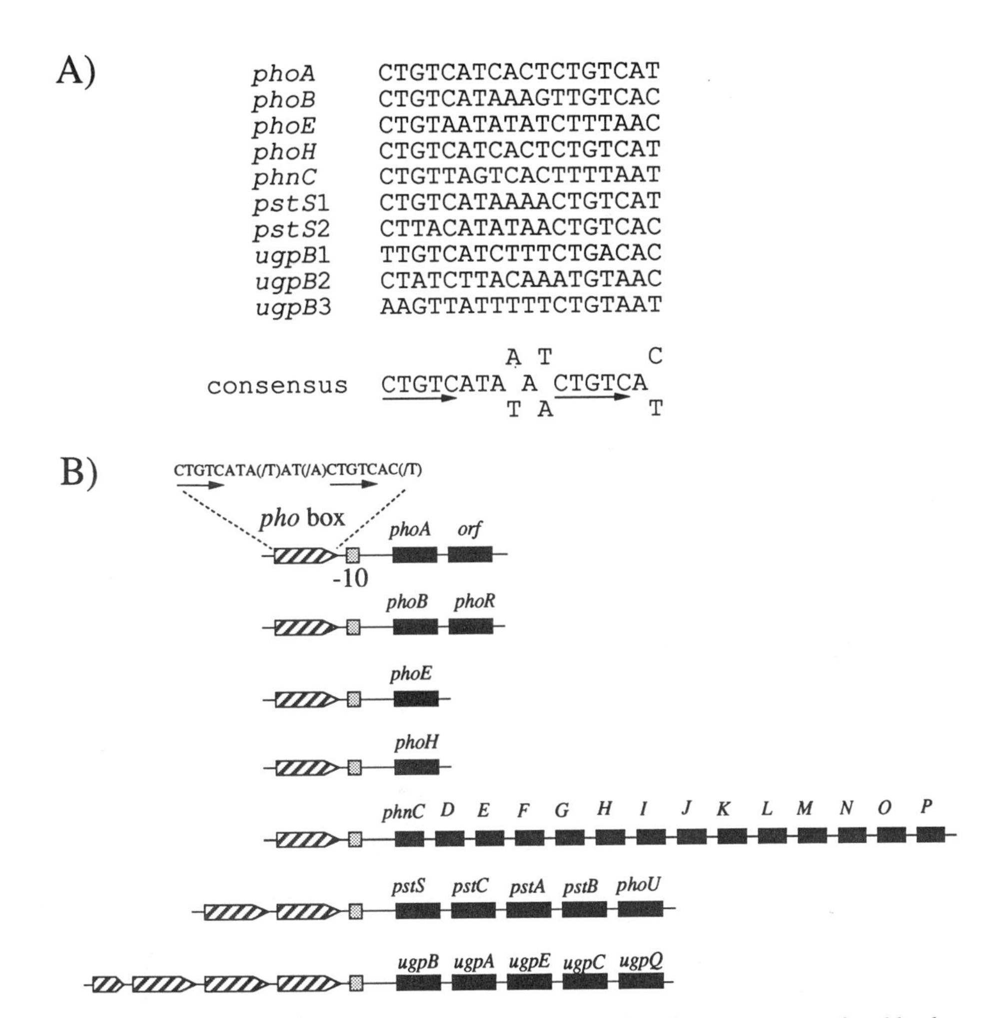

FIG. 3. Schematic presentation of the *pho* regulon genes. (A) The homologous sequences shared by the genes belonging to the *pho* regulon (the *pho* boxes) are shown. (B) The *pho* boxes, −10 sequences, and coding regions are illustrated as indicated. The *ugp* promoter contains 3.5 *pho* boxes.

assay with the truncated promoter fragments and by DNase I footprinting. The results of these in vitro experiments also indicate that PhoB binds to each *pho* box of the *pstS* promoter. The *pho* box is likely to consist of two direct repeats of 5′-CTGTC that flank the A+T-rich 6-bp spacer. Single-point mutations in any of these sequences in a truncated *pstS* promoter that contains only the distal *pho* box caused reduction in transcriptional activity, but other mutations in the *pho* box affected this activity only marginally at most, indicating that these direct repeats play an important role in the interaction with PhoB. We also examined the binding of PhoB to the regulatory regions of the *pho* genes by methylation protection experiments (Makino et al., 1988; Kasahara et al., 1991; Kim et al., 1993). In most cases, methylation of Gs at nucleotides 5′-CT<u>G</u>TC in the upper strands and 5′-<u>G</u>ACAG in the lower strands was reduced the most by PhoB. Therefore, these Gs are very important for the interaction with PhoB, probably by directly forming hydrogen bonds with PhoB.

Mutational Analysis of PhoB

The transcriptional activator PhoB, composed of 229 amino acids, is postulated to contain at least three functional domains (Makino et al., 1989) (Fig. 1): domain I, a domain for phosphorylation; domain II, a domain for DNA (the *pho* box) binding; and domain III, a domain for interaction with the RNA polymerase holoenzyme. To identify these domains in PhoB, we constructed a series of *phoB* mutants, including those that encode amino (N)-terminal or carboxy (C)-terminal

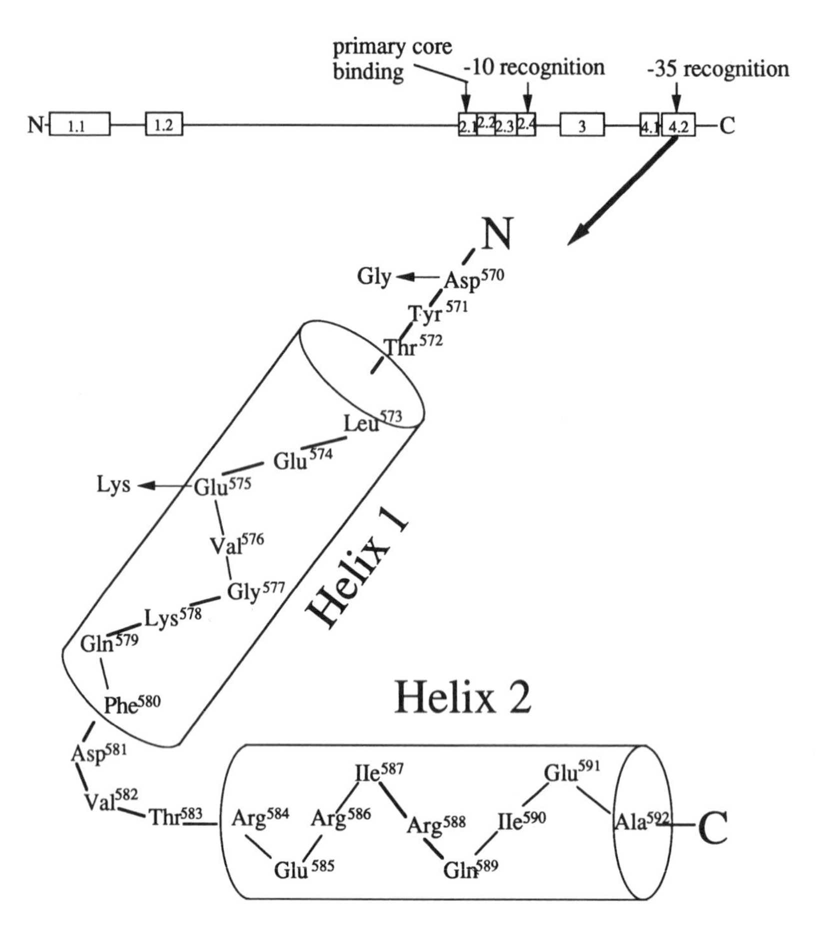

FIG. 4. Domain structure of σ^{70} and amino acid substitutions in the mutant σ^{70} subunits that specifically decrease transcription from the *pho* promoters. In σ^{70}, four conserved regions (regions 1 through 4) among many σ factors, their subregions such as 1.1, and the functions of the subregions are shown at the top as reported by Helmann and Chamberlin (1988). A schematic diagram of the putative HTH structure of region 4.2 of σ^{70} of RNA polymerase is drawn at the bottom. The positions of two amino acid substitutions in the mutant σ^{70} subunits are indicated by arrows. The second helix (helix 2) has been proposed for the direct recognition helix of the −35 sequence (Helmann and Chamberlin, 1988). N and C indicate amino- and carboxy-terminal regions, respectively, of the σ^{70} subunit.

truncated PhoB and point mutants, and analyzed transcription-enhancing, phosphate-accepting, and DNA-binding activities (details to be published elsewhere). The truncated proteins, containing at least the N-terminal 127 amino acids (domain I), were fully competent for accepting phosphate from phospho-PhoR1084, which lacks the N-terminal 83 residues of PhoR and constitutively promotes phosphorylation of intact PhoB (Makino et al., 1989). Biochemical and mutational analysis revealed that Asp-53 of PhoB is the phosphate-accepting residue from phospho-PhoR and that Thr-83 plays an important role in the phosphate transfer reaction. Domains II and III were contained in the C-terminal 90 amino acids of PhoB. This 90-amino-acid PhoB constitutively activated transcription from the *pho* genes. Therefore, we propose that the N-terminal domain physically blocks the C-terminal activator domain and that the phosphorylation of Asp-53 releases this block by bringing about the conformational change in domain I. When any one of three Arg, two Thr, or two Gly residues in the C-terminal region was replaced by other amino acids, the DNA-binding activity of PhoB was abolished; therefore, they should play important roles in the interaction with DNA (domain II). A mutant PhoB with an alteration of Glu-77 to Lys did not activate transcription, but it had normal DNA-binding and phosphate-accepting activities. The region containing Glu-177 of PhoB may be related to the interaction with RNA polymerase (domain III). We do not know whether domain III is physically separable from domain II or overlaps it.

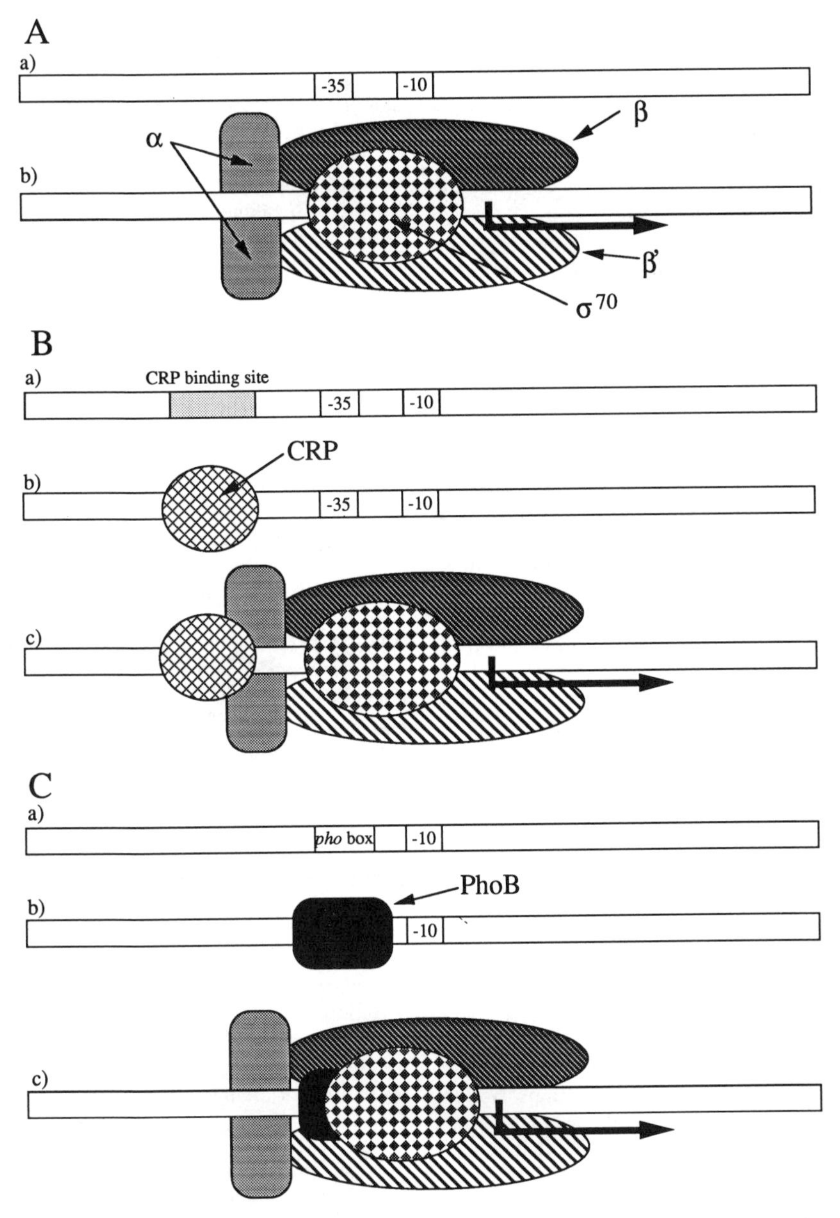

FIG. 5. Schematic presentation of transcription activation by RNA polymerase at three kinds of promoters. The −10 and −35 signals, CRP-binding sites, and PhoB-binding sites (*pho* box) in the three types of promoters are drawn. The heavy arrow represents transcription into mRNA when the system is activated. The subunits of RNA polymerase, CRP, and PhoB are illustrated as indicated. (A) Activator-independent promoters: empty (a) or bound by RNA polymerase (b). The σ^{70} subunit mediates interaction between RNA polymerase and the promoter by making direct contact with the −10 and −35 sites. (B) CRP-dependent promoters that contain the CRP binding sites located upstream from the −35 sites: empty (a), bound by CRP (b), or bound by CRP and RNA polymerase (c). CRP bound to the binding site facilitates the interaction of RNA polymerase with the promoter via direct contact with the C-terminal part of the α subunit. Although the σ^{70} subunit interactions with the −10 and −35 sites are required, they are not enough for the formation of the stable complex with the promoter. Interaction between CRP and RNA polymerase via the α subunit mutually increases their interactions with the promoter. This model has been proposed by Igarashi and Ishihama (1990). (C) PhoB-dependent promoters: empty (a), bound by PhoB (b), or bound by PhoB and RNA polymerase (c). PhoB bound to the *pho* box promotes interaction of RNA polymerase with the promoter via direct contact with the σ^{70} subunit. Although σ^{70} interaction with the −10 site of the *pho* promoter is required, it alone is not enough for the formation of stable complex with the promoter. The process requires the function of PhoB bound to the *pho* box. Modified from Makino et al. (1993) with permission.

Role of the σ^{70} Subunit of RNA Polymerase in Transcriptional Activation by PhoB

Using a series of C-terminally truncated α subunits of RNA polymerase, Ishihama and coworkers have demonstrated that some activators such as CRP, which binds to the upstream region from the promoter, make direct contact with the C-terminal region of the α subunit in transcriptional activation (Igarashi and Ishihama, 1990; Igarashi et al., 1991). Genetic studies also supported the idea that activators such as CRP, Fnr, and OmpR make direct contact with the C-terminal region of the α subunit in transcriptional activation (Zou et al., 1992; Lombardo et al., 1991; Slauch et al., 1991). However, in vitro transcription analysis with PhoB indicated that the C-terminal region of the α subunit of RNA polymerase is dispensable for the activation of transcription from the PhoB-dependent promoters (Igarashi et al., 1991). From these results, we postulated that PhoB specifically interacts with other subunits of RNA polymerase in transcriptional activation. To identify the subunit of RNA polymerase involved in the specific interaction with PhoB, we attempted by localized mutagenesis on the chromosome to isolate *rpo* mutants which are specifically defective in the expression of the *pho* genes. We isolated two mutants with such properties, and they had mutations in the *rpoD* gene encoding the σ^{70} subunit of RNA polymerase (Makino et al., 1993). The *rpoD* mutations altered amino acids within and near the first helix of the putative helix-turn-helix (HTH) motif in the C-terminal region (region 4.2) of σ^{70} (Fig. 4). Transcription from the *pho* promoters in vivo was greatly reduced in these mutants, whereas transcription from the PhoB-independent promoters was affected only marginally at most. The reconstituted RNA polymerase holoenzymes containing the mutant σ^{70} proteins were very defective in the transcription from the *pstS* promoter in vitro, whereas they were proficient in transcribing the PhoB-independent promoters. Phosphorylated PhoB mediated the specific binding of the wild-type holoenzyme to the *pstS* promoter, but it did not mediate the binding of the mutant holoenzymes. These results suggest that PhoB promotes specific interaction between RNA polymerase and the *pho* promoters for the transcriptional activation and that the first helix of the putative HTH motif of σ^{70} plays an essential role in the interaction.

To further study the role of the first helix of region 4.2 of σ^{70} in transcriptional activation by PhoB, we made a series of site-directed mutations to alter amino acids in the motif and purified the mutant σ^{70} proteins (unpublished data). Most of the reconstituted RNA polymerases containing the mutant σ^{70} proteins reduced the in vitro transcription from the *pstS* promoter, whereas they transcribed the *tac* promoter normally. Kumar et al. (1994) also have demonstrated that RNA polymerase containing a C-terminally truncated σ^{70} with a deletion up to the second helix of the putative HTH motif, which is proposed to be the direct recognition helix for the -35 sequence (Helmann and Chamberlin, 1988), could not transcribe most promoters but could transcribe the *pstS* promoter in the presence of PhoB in vitro. However, RNA polymerases with further deletion in σ^{70} could not transcribe *pstS* which does not possess the -35 sequence. These results are consistent with the hypothesis that PhoB directly interacts with the first helix of region 4.2 of σ^{70} to introduce the RNA polymerase into the *pho* promoters for transcriptional activation and that the second helix is involved in the interaction with the -35 sequence (Fig. 5).

Conclusions

The pathway of signal transduction in the Pho regulon involves PstS, PstC, PstA, and PstB (phosphate transporter), PhoU, PhoR (sensor), PhoB (activator), and RNA polymerase. An external phosphate signal received by PstS, PstC, PstA, PstB, and PhoU is transmitted to PhoR. With limited phosphate, PhoR autophosphorylates by using ATP and phosphorylates PhoB. The activated PhoB (phospho-PhoB) binds to the *pho* boxes of the *pho* promoters; interacts with RNA polymerase, probably by making direct contact with the first helix of the HTH motif of σ^{70}; and permits the RNA polymerase to enter the promoters for the initiation of RNA synthesis. With excess phosphate, PhoR facilitates the dephosphorylation of phospho-PhoB, and the transcription from the *pho* promoters is abolished. The chemical or physical nature of the signal transmitted from PstS, PstC, PstA, PstB, and PhoU to PhoR is not known at present.

This work was supported by a Grant-in-Aid for Scientific Research from the Ministry of Education, Science, and Culture of Japan and by a travel grant to K.M. from the Yamada Foundation for the Promotion of Science.

REFERENCES

Amemura, M., K. Makino, H. Shinagawa, A. Kobayashi, and A. Nakata. 1985. Nucleotide sequence of the genes involved in phosphate transport and regulation of the phosphate regulon in *Escherichia coli*. *J. Mol. Biol.* **184:**241–250.

Amemura, M., K. Makino, H. Shinagawa, and A. Nakata. 1986. Nucleotide sequence of the *phoM* region of *Escherichia coli:* four open reading frames that may constitute an operon. *J. Bacteriol.* **168:**294–302.

Amemura, M., K. Makino, H. Shinagawa, and A. Nakata. 1990. Cross talk to the phosphate regulon of *Escherichia coli* by PhoM protein: PhoM is a histidine protein kinase and catalyzes phosphorylation of PhoB and PhoM open reading frame 2. *J. Bacteriol.* **172:**6300–6307.

Chen, C.-M., Q.-Z. Ye, Z. Zhu, B. L. Wanner, and C. T. Walsh. 1990. Molecular biology of carbon-phosphorus bond cleavage: cloning and sequencing of the *phn* (*psiD*) genes involved in alkylphosphate uptake and C-P lyase activity in *Escherichia coli* B. *J. Biol. Chem.* **265:**4461–4471.

Helmann, J. D., and M. J. Chamberlin. 1988. Structure and function of bacterial sigma factors. *Annu. Rev. Biochem.* **57:**839–872.

Igarashi, K., A. Hanamura, K. Makino, H. Aiba, H. Aiba, T. Mizuno, A. Nakata, and A. Ishihama. 1991. Functional map of the α subunit of *Escherichia coli* RNA polymerase: two modes of transcription activation by positive factors. *Proc. Natl. Acad. Sci. USA* **88:**8958–8962.

Igarashi, K., and A. Ishihama. 1990. Bipartite functional map of the *E. coli* RNA polymerase α subunit: involvement of the C-terminal region in transcriptional activation by cAMP-CRP. *Cell* **65:**1015–1022.

Kasahara, M., K. Makino, M. Amemura, A. Nakata, and H. Shinagawa. 1991. Dual regulation of the *ugp* operon by phosphate and carbon starvation at two interspaced promoters. *J. Bacteriol.* **173:**549–558.

Kikuchi, Y., K. Yoda, M. Yamasaki, and G. Tamura. 1981. The nucleotide sequence of the promoter and the amino-terminal region of alkaline phosphatase structural gene (*phoA*) of *Escherichia coli. Nucleic Acids Res.* **9:**5671–5678.

Kim, S.-K., K. Makino, M. Amemura, H. Shinagawa, and A. Nakata. 1993. Molecular analysis of the *phoH* gene, belonging to the phosphate regulon in *Escherichia coli. J. Bacteriol.* **175:**1316–1324.

Kimura, S., K. Makino, H. Shinagawa, M. Amemura, and A. Nakata. 1989. Regulation of the phosphate regulon of *Escherichia coli:* characterization of the promoter of the *pstS* gene. *Mol. Gen. Genet.* **215:**374–380.

Kumar, A., B. Grimes, N. Fujita, K. Makino, R. A. Malloch, R. S. Hayward, and A. Ishihama. 1994. Role of the sigma^{70} subunit of *Escherichia coli* RNA polymerase in transcription activation. *J. Mol. Biol.* **235:**405–413.

Lombardo, M.-J., D. Bagga, and C. G. Miller. 1991. Mutations in *rpoA* affect expression of anaerobically regulated genes in *Salmonella typhimurium. J. Bacteriol.* **173:**7511–7518.

Magota, K., N. Otuji, T. Miki, T. Horiuchi, S. Tsunasawa, J. Kondo, F. Sakiyama, M. Amemura, T. Morita, H. Shinagawa, and A. Nakata. 1984. Nucleotide sequence of the *phoS* gene, the structural gene for the phosphate-binding protein of *Escherichia coli* K-12. *J. Bacteriol.* **157:**909–917.

Makino, K., M. Amemura, S.-K. Kim, A. Nakata, and H. Shinagawa. 1993. Role of the σ^{70} subunit of RNA polymerase in transcriptional activation by activator protein PhoB in *Escherichia coli. Genes Dev.* **7:**149–160.

Makino, K., M. Amemura, S.-K. Kim, H. Shinagawa, and A. Nakata. 1992. Signal transduction of the phosphate regulon in *Escherichia coli* mediated by phosphorylation, p. 191–200. *In* S. Papa, A. Azzi, and J. M. Tager (ed.), *Adenine Nucleotides in Cellular Energy Transfer and Signal Transduction,* Birkhauser Verlag, Basel.

Makino, K., S.-K. Kim, H. Shinagawa, M. Amemura, and A. Nakata. 1991. Molecular analysis of the cryptic and functional *phn* operons for phosphonate use in *Escherichia coli* K-12. *J. Bacteriol.* **173:**2665–2672.

Makino, K., H. Shinagawa, M. Amemura, T. Kawamoto, M. Yamada, and A. Nakata. 1989. Signal transduction in the phosphate regulon of *Escherichia coli* involves phosphotransfer between PhoR and PhoB proteins. *J. Mol. Biol.* **210:**551–559.

Makino, K., H. Shinagawa, M. Amemura, S. Kimura, A. Nakata, and A. Ishihama. 1988. Regulation of the phosphate regulon of *Escherichia coli:* activation of *pstS* transcription by PhoB protein in vitro. *J. Mol. Biol.* **203:**85–95.

Makino, K., H. Shinagawa, M. Amemura, and A. Nakata. 1986a. Nucleotide sequence of the *phoB* gene, the positive regulatory gene for the phosphate regulon of *Escherichia coli* K 12. *J. Mol. Biol.* **190:**37–44.

Makino, K., H. Shinagawa, M. Amemura, and A. Nakata. 1986b. Nucleotide sequence of the *phoR* gene, a regulatory gene for the phosphate regulon of *Escherichia coli. J. Mol. Biol.* **192:**549–556.

Makino, K., H. Shinagawa, and A. Nakata. 1982. Cloning and characterization of the alkaline phosphatase positive regulator gene (*phoB*) of *Escherichia coli. Mol. Gen. Genet.* **187:**181–186.

Makino, K., H. Shinagawa, and A. Nakata. 1984. Cloning and characterization of the alkaline phosphatase positive regulatory gene (*phoM*) of *Escherichia coli. Mol. Gen. Genet.* **195:**381–390.

Makino, K., H. Shinagawa, and A. Nakata. 1985. Regulation of the phosphate regulon of *Escherichia coli* K 12. Regulation and role of the regulatory gene *phoR. J. Mol. Biol.* **184:**231–240.

Metcalf, W. W., P. M. Steed, and B. L. Wanner. 1990. Identification of phosphate starvation-inducible genes in *Escherichia coli* K-12 by DNA sequence analysis of *psi::lacZ*(Mu d1) transcriptional fusions. *J. Bacteriol.* **172:**3191–3200.

Metcalf, W. W., and B. L. Wanner. 1993. Mutational analysis of an *Escherichia coli* fourteen-gene operon for phosphonate degradation, using Tn*phoA'* elements. *J. Bacteriol.* **175:**3430–3442.

Nakata, A., M. Amemura, K. Makino, and H. Shinagawa. 1987. Genetic and biological analysis of the phosphate-specific transport system in *Escherichia coli,* p. 150–155. *In* A. Torriani-Gorini, F. G. Rothman, S. Silver, A. Wright, and E. Yagil (ed.), *Phosphate Metabolism and Cellular Regulation in Microorganisms.* American Society for Microbiology, Washington, D.C.

Overbeeke, N., H. Bergmans, F. Van Mansfeld, and B. Lugtenberg. 1983. Complete nucleotide sequence of *phoE,* the structural gene for the phosphate limitation inducible outer membrane pore protein of *Escherichia coli* K12. *J. Mol. Biol.* **163:**513–532.

Overduin, P., W. Boos, and J. Tommassen. 1988. Nucleotide sequence of the *ugp* genes of *Escherichia coli:* homology to the maltose system. *Mol. Microbiol.* **2:**767–775.

Parkinson, J. S. 1993. Signal transduction schemes of bacteria. *Cell* **73:**857–871.

Rudd, K. E. 1992. Alignment of *E. coli* DNA sequences to a revised, integrated genomic restriction map, section 2A. *In* J. H. Miller (ed.), *A Short Course in Bacterial Genetics.* Cold Spring Harbor Laboratory Press, Cold Spring Harbor, N.Y.

Shinagawa, H., K. Makino, M. Amemura, and A. Nakata. 1987. Structure and function of the regulatory genes for the phosphate regulon in *Escherichia coli,* p. 20–25. *In* A. Torriani-Gorini, F. G. Rothman, S. Silver, A. Wright, and E. Yagil (ed.), *Phosphate Metabolism and Cellular Regulation in Microorganisms.* American Society for Microbiology, Washington, D.C.

Slauch, J. M., F. D. Russo, and T. J. Silhavy. 1991. Suppressor mutations in *rpoA* suggest that OmpR controls transcription by direct interaction with the α subunit of RNA polymerase. *J. Bacteriol.* **173:**7501–7510.

Stock, J. B., A. Ninfa, and A. M. Stock. 1989. Protein phosphorylation and regulation of adaptive responses in bacteria. *Microbiol. Rev.* **53:**450–490.

Surin, B. P., H. Rosenberg, and G. B. Cox. 1985. Phosphate-specific transport system of *Escherichia coli:* nucleotide sequence and gene-polypeptide relationships. *J. Bacteriol.* **161:**189–198.

Wackett, L. P., B. L. Wanner, C. P. Venditti, and C. T. Walsh. 1987. Involvement of the phosphate regulon and *psiD* locus in carbon-phosphorus lyase activity of *Escherichia coli* K-12. *J. Bacteriol.* **169:**1753–1756.

Wanner, B. L. 1983. Overlapping and separate controls on the phosphate regulon in *Escherichia coli* K-12. *J. Mol. Biol.* **166:**283–308.

Wanner, B. L. 1992. Is cross-regulation by phosphorylation of two-component response regulator proteins important in bacteria? *J. Bacteriol.* **174:**2053–2058.

Wanner, B. L. 1993. Gene regulation by phosphate in enteric bacteria. *J. Cell. Biochem.* **51:**47–54.

Wanner, B. L., and J. A. Boline. 1990. Mapping and molecular cloning of the *phn* (*psiD*) locus for phosphonate utilization in *Escherichia coli. J. Bacteriol.* **172:**1186–1196.

Wanner, B. L., and R. McSharry. 1982. Phosphate-controlled gene expression in *Escherichia coli* using Mu*dl*-directed *lacZ* fusions. *J. Mol. Biol.* **158:**347–363.

Yamada, M., K. Makino, H. Shinagawa, and A. Nakata. 1990. Regulation of the phosphate regulon of *Escherichia coli:* properties of *phoR* deletion mutants and subcellular location of PhoR protein. *Mol. Gen. Genet.* **220:**366–372.

Zou, C., N. Fujita, K. Igarashi, and A. Ishihama. 1992. Mapping the cAMP receptor protein contact site on the α subunit of *Escherichia coli* RNA polymerase. *Mol. Microbiol.* **6:**2599–2605.

Chapter 3

Multiple Controls of the *Escherichia coli* Pho Regulon by the P_i Sensor PhoR, the Catabolite Regulatory Sensor CreC, and Acetyl Phosphate

BARRY L. WANNER

Department of Biological Sciences, Purdue University, West Lafayette, Indiana 47907

The *Escherichia coli* phosphate regulon (Pho regulon) is a paradigm for a signal transduction pathway (Wanner, 1993) in which a cell surface protein (s) regulates gene expression by phosphorylation and dephosphorylation of a regulatory protein (the PhoB protein). The PhoB protein is a DNA-binding protein and transcriptional activator that specifically recognizes Pho regulon promoters and activates transcription only when phosphorylated (Makino et al., 1989). The primary control of the Pho regulon involves detection of the extracellular P_i level by the phosphate-specific transport (Pst) system, a cell surface receptor complex for P_i uptake. The Pst system, together with a protein called PhoU, (somehow) regulates the activity of the PhoR protein (the P_i sensor), which in turn activates the PhoB protein by phosphorylation under conditions of P_i limitation and inactivates the PhoB protein (presumably by dephosphorylation) when P_i is in excess. In addition, the Pho regulon may be subject to cross regulation by two other cell-signaling pathways (Wanner, 1992). Input signals from controls involving carbon and energy metabolism lead to phosphorylation of the PhoB protein by a membrane protein called CreC (the catabolite sensor, formerly called PhoM) and by acetyl phosphate (a key intermediate in P_i, C, and energy metabolism), which may also be important as an effector of gene regulation. While the integration of multiple input signals may be especially important for the overall (global) control of cell growth, these controls by the CreC protein and acetyl phosphate are clearly evident only in the absence of the PhoR protein.

Studies on the Pho regulon emanated from the early observations of Horiuchi et al. (1959) and Torriani (1959, 1960) that the synthesis of bacterial alkaline phosphatase (Bap) was induced many hundred-fold by P_i limitation. Shortly thereafter, Echols and coworkers in C. Levinthal's laboratory laid the foundation for understanding this control of Bap synthesis by isolating regulatory mutants (Echols et al., 1961; Garen and Echols, 1962a; Adhya, 1993). Among other things, this early work provided the first evidence that a genetic control could be positive (Garen and Echols, 1962b), at a time when the operon model of Jacob and Monod was favored (which provided an explanation only for how a genetic control could be negative). These early studies also showed that the Pho regulon was subject to negative control. Subsequent studies showed that it was subject to a complex control mechanism(s) involving multiple positive (as well as negative) controls (Wanner and Latterell, 1980). Continued genetic studies have shed new light on these controls (Wanner, 1986a; Wanner et al., 1988b; Wanner and Wilmes-Riesenberg, 1992). Such work has recently provided the first evidence that the old and yet mysterious molecule acetyl phosphate (Lipmann, 1944) may be an effector of gene regulation (Wanner and Wilmes-Riesenberg, 1992; Wanner, 1992). My own view on how these multiple controls involving P_i limitation, the catabolite regulatory sensor CreC, and acetyl phosphate may regulate the Pho regulon is described in this chapter.

Phosphate Starvation-Inducible Genes and P Assimilation

It is now known that more than 30 phosphate starvation-inducible (*psi*) genes are coregulated with the *phoA* gene as members of the Pho regulon (Wanner, 1987a, 1990, 1993). All of these genes probably have a role in P assimilation because they are turned on when P_i (the preferred P source) is limited (Fig. 1). They are arranged in eight transcriptional units. Each gene or operon is preceded by a promoter containing an upstream activation site with a consensus "Pho box" sequence for activation by the PhoB protein (Makino et al., 1986). The *ugpBAECQ* operon is preceded by two promoters, including one *psi* promoter and another that is C regulated (Wanner and McSharry, 1982; Kasahara et al., 1991; Su et al., 1991). With the exception of the *psiE* gene, these

Gene/operon	Map	Description
phoA (*psiA*)-*psiF* (Orf-106)	8.7'	Bap, Unknown
phoBR	9.0'	Regulator, P_i sensor
phoE	5.8'	Polyanion porin
phoH(*psiH*)	23.6'	ATPase (unknown)
phnCDEFGHIJKLMNOP(*psiD*)	92.8'	Phosphonate utilization
pstSCAB-phoU	83.7'	P_i uptake and repression
ugpBAECQ(*psiB, C*)	75.7'	Uptake *sn*-Glycerol-3-Phosphate
psiE(Orf-136)	91.4'	Unknown
> 10 other *psi* promoters	Various	Unknown

FIG. 1. Sequenced *psi* genes of the *E. coli* Pho regulon. Rectangles preceding a *psi* gene represent occurrences of Pho boxes. As shown, the *phoE* gene, *pstSCAB-PhoU* operon, and the *ugpBAECQ* operon contain multiple Pho box sequences. Genes or operons corresponding to particular *psi*::*lacZ* (Mu d1) fusion mutant(s) are identified by the respective *psi* gene name in parentheses (Wanner et al., 1981; Wanner and McSharry, 1982). Recent studies on the 14-gene *phnC*-to-*phnP* operon (Metcalf and Wanner, 1993) are described in chapter 36 and elsewhere (Wanner, in press).

psi genes appear to be regulated similarly by the PhoB, PhoR, and CreC proteins and by acetyl phosphate. Despite the presence of a sequence preceding the *psiE* gene with weak similarity to the Pho box, earlier studies showed that the *psiE* gene appears to be regulated differently by the CreC protein (Wanner and McSharry, 1982; Wanner, 1983, 1986a). Also, it should be mentioned that studies on gene regulation by acetyl phosphate have concerned primarily the *phoA* gene (Wanner and Wilmes-Riesenberg, 1992; Wanner, 1992).

The Pho regulon is subject to multiple positive controls. Its primary control is by P_i limitation. However, two P_i-independent controls are probably important as well. To understand why the Pho regulon may be subject to multiple controls, it is important to recognize that the process of P assimilation involves two (or more) steps (Fig. 2). First, an environmental P source is taken up by a transport system, and then P_i (or a phosphoryl group) is incorporated into ATP (the primary phosphoryl donor in metabolism) via one of several central pathways of C, energy, and P_i metabolism. Eventually, P is incorporated into membrane lipids, complex carbohydrates, nucleic acids, and phosphoproteins. Therefore, it is to be expected that Pho regulon control may involve a regulatory coupling(s) to different steps in P_i metabolism. Accordingly, P_i control is coupled to a P_i transporter (the Pst system), and the P_i-independent controls may be forms of cross regulation that are coupled to subsequent steps in P_i metabolism (Wanner, 1992), which are also steps in C and energy metabolism.

P_i Control of the Pho Regulon and Transmembrane Signal Transduction

P_i control of the Pho regulon is a form of transmembrane signal transduction. It detects the extracellular P_i concentration and involves both negative control and positive control (activation). The PhoR protein has a role in both processes (as predicted on the basis of genetic studies). Hence the PhoR protein probably exists in two forms, repressor ($PhoR^R$) and activator ($PhoR^A$) forms (Wanner and Latterell, 1980; Wanner, 1993). However, the $PhoR^R$ or $PhoR^A$ protein does not act as a classical repressor or activator protein since the PhoR protein is not a DNA-binding protein. Instead, the PhoR protein acts by modifica-

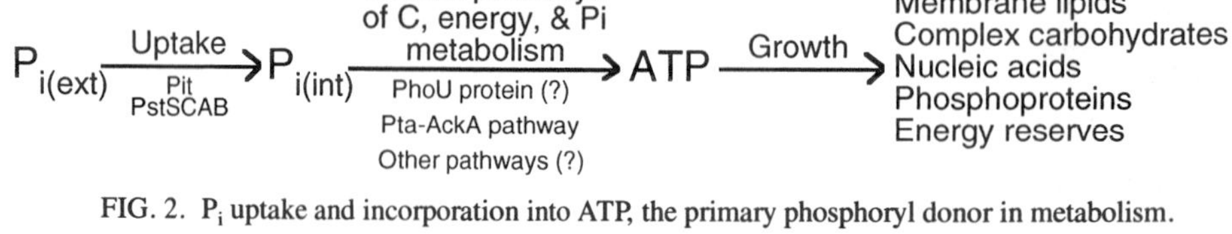

$P_{i(ext)}$ —Uptake (Pit, PstSCAB)→ $P_{i(int)}$ —Central pathways of C, energy, & Pi metabolism (PhoU protein (?), Pta-AckA pathway, Other pathways (?))→ ATP —Growth→ Membrane lipids, Complex carbohydrates, Nucleic acids, Phosphoproteins, Energy reserves

FIG. 2. P_i uptake and incorporation into ATP, the primary phosphoryl donor in metabolism.

tion of the activator protein PhoB. The formation of the $PhoR^R$ protein requires excess P_i, the PhoU protein, and an intact (although not necessarily functional) Pst system (composed of the PstA, PstB, PstC, and PstS proteins). Activation by P_i limitation requires the $PhoR^A$ protein and the PhoB protein (Wanner, 1990). Derepression due to P_i limitation occurs when the environmental P_i concentration falls below about 4 μM. No concomitant decrease in the intracellular P_i concentration occurs before derepression of the Pho regulon (Willsky et al., 1973). Also, even though repression is coupled to the Pst system, it is independent of transport per se. A missense change in the PstA protein abolishes transport without affecting P_i repression (Cox et al., 1988). Furthermore, recent measurements indicate that the intracellular P_i concentration (of 9 to 13 mM when extracellular P_i is in excess) remains high (about 7 mM) during derepression of the Pho regulon (Rao et al., 1993), thus providing corroborative evidence for control by extracellular P_i.

Evidence for activation by the PhoB protein was provided by the isolation of *phoB* mutants (Bracha and Yagil, 1973; Brickman and Beckwith, 1975), and this was later verified by in vitro studies (Inouye et al., 1977; Makino et al., 1988). An understanding of how the PhoR protein acted originated from the observation that the PhoB and PhoR proteins belong to families of partner proteins (called two-component regulatory systems) that share sequence similarities at the protein level with other members of the same family (Nixon et al., 1986; Ronson et al., 1987). One family includes the PhoB protein, and the other family includes the PhoR protein. The PhoB protein family contains highly conserved response regulatory or receiver domains; the PhoR protein family contains highly conserved sensory or transmitter domains (Kofoid and Parkinson, 1988; Stock et al., 1989). The finding of these sequence similarities was especially important because activation of the NtrC (NR_I) protein (a PhoB homolog) by the NtrB (NR_{II}) protein (a PhoR homolog) in nitrogen control was shown to involve phosphorylation of the NtrC protein by the NtrB protein (Ninfa and Magasanik, 1986; Keener and Kustu, 1988). Likewise, it was shown that the PhoB protein was phosphorylated by the PhoR protein and that phospho-PhoB was an even better transcriptional activator (Makino et al., 1989). Also, some PhoR homologs such as the EnvZ protein were shown to act as a phosphoprotein phosphatase of its cognate receiver protein (Igo et al., 1989).

A model for Pho regulon control by environment P_i is shown in Fig. 3. This model is fully consistent with earlier genetic studies. It is also supported by biochemical studies on the PhoB and PhoR proteins and their homologs. Accordingly, P_i repression would lead to formation of a "repressor complex" containing the PhoR protein, the PhoU protein, and all four proteins of the Pst transporter (the PstA, PstB, PstC, and PstS proteins). Importantly, this complex would form only when the PstS protein is saturated with P_i. Since the PstS protein has a K_d of 0.8 μM, the PstS protein would be nearly completely saturated above an external P_i concentration of 4 μM. Also, it is reasonable to suppose that the Pst transporter would undergo a conformational change when P_i is bound to the PstS protein. This conformational change would be transmitted via protein-protein interactions to the PhoU and PhoR proteins, which would lead to formation of the $PhoR^R$ protein. This would explain the requirement for an intact Pst system (without a requirement for actual P_i uptake) for P_i repression. Protein-protein interactions between the transporter and the PhoR protein or between the PhoU and PhoR proteins would be especially important for maintaining the PhoR protein in its repressor ($PhoR^R$) form. The $PhoR^R$ protein may inactivate the phospho-PhoB protein by dephosphorylation, which may involve an activity of the $PhoR^R$ protein (or an accessory protein such as the PhoU protein).

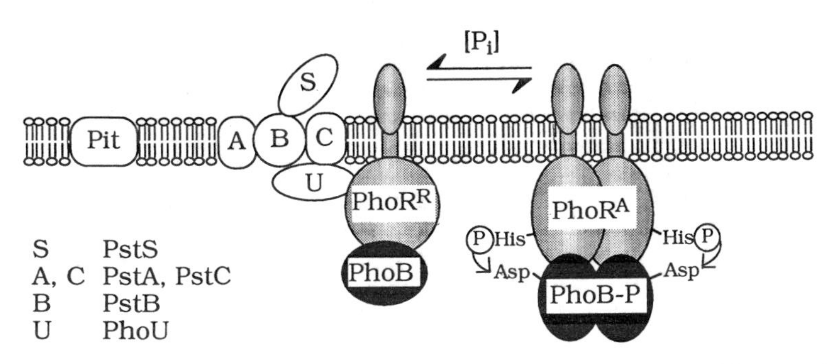

FIG. 3. Transmembrane signal transduction by environmental P_i. PhoB-P, phospho-PhoB protein. Like in nitrogen control (Magasanik, 1989), the $PhoR^A$ protein is believed to be autophosphorylated on a conserved histidine residue, and this phosphoryl group is believed to be transferred to a conserved aspartyl residue on the PhoB protein.

Under conditions of P_i limitation, the PstS protein would no longer be saturated with P_i, and the (proposed) concomitant conformational change in the Pst transporter would lead to release of the PhoR^R protein from the repressor complex. Since no additional factor is necessary for activation, the PhoR^A protein would form spontaneously upon its release. The PhoR^A protein would then be autophosphorylated on a histidinyl residue, and it would phosphorylate the PhoB protein by transfer of this phosphoryl group to an aspartyl residue on the PhoB protein. Upon binding to DNA, the phospho-PhoB protein may form a dimer (or higher oligomer) that would lead to transcriptional activation of Pho regulon *psi* promoters. Oligomerization has been shown to be involved in the mechanism of activation by phosphorylation-dependent PhoB homologs (Nakashima et al., 1991).

A central feature of this model involves the interconversion of the PhoR^R and PhoR^A proteins. Several lines of evidence suggest that this may involve an association and disassociation of monomers and dimers, as shown in Fig. 3. An allele-specific *phoR* mutation that is dominant leads to activation in the presence of a wild-type gene (Chang and Wanner, unpublished). Such dominant alleles are indicative of interactions between the gene products, which would occur if one form of the PhoR protein were multimeric. The mechanism of autophosphorylation of PhoR homologs involves transphosphorylation of one subunit by another subunit (Wolfe and Stewart, 1993; Swanson et al., 1993). This implies that autophosphorylation of the PhoR^A protein may require the formation of a dimer (or higher oligomer). Furthermore, a truncated form of the PhoR protein appears to be always activated (Makino et al., 1989). Presumably, its hydrophobic N terminus is necessary for tethering the PhoR protein to the membrane, which may be necessary for its association in a repressor complex and concomitant formation of the PhoR^R protein, which is required for dephosphorylation of the phospho-PhoB protein.

P_i-Independent Control of the Pho Regulon

Discovery of the catabolite regulatory sensor CreC. It is now clear that two P_i-independent controls act on the Pho regulon in the absence of the PhoR protein. Both are regulated by the C source (in different ways), lead to activation by the PhoB protein, and probably involve phosphorylation of the PhoB protein (Wanner et al., 1988b; Wanner, 1992). One P_i-independent control requires the CreC protein (a PhoR homolog), which has been shown to phosphorylate the PhoB protein (Amemura et al., 1990); and the other P_i-independent control requires acetyl phosphate, which may directly phosphorylate the PhoB protein (Wanner, 1992). Studies that led to the discovery of these P_i-independent controls are summarized below.

Evidence for activation by the CreC protein was provided by the isolation of mutants with lesions in the *creC* gene (formerly called *phoM*), which were fortuitously isolated in a *phoR* mutant (Wanner and Latterell, 1980). It had been known for a long time (Garen and Echols, 1962a) that defective (Def) *phoR* alleles abolished P_i repression and simultaneously led to an induced level of Bap synthesis. However, this level was substantially lower than the amount in *phoR*$^+$ cells during P_i limitation. This lowered level of synthesis was actually the basis for the original proposal that the PhoR protein had a role as an activator, as well as a role as a repressor.

Table 1 shows the effects of P_i control and P_i independent control on the Pho regulon. A wild-type strain shows more than 2,000-fold induction during P_i limitation (lines 1 and 2). A *phoR* (Def) mutant shows more than 400-fold induction (above the repressed level in a wild-type strain) during growth on glucose minimal media (line 3). This synthesis requires the CreC protein because it is abolished in a *phoR* (Def) *creC* (Def) mutant (line 4). Also, this synthesis is C source regulated (Wanner et al., 1988b; Wilmes-Riesenberg and Wanner, unpublished). This C source regulation by the CreC protein had been overlooked in earlier studies because much of that work had been done with a mutant that is defective in C source control by the CreC protein (Wanner, 1987b; Wanner et al., 1988a, 1988b). In addition, the Pho regulon is induced in a *phoR creC* mutant when grown on a high-energy medium (line 5). However, this additional control was initially seen only during growth on agar. It was not observed in broth cultures, except during the late stationary growth phase (Wanner, 1985, 1986b; Wilmes-Riesenberg and Wanner, unpublished). This control is also seen during logarithmic growth in broth cultures containing pyruvate as a C source (Wanner and Wilmes-Riesenberg, 1992).

Mutational studies on P_i-independent Pho regulon control. New Pho regulon regulatory mutants were isolated to study the P_i-independent Pho regulon control(s). Several approaches were taken. Mutants showing an altered Bap phenotype were isolated on indicator agar containing different C sources. Some showed increased Bap synthesis, while others showed decreased Bap synthesis. Mutants were isolated from both *phoR creC*$^+$ and *phoR creC* strains. Some were spontaneous mutants; some were chemical-induced mutants; and some were transposon-induced mutants. Altogether, hundreds of mutants were characterized in a variety of ways, and several dozen of these were identified. Some were identified on the basis of a mutant phenotype; some were identified

TABLE 1. P_i control and P_i-independent controls of the Pho regulon

Genotype[a]	Growth condition[b]	Phenotype (Sp act)[c]
Wild type	Glucose minimal + P_i	Repressed (0.1)
Wild type	Glucose minimal − P_i	Induced, high level (280)
phoR	Glucose minimal ± P_i	Induced, low level (42.7)
phoR creC	Glucose minimal ± P_i	Negative (0.2)
phoR creC	High-energy medium ± P_i	Induced, high level (>250)

[a]Nonsense or deletion *phoR* and *creC* mutations have similar effects.

[b]Glucose minimal is glucose-MOPS medium with excess P_i ($+P_i$) or limiting P_i ($-P_i$). High-energy medium is tryptone-yeast extract agar containing 1% glucose (Wanner, 1985, 1986b).

[c]Numbers in parentheses are Bap specific activities (Wanner and Wilmes-Riesenberg, 1992).

by mapping; and some were identified by cloning and DNA sequencing a segment of the mutated gene. In brief, 16 loci were found in which mutations dramatically affected P_i independent Pho regulon control. Several of these are listed in Table 2. These results clearly indicated that P_i independent Pho regulon control was coupled to central pathways of C and energy metabolism. In addition, numerous double mutants were constructed to test for epistasis between certain mutations. These results showed that mutations affected two separate P_i-independent controls, involving the CreC protein or acetyl phosphate (Wanner et al. unpublished; Wanner and Bernstein, 1982; Wanner et al., 1988b; Wanner and Wilmes-Riesenberg, 1992; Wilmes-Riesenberg and Wanner, unpublished).

Discovery of acetyl phosphate as an effector of gene regulation. Evidence for activation by acetyl phosphate was provided by the isolation of mutants altered in the *ackA* gene (for acetate kinase) that led to an accumulation of acetyl phosphate (Wanner and Wilmes-Riesenberg, 1992). These were isolated as Bap$^+$ revertants of a *phoR creC* mutant. Such Bap$^+$ revertants were first isolated as "suppressor" mutants nearly 15 years ago. Those resulted from experiments aimed toward determining whether the PhoR and CreC proteins had roles in activation of the Pho regulon. Because other results strongly indicated that the PhoR and CreC proteins had such roles, it was surprising to find that *phoR creC* mutants (with missense, nonsense, or deletion mutations) gave rise to frequent Bap$^+$ revertants (Wanner et al., unpublished data; Wanner and Bernstein, 1982). However, revertants with an *ackA* mutation were difficult to study because of their severe growth defect and their propensity to frequently yield Bap$^-$ compensatory mutants (Wanner and Wilmes-Riesenberg, 1992). As a consequence, it was quite some time before we had identified the *ackA* mutants. Later on, a transposon-induced *ackA* mutant was identified by cloning the transposon, sequencing the flanking DNA, and searching GenBank for the mutated gene. In 1989 the *ackA* gene sequence was reported (Matsuyama et al., 1989), and a match was found in a data base search. Mutants blocked in acetyl phosphate synthesis with mutations in the *pta* gene (for phosphotransacetylase) were among the Bap$^-$ compensatory mutants of the slow-growing *phoR creC ackA* mutants, and they were identified by their close linkage to the *ackA* gene and by enzyme assays (Wanner and Wilmes-Riesenberg, 1992).

Our studies on *ackA* and *pta* mutants showed that acetyl phosphate was necessary for activation of the Pho regulon in a *phoR creC* mutant. Also, this work provided the first evidence that acetyl phosphate may be an effector of gene regulation (Wanner and Wilmes-Riesenberg, 1992). Acetyl

TABLE 2. Genes involved in P_i-independent control of the Pho regulon[a]

Genes mutated[b]	Function	Mechanism
creA, creB, creC, creD	Catabolite regulatory (*creABCD*) operon; unknown	Activation by CreC sensor
ackA, pta	Acetate kinase, phosphotransacetylase; Pta-AckA pathway for ATP synthesis	Activation by acetyl phosphate
arcA, cya, crp, icd, mdh, ompR, ops, pstHI, pur	Aerobic respiratory control, cyclic AMP synthesis, catabolite activator, isocitrate dehydrogenase, malate dehydrogenase, osmoregulation, exopolysaccharide, production, phosphohistidinoprotein (HPr) or enzyme I, guanine biosynthesis	Activation by CreC sensor or by acetyl phosphate (depending on the mutation)

[a]Mutations lead to increased or decreased Bap synthesis in a *phoR* strain.

[b]Genes were identified by phenotypic testing and mapping or by sequencing DNA adjacent to an insertion site (for those resulting from transposon mutagenesis) and searching GenBank for the mutated gene.

phosphate is an intermediate of the phosphotransacetylase-acetate kinase (Pta-AckA) pathway. Acetyl phosphate and coenzyme A are made from acetyl coenzyme A and P_i by phosphotransacetylase, and acetyl phosphate and ADP are converted to ATP and acetate by acetate kinase (Fig. 4A). Importantly, mutations or growth conditions that lead to activation of the Pho regulon (Fig. 4B) are expected to lead to an accumulation of acetyl phosphate (Fig. 4C). Conversely, mutations or growth conditions that lead to reduced expression of the Pho regulon (Fig. 4B) are expected to lead to decreased acetyl phosphate (Fig. 4C). Since acetyl phosphate acts as a phosphoryl donor (for example, in the synthesis of ATP by acetate kinase), we proposed that acetyl phosphate might directly activate the Pho regulon by phosphorylation of the PhoB protein (Wanner and Wilmes-Riesenberg, 1992; Wanner, 1992). Following our discovery of acetyl phosphate as an effector of Pho regulon control, acetyl phosphate as well as other small phospho donor molecules were shown to phosphorylate PhoB homologs directly (Deretic et al., 1992; Feng et al., 1992; Holman et al., 1994; Lukat et al., 1992; McCleary et al., 1993; Roggiani and Dubnau, 1993). Whether these other systems are normally regulated by acetyl phosphate is unknown.

Although these effects due to acetyl phosphate are observed in *phoR creC*$^+$ and *phoR creC* mutants, there is circumstantial evidence that acetyl phosphate may have a role in Pho regulon control in a wild-type strain. Wilkins (1972) showed that pyrimidine and purine limitations can cause derepression of Bap synthesis in a *phoR*$^+$ strain when P_i was in excess. Although this led to the hypothesis that an unknown nucleotide was an effector of Pho regulon control (Morris et al., 1974; Wanner, 1987a), no such effector has been identified in spite of deliberate searches (Torriani, 1990). Instead, regulation by these limitations may be entirely due to acetyl phosphate. While some Bap$^+$ revertants of a *phoR creC* mutant carried mutations that affect acetyl phosphate synthesis directly (Wanner and Wilmes-Riesenberg, 1992), others carried mutations that probably affect acetyl phosphate synthesis indirectly. The latter included mutations

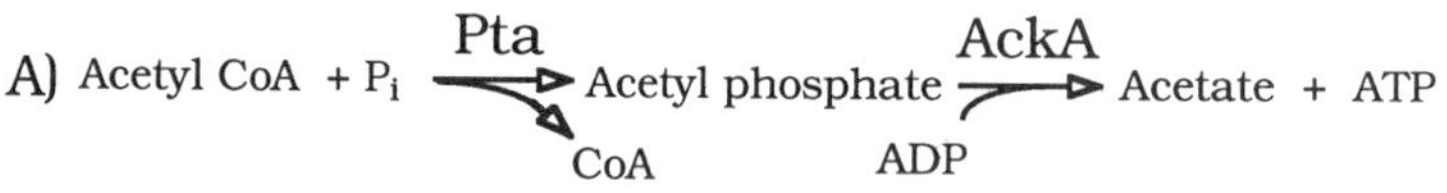

A) Acetyl CoA + P_i —Pta→ Acetyl phosphate (CoA) —AckA→ Acetate + ATP (ADP)

B) Mutational effects on induction of PHO regulon

Genotype	Carbon source		
	Glucose	Pyruvate	Acetate
Wild-type	0.2	104	0.6
AckA$^-$	422	409	1.4
Pta$^-$	0.2	0.2	43.7
AckA$^-$ Pta$^-$	0.2	0.2	0.3

C) Expected effects on level of acetyl phosphate

Genotype	Carbon source		
	Glucose	Pyruvate	Acetate
Wild-type	Low	High	Low
AckA$^-$	Increased	Increased	Decreased
Pta$^-$	Decreased	Decreased	Increased
AckA$^-$ Pta$^-$	Absent	Absent	Absent

FIG. 4. Effects of *pta* and *ackA* mutations on induction of the Pho regulon and expected effects on level of acetyl phosphate. (A) The Pta-AckA pathway. This pathway normally operates in the direction indicated, although both phosphotransacetylase (Pta) and acetate kinase (AckA) are reversible in vivo. (B) Mutational effects on induction of the Pho regulon. Numbers indicate the amount of Bap made in each strain when grown on each compound as the sole carbon source. A *pta* or *ackA* mutant grows on acetate because acetate is also metabolized via acetylcoenzyme A (CoA) synthetase. Data are from Wanner and Wilmes-Riesenberg (1992). (C) Expected effects on level of acetyl phosphate. The amount of Bap synthesis in *phoR creC* mutants parallels the expected level of acetyl phosphate under all conditions.

that led to a guanine auxotrophy (Table 2) (Wanner et al., unpublished).

P_i Control and Cross Regulation of the Pho Regulon

I have used the term cross regulation elsewhere (Wanner, 1992) to describe the basis for the P_i-independent controls of the Pho regulon by the CreC protein and acetyl phosphate. This term refers to the control of a response regulator (the PhoB protein) by a different regulatory system. Also, this term is meant to imply a control(s) of physiological importance. The hypothesis of cross regulation originated from our discovery that acetyl phosphate was an activator of the Pho regulon. It was particularly interesting that an intermediate in P_i metabolism (for incorporation of P_i into ATP) was involved in Pho regulon control. Therefore, I proposed that the control of the Pho regulon involving acetyl phosphate may detect the ratio of the ATP (or other nucleotide) concentration to the acetyl phosphate concentration, with a lowered ratio causing induction (Fig. 5). I suggested this because the control of a pathway is expected to involve the end product of a pathway. By extension, P_i independent control by the CreC protein may be coupled to a different central pathway of C and energy metabolism (for entry of P_i into ATP).

Accordingly, the Pho regulon appears to be subject to three controls, each of which leads to activation of the PhoB protein by phosphorylation (Fig. 5) (Wanner, 1992, 1993). P_i control involves the Pst transporter and the PhoU protein, detection of $P_{i(ext)}$, and activation (and repression) by the sensor PhoR. Control by an unknown central pathway involves detection of an unknown signal and activation by the sensor CreC. Control by the Pta-AckA pathway involves detection of acetyl phosphate (or more probably the ratio of the ATP to acetyl phosphate concentrations) and activation by acetyl phosphate (in a manner that may be inhibited by ATP). These controls are entirely consistent with our current understanding of Pho regulon control. Further, it seems likely that cross regulation of this sort may also be important in the control of other two-component regulatory systems, especially ones associated with key steps in central metabolism.

It should be pointed out that the concept of cross regulation is only a hypothesis. This is because these P_i-independent controls are clearly evident only in a *phoR* mutant. Nevertheless, there are several ways in which they may operate in a wild-type strain (Wanner, 1992). There may be a control(s) of the phospho-PhoB protein phosphatase activity of the $PhoR^R$ (or the PhoU) protein. There may be growth conditions under which the amount of the PhoR protein is so low that it does lead to phospho-PhoB protein dephosphorylation. In this regard, the *phoBR* operon is autogenously regulated (Guan et al., 1983; Wanner and Chang, 1987). When P_i is in excess, it is expressed at a low level as a result of P_i repression. It is induced several hundred-fold under conditions of P_i limitation. Alternatively, the regulation observed in *phoR* mutants may reflect normal controls by the CreC protein and acetyl phosphate on basal-level expression of the Pho regulon. Such controls may be especially important in wild-type strains undergoing shifts of C and energy sources.

Prospectus

Much has been learned about gene regulation by studying the Pho regulon over the past three decades. However, many new questions now await future investigations. How does the PhoR protein act as a P_i sensor? Does it detect P_i directly? Does it detect P_i indirectly via an association(s) with the Pst system? Via the PhoU protein? If protein-protein interactions are involved (as seems likely, though unproved), what are these interactions? Are there specific interactions between the PhoR protein and the Pst system, or between the PhoR and PhoU proteins? Also, recent genetic studies indicate a new role for the PhoU protein (Steed and Wanner, 1993). However, this new role for the PhoU protein is unknown. Is it related to P_i metabolism? If so, what reaction does the PhoU protein catalyze? What is the signal for activation by the CreC protein? Also, it is likely that the CreC protein has a role (together with the PhoB homolog CreB) in the

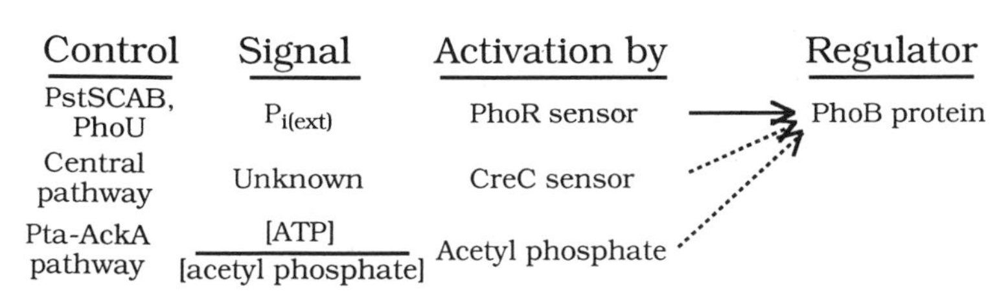

FIG. 5. P_i control and cross regulation of the PHO regulon by controls involving C, energy, and P_i metabolism. Reprinted with permission from Wanner (1992).

control of other (unknown) genes. What are these genes? Are they involved in central metabolism? What is the mechanism of activation by acetyl phosphate? Is activation by acetyl phosphate affected by ATP (or another nucleotide)? Are genes for acetyl phosphate synthesis regulated? If so, what role do those factors play in Pho regulon control and in P assimilation? How important is cross regulation in the Pho regulon and in cell biology in general? Probing for answers to these and other basic questions in new studies of the Pho regulon is likely to continue to entertain many new investigators in the future.

My laboratory is supported by NIH grant GM35392 and NSF grant DMB9108005.

I am indebted to many individuals who contributed to my studies of the Pho regulon over the past 18 years. I am especially grateful to Deepak Agrawal, Bey-Dih Chang, Jisong Cui, Weihong Jiang, Ki-Sung Lee, Bill Metcalf, Connie Schmellik-Sandage, Paul Steed, Mary Wilmes-Riesenberg, and Dennis Young, who (as Ph.D. or postdoctoral students) have contributed to studies in my laboratory since the 1986 phosphate meeting in Concarneau.

Addendum in Proof

Three additional loci that alter P_i-independent Pho regulon control correspond to the *rpiA* gene (encoding ribose phosphate isomerase; GenBank X73026), the *ppk* gene (encoding polyphosphate kinase; GenBank L03719), and an open reading frame of unknown function (GenBank Z19601) which shares sequence similarity to a ribulose 5′-phosphate epimerase gene from another organism (unpublished results). Each of these (like those loci given in Table 2) is connected to P_i, C, and energy metabolism.

REFERENCES

Adhya, S. 1993. Obituary Harrison Echols (1933–1993). *Cell* **73:**833–834.

Amemura, M., K. Makino, H. Shinagawa, and A. Nakata. 1990. Cross talk to the phosphate regulon of *Escherichia coli* by PhoM protein: PhoM is a histidine protein kinase and catalyzes phosphorylation of PhoB and PhoM open reading frame 2. *J. Bacteriol.* **172:**6300–6307.

Bracha, M., and E. Yagil. 1973. A new type of alkaline phosphatase-negative mutants in *Escherichia coli* K12. *Mol. Gen. Genet.* **122:**53–60.

Brickman, E., and J. Beckwith. 1975. Analysis of the regulation of *Escherichia coli* alkaline phosphatase synthesis using deletions and f80 transducing phages. *J. Mol. Biol.* **96:**307–316.

Chang, B.-D., and B. L. Wanner. Unpublished data.

Cox, G. B., D. Webb, J. Godovac-Zimmermann, and H. Rosenberg. 1988. Arg-220 of the PstA protein is required for phosphate transport through the phosphate-specific transport system in *Escherichia coli* but not for alkaline phosphatase repression. *J. Bacteriol.* **170:**2283–2286.

Deretic, V., J. H. J. Leveau, C. D. Mohr, and N. S. Hibler. 1992. *In vitro* phosphorylation of AlgR, a regulator of mucoidy in *Pseudomonas aeruginosa,* by a histidine protein kinase and effects of small phospho-donor molecules. *Mol. Microbiol.* **6:**2761–2767.

Echols, H., A. Garen, S. Garen, and A. Torriani. 1961. Genetic control of repression of alkaline phosphatase in *E. coli. J. Mol. Biol.* **3:**425–438.

Feng, J., M. R. Atkinson, W. McCleary, J. B. Stock, B. L. Wanner, and A. J. Ninfa. 1992. Role of phosphorylated metabolic intermediates in the regulation of glutamine synthetase synthesis in *Escherichia coli. J. Bacteriol.* **174:**6061–6070.

Garen, A., and H. Echols. 1962a. Properties of two regulating genes for alkaline phosphatase. *J. Bacteriol.* **83:**297–300.

Garen, A., and H. Echols. 1962b. Genetic control of induction of alkaline phosphatase synthesis in *E. coli. Proc. Natl. Acad. Sci. USA* **48:**1398–1402.

Guan, C.-D., B. Wanner, and H. Inouye. 1983. Analysis of regulation of *phoB* expression using a *phoB-cat* fusion. *J. Bacteriol.* **156:**710–717.

Holman, T. R., Z. Wu, B. L. Wanner, and C. T. Walsh. 1994. Identification of the DNA-binding site for the phosphorylated VanR protein required for vancomycin resistance in *Enterococcus faecium. Biochemistry* **33:**4625–4631.

Horiuchi, T., S. Horiuchi, and D. Mizuno. 1959. A possible negative feedback phenomenon controlling formation of alkaline phosphomonoesterase in *Escherichia coli. Nature* (London) **183:**1529–1530.

Igo, M. M., A. J. Ninfa, J. B. Stock, and T. J. Silhavy. 1989. Phosphorylation and dephosphorylation of a bacterial transcriptional activator by a transmembrane receptor. *Genes Dev.* **3:**1725–1734.

Inouye, H., C. Pratt, J. Beckwith, and A. Torriani. 1977. Alkaline phosphatase synthesis in a cell-free system using DNA and RNA templates. *J. Mol. Biol.* **110:**75–87.

Kasahara, M., K. Makino, M. Amemura, A. Nakata, and H. Shinagawa. 1991. Dual regulation of the *ugp* operon by phosphate and carbon starvation at two interspaced promoters. *J. Bacteriol.* **173:**549–558.

Keener, J., and S. Kustu. 1988. Protein kinase and phosphoprotein phosphatase activities of nitrogen regulatory proteins NTRB and NTRC of enteric bacteria: roles of the conserved amino-terminal domain of NTRC. *Proc. Natl. Acad. Sci. USA* **85:**4976–4980.

Kofoid, E. C., and J. S. Parkinson. 1988. Transmitter and receiver modules in bacterial signaling proteins. *Proc. Natl. Acad. Sci. USA* **85:**4981–4985.

Lipmann, F. 1944. Enzymatic synthesis of acetyl phosphate. *J. Biol. Chem.* **155:**55–70.

Lukat, G. S., W. R. McCleary, A. M. Stock, and J. B. Stock. 1992. Phosphorylation of bacterial response regulator proteins by low molecular weight phospho-donors. *Proc. Natl. Acad. Sci. USA* **89:**718–722.

Magasanik, B. 1989. Regulation of transcription of the *glnALG* operon of *Escherichia coli* by protein phosphorylation. *Biochimie* **71:**1005–1012.

Makino, K., H. Shinagawa, M. Amemura, T. Kawamoto, M. Yamada, and A. Nakata. 1989. Signal transduction in the phosphate regulon of *Escherichia coli* involves phosphotransfer between PhoR and PhoB proteins. *J. Mol. Biol.* **210:**551–559.

Makino, K., H. Shinagawa, M. Amemura, S. Kimura, A. Nakata, and A. Ishihama. 1988. Regulation of the phosphate regulon of Escherichia coli: activation of pstS transcription by PhoB protein in vitro. *J. Mol. Biol.* **203:**85–95.

Makino, K., H. Shinagawa, M. Amemura, and A. Nakata. 1986. Nucleotide sequence of the phoB gene, the positive regulatory gene for the phosphate regulon of Escherichia coli K-12. *J. Mol. Biol.* **190:**37–44.

Matsuyama, A., H. Yamamoto, and E. Nakano. 1989. Cloning, expression, and nucleotide sequence of the *Escherichia coli* K-12 *ackA* gene. *J. Bacteriol.* **171:**577–580.

McCleary, W. R., J. B. Stock, and A. J. Ninfa. 1993. Is acetyl phosphate a global signal in *Escherichia coli? J. Bacteriol.* **175:**2793–2798.

Metcalf, W. W., and B. L. Wanner. 1993. Mutational analysis of an *Escherichia coli* fourteen-gene operon for phosphonate degradation using Tn*phoA′* elements. *J. Bacteriol.* **175:**3430–3442.

Morris, H., M. J. Schlesinger, M. Bracha, and E. Yagil. 1974. Pleiotropic effects of mutations involved in the regulation of *Escherichia coli* K-12 alkaline phosphatase. *J. Bacteriol.* **119:**583–592.

Nakashima, K., K. Kanamaru, H. Aiba, and T. Mizuno. 1991. Signal transduction and osmoregulation in *Escherichia coli.* A novel type of mutation in the phosphorylation domain of the activator protein, OmpR, results in a defect in its phosphorylation-dependent DNA binding. *J. Biol. Chem.* **266:**10775–10780.

Ninfa, A. J., and B. Magasanik. 1986. Covalent modification of the *glnG* product, NRI, by the *glnL* product, NRII, regulates the transcription of the *glnALG* operon in *Escherichia coli. Proc. Natl. Acad. Sci. USA* **83:**5909–5913.

Nixon, B. T., C. W. Ronson, and F. M. Ausubel. 1986. Two-component regulatory systems responsive to environmental stimuli share strongly conserved domains with the nitrogen assimilation regulatory genes *ntrB* and *ntrC. Proc. Natl. Acad. Sci. USA* **83:**7850–7854.

Rao, N. N., M. F. Roberts, A. Torriani, and J. Yashphe. 1993. Effect of *glpT* and *glpD* mutations on expression of the *phoA* gene in *Escherichia coli. J. Bacteriol.* **175:**74–79.

Roggiani, M., and D. Dubnau. 1993. ComA, a phosphorylated response regulator protein of *Bacillus subtilis,* binds to the promoter region of *srfA. J. Bacteriol.* **175:**3182–3187.

Ronson, C. W., B. T. Nixon, and F. M. Ausubel. 1987. Conserved domains in bacterial regulatory proteins that respond to environmental stimuli. *Cell* **49:**579–581.

Steed, P. M., and B. L. Wanner. 1993. Use of the *rep* technique for allele replacement to construct *pstSCAB-phoU* operon mutants: evidence for a new role for the PhoU protein in the PHO regulon. *J. Bacteriol.* **175:**6797–6809.

Stock, J. B., A. J. Ninfa, and A. M. Stock. 1989. Protein phosphorylation and regulation of adaptive responses in bacteria. *Microbiol. Rev.* **53:**450–490.

Su, T.-Z., H. P. Schweizer, and D. L. Oxender. 1991. Carbon-starvation induction of the *ugp* operon, encoding the binding protein-dependent *sn*-glycerol-3-phosphate transport system in *Escherichia coli. Mol. Gen. Genet.* **230:**28–32.

Swanson, R. V., R. B. Bourret, and M. I. Simon. 1993. Intermolecular complementation of the kinase activity of CheA. *Mol. Microbiol.* **8:**435–441.

Torriani, A. 1959. Effect of inorganic phosphate (P_i) on formation of phosphatases by *E. coli. Fed. Proc.* **18:**339.

Torriani, A. 1960. Influence of inorganic phosphate in the formation of phosphatases by *Escherichia coli. Biochim. Biophys. Acta* **38:**460–469.

Torriani, A. 1990. From cell membrane to nucleotides: the phosphate regulon in *Escherichia coli. Bioessays* **12:**371–376.

Wanner, B. L. 1983. Overlapping and separate controls on the phosphate regulon in *Escherichia coli* K-12. *J. Mol. Biol.* **166:**283–308.

Wanner, B. L. 1985. Phase mutants: evidence of a physiologically regulated "change-in-state" gene system in *Escherichia coli. UCLA Symp. Mol. Cell. Biol. New Ser.* **20:**103–122.

Wanner, B. L. 1986a. Novel regulatory mutants of the phosphate regulon in *Escherichia coli* K-12. *J. Mol. Biol.* **191:**39–58.

Wanner, B. L. 1986b. Bacterial alkaline phosphatase clonal variation in some *Escherichia coli* K-12 *phoR* mutant strains. *J. Bacteriol.* **168:**1366–1371.

Wanner, B. L. 1987a. Phosphate regulation of gene expression in *Escherichia coli,* p. 1326–1333. *In* F. C. Neidhardt, J. L. Ingraham, K. B. Low, B. Magasanik, M. Schaechter, and H. E. Umbarger (ed.), *Escherichia coli and Salmonella typhimurium: Cellular and Molecular Biology,* vol. 2. American Society for Microbiology, Washington, D.C.

Wanner, B. L. 1987b. Control of *phoR*-dependent bacterial alkaline phosphatase clonal variation by the *phoM* region. *J. Bacteriol.* **169:**900–903.

Wanner, B. L. 1990. Phosphorus assimilation and its control of gene expression in *Escherichia coli,* p. 152–163. *In* G. Hauska and R. Thauer (ed.), *The Molecular Basis of Bacterial Metabolism.* Springer-Verlag KG, Heidelberg, Germany.

Wanner, B. L. 1992. Is cross regulation by phosphorylation of two-component response regulator proteins important in bacteria? *J. Bacteriol.* **174:**2053–2058.

Wanner, B. L. 1993. Gene regulation by phosphate in enteric bacteria. *J. Cell. Biochem.* **51:**47–54.

Wanner, B. L. An overview, molecular genetics of carbon-phosphorus bond cleavage in bacteria. *Biodegradation,* in press.

Wanner, B. L., and J. Bernstein. 1982. Determining the *phoM* map location in *Escherichia coli* K-12 by using a nearby transposon Tn*10* insertion. *J. Bacteriol.* **150:**429–432.

Wanner, B. L., J. Bernstein, and M. Lyster. Unpublished data.

Wanner, B. L., and B.-D. Chang. 1987. The *phoBR* operon in *Escherichia coli* K-12. *J. Bacteriol.* **169:**5569–5574.

Wanner, B. L., and B.-D. Chang. Unpublished data.

Wanner, B. L., and P. Latterell. 1980. Mutants affected in alkaline phosphatase expression: evidence for multiple positive regulators of the phosphate regulon in *Escherichia coli. Genetics* **96:**242–266.

Wanner, B. L., and R. McSharry. 1982. Phosphate-controlled gene expression in *Escherichia coli* using Mu*d1*-directed *lacZ* fusions. *J. Mol. Biol.* **158:**347–363.

Wanner, B. L., S. Wieder, and R. McSharry. 1981. Use of bacteriophage transposon Mu *d*1 to determine the orientation for three *proC*-linked phosphate-starvation-inducible (*psi*) genes in *Escherichia coli* K-12. *J. Bacteriol.* **146:**93–101.

Wanner, B. L., M. R. Wilmes, and E. Hunter. 1988a. Molecular cloning of the wild-type *phoM* operon in *Escherichia coli* K-12. *J. Bacteriol.* **170:**279–288.

Wanner, B. L., M. R. Wilmes, and D. C. Young. 1988b. Control of bacterial alkaline phosphatase synthesis and variation in an *Escherichia coli* K-12 *phoR* mutant by adenyl cyclase, the cyclic AMP receptor protein, and the *phoM* operon. *J. Bacteriol.* **170:**1092–1102.

Wanner, B. L., and M. R. Wilmes-Riesenberg. 1992. Involvement of phosphotransacetylase, acetate kinase, and acetyl phosphate synthesis in the control of the phosphate regulon in *Escherichia coli. J. Bacteriol.* **174:**2124–2130.

Wilkins, A. S. 1972. Physiological factors in the regulation of alkaline phosphatase synthesis in *Escherichia coli. J. Bacteriol.* **110:**616–623.

Willsky, G. R., R. L. Bennett, and M. H. Malamy. 1973. Inorganic phosphate transport in *Escherichia coli:* involvement of two genes which play a role in alkaline phosphatase regulation. *J. Bacteriol.* **113:**529–539.

Wilmes-Riesenberg, M. R., and B. L. Wanner. Unpublished data.

Wolfe, A. J., and R. C. Stewart. 1993. The short form of the CheA protein restores kinase activity and chemotactic ability to kinase-deficient mutants. *Proc. Natl. Acad. Sci. USA* **90:**1518–1522.

Chapter 4

Phosphate, Phosphorylated Metabolites, and the Pho Regulon of *Escherichia coli*

NARAYANA N. RAO,[1] ANITA KAR,[2] MARY F. ROBERTS,[3] JACOB YASHPHE,[4] AND ANNAMARIA TORRIANI-GORINI[5]

Department of Biochemistry, Stanford University School of Medicine, Stanford, California 94305[1]; Department of Zoology, University of Pune, Pune 411107, India[2]; Department of Chemistry, Boston College, Chestnut Hill, Massachusetts, 02167[3]; Department of Bacteriology, Hadassah Medical School, The Hebrew University, Jerusalem, Israel[4]; and Department of Biology, Massachusetts Institute of Technology, Cambridge, Massachusetts 02139

In *Escherichia coli* a number of genes and operons are involved in the uptake and the assimilation of P_i and phosphorylated compounds (Rao and Torriani, 1990; Torriani, 1990; Torriani and Ludtke, 1985; Wanner, 1987). The genes of the phosphate (Pho) regulon induced during P_i-limited growth are regulated by a two-component regulatory system consisting of PhoB, the positive response regulator protein; and PhoR, the sensor kinase (Makino et al., 1989; Stock et al., 1989, 1990; Wanner, 1992). An intact phosphate-specific transport (Pst) system is required for the repression of the Pho regulon when the cells are grown in a P_i-rich medium. Therefore, it is likely that the Pst system is involved in signal transduction across the inner membrane, indicating the level of P_i in the medium. The experiments described here suggest that P_i triggers the repression of the regulon from outside the cytoplasmic membrane of the *E. coli* cells. We also investigated the phenomenon of escape synthesis of alkaline phosphatase (AP) during growth. A specific nucleotide (dUTP) was observed to be involved in an escape synthesis of AP during growth of *E. coli* in excess phosphate. Another subject of investigation was the appearance of an adenylylated nucleotide, at the onset of P_i limitation, which may act as an alarmone—a signal molecule which alerts the cell to the onset of a metabolic stress.

Is P_i the Trigger for Repression of the Pho Regulon?

The Pst system comprises five distinct proteins encoded by *pstS, pstA, pstB, pstC,* and *phoU* (Amemura et al., 1985; Rosenberg, 1987). Many mutations in the Pst-encoding genes resulted in the loss of P_i transport and in constitutive synthesis of AP, which belongs to the Pho regulon (Rao and Torriani, 1990; Torriani, 1990; Torriani and Ludtke, 1985; Wanner, 1987). It was therefore assumed that the Pst-mediated transport of P_i was required for repression of the synthesis of the Pho regulon proteins. However, mutations in *pstA* and *pstC* created by site-directed mutagenesis resulted in the loss of P_i uptake but did not affect the repression of AP synthesis (Cox et al., 1988, 1989). In these mutants, P_i can exert repression either from inside, after entering the cell through the alternative and constitutive Pit (phosphate inorganic transport) system, or from outside, by triggering a transmembrane signal. To address this question, phosphate was furnished to the cells in the form of *sn*-glycerol-3-phosphate (*sn*-G3P) and its effect on the synthesis of AP was examined. The *sn*-G3P may enter the cell by either of the following two specific transport systems (Fig. 1): (i) GlpT, which is induced by *sn*-G3P and is geared to the utilization of *sn*-G3P as a carbon source (Larson et al., 1982; Lin, 1976), or (ii) Ugp, which belongs to the Pho regulon and is geared to the utilization of *sn*-G3P as a P_i source (Brzoska et al., 1994; Schwiezer et al., 1982). Under certain conditions, GlpT is also the only route for the exit of P_i from the cell by exchange with external *sn*-G3P (Elvin et al., 1985; Rosenberg, 1987). The mutational inactivation of *glpT* was used to study the effects of intracellular and extracellular P_i (Fig. 1) on the expression of the Pho regulon.

One of the difficulties in the study of AP synthesis with *sn*-G3P as the sole phosphate source is the hydrolysis of this compound in the periplasm of *E. coli* cells by AP to glycerol and P_i. The latter, when present in growth medium at sufficient concentrations ($>$0.1 mM), is known to repress the synthesis of the Pho regulon proteins including AP (Torriani and Ludtke, 1985; Wanner, 1987). To avoid this interference, we used an *E. coli* strain (MPh-45) which carries a chromosomal *phoA-lacZ* fusion (Rao et al., 1993; Sarthy et al., 1981) and whose *phoA*$^+$ chromosomal function was eliminated by ethyl methanesulfonate mutagenesis (strain JY-16).

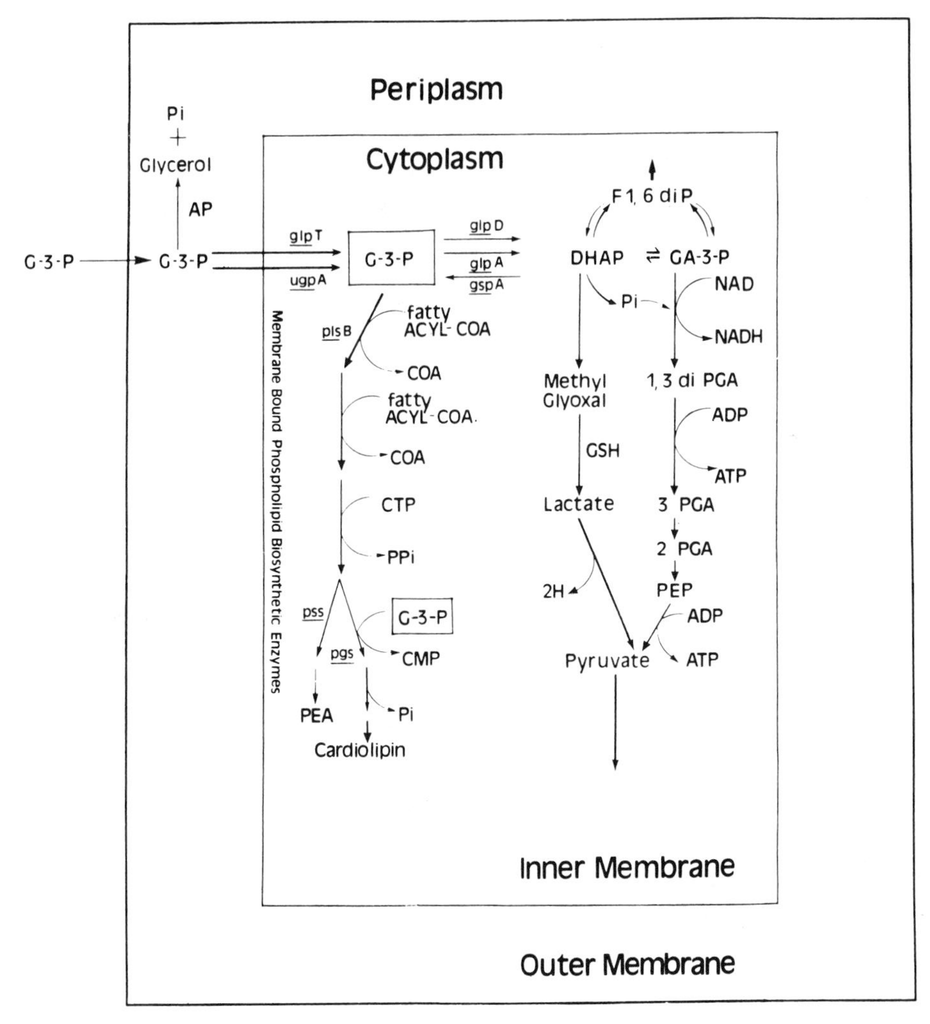

FIG. 1. Biosynthesis of glycerophospholipids in *E. coli.*

The expression of the *phoA-lacZ* fusion as β-galactosidase in cells of strain JY-16 was under P_i control and was also repressed when *sn*-G3P was supplied as the sole phosphate source (Fig. 2). This repression could be due to either *sn*-G3P itself or another compound, including P_i, derived from it metabolically (Fig. 1). The P_i produced can affect the activity of the *phoA* promoter either directly from inside the cell or indirectly from outside after its efflux. To test these possibilities, a *glpT* deletion was introduced into the strain JY-16. It has been shown that the *glpT* permease exchanged cytoplasmic P_i with the *sn*-G3P present in the medium (Ambudkar et al., 1986; Elvin et al., 1985). The *glpT* permease is the only exit route for P_i under the growth conditions used in the experiments. The alternate exit route, which involves the hexose 6-phosphate transport system, is absent because of the lack of its inducer, glucose 6-phosphate, in the growth medium (Kadner et al., 1987). The high differential rate of β-galactosidase synthesis ($\Delta U/\Delta OD_{540}$ [optical density at 540 nm] unit $\approx$ 5,000) found in the $\Delta glpT$ strain (JY-18) (Fig. 3A) grown on *sn*-G3P could be attributed to the absence of external P_i as a result of the inability of the cells to exchange cytoplasmic P_i with the external *sn*-G3P. Thus, the repression observed in the $glpT^+$ strain (JY-16) grown on *sn*-G3P was triggered by external P_i (Fig. 2).

Since it has been demonstrated (Larson et al., 1982; Lin, 1976) that the Glp system is geared to the utilization of *sn*-G3P as a carbon source through the glycolytic pathway (Fig. 1), the possibility that a *glpT* mutation prevents the accumulation of a negative effector in this pathway was examined. A *glpD* mutation, which inactivated

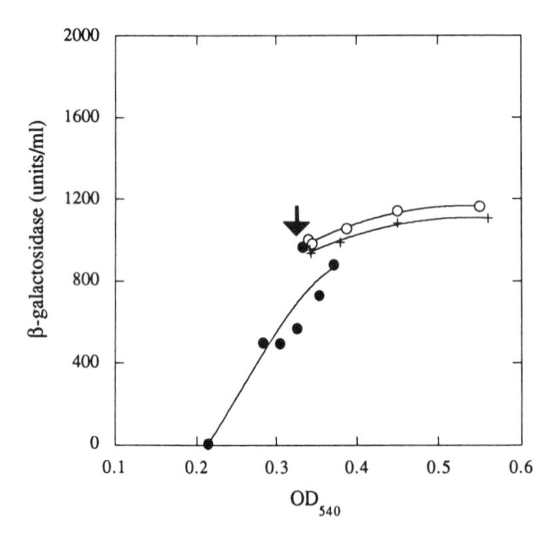

FIG. 2. Expression of *phoA* promoter (β-galactosidase synthesis) in *E. coli* JY-16 (*phoA-lacZ*). The differential rate of enzyme synthesis is expressed as increase in activity (in units per milliliter) versus increase in bacterial mass (OD_{540}). ●, 0.05 mM P_i; ↓, addition of 2 mM P_i (○) or 2 mM *sn*-G3P (+).

the aerobic *sn*-G3P dehydrogenase (Fig. 1), was introduced into strain JY-16 (yielding strain JY-24). This mutation resulted in a fully repressed level of β-galactosidase, eliminating the abovementioned possibility. Further, the introduction of the *glpD* mutation to strain JY-18 Δ*glpT* (yielding strain JY-37) (Fig. 3B) did not significantly affect the high induction observed in the Δ*glpT* parent strain (Fig. 3A). This observation suggests that a positive effector may be produced in the glycerophospholipid biosynthetic pathway rather than in the glycolytic pathway (Fig. 1).

To verify that internal P_i is not involved in the regulation of the *phoA* promoter, its intracellular concentrations in *glpT*$^+$ and *glpT* mutant cells were estimated by ^{31}P nuclear magnetic resonance spectroscopy (202.3 MHz) (Rao et al., 1993). Regardless of the type of cells examined (*glpT*$^+$ or *glpT*), there was always a high level of P_i (9 to 13 mM) in oxygenated cell suspensions (Table 1). The level of intracellular P_i was not significantly different when the cells were grown in excess P_i or in excess *sn*-G3P (Table 1). Hence, although wild-type *E. coli* cells grown on a medium containing more than 0.1 mM P_i are repressed for AP synthesis, the more than 8 mM intracellular P_i in the *glpT* mutant (JY-18) could not repress AP synthesis. The concentration of phosphomonoesters, which were predominantly *sn*-G3P, did not vary significantly between the strains (Table 1).

Nucleotides and the Escape Synthesis of Alkaline Phosphatase

Work from the J. Gallant laboratory, reported over 20 years ago, indicated that disturbances in nucleotide metabolism resulted in partial derepression of AP in media containing excess P_i (Wilkins, 1972). The escape synthesis was small (about 1 to 5% of the induced wild-type levels), and it was suggested to be due to an accumulation of an adenine derivative. The synthesis was interpreted as a possible effect on the activation of PhoR.

Recent results (Lee et al., 1990) suggested that the PhoB protein could be phosphorylated by an

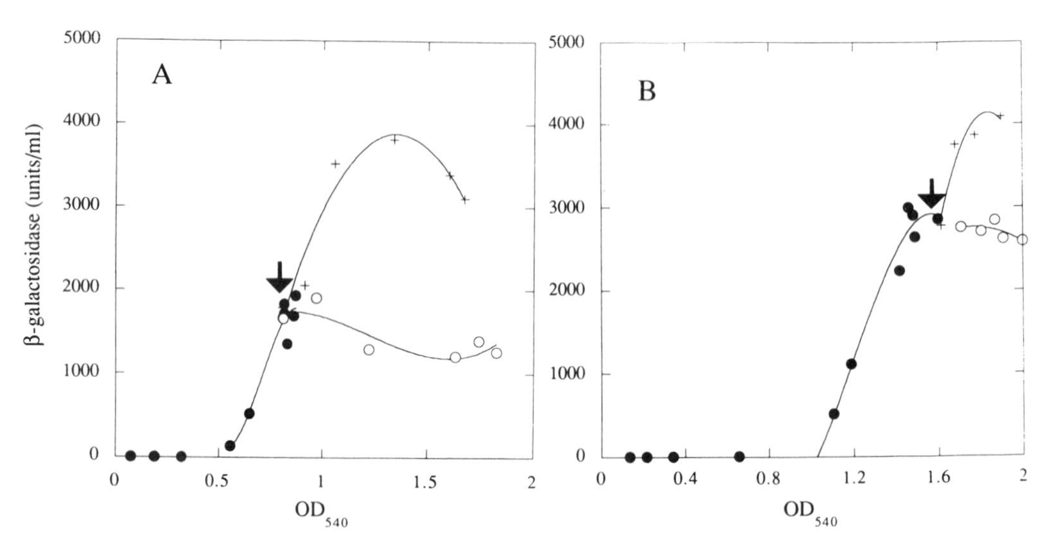

FIG. 3. Expression of *phoA* promoter (β-galactosidase synthesis) in *E. coli* Δ*glpT* JY-18 (A) and Δ*glpT glpD* JY-37 (B) (see the legend to Fig. 2). ●, 0.1 mM P_i; ↓, addition of 2 mM P_i (○) or 2 mM *sn*-G3P (+).

TABLE 1. ^{31}P analysis of intracellular P_i, phosphomonoesters, and ATP of intact *E. coli* cells

Strain	P_i source (2 mM)	Intracellular concn[a] of:		
		P_i[b]	PME[c]	ATP
JY-16 (*phoA-lacZ glpT*+)	P_i	2.7 (11.5)	1.8	3.2
	sn-G3P	2.4 (10.3)	2.8	2.7
JY-18 (*phoA-lacZ* Δ*glpT*)	P_i	3.0 (12.7)	2.0	2.9
	sn-G3P	2.1 (8.9)	2.1	3.5

[a]Concentration (micromoles per 100 mg [dry weight] of cells) was determined from fractional integrated intensity in each resonance of interest and total phosphorus content (errors estimated from integrations, ±25%). Glucose was the carbon source.
[b]Millimolar concentrations in parentheses.
[c]PME (phosphomonoesters) were assumed to be primarily *sn*-G3P.

excess of ATP in the absence of PhoR, as a result of overproduction of an acetate kinase (AckA). This provoked an escape synthesis of AP (ca. 6%) in a *phoR phoM* mutant in which PhoB cannot be phosphorylated by either PhoR kinase or PhoM kinase. These observations implicate nucleotides as playing a role in the regulation of *phoA* expression. Earlier experiments with permeabilized cells clearly demonstrated a variation in the nucleotide pools of P_i-rich and P_i-starved cells (Rao et al., 1986). Two specific nucleotides (S2 and S15) were present either at P_i starvation of the wild-type or at any point of growth in *phoU35*, a mutation of the *pst* operon. These nucleotides, when added to a permeabilized cell system, induced P_i-independent synthesis of AP (Rao et al., 1986; Torriani, 1990) (Table 2). In an attempt to identify a nucleotide which can reproduce the results obtained with S2 and S15, more than 40 nucleotides were tested; only dUTP showed an

TABLE 2. Synthesis of alkaline phosphatase and β-galactosidase in permeabilized cells of *E. coli* K10

Cells or nucleotide	Enzymatic activity[a]	
	Phosphatase	β-galactosidase
Cells		
Viable cells + IPTG[b] + IVM[c]	0.16	733
Permeabilized cells + IVM	0.021	1.93
Permeabilized cells + IPTG + IVM	0.021	56
Nucleotides[d]		
dUTP		
100 μM	0.127	2.93
50 μM	0.109	
25 μM	0.056	
dUDP (100 μM)	0.010	
dUMP (100 μM)	<0.021	
dU (100 μM)	<0.021	
dCTP (100 μM)	<0.021	
dTTP (100 μM)	<0.021	
Nucleotides extracted from *E. coli*		
HPLC fraction 19[e]	0.987	2.93
TLC spots S1, S3, S4, S5	<0.021	
TLC spot S2	1.779	
HPLC fraction 20	1.501	2.93
TLC spot S15	1.979	
HPLC fraction 12	<0.021	

[a]The enzymatic activities are expressed as units per 10^9 cells; 1 U = 1 nmol of substrate hydrolyzed per min at 37°C. The AP activity of the viable K10 cells repressed overnight and resuspended in IVM (i.e., the first entry in the table) has been subtracted from the other values as a background.
[b]IPTG (isopropyl-β-D-thiogalactopyranoside) is a β-galactosidase inducer added at 1.0 mM.
[c]IVM, in vitro mix of Zubay (1973) in which the permeabilized cells are suspended.
[d]This is a partial list of the nucleotides tested. All others gave a negative response, <0.021.
[e]See the legend to Fig. 4.

inducing effect of AP synthesis (Table 2). The observation that the addition dUTP provokes an increased amount of AP was particularly useful because the commercial availability of dUTP allows one to make samples large enough to be checked by immunoblot analysis. In this kind of experiment, the proteins have been labeled with ^{35}S. The results of the immunoblot analysis show that only the phosphatase synthesized in the presence of dUTP and ^{35}S was labeled, whereas the background observed in the absence of dUTP had no ^{35}S-labeled phosphatase. Therefore, this synthesis by the permeabilized cells was de novo above a background from the repressed K10 cells (Table 2). The nucleotide S2 may have some analogies with dUTP since S2 comigrates with dUTP in two-dimensional thin-layer chromatography (TLC) (Fig. 4).

A strain of *E. coli* K10 deficient in deoxyuridine triphosphatase (dUTPase) activity (*dut-1*) (Hochhauser and Weiss, 1978) presented an escape synthesis of AP during the exponential phase of growth (Fig. 5). The differential rate of enzyme synthesis was about sevenfold higher in the *dut-1* mutant (strain BW 285) than in the wild-type cells (strain BW 35) (Table 3). Complementation with a *dut*$^+$ plasmid of the *dut-1* mutant (strain BW 333) abolished the escape synthesis. The AP induced in the P_i-starved *dut-1* mutant (strain BW 285) was about 73% of the activity seen in the corresponding wild-type cells.

The effects of the *dut-1* mutation were similar to those of thymine starvation obtained in a thymine auxotroph (Hochhauser and Weiss, 1978; Taylor et al., 1980). There was an eightfold increase in the differential rate of AP synthesis in the thymine-starved, P_i-repressed B3 (*thyA14*) cells (Table 3). The thymine-supplemented control culture was repressed and did not show any increase in AP synthesis until the cells were starved for phosphate.

P_i Starvation-Induced Alarmones

Insufficient supply of nutrients causes the occurrence of unique phosphorylated metabolites in *E. coli* and *Salmonella typhimurium* (Bochner, 1987; Stephens et al., 1975). These stress-related compounds are termed alarmones; the best known alarmone is guanosine 5′-diphosphate 3′-diphosphate (ppGpp) and cyclic AMP. The nucleotide ppGpp is a positive effector for histidine operon transcription and is a general signal for amino acid deficiency (Stephens et al., 1975).

The internal P_i levels in *E. coli* cells remain relatively constant and high (10.8 $\pm$ 1.3 mM) regardless of the concentration and the type of P_i source in the growth medium (Table 1). Also, many of the *pst* mutants are insensitive to repressible amounts of P_i in the medium. These observations prompted a search for one or more molecules imparting a signal of P_i sufficiency or limitation by comparing the cellular pools of acid-soluble phosphate metabolites during P_i uptake in wild-type and in transport-defective mutants (e.g., *pit*, *pst*, and *pst pit*). A sample of exponentially growing *E. coli* cells ($OD_{540} \approx 0.25$) in a P_i-limited medium was pulse-labeled with ^{32}P for 2 min.

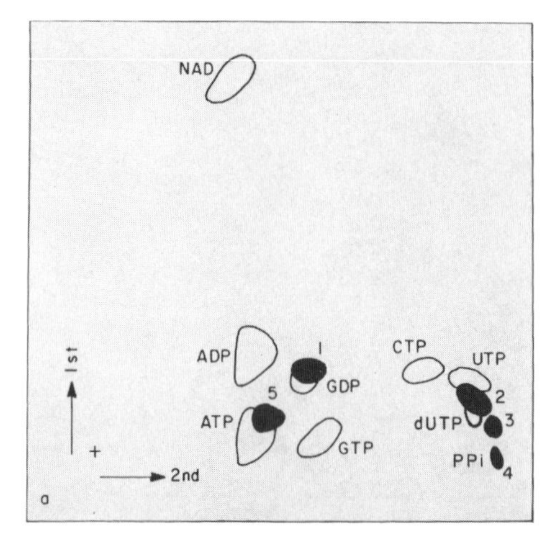

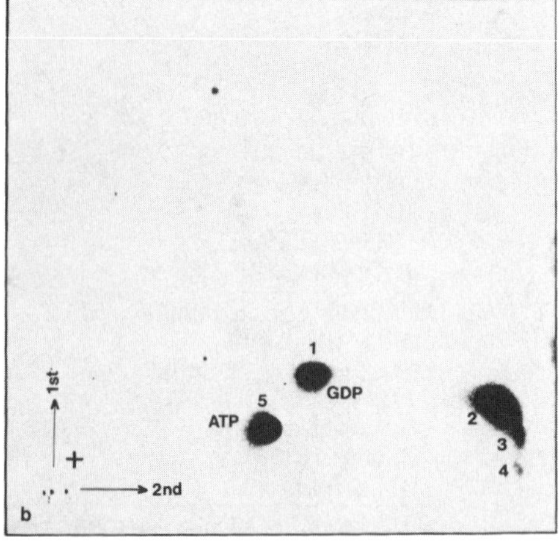

FIG. 4. Two dimensional TLC separation of ^{32}P-labeled nucleotides from high-pressure liquid chromatography (HPLC) fraction 19. Guanidine hydrochloride (0.9 M) was used for the first dimension, and ammonium sulfate solvent [43.5 g of $(NH_4)_2SO_4$, 0.4 g of (NH_4) HSO_4, and 4 g of disodium EDTA in 100 ml of distilled water] was used for the second dimension. (a) Tracing of autoradiogram with identity of metabolites by UV260; (b) photographic reproduction of the autoradiogram.

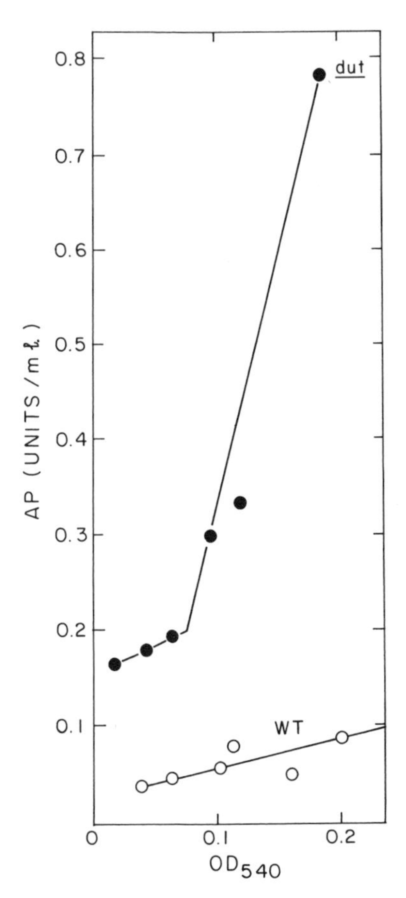

FIG. 5. Synthesis of AP plotted with respect to increase of bacterial mass. OD_{540} = 4 × 10^8 cells per ml. Cells used were *E. coli* BW 35 (dut^+) wild type (WT) (○) and BW 285 (*dut-1*) (●).

When the culture reached the stationary phase (OD_{540} ≈ 1.0), indicating the P_i starved conditions, another sample was also labeled for 2 min. In both cases, after the ^{32}P labeling, samples were extracted with 1 M formic acid as described previously (Bochner and Ames, 1982; Rao et al., 1986). The ^{32}P-labeled phosphate metabolites were resolved in two-dimensional thin-layer chromatography, and the compounds were identified by cochromatographic, chemical, and enzymatic analyses (Bochner and Ames 1982; Lee et al., 1983a, 1983b) (Fig. 6). Of the acid-extractable metabolites which incorporated ^{32}P in the first 2 min of P_i uptake, about 50% were nucleotides. Three unique nucleotides, P1, P2, and P3, were detected in P_i-starved wild-type (pit^+ pst^+) cells (Fig. 6B and D). When a cocktail of compounds which dissipate the proton motive force was added to the P_i-starved cells, the above metabolites could not be detected. The cocktail, which contained the energy transfer inhibitor *N,N′*-dicyclohexylcarbodiimide

TABLE 3. Escape synthesis of AP in *E. coli*

Strain[a]	Escape synthesis[b]
BW 35 (wild type)	0.80
BW 285 (*dut-1*)	6.00
BW 333 (*dut-1* p*dut*$^+$)	0.75
B 3 (*thyA14*) + thymine	0.75
B 3 (*thyA14*), no thymine	6.5

[a]The cells were grown at 30°C. Genotypes: BW 35, KL16 Hfr *relA1 thi-1 spoT1;* BW 285, same as BW 35 but *dut-1;* B 3, F^- *thyA14*.

[b]Escape synthesis is enzyme synthesis during growth at high P_i. It is expressed as differential rate of synthesis, since the cells are growing exponentially: Δenzyme (U/ml)/Δbacterial mass (OD_{540} = 1.0).

(DCCD), the respiration inhibitor NaN_3, and the uncoupler carbonyl cyanide *m*-chlorophenylhydrazone (CCCP), completely inactivated the Pit system (Brucker et al., 1984; Rosenberg et al., 1979). In this experiment, the rate of ^{32}P uptake by the pit^+ pst^+ strain was decreased by about 50% compared with that by a *pit* pst^+ strain, which was not affected. These three spots were also not found in autoradiograms of a *pit* strain. Therefore, the synthesis of those compounds may be related to the function of Pit-mediated transport. One of these nucleotides, P1, was isolated and found to be hydrolyzed to P_i by AP digestion. When P1 was treated with snake venom phosphodiesterase at pH 8.3, it produced PP_i and p5′A3′pp (adenosine 5′-monophosphate 3′-diphosphate). On the basis of chemical and enzymatic analysis (Bochner and Ames, 1982), P1 was tentatively identified as ppp5′A3′pp (adenosine 5′-triphosphate 3′-diphosphate). The concentration of adenylylated nucleotides increases dramatically in *E. coli* and *S. typhimurium* cells subjected to heat shock (Lee et al., 1983a, 1983b). Limitation of certain nutrients required for growth by the above bacteria also triggers the synthesis of acid-soluble, low-molecular-weight nucleotides which are unique for a particular stress (Bochner, 1987; Lee et al., 1983a, 1983b). Highly phosphorylated adenine derivatives were synthesized by sporulating cells of *Bacillus subtilis* during P_i starvation (Rhaese and Groscurth, 1974; Rhaese et al., 1977). In agreement with the hypothesis proposed for prokaryotes that AppppA and ApppG are signal nucleotides or alarmones of oxidative stress (Bochner et al., 1984), we believe that the unique nucleotides detected in P_i-starved *E. coli* cells serve as indicators of P_i availability and the need to induce the regulon.

Conclusion

In summary, the Pho regulon is regulated by the external P_i, not by the cytoplasmic P_i. The cytoplasmic P_i is maintained at a high concentration (10.8 ± 1.3 mM) under all growth conditions.

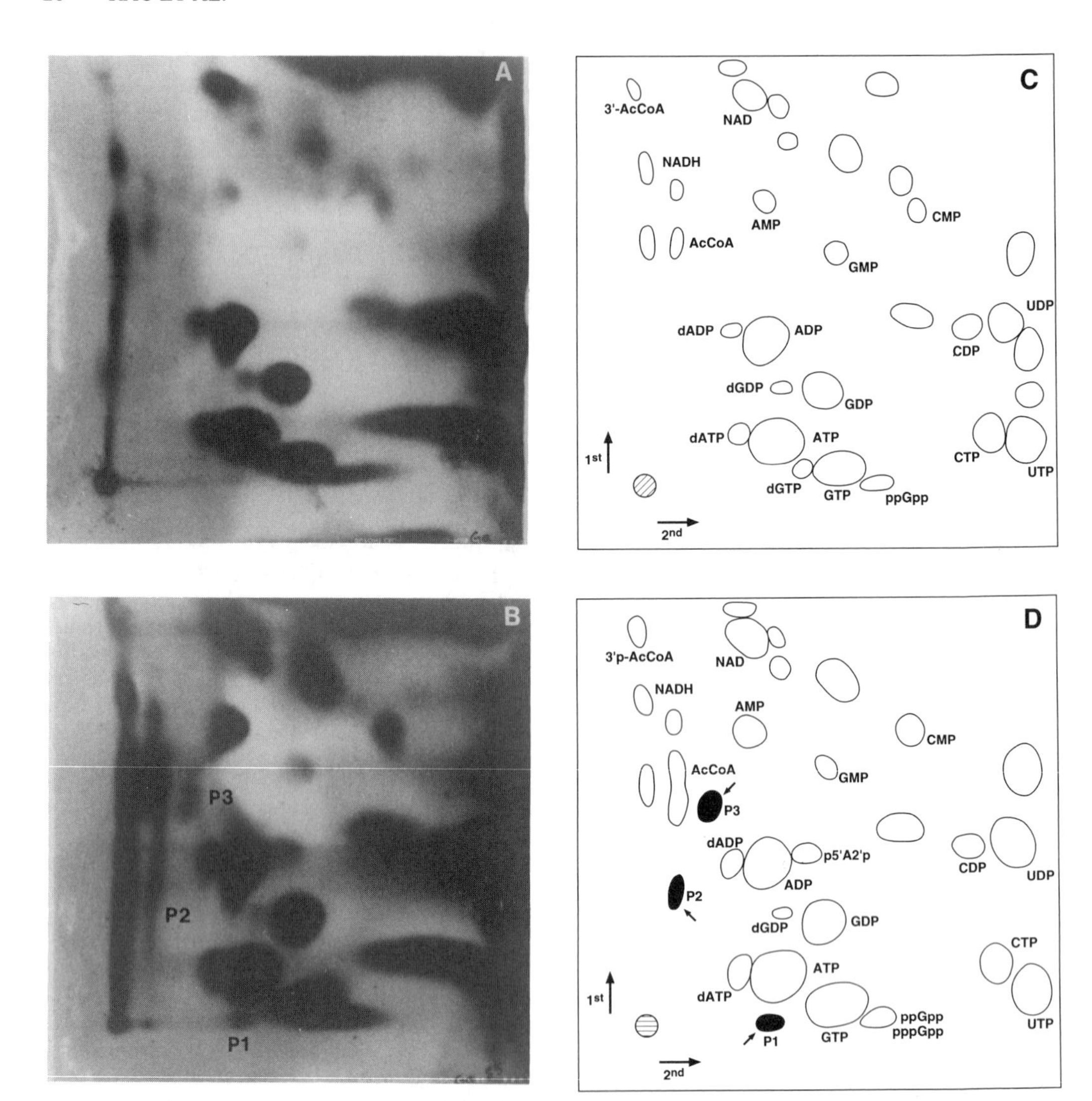

FIG. 6. Autoradiograms exposed from two-dimensional thin-layer chromatograms developed with guanidine hydrochloride solvent followed by ammonium sulfate solvent (see the legend to Fig. 4), as described previously (Rao et al., 1986). (A and B) Autoradiograms showing ^{32}P-labeled pools from *E. coli* K10 wild-type cells pulse-labeled for 2 min during exponential (A) and stationary (B) phases of growth. (C and D) Diagrammatic representations showing the identities of the metabolites resolved by the autoradiograms in panels A and B, respectively. The spots P1, P2, and P3, indicated by arrows, are the metabolites accumulated in P_i-starved cells (corresponding to the stationary phase).

A mutant (*dut-1*) with a highly reduced dUPTase activity (less than 1% of the activity in the wild type) displayed an escape synthesis of AP during growth in a medium containing excess phosphate. AP synthesis was induced in the presence of dUTP in permeabilized *E. coli* cells grown in P_i-rich medium (Table 3; Fig. 5). Hence, the escape synthesis observed in the *dut-1* mutant was attributed to the dUTP accumulated during growth. Since no specific nucleotides are required when AP synthesis is performed in a cell-free system with *phoA* DNA (Inouye et al., 1977), the requirement for nucleotides (S2, S15, and dUTP) (Table 2) for the expression of *phoA* in permeabilized cells is probably due to the activation of a step in the regulatory mechanism.

Unique phosphorylated metabolites (for example, pppApp) appeared during stress-induced growth of wild-type *E. coli* cells starved for phosphate (Fig. 6) but were not detected in exponentially growing cells. Dissipation of the proton motive force, the driving force for the operation of the Pit system, also led to the disappearance of these phosphorylated metabolites. We propose that one of the nucleotides characterized, pppApp, or a related adenylylated nucleotide may serve as an alarmone signaling the onset of P_i starvation.

REFERENCES

Ambudkar, S. V., T. J. Larson, and P. C. Maloney. 1986. Reconstitution of sugar transport systems of *Escherichia coli. J. Biol. Chem.* **261:**9083–9086.

Amemura, M., K. Makino, H. Shinagawa, A. Kobayashi, and A. Nakata. 1985. Nucleotide sequence of the genes involved in phosphate transport and regulation of the phosphate regulon in *Escherichia coli. J. Mol. Biol.* **184:**241–259.

Bochner, B. R. 1987. Global analysis of phosphorylated metabolites to search for alarmones, p. 289–294. *In* A. Torriani-Gorini, F. Rothman, S. Silver, A. Wright, and E. Yagil (ed.), *Phosphate Metabolism and Cellular Regulation in Microorganisms.* American Society for Microbiology, Washington, D.C.

Bochner, B. R., and B. N. Ames. 1982. Complete analysis of cellular nucleotides by two-dimensional thin layer chromatography. *J. Biol. Chem.* **257:**9759–9769.

Bochner, B. R., P. C. Lee, S. W. Wilson, C. W. Cutler, and B. N. Ames. 1984. AppppA and related adenylylated nucleotides are synthesized as a consequence of oxidation stress. *Cell* **37:**225–232.

Brucker, R., R. Levitz, E. Yagil, and I. Friedberg. 1984. Complementation tests between mutations in the phosphate specific transport region of *Escherichia coli. Curr. Microbiol.* **10:**303–308.

Brzoska, P., M. Rimmele, K. Brzostek, and W. Boos. 1994. The pho regulon-dependent Ugp uptake system for glycerol-3-phosphate in *Escherichia coli* is *trans* inhibited by P_i. *J. Bacteriol.* **176:**15–20.

Cox, G. B., D. Webb, J. Godovac-Zimmerman, and H. Rosenberg. 1988. Arg-220 of the PstA protein is required for phosphate transport through the phosphate-specific transport system in *Escherichia coli* but not for alkaline phosphatase repression. *J. Bacteriol.* **170:**2283–2286.

Cox, G. B., D. Webb, and H. Rosenberg. 1989. Specific amino acid residues in both the PstB and PstC proteins are required for phosphate transport by the *Escherichia coli* Pst system. *J. Bacteriol.* **171:**1531–1534.

Elvin, C. M., C. M. Hardy, and H. Rosenberg. 1985. P_i exchange mediated by the GlpT-dependent *sn*-glycerol-3-phosphate transport system in *Escherichia coli. J. Bacteriol.* **161:**1054–1058.

Hochhauser, S. J., and B. Weiss. 1978. *Escherichia coli* mutants deficient in deoxyuridine triphosphatase. *J. Bacteriol.* **134:**157–166.

Inouye, H., C. Pratt, J. Beckwith, and A. Torriani. 1977. Alkaline phosphatase synthesis in a cell-free system using DNA and RNA templates. *J. Mol. Biol.* **110:**75–87.

Kadner, R. J., D. M. Shattuck-Eidens, and L. A. Weston. 1987. Exogenous induction of the *uhp* sugar phosphate transport system, p. 178–183. *In* A. Torriani-Gorini, F. Rothman, S. Silver, A. Wright, and E. Yagil (ed.), *Phosphate Metabolism and Cellular Regulation in Microorganisms.* American Society for Microbiology, Washington, D.C.

Larson, T. J., G. Schumacher, and W. Boos. 1982. Identification of the *glpT*-encoded *sn*-glycerol-3-phosphate permease of *Escherichia coli,* an oligomeric integral membrane protein. *J. Bacteriol.* **152:**1008–1021.

Lee, P. C., B. R. Bochner, and B. N. Ames. 1983a. AppppA, heat shock stress, and cell oxidation. *Proc. Natl. Acad. Sci. USA* **80:**7496–7500.

Lee, P. C., B. R. Bochner, and B. N. Ames. 1983b. Diadenosine 5′,5‴-P^1,P^4-tetraphosphate and related adenylylated nucleotides in *Salmonella typhimurium. J. Biol. Chem.* **258:**6827–6834.

Lee, T.-Y., K. Makino, H. Shinagawa, and A. Nakata. 1990. Overproduction of acetate kinase activates the phosphate regulon in the absence of the *phoR* and *phoM* functions of *Escherichia coli. J. Bacteriol.* **172:**2245–2249.

Lin, E. C. C. 1976. Glycerol dissimilation and its regulation in bacteria. *Annu. Rev. Microbiol.* **30:**535–578.

Makino, K., H. Shinagawa, M. Amemura, T. Kawamoto, M. Yamada, and A. Nakata. 1989. Signal transduction in the phosphate regulon of *Escherichia coli* involves phosphate transfer between PhoR and PhoB proteins. *J. Mol. Biol.* **210:**551–559.

Rao, N. N., M. F. Roberts, A. Torriani, and J. Yashphe. 1993. Effect of *glpT* and *glpD* mutations on expression of the *phoA* gene in *Escherichia coli. J. Bacteriol.* **175:**74–79.

Rao, N. N., and A. Torriani. 1990. Molecular aspects of phosphate transport in *Escherichia coli. Mol. Microbiol.* **4:**1083–1090.

Rao, N. N., E. Wang, J. Yashphe, and A. Torriani. 1986. Nucleotide pool in Pho regulon mutants and alkaline phosphate synthesis in *Escherichia coli. J. Bacteriol.* **166:**205–211.

Rhaese, H. J., and R. Groscurth. 1974. Studies on the control of development. *In vitro* synthesis of HPN and MS nucleotides by ribosomes from either sporulating or vegetative cells of *Bacillus subtilus. FEBS Lett.* **44:**87–93.

Rhaese, H. J., J. A. Hoch, and R. Groscurth. 1977. Studies on the control of development: isolation of *Bacillus subtilis* mutants blocked early in sporulation and defective in synthesis of highly phosphorylated nucleotides. *Proc. Natl. Acad. Sci. USA* **74:**1125–1129.

Rosenberg, H. 1987. Phosphate transport in prokaryotes, p. 205–248. *In* B. P. Rosen and S. Silver (ed.), *Ion Transport in Prokaryotes.* Academic Press, Inc., New York.

Rosenberg, H., R. G. Gerdes, and F. M. Harold. 1979. Energy coupling to the transport of inorganic phosphate in *Escherichia coli* K-12. *Biochem. J.* **178:**133–137.

Sarthy, A., S. Michaelis, and J. Beckwith. 1981. Use of gene fusions to determine the orientation of gene *phoA* on the *Escherichia coli* chromosome. *J. Bacteriol.* **145:**293–298.

Schwiezer, H., M. Argast, and W. Boos. 1982. Characteristics of a binding protein-dependent transport system for *sn*-glycerol-3-phosphate in *Escherichia coli* that is part of the Pho regulon. *J. Bacteriol.* **150:**1154–1163.

Stephens, J. C., S. W. Artz, and B. N. Ames. 1975. Guanosine 5′-diphosphate 3′-diphosphate (ppGpp): positive effector for histidine operon transcription and general signal for amino acid deficiency. *Proc. Natl. Acad. Sci. USA* **72:**4389–4393.

Stock, J. B., A. J. Ninfa, and A. M. Stock. 1989. Protein phosphorylation and regulation of adaptive responses in bacteria. *Microbiol. Rev.* **53:**450–490.

Stock, J. B., A. M. Stock, and J. M. Mottonen. 1990. Signal transduction in bacteria. *Nature* (London) **344:**395–400.

Taylor, A. F., P. G. Siliciano, and B. Weiss. 1980. Cloning of the *dut* (deoxyuridine triphosphatase) gene of *Escherichia coli. Gene* **9:**321–336.

Torriani, A. 1990. From cell membrane to nucleotides: the phosphate regulon in *Escherichia coli. Bioessays* **12:**371–376.

Torriani, A., and D. N. Ludtke. 1985. The Pho regulon of *Escherichia coli,* p. 224–242. *In* M. Schaechter, F. C. Neidhardt, J. Ingraham, and N. O. Kjeldgaard (ed.), *The Molecular Biology of Bacterial Growth.* Jones and Bartlett Publishers, Boston.

Wanner, B. L. 1987. Phosphate regulation of gene expression in *Escherichia coli,* p. 1326–1333. *In* F. C. Neidhardt, J. L. Ingraham, K. B. Low, B. Magasanik, M. Schaechter, and H. E. Umbarger (ed.), *Escherichia coli and Salmonella typhimurium: Cellular and Molecular Biology,* vol. 2. American Society for Microbiology, Washington, D.C.

Wanner, B. L. 1992. Is cross regulation by phosphorylation of two-component response regulator proteins important in bacteria? *J. Bacteriol.* **174:**2053–2058.

Wilkins, A. 1972. Physiological factors in the regulation of alkaline phosphatase synthesis in *Escherichia coli. J. Bacteriol.* **110:**616–623.

Zubay, G. 1973. *In vitro* synthesis of protein in microbial systems. *Annu. Rev. Genet.* **7:**267–287.

Chapter 5

The Ugp Paradox: the Phenomenon That Glycerol-3-Phosphate, Exclusively Transported by the *Escherichia coli* Ugp System, Can Serve as a Sole Source of Phosphate but Not as a Sole Source of Carbon Is Due to *trans* Inhibition of Ugp-Mediated Transport by Phosphate

PIUS BRZOSKA,[1,2] MARTINA RIMMELE,[1] KATARZYNA BRZOSTEK,[1,3] AND WINFRIED BOOS[1]

Department of Biology, University of Konstanz, 78434 Konstanz, Germany[1]; Radiation Oncology Research Laboratory, Department of Radiation Oncology, University of San Francisco, San Francisco, California 94103-0806[2]; and Institute of Microbiology, Warsaw University, Nowy Swiat 67, PL-00046 Warsaw, Poland[3]

The Ugp (*u*ptake of *g*lycerol *p*hosphate) system is a typical periplasmic binding protein-dependent multicomponent transport system specific for glycerol 3-phosphate (G3P) and glyceryl phosphoryl phosphodiesters, the diacylation products of phospholipids (Brzoska and Boos, 1988; Schweizer et al., 1982). The genes *ugpB*, *ugpA*, *ugpE*, *ugpC*, and *ugpQ* (Overduin et al., 1988) form an operon, located at min 75 on the *Escherichia coli* chromosome, encoding the specific binding protein (*ugpB*) (Argast and Boos, 1979), the two membrane-bound components (*ugpA* and *ugpE*), and the ATP-binding fold-containing subunit (*ugpC*), supposedly the energy module of the system. The last gene in the operon, *ugpQ*, encodes a peculiar enzyme, a glyceryl phosphoryl phosphodiesterase, which hydrolyzes only diesters that are in the process of being transported by the system and appear at the inner surface of the membrane. Internal phosphodiesters are not hydrolyzed (Brzoska and Boos, 1988; Tommassen et al., 1991). UgpQ is not necessary for the transport of G3P or for the transport of glyceryl phosphoryl phosphodiesters.

The *ugp* operon is under the control of PhoB, the central gene activator of the Pho regulon (Argast and Boos, 1980; Schweizer and Boos, 1985). The Pho control circuit consists of a typical two-component system, with PhoB as the response regulator and PhoR, the histidine kinase, as the membrane-bound sensor (Scholten and Tommassen, 1993; Wanner, 1993). The signal of phosphate surplus (repression) or phosphate limitation (derepression) is mediated via the recognition or transport of P_i by the binding protein-dependent and multicomponent uptake system Pst (Webb et al., 1992), which is also dependent in its synthesis on PhoB. The product of *phoU* (Muda et al., 1992), the last gene in the *pst* operon, plays a key role in the transmission of the signal even though neither the signal nor the mechanism of its transmission is understood (Steed and Wanner, 1993).

The promoter of the *ugp* operon contains multiple copies of the typical *phoB* box (Kasahara et al., 1991; Overduin et al., 1988) for the binding of the gene activator PhoB; it is therefore a Pho-controlled operon. Recently it has been reported that *ugp* is also controlled by the cyclic AMP/catabolite gene activator protein (CAP) system (Kasahara et al., 1991; Su et al., 1991).

Aside from extended homologies in the polypeptide carrying the ATP-binding fold that has been observed in all binding-protein-dependent transport systems (Higgins, 1992), the remaining components of these systems (Köster and Bohm, 1992), in particular the binding proteins (Tam and Saier, 1993), show only marginal homology. In contrast, the sequence analysis of the *ugp* genes revealed a surprising homology on the protein level to the corresponding genes of the binding-protein-depending transport system for maltose and maltodextrins (Overduin et al., 1988). This homology can be seen in all polypeptides, but it is particularly evident between MalK and UgpC, the ATP-binding cassettes. Both proteins show an unusually long C terminus, which in MalK has been identified as the target for regulation of transport activity by enzyme III^{Glc} of the phosphotransferase system, as well as being a regulating domain for *mal* expression (Dean et al., 1990; Kühnau et al., 1991). No such function has been found in the Ugp system; in particular, it is clear that strains lacking *ugpC* (or any other *ugp* gene) are not *pho* constitutive, whereas strains lacking MalK function are constitutive for *mal* gene expression (Bukau et al., 1986; Reyes and Shuman, 1988). In a wild-type strain growing on G3P, the Ugp system is not induced because of the repression of the Pho regulon by P_i released from the cell after the uptake of G3P by GlpT permease, the major G3P transport system of *E. coli* (Rao et al., 1993).

The GlpT transport system (Ambudkar et al., 1986; Hayashi et al., 1964; Larson et al., 1982) is part of the *glp* system mediating uptake and metabolism of glycerol, G3P, and glycerol phosphoryl phosphodiesters (Lin, 1987). GlpT and all the other *glp*-encoded proteins are under the control of GlpR (Larson et al., 1987), the repressor of the system, the inducer being G3P (Lin and Iuchi, 1991). Growth on glycerol or glycerol phosphoryl phosphodiesters induces the system after metabolism of these compounds to G3P. The GlpT transport system can function in two modes. In the exchange mode G3P is taken up in exchange for P_i (Ambudkar et al., 1986; Elvin et al., 1985). Net uptake of G3P occurs essentially by proton symport. The GlpT permease is a tightly membrane-bound oligomeric complex of identical polypeptide subunits (Larson et al., 1982). Comparing the kinetic properties of the Ugp and the GlpT transport systems in their capabilities to transport G3P, one observes that the maximal velocities of both systems are of the same order of magnitude (around 10 nmol/min/10^9 cells), whereas the apparent affinity of the Ugp system for its substrate G3P is higher than that of the GlpT system (1 to 2 μM versus 12 to 20 μM) (Schweizer et al., 1982). Therefore, one would assume that both systems were able to supply enough G3P for growth. However, whereas G3P transported exclusively via the GlpT system can serve as the sole source of both carbon and phosphate, G3P transported exclusively via the Ugp system can serve as the sole source of phosphate but not of carbon. In both cases G3P enters the same metabolic pathway, oxidation by G3P dehydrogenase being the first step (Schweizer et al., 1982). Here we demonstrate that this phenomenon can be explained by a *trans* inhibition of the Ugp transport system by P_i, the metabolic product of G3P.

Observations Relevant to the Ugp Paradox

Incorporation of G3P into cellular material is slowed when G3P is taken up by the Ugp system but not when it is taken up by the GlpT system. When G3P was given to cells growing on maltose and transporting G3P exclusively by either the GlpT system or the Ugp system, the cells incorporated [^{14}C]G3P at the same rate but only during the first few minutes. Although the rate of incorporation of G3P mediated via the GlpT system remained high, the incorporation of G3P mediated via the Ugp system rapidly declined and the specific incorporation (normalized per cell) remained low over the entire period of several hours. This is shown in Fig. 1. In this experiment both strains lacked Pst, the high-affinity transport system for P_i, and were therefore constitutive for the *pho* regulon (Schweizer et al., 1982). This observation is the basis for the Ugp paradox, the phenomenon that the uptake of G3P through Ugp in the long run is insufficient to maintain growth.

Transport or recognition of substrate by the Ugp system does not elicit a PhoR/PhoB-mediated repression of the Pho regulon. We considered the possibility that G3P, as a result of being transported through the Ugp system, represses the Pho regulon and thus the expression of *ugp* by interacting with the PhoR/PhoB regulatory signal transduction. This would be analogous to repres-

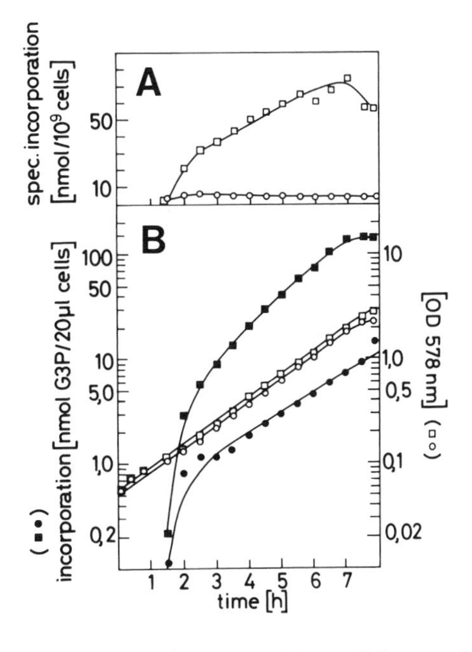

FIG. 1. Incorporation of ^{14}C into cellular material after uptake of [^{14}C]G3P mediated by either the GlpT or the Ugp transport system. Strain Brz70 (*ugp*$^+$ *glpT*) (circles) and strain Brz80 (*glpT*$^+$ *ugp*) (squares) were grown to logarithmic phase in G+L medium to 1 mM P_i and with 0.2% maltose as the carbon source. The initial rate of incorporation is the same in both strains. It rapidly declines when G3P is transported by the Ugp system. Thus, the specific incorporation (A) remains much lower in the Ugp$^+$ strain than in the GlpT$^+$ strain. In panel B, open symbols indicate the optical density at 578 nm (OD_{578}) of the culture and the solid symbols indicate the amount of radioactivity associated with the cells. Reprinted from Brzoska et al. (1994) with permission.

sion of the Pho regulon by P_i during transport through the Pst system. To test this possibility, we used a *ugp*$^+$ *glpT pst* strain, lysogenic with a *ugpA-lacZ* fusion, which allowed us to monitor *ugp* expression by measuring β-galactosidase activity. The strain was grown on maltose in G+L medium (Garen and Levinthal, 1960) with 1 mM P_i. The addition of G3P did not reduce the high specific activity of β-galactosidase (data not shown). Thus, it is clear that G3P taken up by the Ugp system does not repress the Pho regulon, because it does not affect the expression of *ugp* (Brzoska et al., 1994).

The above-mentioned *ugp*$^+$ *glpT pst* strain, in which the effect of Ugp-mediated G3P transport on the expression of the Pho system was measured, is *pst* and therefore Pho constitutive. One could therefore argue that *pst*-derived Pho constitutivity could be dominant over a repression brought about by Ugp-mediated transport of G3P. To test this possibility, we used a *pst*$^+$ strain. To avoid interference with P_i, liberated from G3P, that would result in Pst-mediated repression, we used glyceryl phosphorylcholine (GPC) as a substrate of the Ugp system and alkaline phosphatase activity as a measure of *pho* expression. The results are shown in Fig. 2. To observe the possible *pho* repression by transport of GPC via Ugp in a *pst*$^+$ strain, we first had to derepress by phosphate starvation. Thus, as seen in Fig. 2, all cultures stopped growing after the external P_i pool was exhausted. As a consequence, alkaline phosphatase activity was strongly induced. The addition of GPC that was exclusively transported by the Ugp system did relieve phosphate starvation (as seen by the resumption of growth) but did not result in the repression of the Pho regulon. In contrast, the addition of 1 mM P_i was very effective in repression. In the control experiment with a *ugp* strain and GPC as the phosphate source, growth did not resume. Thus, it is clear that transport or recognition of substrate by the Ugp system does not lead to repression of the Pho regulon, as does transport of P_i through the Pst system (Brzoska et al., 1994).

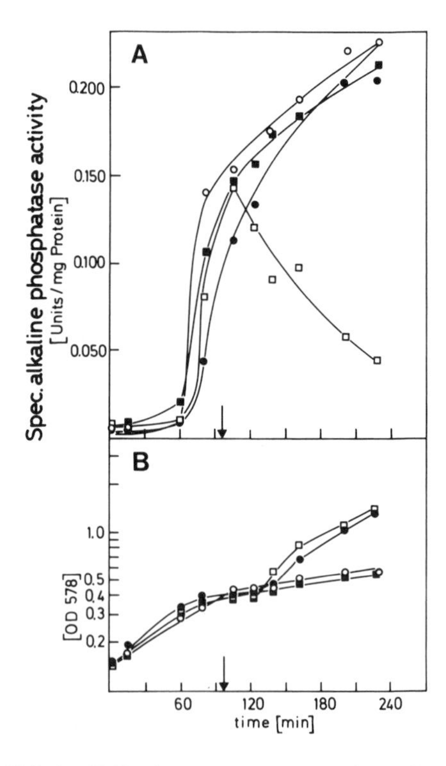

FIG. 2. GPC, when transported through the Ugp system, does not repress *phoA*, in contrast to P_i when transported through the Pst system. Strain Brz332 (*ugp glpT glpQ*) (squares) and strain TL45 (*ugp*$^+$ *glpT glpQ*) (circles) were grown in G+L medium–0.1 mM P_i and with 0.2% glucose as the carbon source. When the growth of the cultures had slowed as a result of P_i limitation (105 min in the figure), 1 mM GPC was added to Brz332 (solid squares) and to TL45 (solid circles). At the same time, 1 mM P_i was added to another culture of Brz332 (open squares) and nothing was added to another culture of TL45 (open circles). Alkaline phosphatase was assayed (A) and the optical density at 578 nm (OD_{578}) was monitored (B). Ugp-mediated uptake of GPC allowed resumption of growth of strain TL45 but did not result in repression of alkaline phosphatase. In contrast, uptake of P_i in strain Brz332 allowed resumption of growth and resulted in repression of alkaline phosphatase. Reprinted from Brzoska et al. (1994) with permission.

Intracellular G3P does not inhibit Ugp-mediated uptake of G3P in *trans*. Since the internal accumulation of G3P did not reduce the expression of *ugp* but the ability of cells to incorporate G3P into cellular material declined after the addition of G3P, it was possible that internal G3P itself would inhibit the further Ugp-mediated uptake of [^{14}C]G3P from the external medium by inhibiting transport activity in *trans*. To test this possibility, we wanted to create a situation in which G3P would be synthesized internally by phosphorylation of glycerol after uptake of external glycerol. For this purpose we constructed a mutant that was *phoA* (to avoid hydrolysis of [^{14}C]G3P), *glpT* (to block entry of G3P via the GlpT system), *glpD* (lacking G3P dehydrogenase activity to prevent the fast degradation of internal G3P), *glpR* (encoding the *glp* repressor to ensure high activity of the glycerol uptake system as well as glycerol kinase), and, finally, *pst* (to ensure constitutive expression of the Pho regulon, including *ugp*). The strain was grown on maltose, and its ability to take up [^{14}C]G3P by the Ugp system was monitored before and after the addition of 1 mM unlabeled glycerol. The addition of glycerol reduced the growth rate of the strain by 52%, a known phenomenon that is due to the accumulation of internal G3P (Cozzarelli et al., 1965). However, as shown in Table 1, the ability of the culture to

TABLE 1. Glycerol, after internal phosphorylation to G3P, does not inhibit Ugp-mediated uptake [^{14}C]G3P[a]

Time (min) after addition of glycerol[b]	OD_{578}[c]	Specific uptake of G3P[d]
0	0.5	68.6
10	0.55	76.2
15	0.58	76.8
20	0.6	84.0
75	0.7	76.4
75 (control, no glycerol)	0.88	65.2

[a]Reprinted from Brzoska et al. (1994) with permission.
[b]The culture was grown in MMA (Miller, 1972) containing 0.2% maltose.
[c]OD_{578}, optical density at 578 nm.
[d]Given in picomoles of G3P taken up by 10^9 cells per minute at a substrate concentration of 0.5 μM.

take up [^{14}C]G3P did not decline even after 1 h of growth in the presence of glycerol. This demonstrated that it was not G3P that inhibited the Ugp system from inside and that metabolism of G3P was necessary for inhibition to take place.

Phosphate contained in G3P is released when G3P is taken up by the GlpT system, but much less is released when it is taken up by the Ugp system. The *ugp*-encoded transport system represents a typical multicomponent and binding-protein-dependent system in which the transport of substrate is coupled to the hydrolysis of ATP. Therefore, it represents a primary pump that should function in an essentially unidirectional fashion. In contrast, the *glpT*-encoded system functions primarily as a G3P/P_i antiporter. To monitor simultaneously the incorporation of G3P into phospholipids (the ^{3}H label is located at the C-2 position of glycerol) and the release of P_i from G3P into the medium, we added a mixture of [^{32}P]G3P and [^{3}H]G3P to cells growing on maltose. We used two strains, one lacking the Ugp system but containing the GlpT system, and the other lacking the GlpT transport system but expressing the Ugp system constitutively as a result of a *pst* mutation. The results are shown in Fig. 3. Both strains were growing exponentially at a generation time of 75 min in Tris medium–0.1 mM P_i with 1% Casamino Acids as the carbon source. When G3P was taken up by the Ugp system, the ^{3}H label of G3P was initially quickly incorporated into phospholipids but little P_i was released into the medium. After about 40 min the incorporation stopped, and it resumed at a low rate only after more than 2 h. The ratio of internal ^{32}P to external $^{32}P_i$ (released from the inside) remained high throughout the experiment. The ratio was 3.6 after 160 min. In contrast, when G3P was transported exclusively by the GlpT system, incorporation of ^{3}H label into phospholipids increased exponentially and massive release of $^{32}P_i$ could be observed after an initial lag period. The ratio of internal ^{32}P to external $^{32}P_i$ (released from the

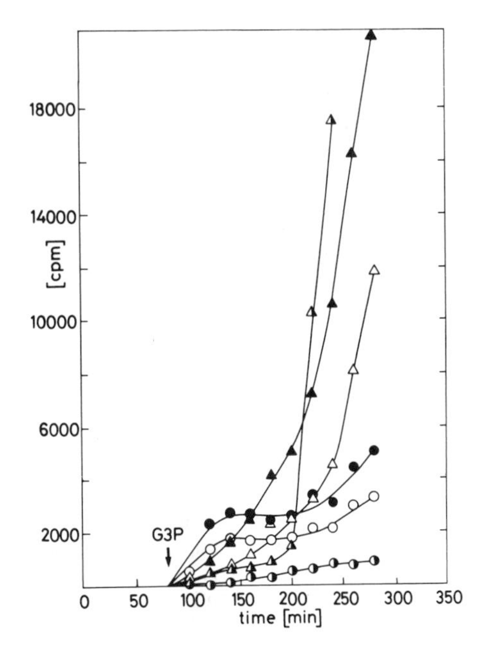

FIG. 3. Release of $^{32}P_i$ from the cells after GlpT- or Ugp-mediated uptake of [^{32}P]G3P. Strains AC105 (*ugp*$^+$ *glpT*) (circles) and AC109 (*glpT*$^+$ *ugp*) (triangles) were grown to logarithmic phase in G+L medium–0.1 mM P_i with 1% Casamino Acids as the carbon source. At an optical density of 578 nm of 0.06 (80 min), 5 mM G3P containing (per filtered sample) 1.2 × 10^5 cpm of [^{32}P]G3P and 3.2 × 10^5 cpm of 2-[^{3}H]G3P was added. At the times indicated, samples were withdrawn and filtered. Open and solid symbols indicate the amount of ^{32}P and ^{3}H, respectively, incorporated exclusively in lipids and membrane-derived oligosaccharides. Half-filled symbols indicate the amount of ^{32}P released as P_i into the supernatant. Both strains grew with a generation time of 75 min. At 160 min after the addition of G3P, the ratio of ^{32}P incorporated into cellular material to $^{32}P_i$ released from internal [^{32}]G3P was 3.6 when G3P was transported by the Ugp system and 0.25 when G3P was transported by the GlpT system. Reprinted from Brzoska et al. (1994) with permission.

inside) in this case was 0.25 (160 min after the addition of G3P). This demonstrated that P_i could not be released easily after G3P had been taken up exclusively by the Ugp system and had been degraded to P_i (Brzoska et al., 1994). Therefore, it appeared likely that the Ugp-mediated uptake of G3P was controlled mainly by the ability of the cells to accumulate, incorporate, or release P_i. The biphasic kinetics of ^{32}P and ^{3}H incorporation into the Ugp$^+$ GlpT$^-$ strain can be explained in the following way. The first phase represents the uninhibited G3P uptake by the Ugp system. Then G3P becomes metabolized, and its metabolic product, most probably P_i, accumulates and blocks further uptake of G3P. The medium contains a low concentration (0.1 mM) of P_i. After some time the cells experience P_i limitations, and the inhibition of Ugp by internal P_i will be relieved to a certain extent, allowing the entry of G3P to satisfy the demand for P_i. Therefore, Ugp-mediated G3P incorporation will resume at a low level (Brzoska et al., 1994).

Internal P_i controls the activity of the Ugp-mediated uptake of G3P. To test whether internal P_i was inhibiting the Ugp transport system in *trans,* we grew strain Brz503, a *glpT phoA* mutant, on glucose in the presence of an initial P_i concentration of 0.1 mM. After the cells had stopped growing as a result of the limitation in P_i, the *pho* system (including *pst* as well as *ugp*) was fully derepressed. The cells were washed with Tris buffer. Different samples of the culture were incubated with various amounts of P_i for 15 min, and the uptake of 0.14 μM [^{14}C]G3P was determined. The results are shown in Table 2. P_i concentrations as low as 0.5 mM inhibited the uptake of G3P about five- to sixfold. The inhibition of G3P uptake by preincubation with P_i was time-dependent (Table 3). Substantial inhibition had occurred after 1 min of incubation, but full inhibition required up to 15 min. Since Pst-mediated P_i transport is expected to be fast, one might also expect that the inhibition of Ugp-mediated uptake by P_i entering via the Pst system would be complete within a short time. This is clearly not the case. There is a fast inhibition followed by a slower inhibition. Our interpretation of this phenomenon is that uptake of P_i via the Pst system will also be inhibited by increasing levels of internal P_i and hence the level of inhibition of the Ugp-mediated G3P uptake will reflect the characteristics of the unidirectional Pst-mediated influx of P_i.

When 10 μM unlabeled G3P was added to the transport assay mixture containing 0.14 μM [^{14}C]G3P, the rate of uptake of the labeled compound was reduced about sevenfold (Table 4). The same degree of inhibition by unlabeled G3P was observed after the cells had been preincubated with P_i. This demonstrates that the K_m of G3P uptake was unaffected by preincubation with P_i whereas the V_{max} was reduced.

The inhibition of Ugp-mediated G3P uptake by incubation with P_i prior to the transport assay was variable. One determining factor was the "growth history" of the bacterial culture. To obtain strong inhibition (up to eightfold), the culture had to be starved extensively for phosphate during growth. Apparently only the complete removal of P_i from the medium (and consequently high levels of the Pst transport system) allowed the full inhibition by P_i to be seen. Strains lacking the Pst system and being fully active for the Ugp system were not inhibited by the addition of P_i (Argast et al., 1978).

TABLE 3. Time dependence of the inhibition of Ugp-mediated G3P uptake by 10 mM P_i

Time (min) after addition of 10 mM P_i	Rate of G3P uptake[a]
0	87.5
1	51.5
2	46.1
3	32.0
5	32.1
8	21.3
10	22.0
30	16.3

[a]The rate is given in picomoles of G3P taken up per minute per 10^9 cells at a [^{14}C]G3P concentration of 0.14 μM.

Discussion

We conclude that Ugp-dependent transport activity is feedback inhibited by internal phosphate. The experiment showing this inhibition involved allowing P_i to enter via Pst, the high-affinity P_i transport system, which most probably operates undirectionally. It is clear that inhibition is not due to external P_i, since Ugp-mediated uptake of G3P in a *pst* mutant is not inhibited even at 50 mM external P_i (Argast et al., 1978). It is possible that a metabolic product of P_i rather than P_i itself is inhibiting G3P uptake. We believe this to be unlikely since the initial inhibition of G3P uptake by P_i uptake was fast. The time dependence of inhibition showed a second, slower phase (up to 15 min); we suggest that this is caused by a feedback inhibition of Pst-mediated P_i transport by internal P_i itself.

We postulate that inhibition of Ugp-mediated uptake of G3P (or other substrates of the Ugp

TABLE 2. Internal P_i inhibits Ugp-mediated uptake of G3P[a]

P_i addition (mM)	Rate of G3P uptake[b]
None	61.9
0.5	16.9
1	11.6
10	11.5

[a]The cells were derepressed for the *pho* system and resuspended in G+L medium.

[b]The rate is given in picomoles of G3P taken up per minute per 10^9 cells at a [^{14}C]G3P concentration of 0.14 μM.

TABLE 4. Competitive inhibition by unlabeled G3P of Ugp-mediated uptake of [^{14}C]G3P in the presence and absence of P_i

Addition of 10 μM G3P	Presence of 10 mM P_i	Rate of [^{14}C]G3P uptake[a]
−	−	75.4
+	−	10.5
−	+	15.9
+	+	1.9

[a]The rate is given in picomoles of [^{14}C]G3P taken up per minute per 10^9 cells at a [^{14}C]G3P concentration of 0.14 μM.

system) by internal P_i is the reason why G3P, transported exclusively by the Ugp system, can be used as the sole source of phosphate but not of carbon. We envision that G3P as the carbon source contains, per carbon, more phosphate than is needed for cellular growth. Thus, after the degradation of G3P, when the excess P_i can no longer be secreted, it will accumulate and inhibit further uptake of the carbon source, G3P, resulting in cessation of growth. Inhibition of Ugp-mediated uptake of G3P by P_i transported via Pst is not 100%. On the other hand, strains relying exclusively on the Ugp system for the transport of G3P are completely unable to grow. Therefore, we conclude that the internal P_i levels reached after Ugp-mediated uptake of G3P and its subsequent degradation to P_i must be higher than the final levels of P_i obtained by Pst-mediated P_i transport.

The function of the Ugp system in a wild-type strain is obviously exclusively geared for the utilization of low concentrations of G3P and glyceryl phosphoryl diesters when P_i becomes limiting. Under these conditions and in the presence of an alternative carbon source, the internal concentration of P_i will control the activity of the Ugp system to allow just enough G3P to enter in order to maintain constant P_i levels. So far, the level to which the internal P_i concentration will rise after Ugp-mediated uptake of G3P has not been measured. In vivo nuclear magnetic resonance studies by Rao et al. (1993) have indicated that the level of internal P_i is rather high (around 10 mM) and independent of the presence or absence of Pst, the major P_i transport system. Also the GlpT-mediated transport and subsequent metabolism of G3P did not result in an increase in the internal P_i level. Of course, this is to be expected since the GlpT system has been described as an G3P-P_i exchange system (Ambudkar et al., 1986; Elvin et al., 1985). Another factor influencing the concentration of internal P_i is oxygen. Unoxygenized, resting cells contain a much higher concentration of internal P_i (ca. 50 mM) (Rao et al., 1993). The influence of aeration on the activity of the Ugp transport system has so far not been tested.

Another puzzle is demonstrated by the second major P_i transport system, Pit (Elvin et al., 1986; Rosenberg et al., 1977). This system is supposedly a proton motive force-driven system (Rosenberg et al., 1979) and should be able to maintain a constant internal P_i concentration by catalyzing entry as well as exit. However, apparently neither Pst nor Pit does allow exit of P_i (Elvin et al., 1985; Willsky and Malamy, 1980). Therefore, the major P_i transport systems should also be feedback inhibited by internal P_i to explain the constant internal P_i concentration.

Together with alkaline phosphatase, the Ugp system is the second example of a member of the *pho* proteins to be controlled by the same signal (P_i) not only on the level of expression (external P_i transported via the Pst transport system) but also on the level of activity (internal P_i). The effect of internal P_i on the activity of the Ugp system may explain the notorious variations encountered in the uptake measurements of this system (Argast et al., 1978). Thus, depending on the state and activity of P_i transport and the supply of external P_i, not only the expression of the *ugp* genes but also the activity of the Ugp system will vary.

The Ugp system belongs to the class of ABC transporters, multicomponent systems driven by an ATP-hydrolyzing subunit (Higgins, 1992). This subunit, UgpC in the Ugp system, is associated with the tightly membrane-bound pore proteins UgpA and UgpE. It is unclear where the inhibition of transport activity by P_i is being exerted. From the analogy of the Ugp system to the equivalent maltose system, in which the ATP-binding site-carrying subunit, MalK, is the target for regulation of transport activity (Dean et al., 1990; Kühnau et al., 1991), one would assume that inhibition of Ugp by P_i is exerted via the UgpC subunit. The fact that the ATP-binding subunits of the maltose and Ugp transport systems can functionally be exchanged (Hekstra and Tommassen, 1993) allows us to test whether UgpC is the target of the *trans* inhibition by P_i. The observation that V_{max} rather than K_m is affected by P_i inhibition would also be consistent with the idea of the nonspecific subunit UgpC being the target.

REFERENCES

Ambudkar, S. V., T. H. Larson, and P. C. Maloney. 1986. Reconstitution of sugar phosphate transport systems of *Escherichia coli*. *J. Biol. Chem.* **261:**9083–9086.

Argast, M., and W. Boos. 1979. Purification and properties of sn-glycerol 3-phosphate-binding protein of *Escherichia coli*. *J. Biol. Chem.* **254:**10931–10935.

Argast, M., and W. Boos. 1980. Coregulation in *Escherichia coli* of a novel transport system for *sn*-glycerol-3-phosphate and outer membrane protein Ic (e,E) with alkaline phosphatase and phosphate-binding protein. *J. Bacteriol.* **143:**142–150.

Argast, M., D. Ludtke, T. J. Silhavy, and W. Boos. 1978. A second transport system for *sn*-glycerol-3-phosphate in *Escherichia coli*. *J. Bacteriol.* **136:**1070–1083.

Brzoska, P., and W. Boos. 1988. Characteristics of a *ugp*-encoded and *phoB*-dependent glycerolphosphoryl diester phosphodiesterase which is physically dependent on the Ugp transport system of *Escherichia coli*. *J. Bacteriol.* **170:**4125–4135.

Brzoska, P., M. Rimmele, K. Brzostek, and W. Boos. 1994. The *pho* regulon-dependent Ugp uptake system for glycerol-3-phosphate in *Escherichia coli* is *trans*-inhibited by P_i. *J. Bacteriol.* **176:**15–20.

Bukau, B., M. Ehrmann, and W. Boos. 1986. Osmoregulation of the maltose regulon in *Escherichia coli*. *J. Bacteriol.* **166:**884–891.

Cozzarelli, N. R., J. P. Koch, S. Hayashi, and E. C. C. Lin. 1965. Growth stasis by accumulated L-glycerophosphate in *Escherichia coli*. *J. Bacteriol.* **90:**1325–1329.

Dean, D. A., J. Reizer, H. Nikaido, and M. H. Saier. 1990. Regulation of the maltose transport system of *Escherichia coli* by the glucose-specific enzyme III of the phosphoenolpyruvate-sugar phosphotransferase system: characterization of inducer exclusion-resistant mutants and reconstitution of inducer exclusion in proteoliposomes. *J. Biol. Chem.* **265:**21005–21010.

Elvin, C. M., N. E. Dixon, and H. Rosenberg. 1986. Molecular cloning of the phosphate (inorganic) transport (*pit*) gene of *Escherichia coli* K12; identification of the pit^+ gene product and physical mapping of the *pit-gor* region of the chromosome. *Mol. Gen. Genet.* **204:**477–484.

Elvin, C. M., C. M. Hardy, and H. Rosenberg. 1985. P_i exchange mediated by the GlpT-dependent *sn* glycerol-3-phosphate transport system in *Escherichia coli. J. Bacteriol.* **161:**1054–1058.

Garen, A., and C. Levinthal. 1960. A fine structure genetic and chemical study of the enzyme alkaline phosphatase of *Escherichia coli. Biochim. Biophys. Acta* **38:**470–483.

Hayashi, S., J. P. Koch, and E. C. C. Lin. 1964. Active transport of L-alpha-glycerophosphate in *Escherichia coli. J. Biol. Chem.* **239:**3098–3105.

Hekstra, D., and J. Tommassen. 1993. Functional exchangeability of the ABC proteins of the periplasmic binding protein-dependent transport systems Ugp and Mal of *Escherichia coli. J. Bacteriol.* **175:**6546–6552.

Higgins, C. F. 1992. ABC transporters—from microorganisms to man. *Annu. Rev. Cell Biol.* **8:**67–113.

Kasahara, M., K. Makino, M. Amemura, A. Nakata, and H. Shinagawa. 1991. Dual regulation of the *ugp* operon by phosphate and carbon starvation at two interpaced promoters. *J. Bacteriol.* **173:**549–558.

Köster, W., and B. Bohm. 1992. Point mutations in 2 conserved glycine residues within the integral membrane protein FhuB affect iron(III) hydroxamate transport. *Mol. Gen. Genet.* **232:**399–407.

Kühnau, S., M. Reyes, A. Sievertsen, H. A. Shuman, and W. Boos. 1991. The activities of the *Escherichia coli* MalK protein in maltose transport, regulation, and inducer exclusion can be separated by mutations. *J. Bacteriol.* **173:**2180–2186.

Larson, T. J., G. Schumacher, and W. Boos. 1982. Identification of the *glpT*-encoded *sn*-glycerol-3-phosphate permease of *Escherichia coli*, an oligomeric integral membrane protein. *J. Bacteriol.* **152:**1008–1021.

Larson, T. J., S. Ye, D. L. Weissenborn, H. J. Hoffmann, and H. Schweizer. 1987. Purification and characterization of the repressor for the *sn*-glycerol 3-phosphate regulon of *Escherichia coli. J. Biol. Chem.* **262:**15869–15874.

Lin, E. C. C. 1987. Dissimilatory pathways for sugars, polyols, and carboxylates, p. 244–284. *In* F. C. Neidhardt, J. L. Ingraham, K. B. Low, B. Magasanik, M. Schaechter, and H. E. Umbarger (ed.), *Escherichia coli and Salmonella typhimurium: Cellular and Molecular Biology.* American Society for Microbiology, Washington, D.C.

Lin, E. C. C., and S. Iuchi. 1991. Regulation of gene expression in fermentative and respiratory systems in *Escherichia coli* and related bacteria. *Annu. Rev. Genet.* **25:**361–387.

Miller, J. H. 1972. *Experiments in Molecular Genetics.* Cold Spring Harbor Laboratory, Cold Spring Harbor, N.Y.

Muda, M., N. N. Rao, and A. Torriani. 1992. Role of PhoU in phosphate transport and alkaline phosphatase regulation. *J. Bacteriol.* **174:**8057–8064.

Overduin, P., W. Boos, and J. Tommassen. 1988. Nucleotide sequence of the *ugp* genes of *Escherichia coli* K-12: homology to the maltose system. *Mol. Microbiol.* **2:**767–775.

Rao, N. N., M. F. Roberts, A. Torriani, and J. Yashphe. 1993. Effect of *glpT* and *glpD* mutations on expression of the *phoA* gene in *Escherichia coli. J. Bacteriol.* **175:**74–79.

Reyes, M., and H. A. Shuman. 1988. Overproduction of MalK protein prevents expression of the *Escherichia coli mal* regulon. *J. Bacteriol.* **170:**4598–4602.

Rosenberg, H., R. G. Gerdes, and K. Chegwidden. 1977. Two systems for the uptake of phosphate in *Escherichia coli. J. Bacteriol.* **131:**505–511.

Rosenberg, H., R. G. Gerdes, and F. M. Harold. 1979. Energy coupling to the transport of inorganic phosphate in *Escherichia coli* K12. *Biochem. J.* **178:**133–137.

Scholten, M., and J. Tommassen. 1993. Topology of the PhoR protein of *Escherichia coli* and functional analysis of internal deletion mutants. *Mol. Microbiol.* **8:**269–275.

Schweizer, H., M. Argast, and W. Boos. 1982. Characteristics of a binding protein-dependent transport system for *sn*-glycerol-3-phosphate in *Escherichia coli* that is part of the *pho* regulon. *J. Bacteriol.* **150:**1154–1163.

Schweizer, H., and W. Boos. 1985. Regulation of *ugp*, the *sn*-glycerol-3-phosphate transport system of *Escherichia coli* K-12 that is part of the *pho* regulon. *J. Bacteriol.* **163:**392–394.

Steed, P. M., and B. L. Wanner. 1993. Use of the *rep* technique for allele replacement to construct mutants with deletions of the *pstSCAB-phoU* operon: evidence of a new role for the PhoU protein in the phosphate regulon. *J. Bacteriol.* **175:**6797–6809.

Su, T. Z., H. P. Schweizer, and D. L. Oxender. 1991. Carbon-starvation induction of the *ugp* operon, encoding the binding protein-dependent *sn*-glycerol-3-phosphate transport system in *Escherichia coli. Mol. Gen. Genet.* **230:**28–32.

Tam, R., and M. H. Saier. 1993. Structural, functional, and evolutionary relationships among extracellular solute-binding receptors of bacteria. *Microbiol. Rev.* **57:**320–346.

Tommassen, J., K. Eiglmeier, S. T. Cole, P. Overduin, T. J. Larson, and W. Boos. 1991. Characterization of two genes, *glpQ* and *ugpQ*, encoding glycerophosphoryl diester phosphodiesterases of *Escherichia coli. Mol. Gen. Genet.* **226:**321–327.

Wanner, B. L. 1993. Gene regulation by phosphate in enteric bacteria. *J. Cell. Biochem.* **51:**47–54.

Webb, D. C., H. Rosenberg, and G. B. Cox. 1992. Mutational analysis of the *Escherichia coli* phosphate-specific transport system, a member of the traffic ATPase (or ABC) family of membrane transporters. *J. Biol. Chem.* **267:**24661–24668.

Willsky, G. R., and M. H. Malamy. 1980. Characterization of two genetically separable inorganic phosphate transport systems in *Escherichia coli. J. Bacteriol.* **144:**356–365.

Chapter 6

Proposed Mechanism for Phosphate Translocation by the Phosphate-Specific Transport (Pst) System and Role of the Pst System in Phosphate Regulation

DIANNE C. WEBB AND GRAEME B. COX

Division of Biochemistry and Molecular Biology, John Curtin School of Medical Research, The Australian National University, P. O. Box 334, Canberra, ACT 2601, Australia

There are two phosphate transport systems present in *Escherichia coli*, the phosphate inorganic transporter (Pit) and the phosphate-specific transporter (Pst) (Rosenberg et al., 1977). The Pit system appears to be a proton/phosphate symporter, whereas the Pst system is a typical periplasmic permease that comprises one periplasmic substrate-binding protein and three membrane-bound components (Ames, 1988) and belongs to the large family of ABC transporters (Hyde et al., 1990) or traffic ATPases (Ames and Joshi, 1990). The Pit system is made constitutively and has a K_t for phosphate of about 25 μM (Rosenberg et al., 1977). The Pst system has a higher affinity for phosphate, with a K_t of about 0.2 μM (Rosenberg et al., 1977). The synthesis of the Pst system is regulated by the concentration of phosphate in the growth medium and has the somewhat unusual property of being a sensory adjunct to a two-component (PhoR, PhoB) regulatory system that controls the promoters of the Pho regulon (see Wanner, 1993). However, the PhoR sensor does have, comparatively, a much reduced periplasmic domain (Scholten and Tommassen, 1993; see also Wanner [1993]), and it is the Pst system that appears to carry out the sensory role. The PhoU protein may provide the link between the transporter and the PhoR. The arrangement of the seven component proteins involved in the transporter-regulator interactions are depicted in Fig. 1, essentially as proposed by Wanner (1993).

Role of the PhoU Protein

The PhoU protein is a peripheral cytoplasmic membrane protein which is solubilized when membranes are suspended in low-ionic-strength buffer (Surin et al., 1985). The only missense mutation reported in the *phoU* gene is the *phoU35* allele (Torriani and Rothman, 1961). This mutation does not affect transport through the Pst system but does cause derepression of the Pho regulon (Surin et al., 1985). The *phoU35* allele has been sequenced, and the only difference detected between the mutant allele and the wild type was a C-to-A transversion resulting in alanine at position 147 being replaced by glutamate. This residue is located in a relatively long region of a predicted α-helix, which is likely to be stabilized by the presence of oppositely charged residues on one face of the putative α-helix (Fig. 2). It is tempting to speculate that this region interacts with a similar region on the PhoR protein (Fig. 2), but the necessary experimental support is lacking. The observation that transport through the Pst system is normal in the presence of the *phoU35* allele does not necessarily indicate that the PhoU protein is not required for Pst function. Two laboratories have addressed this question through the isolation of *phoU* deletion mutants (Muda et al., 1992; Steed and Wanner, 1993). The *phoU* deletion mutant isolated by Steed and Wanner (1993) grew extremely poorly, although phosphate uptake by this strain was similar to that by the wild type. The poor growth could be reversed by combining the *phoU* deletion with mutations affecting Pst function. Furthermore, compensatory mutations, affecting the function of the Pst complex, arose during growth of the Δ*phoU* mutant. Steed and Wanner (1993) have proposed that, in addition to its role as a negative regulator, PhoU has a second role, i.e., that of an enzyme involved in intracellular metabolism. The *phoU* deletion mutant isolated by Muda et al. (1992) had somewhat different properties from those of the mutant isolated by Steed and Wanner (1993), but this may be due to the presence of additional compensatory mutations in the mutant isolated by Muda et al.

Structure-Function Relationships in the PstA and PstC Proteins

The PstA and PstC proteins are integral membrane proteins which presumably mediate the movement of phosphate from outside to inside the cell. We have used a modeling, site-directed mu-

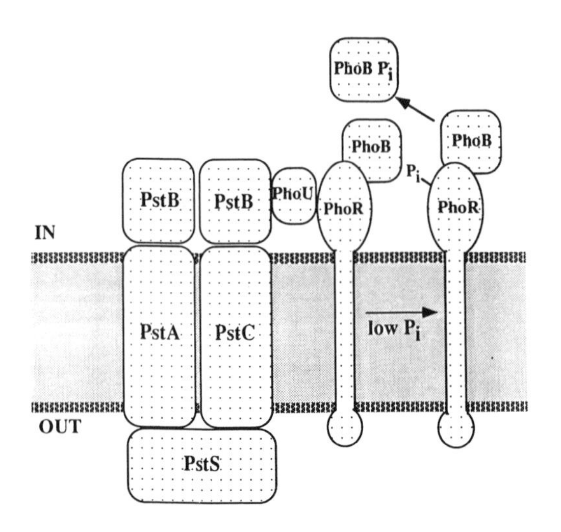

FIG. 1. Diagram of the location in the membrane and interactions of the seven proteins involved in phosphate regulation, based on the proposal of Wanner (1993). The four Pst proteins are also involved in phosphate transport. By analogy with other traffic ATPases, the PstB protein probably functions as a dimer.

```
      234
PhoR  LEGAVAEKALHTMREQTQRME

PhoU  HTIQMLHDVLDAFARMDIDEA
      134           ↓
                    E
```

FIG. 2. The predicted α-helical sequence in which the substitution (A147E) resulting from the *PhoU35* mutation occurs. The charged residues underlined occur at an α-helical periodicity. The A147E substitution is both underlined and indicated by an arrow. The charged residues in a postulated interacting sequence in the PhoR protein are also underlined.

tagenesis approach in an attempt to understand structure-function relationships and gain insight into the molecular mechanism. From hydropathy plots (Engelman et al., 1986) and use of the "positive inside" rule (von Heijne, 1986), models of both the A and C subunits have been developed (Webb et al., 1992). Initially a number of charged residues in both subunits were targeted, and three were found to be essential for phosphate transport (Fig. 3): two residues, R237 and E241, in the PstC protein, and one residue, R220, in the PstA

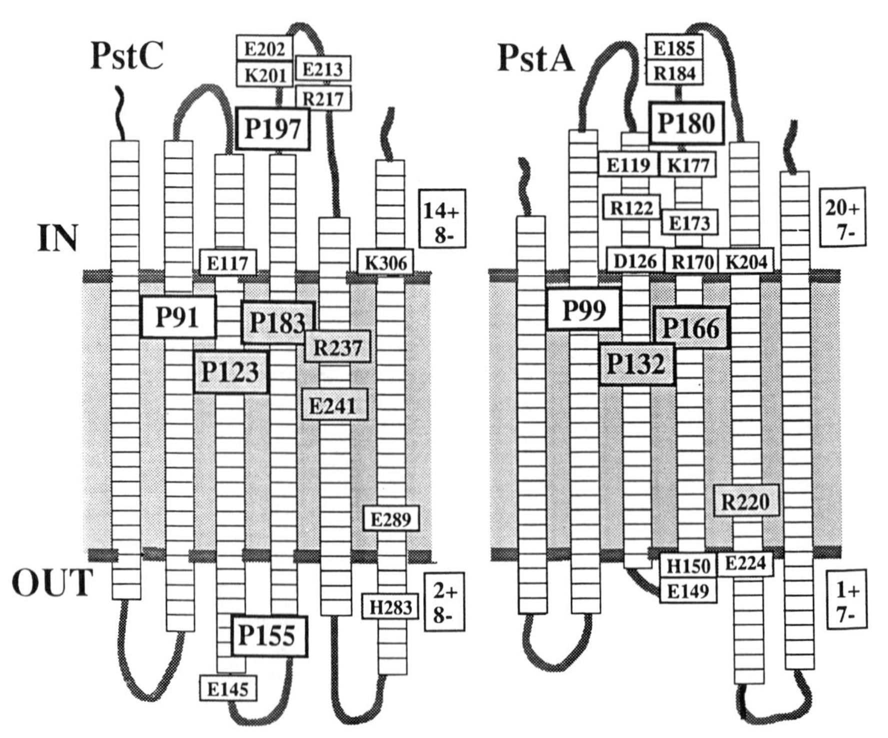

FIG. 3. Putative transmembrane helices of the PstC and PstA proteins, based on structures proposed by Webb et al. (1992). All labeled residues have been altered by site-directed mutagenesis, and shaded boxes indicate residues important for function. The distribution of charged residues on either side of the membrane is indicated. Transmembrane helices referred to in the text are numbered 1 to 6 from the N-terminal end of each subunit.

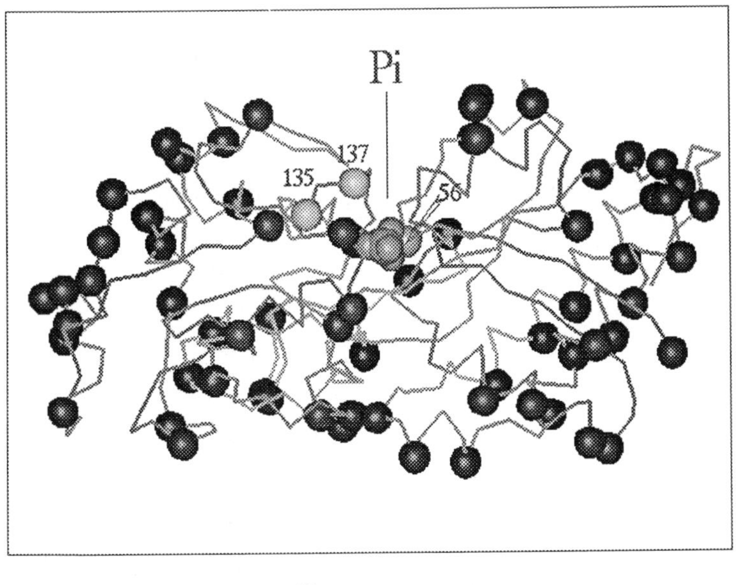

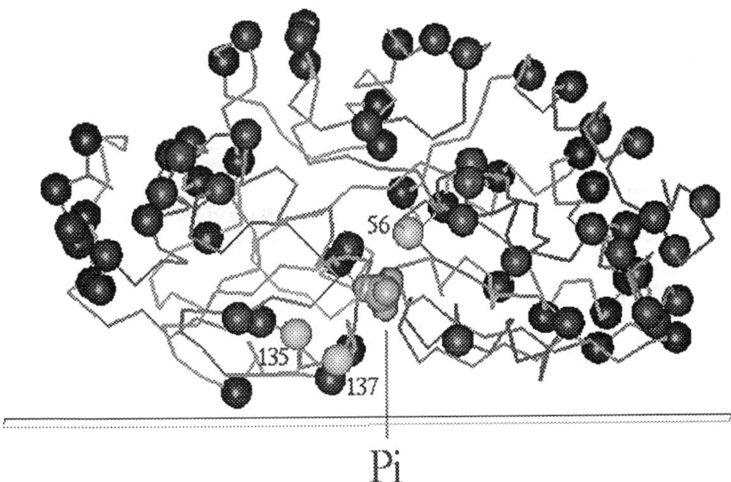

FIG. 4. Crystal structure of the phosphate-loaded PstS protein (Luecke and Quiocho, 1992). Only the α-carbon backbone is shown, and residues carrying charged side chains are depicted as spheres. One structure has been rotated through 90° with respect to the other. The lower structure is suggested to be side-on looking across the membrane surface. The phosphate anion is indicated, and charged residues involved in hydrogen bonding to the phosphate are numbered.

protein. Even though these mutations prevented phosphate transport, alkaline phosphatase remained repressed at high phosphate concentrations in the medium. It appears that a structurally intact Pst system, but not necessarily a functional one, is sufficient to retain its sensory role.

There were two surprising results in the mutagenesis experiments, i.e., that the two mutations E289Q (PstC) and E224Q (PstA) had no effect either on the transport of phosphate or on alkaline phosphatase repression. The E289 residue is located in the membrane in the proposed model, and, given the energy cost of burying a charged residue in a hydrophobic environment, it would be expected that this residue would be important either structurally or functionally. Alternatively, the model may be incorrect. The E224 residue is one turn of an α-helix away from the essential R220 residue, and in an α-helical structure the two residues would form an ion pair. It was surprising that only one residue of the ion pair was found to be essential for function. The double mutant E224Q (PstA) E289L (PstC) was constructed and was found to have lost phosphate transport activity,

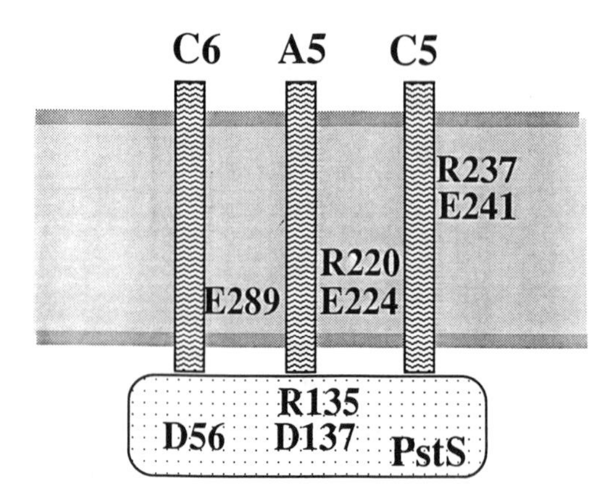

FIG. 5. Proposed interactions between transmembrane helices C6, A5, and C5 and the PstS protein. The charged residues involved in phosphate binding and translocation are indicated. See Fig. 3 and 4 for detailed structures.

whereas alkaline phosphatase remained repressed (Webb, unpublished). There are therefore two salt-bridged arginine residues, R237 (PstC) and R220 (PstA), probably located in the bilayer, that are essential for phosphate movement through the membrane. The residue R237 (PstC) forms an ion pair with the essential residue E241 (PstC). The essential R220 (PstA) forms an ion pair with E224 (PstA), which is essential only in the absence of E289 (PstC). It seems likely, therefore, that the PstA residue R220 is able to interact with either E224 (PstA) or E289 (PstC). These three residues are placed on the periplasmic side of the bilayer (Fig. 3) and might be expected to interact with the PstS protein. The putative helix 6 of the PstC protein would therefore be in close proximity to helix 5 of the PstA protein.

Structure of the PstS Protein

The crystal structure of the PstS protein with bound P_i has been solved in Quiocho's laboratory (Fig. 4). The anhydrous phosphate (as either the dianion or monoanion) is bound without counterions via 12 hydrogen bonds. There are three charged residues involved in the hydrogen bonding, an arginine residue (R135), salt bridged to an aspartate residue (D137), and an acidic residue (D56). Luecke and Quiocho (1990) suggest that D56 plays a key role in specificity, enabling the protein to distinguish between the protonated phosphate and the sulfate oxyanion. As far as charged residues are concerned, the environment of the phosphate in the PstS protein is mirrored by the charged residues on the periplasmic side of helices 6 (PstC) and 5 (PstA) (Fig. 5). It is worth noting that the sulfate-binding protein lacks the equivalent of D56 (Luecke and Quiocho, 1990) and that the subunits of the sulfate transport system equivalent to PstA and PstC also lack an

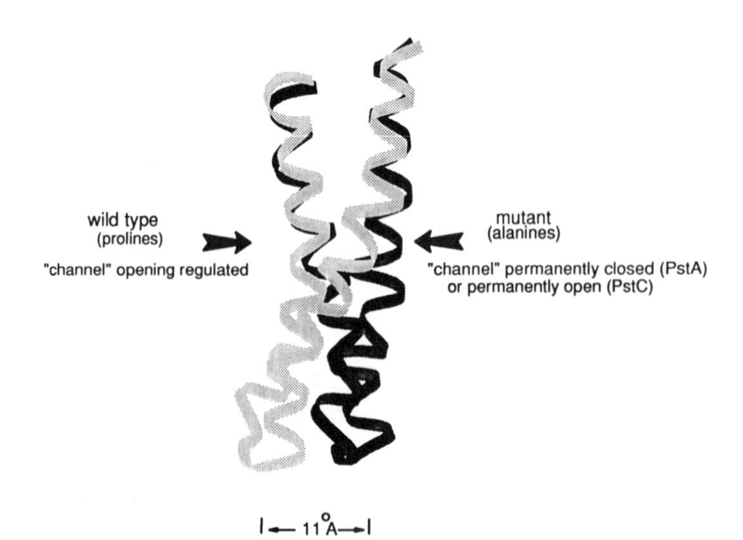

FIG. 6. Helices 3 and 4 of the PstA subunit were modeled by using the Biosym molecular modeling programs. The two models were generated by replacing the wild-type prolines (light shading) with alanines (dark shading). The substitution results in a shift of the "turn" end of the helical hairpin of about 11 Å. If residues P132 and P166 in the PstA protein are both replaced by alanine, transport activity is lost, possibly because the channel is in a permanently closed state. If residues P123 and P183 in the PstC protein are both replaced by alanine, the channel appears to be in a permanently open state. Only PstA helices 3 and 4 were modeled.

acidic residue corresponding to E289 of PstC (Webb, unpublished). However, the salt-bridged arginine residues R220 (PstA) and R237 (PstC) are conserved. The phosphate is bound close to the flat surface of the binding protein and to one side (Fig. 4), and it is therefore proposed that helices 5 (PstA) and 6 (PstC) interact with the PstS protein in the region of the phosphate-binding site (Fig. 5).

Role of Proline Residues in the PstA and PstC Subunits

Brandl and Deber (1986) suggested that the *cis-trans* isomerization of a peptidyl-prolyl bond buried in the membrane may provide the reversible conformational change required for the opening and closing of a transport channel. In the proposed models for PstA and PstC (Fig. 3), there are three buried proline residues in each subunit. These residues plus P180 of PstA and P155 and P197 of PstC were substituted by leucine (Webb et al., 1992). Such substitutions of the buried proline residues in helices 3 and 4 of the proposed model (Fig. 3) caused loss of transport activity, whereas alteration of the other proline residues had little or no effect. However, in contrast to the effect of leucine substitution, if P132 or P166 of PstA or P123 or P183 of PstC was replaced by alanine, only a partial or no effect on phosphate transport activity ensued. The double mutants P132A P166A (PstA) and P123A P183A (PstC) were prepared, and whereas the PstA double mutant had lost phosphate transport activity, the growth of the PstC double mutant was inhibited by 100 mM phosphate. The P123A P183A (PstC) double mutant had a very high phenotypic reversion frequency, and it was not possible to grow a pure culture to carry out phosphate transport activity measurements. The phosphate inhibition of growth was largely removed when the R237Q E241Q double mutation in PstC was introduced into the proline double-mutant strain. It was concluded that the double proline mutation caused the phosphate transporter to be permanently "open" (Webb et al., 1992). The putative helical hairpin formed by helices 3 and 4 of the PstA protein was modeled by Frank Gibson by using the Biosym molecular modeling programs (Biosym Technologies, San Diego, Calif.) and energy minimized (Fig. 6). Proline residues 132 and 166 were replaced by alanine, and the mutant helices were also energy minimized (Fig. 6). When the open ends of the helices are aligned, the closed loop has shifted about 11 Å as a result of substituting the alanine residues for the proline residues.

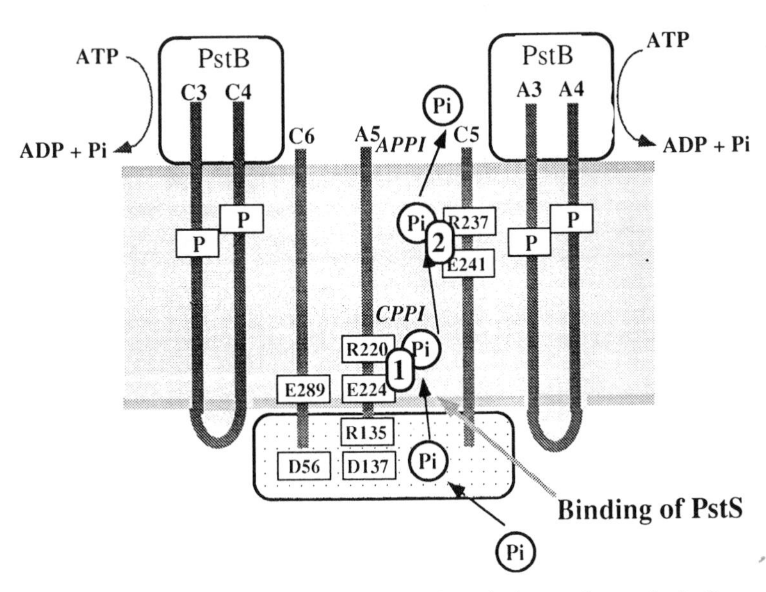

FIG. 7. Proposed mechanism for phosphate translocation through the membrane via the Pst system. Phosphate bound to the PstS protein is transferred to the first membrane-binding site (site 1) on interaction of PstS with PstA and C. The two subsequent movements of phosphate, from membrane-binding site 1 to membrane-binding site 2 and then from 2 to the cytoplasm, occur as a result of conformational changes brought about by *cis-trans* isomerization changes at peptidyl-prolyl bonds energized by ATP hydrolysis on the two PstB subunits. *APPI* indicates the peptidyl-prolyl isomerase activity of the PstB subunit with the PstA subunit as substrate. *CPPI* indicates a similar activity but with the PstC subunit as substrate.

Proposed Mechanism of Phosphate Transport

The mechanism of phosphate transport proposed is summarized in Fig. 7. The phosphate dianion in the periplasmic space is sequestered by the PstS protein. There are three charged residues involved in hydrogen bonding of the phosphate (D56, R135, and D137), with D56 thought to play a key role in the specificity determination (Luecke and Quiocho, 1990). The interaction of PstS with the PstA and PstC subunits results in the transfer of phosphate to a binding site that also involves three charged residues (R220, E224, and E289) on helices A5 and C6, with E289 proposed to play a role in specificity. The conformational changes required to remove the PstS protein and to bring the sequestered phosphate into contact with the second membrane site (R237, E241) are proposed to be energized by ATP hydrolysis catalyzed by the PstB subunit and is achieved by peptidyl-prolyl isomerization on helices C3 and C4 (or A3 and A4). The separation of membrane sites 1 and 2 and the release of phosphate to the inside of the cell are achieved by a similar sequence of events but involving helices A3 and A4 (or C3 and C4). The movement of phosphate is thought to occur in a groove at the protein-phospholipid interface (Webb et al., 1992).

We thank Frank Gibson for computing.

We thank the following for their generous support: John Proud, The Raymond E. Purves Foundation, The James N. Kirby Foundation, C. H. Warman, The Bruce and Joy Reid Foundation, and Rutherford Robertson.

REFERENCES

Ames, G. F.-L. 1988. Structure and mechanism of bacterial periphasmic transport systems. *J. Bioenerg. Biomembr.* **20:**1–18.

Ames, G. F.-L., and A. Joshi. 1990. Energy coupling in bacterial periplasmic permeases. *J. Bacteriol.* **172:**4133–4137.

Brandl, C. J., and C. M. Deber. 1986. Hypothesis about the function of membrane-buried proline residues in transport proteins. *Proc. Natl. Acad. Sci. USA* **83:**917–921.

Engelman, D. M., T. A. Steitz, and A. Goldman. 1986. Identifying non-polar transbilayer helices in amino acid sequences of membrane proteins. *Annu. Rev. Biophys. Biophys. Chem.* **15:**321–353.

Hyde, S. C., P. Emsley, M. M. Hartshorn, U. Gileadi, S. R. Pearce, M. P. Gallagher, D. R. Gill, R. E. Hubbard, and C. F. Higgins. 1990. Structural model of ATP-binding proteins associated with cystic fibrosis, multidrug resistance and bacterial transport. *Nature* (London) **346:**362–365.

Luecke, H., and F. A. Quiocho. 1990. High specificity of a phosphate transport protein determined by hydrogen bonds. *Nature* (London) **347:**4092–4096.

Luecke, H., and F. A. Quiocho. 1992. *Brookhaven Protein Data Bank, pdb1abh.ent.* Brookhaven National Laboratory, Brookhaven, N.Y.

Muda, M., N. N. Rao, and A. Torriani. 1992. Role of PhoU in phosphate transport and alkaline phosphatase regulation. *J. Bacteriol.* **174:**8057–8064.

Rosenberg, H., R. G. Gerdes, and K. Chegwidden. 1977. Two systems for the uptake of phosphate in *Escherichia coli.* *J. Bacteriol.* **131:**505–511.

Scholten, M., and J. Tommassen. 1993. Topology of the PhoR protein of *Escherichia coli* and functional analysis of internal deletion mutants. *Mol. Microbiol.* **8:**269–275.

Steed, P. M., and B. L. Wanner. 1993. Use of the rep technique for allele replacement to construct mutants with deletions of the pstSCAB-phoU operon: evidence of a new role for the phoU protein in the phosphate regulon. *J. Bacteriol.* **175:**6797–6809.

Surin, B. P., H. Rosenberg, and G. B. Cox. 1985. Phosphate-specific transport system of *Escherichia coli:* nucleotide sequence and gene-polypeptide relationships. *J. Bacteriol.* **161:**189–198.

Torriani, A., and F. Rothman. 1961. Mutants of *Escherichia coli* constitutive for alkaline phosphatase. *J. Bacteriol.* **81:**835–836.

von Heijne, G. 1986. The distribution of positively charged residues in bacterial inner membrane proteins correlates with the trans-membrane topology. *EMBO J.* **5:**3021–3027.

Wanner, B. L. 1993. Gene regulation by phosphate in enteric bacteria. *J. Cell. Biochem.* **51:**47–54.

Webb, D. C. Unpublished observation.

Webb, D. C., H. Rosenberg, and G. B. Cox. 1992. Mutational analysis of the *Escherichia coli* phosphate-specific transport system, a member of the traffic ATPase (or ABC) family of membrane transporters. *J. Biol. Chem.* **267:**24661–24668.

Chapter 7

Phosphate Inorganic Transport (Pit) System in *Escherichia coli* and *Acinetobacter johnsonii*

HENDRIK W. VAN VEEN,[1] TJAKKO ABEE,[2] GERARD J. J. KORTSTEE,[3] WIL N. KONINGS,[1] AND ALEXANDER J. B. ZEHNDER[4]

Department of Microbiology, University of Groningen, P. O. Box 14, NL-9750 AA Haren,[1] Department of Food Science, Agricultural University Wageningen, NL-6703 HD Wageningen,[2] and Department of Microbiology, Agricultural University Wageningen, NL-6703 CT Wageningen,[3] The Netherlands, and Swiss Federal Institute for Environmental Science and Technology (EAWAG)/Swiss Federal Institute of Technology (ETH), CH-8600 Dübendorf, Switzerland[4]

Although the phosphate inorganic transport (Pit) system is the major secondary uptake system for P_i in *Escherichia coli*, its transport mechanism has been studied less extensively than that of the P_i-linked antiporters for *sn*-glycerol 3-phosphate (GlpT) and glucose 6-phosphate (UhpT) (for a review, see Maloney et al. [1990]) or the binding-protein-dependent phosphate-specific transport system (Pst) (for a review, see Silver and Walderhaug [1992]). The transport of the phosphate anion via these four systems has generally been interpreted in terms of the translocation of mono- and/or dibasic phosphate. Thus, the periplasmic binding protein which is the initial receptor in the Pst system, shows affinity for $H_2PO_4^-$ and HPO_4^{2-} (Kubena et al., 1986; Luecke and Quiocho, 1990). GlpT and UhpT belong to a family of P_i-linked antiporters which mediate electroneutral exchange of $H_2PO_4^-$, organic phosphate anions, or both (Maloney et al., 1990). In addition, it is generally assumed that HPO_4^{2-} is the phosphate ion species which is transported via Pit (Rosenberg et al., 1984; Rosenberg, 1987). However, recent work on the mechanism and energetics of P_i transport via Pit in *E. coli* and *Acinetobacter johnsonii* suggests that this system may represent a new class of bacterial porters whose operation involves the transport of a soluble, neutral metal phosphate ($MeHPO_4$) complex rather than P_i (Van Veen et al., 1993b, 1994a, b). This surprising aspect of the mechanism of Pit is not immediately evident. Instead, there is a successful masquerade because in many P_i transport studies in the past, divalent metal ions and P_i were simultaneously present under the experimental conditions. For this reason, the cotransport across biomembranes of divalent metal ions and P_i via metal phosphate chelates may be a more common phenomenon in prokaryotic and eukaryotic cells. A study of the Pit system may therefore have wide significance for the general field of transport.

Accordingly, three issues will be dealt with in this review. First, a concise description of the basic properties of the Pit system in *E. coli* which have been brought to light during the past 20 years of research will be presented. The reader is referred to the excellent review by Rosenberg (1987) for a more detailed account of the early studies on phosphate transport via Pit. Second, the current evidence concerning the substrate specificity and mechanism of energy coupling ($MeHPO_4$/H^+ symport) of Pit will be summarized. Third, some implications for other fields of study will be discussed. Clearly, all the relevant areas cannot be treated in the detail they deserve. We do hope, however, that the topics emphasized will offer perspectives for future research.

Basic Properties of Pit

The *E. coli* Pit system was originally described by Bennett and Malamy (1970), Medveczky and Rosenberg (1971), and Willsky et al. (1973). It was differentiated from the Pst system on the basis of specificity for P_i and its toxic analog arsenate. Pit shows a relatively low substrate specificity, with a K_m of about 25 μM for P_i and arsenate, whereas the Pst system has a 100-fold-higher affinity for P_i (K_m, 0.25 μM P_i) than for arsenate (Rosenberg et al., 1977; Willsky and Malamy, 1980a, 1980b; Silver et al., 1981). Pit function is insensitive to a cold osmotic shock or spheroplast formation (Medveczky and Rosenberg, 1970; Rae and Strickland, 1976a; Rosenberg et al., 1977). Hence, Pit does not include a periplasmic phosphate-binding protein, as was shown for the Pst system.

Pst-deficient mutants of *E. coli* were used for a further study of Pit (Rosenberg et al., 1977; Willsky and Malamy, 1980a). P_i transport in these mutants was not affected by P_i deprivation, showing that Pit is constitutively expressed. P_i uptake via Pit was completely abolished by the protonophore carbonyl cyanide 3-chlorophenylhydrazone (CCCP), indicating a role for the proton motive force as the driving force for Pit. Additional evidence supporting this suggestion came from a study of P_i uptake in *E. coli* strains carrying the Pit system in a normal or H^+-translocating F_0F_1 ATPase-defective background (Rosenberg et al., 1979). The presence of an inactive F_0F_1 ATPase did not affect glucose-energized P_i uptake via Pit under aerobic conditions, but strongly reduced P_i uptake under anaerobic conditions. Apparently, the generation of a proton motive force through proton pumping via the H^+-ATPase is essential for energization of Pit under anaerobic conditions. In the presence of oxygen, Pit can be energized by a proton motive force generated via redox-reaction-coupled primary H^+ translocation by the respiratory chain. Further insight into energy coupling to Pit came from the work of Konings and Rosenberg (1978), who studied the effect of ionophores on P_i transport in membrane vesicles from *E. coli* energized by the oxidation of D-lactate or ascorbate/phenazine methosulfate. Although in these experiments, P_i transport was biased toward the ΔpH component of the proton motive force, the transport process appeared to be electrogenic. This latter property seems to be restricted to Pit since GlpT and UhpT mediate an electroneutral P_i transport mechanism (Maloney et al., 1990). The absence of proton motive force-driven uptake of P_i in membrane vesicles prepared from the Pit-deficient *E. coli* K-10 is consistent with this notion (Konings and Rosenberg, 1978).

Like other secondary transport systems (e.g., LacY [Kaback, 1990]), Pit is probably composed of a single polypeptide. All known *pit* mutations map within the same locus (77 min) of the *E. coli* chromosome and are complemented by transformation with a plasmid carrying a 2.2-kb chromosomal *Sal*I-*Ava*I fragment. This fragment was sequenced, and an open reading frame comprising 1,287 bp was found. The deduced polypeptide contains 429 amino acids, corresponding to a molecular mass of 46.2 kDa (Elvin et al., 1986, 1987). The actual sequence data have not yet been published.

Substrate Specificity

Since the idea of $MeHPO_4/H^+$ symport arose during studies of P_i transport by the polyphosphate-accumulating *A. johnsonii* 210A, it is useful to summarize relevant information from that work. Transport studies with whole cells gave kinetic evidence for the presence of two phosphate transport systems which show a strong analogy to the Pst and Pit systems in *E. coli* (Bonting et al., 1992; Van Veen et al., 1993a). Pit in *A. johnsonii* 210A is a constitutive phosphate/arsenate transport system with a K_m for P_i of about 9 μM and a K_i for arsenate of about 11 μM at pH 7.0. P_i transport via *A. johnsonii* Pit is strongly inhibited by the uncoupler CCCP (Van Veen et al., 1993a, 1993b).

A. johnsonii Pit has been characterized in membrane vesicles and proteoliposomes in which the carrier protein was reconstituted (Van Veen et al., 1993b). During proton motive force-driven uptake, the apparent K_m for P_i increased from about 5 μM at pH 8.0 to 24 μM at pH 6.0. Although this result would be consistent with a specificity for HPO_4^{2-}, further studies pointed to the translocation of a neutral $MeHPO_4$ complex. Thus, P_i uptake via *A. johnsonii* Pit was strictly dependent on the presence of divalent cations, such as Ca^{2+}, Mg^{2+}, Co^{2+}, or Mn^{2+} (Fig. 1). Similar observations were reported for a *Bacillus subtilis* citrate permease mediating the transport of a Me^{2+}-citrate complex (Bergsma and Konings, 1983; Willecke et al., 1973). Calculation of the concentrations of several P_i species under the experimental conditions indicated that at pH 7.0, 31% (Ca^{2+}) to 87% (Mn^{2+}) of P_i was present as a neutral $MeHPO_4$ complex (Van Veen et al., 1993b,

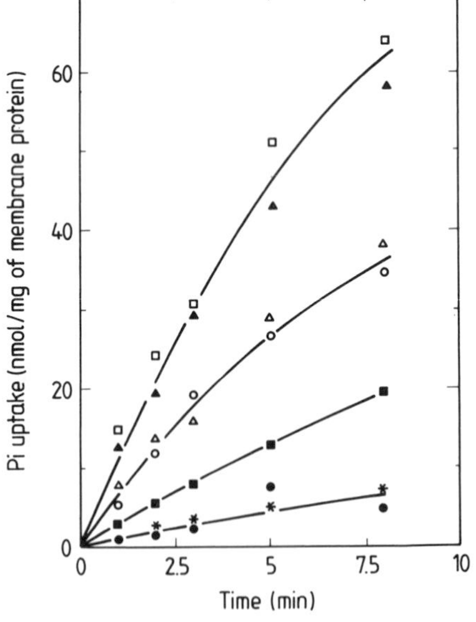

FIG. 1. P_i transport via *A. johnsonii* Pit is dependent on the presence of divalent cations. Imposed proton motive force-driven uptake of 50 μM $^{32}P_i$ was measured in proteoliposomes in an *N*-methyl-*o*-glucamine–piperazine-*N*,*N'*-bis(2-ethanesulfonic acid) (PIPES)-based buffer (pH 7.0) in the presence of 2 mM Mg^{2+} (○), 2 mM Ca^{2+} (△), 2 mM Co^{2+} (▲), 2 mM Mn^{2+} (□), or 0.5 mM EDTA (●) or in the absence of added divalent cations (■) or an artificially imposed Δp of −240 mV (*). Reprinted from Van Veen et al. (1993b) with permission.

1994a). The stimulation of P_i uptake by the divalent metal ions in experiments whose results are shown in Fig. 1 correlated well with the concentrations of $MeHPO_4$ in the incubation mixtures. Moreover, a reevaluation of the kinetic data for *A. johnsonii* Pit revealed a pH-independent K_m of 7.9 μM $MeHPO_4$ (Van Veen et al., 1993b). Finally, control measurements excluded other possible explanations for the metal dependence of P_i transport via *A. johnsonii* Pit, such as binding of $MeHPO_4$ to membranes, internal precipitation of $MeHPO_4$ as a result of solute accumulation, or secondary effects of divalent cations on the magnitude or stability of the artificially imposed proton motive force in proteoliposomes (Van Veen et al., 1994a).

Additional experiments have been carried out to elucidate the substrate specificity and transport mechanism of *A. johnsonii* Pit (Van Veen et al., 1993b, 1994a). However, we will now return to the work on the Pit system of *E. coli* because it appears that *A. johnsonii* Pit operates in a similar way to *E. coli* Pit (Van Veen et al., 1994b). Proton motive force-driven P_i transport in proteoliposomes in which Pit carrier protein from *E. coli* was reconstituted was found to be strictly divalent-cation dependent. The K_m for $MeHPO_4$ of 8.8 μM is very similar to the one obtained for *A. johnsonii* Pit. Specificity for $MeHPO_4$ was tested further by studying the proton motive force-driven transport of Mg^{2+} and Ca^{2+} in proteoliposomes in the presence and absence of P_i. Besides a divalent-cation-dependent uptake of P_i, such studies demonstrated (i) the P_i-dependent uptake of Mg^{2+} and Ca^{2+}, (ii) the inhibition by Mg^{2+} of Ca^{2+} uptake in the presence of P_i but not of P_i uptake in the presence of Ca^{2+}, and (iii) the equimolar transport of Ca^{2+} and P_i. Verification that $MeHPO_4$ but not P_i is the authentic substrate of Pit came from transport experiments performed in the absence of a proton motive force. Measurements of solute transport via exchange and efflux reactions allowed an easy experimental control over *cis* and *trans* compartments. Under these conditions, *E. coli* Pit could mediate the efflux and homologous exchange of $MeHPO_4$ but not the heterologous exchange of $MeHPO_4$ and the substrates for the GlpT and UhpT system: P_i, glycerol 3-phosphate, or glucose 6-phosphate.

At present the evidence is most simply interpreted by the translocation of $MeHPO_4$ but not of P_i via the Pit system in both *E. coli* and *A. johnsonii.* Because of the almost identical pH dependence of $MeHPO_4$ and HPO_4^{2-} in aqueous solutions (Van Veen et al., 1994a), it is now understandable why earlier studies had pointed out HPO_4^{2-} as the physiological substrate of Pit (Rosenberg et al., 1984; Rosenberg, 1987). With hindsight, we could say that the previously observed Mg^{2+} dependence of Pit function in cells (Medveczky and Rosenberg, 1971; Rae and Strickland, 1976b) and membrane vesicles (Konings and Rosenberg, 1978) reflected the translocation of $MgHPO_4$ via this system.

Transport Mechanism

Artificial imposition of ion diffusion gradients in proteoliposomes containing Pit protein confirmed previous work with membrane vesicles of *E. coli* (Konings and Rosenberg, 1978). Both a membrane potential and a pH gradient can drive $MeHPO_4$ transport through this system. A detailed analysis of the steady-state accumulation level of membrane potential-driven uptake of $MeHPO_4$ indicated the translocation of a (neutral) $MeHPO_4$ complex in symport with one proton (Van Veen et al., 1994b).

The mechanism of $MeHPO_4/H^+$ symport via Pit has been deduced from the pH and proton motive force dependence of $MeHPO_4$ uptake, efflux, counterflow, and equilibrium exchange (Van Veen et al., 1993b, 1994b). In the present discussion these facts will not be recited but will be incorporated into two models. In the cellular model, the overall transport reactions are conveniently summarized (Fig. 2). The uptake of a soluble, neutral $MeHPO_4$ complex occurs in symport with that of a proton (Fig. 2A). During $MeHPO_4$ efflux the transport reaction is reversed, resulting in the generation of a proton motive force. In homologous exchange of $MeHPO_4$ (Fig. 2B), no net translocation of protons takes place. In the kinetic model (Fig. 3), the vectorial translocation of $MeHPO_4$ across the cytoplasmic membrane can be considered a cyclic process in which binding and dissociation of $MeHPO_4$ and a proton on the outer and inner surfaces of the membrane occur via an ordered mechanism. Thus, the process of efflux involves protonation of the carrier protein on the inner surface of the membrane followed by binding of $MeHPO_4$. The loaded carrier protein reorients the binding sites to the outer surface of the membrane, after which $MeHPO_4$ and then the catalytic proton are released from the carrier. A conformational change of the empty carrier restores the initial orientation of the binding sites. During homologous $MeHPO_4$ exchange, the carrier recycles without being deprotonated (Fig. 3).

The dissociation of the catalytic proton from the carrier protein appears to be rate limiting for $MeHPO_4$ transport. As a result, $MeHPO_4$ uptake via Pit is strongly inhibited by a low internal pH. For optimal function of Pit in cells it is therefore essential to maintain a constant alkaline pH in the cytosol. In bacteria, electrogenic uptake of K^+ accompanied by expulsion of H^+ from the cytoplasm is an important mechanism for alkalinization of the cellular interior (Bakker, 1993). Evidence confirming the relevance of this process for

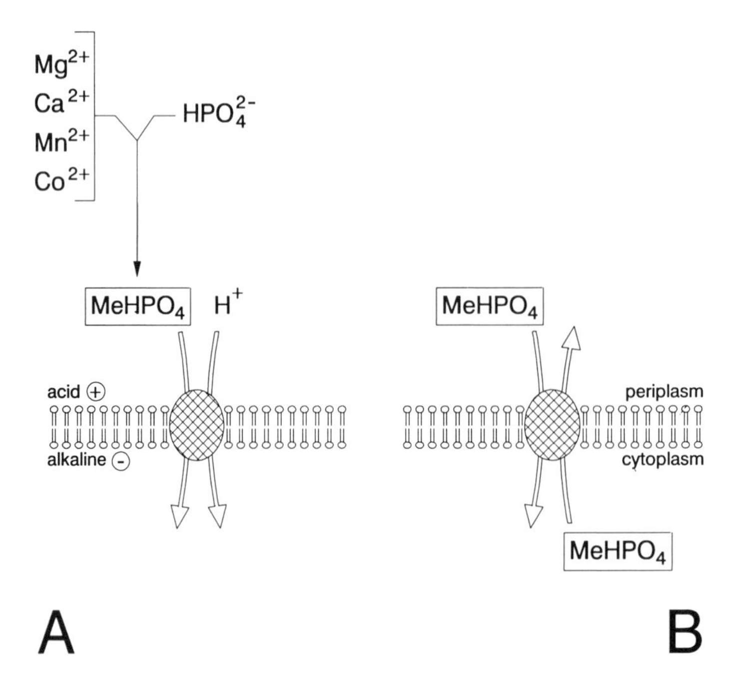

FIG. 2. Cellular model showing overall transport reactions via Pit. (A) Proton motive force-driven uptake of $MeHPO_4$, which is formed by complexation of Me^{2+} and HPO_4^{2-}. (B) Homologous $MeHPO_4$ exchange.

phosphate transport via Pit came from the work of Russell and Rosenberg (1979, 1980), who demonstrated that although potassium ions greatly stimulate Pit function in *E. coli* cells, K^+ transport and phosphate transport are linked indirectly via proton circulation.

Pit catalyzes completely reversible transport reactions under deenergized conditions. However, unlike the exchange reactions mediated via other proton symporters (Driessen et al., 1987; Kaback, 1990) or the P_i-linked antiporters UhpT and GlpT (Maloney et al., 1990), the homologous exchange of $MeHPO_4$ via Pit is inhibited by the membrane potential (Fig. 3). This difference in behavior may play an important role in the intriguing paradox that Pit, being a secondary phosphate transporter, mediates an apparent unidirectional uptake of phosphate in cells under physiological conditions (Willsky and Malamy, 1980a; Elvin et al., 1985; Rosenberg, 1987). In addition, the exchange reaction catalyzed by Pit may have remained unnoticed in previous studies (Elvin et al., 1985; Ambudkar et al., 1986; Rosenberg, 1987; Sonna et al., 1988) because (i) the substrate is $MeHPO_4$ rather than P_i and (ii) the maximal velocity of metal-phosphate exchange is at least two orders of magnitude lower than that of homologous P_i exchange via the UhpT or GlpT system in fully induced cells or in its membrane vesicles (Maloney et al., 1990; Van Veen et al., 1993b, 1994b).

The kinetic model depicted in Fig. 3 provides a useful framework for discussing the energetics and mechanism of $MeHPO_4/H^+$ symport via Pit but does not represent the actual molecular translocation pathway. Thus, the recycling of the Pit carrier should not be interpreted in terms of diffusion of the whole transport protein within the membrane. This would clearly be incorrect on thermodynamic grounds and would be incompatible with current structural models of secondary transport proteins (Henderson, 1991). The alternate exposure of the binding sites of the carrier protein to the outer and inner surfaces of the membrane could, for example, be envisaged as a small conformational change which would realign the substrate-binding sites within a pore-like translocation pathway formed by the transmembrane domains of the carrier protein. A thorough understanding of the translocation mechanism at the molecular level should involve knowledge of the identity of specific amino acid residues which play a role in the catalytic process.

Broader Implications

The finding of $MeHPO_4/H^+$ symport in *E. coli* and the polyphosphate-accumulating *A. johnsonii* 210A suggests that this reaction may be a general mechanism for the transport of divalent metal ions and P_i (plus arsenate) in bacteria. P_i transport in general or via a Pit-like system in particular in *Micrococcus lysodeikticus* (Friedberg, 1977), *Acinetobacter lwoffii* (Yashphe et al., 1992), *Pseudomonas aeruginosa* (Lacoste et al., 1981), and

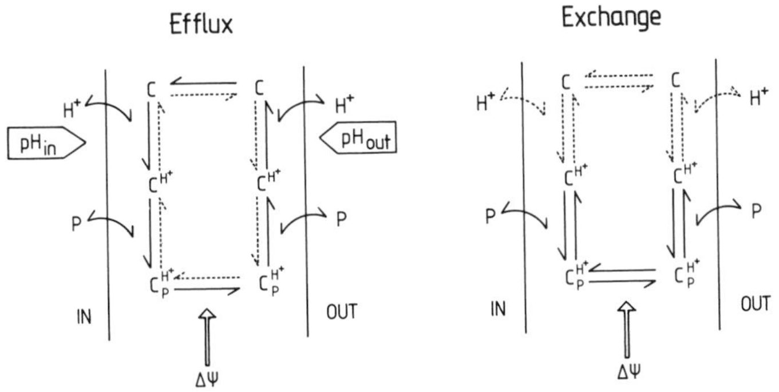

FIG. 3. Kinetic model for the transport cycle of the Pit system catalyzing $MeHPO_4$ efflux and exchange. The model consists of a single transport loop linking six discrete states of the Pit carrier. The state transitions include one transmembrane charge transport step and one step each for binding of $MeHPO_4$ and proton at each side of the membrane. In the model an ordered mechanism for association and dissociation of $MeHPO_4$ and proton to and from the carrier protein is suggested. Abbreviations: C, carrier protein; P, $MeHPO_4$; H^+, protons; IN and OUT, inside and outside of the cytoplasmic membrane, respectively. Solid and dashed arrows indicate the major and minor steps (in terms of rates), respectively, involved in the efflux and exchange reactions. The internal and external pH have influence upon the (de)protonation of the carrier protein. The reorientation of the binding sites of the positively charged ternary carrier-H^+-$MeHPO_4$ complex is affected by the membrane potential ($\Delta\psi$). Reprinted from Van Veen et al. (1993b) with permission.

Bacillus cereus (Rosenberg et al., 1969) was reported to be stimulated by Mg^{2+}. In addition, some investigators observed a stimulation by P_i of Me^{2+} transport, e.g., the uptake of Mn^{2+} in the manganese polyphosphate-accumulating *Lactobacillus plantarum* (Archibald and Duong, 1984) and of Mg^{2+}, Ca^{2+}, Mn^{2+}, and Co^{2+} via a general divalent cation transport system in *Bacillus subtilis* (Kay and Ghei, 1981). Interestingly, a Pit mutant of *B. subtilis* was strongly impaired in the transport of Ca^{2+} and Co^{2+}. The mutant still elicited significant Mn^{2+} transport as a result of uptake via a second Mn^{2+}-specific high-affinity uptake system (Kay and Ghei, 1981).

In *A. johnsonii* 210A, transport of $MeHPO_4$ is closely related to the metabolism of cytoplasmic polyphosphate granules in which P_i and divalent metal ions are accumulated (Van Veen et al., 1994a). When oxidative phosphorylation is impaired (e.g., under anaerobic conditions), this strictly aerobic organism degrades its metal polyphosphate, resulting in the excretion of $MeHPO_4$ and H^+ via Pit. Recent results indicate that during this efflux the components of the proton motive force are generated (Van Veen et al., 1993b, submitted). The $MeHPO_4$ efflux-induced proton motive force can be coupled to energy consuming processes such as the uptake of solutes and the synthesis of ATP via the H^+-ATPase (Van Veen et al., submitted). Thus, metabolic energy from polyphosphate degradation can be conserved in a similar way to that proposed by Michels et al. (1979) in their "energy recycling model."

Research on prokaryotic calcium transport systems lags behind that on eukaryotes. Although information on bacterial Ca^{2+} efflux systems is available (for a review, see Ambudkar and Rosen [1990]), the mechanisms for Ca^{2+} entry into bacteria are unclear (Lynn and Rosen, 1987). The transport of $MeHPO_4$, including the calcium phosphate complex, via Pit provides *E. coli* with such a mechanism. Pit may be functionally linked to the Ca^{2+}/nH^+ (where n is the number of protons translocated in antiport with Ca^{2+} [$n \geq 3$]) antiporter of this organism. Thus, a chemiosmotic circuit for divalent cations can be envisaged in which Pit mediates the entrance of P_i and divalent cations, whereas the antiporter catalyzes the proton motive force-driven extrusion of Ca^{2+} and other divalent cations such as Mn^{2+}, Sr^{2+}, and Ba^{2+} (Brey and Rosen, 1979) in order to maintain low concentrations of these ions in the cytosol.

Transport of metal phosphate chelates may also be encountered in eukaryotic cells. In many biological systems P_i transport is linked to the cellular metabolism and transport of divalent cations. Three such examples might be considered. First, divalent cations stimulate the uptake of P_i across the plasma and vacuolar membranes in *Saccharomyces cerevisiae* and other lower eukaryotes (Klionsky et al., 1990; Kotyk and Horák, 1981; Nieuwenhuis, 1982). Second, P_i is known to have a large influence on Ca^{2+} transport by mitochondria isolated from a range of tissues and species (Lehninger, 1970; Pressman, 1970; Meisner et al., 1972; Wohlrab, 1986; Breitbart et al., 1990; Bygrave et al., 1990). Third, the influx of Mg^{2+} in rat hepatocytes has recently been suggested to occur by sodium motive force-driven Mg^{2+}/P_i cotransport (Günther and Höllriegl,

1993). Clearly, the weight of the current evidence necessitates the continued consideration of the impact of divalent metal ions on P_i transport processes and vice versa. In addition, it provides an adequate basis on which to encourage further experimentation on the topic.

This work was financially supported by the Netherlands Organization for Scientific Research (NWO) and the Technology Foundation (STW).

REFERENCES

Ambudkar, S. V., T. J. Larson, and P. C. Maloney. 1986. Reconstitution of sugar phosphate transport systems of *Escherichia coli. J. Biol. Chem.* **261:**9083–9086.

Ambudkar, S. V., and B. P. Rosen. 1990. Ion-exchange systems in prokaryotes, p. 247–271. *In* T. A. Krulwhich (ed.), *The Bacteria,* vol. 12. Academic Press, Inc., New York.

Archibald, F. S., and M.-N. Duong. 1984. Manganese acquisition by *Lactobacillus plantarum. J. Bacteriol.* **158:**1–8.

Bakker, E. P. 1993. Cell K^+ and K^+ transport systems in prokaryotes, p. 205–224. *In* E. P. Bakker (ed.), *Alkali Cation Transport Systems in Prokaryotes,* CRC Press, Inc., Boca Raton, Fla.

Bennett, R. L., and M. H. Malamy. 1970. Arsenate resistant mutants of *Escherichia coli. Biochem. Biophys. Res. Commun.* **40:**496–503.

Bergsma, J., and W. N. Konings. 1983. The properties of citrate transport in membrane vesicles from *Bacillus subtilis. Eur. J. Biochem.* **134:**151–156.

Bonting, C. F. C., H. W. Van Veen, A. Taverne, G. J. J. Kortstee, and A. J. B. Zehnder. 1992. Regulation of polyphosphate metabolism in *Acinetobacter* strain 210A grown in carbon- and phosphate-limited continuous cultures. *Arch. Microbiol.* **158:**139–144.

Breitbart, H., R. Wehbie, and H. A. Lardy. 1990. Calcium transport in bovine sperm mitochondria: effect of substrates and phosphate. *Biochim. Biophys. Acta* **1026:**57–63.

Brey, R. N., and B. P. Rosen. 1979. Cation/proton antiport systems in *Escherichia coli. J. Biol. Chem.* **254:**1957–1963.

Bygrave, F. L., L. Lenton, J. G. Altin, B. A. Setchell, and A. Karjalainen. 1990. Phosphate and calcium uptake by mitochondria and by perfused rat liver induced by the synergistic action of glucagon and vasopressin. *Biochem. J.* **267:**69–73.

Driessen, A. J. M., S. De Jong, and W. N. Konings. 1987. Transport of branched chain amino acids in membrane vesicles of *Streptococcus cremoris. J. Bacteriol.* **169:**5193–5200.

Elvin, C. M., N. E. Dixon, and H. Rosenberg. 1986. Molecular cloning of the phosphate (inorganic) transport (*pit*) gene of *Escherichia coli* K-12. Identification of the *pit*$^+$ gene product and physical mapping of the *pit-gor* region of the chromosome. *Mol. Gen. Genet.* **204:**477–484.

Elvin, C. M., C. M. Hardy, and H. Rosenberg. 1985. P_i exchange mediated by the GlpT-dependent *sn*-glycerol 3-phosphate transport system in *Escherichia coli. J. Bacteriol.* **161:**1054–1058.

Elvin, C. M., C. M. Hardy, and H. Rosenberg. 1987. Molecular studies on the phosphate inorganic transport system of *Escherichia coli,* p. 156–158. *In* A. Torriani-Gorini, F. G. Rothman, S. Silver, A. Wright, and E. Yagil (ed.), *Phosphate Metabolism and Cellular Regulation in Microorganisms.* American Society for Microbiology, Washington, D.C.

Friedberg, I. 1977. Phosphate transport in *Micrococcus lysodeikticus. Biochim. Biophys. Acta* **466:**451–460.

Günther, T., and V. Höllriegl. 1993. Na^+- and anion-dependent Mg^{2+} influx in isolated hepatocytes. *Biochim. Biophys. Acta* **1149:**49–54.

Henderson, P. J. F. 1991. Studies of translocation catalysis. *Biosci. Rep.* **11:**477–538.

Kaback, H. R. 1990. Active transport: membrane vesicles, bioenergetics, molecules and mechanisms, p. 151–202. *In* T. A. Krulwhich (ed.), *The Bacteria,* vol. 12. Academic Press, Inc., New York.

Kay, W. W., and O. K. Ghei. 1981. Inorganic cation transport and the effects on C_4 dicarboxylate transport in *Bacillus subtilis. Can. J. Microbiol.* **27:**1194–1201.

Klionsky, D. J., P. K. Herman, and S. D. Emr. 1990. The fungal vacuole: composition, function and biogenesis. *Microbiol. Rev.* **54:**266–292.

Konings, W. N., and H. Rosenberg. 1978. Phosphate transport in membrane vesicles from *Escherichia coli. Biochim. Biophys. Acta* **508:**370–378.

Kotyk, A., and J. Horák. 1981. Transport processes in the plasma membrane, p. 49–64. *In* W. N. Arnold (ed.), *Yeast Cell Envelopes: Biochemistry, Biophysics and Ultrastructure,* vol. 1. CRC Press, Inc., Boca Raton, Fla.

Kubena, B. D., H. Luecke, H. Rosenberg, and F. A. Quiocho. 1986. Crystallization and X-ray diffraction studies of a phosphate-binding protein involved in active transport in *Escherichia coli. J. Biol. Chem.* **261:**7995–7996.

Lacoste, A.-M., A. Cassaigne, and E. Neuzil. 1981. Transport of inorganic phosphate in *Pseudomonas aeruginosa. Curr. Microbiol.* **6:**115–120.

Lehninger, A. L. 1970. Mitochondria and calcium transport. *Biochem. J.* **119:**129–138.

Luecke, H., and F. A. Quiocho. 1990. High specificity of a phosphate transport protein determined by hydrogen bonds. *Nature* (London) **347:**402–406.

Lynn, A. R., and B. P. Rosen. 1987. Calcium transport in prokaryotes, p. 181–201. *In* B. P. Rosen, and S. Silver (ed.), *Ion Transport in Prokaryotes.* Academic Press, Inc., New York.

Maloney, P. C., S. V. Ambudkar, V. Anantharam, L. A. Sonna, and A. Varadhachary. 1990. Anion-exchange mechanisms in bacteria. *Microbiol. Rev.* **54:**1–17.

Medveczky, N., and H. Rosenberg. 1970. The phosphate-binding protein of *Escherichia coli. Biochim. Biophys. Acta* **211:**158–168.

Medveczky, N., and H. Rosenberg, 1971. Phosphate transport in *Escherichia coli. Biochim. Biophys. Acta* **241:**494–506.

Meisner, H., F. Palmieri, and E. Quagliariello. 1972. Effect of cations and protons on the kinetics of substrate uptake in rat liver mitochondria. *Biochemistry* **11:**949–955.

Michels, P. A. M., J. P. J. Michels, J. Boonstra, and W. N. Konings. 1979. Generation of an electrochemical proton gradient in bacteria by the excretion of metabolic end products. *FEMS Microbiol. Lett.* **5:**357–364.

Nieuwenhuis, B. J. W. M. 1982. *Phosphate and Divalent Cation Uptake in Yeast.* Ph.D. thesis. University of Nijmegen, Nijmegen, The Netherlands.

Pressman, B. C. 1970. Energy-linked transport in mitochondria, p. 213–250. *In* E. Racker (ed.), *Membranes of Mitochondria and Chloroplasts.* Van Nostrand-Reinhold, New York.

Rae, A. S., and K. P. Strickland. 1976a. Studies on phosphate transport in *Escherichia coli.* I. Reexamination of the effect of osmotic and cold shock on phosphate uptake and some attempts to restore uptake with phosphate binding protein. *Biochim. Biophys. Acta* **433:**555–563.

Rae, A. S., and K. P. Strickland. 1976b. Studies on phosphate transport in *Escherichia coli.* II. Effects of metabolic inhibitors and divalent cations. *Biochim. Biophys. Acta* **433:**564–582.

Rosenberg, H. 1987. Phosphate transport in prokaryotes, p. 205–248. *In* B. P. Rosen, and S. Silver (ed.), *Ion Transport in Prokaryotes.* Academic Press, Inc., New York.

Rosenberg, H., R. G. Gerdes, and K. Chegwidden. 1977. Two systems for the uptake of phosphate in *Escherichia coli. J. Bacteriol.* **131:**505–511.

Rosenberg, H., R. G. Gerdes, and F. M. Harold. 1979. Energy coupling to the transport of inorganic phosphate in *Escherichia coli* K-12. *Biochem. J.* **178:**133–137.

Rosenberg, H., C. M. Hardy, and B. P. Surin. 1984. Energy coupling to phosphate transport in *Escherichia coli,* p. 50–52. *In* L. Leive and D. Schlessinger (ed.), *Microbiology—*

1984. American Society for Microbiology, Washington, D.C.

Rosenberg, H., N. Medveczky, and J. M. La Nauze. 1969. Phosphate transport in *Bacillus cereus*. *Biochim. Biophys. Acta* **193:**159–167.

Russell, L. M., and H. Rosenberg. 1979. Linked transport of phosphate, potassium ions and protons in *Escherichia coli*. *Biochem. J.* **184:**13–21.

Russell, L. M., and H. Rosenberg. 1980. The nature of the link between potassium transport and phosphate transport in *Escherichia coli*. *Biochem. J.* **188:**715–723.

Silver, S., K. Budd, K. M. Leahy, W. V. Shaw, D. Hammond, R. P. Novick, G. R. Willsky, M. H. Malamy, and H. Rosenberg. 1981. Inducible plasmid-determined resistance to arsenate, arsenite, and antimony(III) in *Escherichia coli* and *Staphylococcus aureus*. *J. Bacteriol.* **146:**983–996.

Silver, S., and M. Walderhaug. 1992. Gene regulation of plasmid- and chromosome-determined inorganic ion transport in bacteria. *Microbiol. Rev.* **56:**195–228.

Sonna, L. A., S. V. Ambudkar, and P. C. Maloney. 1988. The mechanism of glucose 6-phosphate transport by *Escherichia coli*. *J. Biol. Chem.* **263:**6625–6630.

Van Veen, H. W., T. Abee, G. J. J. Kortstee, W. N. Konings, and A. J. B. Zehnder. 1993a. Characterization of two phosphate transport systems in *Acinetobacter johnsonii* 210A. *J. Bacteriol.* **175:**200–206.

Van Veen, H. W., T. Abee, G. J. J. Kortstee, W. N. Konings, and A. J. B. Zehnder. 1993b. Mechanism and energetics of the secondary phosphate transport system of *Acinetobacter johnsonii* 210A. *J. Biol. Chem.* **268:**19377–19383.

Van Veen, H. W., T. Abee, G. J. J. Kortstee, W. N. Konings, and A. J. B. Zehnder. 1994a. Substrate specificity of the two phosphate transport systems of *Acinetobacter johnsonii* 210A in relation to P_i speciation in its aquatic environment. *J. Biol. Chem.* **269:**16212–16216.

Van Veen, H. W., T. Abee, G. J. J. Kortstee, W. N. Konings, and A. J. B. Zehnder. 1994b. Translocation of metal phosphate via the phosphate inorganic transport (Pit) system of *Escherichia coli*. *Biochemistry* **33:**1766–1770.

Van Veen, H. W., T. Abee, G. J. J. Kortstee, H. Pereira, W. N. Konings, and A. J. B. Zehnder. Generation of a proton motive force by the excretion of metal phosphate in the polyphosphate-accumulating *Acinetobacter johnsonii* strain 210A. Submitted for publication.

Willecke, K., E.-M. Grier, and P. Oehr. 1973. Coupled transport of citrate and magnesium in *Bacillus subtilis*. *J. Bacteriol.* **144:**366–374.

Willsky, G. R., R. L. Bennett, and M. H. Malamy. 1973. Inorganic phosphate transport in *Escherichia coli:* involvement of two genes which play a role in alkaline phosphatase regulation. *J. Bacteriol.* **113:**529–539.

Willsky, G. R., and M. H. Malamy. 1980a. Characterization of two genetically separable inorganic phosphate transport systems in *Escherichia coli*. *J. Bacteriol.* **144:**356–365.

Willsky, G. R., and M. H. Malamy. 1980b. Effect of arsenate on inorganic phosphate transport in *Escherichia coli*. *J. Bacteriol.* **144:**366–374.

Wohlrab, H. 1986. Molecular aspects of inorganic phosphate transport in mitochondria. *Biochim. Biophys. Acta* **853:**115–134.

Yashphe, J., H. Chikarmane, M. Iranzo, and H. O. Halvorson. 1992. Inorganic phosphate transport in *Acinetobacter lwoffi*. *Curr. Microbiol.* **24:**275–280.

Chapter 8

The Pho Regulon of *Bacillus subtilis* Is Regulated by Sequential Action of Two Genetic Switches

F. MARION HULETT, GUOFU SUN, AND WEI LIU

Department of Biological Sciences, Laboratory for Molecular Biology, University of Illinois at Chicago, Chicago, Illinois 60607-7020

Depletion of nutrients is a form of stress often encountered by a bacterial cell; it results in the cessation of exponential growth. The transition from exponential growth to stationary phase is a gradual process in which expression of genes which are normally silent during exponential growth yields gene products required for survival under the particular stress condition encountered; additionally, many growth-related genes continue to be expressed. Faced with nutrient depletion, *Bacillus subtilis* can, depending on the environmental and metabolic conditions, choose one of two options, either a nondividing stationary period of low metabolic activity or a developmental program culminating in the formation of a free endospore. If the gene products expressed in response to the stress replenish the depleted nutrient before the commitment to sporulation is made, the cell can resume vegetative growth.

In soil, the natural environment of *B. subtilis,* P_i is the major limiting nutrient for biological growth and is often present at levels 2 to 3 orders of magnitude lower than those of other required ions. Soil bacteria, including *B. subtilis,* have evolved complex regulatory systems for utilizing this limited nutrient efficiently. Studies of regulatory mechanisms governing alkaline phosphatase (APase) expression, during both phosphate deprivation and the developmental process of sporulation, yield insight into signal transduction and regulatory mechanisms which allow cells to respond in various developmental ways depending on the stimulus.

The study of phosphate metabolism in *Bacillus* species in general and *B. subtilis* in particular has been complicated by a fact that increases the potential importance of this process: APase, the usual enzyme of choice as a reporter of phosphate metabolism, is encoded by a multigene family (Bookstein et al., 1990; Hulett et al., 1990; Kapp et al., 1990). Two unlinked APase structural genes, *phoA* (formerly *phoAIV*) and *phoB* (formerly *phoAIII*), have been cloned by using reverse genetics (Hulett et al., 1991). Characterization of the genes showed that the sequences encoding the mature proteins are 64% identical and that the deduced protein sequences are 63% identical. Both genes are expressed during phosphate-limited growth (Bookstein et al., 1990; Kapp et al., 1990) and during the developmental process leading to sporulation (Birkey et al., unpublished; Chesnut et al., 1991). APase transcriptional regulation during phosphate-limited stress and during spore development appears to be controlled by different signal transduction regimens (Chesnut et al., 1991; Piggot and Taylor, 1977).

During phosphate starvation of *B. subtilis,* at least three *trans*-acting regulators are involved in controlling the synthesis of the APase gene family. Two genes, *phoP* and *phoR*, encoding proteins which show similarity in sequence to procaryotic two-component signal transduction regulators (Lee and Hulett, 1992; Miki et al., 1965; Seki et al., 1987) are required for phosphate starvation-mediated induction of APases (Chesnut et al., 1991; Miki et al., 1965; Piggot and Taylor, 1977). PhoP is 40% identical to PhoB of *Escherichia coli,* the transcription activator protein for the Pho regulon of *E. coli* (Makino et al., 1988; Seki et al., 1988; Yamada et al., 1989). The kinase for PhoB of *E. coli,* PhoR, and the carboxyl-terminal three-fourths of PhoR from *B. subtilis* show significant similarity (Seki et al., 1987). The *B. subtilis* PhoR has an amino-terminal extension of 137 amino acids that is not found in the *E. coli* PhoR. A third gene, *spo0A,* which is involved in regulation of gene expression during the transition from exponential to stationary growth and is essential for sporulation (Ferrari et al., 1985), also influences phosphate starvation-induced APase production (Jensen et al., 1993). A mutation in the *spo0A* gene results in hyperinduction of total APase activity, suggesting that Spo0A acts to repress or turn off APase expression in a wild-type cell.

PhoA and PhoB are the major APase proteins expressed during phosphate limitation, accounting for 98% of that activity. Both members of the signal transduction switch that responds to limiting phosphate, PhoP (response regulator) and

PhoR (histidine kinase), are equally required for the transcription of either APase gene during phosphate limitation. This indicates that no P_i-independent controls activate PhoP in the absence of its cognate sensor, PhoR, at least under the culture conditions used in the reported study (Hulett et al., 1994). In *E. coli,* a mutation in *phoR*, the gene encoding the histidine kinase, results in constitutive expression of the Pho regulon, removing both repression and induction of the regulon. This constitutive expression is believed to involve cross regulation by CreC (formerly PhoM), a sensor kinase induced by growth on glucose, or to involve acetyl phosphate during growth on pyruvate (Wanner, 1992; Wanner and Wilmes-Riesenberg, 1992).

Comparison of the APase promoters used during phosphate limitation shows that both promoters have −10 regions which are similar (*phoA*) or identical (*phoB*) to the consensus sequence for a sigma A promoter (Fig. 1). However, there is no similarity at the −35 region to any sigma consensus sequence or to each other. Interestingly, both APase promoters have an identical 6-bp sequence TTAACA positioned 9 bp 5′ of the −10 region. The significance of this sequence, if any, is unknown. No sequence similar to the Pho box consensus sequence (Makino et al., 1988), for PhoB binding in *E. coli* (or presumably PhoP in *B. subtilis*), was found in either APase gene promoter used during phosphate starvation induction. The 76-bp sequence upstream of the transcriptional start site for the *phoB* promoter used during phosphate starvation induction (the P_v promotes) is sufficient for full promoter activity (Chesnut et al., 1991). Deletion of a further 32 bp, leaving 44 bp upstream of the transcriptional start of the P_v promoter, eliminated all promoter activity. This 32-bp sequence, essential for P_v promoter activity of *phoB*, does show some similarity to a sequence upstream of the P_v promoter transcription start site of *phoA*, as shown in Fig. 1. Again, the significance of this sequence similarity is unknown. A sequence with similarity to the *E. coli* "Pho box" sequence, which was noted previously (Bookstein et al., 1990) in the *phoB* (APase B) 5′ region, could be deleted without changing the P_v promoter activity (Chesnut et al., 1991). This suggests that if PhoP binds directly to the APase promoters, the binding sequence is different from that for PhoB of *E. coli* (Chesnut et al., 1991). Preliminary evidence, obtained by using partially purified preparations of PhoP for gel retardation studies, suggests that PhoP may bind directly to the *phoA* promoter region (Liu and Hulett, unpublished).

Spo0A, a member of the regulator class of two-component systems (Burbulys et al., 1991; Ferrari et al., 1988), is active when phosphorylated by either of two kinases, KinA or KinB (and possibly other, as yet unidentified kinases), via a phosphorelay involving several of the *spo0* gene products (Burbulys et al., 1991). A major role of Spo0A ~ P is to down-regulate *abrB* transcription (Strauch et al., 1990), thereby reducing the concentration of AbrB, the growth phase repressor of many post-exponentially expressed genes (Perego et al., 1991; Strauch et al., 1989). The result of Spo0A phosphorylation is to allow the expression of protease (Ferrari et al., 1986, 1988), antibiotics (Marahiel et al., 1987; Robertson et al., 1989), and other genes (Dabnau, 1991) that are repressed by AbrB during exponential growth.

Current evidence suggests that the mechanism by which *spo0A* regulates the synthesis of APases is not via repression of *abrB*, since *spo0A* mutants show an APase hyperinduction phenotype (Hulett et al., 1994; Jensen et al., 1993) and an *abrB* mutation in a *spo0A* background has little further effect on APase transcription. The initiation of APase induction at approximately 0.08 to 0.1 mM P_i (Hulett and Jensen, 1988) is not changed in a *spo0A* strain, but the induction period is extended for 2 to 3 h rather than being turned off as it is in the spo^+ parent strain (Hulett et al., 1994). Mutations in *phoP* or *phoR* are epistatic to *spo0A* mutations; *phoP spo0A* double mutants produce APase enzyme levels equal to those produced by *phoP* or *phoR* mutations alone (<5% of wild-type levels [Sun and Hulett, unpublished]). Interestingly, mutations in *spo0A* cause increased expression of the *phoPR* operon during phosphate-limited growth, which is similar to the transcriptional effect on

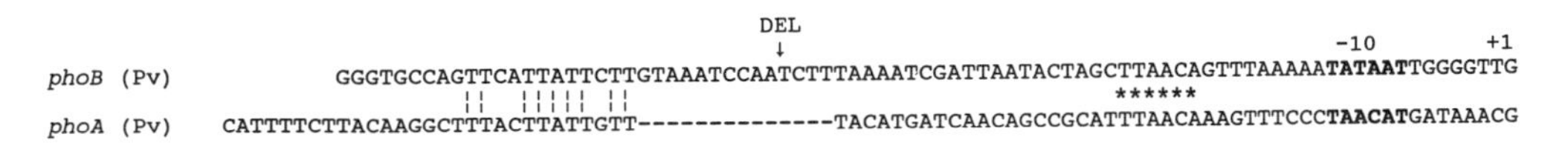

FIG. 1. Comparison of *phoA* and *phoB* phosphate starvation-inducible (PSI) promoters. The sequences are aligned at the transcription start site, +1. The −10 region is marked, and the sigma A-like consensus sequence in each promoter is in boldface type. Asterisks mark an identical 6-bp sequence found 9 bp upstream of the −10 sequence in both promoters. The PhoB promoter sequence is the complete sequence necessary for phosphate starvation induction of *phoB*. The arrow marks the position of a deletion which destroys all *phoB* promoter (PSI) function. || identify sequences within the deleted region of the *phoB* promoter shown to be necessary for PSI function and a sequence in the 5′ region of the *phoA* promoter. -- identifies spacing required for alignment.

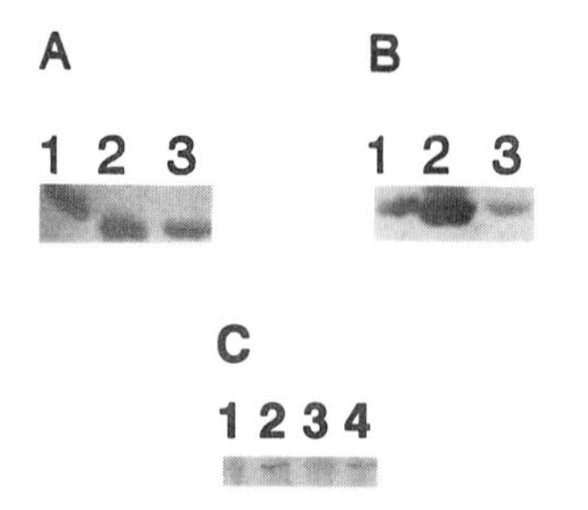

FIG. 2. Concentrations of PhoA (APase A), PhoB (APase B), and PhoP (response regulator) are all elevated in a *spo0A* strain cultured under phosphate-limited conditions. (A) Western blot analysis with PhoB polyclonal antibody. Lanes: A1, marker (purified PhoB protein); A2, 200 μg of protein from *spo0A* strain culture lysate; A3, 200 μg of protein from *spo*+ parental strain. (B) Same as panel A but with PhoA polyclonal antibody and purified PhoA protein in lane B1. Culture samples in panels A and B were taken after 11 h of growth in LPDM medium (Jensen et al., 1993). (C) Western blot with polyclonal antibody to a PhoP synthetic peptide. Lanes: C1 and C3, 20 μg of a parental spo^+ strain cell lysate taken after 12 h (lane C1) or 14 h (lane C2) growth in LPDM medium; C2 and C4, 20 μg of protein from a *spo0A* strain cell lysate taken after 12 h (lane C2) or 14 h (lane C4) in LPDM medium.

phoA or *phoB* (Hulett et al., 1994). We believe that the action of Spo0A on the *phoPR* promoter is indirect since gel retardation of or footprinting on the *phoB* or *phoPR* promoter by Spo0A was not observed (Strauch et al., unpublished).

To determine if increased transcription of *phoA*, *phoB*, and *phoPR* was translated into increased amounts of protein, we compared Western immunoblot analysis of equal amounts of protein from an induced *spo0A* strain and its isogenic parent strain (Fig. 2). PhoA (APase A), PhoB (APase B), and PhoP were all present in higher concentrations in a *spo0A* strain than in the parent strain, although the increased production of PhoA was most dramatic. Although rigorous studies of protein turnover, etc., have not been conducted, these data are consistent with the hypothesis that increased transcription does result in increased levels of the individual proteins in a *spo0A* strain.

It was of interest to determine if the induction of the *phoPR* operon during phosphate-limited growth was dependent on PhoP and PhoR. A *phoPR-lacZ* promoter fusion was used to compare *phoPR* transcription in strains containing either an in-frame *phoP* deletion or a *phoR* deletion with that in an isogenic wild-type strain (Fig. 3). APase and β-galactosidase from the *phoP* promoter fusion showed that the kinetics of induction of APase (Fig. 3A) and of the *phoPR* operon (Fig. 3B) are similar in a wild-type strain. The specific activities of both enzymes declined during exponential growth (1 to 5 h of growth [data not shown]) because of an increase in culture density without new enzyme synthesis. When growth slowed (after 5 h) because of phosphate depletion, APase and *phoPR* transcription were induced, essentially in parallel; this continued for several hours before being turned off. (Spo0A is responsible for repressing APase and *phoPR* transcription

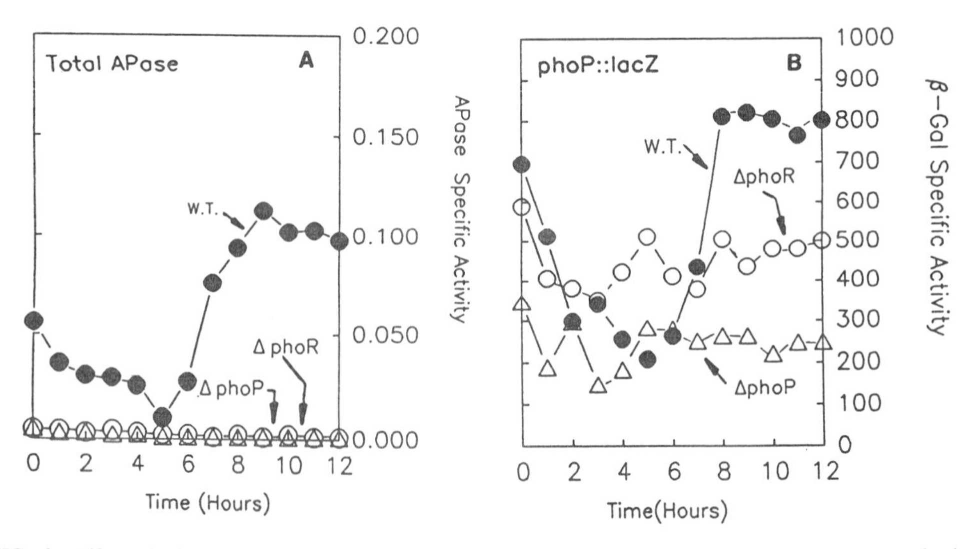

FIG. 3. Effect of *phoP* or *phoR* mutations on *phoPR* transcription and APase synthesis. (A) The abscissa indicates the time of growth in LPDM medium (Jensen et al., 1993). The ordinate indicates APase specific activity. (B) β-Galactosidase specific activity from a *phoP-lacZ* promoter fusion. Solid circles, spo^+ parental strain; open circles, isogenic *phoR* deletion mutant; open triangles, isogenic *phoP* deletion (in-frame) mutant.

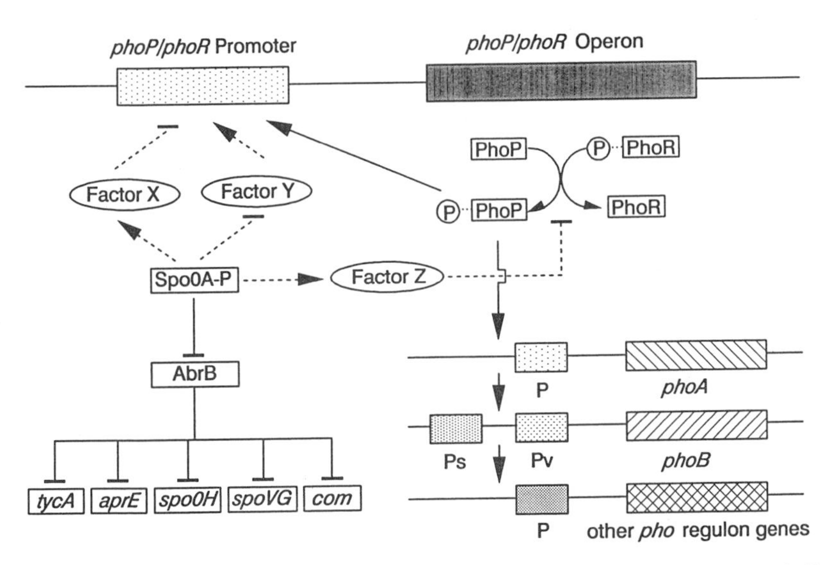

FIG. 4. Model for the roles of PhoP, PhoR, and Spo0A in Pho regulon regulation. Arrowheads indicate positive regulation. Horizontal bars indicate negative regulation. Factor X, factor Y, and factor Z are hypothetical intermediates, indicating (via dashed lines) that Spo0A ~ P regulation on *phoPR* transcription may be indirect or may affect phosphorylation of PhoP (factor Z). Arrows with solid lines identify promoters believed to require PhoP ~ P for induction. Solid lines with bars identify negative regulation by Spo0A or AbrB. PhoP, response regulator; PhoR, histidine kinase; Spo0A, response regulator; AbrB, repressor of many postexponentially expressed genes. Genes regulated by Spo0A and/or AbrB are not limited to those shown.

after the induction period [Hulett et al., 1994; Jensen et al., 1993].) There is low-level constitutive *phoPR* transcription in a *pho*P deletion strain (Fig. 3B) and no APase expression (Fig. 3A). Therefore, PhoP is required for autoinduction of the *phoPR* operon, which coincides with Pho regulon gene expression during phosphate limitation. A higher-level constitutive expression of *phoPR* transcription occurs in a *phoR* mutant strain than in the *phoP* mutant (Fig. 3B). It is unclear whether this increase in *phoPR* constitutive expression over that seen in the *phoP* mutant is due to a function of unphosphorylated PhoP or to PhoP phosphorylated via cross regulation. However, it is clear that PhoP, in a *phoR* mutant strain, is unable to activate transcription of APase genes (Fig. 3A) and other PhoP regulon genes (Sun and Hulett, unpublished) and is not capable of autoinduction of the *phoPR* operon during phosphate-limited growth (Fig. 3B).

Our current understanding of the regulation of APase and other Pho regulon genes during phosphate limitation is diagramed in Fig. 4. The model is based on data described here and on general properties conserved among two-component signal transduction systems. When *B. subtilis* experiences phosphate depletion, APases, which are dependent on PhoP and PhoR for expression, are synthesized. Since mutations in PhoR are as deleterious to APase production as are mutations in PhoP, we propose that (under the experimental conditions used) PhoR alone is responsible for activation of PhoP, presumably by phosphorylation (Makino et al., 1989). The activated PhoP acts as a transcription activator of Pho regulon genes including *phoA*, *phoB*, and *phoPR*. Induction continues for several hours. Under these culture conditions, no phosphorylated substrates are available, and therefore the phosphate response system fails to provide the limiting nutrient, P_i. The cell then abandons the phosphate response. The genes encoding the phosphate-sensing genetic switch, the *phoPR* operon, are repressed by a second genetic switch (Burbulys et al., 1991; Ferrari et al., 1985), Spo0A ~ P, which concomitantly releases other adaptive response systems from repression by AbrB (Strauch et al., 1990) or enables the cell to sporulate (Burbulys et al., 1991). The Spo0A ~ P repression of the *phoPR* operon is apparently indirect (Strauch et al., unpublished). Spo0A ~ P may regulate genes whose products regulate *phoPR* transcription through a pathway independent of PhoP and PhoR or, conversely, through a pathway affecting the phosphorylation state of PhoP which, in turn, affects *phoPR* transcription.

We thank Lei Shi for help in preparation of the manuscript.
This work was supported by Public Health Service research grant GM33471 from the National Institutes of Health.

REFERENCES

Birkey, S., L. Shi, and F. M. Hulett. Unpublished data.

Bookstein, C., C. W. Edwards, N. V. Kapp, and F. M. Hulett. 1990. The *Bacillus subtilis* 168 alkaline phosphatase III gene: impact of a *phoAIII* mutation on total alkaline phosphatase synthesis. *J. Bacteriol.* **172:**3730–3737.

Burbulys, D., K. A. Trach, and J. A. Hoch. 1991. Initiation of sporulation in *B. subtilis* is controlled by a multicomponent phosphorelay. *Cell* **64:**545–552.

Chesnut, R. S., C. Bookstein, and F. M. Hulett. 1991. Separate promoters direct expression of *phoAIII*, a member of the *Bacillus subtilis* multigene family, during phosphate starvation and sporulation. *Mol. Microbiol.* **5:**2181–2190.

Dubnau, D. 1991. Genetic competence in *Bacillus subtilis. Microbiol. Rev.* **55:**395–424.

Ferrari, E., D. Henner, M. Perego, and J. A. Hoch. 1988. Transcription of *Bacillus subtilis* subtilisin and expression of subtilisin in sporulation mutants. *J. Bacteriol.* **170:**289–295.

Ferrari, E., S. Howard, and J. A. Hoch. 1986. Effect of stage 0 sporulation mutations on subtilisin expression. *J. Bacteriol.* **166:**173–179.

Ferrari, F. A., K. Trach, D. LeCoq, J. Spence, E. Ferrari, and J. A. Hoch. 1985. Characterization of the *spo0A* locus and its deduced product. *J. Bacteriol.* **82:**2647–2651.

Hulett, F. M., C. Bookstein, and K. Jensen. 1990. Evidence for two structural genes for alkaline phosphatase in *Bacillus subtilis. J. Bacteriol.* **172:**735–740.

Hulett, F. M., and K. Jensen. 1988. Critical roles of *spo0A* and *spo0H* in vegetative alkaline phosphatase production in *Bacillus subtilis. J. Bacteriol.* **170:**3765–3768.

Hulett, F. M., E. E. Kim, C. Bookstein, N. V. Kapp, C. W. Edwards, and H. W. Wyckoff. 1991. *Bacillus subtilis* alkaline phosphatases III and IV. Cloning, sequencing, and comparisons of deduced amino acid sequence with *Escherichia coli* alkaline phosphatase three-dimensional structure. *J. Biol. Chem.* **266:**1077–1084.

Hulett, F. M., J. W. Lee, L. Shi, G. Sun, R. Chesnut, E. Sharkava, M. F. Duggan, and N. Kapp. 1994. Sequential action of two-component genetic switches regulates the PHO regulon in *Bacillus subtilis. J. Bacteriol.* **176:**1348–1358.

Jensen, K. K., E. Sharkova, M. F. Duggan, Y. Qi, A. Koide, J. A. Hoch, and F. M. Hulett. 1993. *Bacillus subtilis* transcription regulator, Spo0A, decreases alkaline phosphatase levels induced by phosphate starvation. *J. Bacteriol.* **175:**3749–3756.

Kapp, N. V., C. W. Edwards, R. S. Chesnut, and F. M. Hulett. 1990. The *Bacillus subtilis phoAIV* gene: effects of in vitro inactivation on total alkaline phosphatase production. *Gene* **96:**95–100.

Lee, J., and F. M. Hulett. 1992. Nucleotide sequence of the *phoP* gene encoding PhoP, the response regulator of the phosphate regulon of *Bacillus subtilis. Nucleic Acids Res.* **19:**5848.

Liu, W., and F. M. Hulett. Unpublished data.

Makino, K., H. Shinagawa, M. Amemura, T. Kawamoto, M. Yamada, and A. Nakata. 1989. Signal transduction in the phosphate regulon of *Escherichia coli* involves phosphotransfer between PhoR and PhoB proteins. *J. Mol. Biol.* **210:**551–559.

Makino, K., H. Shinagawa, M. Amemura, K. Kimura, and A. Nakata. 1988. Regulation of the phosphate regulon of *Escherichia coli.* Activation of *pstS* transcription by PhoB protein *in vitro. J. Mol. Biol.* **203:**85–95.

Marahiel, M. A., P. Zuber, G. Czekay, and R. Losick. 1987. Identification of the promoter for a peptide antibiotic biosynthesis gene from *Bacillus brevis* and its regulation in *Bacillus subtilis. J. Bacteriol.* **169:**2215–2222.

Miki, T., A. Minami, and Y. Ikeda. 1965. The genetics of alkaline phosphatase formation in *Bacillus subtilis. Genetics* **52:**1093–1100.

Perego, M., J.-J. Wu, G. B. Spiegelman, and J. A. Hoch. 1991. Mutational dissociation of the positive and negative regulatory properties of the Spo0A sporulation transcription factor of *Bacillus subtilis. Gene* **100:**207–212.

Piggot, P. J., and S. Y. Taylor. 1977. New types of mutations affecting formation of alkaline phosphatase by *Bacillus subtilis* in sporulation conditions. *J. Gen. Microbiol.* **128:**663–669.

Robertson, J. B., M. Gocht, M. A. Marahiel, and P. Zuber. 1989. AbrB, a regulator of gene expression in *Bacillus*, interacts with the transcription initiation regions of a sporulation gene and an antibiotic biosynthesis gene. *Proc. Natl. Acad. Sci. USA* **86:**8457–8461.

Seki, T., H. Yoshikawa, H. Takahashi, and H. Saito. 1987. Cloning and nucleotide sequence of *phoP*, the regulatory gene for alkaline phosphatase and phosphodiesterase in *Bacillus subtilis. J. Bacteriol.* **169:**2913–2916.

Seki, T., H. Yoshikawa, H. Takahashi, and H. Saito. 1988. Musleotide sequence of the *Bacillus subtilis phoP* gene. *J. Bacteriol.* **170:**5935–5938.

Strauch, M. A., J. A. Hoch, and F. M. Hulett. Unpublished data.

Strauch, M. A., G. B. Spiegelman, M. Perego, W. C. Johnson, D. Burbulys, and J. A. Hoch. 1989. The transition state transcription regulator *abrB* of *Bacillus subtilis* is a DNA binding protein. *EMBO J.* **8:**1615–1621.

Strauch, M., V. Webb, G. Spiegelman, and J. A. Hoch. 1990. The Spo0A protein of *Bacillus subtilis* is a repressor of the *abrB* gene. *Proc. Natl. Acad. Sci. USA* **87:**1801–1805.

Sun, G., and F. M. Hulett. Unpublished data.

Wanner, B. L. 1992. Is cross regulation by phosphorylation of two-component response regulator proteins important to bacteria? *J. Bacteriol.* **174:**2053–2058.

Wanner, B. L., and M. R. Wilmes-Riesenberg. 1992. Involvement of phosphotransacetylase, acetate kinase, and acetyl phosphate synthesis in control of the phosphate regulon in *Escherichia coli. J. Bacteriol.* **174:**2124–2130.

Yamada, M., K. Makino, M. Amemura, H. Shinagawa, and A. Nakata. 1989. Regulation of the phosphate regulon of *Escherichia coli:* analysis of mutant *phoB* and *phoR* genes causing different phenotypes. *J. Bacteriol.* **171:**5601–606.

II. REGULATION OF PHOSPHATE METABOLISM IN *SACCHAROMYCES CEREVISIAE*

Chapter 9

Introduction: Regulation of Phosphate Metabolism in *Saccharomyces cerevisiae*

YASUJI OSHIMA[1] AND HARLYN HALVORSON[2]

Department of Biotechnology, Osaka University, Yamada-kami, Suita-shi, Osaka 565, Japan,[1] and Department of Biology, University of Massachusetts—Dartmouth, N. Dartmouth, Massachusetts 02747[2]

By the early 1950s, *Saccharomyces cerevisiae* was established as a major model system for studies of the adaptive response of microorganisms to changes in the medium. It was recognized early that, in response to various sugars in the medium, yeast cells modify their enzymatic composition (induced enzyme biosynthesis). The distribution of acid and alkaline phosphatases in various yeast cell fractions is dependent on the type and concentration of monosaccharides in the medium.

By the 1970s, the genetic control of phosphate regulation began to emerge. Ezra Yagil identified *phoA* as a gene controlling phosphatase activity. Shortly thereafter, Oshima found a second gene (*phoB*) which also regulated yeast phosphatase, and he went on to define genetically the regulatory system. By the time of the Concarneau meeting (Torriani-Gorini et al., 1987), the genetic control of phosphate regulation had been largely described and the transition to molecular studies had begun. In this section, the current status of the molecular biology of regulation of phosphate metabolism in *S. cerevisiae* is presented.

In chapter 10, Fig. 1 outlines our understanding of the molecular regulation of phosphate metabolism; in that chapter Ogawa et al. discuss upstream activation sites (UAS) regulating the phosphatase genes *PHO5* (encoding repressible acid phosphatase), *PHO8* (encoding alkaline phosphatase by *PHO84* P_i transport), and *PHO81* (one of the regulatory genes in phosphate regulation). In chapter 11, Parent et al. present data on protein-protein and protein-DNA interactions at distinct UAS elements during UAS-dependent activation of *PHO5*. In chapter 12, Fujino et al. describe the structure of the protein kinase and how it is encoded by *PHO85*. In this case it is unlikely that the kinase activity is directly regulated by the phosphate concentration. As shown in Chapter 10, Fig. 1, it is still possible that phosphate acts instead through the *PHO81p* mediator.

REFERENCES

Torriani-Gorini, A., F. G. Rothman, S. Silver, A. Wright, and E. Yagil (ed.). 1987. *Phosphate Metabolism and Cellular Regulation in Microorganisms.* American Society for Microbiology, Washington, D.C.

Chapter 10

Regulatory Circuit for Phosphatase Genes in *Saccharomyces cerevisiae*: Specific *cis*-Acting Sites in *PHO* Promoters for Binding the Positive Regulator Pho4p

NOBUO OGAWA, NAOYUKI HAYASHI, HIROYUKI SAÏTO, KEN-ICHI NOGUCHI, YASUJI YAMASHITA, AND YASUJI OSHIMA

Department of Biotechnology, Faculty of Engineering, Osaka University, 2-1 Yamadaoka, Suita-shi, Osaka 565, Japan

The transcriptions of the genes encoding three isozymes (p60, p58, and p56) of repressible acid phosphatase (EC 3.1.3.2; rAPase), a repressible alkaline phosphatase (EC 3.1.3.1; rALPase), and a P_i transporter in *Saccharomyces cerevisiae* are coordinately repressed by P_i in the medium (for review, see Oshima [1991] and Johnston and Carlson [1992]). This phosphatase (*PHO*) system has been studied extensively with numerous technical advantages (Yoshida et al., 1987). The genes under the regulation of P_i are *PHO5* (encoding p60, a major fraction of rAPase), *PHO10* (p58), *PHO11* (p56), *PHO8* (rALPase), and *PHO84* (P_i transporter). The P_i signals are conveyed to these genes by a system consisting of products of at least five genes, *PHO2* (= *BAS2/GRF10*), *PHO4*, *PHO80*, *PHO81*, and *PHO85*. In a current model for phosphatase regulation (Fig. 1), a positive regulatory factor (or transcriptional activator), Pho4p, encoded by *PHO4* is proposed to be indispensable for transcription of the *PHO* structural genes for the enzymes. The Pho2p protein is an additional transcriptional activator of all the *PHO* structural genes except *PHO8* but does not have a direct function in transmission of P_i signals (Yoshida et al., 1989b). Pho2p, also known as Bas2p or Grf10p (Yoshida et al., 1989a), activates the transcriptions of *HIS4* (Arndt et al., 1987; Tice-Baldwin et al., 1989) and *TRP4* (Braus et al., 1989) and adenine biosynthesis (Arndt et al., 1987). In high-P_i medium, the Pho85p protein, a homolog of the protein kinases encoded by *PHO85* (Toh-e et al., 1988), may activate Pho80p (Uesono et al., 1992), a negative regulator, and the Pho80p protein may interact with Pho4p, inhibiting its function; therefore, Pho4p does not activate the transcriptions of *PHO* genes. When the P_i concentration in the medium is sufficiently low, Pho81p inhibits the function of Pho80p or eliminates P_i (or a derivative of it) that activates Pho80p (or Pho85p), thus allowing Pho4p (with Pho2p) to transcribe the *PHO* genes (Yoshida et al., 1989b). The *PHO4* (Yoshida et al., 1989a), *PHO80* (Yoshida et al., 1989b; Madden et al., 1990), and *PHO85* (Madden et al., 1990; Uesono et al., 1992) genes are transcribed constitutively at low levels, whereas the level of *PHO2* transcription is low but is self-regulated (Yoshida et al., 1989a). Although the *PHO81* gene produces a regulatory factor, as described above, its transcription is under regulation by Pho4p, indicating that the regulatory system forms a closed circuit (Yoshida et al., 1987, 1989b; Ogawa et al., 1993).

On the basis of the epistatic relationship of the *PHO2* and *PHO4* genes with all the other regulatory genes in the *PHO* system, it was suggested that Pho2p and Pho4p bind or interact with the regulatory site located in the upstream regions of *PHO* structural genes. In fact, such sequences, termed upstream activation sites (UASs), of the *PHO5* gene have been proposed (Bergman et al., 1986; Nakao et al., 1986; Rudolph and Hinnen, 1987). These proposed sites are, however, not consistent with each other. A region from positions −296 to −277 in the *PHO5* promoter was suggested to act as a binding site of Pho2p (Vogel et al., 1989). However, deletion of this site was found to have no effect on *PHO5* expression (Rudolph and Hinnen, 1987), and hybrid promoters containing only the suggested Pho4p-binding site but not the Pho2p-binding site could not derepress rAPase in a *pho2* mutant (Sengstag and Hinnen, 1988). Thus, the sites suggested by these authors for binding of Pho2p and Pho4p and their functions were conflicting.

To clarify the Pho4p-binding site, we first investigated the specific *cis*-acting regulatory sequence in the upstream region of *PHO8*, because Pho4p but not Pho2p is required for transcription. We found that a specific 6-bp sequence, CACGTG, is the Pho4p-binding site in the *PHO8* pro-

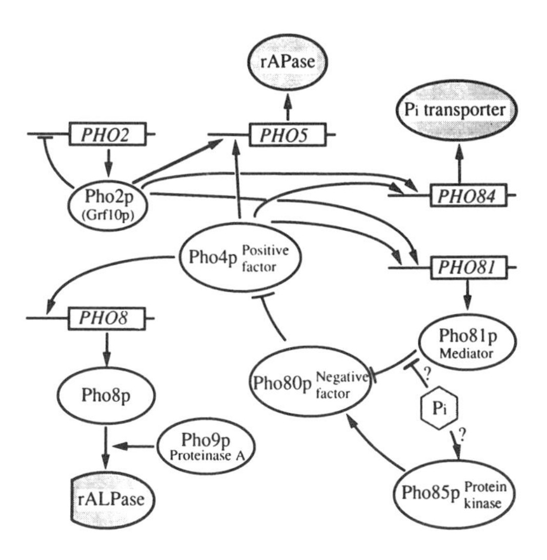

FIG. 1. Regulatory circuit for phosphatase genes. The protein factors, enzymes, and enzyme precursors in the regulatory circuit are shown in open ovals, and the genes under regulation by these proteins are shown in boxes. The factors in the shaded hexagon and ovals, P_i, rAPase, rALPase, and P_i transporter, are the input signal and outputs of the regulatory circuit, respectively. The regulatory factor, for which the corresponding gene is not shown, is known to be produced constitutively in all cases, except *PHO9* (= *PEP4*), whose regulation is unknown. The arrows in the regulatory circuit indicate a positive or stimulating function of the factors, and the bars indicate repressive or inhibitory functions. The Pho9p (Pep4p) protein, proteinase A, is needed for processing the Pho8p polypeptide.

moter (Hayashi and Oshima, 1991), as suggested for the *PHO5* promoter by Vogel et al. (1989) and confirmed by Fisher et al. (1991). Then we examined the 6-bp motif of the UASs of *PHO5*, *PHO81*, and *PHO84*. In this article, we summarize these analyses and suggest that the 8-bp motifs GCACGTGG and GCACGTTT are binding sites of Pho4p and the Pho4p-Pho2p complex, respectively, for transcriptional regulation of *PHO* genes.

Structures of Transcriptional Activators Pho4p and Pho2p

The open reading frame of the positive regulatory gene *PHO4* suggests that Pho4p consists of 312 amino acids (aa) (Yoshida et al., 1989a). In a previous study (Ogawa and Oshima, 1990), we found that the 85-aa C terminus of Pho4p has a basic region and an amphipathic helix-loop-helix structure side by side and that these structures function as a DNA-binding domain. A site for interaction with the negative regulator Pho80p appears to be located in a region from aa 163 to 202. The N-terminal region (aa 1 to 109), which is rich in acidic amino acids, was suggested to be the transcriptional activation domain, and the region from aa 203 to 227 was suggested to be involved in oligomerization of the protein.

The Pho2p protein, deduced from the nucleotide sequence, consists of 519 aa (Yoshida et al., 1989a) or 559 aa (Sengstag and Hinnen, 1987), depending on the yeast strain, and contains a homeo-box from aa 76 to 135 or from aa 77 to 136 (Bürglin, 1988). A portion of the homeo-box of Pho2p has similarity to a short stretch of the amino acid sequence of Pho4p (Bürglin, 1988). It also has a stretch of glutamine residues (14 of 18 aa) on the N-terminal side of the homeo-box.

Specific Nucleotide Sequences for Pho4p Binding

Specific *cis*-acting sequence in the *PHO8* promoter. To identify the Pho4p-binding sequence, we first investigated the specific *cis*-acting regulatory sequence in the upstream region of the *PHO8* gene encoding rALPase, because Pho4p but not Pho2p is thought to interact with its promoter (Hayashi and Oshima, 1991). The rALPase activity of cells was enhanced two- to threefold more in low-P_i medium than in high-P_i medium as a result of transcription of *PHO8*. Deletion analysis of the *PHO8* promoter revealed two separate regulatory regions for derepression of rALPase at nucleotides −704 to −661 (distal region) and −548 to −502 (proximal region) and an inhibitory region at −421 to −289 relative to the translation initiation codon (Fig. 2). Gel retardation experiments with a β-galactosidase–Pho4p fusion protein suggested that Pho4p binds to a 132-bp fragment of the *PHO8* DNA from positions −615 to −484 bearing the proximal region but not to a 226-bp fragment from positions −841 to −616 bearing the distal region. The fusion protein also binds to a 19-bp synthetic oligonucleotide having the same 12-bp nucleotide sequence as the *PHO8* DNA from positions −536 to −525. No similarities of the nucleotide sequences in the proximal and distal regions of *PHO8* with the published UASs of *PHO5* (Bergman et al., 1986; Nakao et al., 1986; Rudolph and Hinnen, 1987) were detected, except for a CACGT sequence in the proximal region of *PHO8* and in the UASs of *PHO5* (Vogel et al., 1989).

The 132-bp fragment of the *PHO8* DNA could sense P_i signals in vivo when it was connected at position −281 in the 5′ upstream region of a *HIS5′-′lacZ* fused gene (Hayashi and Oshima, 1991). A 29-bp synthetic oligonucleotide having the same 20-bp sequence as the *PHO8* DNA from positions −544 to −525, bearing the CACGT motif, could bind with the β-galactosidase–Pho4p fusion protein in vitro but could not sense the P_i signals in vivo. Since the CACGT sequence at this

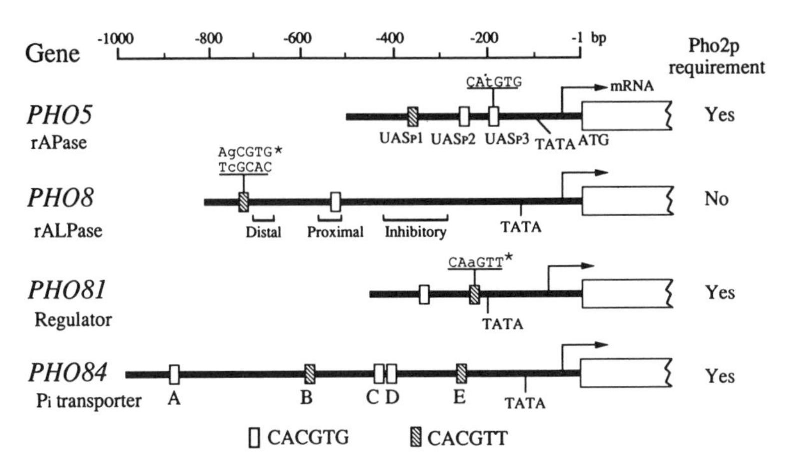

FIG. 2. Distribution of the 6-bp motif, CACGT(G/T), in the promoter regions of the *PHO* structural genes. The open box at the right end of individual genes represents the open reading frame. Short open and hatched boxes indicate CACGTG and CACGTT motifs, respectively. Three 6-bp sequences, analogous to the 6-bp motif, shown in the *PHO5*, *PHO8*, and *PHO81* promoters are indicated by their sequences (lowercase letters indicate deviation from the 6-bp motifs). The function and Pho4p-binding ability of the sequence analogous to the 6-bp motif in the *PHO8* promoter (marked with an asterisk) were described by Barbarić et al. (1992), but those of the *PHO81* promoter have not yet been examined.

position overlaps the *Pma*CI restriction sequence CACGTG, we inserted various linkers into this blunt-ended restriction site in the 132-bp fragment. The 5-bp CACGT sequence from positions −535 to −531 (Fig. 2) was found by gel shift experiments to be essential for binding the β-galactosidase–Pho4p fusion protein and was also necessary for derepression of rALPase in vivo (Hayashi and Oshima, 1991). The effects of linker insertions at the *Pma*CI site also suggested that the CACGTG sequence is more efficient than the CACGTT sequence for *PHO8* gene expression. We confirmed that this region actually binds with Pho4p by DNase I footprinting in vitro with a purified Pho4p preparation (Fig. 3). The Pho4p protein protected a 22-bp region from nucleotides −541 to −520. This region contains a 6-bp CACGTG sequence in its center.

UAS of *PHO81*. The *PHO81* gene encodes one of the regulators for the *PHO* regulon. The nucleotide sequence of this gene suggests that it encodes a 134-kDa protein, consisting of 1,179 aa with six repeats of a 33-aa sequence homologous to the ankyrin repeat (Lux et al., 1990) at aa 424 to 656 and an asparagine-rich region (26 of 33 aa) from aa 222 to 254 (Ogawa et al., 1993). Transcription of *PHO81* is activated by Pho4p in cooperation with Pho2p. Deletion analysis showed that a 95-bp region from nucleotides −385 to −291 is essential for the response to P_i signals. The purified preparation of the Pho4p protein protected a 19-bp region from positions −350 to −332 in the 95-bp fragment from DNase I digestion in vitro (Fig. 3). This protected region contains a CACGTG sequence at positions −344 to −339 (Fig. 2). This 6-bp sequence is thought to be the only copy of the 6-bp CACGTG motif in the promoter from position −1231 to the ATG codon. Although the transcription of the *PHO81* gene was severely reduced in a *pho2* mutant (Yoshida et al., 1989b), transcription of the *lacZ* gene connected to the 95-bp fragment was regulated by P_i but was independent of the *pho2* deletion in the host cells (Ogawa et al., 1993). These findings suggest that the 95-bp region may not have the complete sequence necessary for full regulation of the *PHO81* gene. When a 1.1-kb *PHO81* fragment from positions −1010 to +103 was connected to the *lacZ* gene and the construct was ligated into YCp50, expression of the *PHO81-lacZ* fusion gene in low-P_i medium was significantly reduced in the *pho2*-disrupted host as judged by much lower β-galactosidase activity than in the *PHO2*$^+$ host (Ogawa et al., 1993). This *PHO2*-dependent expression of *PHO81* was also observed with a shorter *PHO81* fragment, a 488-bp fragment from nucleotides −385 to +103 connected to the same *lacZ* fragment. These results strongly suggest that a specific site that interacts with Pho2p is located in the region downstream of position −290.

Distribution of 6-bp motifs in the *PHO84* promoter. According to the nucleotide sequence of the cloned DNA fragment of *PHO84* (Bun-ya et al., 1991), five copies of the 6-bp CACGT(G/T) motif are present at nucleotides −880 (site A), −587 (site B), −436 (site C), −414 (site D), and −262 (site E) (Fig. 2). Deletion analyses of the promoter region and base substitutions in the 6-bp

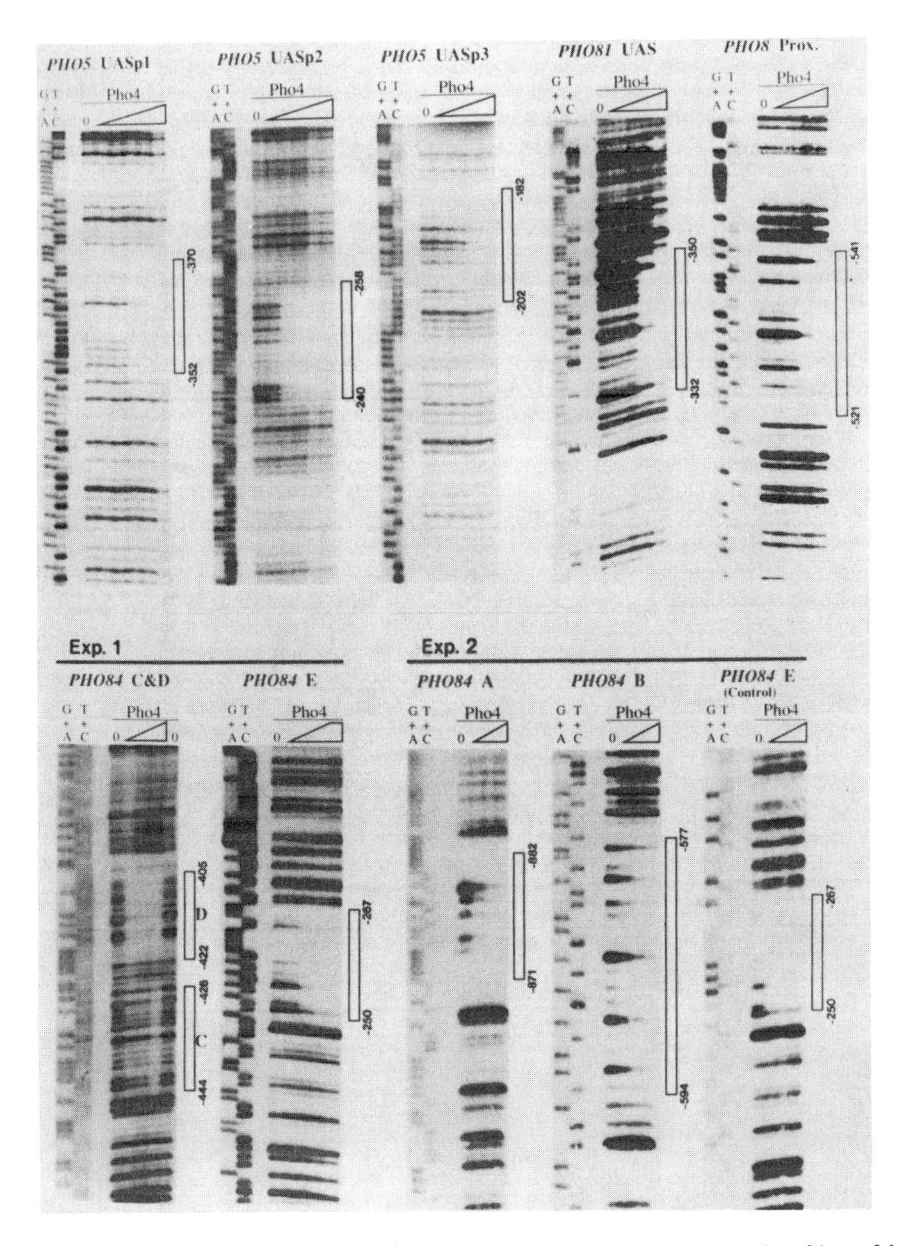

FIG. 3. DNase I footprinting analysis of Pho4p-binding sites in the *PHO* promoters. Samples of 1 ng of the short fragment of the promoter DNA (80 to 177 bp) labeled with ^{32}P at the 3′ end were mixed with 1.0 μg of calf thymus DNA and treated with various amounts (0 to 500 ng) of purified Pho4p in 20 μl of reaction mixture at 25°C for 30 min. Then 1 μl of DNase I solution (1 ng/μl) was added, and incubation was continued for 7 min. Other procedures were as described by Ogawa et al. (1993). Lanes G+A and T+C are the Maxam-Gilbert sequencing ladders of the same 3′-end-labeled fragments used as position markers. Protected regions are indicated by boxes, and the numbers indicate nucleotide positions at the protected ends. Protection profiles of the *PHO84* promoter were examined by two independent experiments, one for sites C, D, and E (experiment 1) and the other for sites A and B with E as the control (experiment 2).

motif revealed that either site C or site D and site E are essential and sufficient for full regulation of *PHO84* transcription but that the distal motifs are dispensable (unpublished data). These five copies of the 6-bp motif were also examined by DNase I footprinting of the purified Pho4p preparation.

For this, short (80- to 177-bp) fragments prepared by cutting the promoter DNA of *PHO84* with appropriate restriction enzymes to have one or two copies of the 6-bp motif were propagated in *Escherichia coli.* The 3′ ends of individual short promoter fragments of *PHO84* were labeled with ^{32}P and subjected to DNase I footprinting as described by Ogawa et al. (1993). The 18-base region from nucleotides −421 to −402 covering site D was clearly protected from DNase I digestion, whereas the 19-base region from nucleotides −444 to −426 covering site C was protected progressively with increased amounts of Pho4p, threefold more Pho4p being required to give the same level of protection as with site D (Fig. 3, experiment 1). Similar protection profiles were obtained for the 18-base region from nucleotides −271 to −251 covering site E but with twofold more Pho4p protein than for protection of site D. Sites A and B were also protected but with two- to threefold more Pho4p protein than for site E (Fig. 3, experiment 2). These protection profiles suggest that the proximal two sites, sites D and E, have the strongest affinities to Pho4p and that the distal three, sites A, B, and C, have lower affinities than do sites D and E. These facts are in agreement with the results of deletion and base substitution experiments showing that sites D and E are essential for *PHO84* regulation and that site C can replace an inactivated site D but that the distal motifs are dispensable (unpublished data).

UAS_p of the *PHO5* gene also contains a 6-bp motif. Rudolph and Hinnen (1987) suggested that transcriptional control of *PHO5* requires three copies of a specific 19-bp dyad sequence, a different sequence from those determined by Bergman et al. (1986) and Nakao et al. (1986) by deletion analysis of the *PHO5* promoter and by comparison with the *PHO11* DNA. The in vitro binding of Pho4p with two of these sequences, UAS_p1 located in the region from −373 to −347 and UAS_p2 located from −262 to −239, was demonstrated by Vogel et al. (1989). These findings were confirmed by our results (Fig. 3). These two UAS_p sites also have one copy each of the CACGT(G/T) sequence in the middle (Fig. 2; also see Fig. 4). We also observed that another 21-bp region in the *PHO5* promoter from nucleotides −202 to −182, which almost completely overlaps the third element detected by Rudolph and Hinnen (1987), was weakly protected by Pho4p from DNase I digestion (Fig. 3). This protected region contains a similar 6-bp motif, CATGTG or CACATG. Since the *PHO5* promoter with a deletion of a region containing this 21-bp sequence was reported to confer 30% of the activity of the wild-type *PHO5* promoter (Rudolph and Hinnen, 1987), this sequence, designated as UAS_p3, might also be involved in *PHO5* regulation.

Discussion

All the 6-bp motifs in the promoter regions of *PHO5*, *PHO8*, *PHO81*, and *PHO84* (Fig. 2), except CAaGTT of *PHO81* (described below), which has not yet been examined, were found to be protected by Pho4p from DNase I digestion in vitro (Fig. 3). The nucleotide sequences of these protected regions, including UAS_p3 of *PHO5*, are compiled in Fig. 4, together with that of the distal site of *PHO8* described by Barbarić et al. (1992). So far 11 footprint profiles have been examined, including 6 with the CACGTG motif (group 1) and 3 with the CACGTT motif (group 2) in the center of the protected regions. The other two, UAS_p3 of *PHO5*, CAtGTG (the lowercase letter indicates a deviation from CACGTG), and the distal one of *PHO8*, CACGcT, have similar sequences. Their UAS functions and relative affinities to Pho4p estimated from the footprint profiles (Fig. 3) are also listed in Fig. 4. Since some of the Pho4p-binding sites are ineffective for regulation, the consensus sequences seem to be GCACGTGGG for group 1 and GCACGTTTT for group 2.

The 6-bp CACGTG core sequence has similarity with a highly conserved octanucleotide, RTCACRTG (where R is a purine residue), called CDEI, in centromeres and in various promoters of *S. cerevisiae* for binding with the CP1/CBF1/CPF1 (Cpf1p) protein, which has a basic region and a helix-loop-helix structure in its DNA-binding domain (Baker and Masison, 1990; Cai and Davis, 1990; Mellor et al., 1990). In this context, it is possible to speculate that the G residue flanking the 5′ end of the 6-bp CACGT(G/T) motif and the other residues flanking the 3′ end have important functions for discrimination of the Pho4p binding sites in *PHO* promoters from the CDEI elements. This view was supported by the report by Fisher and Goding (1992) that the presence of T residues flanking the 5′ end of the CACGTG motif inhibits Pho4p binding but not Cpf1p binding. Overexpression of *PHO4*, however, could suppress the methionine auxotrophy of a *cep1* mutant, whereas *CEP1* overexpression failed to suppress the phenotype of a *pho4* mutant (O'Connell and Baker, 1992).

The transcriptions of all these *PHO* structural genes, except *PHO8*, depend on the function of Pho2p. Several attempts to detect a Pho2p-binding site in the *PHO5* promoter have been unsuccessful. There is a report, however, that a 31-bp fragment of the *PHO5* promoter from nucleotides −381 to −351 bearing UAS_p1 (GCACGTTTT) could respond to a *pho2* mutation of host cells (Vogel et al., 1989). Transcription of the *PHO81* gene in the *pho2* mutant in low-P_i medium was leaky (Yoshida et al., 1989b), and a specific region for the Pho2p-dependent expression of *PHO81* was suggested to be located downstream

Site	UAS activity	Nucleotide sequence	Affinity to Pho4p
Group 1			
PHO5 UASp2	+	-258 A C T C A C A C A C G T G G G A C T A G C -240	High
PHO84 Site D	+	-421 T T T C C A G C A C G T G G G G C G G A -402	High
PHO81 UAS	+	-350 T T A T G G C A C G T G C G A A T A A -332	High
PHO8 Proximal	+	-520 G T G A T C G C T G C A C G T G G C C C G A -541	High
PHO5 UASp3	±	-182 T A A T T T G G C A T G T G C G A T C T C -202	Low
PHO84 Site A	-	-870 T T T A T C A C G T G A C A C T T T T T -889	Low
PHO84 Site C	±	-426 A C G T C C A C G T G G A A C T A T T -444	Low
Consensus 1		t t - - - G C A C G T G G G - c - a	
Group 2			
PHO5 UASp1	+	-370 T A A A T T A G C A C G T T T T C G C -352	Medium
PHO84 Site E	+	-251 A A T A C G C A C G T T T T T A A T C T A -271	Medium
PHO84 Site B	-	-592 T T A C G C A C G T T G G T G C T G -573	Low
PHO8 Distal	±	-719 T T A C C C G C A C G C T T A A T A T -737	Low
Consensus 2		a a t - - G C A C G T T T T	

FIG. 4. Regions in the *PHO* promoters protected from DNase I digestion in vitro by Pho4p. The protected sequences shown in Fig. 3 were compiled for strands having sequences conforming most closely to GCACGTG or GCACGTT along with the data of Barbarić et al. (1992) for that flanking the distal region of the *PHO8* promoter. These sequences were classified into two classes depending on whether they contained the CACGTG (group 1) or CACGTT (group 2) sequence in the core region. The relative affinity of each sequence to Pho4p protein was estimated from the amount of Pho4p protein required for protection.

of nucleotide −290 (Ogawa et al., 1993). A putative GCACGTTTT sequence, GCAaGTTTT, is found at position −234 close to the suggested TATA box in the *PHO81* promoter (Fig. 2). The E site of the *PHO84* promoter has a GCACGTTTT sequence, whereas the *PHO8* gene has no such sequence but has the sequence GCACGcTTa from nucleotides −725 to −733 directly flanking the distal region (Fig. 2). This 9-bp sequence is located in the region weakly protected by Pho4p (Barbarić et al., 1992). The CACGTG sequence is suggested to be more efficient than CACGTT for binding Pho4p (Fisher et al., 1991) and *PHO8* expression (Hayashi and Oshima, 1991). Pho4p has a small stretch of amino acid sequence with similarity to an amino acid sequence in the Pho2p homeo-box (Bürglin, 1988). These facts suggest the possibility that the factor actually binding to the GCACGTTTT sequence is a Pho4p-Pho2p complex. This possibility requires examination.

This study was supported by grants to Y.O. from the Ministry of Education, Science and Culture of Japan and to N.O. from the Inamori Foundation.

REFERENCES

Arndt, K. T., C. Styles, and G. R. Fink. 1987. Multiple global regulators control *HIS4* transcription in yeast. *Science* **237:**874–880.

Baker, R. E., and D. C. Masison. 1990. Isolation of the gene encoding the *Saccharomyces cerevisiae* centromere-binding protein CP1. *Mol. Cell. Biol.* **10:**2458–2467.

Barbarić, S., K.-D. Fascher, and W. Hörz. 1992. Activation of the weakly regulated *PHO8* promoter in *S. cerevisiae:* chromatin transition and binding sites for the positive regulatory protein PHO4. *Nucleic Acids Res.* **20:**1031–1038.

Bergman, L. W., D. C. McClinton, S. L. Madden, and L. H. Preis. 1986. Molecular analysis of the DNA sequences involved in the transcriptional regulation of the phosphate-repressible acid phosphatase gene (*PHO5*) of *Saccharomyces cerevisiae. Proc. Natl. Acad. Sci. USA* **83:**6070–6074.

Braus, G., H.-U. Mösch, K. Vogel, A. Hinnen, and R. Hütter. 1989. Interpathway regulation of the *TRP4* gene of yeast. *EMBO J.* **8:**939–945.

Bun-ya, M., M. Nishimura, S. Harashima, and Y. Oshima. 1991. The *PHO84* gene of *Saccharomyces cerevisiae* encodes an inorganic phosphate transporter. *Mol. Cell. Biol.* **11:**3229–3238.

Bürglin, T. R. 1988. The yeast regulatory gene *PHO2* encodes a homeo box. *Cell* **53:**339–340.

Cai, M., and R. W. Davis. 1990. Yeast centromere binding protein CBF1, of the helix-loop-helix protein family, is required for chromosome stability and methionine prototrophy. *Cell* **61:**437–446.

Fisher, F., and C. R. Goding. 1992. Single amino acid substitutions alter helix-loop-helix protein specificity for bases flanking the core CANNTG motif. *EMBO J.* **11:**4103–4109.

Fisher, F., P.-S. Jayaraman, and C. R. Goding. 1991. C-Myc and the yeast transcription factor PHO4 share a common CACGTG-binding motif. *Oncogene* **6:**1099–1104.

Hayashi, N., and Y. Oshima. 1991. Specific *cis*-acting sequence for *PHO8* expression interacts with PHO4 protein, a positive regulatory factor, in *Saccharomyces cerevisiae. Mol. Cell. Biol.* **11:**785–794.

Johnston, M., and M. Carlson. 1992. Regulation of carbon and phosphate utilization, p. 193–281. *In* E. W. Jones, J. R. Pringle, and J. R. Broach (ed.), *The Molecular and Cellular Biology of the Yeast Saccharomyces: Gene Expression.* Cold Spring Harbor, N.Y.

Lux, S. E., K. M. John, and V. Bennett. 1990. Analysis of cDNA for human erythrocyte ankyrin indicates a repeated structure with homology to tissue-differentiation and cell-cycle control proteins. *Nature* (London) **344:**36–42.

Madden, S. L., D. L. Johnson, and L. W. Bergman. 1990. Molecular and expression analysis of the negative regulators involved in the transcriptional regulation of acid phosphatase production in *Saccharomyces cerevisiae. Mol. Cell. Biol.* **10:**5950–5957.

Mellor, J., W. Jiang, M. Funk, J. Rathjen, C. A. Barnes, T. Hinz, J. H. Hegemann, and P. Philippsen. 1990. CPF1, a yeast protein which functions in centromeres and promoters. *EMBO J.* **9:**4017–4026.

Nakao, J., A. Miyanohara, A. Toh-e, and K. Matsubara. 1986. *Saccharomyces cerevisiae PHO5* promoter region: location and function of the upstream activation site. *Mol. Cell. Biol.* **6**:2613–2623.

O'Connell, K. F., and R. E. Baker. 1992. Possible cross-regulation of phosphate and sulfate metabolism in *Saccharomyces cerevisiae*. *Genetics* **132**:63–75.

Ogawa, N., K. Noguchi, Y. Yamashita, T. Yasuhara, N. Hayashi, K. Yoshida, and Y. Oshima. 1993. Promoter analysis of the *PHO81* gene encoding a 134 kDa protein bearing ankyrin repeats in the phosphatase regulon of *Saccharomyces cerevisiae*. *Mol. Gen. Genet.* **238**:444–454.

Ogawa, N., and Y. Oshima. 1990. Functional domains of a positive regulatory protein, PHO4, for transcriptional control of the phosphatase regulon in *Saccharomyces cerevisiae*. *Mol. Cell. Biol.* **10**:2224–2236.

Oshima, Y. 1991. Impact of the Douglas-Hawthorne model as a paradigm for elucidating cellular regulatory mechanisms in fungi. *Genetics* **128**:195–201.

Rudolph, H., and A. Hinnen. 1987. The yeast *PHO5* promoter: phosphate-control elements and sequences mediating mRNA start-site selection. *Proc. Natl. Acad. Sci. USA* **84**:1340–1344.

Sengstag, C., and A. Hinnen. 1987. The sequence of the *Saccharomyces cerevisiae* gene *PHO2* codes for a regulatory protein with unusual aminoacid composition. *Nucleic Acids Res.* **15**:233–246.

Sengstag, C., and A. Hinnen. 1988. A 28-bp segment of the *Saccharomyces cerevisiae PHO5* upstream activator sequence confers phosphate control to the *CYC-lacZ* gene fusion. *Gene* **67**:223–228.

Tice-Baldwin, K., G. R. Fink, and K. T. Arndt. 1989. BAS1 has a Myb motif and activates *HIS4* transcription only in combination with BAS2. *Science* **246**:931–935.

Toh-e, A., K. Tanaka, Y. Uesono, and R. B. Wickner. 1988. *PHO85*, a negative regulator of the PHO system, is a homolog of the protein kinase gene, *CDC28*, of *Saccharomyces cerevisiae*. *Mol. Gen. Genet.* **214**:162–164.

Uesono, Y., M. Tokai, K. Tanaka, and A. Toh-e. 1992. Negative regulators of the *PHO* system in *Saccharomyces cerevisiae:* characterization of *PHO80* and *PHO85*. *Mol. Gen. Genet.* **231**:426–432.

Vogel, K., W. Hörz, and A. Hinnen. 1989. The two positively acting regulatory proteins PHO2 and PHO4 physically interact with *PHO5* upstream activation regions. *Mol. Cell. Biol.* **9**:2050–2057.

Yoshida, K., Z. Kuromitsu, N. Ogawa, K. Ogawa, and Y. Oshima. 1987. Regulatory circuit for phosphatase synthesis in *Saccharomyces cerevisiae*, p. 49–55. *In* A. Torriani-Gorini, F. G. Rothman, S. Silver, A. Wright, and E. Yagil (ed.), *Metabolism and Cellular Regulation in Microorganisms.* American Society for Microbiology, Washington, D.C.

Yoshida, K., Z. Kuromitsu, N. Ogawa, and Y. Oshima. 1989a. Mode of expression of the positive regulatory genes *PHO2* and *PHO4* of the phosphatase regulon in *Saccharomyces cerevisiae*. *Mol. Gen. Genet.* **217**:31–39.

Yoshida, K., N. Ogawa, and Y. Oshima. 1989b. Function of the *PHO* regulatory genes for repressible acid phosphatase synthesis in *Saccharomyces cerevisiae*. *Mol. Gen. Genet.* **217**:40–46.

Chapter 11

Protein-DNA and Protein-Protein Interactions Regulating the Phosphatase Multigene Family of *Saccharomyces cerevisiae*

STEPHEN A. PARENT,[1] MICHAEL C. JUSTICE,[1] LING-WEN YUAN,[2] JAMES E. HOPPER,[2] AND KEITH A. BOSTIAN[3]

Merck Research Laboratories, P.O. Box 2000, Rahway, New Jersey 07065[1]; Pennsylvania State University College of Medicine, Hershey, Pennsylvania 17033[2]; and Microcide Pharmaceuticals, 4040 Campbell Avenue, Menlo Park, California 94025[3]

Many eukaryotic genes are regulated by proteins which control transcription initiation in response to specific intra- and extracellular signals. In yeast cells, transcription is regulated in response to diverse stimuli including nutritional changes in the environment, the pheromone responsiveness of the cell, and its cell type. We are studying a family of regulatory proteins which control expression of enzymes involved in the acquisition and metabolic integration of P_i in *Saccharomyces cerevisiae* (Bostian et al., 1983; Dawes and Senior, 1973). Several enzymes involved in this process include acid phosphatase (APase) and alkaline phosphatase (AlkPase), a phosphate permease, a polyphosphate kinase, and a polyphosphatase. In cells grown in P_i-rich media, these enzymes are present in basal amounts. Upon P_i starvation, the enzymes are derepressed and metabolism becomes dependent upon exocellular APase. When P_i is replenished, APase and AlkPase are repressed by a coordinated mechanism.

Regulation of the repressible APases (rAPases; encoded by *PHO5*, *PHO10*, and *PHO11*) (Bostian et al., 1980; Andersen et al., 1983; Lemire et al., 1985), alkaline phosphatase (*PHO8*) (Kaneko et al., 1985, 1987), and an inducible phosphate permease (*PHO84*) (Bun-Ya et al., 1991) is predominantly transcriptional. The genes encoding these enzymes are transcribed when external P_i is limiting and repressed when P_i is in excess. In contrast, the *PHO3* gene, encoding a minor APase, is expressed predominantly during high-P_i growth (Tait-Kamradt et al., 1986) and repressed by thiamine (Schweingruber et al., 1986). The *PHO5*, *PHO8*, *PHO10*, *PHO11*, and *PHO84* genes are controlled by several genetically defined regulatory proteins. Recessive *pho4* and *pho81* mutations block derepression of all these genes (Toh-e et al., 1973), whereas *pho80* or *pho85* mutations lead to constitutive synthesis of the enzymes (Toh-e et al., 1973; Ueda et al., 1975; Tamai et al., 1985). In *pho2* mutants, genes encoding rAPase and the repressible permease do not respond to P_i starvation (Toh-e et al., 1973; Tamai et al., 1985).

A Working Model Explaining *PHO* Regulation

Current models explaining regulation of this gene family are based on genetic and biochemical data (Yoshida et al., 1989b) (Fig. 1). In the presence of corepressor (P_i excess), the negative factors (*PHO80* and *PHO85*) compete with *cis* promoter sequences (at an upstream activation site) for binding of the positive transcriptional factor *PHO4*. When P_i is limiting, the product of the *PHO81* gene metabolizes an unknown corepressor, relieving Pho4 from the repressive action of Pho80 and Pho85. In the derepressed state, Pho4 binds to DNA sequences in the regulated promoters and interacts with the positive factor Pho2 to activate transcription.

Results from many genetic experiments including epistasis and hypostasis tests support this model. In addition, the phenotypes of *PHO4*^{c} mutations (formerly called *PHO82*^{c}) suggest that interactions between the regulatory proteins control transcription in this system. These mutations confer weakly dominant constitutive phenotypes, map within *PHO4* coding sequences, and have been proposed to disrupt the site of interaction with negative regulators (Toh-e and Oshima, 1974; Toh-e et al., 1981). The recent discovery that a *PHO80-2* mutation suppresses specific *PHO4*^{c} mutations supports this (Okada and Toh-e, 1992).

Molecular data from experiments in several laboratories provide evidence to support this model. First, coordinate control of the rAPases, AlkPase, and the phosphate permease is exerted at the level of mRNA synthesis through the *PHO* regulatory genes. Nonderepressible *pho4* and *pho81* mutants fail to transcribe *PHO5*, *PHO8*, and *PHO84* under derepressing conditions, whereas *pho80* and *pho85* mutants express these RNAs constitutively (Lemire et al., 1985; Kaneko et al., 1985; Bun-

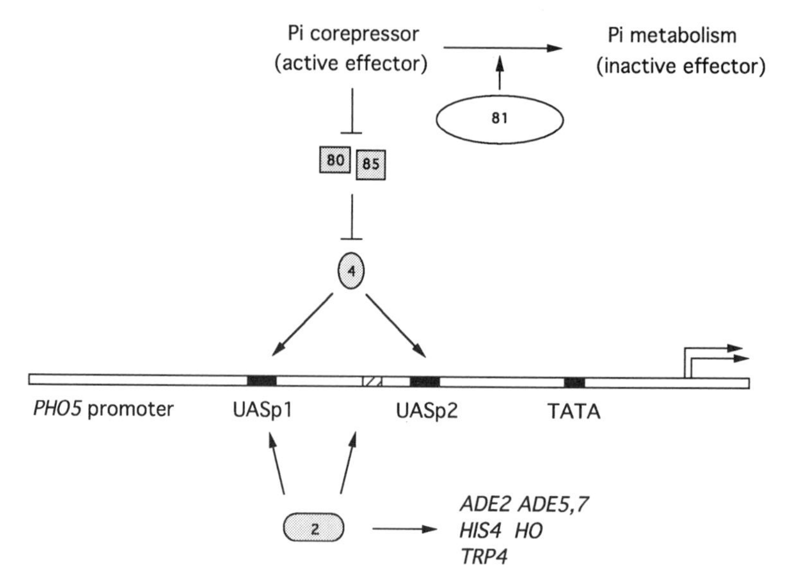

FIG. 1. Regulatory interactions controlling expression of the *PHO5* promoter. Oblongs depict the products of the positive-acting *PHO2*, *PHO4*, and *PHO81* genes. The products of the negative-acting *PHO80* and *PHO85* gene products are depicted by squares. *PHO2*, *PHO4*, *PHO80*, and *PHO85* RNAs are expressed constitutively. The hatched region in the *PHO5* promoter depicts the Pho2-binding site, which does not affect promoter function (Rudolph and Hinnen, 1987; Vogel et al., 1989). The Pho2-UASp1 interaction is described in the text. *PHO4* also regulates genes encoding minor rAPases (*PHO10* and *PHO11*), alkaline phosphatase (*PHO8*), a phosphate permease (*PHO84*), and the positive regulator Pho81. Promoters of the *PHO8* and *PHO81* genes also contain UASp elements to which Pho4 protein binds. The sequences of the *PHO5* UASp elements are also conserved in the *PHO11* promoter (Hinnen et al., 1987; Tait-Kamradt et al., 1986). Although *PHO2* regulates expression of *PHO81*, a Pho2-binding site has not been identified in the promoter of this gene (Ogawa et al., 1993). *PHO2* is not required for *PHO8* expression.

Ya et al., 1991). The *PHO2* gene product is required for transcription of the *PHO5* and *PHO84* genes. Second, repression and derepression of rAPase mRNAs occurs in the absence of protein synthesis (Lemire et al., 1985). Lastly, in wild-type cells, *PHO2*, *PHO4*, *PHO80*, and *PHO85* RNAs are expressed constitutively in a P_i-independent manner (Koren et al., 1986; Legrain et al., 1986; Berben et al., 1988; Madden et al., 1988; Yoshida et al., 1989a; Gilliquet et al., 1990). Although *PHO2* RNA is expressed constitutively, the gene appears to be posttranscriptionally repressed by P_i in the stationary phase and autoregulated.

Little is known about what metabolic signal serves as corepressor in this system. Expression of rAPase is influenced by endogenous phosphate pools (Bostian et al., 1983). In the same study, the finding that changes in intracellular P_i levels do not correlate with rAPase derepression suggests that P_i may not serve as corepressor. An inverse correlation of rAPase activity with acid-soluble polyphosphate suggests that PP_i or a low-molecular-weight polyphosphate may serve as a metabolic regulator controlling APase expression (Bostian et al., 1983).

Important Protein-DNA and Protein-Protein Interactions in the *PHO* System

Roles of the regulatory proteins have been elucidated by analyses of the *PHO5* and *PHO8* promoters (Rudolph and Hinnen, 1987; Hayashi and Oshima, 1991), the isolation of the *PHO* regulatory genes (Koren et al., 1986; Legrain et al., 1986; Toh-e and Shimauchi, 1986; Sengstag and Hinnen, 1987; Uesono et al., 1987), and the characterization of their products through molecular genetic and biochemical techniques. The model in Fig. 1 suggests that balanced levels of the regulatory proteins are required to maintain P_i-sensitive regulation. To test this, we overexpressed Pho4 protein in wild-type cells and various *pho* mutants and examined rAPase expression (Parent et al., 1987). In a wild-type cell, overexpression of *PHO4* results in high levels of rAPase in low-P_i media and constitutive expression under normally repressing conditions. Increasing the *PHO4* gene dosage also suppresses the nonderepressible phenotypes of recessive *pho2-1* and *pho81-1* mutants. These findings support the genetic epistasis-hypostasis relationships in this system and have been confirmed and extended (Yoshida et al., 1989b;

Berben et al., 1988). To explain these phenotypes, we proposed that Pho4 was a limiting factor determining derepressed levels of rAPase and that its concentration, relative to the negative factors, was crucial for maintaining repression. The ability of *PHO4* overexpression to suppress the *pho2-1* mutation might be explained if Pho4 and Pho2 function cooperatively at the *PHO5* promoter. Sequence analysis of the *PHO4* gene indicates that it encodes a member of the basic helix-loop-helix (bHLH) family of sequence-specific DNA-binding proteins (Ogawa and Oshima, 1990), including c-*myc* and MyoD. The bHLH region of Pho4 also shares sequence similarity with the transcriptional activator *NUC-1*, which regulates genes involved in phosphate metabolism in *Neurospora crassa* (Kang and Metzenberg, 1990). Functional domains of Pho4 include a COOH-terminal bHLH region required for DNA binding, an NH_2-terminal domain which is necessary to activate transcription, regions that might interact with Pho80, and domains involved in multimerization (Ogawa and Oshima, 1990).

Pho2 (Bas2/Grf10) belongs to the homeodomain family of DNA-binding proteins which recognize $(TAAT)_n$ sequences (Arndt et al., 1987; Berben et al., 1988), and, in addition to being essential for *PHO5* and *PHO84* expression, it regulates transcription of genes involved in diverse biochemical and cellular pathways (Arndt et al., 1987; Tice-Baldwin et al., 1989; Braus et al., 1989; Daignan-Fornier and Fink, 1992; Brazas and Stillman, 1993a, 1993b). Recombinant Pho2 protein expressed in vitro or in *Escherichia coli* binds to DNA sequences in promoters of the *ADE2*, *ADE5,7*, *HIS4*, *HO*, *PHO5*, and *TRP4* genes. In cooperation with the sequence-specific DNA-binding transcription factor Bas1, Pho2 controls basal expression of the *HIS4* gene and the basal and induced expression of *ADE1*, *ADE2*, *ADE5,7*, and *ADE8*. At the *TRP4* promoter, Pho2 binds upstream activation sequence 1 (UAS1), the major *GCN4*-dependent activation element, and antagonizes *GCN4*-mediated general amino acid control of the gene when P_i is limiting. This pleiotropic regulator is also necessary for full expression of the *HO* gene, encoding the endonuclease which initiates mating-type switching. The function of a *SWI*-responsive *HO* UAS element requires cooperative DNA-binding interactions between Pho2 and the zinc finger protein Swi5. Surprisingly, deletion of the Pho2-binding site at positions -277 to -296 in the *PHO5* promoter has little effect on promoter function (Rudolph and Hinnen, 1987).

Deletion analyses of several phosphate-regulated promoters have identified *cis*-acting UASs (UASp) required for P_i-responsive transcription (Rudolph and Hinnen, 1987; Hayashi and Oshima, 1991). Many of these sites confer Pho4-dependent P_i-regulated expression upon heterologous promoters, contain the core consensus sequence (CANNTG) recognized by bHLH proteins, and bind recombinant Pho4 protein in vitro (Vogel et al., 1989; Hayashi and Oshima, 1991; Barbaric et al., 1992). Analyses of the chromatin structures of the *PHO5* and *PHO8* promoters indicate the presence of highly ordered structures which change significantly upon derepression, resulting in large regions which are nucleosome free (Almer et al., 1986; Barbaric et al., 1992). The findings that these transitions are dependent upon *PHO4* (Fascher et al., 1990) and that several Pho4-binding sites in the *PHO5* and *PHO8* promoters are within regions hypersensitive to various nucleases suggest that Pho4 interactions at some of these sites play a role in initiating the chromatin transitions associated with derepression. Although binding of Pho2 to the UASp elements has not been demonstrated, the activities of the *PHO5* UASp1 (Sengstag and Hinnen, 1988) and UASp2 (Justice et al., unpublished) elements in heterologous promoters are Pho2 dependent, suggesting that this protein also contributes to UASp function in vivo.

The precise roles of the *PHO80*, *PHO81*, and *PHO85* genes are unknown. Molecular analysis of the *PHO81* gene indicates that, like *PHO5*, its transcription is also controlled by P_i and the *PHO* regulatory genes (Yoshida et al., 1989b), through a P_i responsive UASp element which binds Pho4 (Ogawa et al., 1993). These observations have led to the model that Pho81 might metabolize an effector molecule to an inactive state (Yoshida et al., 1989b). Although the predicted Pho81 amino acid sequence provides no clues about its function, the protein does contain several ankyrin repeats which have been suggested to be important for promoting protein-protein interactions (Ogawa et al., 1993). Pho80 shares strong sequence similarity to the *N. crassa* PREG regulatory protein, which, in conjunction with PGOV, antagonizes the positive-acting NUC-1 transcriptional activator involved in P_i metabolism (Kang and Metzenberg, 1993). Increasing the *PHO80* gene dosage represses APase expression under normally derepressing conditions (Madden et al., 1988; Yoshida et al., 1989b), and this phenotype can be abrogated by overexpressing Pho4, suggesting that the ratio of these proteins is crucial for maintaining regulation. Pho80 overexpression also suppresses the constitutive phenotypes of *pho85*Δ and *PHO81*^{c} mutants (Madden et al., 1990; Creasy et al., 1993), suggesting that Pho80 is a key regulator controlling APase repression. The finding that increasing the dosage of Pho85 specifically suppresses the constitutive phenotype of a *pho80-1* mutation also suggests that these proteins interact (Gilliquet et al., 1990). The predicted Pho85 amino acid sequence shares significant similarity with the protein kinase Cdc28 (Toh-e et al., 1988). Although the substrate(s) of Pho85 is unknown, casein is phosphorylated in vitro with whole-cell immunoprecipitates containing Pho85, suggesting that Pho85 possesses kinase activity (Uesono et al., 1992). In the same study, Pho80 and at least two other proteins were phosphorylated in derepressed cells in a *PHO85*-dependent manner.

Important Pho4 and Pho2 Interactions with the *PHO5* Promoter

To understand how Pho4 overexpression suppresses *pho2* mutations (Parent et al., 1987; Berben et al., 1988; Parent and Bostian, unpublished), we examined Pho2- and Pho4-DNA interactions by using *PHO5* UASp elements. Mutations in the *PHO5* UASp1 core sequence recognized by bHLH proteins reduce the ability of recombinant Pho4 to bind this element (Fisher et al., 1991). To examine the importance of this interaction, we tested the function of UASp1 and UASp2 in vitro and in vivo. Pho4-DNA interactions were examined in electrophoretic mobility shift assays with radiolabeled UASp DNA probes and yeast cell extracts. Protein-DNA complexes were detected with UASp probes and extracts that were prepared from cells grown in high- and low-P_i media (Fig. 2A and B, lanes 2 and 3). Two unpublished observations suggest that the complexes indicated by the solid circles contain Pho4: (i) the complexes are absent in reactions with *pho4Δ* extracts, and (ii) their mobilities are altered upon addition of Pho4 antiserum. Interestingly, recombinant (Vogel et al., 1989; Fisher et al., 1991) and native Pho4 proteins (Fig. 2A and B) bind less efficiently to UASp1 than to UASp2. Competition electrophoretic mobility shift assays with radiolabeled UASp2 probe and unlabeled UASp DNAs support these findings (unpublished observations). Although several studies indicate that Pho4 interacts with the UASp elements with different efficiencies, 20- to 80-bp deletions encompassing either site reduce *PHO5* promoter activity by 90% (Rudolph and Hinnen, 1987). Either element confers P_i-responsive expression to a *CYC1-lacZ* reporter in vivo (Table 1). However, the site (UASp1) which is most active when cells are starved for P_i in vivo binds Pho4 less efficiently. To investigate these elements in the native promoter, single-base mutations were introduced into the bHLH core-binding sites within the *PHO5* gene and assayed in vivo (Table 2). As expected, the wild-type promoter is regulated by extracellular P_i. UASp1 and UASp2 T-to-C mutations at the +2 position of the core consensus site inhibited Pho4 binding (unpublished observations) and lowered promoter activity to similar levels when cells

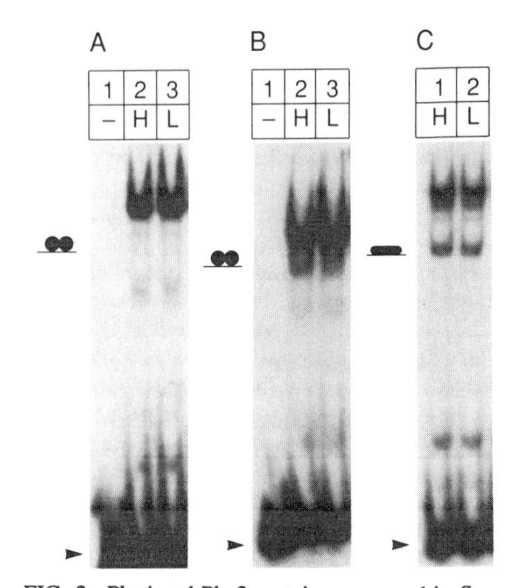

FIG. 2. Pho4 and Pho2 proteins expressed in *S. cerevisiae* interact with UASp elements of the *PHO5* promoter. Wild-type cells were grown to an optical density at 660 nm (OD_{660}) of 1.5 in low- or high-P_i medium and harvested by centrifugation. Whole-cell extracts prepared by breaking cells with glass beads were assayed in electrophoretic mobility shift assays with radiolabeled UASp DNA probes (Justice et al., unpublished observations). Nonspecific competitor DNAs [poly(dA-dT)-poly(dA-dT) or poly(dI-dC)-poly(dI-dC)] were included in the reactions and enabled the Pho2-UASp1 and Pho4-UASp1 complexes to be distinguished (data not shown). (A and B) Reaction mixtures (24 μl) contained approximately 1 ng of UASp1 or UASp2 DNA probes, respectively, and 0.1 μg of nonspecific competitor poly(dA-dT)-poly(dA-dT). Lanes: 1, no extract; 2, 3 μg of wild-type extract from high-P_i-grown cells; 3, 3 μg of wild-type extract from low-P_i-grown cells. Complementary oligonucleotides were synthesized, annealed, and radiolabeled with [α-^{32}P]dCTP (Sambrook et al., 1989). The *PHO5* promoter sequence is indicated by boldface type. Annealed oligonucleotides contain a 4-base *Xho*I overhang. Sequences are as follows: UASp1 top strand, 5′-TCGAGT**ATTAAATTAGCACGTTTTCGCATAGAC**-3′; UASp1 bottom strand, 5′-TCGAG**TCTATGCGAAAACGTGCTAATTTAATAC**-3′; UASp2 top strand, 5′-TCGAG**TGGCACTCACACGTGGGACTAGCAC**-3′; UASp2 bottom strand, 5′-TCGAG**TGCTAGTCCCACGTGTGAGTGCCAC**-3′. (C) Reaction mixtures (24 μl) contain approximately 1 ng of the UASp1 DNA probe used in panel A and 0.1 μg of the nonspecific competitor poly(dI-dC)-poly(dI-dC). Lanes: 1, 3 μg of wild-type extract from high-P_i grown cells; 2, 3 μg of wild-type extract from low-P_i grown cells. The arrowheads represent the positions of the unbound DNA probes. Solid circles on thin line represent the positions of the Pho4-DNA complexes. These complexes were absent in *pho4Δ* extracts, and their migration is altered on the addition of Pho4 antiserum to reactions with wild-type extracts (unpublished observations). Solid bar on thin line represents the position of the Pho2-DNA complexes. This complex was not detected in *pho2Δ* extracts, and the addition of Pho2 antiserum to reactions with wild-type extracts prevented complex formation (unpublished observations).

TABLE 1. UASp1 and UASp2 confer phosphate-regulated expression upon a *CYC1* basal promoter

UAS[a]	Sequence[b]	β-Galactosidase activity (U/OD_{660})		L/H ratio
		High P_i (H)	Low P_i (L)	
—	—	4.7	4.5	1
UASp1	TATTAAATTAGCACGTTTTCGCATAGA	5.8	48.3	8
UASp2	TGGCACTCACACGTGGGACTAGCA	5.1	12.2	2

[a]Double-stranded oligonucleotides containing UASp1 and UASp2 (see the legend to Fig. 2) were cloned into the *Sal*I site upstream of a *CYC1-lacZ* reporter gene in the 2μm plasmid pLG-ΔBS. Wild-type transformants containing the reporter plasmids were grown in high- and low-P_i media and assayed for β-galactosidase activity under standard conditions (Guarente, 1983). Data in the first row are activities expressed by transformants containing pLG-ΔBS. At an OD_{660} of 1, cells were placed on ice after the addition of cycloheximide to a final concentration of 0.1 mg/ml, harvested by centrifugation, resuspended in Z buffer, and assayed with *o*-nitrophenyl-β-D-galactoside (ONPG) after permeabilization with chloroform and SDS. Units were calculated by the method of Miller (1972) with the OD_{660} of the culture for normalization.

[b]The sequence of the UASp elements is shown with the Pho4 core-binding site in large font. These sequences are protected from DNase I digestion by Pho4 protein (Vogel et al., 1989).

were grown in low-P_i medium (Table 2, UASp1-M8 and UASp2-M19).

To begin to understand why the relative binding of Pho4 to the UASp elements does not correlate with the activities of the regulatory sites in the native and heterologous promoters, we reinvestigated Pho2-UASp interactions. Pho2 protein from yeast cells grown in high- and low-P_i media formed a protein-DNA complex with UASp1 in electrophoretic mobility shift assays (Fig. 2C; solid bar indicates the position of the complex). This complex was not detected with *pho2*Δ extracts, and incubation of Pho2 antibody with wild-type extracts prevented its formation (unpublished observations), suggesting that Pho2 binds to UASp1. Specific binding to UASp2 was not detected under these conditions (unpublished observations). We compared the DNA sequence of UASp1 with that of several Pho2(Bas2)-binding sites and generated the following consensus sequence: (T/C)TAATT(G/T)A(A/T)T. Matches to this sequence include sites in the promoters of the *HIS4* (TTAATTAATT at −226 to −217), *HO* (TAAATTGATG at −1317 to −1326), *TRP4* (CTAATATTAT at −242 to −233), and *PHO5* (CTAAATGAAT at −283 to −274 and CTAATTTAAT at −363 to −372 in UASp1) genes. A mutation altering the TAAT sequences in the UASp1 site (CCGGCTCGGC) lowered *PHO5* promoter activity in vivo (Table 2), suggesting that Pho2 interacts with this sequence to activate transcription.

Distinct Regulatory Interactions within the *PHO5* Promoter

Although UASp1 and UASp2 contribute significantly to *PHO5* promoter function, several experiments indicate that these elements differ in at

TABLE 2. Effects of *PHO4* and *PHO2* binding-site mutations on *PHO5* promoter function

Plasmid[a]	*PHO5* promoter	APase activity[b] (U/OD_{660})		L/H ratio
		High P_i (H)	Low P_i (L)	
YCplac113	—	1.5	8.0	5
MJ5	Wild type	5.4	93.0	17
MJ58	UASp1-M8 (*pho4*)	5.9	24.8	4
MJ54	UASp2-M19 (*pho4*)	6.0	30.8	5
MJ59	UASp1-M20 (*pho2*)	4.8	33.8	7

[a]Wild-type and mutant *PHO5* genes were cloned into the single-copy vector YCplac113 (Gietz and Sugino, 1988). UASp mutations were generated by site-directed mutagenesis with the Bio-Rad Muta-Gene kit and M13mp19 templates containing the *PHO5* promoter.

[b]GG100-14D (*pho5-1 pho3-1*) transformants containing the plasmids were grown in high- and low-P_i media and assayed for APase activity using published procedures (Bostian et al., 1980; Tait-Kamradt et al., 1986). Briefly, cells were harvested at an OD_{660} of 1 after the addition of cycloheximide to a final concentration of 0.1 mg/ml, frozen on dry ice, thawed, and assayed with *p*-nitrophenyl phosphate as the substrate. One unit of enzyme activity is the amount producing 1 nmol of *p*-nitrophenol per min. Activity was normalized to culture density measured at 660 nm.

least two important aspects. First, recombinant and native Pho4 bind more efficiently to UASp2 than to UASp1. Second, Pho2 binds to UASp1 in vitro. A mutation in UASp1 altering the TAAT sequences which are recognized by many homeodomain proteins reduces promoter activity, suggesting a direct molecular role for Pho2 in regulating *PHO5* expression. The proximity of the Pho4- and Pho2-binding sites in UASp1 suggests that these proteins cooperate or synergize at this element to activate *PHO5* transcription. We propose that the proteins bind cooperatively to DNA or present a composite activation domain which is necessary for efficient transcription. In both models, overexpression of Pho4 might suppress a loss of Pho2 function.

Pho4 and Pho80 Physically Interact In Vivo

Results from several experiments suggest that Pho4 and Pho80 interact in vivo. To address this model, we used the two-hybrid system (Fields and Song, 1989). In this system, the Gal4 DNA-binding domain is fused to one protein and the Gal4 transcription activation domain is fused to the second protein. If the two proteins physically interact and reestablish the proximity of the Gal4 domains in vivo, transcription of genes regulated by *GAL* UAS elements is activated in a *gal4Δ gal80Δ* strain. We fused Pho80 to the Gal4 DNA-binding domain and Pho4 to the Gal4 activation domain and tested the ability of these fusion proteins to activate transcription of a *GAL1-lacZ* reporter. Transformants expressing the Gal4-Pho80 and Gal4-Pho4 fusion proteins expressed β-galactosidase activity (Table 3). Furthermore, these interactions occur in cells grown in high- and low-P_i media (unpublished observations). Although interpretations of these results require several assumptions, they suggest that Pho4 and Pho80 interact in vivo and that regulation is not achieved by controlling the ability of Pho80 to interact with and inhibit Pho4 function. Similar results have been observed in the *GAL* system (Leuther and Johnston, 1992). We are testing this hypothesis by examining physical interactions between these proteins in wild-type cells.

TABLE 3. Pho4 and Pho80 physically interact in vivo

Plasmids[a]	β-Galactosidase activity[b] (U/OD_{660})
Gal4(DBD) + Gal4(AD)	0.7
Gal4(DBD)-Pho80 + Gal4(AD)	0.6
Gal4(DBD) + Gal4(AD)-Pho4	1.3
Gal4(DBD)-Pho80 + Gal4(AD)-Pho4	37.0
Gal4	191.9

[a]The Gal4(DBD)-Pho80 and Gal4(AD)-Pho4 fusion genes were constructed in pMA424 and pGAD-2F (Fields and Song, 1989) by standard methods. Gal4(DBD) is the Gal4 DNA binding domain, and Gal4(AD) is the Gal4 activation domain.

[b]GGY1:171 (*gal4Δ gal80Δ GAL1*-LacZ) transformants grown in selective media containing galactose, glycerol, and lactic acid were assayed for β-galactosidase activity under standard conditions as described in Table 1.

REFERENCES

Almer, A., H. Rudolph, A. Hinnen, and W. Hörz. 1986. Removal of positioned nucleosomes from the yeast *PHO5* promoter upon *PHO5* induction releases additional upstream activating DNA elements. *EMBO* **5:**2689–2696.

Andersen, N., G. P. Thill, and R. A. Kramer. 1983. RNA and homology mapping of two DNA fragments with repressible acid phosphatase genes from *Saccharomyces cerevisiae. Mol. Cell. Biol.* **3:**562–569.

Arndt, K. T., C. Styles, and G. R. Fink. 1987. Multiple global regulators control *HIS4* transcription in yeast. *Science* **237:**874–880.

Barbaric, S., K.-D. Fascher, and W. Hörz. 1992. Activation of the weakly regulated *PHO8* promoter in *S. cerevisiae:* chromatin transition and binding sites for the positive regulatory protein *PHO4. Nucleic Acids. Res.* **20:**1031–1038.

Berben, G., M. Legrain, and F. Hilger. 1988. Studies on the structure, expression and function of the yeast regulatory gene *PHO2. Gene* **66:**307–312.

Bostian, K. A., J. M. Lemire, L. E. Cannon, and H. O. Halvorson. 1980. In vitro synthesis of repressible yeast acid phosphatase: identification of multiple mRNAs and products. *Proc. Natl. Acad. Sci. USA* **77:**4504–4508.

Bostian, K. A., J. M. Lemire, and H. O. Halvorson. 1983. Physiological control of repressible acid phosphatase gene transcripts in *Saccharomyces cerevisiae. Mol. Cell. Biol.* **3:**839–853.

Braus, G., H.-U. Mösch, K. Vogel, A. Hinnen, and R. Hütter. 1989. Interpathway regulation of the *TRP4* gene of yeast. *EMBO J* **8:**939–945.

Brazas, R. M., and D. J. Stillman. 1993a. Identification and purification of a protein that binds DNA cooperatively with the yeast SWI5 protein. *Mol. Cell. Biol.* **13:**5524–5537.

Brazas, R. M., and D. J. Stillman. 1993b. The Swi5 zinc-finger and Grf10 homeodomain proteins bind DNA cooperatively at the yeast *HO* promoter. *Proc. Natl. Acad. Sci. USA* **90:**11237–11241.

Bun-Ya, M., M. Nishimura, S. Harashima, and Y. Oshima. 1991. The *PHO84* gene of *Saccharomyces cerevisiae* encodes an inorganic phosphate transporter. *Mol. Cell. Biol.* **11:**3229–3238.

Creasy, C. L., S. L. Madden, and L. W. Bergman. 1993. Molecular analysis of the PHO81 gene of *Saccharomyces cerevisiae. Nucleic Acids Res.* **21:**1975–1982.

Daignan-Fornier, B., and G. R. Fink. 1992. Coregulation of purine and histidine biosynthesis by the transcriptional activators BAS1 and BAS2. *Proc. Natl. Acad. Sci. USA* **89:**6746–6750.

Dawes, E. A., and P. J. Senior. 1973. The role and regulation of energy and reserve polymers in micro-organisms. *Adv. Microb. Physiol.* **10:**135–166.

Fascher, K.-D., J. Schmitz, and W. Hörz. 1990. Role of *trans*-activating proteins in the generation of active chromatin at the *PHO5* promoter in *S. cerevisiae. EMBO J.* **9:**2523–2528.

Fields, S., and O.-K. Song. 1989. A novel genetic system to detect protein-protein interactions. *Nature* (London) **340:**245–246.

Fisher, F., P.-S. Jayaraman, and C. R. Goding. 1991. C-myc and the yeast transcription factor PHO4 share a common CACGTG-binding motif. *Oncogene* **6:**1099–1104.

Gietz, R. D., and A. Sugino. 1988. New yeast-*Escherichia coli* shuttle vectors constructed with in vitro mutagenized yeast genes lacking six-base pair restrictions sites. *Gene* **74:**527–534.

Gilliquet, V., M. Legrain, G. Berben, and F. Hilger. 1990. Negative regulatory elements of the *Saccharomyces cere-*

visiae PHO system: interaction between PHO80 and PHO85 proteins. *Gene* **96:**181–188.

Guarente, L. 1983. Yeast promoters and *lacZ* fusions designed to study expression of cloned genes in yeast. *Methods Enzymol.* **101:**181–191.

Hayashi, N., and Y. Oshima. 1991. Specific *cis*-acting sequence for PHO8 expression interacts with PHO4 protein, a positive regulatory factor, in *Saccharomyces cerevisiae. Mol. Cell. Biol.* **11:**785–794.

Hinnen, A., W. Bajwa, B. Meyhack, and H. Rudolph. 1987. Molecular aspects of acid phosphatase synthesis in *Saccharomyces cerevisiae*, p. 56–62. *In* A. Torriani-Gorini, F. G. Rothman, S. Silver, A. Wright, and E. Yagil (ed.), *Phosphate Metabolism and Cellular Regulation in Microorganisms.* American Society for Microbiology, Washington, D.C.

Justice, M., et al. Unpublished observations.

Kaneko, Y., N. Hayashi, A. Toh-e, I. Banno, and Y. Oshima. 1987. Structural characteristics of the *PHO8* gene encoding repressible alkaline phosphatase in *Saccharomyces cerevisiae. Gene* **58:**137–148.

Kaneko, Y., Y. Tamai, A. Toh-e, and Y. Oshima. 1985. Transcriptional and post-transcriptional control of *PHO8* expression by *PHO* regulatory genes in *Saccharomyces cerevisiae. Mol. Cell. Biol.* **5:**248–252.

Kang, S., and R. L. Metzenberg. 1990. Molecular analysis of *nuc-1*$^+$, a gene controlling phosphorus acquisition in *Neurospora crassa. Mol. Cell. Biol.* **10:**5839–5848.

Kang, S., and R. L. Metzenberg. 1993. Insertional mutagenesis in *Neurospora crassa:* cloning and molecular analysis of the *preg*$^+$ gene controlling the activity of the transcriptional activator NUC-1. *Genetics* **133:**193–202.

Koren, R., J. LeVitre, and K. A. Bostian. 1986. Isolation of the positive-acting regulatory gene *PHO4* from *Saccharomyces cerevisiae. Gene* **41:**271–280.

Legrain, M., M. DeWilde, and F. Hilger. 1986. Isolation, physical characterization and expression analysis of the *Saccharomyces cerevisiae* positive regulatory gene *PHO4. Nucleic Acids Res.* **14:**3059–3073.

Lemire, J. M., T. Willcocks, H. O. Halvorson, and K. A. Bostian. 1985. Regulation of repressible acid phosphatase gene transcription in *Saccharomyces cerevisiae. Mol. Cell. Biol.* **5:**2131–2141.

Leuther, K. K., and S. A. Johnston. 1992. Nondissociation of GAL4 and GAL80 in vivo after galactose induction. *Science* **256:**1333–1335.

Madden, S. L., C. L. Creasy, V. Srinivas, W. Fawcett, and L. W. Bergman. 1988. Structure and expression of the *PHO80* gene of *Saccharomyces cerevisiae. Nucleic Acids Res.* **16:**2625–2637.

Madden, S. L., D. L. Johnson, and L. W. Bergman. 1990. Molecular and expression analysis of the negative regulators involved in the transcriptional regulation of acid phosphatase production in *Saccharomyces cerevisiae. Mol. Cell. Biol.* **10:**5950–5957.

Miller, J. H. 1972. *Experiments in Molecular Genetics.* Cold Spring Harbor Laboratory, Cold Spring Harbor, N.Y.

Ogawa, N., K. Noguchi, Y. Yamashita, T. Yasuhara, N. Hayashi, K. Yoshida, and Y. Oshima. 1993. Promoter analysis of the *PHO81* gene encoding a 134 kDa protein bearing ankyrin repeats in the phosphatase regulon of *Saccharomyces cerevisiae. Mol. Gen. Genet.* **238:**444–454.

Ogawa, N., and Y. Oshima. 1990. Functional domains of a positive regulatory protein, PHO4, for transcriptional control of the phosphatase regulon in *Saccharomyces cerevisiae. Mol. Cell. Biol.* **10:**2224–2236.

Okada, H., and A. Toh-e. 1992. A novel mutation occurring in the *PHO80* gene suppresses the *PHO4*^{c} mutations of *Saccharomyces cerevisiae. Curr. Genet.* **21:**95–99.

Parent, S. A., and K. A. Bostian. Unpublished results.

Parent, S. A., A. G. Tait-Kamradt, J. LeVitre, O. Lifanova, and K. A. Bostian. 1987. Regulation of the phosphatase multigene family of *Saccharomyces cerevisiae,* p. 63–70. In A. Torriani-Gorini, F. G. Rothman, S. Silver, A. Wright, and E. Yagil (ed.), *Phosphate Metabolism and Cellular Regulation in Microorganisms.* American Society for Microbiology, Washington, D.C.

Rudolph, H., and A. Hinnen. 1987. The yeast *PHO5* promoter: phosphate control elements and sequences mediating mRNA start-site selection. *Proc. Natl. Acad. Sci. USA* **84:**1340–1344.

Sambrook, J., E. F. Fritsch, and T. Maniatis. 1989. *Molecular Cloning: A Laboratory Manual,* 2nd ed. Cold Spring Harbor Laboratory, Cold Spring Harbor, N.Y.

Schweingruber, M. E., R. Fluri, K. Maundrell, A.-M. Schweingruber, and E. Dumermuth. 1986. Identification and characterization of thiamin repressible acid phosphatase in yeast. *J. Biol. Chem.* **261:**15877–15882.

Sengstag, C., and A. Hinnen. 1987. The sequence of the *Saccharomyces cerevisiae* gene *PHO2* codes for a regulatory protein with unusual amino acid composition. *Nucleic Acids Res.* **15:**233–246.

Sengstag, C., and A. Hinnen. 1988. A 28-bp segment of the *Saccharomyces cerevisiae PHO5* upstream activator sequence confers phosphate control to the *CYC1-lacZ* gene fusion. *Gene* **67:**223–228.

Tait-Kamradt, A. G., K. J. Turner, R. A. Kramer, Q. D. Elliott, S. J. Bostian, G. P. Thill, D. T. Rogers, and K. A. Bostian. 1986. Reciprocal regulation of the tandemly duplicated *PHO5/PHO3* gene cluster within the acid phosphatase multigene family of *Saccharomyces cerevisiae. Mol. Cell. Biol.* **6:**1855–1865.

Tamai, Y., A. Toh-e, and Y. Oshima. 1985. Regulation of inorganic phosphate transport systems in *Saccharomyces cerevisiae. J. Bacteriol.* **164:**964–968.

Tice-Baldwin, K., G. R. Fink, and K. T. Arndt. 1989. BAS1 has a Myb motif and activates *HIS4* transcription only in combination with BAS2. *Science* **246:**931–935.

Toh-e, A., S. Inouye, and Y. Oshima. 1981. Structure and function of the *PHO82-pho4* locus controlling the synthesis of repressible acid phosphatase of *Saccharomyces cerevisiae. J. Bacteriol.* **145:**221–232.

Toh-e, A., and Y. Oshima. 1974. Characterization of a dominant, constitutive mutation, PHOO, for the repressible acid phosphatase synthesis in *Saccharomyces cerevisiae. J. Bacteriol.* **120:**608–617.

Toh-e, A., and T. Shimauchi. 1986. Cloning and sequencing of the *PHO80* gene and *CEN15* of *Saccharomyces cerevisiae. Yeast* **2:**129–139.

Toh-e, A., K. Tanaka, Y. Uesono, and R. B. Wickner. 1988. *PHO85*, a negative regulator of the PHO system, is a homolog of the protein kinase gene, *CDC28*, of *Saccharomyces cerevisiae. Mol. Gen. Genet.* **214:**162–164.

Toh-e, A., Y. Ueda, S. I. Kakimoto, and Y. Oshima. 1973. Isolation and characterization of acid phosphatase mutants in *Saccharomyces cerevisiae. J. Bacteriol.* **113:**727–738.

Ueda, Y., A. Toh-e, and Y. Oshima. 1975. Isolation and characterization of recessive, constitutive mutations for repressible acid phosphatase synthesis in *Saccharomyces cerevisiae. J. Bacteriol.* **122:**911–922.

Uesono, Y., K. Tanaka, and A. Toh-e. 1987. Negative regulators of the PHO system in *Saccharomyces cerevisiae:* isolation and structural characterization of *PHO85. Nucleic Acids Res.* **15:**10299–10309.

Uesono, Y., M. Tokai, K. Tanaka, and A. Toh-e. 1992. Negative regulators of the PHO system of *Saccharomyces cerevisiae:* characterization of *PHO80* and *PHO85. Mol. Gen. Genet.* **231:**426–432.

Vogel, K., W. Hörz, and A. Hinnen. 1989. The two positively acting regulatory proteins PHO2 and PHO4 physically interact with *PHO5* upstream activation regions. *Mol. Cell. Biol.* **9:**2050–2057.

Yoshida, K., Z. Kuromitsu, N. Ogawa, and Y. Oshima. 1989a. Mode of expression of the positive regulatory genes *PHO2* and *PHO4* of the phosphatase regulon of *Saccharomyces cerevisiae. Mol. Gen. Genet.* **217:**31–39.

Yoshida, K., N. Ogawa, and Y. Oshima. 1989b. Function of the *PHO* regulatory genes for repressible acid phosphatase synthesis in *Saccharomyces cerevisiae. Mol. Gen. Genet.* **217:**40–46.

Chapter 12

Characterization of Pho85 Kinase of *Saccharomyces cerevisiae*

MARIE FUJINO,[1] MASAFUMI NISHIZAWA,[2] S.-J. YOON,[1] TOMOKO OGUCHI,[1] AND AKIO TOH-E[1]

Department of Biology, Faculty of Science, University of Tokyo, 7-3-1 Hongo, Bunkyo-ku, Tokyo 113,[1] and Department of Microbiology, School of Medicine, Keio University, 35 Shinanomachi, Shinjuku-ku, Tokyo 160,[2] Japan

Responding to the change of concentration of P_i in the medium, *Saccharomyces cerevisiae* cells regulate expression of a group of genes; acid and alkaline phosphatases, for example, are derepressed (or induced) when cells are growing under phosphate-limiting conditions (Schurr and Yagil, 1971; Toh-e et al., 1973). This regulatory system has been designated the PHO system. A marker enzyme of this system is repressible acid phosphatase (rAPase),which is encoded by the *PHO5* gene. Because rAPase localizes in the periplasmic space, its activity is easily detected by an in situ staining method (Toh-e and Oshima, 1974). By monitoring the rAPase activity, many mutants involved in rAPase production were isolated and characterized. So far five major regulatory genes have been characterized: two positive (*PHO2* and *PHO4*) and two negative (*PHO80* and *PHO85*) ones, and *PHO81*, which is apparently a positive factor but does not function as a transcription factor like *PHO4* (see below) (Oshima, 1982). A defect in *PHO2*, *PHO4*, or *PHO81* results in a noninducible phenotype, whereas a defect either in *PHO80* or in *PHO85* results in a constitutive phenotype for rAPase production. The genes whose expression is under the control of the PHO system are *PHO5*, *PHO8* (repressible alkaline phosphatase), *PHO84* (phosphate transporter), and *PHO81*. Expression of all these genes is controlled at the transcriptional level (Lemire et al., 1985; Kaneko et al., 1987; Bun-ya et al., 1991; Yoshida et al., 1987).

The genetic model (Oshima, 1982) predicts the following scenario of PHO regulation. The *PHO4* gene product stimulates transcription of *PHO5* and other genes belonging to the PHO system by binding to the Pho4 binding site, which is located in the upstream region of each gene when phosphate is limiting or either of the negative regulators is inactivated. Pho2 is also a DNA-binding protein and is needed for activation of transcription of some genes belonging to the PHO system, such as *PHO5*. Recently, Schmid et al. (1992) discovered changes in chromatin structure of the *PHO5* locus responding to derepression conditions. When a sufficient amount of P_i is available, Pho4 protein is inactivated most probably by making a complex with the negative factor(s) (Okada and Toh-e, 1992), thereby shutting off the transcription of the *PHO5* gene.

Our main interest is to understand how yeast cells transmit information about phosphate availability to the transcription machinery. Since *PHO85* encodes a protein kinase (Uesono et al., 1987; Toh-e et al., 1988), we believe that protein phosphorylation and dephosphorylation exert an important function in signal transduction in the PHO system. Besides, Pho85 kinase is a member of the Cdc28/Cdc2 kinase family; the ATP-binding and PSTAIR domains are highly similar among these kinases. Therefore, Pho85 kinase may affect some other cellular function besides transcriptional regulation of the PHO system. In this communication, we describe experiments to characterize Pho85 kinase activity and its regulation. We also describe the results of in vitro mutagenesis of the specific domains of Pho85 kinase and analyses of these mutants.

Pleiotropic Effects of the *pho85* Deletion

Strains containing a null allele of *pho85* display several apparently unrelated phenotypes as well as constitutive rAPase production (data not shown); they are respiration deficient (Pet^-), Gal^-, and Suc^- and grow slowly. All these phenotypes were suppressed by a *pho4* null mutation. However, since inducibility of *GAL* genes (*GAL10*, *GAL7*, and *GAL2*) was not drastically different in the wild type and a null *pho85* strain (data not shown), the mechanism underlying the Gal^- phenotype encoded by the null *pho85* mutation remains to be solved.

Construction of PHO85 Derivatives

Construction of intronless *PHO85*. The *PHO85* gene contains an intron at codon 6 of the Pho85 protein. To avoid possible trouble in overproducing Pho85 protein in *S. cerevisiae* and *Escherichia coli*, we eliminated the intron by replacing the sequence encompassing the translation-initiating codon ATG to the *Hin*dIII site covering codons 11 and 12 with a synthetic oligonucleotide corresponding to the same sequence without the intron (pSY856). This derivative of *PHO85* was found to be functional, as judged by its ability to complement a *pho85* mutation.

Construction of *GAP-GST-PHO85*. pSY856 DNA was digested with *Xho*I, filled in with the Klenow fragment of *E. coli* DNA polymerase I and four deoxynucleoside triphosphates, and then further digested with *Sal*I. The *PHO85* gene without the intron was recovered as a DNA fragment with one end blunt and *Sal*I at the other end. pKOMI contains the promoter of the *GAP* gene (encoding yeast glyceraldehyde-3-phosphate dehydrogenase [GAP]) and the *GST* gene (encoding glutathione-*S*-transferase [GST]). The DNA fragment containing *PHO85* prepared as described above was ligated with pKOMI digested with *Pvu*II and *Sal*I. That the *GAP-GST-PHO85* gene was properly constructed was confirmed by examining the production of the expected GST-Pho85 fusion protein in *S. cerevisiae* by Western immunoblotting with anti-Pho85 antibody and anti-GST antibody.

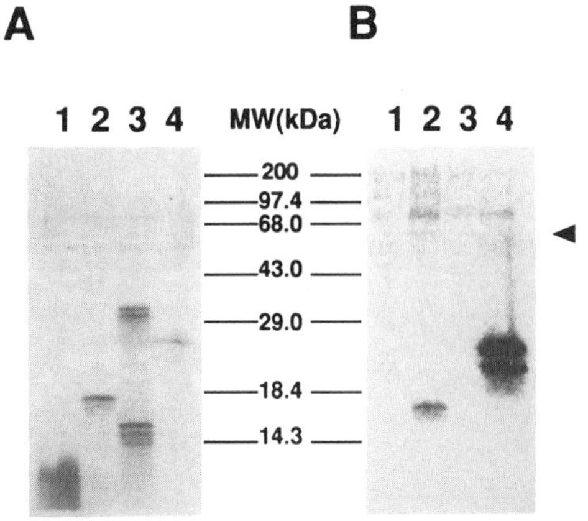

Assay and Substrate Specificity of Pho85 Kinase

Previously we detected Pho85 kinase activity by using an immunoprecipitation assay with anti-Pho85 protein antibody as an enzyme source and casein as substrate (Uesono et al., 1992). In this work we constructed a GST-Pho85 fusion protein and confirmed that the fusion was active by genetic complementation. However, after extensive trials, we found that this protein produced in yeast cells could not be recovered from glutathione beads for unknown reasons. Therefore, we obtained the protein by immunoprecipitation with anti-GST antibody. Affinity-purified anti-GST antibody was found to precipitate the fusion protein from extract satisfactorily, as we expected.

We tested the ability of the GST-Pho85 fusion protein to use protamine, histone H1, κ-casein, and myelin basic protein (MBP) as substrates. As seen in Fig. 1, κ-casein was the best substrate among the proteins tested, MBP was a poor substrate, and protamine and histone H1 were not used as substrates. Furthermore, there is no significant phosphorylation signal at the position of the GST-Pho85 fusion protein (60 kDa) in Fig. 1, indicating that Pho85 kinase is not autophosphorylated.

FIG. 1. Substrate specificity of Pho85 kinase. The *GAP-GST-PHO85* fusion gene was introduced into *pho85* host YAT1534 (*MAT***a** *pho3 pho85 ura3 trp1 his3*), and extract was prepared from the transformant cells grown in high-P_i medium. GST-Pho85 fusion protein was precipitated by adding 1 μg of the anti-GST antibody with 200 μl of extract. The protein was washed three times with RIPA buffer followed by three times with reaction buffer and divided into four equal portions, each of which was mixed with 30 μl of reaction buffer containing 20 mM Tris-HCl (pH 7.5), 10 mM $MgCl_2$, 5 μCi of [γ-^{32}P]ATP, and 5 μg of substrate except protamine (10 μg). After incubation at 30°C for 30 min, the reaction was stopped by adding 15 μl of 5× sodium dodecyl sulfate-polyacrylamide gel electrophoresis (SDS-PAGE) sample buffer; then the protein was separated by SDS-PAGE. An arrowhead indicates the position of GST-Pho85 fusion protein. (A) Coomassie brilliant blue staining. (B) Autoradiography. Lanes: 1, protamine; 2, MBP; 3, histone H1; 4, κ-casein.

Does P_i Regulate Pho85 Kinase Activity?

To investigate whether P_i regulates Pho85 kinase activity, cells containing *GST-PHO85* were grown in high-P_i and low-P_i medium and a portion of cells was withdrawn at intervals and assayed for acid phosphatase and GST-Pho85 kinase activity (Fig. 2). Derepression of rAPase began after 60 min of incubation in low-P_i medium. In contrast, cells grown in high-P_i medium remained phosphatase negative throughout the experiment. A constant amount of immunoprecipitate of GST-Pho85 fusion protein prepared at each time point was subjected to kinase assay. Protein kinase activity remained constant irrespective of the concentration of P_i in the medium. This result indicates that Pho85 kinase activity is not regulated by P_i in the medium.

Mutations in the PSTAIR Sequence Introduced by In Vitro Mutagenesis

Pho85 kinase contains the complete PSTAIR sequence, a hallmark of Cdc28/Cdc2 family kinases. To dissect the function of this region of

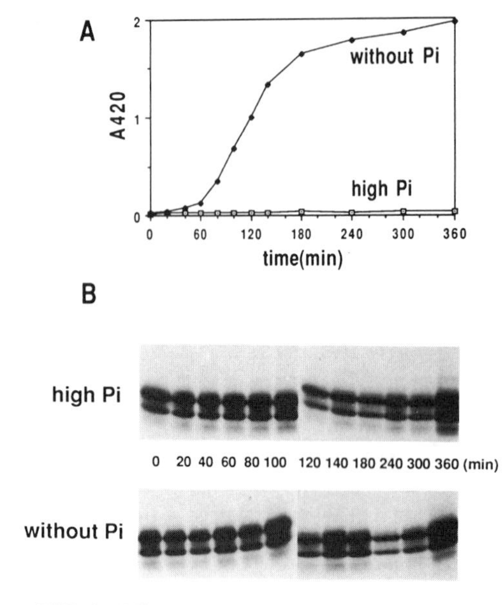

FIG. 2. Effect of P_i concentration in the medium on Pho85 kinase activity. The *pho85* strain YAT1534 (*MAT***a** *pho3 pho85 ura3 trp1 his3*) containing the *GAP-GST-PHO85* fusion gene was grown either in high-P_i or in low-P_i medium. At the indicated time point, cells were harvested, extract was prepared, and GST-Pho85 fusion protein was precipitated with the anti-GST antibody as described in the legend to Fig. 1. Pho85 kinase activity was assayed with κ-casein as the substrate. (A) Time course of acid phosphatase formation. (B) Pho85 kinase assay.

the Pho85 protein, a charged amino acid and some other amino acid in this region were substituted by alanine. Each mutant *pho85* gene thus constructed was introduced into *pho85* or wild-type strains to determine whether a functional change had been introduced by the substitution. Among the *pho85* alleles thus constructed, the *PHO85*E53A and *PHO85*K58A mutant genes showed a clear $Pho85^-$ phenotype (Fig. 3A). Interestingly, the former mutation showed a dominant negative phenotype. This mutant Pho85E53A protein may outcompete the wild-type Pho85 protein for some unknown factor(s). These results indicate that the PSTAIR sequence also plays an important function(s) in Pho85 kinase.

Role of Tyrosine Residue in the ATP-Binding Domain

Figure 3B highlights the amino acid sequence of the ATP-binding domain of Cdc28/Cdc2 kinases. Two tyrosine residues (15Y and 19Y in Cdc2 kinase) are conserved in these three kinases. The role of 15Y in Cdc2 kinase is well documented (Gould et al., 1989, 1990). This Y residue was shown to play a pivotal role in the regulation of Cdc2 kinase activity; the form of Cdc2 kinase phosphorylated at 15Y is inactive, and dephosphorylation of 15Y-P activates Cdc2 kinase. Therefore, phenylalanine-substituted Cdc2 protein, Cdc2Y15F, is constitutively active, deregulating progression of the cell cycle of *Schizosac-*

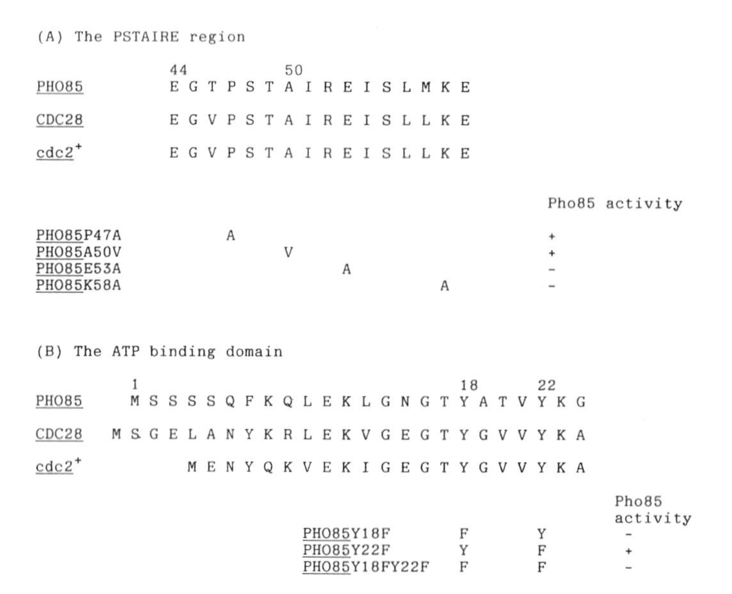

FIG. 3. Site-directed mutagenesis of the PSTAIR region (A) and the ATP-binding region (B). Pho85 activity was assayed by complementation of a *pho85* mutation. Numbering of amino acid is that of Pho85 protein.

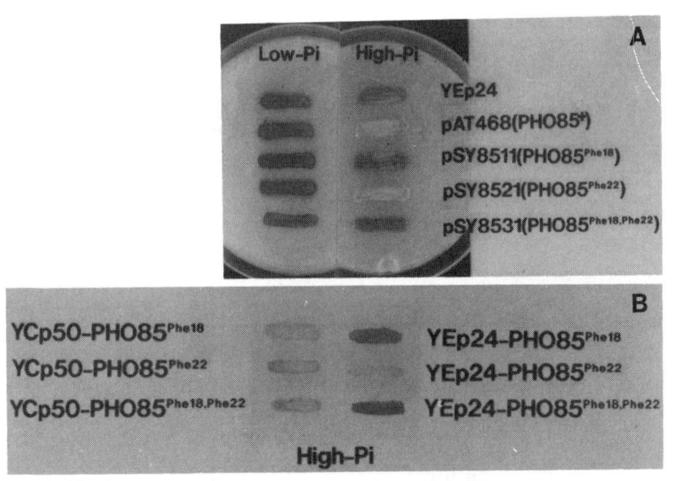

FIG. 4. Phenotype shown by the phenylalanine-substituted *PHO85* genes. (A) YAT1534 (*MAT***a**, *pho3 pho85 ura3 trp1 his3*) containing one of the *PHO85* mutant gene on a low-copy-number vector was brushed onto high-P_i or low-P_i medium. (B) Wild-type strain YAT1651 (*MAT*α *pho3 leu2 his3 trp1 ura3*) containing one of the *PHO85* mutant genes either on a multicopy vector or a single-copy vector was brushed on high-P_i medium. After incubation for 3 days at 30°C, acid phosphatase formation was diagnosed by the staining method.

charomyces pombe. In contrast to the role of 15Y of Cdc2 kinase, 19Y of Cdc28 kinase, which corresponds to 15Y of Cdc2 kinase, does not seem to play any role in Cdc28 kinase, although the phosphorylation-dephosphorylation cycle at this Y residue is confirmed (Sorger and Murray, 1992; Amon et al., 1992; Atherton-Fessler et al., 1993). Since these kinases can replace each other, it is of interest to know what makes this difference. Pho85 kinase also contains the conserved Y residue (18Y). We are interested in examining whether 18Y of Pho85 kinase has any effect on Pho85 kinase activity or its regulation. To address this question, we constructed three mutants, containing *PHO85*Y18F, *PHO85*Y22F, or *PHO85*Y18FY22F. Each of the mutant *PHO85* genes on a single-copy YCp vector was introduced into a *pho85* mutant to test whether it has *PHO85* activity. Each transformant was grown on a plate containing either high-P_i or low-P_i medium and then assayed for acid phosphatase activity (Fig. 4). All transformants were acid phosphatase positive on the low-P_i plate. On high-P_i medium, the *PHO85*Y22F gene complemented the *pho85* defect but neither *PHO85*Y18F nor *PHO85*Y18FY22F gene did so, indicating that the last two mutant *PHO85* genes resulted in a loss of function of Pho85 kinase. Each of the mutant *PHO85* genes was introduced into the wild-type strain, and acid phosphatase production by each of the transformants was assayed by the staining method. When *PHO85* derivative gene was on a low-copy-number plasmid, all the transformants showed the wild-type phenotype, whereas when they were on a multicopy plasmid, *PHO85*Y18F and *PHO85*Y18FY22F, but not *PHO85*Y22F, gave rise to a constitutive phenotype. Overexpression of *PHO85*Y18F or *PHO85*Y18FY22F resulted in a dominant negative phenotype.

To explore the in vivo function of *PHO85*Y18F further, we compared the consequence of a *pho85* or *PHO85*Y18F mutation by testing a double mutant containing either of these mutations and a *pho81* mutation. A *pho81* mutation alone resulted in a negative phenotype. The phenotype of the *pho81* strain is explained as follows. Pho81 protein activates expression of *PHO5* by inactivating a negative factor(s) such as Pho80 or Pho85 or both. In the *pho81* mutant strain the negative factors are not inactivated and the repression conditions persist. A *pho85 pho81* double mutant showed a constitutive phenotype (Table 1). This is because, in this strain, the negative factor to be inactivated by the action of *PHO81* has already been lost. A similar experiment was conducted with a *PHO85*Y18F strain. Unexpectedly, a *PHO85*Y18F *pho81* double mutant showed a negative phenotype similar to that of the *pho81* strain (Table 1). This result indicates that the mechanism of the constitutive phenotype encoded by *PHO85*Y18F is different from that exhibited by *pho85*. The constitutive *pho84* mutation shows a similar genetic behavior to that of *PHO85*Y18F. In *pho84* mutants, phosphate transport is impaired. The phenotype of the *PHO85*Y18F strain can be explained by assuming that Pho85Y18F protein somehow blocks the function of Pho84 protein.

TABLE 1. Genetic characteristics of the *PHO85*Y18F mutation

Strain	Acid phosphatase activity in:	
	High-P_i medium	Low P_i medium
pho85	+	+
pho81	−	−
pho85 pho81	+	+
*PHO85*Y18F	+	+
*PHO85*Y18F *pho81*	−	−

Since *PHO85*Y18F and *PHO85*Y18FY22F mutants displayed the same genetic behavior, we conducted a biochemical study with a strain containing the latter mutant gene, as described below. GST-Pho85Y18FY22F and GST-Pho85 fusion proteins were immunoprecipitated, and a similar amount of immunoprecipitate was used for assaying protein kinase activity with κ-casein as substrate. During incubation of the reaction mixture, a portion of it was withdrawn at the indicated time point and the amount of ^{32}P incorporated into κ-casein was quantified by using an imaging analyzer (Fuji Film Co.). The results (Fig. 5) clearly show that the GST-Pho85Y18FY22F fusion protein was defective in kinase activity, even though the Y15F substitution in Cdc2 kinase resulted in a constitutively high kinase activity and the Y19F substitution in Cdc28 kinase neither affected enzyme activity nor induced phenotypic changes. Therefore, this is the third case of the effect of substitution of Y to F in the ATP-binding domain of Cdc28/Cdc2 family kinases.

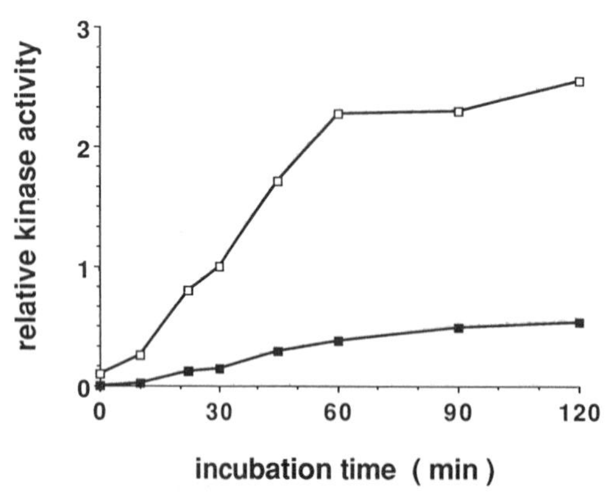

FIG. 5. Pho85Y18FY22F protein is defective in Pho85 kinase activity. The *GAP-GST-PHO85* gene or *GAP-GST-PHO85F*Y18FY22F gene was introduced into YAT1534 (*MATa pho3 pho85 ura3 trp1 his3*). GST-Pho85 and GST-Pho85Y18FY22F fusion proteins were prepared as described in the legend to Fig. 1. A comparable amount of each immunoprecipitate was subjected to Pho85 kinase assay with κ-casein as the substrate. Symbols: □, GST-Pho85; ■, GST-Pho85Y18FY22F.

Conclusions

(i) Among amino acid substitutions in the PSTAIR region, *PHO85*E53A resulted in a clear dominant negative Pho85$^-$ phenotype, probably by outcompeting some factor(s) from wild-type Pho85 kinase.

(ii) When the Y residues in the conserved ATP-binding domain of Pho85 protein were substituted by F, the *PHO85*Y22F mutant behaved just like a wild-type strain, whereas the *PHO85*Y18F and *PHO85*Y18FY22F mutants showed a loss of function, indicating that the Y-to-F substitution in Pho85, similar to corresponding substitutions in Cdc2 and Cdc28, does not affect enzyme activity. Pho85Y18FY22F protein showed a reduced activity of Pho85 kinase. When either the *PHO85*Y18F or *PHO85*Y18FY22F gene was expressed on a multicopy plasmid, wild-type cells containing this gene showed a constitutive PHO phenotype. In spite of the resemblance of the phenotype between *pho85* and *PHO85*Y18F strains, the consequence of each mutation is different; the genetic behavior of the *PHO85*Y18F mutation is similar to that of a *pho84* mutation. We propose that Pho85Y18F protein somehow interferes with Pho84 function, implying a direct interaction between Pho84 protein and Pho85 protein.

(iii) Pho85 kinase activity is not regulated by the concentration of P_i in the medium when kinase activity is assayed with κ-casein as substrate. This may be due to the use of an artificial substrate in the assay, or P_i may exert its function by regulating the accessibility of Pho85 kinase to its substrate in vivo.

REFERENCES

Amon, A., U. Surana, I. Muroff, and K. Nasmyth. 1992. Regulation of p34^{CDC28} tyrosine phosphorylation is not required for entry into mitosis in *Saccharomyces cerevisiae. Nature* (London) **355:**368–371.

Atherton-Fessler, S., L. L. Parker, R. L. Geahlen, and H. Piwnica-Worms. 1993. Mechanisms of p34^{cdc2} regulation. *Mol. Cell. Biol.* **13:**1675–1685.

Bun-ya, M., M. Nishimura, S. Harashima, and Y. Oshima. 1991. The *PHO84* gene of *Saccharomyces cerevisiae* encodes an inorganic phosphate transporter. *Mol. Cell. Biol.* **11:**3229–3238.

Gould, K., and P. Nurse. 1989. Tyrosine phosphorylation of the fission yeast *cdc2*$^+$ protein kinase regulates entry into mitosis. *Nature* (London) **342:**39–44.

Gould, K. L., S. Moreno, N. Tonks, and P. Nurse. 1990. Complementation of the mitotic activator, p80^{cdc25}, by a human protein tyrosine phosphatase. *Science* **250:**1573–1576.

Kaneko, Y., N. Hayashi, A. Toh-e, and Y. Oshima. 1987. Structural characteristics of the *PHO8* gene encoding repressible alkaline phosphatase in *Saccharomyces cerevisiae. Gene* **58:**137–148.

Lemire, J. M., T. Willcocks, H. O. Halvorson, and K. A. Bostian. 1985. Regulation of repressible acid phosphatase gene transcription in *Saccharomyces cerevisiae. Mol. Cell. Biol.* **5:**2131–2141.

Okada, H., and A. Toh-e. 1992. A novel mutation occurring in the *PHO80* gene suppresses the *PHO4*^{c} mutations of *Saccharomyces cerevisiae. Curr. Genet.* **21:**95–99.

Oshima, Y. 1982. Regulatory circuits for gene expression: the metabolism of galactose and of phosphate, p. 159–180. *In*

J. N. Srathern, E. W. Jones, and J. R. Broach (ed.), *The Molecular Biology of the Yeast Saccharomyces: Metabolism and Gene Expression.* Cold Spring Harbor Laboratory, Cold Spring Harbor, N.Y.

Schmid, A., K.-D. Fascher, and W. Horz. 1992. Nucleosome disruption at the yeast *PHO5* promoter upon *PHO5* induction occurs in the absence of DNA replication. *Cell* **71:**853–864.

Schurr, A., and E. Yagil. 1971. Regulation and characterization of acid and alkaline phosphatase in yeast. *J. Gen. Microbiol.* **65:**291–303.

Sorger, P. K., and A. W. Murray. 1992. S-phase feedback control in budding yeast independent of tyrosine phosphorylation of p34^{CDC28}. *Nature* (London) **355:**365–368.

Toh-e, A., and Y. Oshima. 1974. Characterization of a dominant, constitutive mutation, *PHO0*, for the repressible acid phosphatase synthesis in *Saccharomyces cerevisiae. J. Bacteriol.* **120:**608–617.

Toh-e, A., K. Tanaka, Y. Uesono, and R. B. Wickner. 1988. *PHO85*, a negative regulator of the PHO system, is a homolog of the protein kinase gene, *CDC28* of *Saccharomyces cerevisiae. Mol. Gen. Genet.* **214:**162–164.

Toh-e, A., Y. Ueda, S.-I. Kakimoto, and Y. Oshima. 1973. Isolation and characterization of acid phosphatase mutants in *Saccharomyces cerevisiae. J. Bacteriol.* **113:**727–738.

Uesono, Y. K. Tanaka, and A. Toh-e. 1987. Negative regulators of the PHO system in *Saccharomyces cerevisiae:* isolation and structural characterization of *PHO85. Nucleic Acids Res.* **15:**10299–10309.

Uesono, Y., M. Tokai, K. Tanaka, and A. Toh-e. 1992. Negative regulators of the PHO system of *Saccharomyces cerevisiae:* characterization of *PHO80* and *PHO85. Mol. Gen. Genet.* **231:**426–432.

Yoshida, K., Z. Kuromitsu, N. Ogawa, K. Ogawa, and Y. Oshima. 1987. Regulatory circuit for phosphatase synthesis in *Saccharomyces cerevisiae*, p. 49–55. *In* A. Torriani-Gorini, F. G. Rothman, S. Silver, A. Wright, and E. Yagil (ed.), *Phosphate Metabolism and Cellular Regulation in Microorganisms.* American Society for Microbiology, Washington, D.C.

III. TRANSPORT OF PHOSPHORYLATED COMPOUNDS AND OTHER OXYANIONS

Chapter 13

Introduction: Honorary P_i

SIMON SILVER[1] AND PETER MALONEY[2]

Department of Microbiology and Immunology, University of Illinois at Chicago, 835 South Wolcott Avenue, Chicago, Illinois 60612-7344,[1] and Department of Physiology, Johns Hopkins University Medical School, Baltimore, Maryland 21205[2]

In any symposium on phosphate, a line must be arbitrarily drawn. Every cellular function involving ATP or other high-energy phosphate bonds can be considered "fair game." For this section, we were allowed (by space limits) to cover "other" than P_i, hence the euphemism "Honorary P_i." This section contains three papers on transport of organic compounds and three on the metabolism of oxyanions other than phosphate. Yan and Maloney (chapter 15) and Kadner et al. (chapter 14) detail the thorough understanding of the transport of hexose phosphates and regulation of the genes involved. Nikaido et al. (chapter 16) summarize the recent remarkable in vitro reconstitution of the maltose transport system. The maltose transporter qualifies doubly: it is the best understood of a large family of transport ATPases, and the Pst-specific P_i system is a member of this family. Stadtman et al. (chapter 18) discuss the enzymology of the reduction and incorporation of selenium into selenocysteine and tRNA. Rosen et al. (chapter 17) describe the recent surprises with arsenate (a phosphate analog) reduction and an ATPase that effluxes arsenite—a system that is not homologous to other transport ATPases. Finally, Grossman et al. (chapter 19) survey the inorganic nutritional needs (phosphate and sulfate) and results of starvation on cyanobacteria, prokaryotes that synthesize all organic compounds for themselves. We expect that by the next phosphate symposium, the Pho regulon paradigm of *Escherichia coli* will have been extended still further to other compounds and organisms.

Chapter 14

Transmembrane Control of the Uhp Sugar-Phosphate Transport System: the Sensation of Glu6P

ROBERT J. KADNER, MICHAEL D. ISLAND, TOD J. MERKEL, AND CAROL A. WEBBER

Department of Microbiology, University of Virginia School of Medicine, Charlottesville, Virginia 22908

Our genetic studies of the Uhp sugar-phosphate transport system began almost 25 years ago in collaboration with Herb Winkler. For many years, study of the genetic and regulatory properties of this system proceeded slowly, spurred by the novelty of this system as one of the few bacterial processes besides chemotaxis that is regulated by an external signal. This exogenous induction was different from the control of all the other regulatory systems popular for study. During the past 10 years, however, there has been an explosion in the number of systems that respond to external signals, and bacteria are clearly just as likely as their eukaryotic relatives to make use of transmembrane signaling processes regulated by protein phosphorylations. In this paper we review our current understanding of the mechanism of regulation of the transport systems for organophosphate compounds; we find that there is a considerable diversity of mechanisms for regulation of a homologous series of transport proteins.

Transport Systems for Organophosphates

Organophosphate compounds must provide bacteria with an available and nutritious source of carbon, phosphorus, or both, judging from the widespread distribution of a number of different biochemical mechanisms for their transport and utilization. One expects that organophosphates might be available to enteric bacteria, such as *Escherichia coli,* from the release of phosphorylated metabolic intermediates from lysed cells, which the almost daily turnover of the intestinal epithelial cells would shower upon the intestinal bacterial flora. Mono- and diesters of glycerol phosphate are common turnover products of bacterial and host cell phospholipids. Ribose phosphates are released during nucleic acid turnover, and hexose and pentose phosphates can be generated by the action of phosphorylases on plant polysaccharides or glycogen. Organophosphates can be taken up by bacteria following their dephosphorylation by extracellular phosphatases, as is the case for the acquisition of glucose-1-phosphate by *E. coli* (Pradel and Boquet, 1989). However, many bacteria also possess the ability to transport organophosphates in unaltered form. The Uhp sugar-phosphate transport system of *E. coli* was first described by Pogell et al. (1966), who recognized its regulation by external inducer. The inability of glucose-6-phosphate (Glu6P) generated intracellularly to induce synthesis of the transport system was convincingly demonstrated by Winkler (1970). Winkler (1973) demonstrated that inducible transport systems for Glu6P were broadly but erratically scattered among many bacterial types, being present in many enteric bacteria, including *E. coli* and some *Shigella, Salmonella,* and *Enterobacter* species, as well as in *Serratia* species and *Staphylococcus aureus.* This activity was not detected in *Proteus mirabilis, Pseudomonas aeruginosa, Corynebacterium diphtheriae, Bacillus subtilis, Gaffkya tetragena,* or several streptococci.

Five P_i-linked sugar-phosphate transport/exchange systems from four organisms have been extensively characterized by Maloney et al. (1990), namely, hexose phosphate-transporting UhpT and glycerol phosphate-transporting GlpT from *E. coli,* phosphoglycerate-transporting PgtP from *Salmonella typhimurium,* the transporter for both Glu6P and glycerol-3-phosphate (Gly3P) in *Staphylococcus aureus,* and the hexose phosphate transporter from *Steptococcus lactis.* The transport mechanisms for these five systems appear to be very similar. We have a better view of the role in cellular physiology and the regulation of these systems in the enteric bacteria, owing to their more accessible genetic manipulability. In addition to these anion exchange systems, *E. coli* possesses the Ugp system, a periplasmic-binding protein-dependent, ATP-driven transport system capable of accumulating mono- and diesters of glycerol phosphate, such as glyceryl-phosphorylethanolamine (Brzoska et al., 1987). The *ugp* genes are part of the *pho* regulon and are expressed under conditions of phosphate limitation. Ugp allows use of its organophosphate substrates as phosphate sources but not as carbon sources. In contrast, GlpT, UhpT, and PgtP allow use of organophosphates as carbon sources. Although they are capable of providing sources of phosphate, this does not appear to be their major role in the cell, owing to their obligatory exchange activity. A primary function in carbon acquisition is consistent with their regulation by the carbon supply through catabolite repression mediated by the cyclic AMP (cAMP)-dependent transcription activator protein, catabolite activator protein (CAP), and their apparent lack of regulation by the phosphorus supply mediated by the Pho regulon.

UhpT Transport Process

The UhpT transport system has several interesting features. Its mechanism of energy coupling for active transport, which it shares with the related transporters GlpT and PgtP, differs dramatically from that of the familiar ion symporters, such as the proton-coupled lactose permease LacY and the proton- or sodium-coupled melibiose permease MelB, which couple substrate accumulation to electrogenic downhill movement of charge or of an ion concentration gradient (Larson, 1987; Sonna et al., 1988). As elucidated by Maloney's group, the members of the chauvinistically named UhpT family of transporters mediate the electroneutral obligatory exchange of organic phosphates or P_i (reviewed by Maloney et al. [1990]). Amino acid sequence alignments revealed that the UhpT family members, namely, UhpT, GlpT (Eiglmeier et al., 1987), and PgtP (Goldrick et al., 1988), show about 35% identity to each other but are quite distinct from the symporters, although several common sequence motifs can be suggested (Marger and Saier, 1993). Analyses of fusions of UhpT (Lloyd and Kadner, 1990; Island et al., 1992) and GlpT (Gött and Boos, 1988) to the topological reporter PhoA or LacZ suggested that both proteins contain 12 transmembrane segments with both termini in the cytoplasm, which is identical to the topology proposed for LacY, MelB, and many other transport proteins. The process of obligatory exchange performed by the UhpT family does not require the operation of a substantially different transport mechanism from the symporters. Most carriers are thought to mediate the alternating exposure of their substrate-binding site(s) between the two sides of their membrane. If the change in access of the substrate-binding site can occur only when that site is loaded, exchange occurs. It remains to be seen whether symporters and antiporters require substantially different transport mechanisms or are minor variations differing only in rate constants for steps in the transport cycle. The operation of two different transport processes by proteins that are similar in transmembrane topology provides an exciting opportunity to address the relationship of structure to transport function.

Obviously, the uptake of phosphorylated compounds is metabolically preferable to the uptake of a free sugar because it avoids the need to phosphorylate the sugar prior to its metabolism. However, a transport system for phosphorylated compounds could allow wasteful loss of important metabolic intermediates unless it acts in a unidirectional manner, as is the case for the periplasmic permeases. The obligatory exchange mechanism of the UhpT family of transporters has the physiological advantage of reducing the potential for loss of catabolic intermediates by efflux, since release must be coupled to the uptake of another compound, and the substrate affinities favor uptake and retention of the organophosphates, relative to P_i. The effectiveness of this type of transport system is further increased if the gene regulatory system is controlled in such a way that the protein is produced only when its substrates are in the growth medium.

Regulation of Uhp Expression

The organophosphate transport systems are regulated in interesting and diverse manners. The Ugp system is part of the *pho* regulon and is regulated by the *phoB-phoR* two-component regulatory systems in response to the flux of P_i entry through the Pst transport system, as described elsewhere in this volume (see chapters 2, 3, 4, 6, and 48). The GlpT transporter is induced by intracellular Gly3P and is controlled by the GlpR repressor protein, along with the other genes of the *glp* regulon (Schweizer et al., 1985). In contrast, both UhpT and PgtP are regulated in response to inducers in the external medium, Glu6P in the case of UhpT. The rationale for this regulation by an external effector is obvious since *E. coli* normally maintains a high internal pool of Glu6P (in the millimolar range) and induction by internal Glu6P would result in constitutive expression of the transporter, possibly leading to loss of internal metabolites. Our current view of the mechanism of transmembrane control of the UhpT system is presented here and diagrammed in Fig. 1.

All of the genes that are specifically involved in the expression and regulation of the UhpT transporter are located in the *uhp* locus, at min 82.8 on the *E. coli* genetic map (kb3868 to 3873 on EcoMap6). The nucleotide sequences of the *uhp* loci in *E. coli* and *Salmonella typhimurium* have been determined (Friedrich and Kadner, 1987; Island et al., 1992). There are four *uhp* genes (*uhpABCT*) which are very closely packed; e.g., the termination codon of *uhpA* overlaps the initiation codon of *uhpB*, and the termination codon of *uhpC* is part of the *uhpT* promoter. Insertion or in-frame deletion mutations in any one of the *uhp*

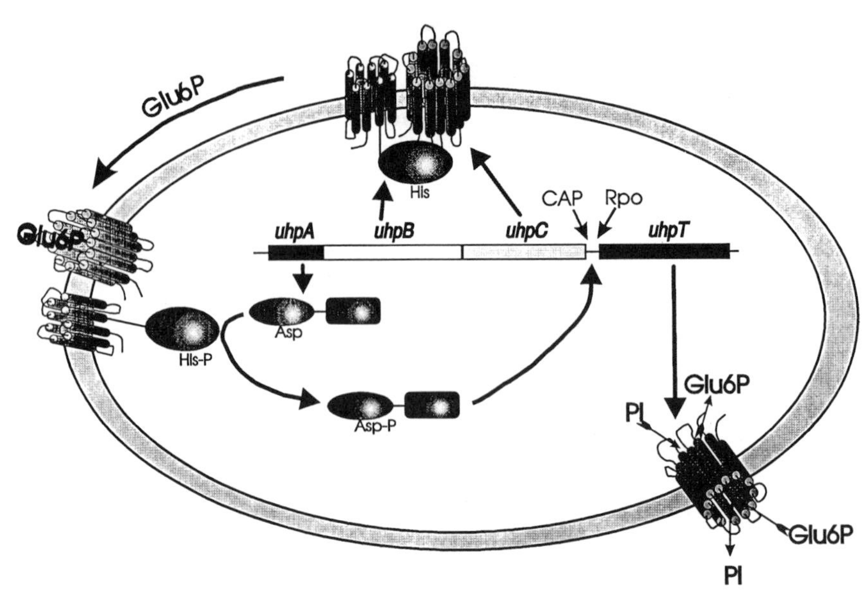

FIG. 1. Representation of the Uhp regulatory system. In this model are indicated the predicted transmembrane topologies of the three membrane-bound Uhp proteins. Binding to UhpC in a complex with UhpB results in activation of the autokinase activity proposed for UhpB, perhaps by relief of an inhibitory interaction with cytoplasmic loops of UhpB and UhpC. Phosphorylation of UhpA by UhpB allows its activation of transcription at the *uhpT* promoter.

genes eliminates UhpT activity. The product of the fourth gene, *uhpT*, was identified as the transporter because its expression from the heterologous *lac* promoter, in a strain deleted for the entire *uhp* region, resulted in production of Uhp transport activity. The *uhpABC* genes regulate the transcription of *uhpT*. Their possible roles in controlling Uhp expression are suggested from sequence comparisons.

Membrane-Associated Regulatory Components

The UhpC protein is similar in length, amino acid sequence, and transmembrane topology to the UhpT family of transport proteins (Island et al., 1992; Kadner et al., 1993). UhpC plays a regulatory role and is not a subunit of the transport system, since Δ*uhpC* strains give rise to Uhp$^+$ variants that express the normal transport system but in a constitutive manner. The sequence similarity of UhpC to the transporters (30 to 35% identity) extends over the full length of these proteins and does not suggest the existence of a distinct signaling module of substantial size. It seems likely that UhpC acts in regulation by providing a binding site for Glu6P in the periplasm and signaling its presence to other components of the regulatory system, perhaps through protein conformation changes.

The UhpA and UhpB proteins are members of the two-component regulatory systems that are being increasingly found to regulate specific bacterial gene expression in response to numerous environmental signals, as discussed elsewhere in this book (chapters 48 to 52) (see other reviews by Parkinson and Kofoid [1992] and Parkinson [1993]). In the bipartite UhpB protein, the cytoplasmic C-terminal half (residues 274 to 500) contains the conserved motifs of the protein kinase module, characteristic of the sensor kinase component. The very nonpolar N-terminal half (residues 1 to 273) is unrelated to other regulatory proteins and is predicted to be extensively embedded in the membrane with eight transmembrane segments. Most membrane-associated sensor kinase proteins have only two transmembrane segments and form a substantial periplasmic domain, which probably contains the ligand-binding sites, as in the case of the chemotaxis transducer proteins. Mutagenesis of the *uhpB* gene showed that numerous random linker insertion mutations in the membrane-embedded N-terminal half result in constitutive Uhp expression, whereas those in the C-terminal half block function (Island and Kadner, 1993). In most cases, the constitutive behavior was dependent on the presence of an active form of UhpC protein, showing that these changes have not bypassed the normal mode of regulation and suggesting that UhpB and UhpC act together in a protein complex. Mutants that allow expression of the Uhp system in the absence of UhpC contain amino acid substitutions in the C-terminal half of UhpB. All of these UhpC-independent isolates caused a change in the charge of the altered residues, most of which were near the conserved kinase motifs. We suspect that the membrane-embedded portion of UhpB acts to negatively regulate the cytoplasmic kinase portion and that this inhibition is relieved when Glu6P binds to a UhpB-UhpC complex in the membrane or when the membrane-embedded portion of UhpB is distorted through mutation or when certain mutations alter the electrostatic surface around the kinase sites. Much more work is needed to assess the changes in kinase function and protein interactions that occur when inducer binding triggers the transmembrane signaling processes.

Transcription Activation by UhpA

The UhpA protein appears to be an activator of transcription at the *uhpT* promoter. UhpA function is absolutely required for any expression of β-galactosidase from a *uhpT-lacZ* transcriptional reporter. Overexpression of UhpA from multicopy plasmid vectors results in constitutive *uhpT* expression that neither requires UhpB or UhpC function nor responds to the presence of inducer (Weston and Kadner, 1988). On the basis of amino acid sequence similarities, UhpA is a member of the large group of bacterial response regulator proteins, most of which are proposed to be transcription activators. These proteins generally contain a characteristic phosphorylation module of 110 to 120 amino acids, usually at the N terminus of the protein, followed by a putative activation module(s). The phosphorylation of an aspartyl residue within the phosphorylation module leads to activation of these proteins for their appropriate function. It is likely that there are several different mechanisms by which these proteins activate transcription. There are at least three discrete phylogenetic families of the response regulator proteins, distinguished partly by subtle sequence differences within the phosphorylation modules and partly by considerable differences in the C-terminal activation regions. The UhpA family of transcription activators is widespread in the bacterial world and includes NarL, NarP, and RcsB from *E. coli*; DegU and ComA from *Bacillus subtilis*; FixJ from *Rhizobium* species; MoxX from *Paracoccus denitrificans*; and an open reading frame from *Streptomyces griseus*. Their 60- to 70-residue C-terminal activation modules are predicted to contain a helix-turn-helix motif and are related in sequence to the DNA-binding regions of transcription activators, such as LuxR from *Vibrio fischeri* and MalT from *E. coli*, which are regulated by direct ligand binding rather than by protein phosphorylation. The phosphorylation and activation modules in the UhpA family are joined by a short and highly variable region likely to serve as a flexible hinge.

The function of the putative modules in UhpA is being investigated by analysis of deletions from the carboxyl terminus. These were generated by unidirectional exonuclease III digestion of the cloned *uhpA* gene followed by ligation to a sequence that provides translation termination signals in all three reading frames (Fig. 2A). When these truncation derivatives were introduced into a Δ*uhpA* strain, most were completely defective in activation of *uhpT-lacZ* expression, unlike the intact gene, which conferred high and constitutive expression. Removal of even seven amino acids obliterated activation. In addition, the truncation derivatives that removed part of the activation module but left the phosphorylation module intact showed a dominant-negative phenotype. They greatly reduced Uhp expression even in the presence of a wild-type *uhpA* gene (Fig. 2B). Deletions that extended into the phosphorylation module did not interfere with UhpA$^+$ function, probably because of the inability of the shorter fragments to form the stable α/β-barrel domain structure as in CheY.

This dominant-negative phenotype could have many explanations. Some of the more obvious ones are that the truncated UhpA competes with the wild-type UhpA for binding to the DNA target site in the *uhpT* promoter, to the UhpB kinase, or to other UhpA monomers if formation of an oligomer is required for UhpA action. Some of these possibilities could be tested from the effect of the UhpA truncations on the constitutive phenotype given by mutations in *uhpA* that bypass the requirement for UhpB and UhpC function or mutations in *uhpB* that bypass the requirement for UhpC function. If the dominant-negative effect of truncated forms of UhpA resulted from formation of inactive mixed oligomers, this phenotype would still be observed with both types of constitutive mutants. If the response were the result of competition for the UhpB kinase, the constitutively active forms of UhpA should be dominant but the constitutively active forms of UhpB should be blocked by the liberated phosphorylation module and show a negative phenotype. This is precisely the behavior that is seen, although we cannot conclude that the dominant-negative effect is the result of competition for the kinase, because more complicated interactions have not been tested.

Transcription Activation by CAP

We think that Uhp may provide a good system with which to study the action of transcription activators, because it appears to have relatively few binding sites for its regulatory proteins. The *uhpT* promoter region extends for 120 bp upstream of the transcription start site and overlaps the end of the *uhpC* gene (Merkel et al., 1992).

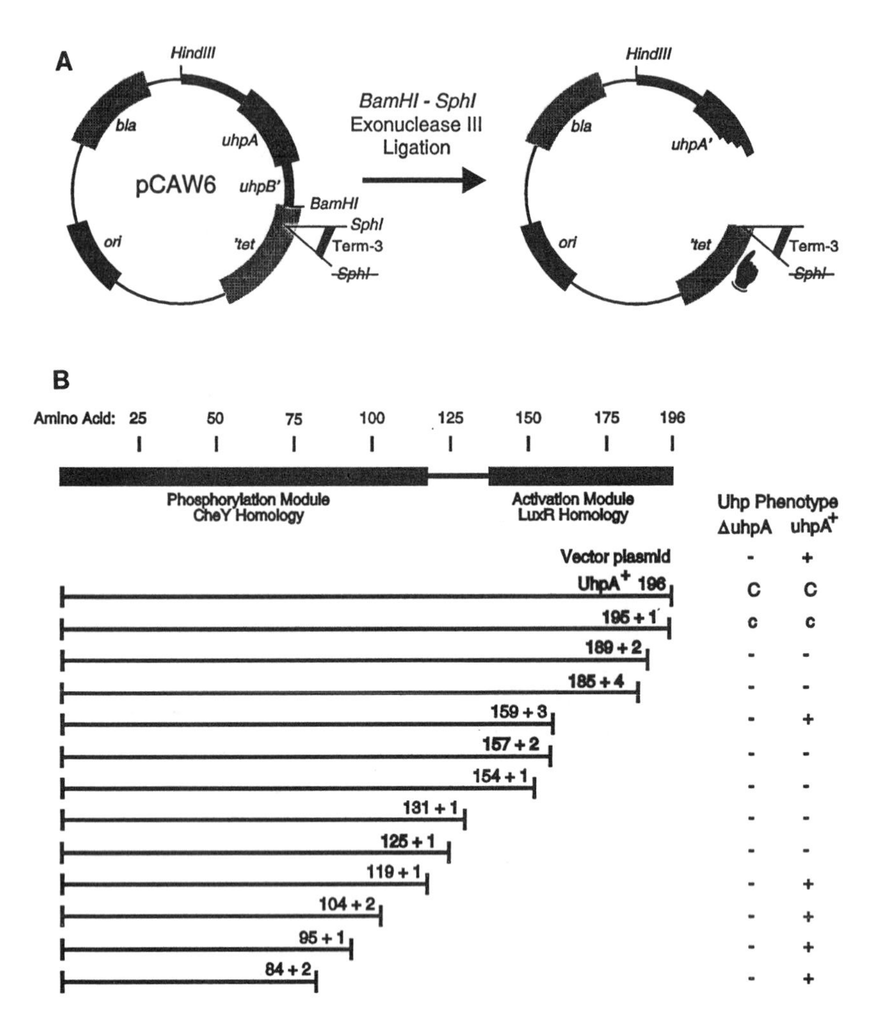

FIG. 2. Construction and phenotypic properties of forms of the UhpA protein truncated from the carboxyl terminus. (A) The *uhpA* gene and flanking sequences were cloned into pBR322 modified by insertion at the unique *Sph*I site of a linker which introduces translational stop codons in all three reading frames. Unidirectional exonuclease III digestion was accomplished after cleavage of the plasmid with *Bam*HI and *Sph*I, followed by ligation. Deletion junctions were determined by using a sequencing primer that hybridized immediately downstream of the termination linker, as indicated by the hand. (B) The extent of the deletion mutations is indicated by the number of amino acid residues from UhpA plus the number of amino acids encoded by the linker before a stop codon is encountered. Plasmids encoding the UhpA truncations were introduced into uhp^+ and a Δ*uhpA* strain and assayed by growth on fructose-6-phosphate and fosfomycin sensitivity. Uhp^- strains are fosfomycin resistant, and Uhp-constitutive strains are sensitive to fosfomycin in the absence or presence of inducing levels of Glu6P.

Two regions within the promoter that are important for transcription activity are a typical −10 region characteristic of σ^{70}-dependent promoters and a palindromic sequence centered at −32, which is unrelated to the usual −35 element present in σ^{70}-dependent promoters. A region of about 40 bp is centered at −64 and appears to be the DNA site for UhpA action, since multiple copies of this region can titrate out the limiting amounts of UhpA protein produced from the chromosomal *uhpA* gene (Weston and Kadner, 1988; Merkel et al., 1992). Centered at −103.5 is a good match to the consensus sequence for the DNA site for binding of CAP, the cAMP-dependent transcription activator protein. This region of the *uhpT* promoter binds purified CAP specifically, and deletion of this region reduces transcription activity by a factor of 7 to 20, comparable to the decrease in *uhpT* transcription seen in *crp* deletion strains unable to make CAP.

CAP action at other CAP-dependent promoters is being extensively studied. CAP binding induces formation of a sharp bend in the DNA, and the CAP-binding site can be placed at various distances from the remainder of the promoter as long as CAP binds on the same face of the DNA helix as RNA polymerase does. CAP action at many promoters is thought to occur through protein contacts and not solely through changes in DNA conformation, since mutational changes affecting a surface loop formed by residues 156 to 162 in CAP cause loss of transcription activation at many CAP-dependent promoters without an effect on the binding or bending of the DNA (Reznikoff, 1991). CAP action at the *uhpT* promoter differs from its action at other CAP-dependent promoters that do not need an additional activator protein. A similar face-of-the-helix periodicity is seen in *uhpT* promoter activity when additional residues are added between the CAP-binding site and the remainder of the promoter. The activation loop of residues 156 to 162 is not important, since mutant forms of CAP that are defective for activation at other CAP-dependent promoters are fully effective at *uhpT.*

Although the mechanism by which CAP activates transcription at the *uhpT* promoter is unknown, the physiological consequences of this effect are apparent as catabolite repression or decreased production of the UhpT transporter in the presence of good carbon sources such as glucose or Glu6P itself. Catabolite repression controls most genes involved in metabolism of potential carbon sources, perhaps to orchestrate the progressive appearance of different metabolic pathways to achieve the most efficient or competitive utilization of the nutrients that a bacterium might encounter. Glucose seems to be the preferred carbon and energy source for *E. coli,* judging from its potency at eliciting catabolite repression. There is a direct mechanism by which glucose entry and phosphorylation inhibit adenylate cyclase activity through enzyme III^{Glc} of the glucose-specific phosphotransferase system, as described in chapters 30 and 31. Glu6P should be an even better, more efficiently metabolized nutrient than glucose, because it can enter the cell already phosphorylated via the UhpT system. It might therefore seem peculiar that Glu6P entry is subject to catabolite repression, as if reflecting that glucose was preferred over Glu6P.

The basis for the catabolite repressibility of UhpT appears to be more than a bacterial sweet tooth in action. This regulatory process operates not only to maintain a hierarchy in the utilization of carbon sources but also to control the flux through central pathways of carbohydrate metabolism. It was shown long ago that Glu6P in combination with cAMP was toxic to wild-type *E. coli* (Ackerman et al., 1974). In a complementary approach, when UhpT was expressed from multicopy plasmids, cells were susceptible to sugar-phosphates even without the addition of cAMP (Kadner et al., 1992). At least two mechanisms of growth inhibition exist, depending on the metabolic fate of the added sugar-phosphate. Elevated transport of UhpT substrates that enter directly into the glycolytic pathway, such as Glu6P, fructose-6-phosphate, or mannose-6-phosphate, resulted in cell killing. Lethality was correlated with the production of the toxic compound, methylglyoxal, which is produced from dihydroxyacetone phosphate by a branch off the glycolytic pathway. Substrates that do not directly enter the glycolytic pathway, such as galactose-6-phosphate, 2-deoxyglucose-6-phosphate, arabinose-5-phosphate, or glucosamine-6-phosphate, elicited a strong but transient growth inhibition, with no loss of cell viability and no production of methylglyoxal. Growth stasis following accumulation of sugar-phosphates is a familiar phenomenon in bacterial metabolism and has been widely used as a selective method in genetic studies of the galactose and other pathways. This growth inhibition could result from the effects of high levels of internal organophosphates on cellular osmolarity or water availability, from depletion of internal P_i pools by the UhpT antiport activity, or from membrane depolarization by increased flux of P_i through the proton-translocating Pit system. Whatever the basis, the unrestricted transport of sugar-phosphates clearly has severe effects on the cell, and the catabolite repressibility of UhpT expression probably exists to limit the entry of material into the catabolic pathways.

Regulation of Organophosphate Transport Systems

An unusual feature of the Uhp system is its use of a third regulatory component that so closely resembles the transport protein whose expression is controlled. Perhaps this process is made more complicated than others to ensure that there is no induction response to the high level of intracellularly generated Glu6P, by using two membrane-embedded components which would both have to be properly oriented to accept only external Glu6P in order for transmission of an inducing signal. Are other organophosphate transport systems similarly regulated? Although it makes sense that catabolic systems for normal metabolites would be made only when that metabolite is present outside the cell, the cell often regulates catabolic systems in response to elevated internal pools, as in the case of GlpT.

The PgtP transporter of *S. typhimurium* is like Uhp in its induction only by external substrate. Studies by Jen-Shiang Hong and colleagues have shown that expression of PgtP is also subject to catabolite repression and is controlled by three regulatory proteins, which are encoded just upstream of the *pgtP* gene but transcribed in the opposite direction. Two of these regulatory proteins, PgtA and PgtB, are related in sequence to the families of two-component regulators, but the similarity to Uhp ends there. The transcription activator, PgtA, is related to the NtrC family of activators, which usually act at σ^{54}-coupled promoters (Yu and Hong, 1986). Neither the probable kinase protein, PgtB, nor the third regulatory protein, PgtC, has the strongly nonpolar character of UhpB and UhpC (Yang et al., 1988), although at least one of them must span the cytoplasmic membrane to be able to participate in transmembrane signaling. It thus seems that the three transport systems for organophosphates are regulated in substantially different manners. Expression of UhpT is regulated by a modification of the familiar two-component regulatory process that uses three components to sense the presence of external inducer. Since the two membrane-bound regulatory proteins appear to function as a complex, we might consider Uhp a 2.5-component regulatory system.

REFERENCES

Ackerman, R. S., N. R. Cozzarelli, and W. Epstein. 1974. Accumulation of toxic concentrations of methylglyoxal by wild-type *Escherichia coli* K-12. *J. Bacteriol.* **119:**357–362.

Brzoska, P., H. Schweizer, M. Argast, and W. Boos. 1987. *ugp*-dependent transport system for *sn*-glycerol-3-phosphate of *Escherichia coli*, p. 170–177. *In* A. Torriani-Gorini, F. G. Rothman, S. Silver, A. Wright, and E. Yagil (ed.), *Phosphate Metabolism and Cellular Regulation in Microorganisms.* American Society for Microbiology, Washington, D.C.

Eiglmeier, K., W. Boos, and S. Cole. 1987. Nucleotide sequence and transcriptional startpoint of the *glpT* gene of *Escherichia coli:* extensive sequence homology of the glycerol-3-phosphate transport protein with components of the hexose-6-phosphate transport system. *Mol. Microbiol.* **1:**251–258.

Friedrich, M. J., and R. J. Kadner. 1987. Nucleotide sequence of the *uhp* region of *Escherichia coli. J. Bacteriol.* **169:**3556–3563.

Goldrick, D., G.-Q. Yu, S.-Q. Jiang, and J.-S. Hong. 1988. Nucleotide sequence and transcription start point of the phosphoglycerate transporter gene of *Salmonella typhimurium. J. Bacteriol.* **170:**3421–3426.

Gött, P., and W. Boos. 1988. The transmembrane topology of the *sn*-glycerol-3-phosphate permease of *Escherichia coli* analysed by *phoA* and *lacZ* protein fusions. *Mol. Microbiol.* **2:**655–663.

Island, M. D., and R. J. Kadner. 1993. Interplay between the membrane-associated UhpB and UhpC regulatory proteins. *J. Bacteriol.* **175:**5028–5034.

Island, M. D., B.-Y. Wei, and R. J. Kadner. 1992. Structure and function of the *uhp* genes for the sugar phosphate transport system in *Escherichia coli* and *Salmonella typhimurium. J. Bacteriol.* **174:**2754–2762.

Kadner, R. J., C. A. Webber, and M. D. Island. 1993. The family of organo-phosphate transport proteins includes a transmembrane regulatory protein. *J. Bioenerg. Biomemb.* **25:**637–645.

Kadner, R. J., G. P. Murphy, and C. M. Stephens. 1992. Two mechanisms for growth inhibition by elevated transport of sugar phosphates in *Escherichia coli. J. Gen. Microbiol.* **138:**2007–2014.

Larson, T. J. 1987. *glpT*-dependent transport of *sn*-glycerol-3-phosphate in *Escherichia coli* K-12, p. 164–169. *In* A. Torriani-Gorini, F. G. Rothman, S. Silver, A. Wright, and E. Yagil (ed.), *Phosphate Metabolism and Cellular Regulation in Microorganisms.* American Society for Microbiology, Washington, D.C.

Lloyd, A. D., and R. J. Kadner. 1990. Topology of the *Escherichia coli uhpT* sugar-phosphate transporter analyzed by using Tn*phoA* fusions. *J. Bacteriol.* **172:**1688–1693.

Maloney, P. C., S. V. Ambudkar, V. Anantharam, L. A. Sonna, and A. Varadhachary. 1990. Anion-exchange mechanisms in bacteria. *Microbiol. Rev.* **54:**1–17.

Marger, M. D., and M. H. Saier, Jr. 1993. A major superfamily of transmembrane facilitators that catalyze uniport, symport, and antiport. *TIBS* **18:**13–20.

Merkel, T. J., D. M. Nelson, C. L. Brauer, and R. J. Kadner. 1992. Promoter elements required for positive control of transcription of the *Escherichia coli uhpT* gene. *J. Bacteriol.* **174:**2763–2770.

Parkinson, J. S. 1993. Signal transduction schemes of bacteria. *Cell* **73:**857–871.

Parkinson, J. S., and E. C. Kofoid. 1992. Communication modules in bacterial signaling proteins. *Annu. Rev. Genet.* **26:**71–112.

Pogell, B. M., B. R. Maity, S. Frumkin, and S. Shapiro. 1966. Induction of an active transport system for glucose-6-phosphate in *Escherichia coli. Arch. Biochem. Biophys.* **116:**406–415.

Pradel, E., and P. L. Boquet. 1989. Mapping of the *Escherichia coli* acid glucose-1-phosphatase gene *agp* and analysis of its expression in vivo by use of an *agp-phoA* protein fusion. *J. Bacteriol.* **171:**3511–3517.

Reznikoff, W. 1991. Catabolite gene activator protein activation of *lac* transcription. *J. Bacteriol.* **174:**655–658.

Schweizer, H., W. Boos, and T. J. Larson. 1985. Repressor for the *sn*-glycerol-3-phosphate regulon of *Escherichia coli* K-12: cloning of the *glpR* gene and identification of its product. *J. Bacteriol.* **161:**563–566.

Sonna, L. A., S. V. Ambudkar, and P. C. Maloney. 1988. The mechanism of glucose-6-phosphate transport by *Escherichia coli. J. Biol. Chem.* **263:**6625–6630.

Weston, L. A., and R. J. Kadner. 1988. Role of *uhp* genes in expression of the *Escherichia coli* sugar-phosphate transport system. *J. Bacteriol.* **170:**3375–3383.

Winkler, H. H. 1970. Compartmentation in the induction of the hexose-6-phosphate transport system of *Escherichia coli. J. Bacteriol.* **101:**470–475.

Winkler, H. H. 1973. Distribution of an inducible hexose-phosphate transport system among various bacteria. *J. Bacteriol.* **116:**1079–1081.

Yang, Y.-L., D. Goldrick, and J.-S. Hong. 1988. Identification of the products and nucleotide sequence of two regulatory genes involved in exogenous induction of phosphoglycerate transport in *Salmonella typhimurium. J. Bacteriol.* **170:**4299–4303.

Yu, G.-Q., and J.-S. Hong. 1986. Identification and nucleotide sequence of the activator gene of the externally induced phosphoglycerate transport system of *Salmonella typhimurium. Gene* **45:**51–57.

Chapter 15

Finding the Hole in UhpT: Applications of Molecular Biology to a Membrane Carrier

RUN-TAO YAN AND PETER C. MALONEY

Department of Physiology, Johns Hopkins University Medical School, Baltimore, Maryland 21205

Carriers related to the protein known as UhpT (named for the uptake of hexose phosphates) catalyze obligatory exchanges among their various substrates (organic phosphate and P_i). Because this involves no associated chemical or photochemical transformations, such exchanges are termed secondary transport events; this distinguishes them from primary events directly linked to such things as ATP hydrolysis or the absorption of light. It is usual to distinguish three reaction mechanisms among the secondary carriers (Maloney, 1987). There are reactions of symport, during which two or more substrates move in the same direction, allowing the downhill movement of one substrate to drive the uphill of another, as in the accumulation of sugar by the H^+/lactose cotransporter (LacY) of *Escherichia coli.* There are also reactions of uniport; in this case, substrate moves alone, independently of other molecules, as in the facilitated diffusion of glucose across the erythrocyte membrane. Finally, there is the category of antiport or exchange. As noted above, this mechanism characterizes UhpT, which catalyzes an anion exchange involving organic phosphates or P_i. UhpT is unusual among antiporters in that its substrates are found as both monovalent and divalent ions under physiological conditions. As a result, these exchanges can become rather complex (Maloney et al., 1990). In the simplest of these reactions, *E. coli* uses UhpT to take up external glucose-6-phosphate (G6P) in exchange for internal phosphate.

As a group, secondary carriers form a widely spread superfamily of membrane proteins, with representatives in all cell types (Maloney, 1990; Marger and Saier, 1993) and with diverse responsibilities in cell biology, ranging from the accumulation of nutrients in bacteria to the cycling of neurotransmitter at synaptic and synaptosomal membranes. The analysis of amino acid (DNA) sequences has now generated several hundred distinct examples, and it has been possible to organize some of these into related groups on the basis of amino acid similarities (Maiden et al., 1987) and to construct evolutionary trees (Marger and Saier, 1993). There is substantial sequence diversity within this superfamily, so it is by no means possible to link all known carriers by arguments of homology. Nevertheless, it seems likely that all carriers follow a common structural imperative and share in a single body plan (Maloney, 1990, 1992; also see below).

The idea that the superfamily of membrane carriers is united by some common structural plan arises from a broad reading of their amino acid sequences. There is now general agreement that carrier protein sequences fall into two main groups (Ambudkar et al., 1990; Maloney, 1990; Marger and Saier, 1993). The larger group, perhaps 80% of the total, includes both prokaryotic and eukaryotic examples, each of which is expected to have 10 to 12 transmembrane segments. By contrast, members in the smaller group (mostly exchange carriers from mitochondria and chloroplasts) are thought to have only five to seven transmembrane segments (Maloney, 1990; Nelson et al., 1993). This surprisingly simple classification is strongly reinforced by biochemical studies, even though these are available for only a few model systems. For example, studies of three examples in the larger group—LacY (Costello et al., 1987) and UhpT (Ambudkar et al., 1990) in *E. coli*, and band 3, the erythrocyte anion-exchange protein (Lindenthal and Schubert, 1991)—agree that monomeric protein functions in transport, suggesting that all carriers having 10 to 12 transmembrane segments may operate as monomers. On the other hand, analysis of examples from mitochondria and chloroplasts, all members of the smaller group, suggests that these work as homodimers (Aquila et al., 1987; Wagner et al., 1989). As a result, it appears that all membrane carriers may resemble each other in having comparable numbers of transmembrane segments in their minimal functioning units. It is this kind of correlation which motivates the idea of a common structural plan.

At this stage of study, further insight into the organization of membrane carriers requires the study of model systems that can be approached by both molecular biology and biochemistry. For this reason, we have broadened the analysis of UhpT to take advantage of scanning mutagenesis, in which a reporter residue is systematically targeted to strategic locations within the molecule. We have chosen to deal first with cysteine as a reporter, because this has proven of considerable value in other systems. Targeted cysteine mutagenesis was important to the early structural models of bacteriorhodopsin (Altenbach et al., 1990), and the strategy is now being used to probe functional regions in the acetylcholine receptor ion channel (Akabas et al., 1992) and a possible H^+-coupling region in LacY (Sahin-Toth and Kaback, 1993). As a prelude to the general use of this method with UhpT, we analyzed the cysteine residues found in the wild-type protein. In doing so, we found that one of these cysteines, C-265, is accessible to a membrane-impermeant, anionic sulfhydral reagent from both membrane surfaces (Yan and Maloney, 1993), implying that we have found a residue (C-265) lining the "hole" through which substrate moves. This opens a new area of study in this field, and we hope to provide direct information about residues comprising the substrate translocation pathway in UhpT and other membrane carriers.

UhpT as a Model Membrane Carrier

The study of UhpT and other P_i-linked antiporters has been under way for some time; indeed, these were the first membrane carriers studied in bacteria. The earliest efforts, from Mitchell's laboratory in the early 1950s, took advantage of the P_i self-exchange reaction in *Staphylococcus aureus* (then called *Micrococcus pyogenes*) to define the basic features of membrane biology in microorganisms (Mitchell, 1953). This work was not continued, in part because Mitchell turned to other areas and in part because *S. aureus* was unsuitable as a genetic system. For these reasons, further study awaited the description of UhpT in *E. coli* (Pogell et al., 1966; Winkler, 1966). UhpT was at that time of special interest for its unusual regulation; expression of this protein is controlled by extracellular but not intracellular G6P. We now appreciate that this reflects a signal transduction event, initiated by occupancy of an external site on a membrane receptor and completed by transcriptional activation of the structural gene. An account of this chain of information transfer, and an outline of how it relates to other "two-component" systems, is given in chapter 14.

That UhpT has surfaced once again as a model for carrier-mediated transport reflects the considerable progress made in the past decade. Biochemical efforts gave strong evidence that UhpT operated as an anion-exchange mechanism (Sonna et al., 1988) and also provided tools for its analysis in solubilized and reconstituted systems (Ambudkar and Maloney, 1986; Sonna et al., 1988; Ambudkar et al., 1990). In parallel, molecular biology provided the UhpT amino acid sequence (Friedrich and Kadner, 1987), and the use of gene fusion technology has now provided convincing evidence that the transmembrane segments identified in hydropathy plots of this and related proteins do, in fact, span the lipid bilayer (Gott and Boos, 1988; Island et al., 1992). Taking these features into account, one may schematically depict UhpT as in Fig. 1, which gives a low-resolution view emphasizing features likely to be found in many (although not all) membrane carriers, i.e., the hydrophobic core with an even number of transmembrane segments (I to XII), N and C termini that are oriented to the cytoplasm, and a relatively long hydrophilic loop, also cytoplasmic, near the center of the molecule.

UhpT Functions as a Monomer

Among the most important of the biochemical studies of UhpT have been those showing that the protein functions as a monomer; thus, all transport functions are represented in the schematic shown in Fig. 1. This conclusion derives from three findings made with solubilized and reconstituted preparations (Ambudkar et al., 1990). (i) The first observation was that solubilized protein could bind substrate. This was proven in two ways, the simpler of which documented that the presence of substrate protected solubilized material against thermal inactivation. Although this is a somewhat unusual use of heat denaturation, quantitative kinetic tests have verified that the method gives

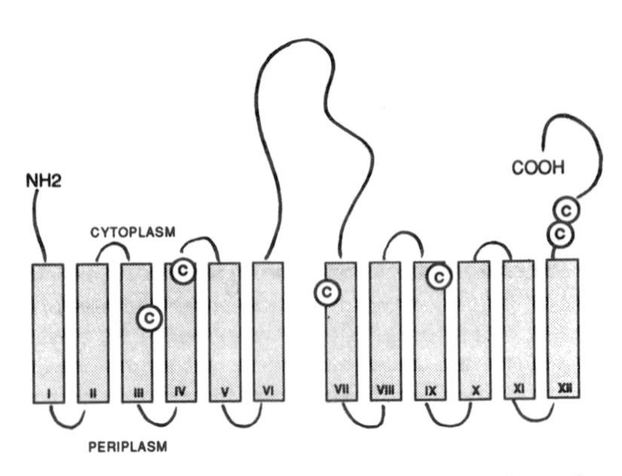

FIG. 1. Low-resolution view of UhpT. The amino acid sequence of UhpT is arranged in two dimensions to show its overall organization in the membrane. Transmembrane segments I to XII are indicated by shaded columns; the six UhpT cysteines (circles) are shown at positions corresponding to residues 108, 143, 265, 331, 436 and 438. This topology differs in minor details from that shown by Island et al. (1992).

binding constants consistent with those measured during membrane transport (Maloney et al., 1992). (ii) Companion experiments then showed that for similar conditions, the solubilized protein was dispersed in monomeric form; that is, size separation chromatography indicated that solubilized UhpT migrated between the 40- and 60-kDa markers, as expected of the 50-kDa monomer. This was true whether or not substrate was present, ruling out the idea that stabilization of activity (above) reflected substrate-induced dimerization or oligomerization. (iii) Finally, it proved possible to extend these conclusions to the operation of UhpT in a lipid bilayer, by performing reconstitution experiments with solubilized (monomeric) protein at a protein-to-lipid ratio so low that single proteoliposomes were unlikely to contain more than a single monomer. Even at this limiting concentration (one monomer per proteoliposome), full activity was retained, proving that monomeric UhpT was fully functional.

Identification of the functional form of the UhpT has set the stage for complementary work that uses molecular biology to characterize individual domains within the molecule. By exploiting targeted mutagenesis of and with cysteine(s), we hope to locate regions that come into close contact with substrate (G6P); progress toward this goal has been encouraging, and, as outlined below, analysis of UhpT variants has allowed us to experimentally define the substrate translocation pathway.

Role of Cysteines in UhpT

The necessary first step to this new line of experiments has been to determine whether the cysteines normally present in UhpT have important functional roles. Therefore, in our initial experiments we used site-directed mutagenesis to replace each of the six endogenous UhpT cysteines with an isosteric serine residue (Yan and Maloney, 1993). This was done in two ways, with equivalent results. On the one hand, we generated each of the six possible Cys→Ser mutants, obtaining strains in which only a single cysteine was changed. In addition, we used these variants along with in vitro restriction enzyme digestion to obtain a second set of derivatives, each containing a single cysteine at one of the normal positions. Assays of transport by both kinds of mutants showed that UhpT function was retained at normal or near-normal levels in all variants, even the mutant lacking cysteine altogether (Table 1). We concluded, therefore, that cysteine residues play no unique or essential role in the antiport functions of UhpT.

LacY is the only other membrane transport protein analyzed in this detail, and there, too, it was found that cysteines are dispensable (van Iwaarden et al., 1991). However, others have

TABLE 1. Cysteine is not required for UhpT function[a]

UhpT derivative[b]	Relative activity (range)[c]
C_6S_0 (wild type)	1.00
C_5S_1	0.36–1.36
C_1S_5	0.48–0.54
C_0S_6 (no cysteine)	0.37

[a]Reprinted with permission from Yan and Maloney (1993).
[b]Variants with single Cys→Ser replacements (C_5S_1), derivatives with a single cysteine (C_1S_5), and the cysteineless mutant (C_0S_6) were studied.
[c]For each set of mutants, the range of G6P transport activity by intact cells is expressed relative to that of the wild type.

found evidence of a disulfide linkage in the membrane sector of the plasma membrane proton pump of *Neurospora crassa* (Rao and Scarborough, 1990), and the character of transport through several mitochondrial carriers is altered by modification of the cysteine residue(s) (Dierks et al., 1990a, 1990b). The null effect of endogenous cysteine(s) in UhpT considerably simplifies the interpretation of our work, since the experimental manipulation of old cysteines or the addition of new ones can now proceed against a neutral background.

A Residue Lining the Translocation Pathway in UhpT

We next examined the set of UhpT variants containing only a single cysteine, with the aim of characterizing the residues responsible for the high sensitivity of UhpT and its relatives to mercury and mercurials (Mitchell, 1953; Maloney et al., 1990). This effort seemed worthwhile, since this information, along with use of permeant and impermeant mercurials, might be relevant to confirming the location of cysteines in UhpT (Fig. 1). By studying the inhibition of G6P transport by *p*-chloromercuribenzoic acid (PCMB), we were able to determine that mercurial sensitivity was attributable to reactions at both C-143 and C-265 (Fig. 2).

PCMB is a weak acid (pK ca. 4), and for this reason it is membrane permeant. Consequently, an attack at C-143 or C-265 could, in principle, occur from either the extracellular or cytoplasmic surface, the latter after nonionic diffusion of the free acid to the inside of the cell. This latter explanation seemed the more likely, since the presumed topology of UhpT places these two residues on the inner aspect of the hydrophobic core, near the cytoplasmic border (Fig. 1). Even so, it was important to test these alternatives in an explicit way, and to do this we used *p*-chloromercuribenzosulfonate (PCMBS), a membrane-impermeant derivative of PCMB. Because of its impermeant nature, we expected that G6P transport by intact

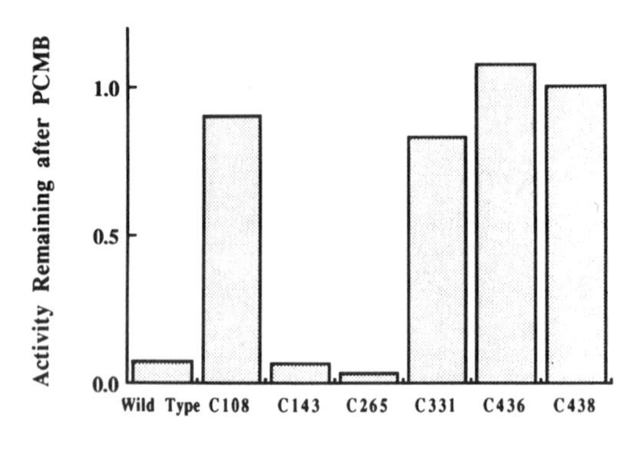

FIG. 2. C-143 and C-265 confer sensitivity to PCMB. The histograms describe the inhibition by PCMB of G6P transport by wild-type UhpT and variants containing only a single cysteine at the residue indicated. Activity is given relative to the untreated wild type cells. Drawn from data reported by Yan and Maloney (1993).

cells would be resistant to inhibition by PCMBS. It also seemed likely that studies of everted membrane vesicles would be informative, since in those cases a latent sensitivity to PCMBS might become evident.

The rationale underlying such experiments was justified by the behavior of C-143 (Table 2). In that case, PCMBS failed to inhibit transport by intact cells, but G6P accumulation by everted vesicles was substantially blocked; this shows directly that C-143 is exposed to the cytoplasm, and only to the cytoplasm, as predicted by the topological model of UhpT (Fig. 1). Unexpectedly, we found no evidence of sidedness on the part of C-265. Instead, C-265 was attacked by PCMBS in tests with both intact cells and everted vesicles (Table 2). Accordingly, we concluded that the water-soluble, membrane-impermeant PCMBS had access to C-265 from both internal and external phases (Yan and Maloney, 1993).

TABLE 2. C265 is accessible from both membrane surfaces[a,b]

UhpT variant	Relative activity in[c]:	
	Intact cells	Everted vesicles
C143	1.00	0.22
C265	0.25	0.15

[a]Reprinted with permission from Yan and Maloney (1993).

[b]These studies used strains expressing UhpT variants having only a single cysteine, at position 143 or 265. G6P transport was measured for intact cells and P_i-loaded everted vesicles incubated for 10 min in the absence or presence of 0.2 mM PCMBS.

[c]Transport activities are given relative to untreated cells or vesicles.

It is self-evident that UhpT contains a pathway through which G6P passes during the process of transport. Thus, our finding that PCMBS comes into contact with C-265 from either membrane surface might be understood if this probe exploits the preexisting substrate translocation pathway. This idea seemed reasonable, in part because G6P and PCMBS resemble each other in molecular size, shape, and charge. To test this possibility in the simplest way, we asked whether G6P could protect C-265 against an attack by PCMBS—one might expect this to occur if the two anions used the same pathway. Indeed, this proved to be the case, and substrate protection could be readily documented, whether the attack by PCMBS was from the extracellular medium or from the cytoplasmic surface (Yan and Maloney, 1993). Such findings strongly suggest that C-265 is one of the residues that lines the translocation pathway in this model membrane carrier.

Engineering a New PCMBS-Sensitive Site in UhpT

Experiments summarized in the preceding section indicate that C-265 may lie directly on the substrate translocation pathway of UhpT. Just as important, they show how one might use a relatively simple technique—targeted cysteine mutagenesis—to reveal significant elements of the structure of a membrane carrier. In an attempt to validate the approach, we have begun to implant new cysteines, hoping to identify other residues on the transport pathway. For example, starting with the cysteineless variant of UhpT, we have engineered a Ser→Cys mutation at position 199. This location was selected for two reasons. First, replacement of serine with cysteine is an isosteric event, and this eliminates one variable that might complicate the interpretation of results. Second, the hydropathy profile of UhpT places position 199 on transmembrane segment VI, within the hydrophobic core of UhpT; we hoped that this would maximize chances of targeting mutagenesis to an informative position.

Some possible outcomes of this work are especially interesting. For example, if the S199C mutant shows PCMBS sensitivity and substrate protection from either membrane surface, we would conclude that position 199 lies on the transport pathway, as does position 265. In this way, one might begin to map the transmembrane segments that delineate this domain. Alternatively, perhaps PCMBS inhibition will display a sidedness, occurring only from one surface. In that case, tests of substrate protection will show whether position 199 lies near a substrate-binding domain (substrate protection) or near some as yet undefined area (no protection).

Although our analysis of the S199C mutant is as yet incomplete, the preliminary data are encouraging. It is clear, for example, that this mutant is inhibited by externally added PCMBS (Table 3), proving that position 199, despite being in the hydrophobic core of UhpT, is exposed (at least transiently) to a water-soluble, membrane-impermeant probe. Moreover, the attack by PCMBS is blocked by substrates such as G6P and P_i (Table 3), further suggesting that position 199 lines the substrate transport pathway (as does position 265) or is within the substrate-binding pocket (these two domains need not overlap, although it is likely that they will share some elements). Studies of everted membrane vesicles are still clearly required, but at present the simplest interpretation of these findings is that position 199, just as position 265, can be used to mark the substrate translocation pathway in UhpT. If this proposition is verified, we will have begun the task of localizing the transport pathway to a specific region within UhpT (Fig. 3).

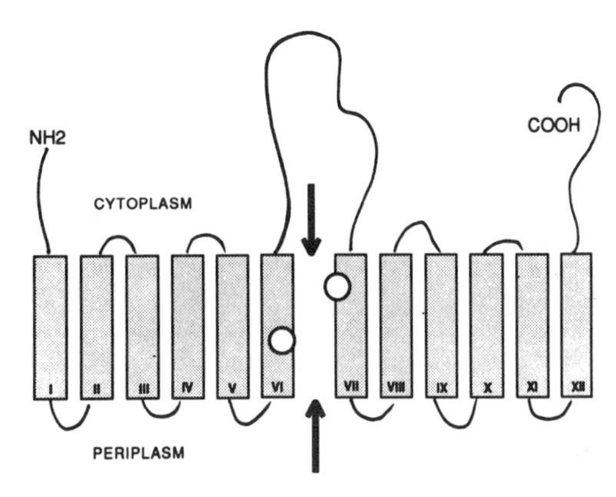

FIG. 3. Substrate translocation pathway of UhpT. This schematic representation of UhpT summarizes the hypothesis that the translocation pathway is composed, in part, of positions 199 and 265 (circles). As a result, it is possible to view substrates as moving between the N-terminal and C-terminal halves of UhpT. This is consistent with the idea (see the text) that UhpT has two functional domains, transmembrane segments I to VI and VII to XII, and that the substrate moves at the interface of these domains.

Implications for Understanding the Structure of Membrane Carriers

Our analysis of cysteines in UhpT sets experimental standards and criteria that define a new region—the translocation pathway—in this membrane carrier (Yan and Maloney, 1993). We now envision the use of cysteine-scanning mutagenesis for each of the 12 transmembrane segments of UhpT. Then, by applying the simple tests developed for C-265 and S199C, we might decide which of these segments do and which do not contribute to the transport pathway. This, in turn, may lead to informed models of the tertiary structure.

Completion of this work may also help in testing a simple argument concerning the general structural principles at work in these membrane proteins. Some time ago, it was suggested that membrane-transporting proteins would be oligomeric structures (Klingenberg, 1979; Kyte, 1980). In part, this suggestion was driven by experimental evidence. However, it also embodied an appreciation that the packing of protein subunits would necessarily enclose a space, perpendicular to the membrane plane, that could serve as a substrate translocation pathway. In this way, substrates would move between, not through, proteins and transporting systems would avoid the difficult problem of designing a catalytic center that, because it is penetrated by substrate, must be assembled and disassembled each transport cycle.

Although this attractive model was consistent with the dimeric structure of the mitochondrial carriers, the proposal lost much of its appeal when monomeric function was described for LacY and other carriers with 10 to 12 transmembrane segments. Perhaps we should now reconsider the general idea. Could LacY, UhpT, and their structural relatives be functional dimers (covalent heterodimers)? If so, the earlier arguments are worth revisiting. In fact, it has been noted that a number of the "monomeric" carriers show homology between the N- and C-terminal halves of the molecule (Maiden et al., 1987). This implies an ancestral homodimer, a twofold symmetry to the original structure, and therefore a pseudo-twofold organization to the contemporary example. Equally interesting, if the translocation pathway was at the interface between ancestral subunits, it is now likely to be found at the intersection of the N- and C-terminal halves of the molecule, a possibility consistent with the properties of positions 199 and 265 in UhpT (Fig. 3). We believe that cysteine-scanning mutagenesis can settle this issue

TABLE 3. Engineering a PCMBS-sensitive site in UhpT

Sample	Relative activity[a]	
	No PCMBS	+ 0.2 mM PCMBS
No additions	1.00	0.22
+ 1 mM G6P	1.01	0.86
+ 100 mM KP_i		0.76
+ 1 mM DTT		0.84

[a]G6P transport activity was measured for intact cells expressing the S199C variant of UhpT. Cells were exposed for 10 min to 0.2 mM PCMBS, in the presence or absence of the UhpT substrate G6P or KP_i. Where shown, dithiothreitol (DTT) was added after the exposure to PCMBS. Activities are expressed relative to the untreated control (no additions).

for UhpT. At the least we are optimistic that within the next several years we will be able to argue in a rational way about critical structural elements in this membrane carrier.

These studies of anion exchange are supported by grants from the Public Health Service (GM24195) and the National Science Foundation (MCB9220823).

We thank our colleagues Robert Kadner and Michael Island, University of Virginia Medical School, for introducing us to the molecular biology of UhpT.

REFERENCES

Akabas, M. H., D. A. Stauffer, M. Xu, and A. Karlin. 1992. Acetylcholine receptor channel structure probed in cysteine-substitution mutants. *Science* **258:**307–310.

Altenbach, C., T. Marti, H. G. Khorana, and W. L. Hubbell. 1990. Transmembrane protein structure: spin labeling of bacteriorhodopsin mutants. *Science* **248:**1088–1092.

Ambudkar, S. V., V. Anantharam, and P. C. Maloney. 1990. UhpT, the sugar phosphate antiporter of *Escherichia coli,* functions as a monomer. *J. Biol. Chem.* **265:**12287–12292.

Ambudkar, S. V., and P. C. Maloney. 1986. Bacterial anion exchange. Use of osmolytes during solubilization and reconstitution of phosphate-linked antiport from *Streptococcus lactis. J. Biol. Chem.* **261:**10079–10086.

Aquila, H., T. Link, and M. Klingenberg. 1987. Solute carriers involved in energy transfer of mitochondria form a homologous protein family. *FEBS Lett.* **212:**109.

Costello, J. J., J. Escaig, K. Matsushita, P. V. Viitanen, D. R. Menick, and H. R. Kaback. 1987. Purified lac permease and cytochrome o oxidase are functional as monomers. *J. Biol. Chem.* **262:**17072–17082.

Dierks, T., A. Salentin, C. Heberger, and R. Kramer. 1990a. The mitochondrial aspartate/glutamate and ADP/ATP carriers switch from obligate counterexchange to unidirectional transport after modification by SH-reagents. *Biochim. Biophys. Acta* **1028:**268–280.

Dierks, T., A. Salentin, and R. Kramer. 1990b. Pore-like and carrier-like properties of the mitochondrial aspartate/glutamate carrier after modification by SH-reagents: evidence for a preformed channel as a structural requirement of carrier-mediated transport. *Biochim. Biophys. Acta* **1028:**281–288.

Friedrich, M. J., and R. J. Kadner. 1987. Nucleotide sequence of the *uhp* region of *Escherichia coli. J. Bacteriol.* **169:**3556–3563.

Gott, P., and W. Boos. 1988. The transmembrane topology of the sn-glycerol 3-phosphate permease of *Escherichia coli* analyzed by phoA and lacZ protein fusions. *Mol. Microbiol.* **2:**655–663.

Island, M. D., B. Y. Wei, and J. J. Kadner. 1992. Structure and function of the *uhp* genes for the sugar phosphate transport system in *Escherichia coli* and *Salmonella typhimurium. J. Bacteriol.* **174:**2754–2762.

Klingenberg, M. 1979. Membrane protein oligomeric structure and transport function. *Nature* (London) **290:**449–454.

Kyte, J. 1980. Molecular considerations relevant to the mechanism of active transport. *Nature* (London) **292:**201–204.

Lindenthal, L., and D. Schubert. 1991. Monomeric erythrocyte band 3 protein transports anion. *Proc. Natl. Acad. Sci. USA* **88:**6540–6544.

Maiden, M. C. J., E. O. Davis, S. A. Baldwin, D. C. M. Moore, and P. J. F. Henderson. 1987. Mammalian and bacterial sugar transport proteins are homologous. *Nature* (London) **325:**641–643.

Maloney, P. C. 1987. Coupling to an energized membrane: role of ion-motive gradients in the transduction of metabolic energy, p. 222–243. *In* F. C. Neidhardt, J. L. Ingraham, K. B. Low, B. Magasanik, M. Schaechter, and H. E. Umbarger (ed.), *Escherichia coli and Salmonella typhimurium: Cellular and Molecular Biology.* American Society for Microbiology, Washington, D.C.

Maloney, P. C. 1990. A consensus structure for membrane transport. *Res. Microbiol.* **141:**374–383.

Maloney, P. C. 1992. The molecular and cell biology of anion transport by bacteria. *Bioessays* **14:**757–762.

Maloney, P. C., S. V. Ambudkar, V. Anantharam, L. A. Sonna, and A. Varadhachary. 1990. Anion-exchange mechanisms in bacteria. *Microbiol. Rev.* **54:**1–17.

Maloney, P. C., V. Anantharam, and M. J. Allison. 1992. Measurement of the substrate dissociation constant of a solubilized membrane carrier. Substrate stabilization of OxlT, the anion exchange protein of *Oxalobacter formigenes. J. Biol. Chem.* **267:**10531–10536.

Marger, M. D., and M. H. Saier. 1993. A major superfamily of transmembrane facilitators that catalyze uniport, symport and antiport. *Trends Biochem. Sci.* **18:**13–20.

Mitchell, P. 1953. Transport of phosphate across the surface of *Micrococcus pyogenes:* nature of the cell 'inorganic phosphate.' *J. Gen. Microbiol.* **9:**273–287.

Nelson, D. R., J. E. Lawson, M. Klingenberg, and M. G. Douglas. 1993. Site-directed mutagenesis of the yeast mitochondrial ADP/ATP translocator. Six arginines and one lysine are essential. *J. Mol. Biol.* **230:**1159–1170.

Pogell, B. M., B. R. Maity, S. Frumkin, and S. Shapiro. 1966. Induction of an active transport system of glucose 6-phosphate in *Escherichia coli. Arch. Biochem. Biophys.* **116:**406–415.

Rao, U. S., and G. A. Scarborough. 1990. Chemical state of the cysteine residues in the *Neurospora crassa* plasma membrane H^+-ATPase. *J. Biol. Chem.* **265:**7227–7235.

Sahin-Toth, M., and H. R. Kaback. 1993. Cysteine scanning mutagenesis of putative transmembrane helices IX and X in the lactose permease of *Escherichia coli. Protein Sci.* **2:**1024–1033.

Sonna, L. A., S. V. Ambudkar, and P. C. Maloney. 1988. The mechanism of glucose 6-phosphate transport by *Escherichia coli. J. Biol. Chem.* **263:**6625–6630.

van Iwaarden, P. R., J. C. Pastore, W. N. Konings, and H. R. Kaback. 1991. Construction of a functional lactose permease devoid of cysteine residues. *Biochemistry* **30:**9595–9600.

Wagner, R., E. C. Apley, G. Gross, and U. I. Flugge. 1989. The rotational diffusion of chloroplast phosphate translocator and of lipid molecules in bilayer membranes. *Eur. J. Biochem.* **182:**165–173.

Winkler, H. H. 1966. A hexose-phosphate transport system in *Escherichia coli. Biochim. Biophys. Acta* **117:**231–240.

Yan, R.-T., and P. C. Maloney. 1993. Identification of a residue in the translocation pathway of a membrane carrier. *Cell* **75:**37–44.

Chapter 16

Maltose Transport System of *Escherichia coli* as a Member of ABC Transporters

HIROSHI NIKAIDO, IRINA D. POKROVSKAYA, LETICIA REYES, ANAND K. GANESAN, AND JASON A. HALL

Department of Molecular and Cell Biology, 229 Stanley Hall, University of California, Berkeley, California 94720

In recent years, as many as 50 membrane transporters in bacteria, yeasts, and animals were found to make up a large family, the ABC (ATP-binding cassette) transporters (Higgins, 1992). An ABC transporter (also called traffic ATPase) is composed of a membrane-associated complex, which typically contains four domains (Fig. 1). Two of these are transmembrane proteins, which presumably form a transport channel and which most frequently appear to contain six membrane-spanning α-helices each. Each of the other two domains contains a nucleotide-binding site, and these domains show significant homology among different members of the ABC transporter family. The various domains frequently exist as independent subunits, especially in bacteria, but in animal cells they are often parts of a larger protein containing more than one domain. An ABC transporter may function either in nutrient uptake or in the export of drugs, peptides, or proteins. This superfamily thus includes not only the transporters functioning in the uptake of small molecules but also the multidrug resistance protein pumping out the anticancer drugs from resistant tumor cells, the presumed chloride channel altered in cystic fibrosis patients (CFTR), and the transporter functioning in the presentation of antigenic peptides on the T-cell surface (Higgins, 1992). Bacterial ABC transporters functioning in nutrient uptake characteristically contain an additional hydrophilic subunit, the binding protein, which resides in the periplasm in gram-negative bacteria or is anchored to the external surface of the cytoplasmic membrane in gram-positive bacteria (Fig. 1).

ABC transporters of *Escherichia coli* include the high-affinity phosphate (PstABCS) (Rao and Torriani, 1990), histidine (Ames, 1986), and maltose (see below) transport systems, among many others. The histidine and maltose systems have been studied particularly intensively in recent years, and here we summarize our current understanding of the maltose system.

The maltose transport system of *E. coli* accumulates both maltose and maltodextrins (up to maltoheptaose) from the external medium. The smaller substrates (e.g., maltose and maltotriose) are able to diffuse through the general-purpose, nonspecific porins of *E. coli*, but the permeation rates are too low to allow the rapid accumulation of these substrates from low external concentrations (Nikaido and Vaara, 1985). Larger members of the maltodextrin series will not be able to diffuse through porin channels at significant rates. For these reasons, *E. coli* produces a specific outer membrane channel, LamB, which contains a specific ligand-binding site within each channel (Luckey and Nikaido, 1980; Gehring et al., 1991a). Once maltose and maltodextrins traverse the outer membrane, they are bound by maltose-binding protein (MBP), the product of the *malE* gene. The ligand-attached MBP then interacts with the membrane-associated transporter, which is made up of one copy each of MalF and MalG, the putative membrane-spanning units, and two copies of MalK, the ATP-binding subunit (Davidson and Nikaido, 1991), and then the ligands are transported across the cytoplasmic membrane, concomitant with the hydrolysis of ATP (Dean et al., 1989; Davidson and Nikaido, 1990). We shall examine the individual steps below, but in a somewhat different order.

Membrane-Associated Transporter

As mentioned above, this complex is composed of two presumed channel proteins, MalF and MalG, together with the nucleotide-binding subunits, MalK. The earliest clue to the existence of the family relationship among what are now known as ABC transporters was the recognition of extensive sequence homology between MalK and the corresponding subunit in the histidine system, HisP (Gilson et al., 1982). This was followed several years later by the even more important discovery that subunits including these two proteins and many others all shared the nucleotide-binding motifs (Walker motifs) (Higgins et al., 1986), a discovery that led to the recognition of the ABC

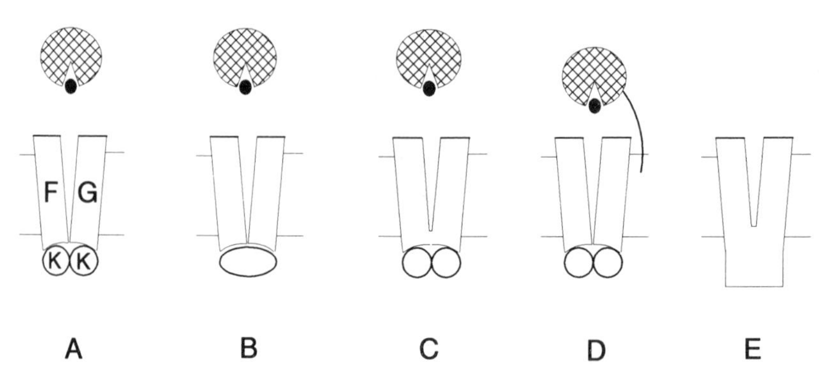

FIG. 1. Domain organization in some examples of ABC transporters. (A) Maltose transport system of *E. coli*, with the two separate membrane-spanning subunits, F and G, associated with the two copies of the ATPase subunit, K. The transporter complex interacts with the ligand-attached binding protein. (B and C) In other examples of bacterial nutrient transport systems, the two copies of the ATPase may be joined together (B), or the two transmembrane domains are joined together to form a single protein (C). (D and E) Finally, in gram-positive bacteria, which do not produce the outer membrane that can confine the binding protein to periplasm, the binding protein may be tethered to the cytoplasmic membrane via a lipid extension (D). In higher animals, all of the domains may be fused into a single protein, as seen in MDR (E).

transporter family. ATP binds to MalK in the absence of MalF and MalG (Ames, 1986), and purified MalK is reported to hydrolyze ATP in the absence of F and G (Walter et al., 1992). The reported rate of hydrolysis, however, is less than 1/10 the rate observed with the full complex (Davidson et al., 1992), and we cannot exclude the possibility that the hydrolysis was catalyzed by improperly folded MalK conformers generated by the denaturation-renaturation process used by Walter et al. (1992).

We have been able to purify the intact maltose transporter complex to about 90% purity from overproducing strains (Davidson and Nikaido, 1991). The transporter is active when incorporated into the bilayers of liposomes, allowing the accumulation of maltose inside the vesicles, driven by hydrolysis of ATP, when the ligand-attached MBP is added from the outside. Although proteoliposome reconstitution of a few other bacterial ABC transporters has been reported (Bishop et al., 1989; Hoshino et al., 1992), to our knowledge this represents the only reported case of a successful purification of the complex. Since our understanding of the mechanism of transport hinges on our knowledge of the structure of the transporter, this gives us some hope that crystallization and crystallographic analysis may become possible one day. The method of Davidson and Nikaido (1991), however, produces a rather poor yield, and is not suitable for such studies. Recently, we have developed a radically different purification protocol, which in preliminary runs produced a yield in excess of 20% (Reyes and Nikaido, unpublished), and efforts to produce both two-dimensional and three-dimensional crystals are in progress.

The maltose transporter complex in an intact, wild-type cell absolutely requires the presence of MBP for transport (Hengge and Boos, 1983). One of the most important contributions to the study of the maltose system was the isolation, by Treptow and Shuman (1985), of *E. coli* mutants that no longer required MBP for maltose transport. These mutants are altered in MalF and MalG subunits, and they transport maltose with a much lower affinity (with a typical K_m of 1 mM rather than 1 μM in wild-type cells) actively and specifically. Why is the presence of MBP no longer required in these mutants? The answer was provided when the transporter complex, containing the mutant MalF (and/or MalG) protein, was reconstituted into proteoliposomes. These mutant transporters hydrolyzed ATP constitutively without the addition of maltose or MBP, whereas ATP hydrolysis by the wild-type transporters occurred only in the presence of both maltose and MBP (Davidson et al., 1992). This unexpected finding led us to the following hypothesis (Fig. 2). It would be wasteful to keep hydrolyzing ATP in the absence of the ligand molecules to be transported. Because the ABC transporters have their ATPase domains (MalK in the maltose transporter) on the inner side of the membrane (Shuman and Silhavy, 1981), a mechanism that tells the ATPase domain of the presence of ligand molecules on the other side of the membrane, and thereby activates the ATPase, is required. We believe that this function is fulfilled by the ligand-attached MBP. The system presumably evolved in such a way to use ligand-attached MBP rather than free maltose as the signaling molecule, because at a low external maltose concentration, practically all of the maltose molecules in the periplasm would exist as mal-

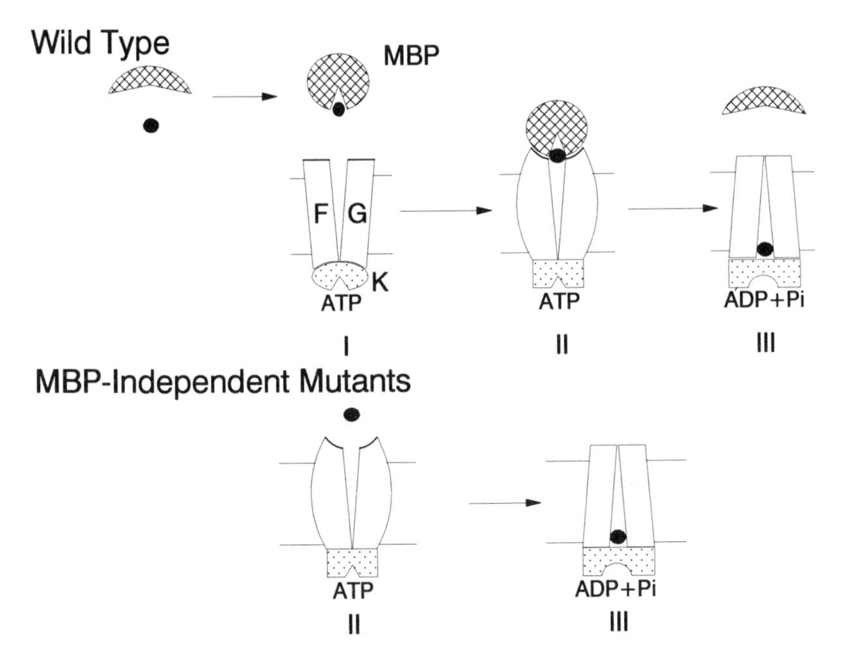

FIG. 2. Interaction between the ligand-attached MBP and the membrane-associated maltose transporter complex. In the wild type, the ATPase subunits, K subunits (here shown as a monomeric entity for clarity), are in an inactive conformation (I), but the binding of the ligand-attached MBP to the outside produces conformational changes that are transmitted via F and G subunits and activate the ATPase (II and III). In the binding-protein-independent mutants, the altered F and G subunits are already in the activated conformation, and therefore the ATPase is also in the active form. ATP is thus constantly hydrolyzed regardless of the presence or absence of ligand-attached MBP. In this figure as well as in Fig. 3, the FG channel in its resting state (conformation I in the wild type and II in the mutant) is shown in an open conformation, but we cannot rule out the possibility that the channel is normally closed (as in conformation III) and requires ATP binding or hydrolysis to become open. Modified from Davidson et al. (1992) with permission.

tose-MBP complex, because of the high affinity of MBP for its ligands. Similarly, we note that *E. coli* uses the maltose-MBP complex as a signal for chemotaxis, whereas free aspartate or serine, for which no binding protein exists, is used as a direct chemotactic signal (Macnab, 1987).

ATP is constitutively hydrolyzed by the mutant $MalFGK_2$ complex, whereas hydrolysis by the wild-type complex requires the binding of ligand-attached MBP (Davidson et al., 1992). This suggests that the conformation of the mutant complex may already resemble that of the wild-type complex to which the ligand-attached MBP has become bound (Fig. 3). This prediction was borne out by measuring the affinity of ligand-attached MBP to membrane vesicles isolated from the wild-type strain and the mutant strains; the apparent dissociation constant (measured as K_m) with the wild-type vesicles was 20 to 50 μM, whereas that with the mutant was only 2.7 μM (Dean et al., 1992).

A similar "signaling" role for ligand-attached binding protein has also been suggested for the histidine transport system (Petronilli and Ames, 1991). In the histidine system, however, a portion of the ATPase subunit, HisP, is reported to be exposed on the external surface of the cytoplasmic membrane (Baichwal et al., 1993), and the "binding-protein-independent" mutations are all located in the HisP subunit. Thus the signaling could conceivably involve the simple, direct interaction between the ligand-attached binding protein and the ATPase subunit. In contrast, in the maltose system, MalK appears to be a true peripheral protein (Shuman and Silhavy, 1981; Walter et al., 1992) and the signal appears to be transmitted across the membrane presumably by conformational changes of the channel subunits, MalF and/or MalG.

This concept that the ABC transporter complex can serve as a transmembrane signaling device may be useful in the study of other systems. Thus the presence of P_i molecules in the periplasm is known to act as a signal to affect the cytoplasmic, PhoU regulatory protein. Furthermore, this signaling process requires the presence of all of the domains of an ABC transporter, the Pst complex, yet does not involve the actual translocation of phosphate molecules into the cytoplasm (Rao and

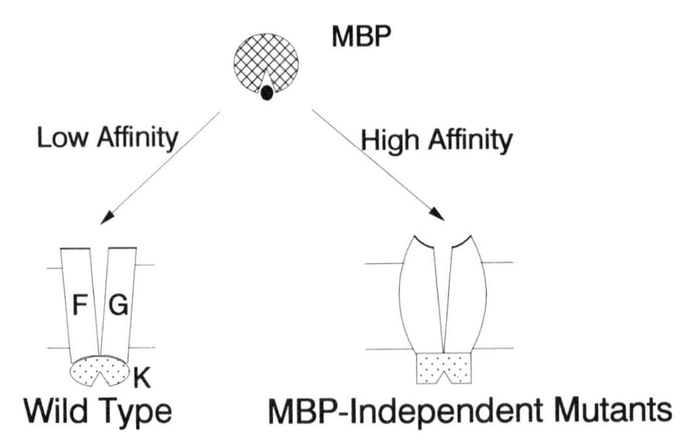

FIG. 3. Affinity between the maltose transporter complex and the ligand-attached MBP. In the binding-protein-independent mutants the external surface of the transporter complex is already in the activated conformation, which would fit snugly with the ligand-attached MBP. Because of this, the mutant transporter is expected to show a higher affinity to the ligand-attached MBP than the wild-type transporter does, a prediction borne out by the experimental results (Dean et al., 1992).

Torriani, 1990). If the presence of ligand-attached phosphate-binding protein acts as a transmembrane signal by altering the conformation of membrane-traversing channel components, PstA and/or PstC, it seems possible that the signal not only activates the PstB ATPase but also affects the PhoU.

Maltose-Binding Protein

When the cells are induced by the presence of maltose in the medium, MBP becomes one of the most abundant periplasmic proteins, with concentrations estimated to reach about 1 mM (Hengge and Boos, 1983). Practically all of the bacterial ABC transporters that function in the uptake of nutrients contain specific binding proteins, and the binding proteins were discovered even in gram-positive bacteria as protein tethered to the cell surface (Higgins, 1992). The ubiquity of the binding proteins suggests that their presence conferred significant improvement to the uptake systems. What was this evolutionary benefit? Clearly, the primary benefit was to increase the affinity of the transport system. An enzyme-like system usually has to pay for an increase in substrate affinity with a decrease in throughput. The use of binding protein, however, allowed cells to accumulate substrates with a very low K_m and still at a reasonably high rate. The maltose system, for example, can transport the substrate with a V_{max} of 40 nmol min^{-1} mg^{-1} (Szmelcman et al., 1976), with a K_m of around 1 μM (Hengge and Boos, 1983). This can be contrasted with proton symporter systems, which have to function without the benefit of binding proteins and which show vastly inferior affinity to the substrates, with reported K_m values in the range of tens to hundreds of micromoles per liter (Henderson, 1990). The evolutionary advantage of the production of binding proteins cannot be ascertained by the analysis of the contemporary functions of these proteins, because contemporary binding proteins carry out many additional functions. (Confusion on this point led some workers to a mistaken notion that the binding proteins do not increase the affinity of transport, simply on the grounds that the mutational loss of the binding proteins in contemporary systems abolishes all transport.) What are the contemporary functions of MBP? We have already mentioned the transmembrane signaling function. Another signaling function is carried out by MBP, as it interacts with the chemotaxis sensor protein Tar. The structure of MBP was solved by X-ray crystallography (Spurlino et al., 1991), and analysis of mutants showed that the face of MBP that interacts with Tar is quite different from the face that interacts with the membrane-associated transporter complex.

Recently we obtained an indication of another contemporary function of MBP. One line of evidence in this story comes from the observation by Treptow and Shuman (1985) that most of their MBP-independent mutants of *E. coli* could transport maltose but not maltodextrin. One interpretation of these data is that MBP is needed for the influx of maltodextrins into the periplasm via the LamB channel (see below) but is not necessary for the influx of maltose via the common porin channels. We have carried out transport studies with the membrane vesicles of these mutants and found that the MBP-independent transport of maltose into these vesicles was not inhibited at all by longer maltodextrins, such as maltoheptaose (Pokrovskaya and Nikaido, unpublished). Since there is no outer membrane barrier in these vesicles, the results suggest that the $MalFGK_2$ transporter from the mutants cannot transport maltodextrins. It thus seems possible that longer maltodextrins must be presented to the transporter complex by MBP, in a correctly aligned position.

In what way are these maltodextrins presented by the MBP? Although MBP binds both maltose and maltodextrins with high affinity, the binding of maltoheptaose produces a blue shift in the fluorescence emission spectrum of MBP, whereas the binding of maltose produces a marked red shift (Szmelcman et al., 1976). The molecular basis of this curious observation became clear recently with the ^3H nuclear magnetic resonance and UV spectroscopic studies of the ligand binding to MBP (Gehring et al., 1991b, 1992). Briefly, the ligands can bind to MBP in one of two alternative ways. In "end-on" binding, the reducing end glucose residue interacts strongly with the binding pocket. Maltose binds predominantly in this manner, and this is the binding mode found in the X-ray crystallographic study of the maltose-MBP complex (Spurlino et al., 1991). In contrast, the longer maltodextrin may also bind via another binding mode, "sideways." In this mode, the oligosaccharide molecule is thought to use its internal glucose residues for binding. The contribution of this binding mode explains how the binding of maltose and of maltodextrins produces different nuclear magnetic resonance signals, fluorescence emission spectra, and UV absorption spectra.

Recent evidence suggests that there are functional differences in the two binding modes. An extensive study of the transport of maltodextrin derivatives was made (Ferenci et al., 1986). These workers found that maltodextrin derivatives in which the reducing terminal glucose residue had been reduced, oxidized, or substituted on the reducing carbon could not be transported by *E. coli*, although most of these derivatives could still be bound tightly by MBP. Why do these derivatives fail to be transported? We found through fluorescence emission spectra and UV absorption spectra that reduced and oxidized maltodextrin derivatives bind to MBP essentially in the sideways mode (Hall and Nikaido, unpublished). Thus, it may be that the sideways binding is a nonproductive mode and that the maltodextrins and their derivatives must be bound end on in order to be delivered properly into the transporter channel. This hypothesis certainly can explain the lack of transport of reduced or oxidized maltodextrins despite their high binding affinity to MBP.

One more line of evidence related to this working hypothesis is the phenotype of maltodextrin-negative mutants of MBP (Wandersman et al., 1979). These mutants can still take up maltose but are incapable of transporting longer maltodextrins. Yet the mutant MBPs show only a marginally lower affinity to maltodextrins, although their affinity to maltose is drastically reduced. Earlier, this phenotype was explained on the basis of physical interaction between MBP and the LamB protein, and results consistent with this notion were noted (see, for example, Bavoil et al. [1983]). However, the influx rate of *p*-nitrophenyl-maltohexaoside was shown by a direct assay to be independent of the presence or absence of MBP (Freundlieb et al., 1988), and thus the MBP-LamB interaction observed earlier was most probably a fortuitous laboratory artifact without physiological relevance. Alternatively, Gehring et al. (1992) suggested that the mutant phenotype could be explained if all ligands bound to the mutant MBP predominantly through the sideways mode. Our preliminary results suggest that this hypothesis is indeed correct (Hall and Nikaido, unpublished). Thus our current working hypothesis assumes that ligands bind to the wild-type MBP via either the end-on or sideways mode. Even free maltose molecules can enter the channel, as shown by the behavior of the binding-protein-independent transporter complex. However, maltodextrins must be presented to the transporter in a correct orientation, and this can apparently occur only with the end-on binding mode. Although more work is necessary, these considerations suggest a way that the structure and functions of the various components of an ABC transporter can be modified to suit the requirement of a particular set of ligands. Indeed, the modular construction of the ABC transporters appears ideal for such modifications, and it is likely that many similar examples will be found in other transporters as we gain more knowledge about other members of this superfamily.

Many of the ideas leading to this work were contributed by K. Gehring. Many experiments described here were carried out by Amy L. Davidson and David A. Dean, who are now located in Baylor College of Medicine and University of California, Los Angeles, respectively.

This study was supported in part by a research grant from U.S. Public Health Service (AI-09644).

REFERENCES

Ames, G. F.-L. 1986. Bacterial periplasmic transport systems: structure, mechanism and evolution. *Annu. Rev. Biochem.* **55:**397–425.

Baichwal, V., D. Liu, and G. F.-L. Ames. 1993. The ATP-binding component of a prokaryotic traffic ATPase is exposed to the periplasmic (external) surface. *Proc. Natl. Acad. Sci. USA* **90:**620–624.

Bavoil, P., C. Wandersman, M. Schwartz, and H. Nikaido. 1983. A mutant form of maltose-binding protein of *Escherichia coli* deficient in its interaction with the bacteriophage lambda receptor protein. *J. Bacteriol.* **155:**919–921.

Bishop, L., R. Agbayani, Jr., S. V. Ambudkar, P. C. Maloney, and G. F.-L. Ames. 1989. Reconstitution of a bacterial periplasmic permease in proteoliposomes and demonstration of ATP hydrolysis concomitant with transport. *Proc. Natl. Acad. Sci. USA* **86:**6953–6957.

Davidson, A. L., and H. Nikaido. 1990. Overproduction, solubilization, and reconstitution of the maltose transport system from *Escherichia coli*. *J. Biol. Chem.* **265:**4254–4260.

Davidson, A. L., and H. Nikaido. 1991. Purification and characterization of the membrane-associated components of the maltose transport system from *Escherichia coli*. *J. Biol. Chem.* **266:**8946–8951.

Davidson, A. L., H. A. Shuman, and H. Nikaido. 1992. Mechanism of maltose transport in *Escherichia coli:* transmembrane signaling by periplasmic binding proteins. *Proc. Natl. Acad. Sci. USA* **89:**2360–2364.

Dean, D. A., A. L. Davidson, and H. Nikaido. 1989. Maltose transport in membrane vesicles of *Escherichia coli* is linked to ATP hydrolysis. *Proc. Natl. Acad. Sci. USA* **86:**9134–9138.

Dean, D. A., L. I. Hor, H. A. Shuman, and H. Nikaido. 1992. Interaction between maltose-binding protein and the membrane-associated maltose transporter complex in *Escherichia coli*. *Mol. Microbiol.* **6:**2033–2040.

Ferenci, T., M. Muir, K.-S. Lee, and D. Maris. 1986. Substrate specificity of the *Escherichia coli* maltodextrin transport system and its component proteins. *Biochim. Biophys. Acta* **860:**44–50.

Freundlieb, S., U. Ehmann, and W. Boos. 1988. Facilitated diffusion of *p*-nitrophenyl-α-D-maltohexaoside through the outer membrane of *Escherichia coli*. Characterization of LamB as a specific and saturable channel for maltooligosaccharides. *J. Biol. Chem.* **263:**314–320.

Gehring, K., K. Bao, and H. Nikaido. 1992. UV difference spectroscopy of ligand binding to maltose-binding protein. *FEBS Lett.* **300:**33–38.

Gehring, K., C.-S. Cheng, H. Nikaido, and B. K. Jap. 1991a. Stoichiometry of maltodextrin-binding sites in LamB, an outer membrane protein from *Escherichia coli*. *J. Bacteriol.* **173:**1873–1878.

Gehring, K., P. G. Williams, J. G. Pelton, H. Morimoto, and D. E. Wemmer. 1991b. Tritium NMR spectroscopy of ligand binding to maltose-binding protein. *Biochemistry* **30:**5524–5531.

Gilson, E., C. F. Higgens, M. Hofnung, G. F.-L. Ames, and H. Nikaido. 1982. Extensive homology between membrane-associated components of histidine and maltose transport systems of *Salmonella typhimurium* and *Escherichia coli*. *J. Biol. Chem.* **257:**9915–9918.

Hall, J. A., and H. Nikaido. Unpublished results.

Henderson, P. J. F. 1990. Proton-linked sugar transport systems in bacteria. *J. Bioenerg. Biomembr.* **22:**525–569.

Hengge, R., and W. Boos. 1983. Maltose and lactose transport in *Escherichia coli*. Examples of two different types of concentrative transport systems. *Biochim. Biophys. Acta* **737:**443–478.

Higgins, C. F. 1992. ABC transporters: from microorganisms to man. *Annu. Rev. Cell Biol.* **8:**67–113.

Higgins, C. F., I. D. Hiles, G. P. C. Salmond, D. R. Gill, J. A. Downie, I. J. Evans, I. B. Holland, L. Gray, S. D. Buckel, A. W. Bell, and M. A. Hermodson. 1986. A family of related ATP-binding subunits coupled to many distinct biological processes in bacteria. *Nature* (London) **323:**448–450.

Hoshino, T., K. Kose-Terai, and K. Sato. 1992. Solubilization and reconstitution of the *Pseudomonas aeruginosa* high affinity branched-chain amino acid transport system. *J. Biol. Chem.* **267:**21313–21318.

Luckey, M., and H. Nikaido. 1980. Diffusion of solutes through channels produced by phage lambda receptor protein of *Escherichia coli:* inhibition by higher oligosaccharides of maltose series. *Biochem. Biophys. Res. Commun.* **93:**166–171.

Macnab, R. M. 1987. Motility and chemotaxis, p. 732–759. *In* F. C. Neidhardt, J. L. Ingraham, K. B. Low, B. Magasanik, M. Schaechter, and H. E. Umbàrger (ed.), *Escherichia coli and Salmonella typhimurium: Cellular and Molecular Biology*, vol. 1. American Society for Microbiology, Washington, D.C.

Nikaido, H., and M. Vaara. 1985. Molecular basis of bacterial outer membrane permeability. *Microbiol. Rev.* **49:**1–32.

Petronilli, V., and G. F.-L. Ames. 1991. Binding protein-independent histidine permease mutants. Uncoupling of ATP hydrolysis from transmembrane signaling. *J. Biol. Chem.* **266:**16293–16296.

Pokrovskaya, I. D., and H. Nikaido. Unpublished results.

Rao, N. N., and A. Torriani. 1990. Molecular aspects of phosphate transport in *Escherichia coli*. *Mol. Microbiol.* **4:**1083–1090.

Reyes, L., and H. Nikaido. Unpublished results.

Shuman, H. A., and T. J. Silhavy. 1981. Identification of the *malK* gene product: a peripheral membrane component of the *E. coli* maltose transport system. *J. Biol. Chem.* **256:**560–562.

Spurlino, J. C., G.-Y. Lu, and F. A. Quiocho. 1991. The 2.3 Å resolution structure of the maltose- or maltodextrin-binding protein, a primary receptor of bacterial active transport and chemotaxis. *J. Biol. Chem.* **266:**5202–5219.

Szmelcman, S., M. Schwartz, T. J. Silhavy, and W. Boos. 1976. Maltose transport in *Escherichia coli* K12. A comparison of transport kinetics in wild-type and λ-resistant mutants with the dissociation constants of the maltose-binding protein as measured by fluorescent quenching. *Eur. J. Biochem.* **65:**13–19.

Treptow, N. A., and H. A. Shuman. 1985. Genetic evidence for substrate and periplasmic-binding-protein recognition by the MalF and MalG proteins, cytoplasmic membrane components of the *Escherichia coli* maltose transport system. *J. Bacteriol.* **163:**654–660.

Walter, C., K. Honer zu Bentrup, and E. Schneider. 1992. Large scale purification, nucleotide binding properties, and ATPase activity of the MalK subunit of *Salmonella typhimurium* maltose transport complex. *J. Biol. Chem.* **267:**8863–8869.

Wandersman, C., M. Schwartz, and T. Ferenci. 1979. *Escherichia coli* mutants impaired in maltodextrin transport. *J. Bacteriol.* **140:**1–13.

Chapter 17

The Arsenite Oxyanion-Translocating ATPase: Bioenergetics, Functions, and Regulation

BARRY P. ROSEN,[1] SIMON SILVER,[2] TATIANA B. GLADYSHEVA,[1] GUANGYONG JI,[2] KRISTINE L. ODEN,[1] SUCHITRA JAGANNATHAN,[2] WEIPING SHI,[1] YAJING CHEN,[2] AND JIANHUA WU[1]

Department of Biochemistry, Wayne State University School of Medicine, Detroit, Michigan 48201,[1] and Department of Microbiology and Immunology, University of Illinois College of Medicine, Chicago, Illinois 60612-7344[2]

The continuing use of herbicidal, insecticidal, and medical arsenicals, as well as release from natural sources, results in the widely disseminated presence of arsenic in the environment (World Health Organization, 1980; Knowles and Benson, 1983). To give just one example, calcium methylarsonate is the sole active ingredient of Ortho Crabgrass Killer, Formula II. Few biological systems can distinguish between arsenate and phosphate. Thus, arsenate acts as a phosphate analog in many enzymatic reactions. The Pst phosphate transport system of *Escherichia coli* is an exception and favors phosphate over arsenate by perhaps 100:1, and the ArsC protein of the bacterial plasmid-borne arsenical resistance (*ars*) operon (see below) is uniquely able to favor arsenate over phosphate (Rosen et al., 1992, in press; Silver et al., 1993a). This is reasonable, since an enzyme without specificity for arsenate could not provide resistance. Proteins that are able to discriminate between arsenate and phosphate should have interesting chemical properties, as already known for the PhoS phosphate-binding protein (Luecke and Quiocho, 1990).

The *ars* operons of plasmids of both gram-negative and gram-positive bacteria confer resistance to arsenate, arsenite, and antimonite. Although As(V) and As(III) both form oxyanions (arsenate and arsenite, respectively), these are chemically different. Arsenite and antimonite, both +3 oxidation state, are oxyanions that are chemically more similar. In solution, arsenate has three ionization states, with arsenic acid (H_3AsO_4) ionizing to $H_2AsO_4^-$ ($pK_1 = 2.25$), to $HAsO_4^{2-}$ ($pK_2 = 6.77$), and to AsO_4^{3-} ($pK_3 = 11.60$). (For convenience we will refer to arsenate as AsO_4^{3-}, although mono- and diprotonated oxyanions are more abundant under physiological conditions and are expected to be the substrates for protein binding.) In contrast, the solution chemistry of arsenite is not as clear; from Raman spectroscopy of arsenite, the predominant form appears to be pyramidal $As(OH)_3$ (Loehr and Plane, 1968; Szymanski et al., 1968), which has a pK of 9.23, ionizing to $As(OH)_2O^-$. Thus, at physiological pH, the anionic species is only a small fraction of the total arsenite. The solution chemistry of antimonite is even less well understood. The form used for the studies described below is antimony potassium tartrate [$K(SbO)C_4H_4O_6$], but it is reasonable to expect to find species common between arsenite and antimonite, since the same proteins produce resistance to both oxyanions. Without knowledge of the detailed chemistry of the substrates, it is difficult to propose precise mechanisms for their interactions with proteins.

The most frequently found mechanism of microbial resistance to heavy metal ions is transport out from the cells by specific membrane carrier proteins (Kaur and Rosen, 1992; Tisa and Rosen, 1990a; Silver and Ji, in press; Silver et al., 1993b). In *Staphylococcus aureus* and *E. coli*, there are plasmids that confer resistance to arsenicals and antimonials (Hedges and Baumberg, 1973; Novick and Roth, 1968). Plasmid-bearing resistant cells accumulate less $^{74}AsO_4^{3-}$ than do sensitive cells (Silver et al., 1981). This was shown to be the result of arsenic efflux systems (Mobley and Rosen, 1982; Silver and Keach, 1982). In *E. coli*, plasmid-mediated arsenic efflux is coupled to ATP energy and not to the electrochemical proton gradient (Mobley and Rosen, 1982; Rosen and Borbolla, 1984); however, in the *S. aureus* system a chemiosmotic system dependent on the membrane potential seems to accomplish the same result (Bröer et al., 1993).

The sequenced *ars* operon from *E. coli* plasmid R773 contains five genes: two regulatory genes (*arsR* and *arsD*) and three structural genes (*arsA*, *arsB*, and *arsC*) (Fig. 1) (Chen et al., 1986; San Francisco et al., 1990; Wu and Rosen, 1991, 1993a, 1993b). The *ars* operons sequenced from plasmids pI258 and pSX267 of different staphylococcal species both contain essentially identical *arsR*, *arsB*, and *arsC* genes (Fig. 1) (Ji and Silver, 1992a; Rosenstein et al., 1992). There have been

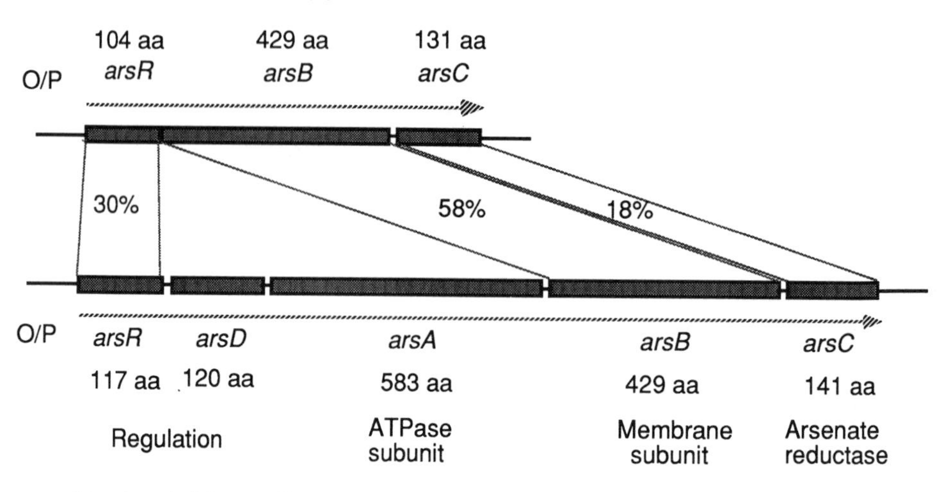

FIG. 1. *ars* operons of gram-negative and gram-positive bacteria. The genes are shown by shaded blocks, with the product lengths in amino acids (aa) above or below. O/P, operator-promoter regions (Fig. 2). Transcriptional direction and length are shown by arrows. Modified from Silver and Walderhaug (1992) with permission.

major surprises and changes in our understanding of the arsenic resistance system in the last 2 years, which will be summarized here. First, the two staphylococcal *ars* operons both lack the *arsA* gene, the determinant of the ATPase subunit of the efflux complex of plasmid R773. It appears that (as will be described below) the *ars* system may function as an ATPase in the presence of the ArsA protein, but with chemiosmotic energy coupling (responding to the membrane potential) in the absence of the ArsA protein. The second unexpected finding is that the product of the *arsC* gene is an arsenate reductase enzyme, converting intracellular As(V) to As(III). This explains why the ArsC product is needed for arsenate resistance but not for arsenite resistance (Chen et al., 1985, 1986) and explains the in vitro specificity of both the ArsA ATPase subunit and the ArsR DNA-binding repressor protein for arsenite.

A third change of our overall thinking concerns how the arsenic resistance ATPase fits with other ATPase efflux systems. It was considered that the Ars system is an "orphan" (Silver et al., 1993a) because of the absence of close homologies to other ATPases, such as the family of P-type ATPases which includes the Cd^{2+} efflux ATPase (Silver et al., 1993b) determined by the same staphylococcal plasmid, pI258, that contains the *ars* arsenic efflux determinant. However, there are no orphan genes or proteins. The lineage of every coding sequence will be traced to a family of related sequences once sufficient sequences are known. The Ars system may be the first well-studied member of a new family of efflux systems. (See Addendum in Proof.)

The ArsR-ArsD Regulatory Circuit

The *ars* operons of plasmids R773 and pI258 are tightly regulated, as are essentially all plasmid metal resistance determinants (Silver and Walderhaug, 1992). In *E. coli,* the plasmid R773 ArsR protein forms a substrate-inducible dimeric repressor that controls the basal level of expression of the operon (Wu and Rosen, 1991, 1993a). In contrast, the ArsD protein apparently serves as a low-affinity substrate-independent repressor that controls the upper level of operon expression (Wu and Rosen, 1993b). The ArsD protein may then prevent overproduction of the ArsB protein, an inner membrane protein (San Francisco et al., 1989) which is toxic when produced in large amounts (Owolabi and Rosen, 1990; Wu and Rosen, 1993b). The concerted action of these two proteins forms a regulatory circuit that provides homeostatic control of the levels of the *ars* gene products (Fig. 2). That being the case, it is surprising that the *arsD* gene and its protein product are missing from the staphylococcal *ars* systems (Ji and Silver, 1992a; Rosenstein et al., 1992) which contain the *arsR* determinant of the transcriptional repressor protein (Rosenstein et al., 1992; Y. Chen and Ji, unpublished). A further difference between the regulation of the *ars* operons of gram-negative and gram-positive bacteria is in the sequence and the positioning of the re-

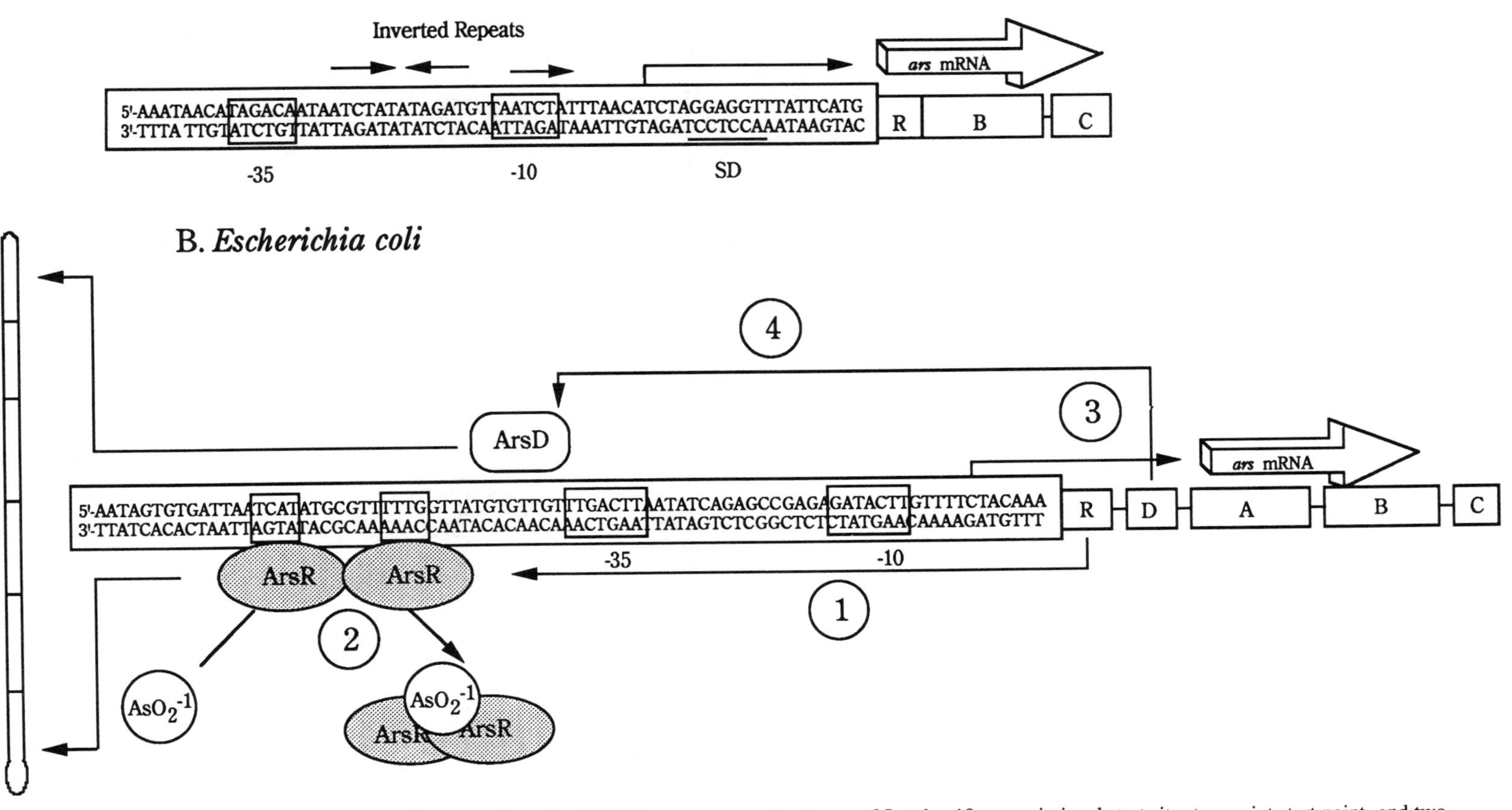

FIG. 2. *ars* operon regulatory circuits. (A) *S. aureus* plasmid pI258. The *ars* operon structure, −35 and −10 transcriptional start site, transcript start point, and two direct repeat hexanucleotides, plus one inverted repeat (presumed binding site for ArsR) are indicated. (B) *E. coli* plasmid R773. The *ars* operon genes and transcription are as in Fig. 1. Constitutive low-level expression of the operon allows synthesis of the small amounts of ArsR protein. The ArsR protein dimerizes to form the active repressor, which binds to the operator; the operator consists of two shaded sequences upstream of the −35 promoter region. Upon induction, the ArsR repressor is released from the operator, allowing transcription. ArsD protein binds to a site within the promoter region, preventing further transcription. Thus, the ArsD protein controls the upper level of *ars* expression.

pressor (operator) site with respect to the initiation site of mRNA synthesis. The operator-promoter regions of the two systems are not recognizably related (Fig. 2), and the ArsR proteins are only 30% identical in their amino acid sequences (Fig. 1 and 3). For the R773 system in *E. coli,* the site of ArsR binding is a nonrepeat tetranucleotide sequence about 20 nucleotides upstream from the −35 region of RNA polymerase binding (Fig. 2B). For the staphylococcal system, an unrelated hexanucleotide triad (consisting of a direct repeat from −27 to −22 and −6 to −11, with an inverted repeat hexanucleotide from −16 to −21, relative to the transcriptional start site, i.e., downstream from the −35 and overlapping the −10 site) is considered a potential binding site for the staphylococcal ArsR (Fig. 2A).

In vivo inducers of the *ars* operon include arsenite As(III), antimonite Sb(III), bismuth Bi(III), and arsenate As(V). In contrast, in vitro ArsR protein-operator interaction was not decreased by addition of arsenate, indicating that arsenate is not a true inducer. As discussed below, arsenate is reduced in vivo to form arsenite, the actual inducer.

The ArsR protein is the best understood member of a newly recognized subfamily of metal-responsive regulatory proteins (Fig. 3). Although ArsR is not closely homologous to the large class of helix-turn-helix regulatory proteins, Bairoch (1993) proposed that it represents a new helix-turn-helix subfamily, with the two cysteine residues in the helix-turn-helix region (Fig. 3) becoming distorted by binding arsenite, leading to release of the protein from its cognate DNA. Shi et al. (submitted) have shown that mutations changing either cysteine of the R773 ArsR protein (Fig. 3) results in loss of inducer

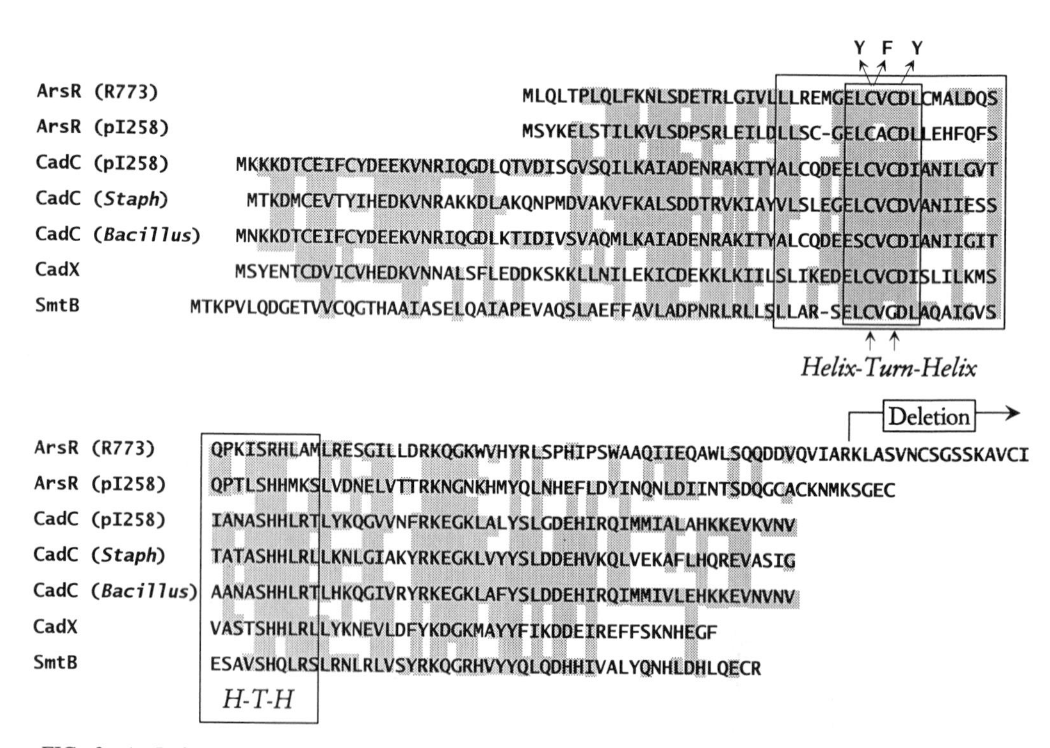

FIG. 3. ArsR family of metalloregulatory proteins. Identities and conservative amino acid replacements are shaded. The inner box shows the highly conserved region with two vicinal cysteines (arrows below) (in all but the SmtB protein). The outer box indicates the presumed helix-turn-helix motif for the DNA-binding region (Bairoch, 1993). Shown are the ArsR proteins of plasmids R773 (San Francisco et al., 1990) and pI258 (Ji and Silver, 1992a), the CadC protein also from plasmid pI258 (Nucifora et al., 1989), the chromosomally encoded CadC proteins from *S. aureus* (Chikramane and Dubin, personal communication) and *B. firmus* (Ivey et al., 1992), the staphylococcal plasmid CadX protein (Dyke, personal communication), and the metallothionein regulatory protein SmtB of *Synechococcus* species (Morby et al., 1993). The ArsR proteins of the gram-positive plasmids pI258 and pSX267 (not shown) (Rosenstein et al., 1992) are identical except for one residue. The nature of three mutations of cysteines (positions 32 and 34) in the R773 ArsR protein plus a deletion from amino acid 102 of R773 (including two C-terminal cysteines) are indicated (Shi et al., submitted).

recognition without affecting ArsR binding to DNA.

The question of the oligomeric form of the ArsR repressor that binds to the *ars* operator-promoter region was addressed by construction of a gene fusion between *arsR* and *blaM*. The chimeric protein contained most of the ArsR protein and all of the mature form of β-lactamase. This chimera was larger than the native ArsR protein, and the dimer formed between the two was observed as a new species intermediate in mass between them (Wu and Rosen, 1993a). Higher-order oligomers were not observed, indicating that the dimer is the active form of the repressor. Heterodimeric ArsR was found in vitro only after coexpression of *arsR* and the *arsR-blaM* fusion gene in the same cell (Wu and Rosen, 1993a). Heterodimers were not formed upon mixing of the wild-type and chimeric ArsR proteins in vitro, indicating that the dimer forms prior to binding to the operator DNA.

From DNase I footprinting analysis, the binding site for R773 ArsR was defined as a region of imperfect dyad symmetry spanning nucleotides (nt) −64 to −40 relative to the transcriptional start of the *ars* mRNA (Wu and Rosen, 1993a), upstream of a canonical −35 region of the promoter site (Fig. 2B). Only two short stretches within that region were protected during hydroxyl radical footprinting, i.e., the 4-bp regions indicated in Fig. 2B. This region was shown to be the target for the ArsR repressor protein in vivo by deletion of nt −105 to −43, which includes most of the ArsR-protected sequence. The cells harboring this deletion plasmid showed constitutive expression of the *ars* operon.

To function, the ArsR protein must have segments involved in DNA binding, inducer binding, and dimer formation. Bairoch (1993) postulated that the first two of these functions will involve the same amino acid segment (indicated as "helix-turn-helix" in Fig. 3). A series of gene fusions was constructed between the *arsR* gene and *blaM*, allowing production of chimeric ArsR–β-lactamase proteins. Chimeras containing 102 or more residues derived from the ArsR protein were fully functional as repressors and retained inducibility (Wu and Rosen, 1991). Chimeras containing 83 or fewer ArsR residues were inactive. A H50Y mutation (in the second helix region of the helix-turn-helix motif) resulted in a constitutive phenotype of expression in the absence of inducer. These results suggest that the N-terminal half of the ArsR protein functions in DNA and inducer binding. A positive selection involving resistance to expression of the toxic ArsB protein was devised for mutant ArsR proteins: those that still formed an active repressor but no longer responded to inducer (Shi et al., 1993). Three mutants involving two of the four cysteinyl residues in the ArsR protein were isolated. Mutants C32Y, C32F, and C34Y (Fig. 3) each had the noninducible phenotype. Each mutant protein was still capable of binding at the operator site on the DNA, but the response to inducer in vitro was greatly reduced. Thus, this vicinal cysteine pair may be the inducer-binding site, as predicted. This cysteine pair is highly conserved in members of the ArsR family (Fig. 3) and has been postulated to be the oxyanion-responsive site of ArsR (Shi et al., submitted). The ArsR protein contains three other cysteinyl residues, C-37, C-108, and C-116. The last two are not present in ArsR–β-lactamase chimeras that are fully functional repressors. Thus, they could not be involved in repressor function. The role of Cys-37 has not been evaluated. The residues involved in interaction with the DNA are not known from direct experiments, although computer analysis suggests that residues 31 to 52 may form a helix-turn-helix.

The second regulatory gene in the *ars* operon, *arsD*, also encodes a regulatory protein (Wu and Rosen, 1993b). The 13-kDa ArsD protein has not been shown to bind to the DNA operator-promoter region, and the protein contains no identifiable DNA-binding motif. However, a frameshift mutation within the *arsD* gene resulted in elevated levels of expression of downstream *ars* genes. When the wild-type *arsD* gene was expressed in *trans* with the operon containing the mutated *arsD* gene, the level of expression of the downstream genes reverted to wild type (Wu and Rosen, 1993b). The *arsD* gene did not, however, affect the low basal level of operon expression set by the *arsR* gene product. Moreover, the *arsD* effect was not affected by inducers of the operon. These results indicate that the ArsD protein is an inducer-independent *trans*-acting regulatory protein that controls the upper level of expression of the *ars* genes. The *arsD* gene was shown to repress a construct in which the *blaM* gene was fused less than 50 bp after the start of transcription, indicating that the site of ArsD action is within the operator-promoter region (Wu and Rosen, 1993b). Moreover, since the effect of *arsD* was observed only after sufficient *ars* mRNA was produced to make sufficient amounts of the ATPase to produce resistance, the ArsD protein is presumed to be a repressor protein with low affinity for the promoter region. The necessity for regulation of high-level expression of the pump proteins can be explained by the toxicity of the ArsB protein when it is overproduced. Thus, the ArsR repressor sets the floor for expression, and the ArsD repressor sets the ceiling.

The Ars Pump

The ATP-coupled efflux pump that is responsible for resistance to arsenite and antimonite in *E. coli* has been discussed in several recent reviews

(Kaur and Rosen, 1992; Silver et al., 1993a; Rosen et al., 1992, in press) and will not be considered in detail. The pump from plasmid R773 is composed of two polypeptides, the ArsA and ArsB proteins (Fig. 4). The ArsA protein is the ATPase subunit and is an arsenite- or antimonite-stimulated ATPase (Hsu et al., 1991; Rosen et al., 1989). ArsA appears to have evolved by duplication and fusion of a gene half the size of the existing *arsA* gene (Chen et al., 1986) and belongs to a family of soluble nucleotide-binding proteins with a variety of intracellular functions. The ATP-binding motif is most closely related to those involved in nitrogen fixation (Chen et al., 1986; Silver et al., 1989). Interestingly, the ArsA protein is the only known member of the family with a transport function. Other homologs include prokaryotic replication proteins (Nishiguchi et al., 1987), the MinD protein involved in septum formation (de Boer et al., 1989), and the P1 phage replication protein ParA (Davis et al., 1992). The most closely related homologs to ArsA are open reading frames (ORFs) (of unknown function) in the nematode *Caenorhabditis elegans* (Sulston et al., 1992) (Fig. 5). On the other hand, the ArsA protein is unrelated in overall sequence to other known transport ATPases.

The ArsA protein is found (as a dimer, as shown in Fig. 4) in a complex with the ArsB protein in the inner membrane of *E. coli* (Tisa and Rosen, 1990b). Only a few homologs of the ArsB

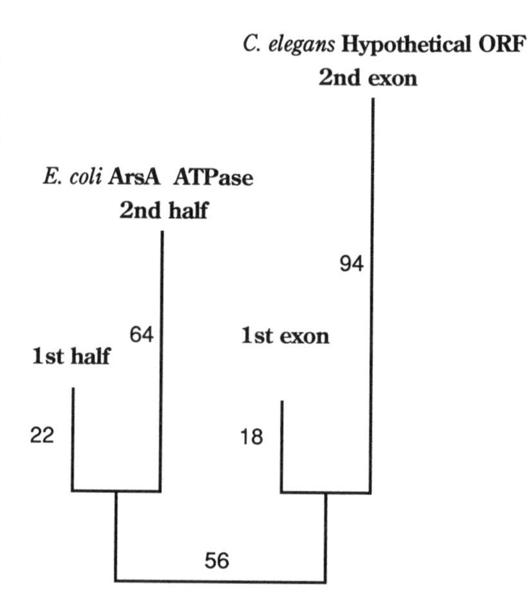

FIG. 5. Relationships among the ArsA family of ATPase proteins (from the "alignment and trees" programs of Feng and Doolittle [1990]). The distances are given in arbitrary units that balance both amino acid identities and substitution by related amino acids. The segments of the ArsA sequence (amino acids 12 through 159 and 331 through 464) that are homologous to the *C. elegans* ORFs are shown separately, as are the two *C. elegans* ORFs of unknown function (Sulston et al., 1992).

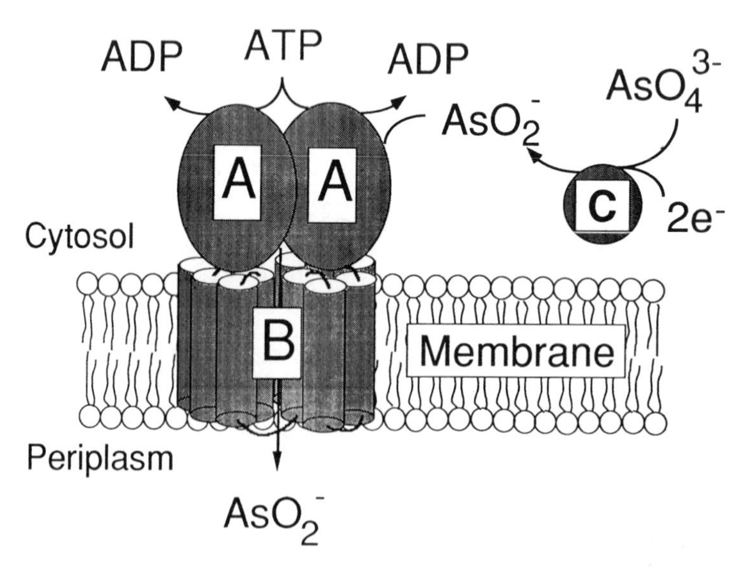

FIG. 4. Plasmid-encoded oxyanion pump. The complex of the ArsA and ArsB proteins forms an oxyanion-translocating ATPase that catalyzes extrusion of oxyanions of arsenite. Two subunits of the ArsA protein, the oxyanion-stimulated ATPase, are indicated to reflect its structure in solution, but the stoichiometry of the ArsA and ArsB proteins in the complex has not been determined. The inner membrane ArsB protein serves both as the membrane anchor for the ArsA protein and as the anion-conducting subunit of the pump. The ArsC protein is the arsenate reductase enzyme.

protein have been identified, and none is very close (Fig. 6). The closest homologs of R773 ArsB are the ArsB proteins of plasmids pI258 and pSX267 of staphylococci (58% identical amino acid positions [Fig. 1]), but several related reading frames of unknown function have recently been identified. One potentially encodes a 45-kDa hydrophobic protein in *Mycobacterium leprae* (21% identical amino acids to ArsB [GenBank accession number L10660]) (Oskam et al., unpublished), and the others are involved in the disease type II oculocutaneous albinism in humans and mice (17% identical positions) (Rinchik et al., 1993). Another new ORF product from *Bacillus subtilis* is 32 to 35% identical with ArsB proteins for the first 177 amino acids of sequence (the DNA sequence corresponding to the remaining 250 amino acids has not been determined). This ORF was recognized by Baum (personal communication) in the 5′ end (the first 532 nt) of the sequence of Renna et al. (1993). Assuming two frameshift errors in the published sequence (Baum, personal communication), this closely homologous ORF starts with a canonical ribosome-binding site 185 nt upstream from the end of *aslR*. The remainder of the presumed *arsB* homolog ORF would have been disrupted by the transposon mutagenesis used to clone the *alsSDR* genes (Renna et al., 1993).

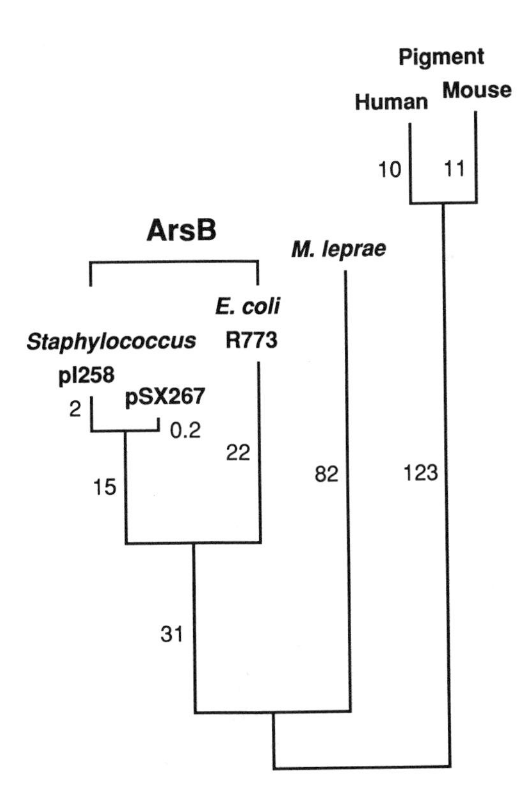

FIG. 6. Relationships between the ArsB membrane carrier proteins and related proteins (see the legend to Fig. 5). The three ArsB proteins form a close cluster with 58% amino acid identities between the R773 ArsB and those of staphylococcal plasmids pI258 and pSX267. The postulated *M. leprae* protein of unknown function (GenBank no. L10660) and the mammalian pigment-related proteins (PIR no. S28911) are also only postulated to be transport proteins.

Topological analysis of the ArsB protein by using gene fusions to three different reporter genes demonstrated that the protein has 12 membrane-spanning α-helices, with 6 periplasmic and 5 cytoplasmic loop regions and with the N and C termini both oriented to the cytoplasmic side of the inner membrane (Wu et al., 1992). Although the ArsB protein of plasmid R773 functions as a subunit of a primary pump in the oxyanion-translocating ATPase, its structure is more similar to that of secondary chemiosmotic porters (Marger and Saier, 1993) than to that of other transport ATPases. The ArsB protein of the staphylococcal *ars* operon functions in resistance without an ArsA subunit and as a chemiosmotic efflux system (Bröer et al., 1993). Therefore, we have postulated that the Ars efflux ATPase arose from the parallel, independent evolution of genes for a soluble ATPase and for a chemiosmotic membrane protein and that the two protein products combined in gram-negative bacteria (but not in gram-positive *ars* systems) to form the R773 ATPase efflux pump (Rosen et al., 1992, in press; Silver et al., 1993a). It is intuitively reasonable that transport systems composed of multiple subunits or even multiple domains on a single polypeptide (as mammalian multidrug resistance) did not emerge de novo. Ancestral genes encoded proteins whose physiological roles could have been different from their present transport functions. The ancestor of the ArsA protein probably was a soluble ATPase involved in a cytosolic process, perhaps in cell division, which requires control over membrane structure but not necessarily over membrane transport. The ancestor of the R773 ArsB protein was probably a chemiosmotic anion carrier coupled to the membrane potential, as are the present-day ArsB proteins from staphylococcal plasmids. Joining of these two proteins provides two advantages. First, ATP binding by ArsA appears to be cooperatively controlled by binding of the oxyanion substrate (Hsu et al., 1991; Kaur and Rosen, unpublished), providing metabolic control to the pump. Secondly, primary ATPase pumps are more efficient mechanisms for creating large concentration gradients than are chemiosmotic secondary transport systems (Rosen and Kashket, 1978). Since ATP levels fall under depolarizing conditions more slowly than the membrane potential does, an ATPase would be less sensitive to the uncoupling action of arsenicals. The addition of the R773 ArsA subunit to the staphylococcal ArsB protein increases the level of arsenite resistance (Bröer et al., 1993), consistent with this hypothesized evolutionary pathway. Furthermore, the R773 ArsB protein (which functions as an obligatory ATP-coupled pump when the ArsA protein is bound) works as a potential-driven chemiosmotic carrier in the absence of the ArsA protein (Dey and Rosen, unpublished). These results all suggest a model for the oxyanion transporter, which can function either as a primary pump or as a secondary carrier, depending on the presence or absence of the ATPase subunit.

The ArsC Arsenate Reductase Enzyme

The ArsC protein is required for resistance to arsenate but not to arsenite or antimonite (Chen et al., 1985, 1986; Ji and Silver, 1992a). Although the ArsC protein was originally suggested to be a pump subunit that altered its substrate specificity, Ji and Silver (1992b) showed that it is an enzyme that converts arsenate (a nonsubstrate of the *ars* membrane pump) to the actual substrate arsenite.

The metabolic coupling of the pI258 and R773 ArsC proteins (which share only 18% identical amino acids [Fig. 1]) to arsenate reduction is not the same. Figure 7 shows our current model, and Fig. 8 presents typical data supporting the hypothesis that ArsC couples to redox energy by thioredoxin with the staphylococcal ArsC (Ji and Silver, 1992b; Ji et al., 1994) or with glutaredoxin with the *E. coli* plasmid R773 ArsC (Gladysheva et al., 1994). With purified R773 ArsC, arsenate-dependent ArsC protein-dependent oxidation of NADPH was coupled through glutathione and glutaredoxin (Fig. 8A) (Gladysheva et al., 1994). In parallel experiments with the pI258 ArsC protein, coupling with glutathione and glutaredoxin did not occur (Fig. 8A), but, rather, arsenate reduction was coupled to NADPH oxidation via thioredoxin and thioredoxin reductase (Fig. 8B) (Ji et al., 1994). In parallel assays of the reduction of $^{73}AsO_4^{3-}$ to $^{73}AsO_2^{-}$, there was good agreement between the oxidation of NADPH and reduction of arsenate by R773 ArsC. The coupled reaction required glutathione reductase and glutaredoxin (R773) or thioredoxin and thioredoxin reductase (pI258) as well as the ArsC protein. It is of interest here that glutathione, which is the major intracellular thiol compound of gram-negative bacteria and eukaryotic cells, is not found in gram-positive bacteria (Newton and Fahey, 1990), such as the staphylococcal host of plasmid pI258. The ArsC proteins of gram-positive and gram-negative bacteria are sufficiently different in sequence that their energy coupling need not be the same in detail.

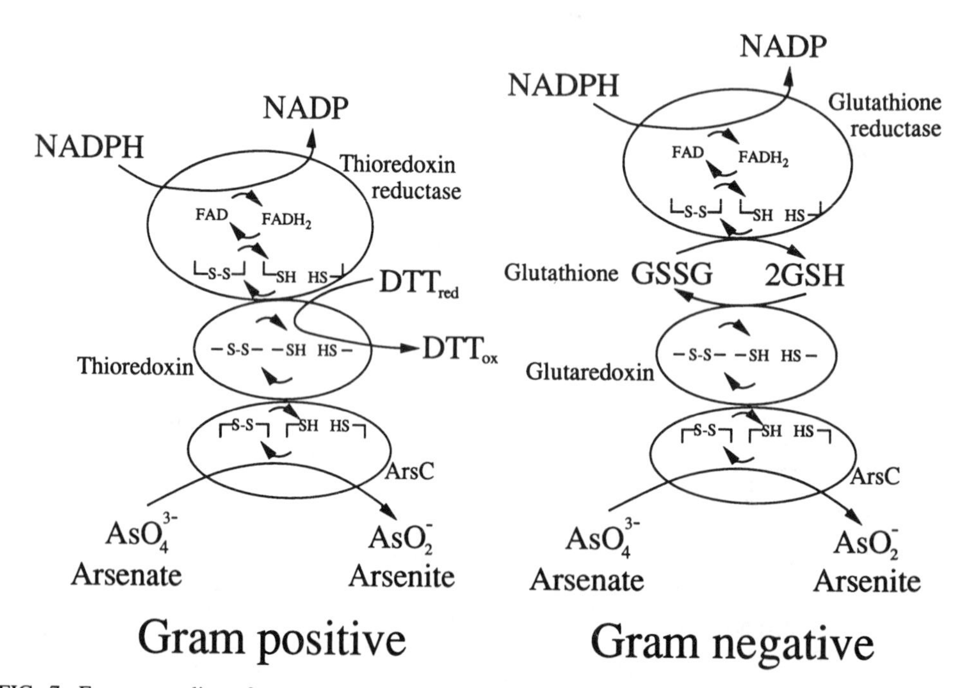

FIG. 7. Energy coupling of gram-positive and gram-negative arsenate reductases to oxidative (NADPH) and reduced dithiol (SH) energy. DTT, reduced or oxidized dithiothreitol. GSH and GSSG, reduced and oxidized glutathione, respectively.

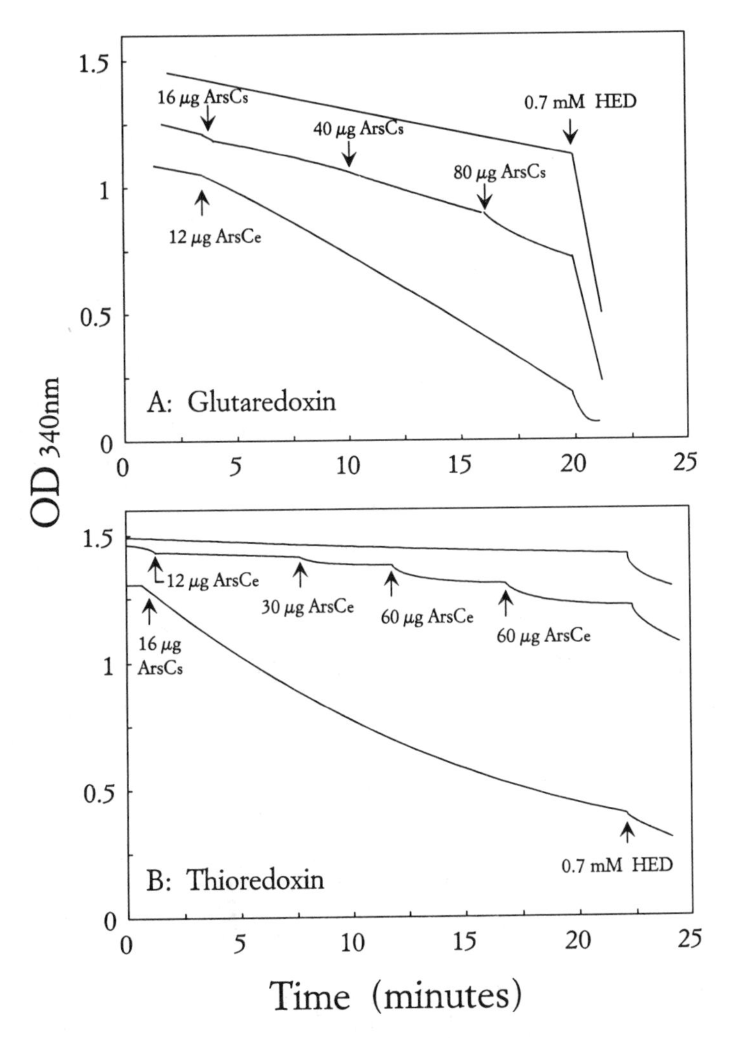

FIG. 8. Oxidation of NADPH coupled to arsenate reduction by purified proteins (Gladysheva and Ji, unpublished). Assay mixtures were monitored by measuring the optical density at 340 nm at 37°C with added 250 μM NADPH and 10 mM arsenate. The upper curves in panels A and B lacked ArsC protein. Purified *E. coli* (R773; ArsCe) or *S. aureus* (pI258; ArsCs) ArsC protein was added as indicated by the arrows, as well as 5 mM reduced glutathione, 10 μg of glutaredoxin, and 50 nM glutaredoxin reductase (A) or 15 μg of thioredoxin and 10 μg of thioredoxin reductase (B). At the end of each assay, 0.7 mM 2-hydroxyethyl disulfide (HED) was added. See Gladysheva et al. (1994) and Ji et al. (1994) for further details.

As expected, phosphate is not an inhibitor of arsenate reduction by ArsC (Ji et al., 1994; Gladysheva et al., 1994). Thus, the ArsC protein is uniquely capable of discriminating between arsenate and phosphate, even more so than the *E. coli* Pst system. While the biochemical properties of the arsenate reductase from bacterial systems are being carefully characterized in our laboratories, it is worth noting that similar activities have been found in mammalian livers. Rabbit liver arsenate reductase has recently been purified 10^3-fold (Aposhian, personal communication).

Potential proteins related to the ArsC protein have been identified through DNA sequence analysis. The closest homolog of the *E. coli* R773 ArsC protein is an ORF near the *Streptomyces* determinant of glutamine synthetase that also confers resistance to the antibiotic bialaphos (Behrmann et al., 1990). The unidentified product of this gene would be 31% identical in sequence to the R773 ArsC (Fig. 9). Since *Streptomyces* species are the major antibiotic producers and have genes for resistance to their own antibiotics, it is tempting to speculate that the ORF is part of an

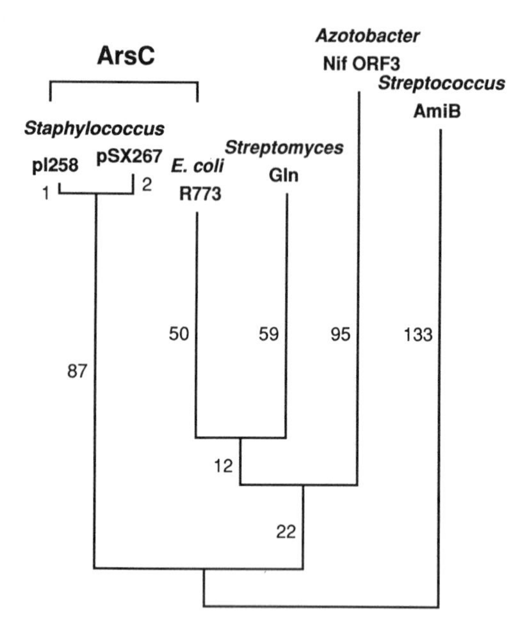

FIG. 9. ArsC family of proteins. The staphylococcal arsenate reductases are closely clustered (97% identical), whereas the R773 arsenate reductase is more closely homologous to *Streptomyces* and *Azotobacter* proteins of unknown function.

operon related to the evolutionary origin of the *ars* operon. This ORF (of unknown function) is related to the R773 ArsC but is only 14% identical with the staphylococcal ArsC proteins in amino acid sequence (although *Streptomyces* and *Staphylococcus* are both gram-positive genera). The ArsC proteins are also weakly related to an ORF of unknown function in a *nif* gene cluster from *Azotobacter vinelandii* (Joerger and Bishop, 1988) and *Rhodospirillum rubrum* (Fitzmaurice et al., 1989). Interestingly, the R773 ArsC protein is less closely related to the pI258 ArsC protein than to these other homologs, even though the two ArsC proteins have the same function.

The *E. coli* ArsC protein has been purified and crystallized (Rosen et al., 1991). It contains no prosthetic groups for electron transfer. Thus, cysteinyl residues are the logical candidates for participants in redox reactions. However, there is a surprising lack of consensus positioning of cysteine residues between the R773 and pI258 ArsC proteins. The pI258 (and pSX267) ArsC sequences contain four cysteinyl residues, whereas the R773 ArsC protein contains only two cysteines, one of which is conserved in both versions, Cys-12 in the R773 numbering. The homologous position in the pI258 ArsC protein contains C-10–TGNSC-15 as a vicinal disulfide pair consistent with the roles pictured in Fig. 7. The second cysteine residue in the R773 ArsC sequence, Cys-106, is positioned approximately in the same region as Cys-82 of the somewhat shorter ArsC sequence of plasmids pI258 and pSX267. If these two cysteinyl residues do undergo a dithiol oxidation-reduction cycle, mutants altered in these positions should be informative. C. M. Chen and Ji (unpublished) have generated Cys-to-Ala mutations in the Cys-82 and Cys-89 positions of the pI258 ArsC and tested in vitro arsenate reductase activity. The C82A mutation eliminated more than 90% (perhaps 100%) of arsenate reductase activity, whereas mutation C89A retained more than 80% of wild-type ArsC in vitro activity. Further mutational analysis, especially of Cys-10 and Cys-15, will be essential for understanding of the models in Fig. 7. However, it has not been directly demonstrated yet whether any cysteinyl residue participates in catalysis, as proposed.

The work in our laboratories was supported by grants from the U.S. Public Health Service (CA54141), the National Science Foundation (DCB86-04781), and the Department of Energy (91ER20056).

Addendum in Proof

A remarkable example of how the arsenic resistance system is moving from orphan to patriarch occurred while this book was in production. The release of 225,419 bp of the *E. coli* chromosome (from minutes 76 to 81.5 on the genetic map; GenBank accession U00039; locus ECOUW76) includes three open reading frames, tentatively labelled *arsE*, *arsF*, and *arsG* (continuing after *arsD*). The predicted protein products of these open reading frames are 74, 91, and 94%, respectively, identical to ArsR, ArsB, and ArsC of the *E. coli* plasmid R773! The as-yet-untested *E. coli* chromosomal version of the *ars* operon is missing the *arsA* gene, as is the staphylococcal version in Fig. 1. The occurrence of arsenic resistance systems with and without the ATPase subunit may mean that the Ars systems are in a sense an intermediate link between the ATP-driven transporters (for example Pst system) and proton motive force-driven transporters.

REFERENCES

Aposhian, V. Personal communication.

Bairoch, A. 1993. A possible mechanism for metal-ion induced DNA-protein dissociation in a family of prokaryotic transcriptional regulators. *Nucleic Acids Res.* **21:**2515.

Baum, B. Personal communication.

Behrmann, I., D. Hillemann, A. Puhler, E. Strauch, and W. Wohlleben. 1990. Overexpression of a *Streptomyces viridochromogenes* gene (*glnII*) encoding a glutamine synthase similar to those of eukaryotes confers resistance against the antibiotic phosphinothricyl-alanyl-alanine. *J. Bacteriol.* **172:**5326–5334.

Bröer, S., G. Ji, A. Bröer, and S. Silver. 1993. Arsenic efflux governed by the arsenic resistance determinant of *Staphylococcus aureus* plasmid pI258. *J. Bacteriol.* **175:**3840–3845.

Chen, C. M., and G. Ji. Unpublished results.

Chen, C. M., H. L. T. Mobley, and B. P. Rosen. 1985. Separate resistances to arsenate and arsenite (antimonite) encoded by the arsenical resistance operon of R-factor R773. *J. Bacteriol.* **161:**758–763.

Chen, C. M., T. Misra, S. Silver, and B. P. Rosen. 1986. Nucleotide sequence of the structural genes for an anion pump: the plasmid-encoded arsenical resistance operon. *J. Biol. Chem.* **261**:15030–15038.

Chen, Y., and G. Ji. Unpublished results.

Chikramane, S. G., and D. T. Dubin. Personal communication.

Davis, M. A., K. A. Martin, and S. J. Austin. 1992. Biochemical activities of the ParA partition protein of P1 plasmid. *Mol. Microbiol.* **6**:1141–1147.

de Boer, P. A. J., R. E. Crossley, and L. I. Rothfield. 1989. A division inhibitor and a topological specificity factor encoded for by the minicell locus determine proper placement of the division septum in *E. coli*. *Cell* **56**:641–649.

Dey, S., and B. P. Rosen. Unpublished results.

Dyke, K. Personal communication.

Feng, D.-F., and R. L. Doolittle. 1990. Progressive alignment and phylogenetic tree construction of protein sequences. *Methods Enzymol.* **183**:375–387.

Fitzmaurice, W. P., L. L. Saari, R. G. Lowery, P. W. Ludden, and G. P. Roberts. 1989. Genes coding for the reversible ADP-ribosylation system of dinitrogenase reductase from *Rhodospirillum rubrum*. *Mol. Gen. Genet.* **218**:340–347.

Gladysheva, T. B., and G. Ji. Unpublished results.

Gladysheva, T. B., K. L. Oden, and B. P. Rosen. 1994. Properties of arsenate reductase from plasmid R773. *Biochemistry* **33**:7288–7293.

Hedges, R. W., and S. Baumberg. 1973. Resistance to arsenic compounds conferred by a plasmid transmissible between strains of *Escherichia coli*. *J. Bacteriol.* **115**:459–460.

Hsu, C. M., P. Kaur, C. E. Karkaria, R. F. Steiner, and B. P. Rosen. 1991. Substrate-induced dimerization of the ArsA protein, the catalytic component of an anion-translocating ATPase. *J. Biol. Chem.* **266**:2327–2332.

Ivey, D. M., A. A. Guffanti, Z. Shen, N. Kudyan, and T. A. Krulwich. 1992. The *cadC* gene product of alkaliphilic *Bacillus firmus* ORF4 partially restores Na^+ resistance to an *Escherichia coli* strain lacking an Na^+/H^+ antiporter (NhaA). *J. Bacteriol.* **174**:4878–4884.

Ji, G., E. A. Garber, L. G. Armes, C.-M. Chen, J. A. Fuchs, and S. Silver. 1994. Arsenate reductase of *Staphylococcus aureus* plasmid pI258. *Biochemistry* **33**:7294–7299.

Ji, G., and S. Silver. 1992a. Regulation and expression of the arsenic resistance operon from *Staphylococcus aureus* plasmid pI258. *J. Bacteriol.* **174**:3684–3694.

Ji, G., and S. Silver. 1992b. Reduction of arsenate to arsenite by the ArsC protein of the arsenic resistance operon of the *Staphylococcus aureus* plasmid pI258. *Proc. Natl. Acad. Sci. USA* **89**:9474–9478.

Joerger, R. D., and P. E. Bishop. 1988. Nucleotide sequence and genetic analysis of the *nifB-nifQ* region from *Azotobacter vinelandii*. *J. Bacteriol.* **170**:1475–1487.

Kaur, P., and B. P. Rosen. 1992. Plasmid-encoded resistance to arsenic and antimony. *Plasmid* **27**:29–40.

Kaur, P., and B. P. Rosen. Unpublished results.

Knowles, F. C., and A. A. Benson. 1983. The biochemistry of arsenic. *Trends Biochem. Sci.* **8**:178–180.

Loehr, T. M., and R. A. Plane. 1968. Raman spectra and structures of arsenious acid and arsenites in aqueous solution. *Inorg. Chem.* **7**:1708–1714.

Luecke, H., and F. A. Quiocho. 1990. High specificity of a phosphate transport protein determined by hydrogen bonds. *Nature* (London) **347**:402–406.

Marger, M. D., and M. H. Saier, Jr. 1993. A major superfamily of transmembrane facilitators that catalyze uniport, symport and antiport. *Trends Biochem. Sci.* **18**:13–20.

Mobley, H. L. T., and B. P. Rosen. 1982. Energetics of plasmid-mediated arsenate resistance in *Escherichia coli*. *Proc. Natl. Acad. Sci. USA* **79**:6119–6122.

Morby, A. P., J. S. Turner, J. W. Huckle, and N. J. Robinson. 1993. SmtB is a metal-dependent repressor of the cyanobacterial metallothionein gene *smtA*: identification of a Zn inhibited DNA-protein complex. *Nucleic Acids Res.* **21**:921–925.

Newton, G. L., and R. C. Fahey. 1990. Glutathione in prokaryotes, p. 69–77. *In* J. Vina (ed.), *Glutathione: Metabolism and Physiological Functions*. CRC Press, Inc., Boca Raton, Fla.

Nishiguchi, R., M. Takanami, and A. Oka. 1987. Characterization and sequence determination of the replicator region in the hairy-root-inducing plasmid pRiA4b. *Mol. Gen. Genet.* **206**:1–8.

Novick, R. P., and C. Roth. 1968. Plasmid-linked resistance to inorganic salts in *Staphylococcus aureus*. *J. Bacteriol.* **95**:1335–1342.

Nucifora, G., L. Chu, T. K. Misra, and S. Silver. 1989. Cadmium resistance from *Staphylococcus aureus* plasmid pI258 *cadA* gene results from a cadmium efflux ATPase. *Proc. Natl. Acad. Sci. USA* **86**:3544–3548.

Oskam, L., C. Hermans, G. Jarings, P. R. Klatser, and R. A. Hartskeerl. Unpublished data.

Owolabi, J. B., and B. P. Rosen. 1990. Differential mRNA stability controls relative gene expression within the plasmid-encoded arsenical resistance operon. *J. Bacteriol.* **172**:2367–2371.

Renna, M. C., N. Najimudin, L. R. Winik, and S. A. Zahler. 1993. Regulation of the *Bacillus subtilis alsS*, *alsD*, and *alsR* genes involved in post-exponential-phase production of acetoin. *J. Bacteriol.* **175**:3863–3875.

Rinchik, E. M., S. J. Bultman, B. Horsthemke, S. T. Lee, K. M. Strunk, R. A. Spritz, K. M. Avidano, M. T. Jong, and R. D. Nicholls. 1993. A gene for the mouse pink-eyed dilution locus and for human type II oculocutaneous albinism. *Nature* (London) **361**:72–76.

Rosen, B. P., and M. G. Borbolla. 1984. A plasmid-encoded arsenite pump produces arsenite resistance in *Escherichia coli*. *Biochem. Biophys. Res. Commun.* **124**:760–765.

Rosen, B. P., S. Dey, and D. Dou. Anion-translocating ATPases. *Biomembranes*, in press.

Rosen, B. P., S. Dey, D. Dou, G. Ji, P. Kaur, M. Ksenzenko, S. Silver, and J. Wu. 1992. Evolution of an ion-translocating ATPase. *Ann. N. Y. Acad. Sci.* **671**:257–272.

Rosen, B. P., and E. R. Kashket. 1978. Energetics of active transport, p. 559–620. *In* B. P. Rosen (ed.), *Bacterial Transport*. Marcel Dekker, Inc., New York.

Rosen, B. P., U. Weigel, R. A. Monticello, and B. P. F. Edwards. 1991. Molecular characterization of an anion pump: purification of the ArsC protein. *Arch. Biochem. Biophys.* **284**:381–385.

Rosen, B. P., U. Weigel, C. Karkaria, and P. Gangola. 1988. Molecular characterization of an anion pump. The *arsA* gene product is an arsenite (antimonate)-stimulated ATPase. *J. Biol. Chem.* **263**:3067–3070.

Rosenstein, R., P. Peschel, B. Wieland, and F. Götz. 1992. Expression and regulation of the *Staphylococcus xylosus* antimonite, arsenite and arsenate resistance operon. *J. Bacteriol.* **174**:3676–3683.

San Francisco, M. J. D., C. L. Hope, J. B. Owolabi, L. S. Tisa, and B. P. Rosen. 1990. Identification of the metalloregulatory element of the plasmid-encoded arsenical resistance operon. *Nucleic Acids Res.* **18**:619–624.

San Francisco, M. J. D., L. S. Tisa, and B. P. Rosen. 1989. Identification of the membrane component of the anion pump encoded by the arsenical resistance operon of R-factor R773. *Mol. Microbiol.* **3**:15–21.

Shi, W., J. Wu, and B. P. Rosen. Isolation of a mutant ArsR repressor protein defective in interaction with inducers. Submitted for publication.

Silver, S., K. Budd, K. M. Leahy, W. V. Shaw, D. Hammond, R. P. Novick, G. R. Willsky, M. H. Malamy, and H. Rosenberg. 1981. Inducible plasmid-determined resistance to arsenate, arsenite, and antimony (III) in *Escherichia coli* and *Staphylococcus aureus*. *J. Bacteriol.* **146**:983–996.

Silver, S., and G. Ji. Newer systems for bacterial resistances to toxic heavy metals. *Environ. Health Perspect.*, in press.

Silver, S., G. Ji, S. Broer, S. Dey, D. Dou, and B. P. Rosen. 1993a. Orphan enzyme or patriarch of a new tribe: the arsenic resistance ATPase of bacterial plasmids. *Mol. Microbiol.* **8**:637–642.

Silver, S., and D. Keach. 1982. Energy-dependent arsenate efflux: the mechanism of plasmid mediated resistance. *Proc. Natl. Acad. Sci. USA* **79**:6114–6118.

Silver, S., B. T. O. Lee, N. L. Brown, and D. A. Cooksey. 1993b. Bacterial plasmid resistances to copper, cadmium, and zinc, p. 38–53. *In* A. J. Welch and S. K. Chapman (ed.), *The Chemistry of the Copper and Zinc Triads.* Royal Society of Chemistry, London.

Silver, S., G. Nucifora, L. Chu, and T. K. Misra. 1989. Bacterial resistance ATPases: primary pumps for exporting toxic cations and anions. *Trends Biochem. Sci.* **14:**76–80.

Silver, S., and M. Walderhaug. 1992. Gene regulation of plasmid- and chromosome-determined inorganic ion transport in bacteria. *Microbiol. Rev.* **56:**195–228.

Sulston, J., Z. Du, K. Thomas, R. Wilson, L. Hiller, R. Staden, N. Halloran, P. Green, J. Thierry-Mieg, L. Qiu, S. Dear, A. Coulson, M. Craxton, R. Durbin, M. Berks, M. Metzstein, T. Hawkins, R. Ainscough, and R. Waterson. 1992. The *C. elegans* genome sequencing project: a beginning. *Nature* (London) **356:**37–41.

Szymanski, H. A., L. Marabella, J. Hoke, and J. Harter. 1968. Infrared and raman studies of arsenic compounds. *Appl. Spectrosc.* **22:**297–304.

Tisa, L. S., and B. P. Rosen. 1990a. Plasmid-encoded transport mechanisms. *J. Bioenerg. Biomembr.* **22:**493–507.

Tisa, L. S., and B. P. Rosen. 1990b. Molecular characterization of an anion pump: the ArsB protein is the membrane anchor for the ArsA protein. *J. Biol. Chem.* **265:**190–194.

World Health Organization. 1980. Arsenic. *Environmental Health Criteria,* 18, World Health Organization, Geneva.

Wu, J. H., and B. P. Rosen. 1991. Regulation of the *ars* operon: the *arsR* gene product is a negative regulatory protein. *Mol. Microbiol.* **5:**1331–1336.

Wu, J. H., and B. P. Rosen. 1993a. Metalloregulated expression of the *ars* operon. *J. Biol. Chem.* **268:**52–58.

Wu, J. H., and B. P. Rosen. 1993b. The *arsD* gene encodes a second trans-acting regulatory protein of the plasmid-encoded arsenical resistance operon. *Mol. Microbiol.* **8:**615–623.

Wu, J. H., L. S. Tisa, and B. P. Rosen. 1992. Membrane topology of the ArsB protein, the membrane component of an anion-translocating ATPase. *J. Biol. Chem.* **267:**12570–12576.

Chapter 18

Selenophosphate: Synthesis, Properties, and Role as Biological Selenium Donor

THRESSA C. STADTMAN, ZSUZSA VERES, AND ICK YOUNG KIM

Laboratory of Biochemistry, National Heart, Lung, and Blood Institute, National Institutes of Health, Building 3, Room 108, Bethesda, Maryland 20892

Selenium is known to occur as a specific component of two types of biological macromolecules, proteins, and tRNAs. Certain selenium-dependent enzymes in both prokaryotes and eukaryotes contain selenocysteine residues that are incorporated cotranslationally as directed by UGA codons. In tRNAs, e.g., lysine, glutamate, and glutamine tRNAs of several bacterial species, selenium is present in 2-selenouridine residues located in the "wobble position" of the anticodons. A highly reactive oxygen-labile selenium compound is the required donor for synthesis of these two different types of molecules. Recently, we have shown that this biological selenium donor is monoselenophosphate, a compound in which selenium is bonded directly to phosphorus (Veres et al., 1992; Glass et al., 1993). The unique chemical properties of selenophosphate, particularly the relative weakness of the P—Se bond compared with the P—S bond and the even stronger P—O bond, make it especially suited for its role as selenium donor. It participates in an addition reaction (Fig. 1) converting 2,3-aminoacrylyl-tRNA to selenocystyl-tRNA (Forchhammer and Böck, 1991; Ehrenreich et al., 1992); it also functions in a substitution reaction (Fig. 2) in which the sulfur of 2-thiouridine residues in tRNAs is replaced with selenium (Wittwer and Stadtman, 1986; Veres et al., 1992).

Biosynthesis of Selenophosphate

The reactive selenium donor compound, monoselenophosphate ($HSePO_3H_2$), is formed from ATP and selenide by selenophosphate synthetase as shown in reaction 1 (Veres et al., 1992; Glass et al., 1993; Veres et al., 1994).

$$ATP + H_2Se \rightarrow HSePO_3H_2 + H_3PO_4 + AMP \quad (1)$$

Selenide prepared by reduction of elemental selenium with borohydride or selenite with excess dithiothreitol is the only selenium compound known to be used by the enzyme as substrate and is not replaceable with sulfide. Other common nucleoside triphosphates do not substitute for ATP in the reaction (Kim et al., 1993). The in vitro reaction is carried out in the presence of dithiothreitol under strictly anaerobic conditions. The selenide-dependent formation of radioactive AMP from [^{14}C]ATP is monitored to determine the course of the reaction. Alternatively, the formation of selenophosphate can be monitored by ^{31}P nuclear magnetic resonance spectroscopy. In this case, the reaction is carried out in a nuclear magnetic resonance spectroscropy tube, and a new ^{31}P resonance is detected at 23.2 ppm. This new signal is well separated from those of ATP, AMP, and P_i (Veres et al., 1992). Substitution of [^{77}Se]selenide for normal isotope abundance Se resulted in splitting of the 23.2-ppm signal, demonstrating that selenium is bonded directly to phosphorus in the new selenophosphate compound. A method for chemical synthesis of the highly oxygen-labile monoselenophosphate was devised, and its structure was established by compositional, spectroscopic, and mass-spectral analyses (Glass et al., 1993). The enzyme product was purified and shown to be indistinguishable from chemically synthesized monoselenophosphate (shown below) by ^{31}P nuclear magnetic resonance spectroscopy and by ion-pairing high-pressure liquid chromatography.

```
        O
        ‖
  HSe—P—OH
        |
        OH
```

Properties of Selenophosphate Synthetase

The product of the *selD* gene from *Escherichia coli* and *Salmonella typhimurium* (Leinfelder et al., 1988; Kramer and Ames, 1988; Stadtman et al., 1989) is a 37-kDa protein (Leinfelder et al., 1990) that forms a labile selenium donor com-

Seryl-$tRNA_{UCA}$ $\xrightarrow{1}$ [PLP-2,3-aminoacrylyl-$tRNA_{UCA}$] $\xrightarrow{2}$ Selenocysteyl-$tRNA_{UCA}$

FIG. 1. Conversion of seryl-$tRNA_{UCA}$ to selenocysteyl-$tRNA_{UCA}$ in prokaryotes. This overall reaction is catalyzed by the *selA* gene product, selenocysteine synthase, a pyridoxal phosphate (PLP)-dependent enzyme. In step 1, the esterified serine residue forms a Schiff base intermediate with enzyme-bound pyridoxal phosphate. A β-elimination of the hydroxyl group forms the 2,3-amino acrylyl-tRNA. In step 2, an activated selenium derivative formed from H_2Se and ATP by the SelD protein is added across the double bond to form selenocysteyl-tRNA. The compound added in step 2 is selenophosphate, $HSePO_3H_2$.

pound required for synthesis of formate dehydrogenase and seleno-tRNAs in these organisms. This selenium donor is essential for synthesis of selenocystyl-tRNA from 2,3-aminoacrylyl-tRNA (Forchhammer and Böck, 1991) and for conversion of 2-thiouridine in tRNAs to 2-selenouridine (Veres et al., 1992). The identity of this compound has been established to be monoselenophosphate (Glass et al., 1993), and thus SELD is named selenophosphate synthetase (Kim et al., 1993). The enzyme was purified from an overproducing strain of *E. coli* (Leinfelder et al., 1990). Apparently homogeneous enzyme preparations can be obtained by using additional phenyl-Sepharose and ATP-agarose affinity chromatographic steps (Veres et al., 1994). The products of the reaction of ATP and selenide catalyzed by highly purified enzyme are a 1:1:1 mixture of selenophosphate, AMP, and P_i (Veres and Stadtman, in press).

Studies with Selenophosphate Synthetase Mutants

There are seven cysteine residues in wild-type selenophosphate synthetase (Leinfelder et al., 1990), and none of these occur in disulfide linkage (Kim et al., 1992). At least one is essential for catalytic activity as shown by sensitivity of the enzyme to alkylating agents. Substitution of serine residues for two cysteine residues located in the amino-terminal region of the protein by site-specific mutagenesis (Kim et al., 1992) showed Cys-17 but not Cys-19 to be essential. A glycine-rich sequence in this region suggestive of nucleotide-binding motifs was subjected to further mutation (Kim et al., 1993). Substitution of Lys-20 with arginine or with glutamine resulted in almost complete or complete loss of activity, respectively. However, binding studies with [γ-^{32}P]8-azido-ATP and manganese-ATP indicated that Lys-20 is not essential for nucleotide binding, and thus it must be required in another step of the reaction sequence. Substitution of Gly-18 with valine resulted in an increase in the ATP K_m value and a 70% decrease in catalytic activity. The mutant enzyme containing asparagine in place of His-13 was indistinguishable from the wild-type enzyme in catalytic activity. Azido-ATP behaved as a competitive inhibitor of ATP in the catalytic reaction. Enzymes containing both Cys-17 and Cys-19 were labeled with [γ-^{32}P]8-azido-ATP following photolysis, but mutant enzymes containing serine residues in place of these cysteine residues were not labeled. Since ATP decreased the incorporation of [γ-^{32}P]8-azido-ATP in these experiments, the normal ATP-binding site appears to be altered by cysteine-to-serine mutations in this amino-terminal portion of the protein.

Role of Selenophosphate in 2-Selenouridine Synthesis

A mutant *Salmonella* strain unable to synthesize seleno-tRNAs (Kramer and Ames, 1988) as a result of a defective *selD* gene (Stadtman et al., 1989) was used as the source of the complementary enzyme(s) required for 2-selenouridine for-

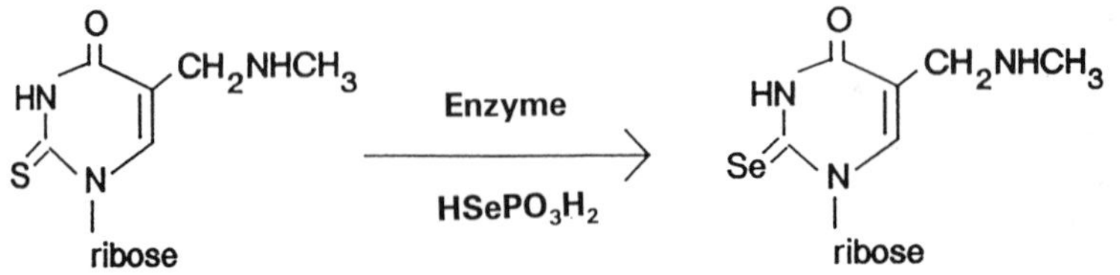

FIG. 2. A 2-thiouridine in the "wobble position" of $tRNA^{Glu}$, $tRNA^{Lys}$, or $tRNA^{Gln}$ is converted to a 2-selenouridine. A thiouridine in the wobble position of the anticodon of an intact tRNA molecule is the substrate, not the free nucleoside. The enzyme that catalyzes this substitution reaction is not characterized. The form of sulfur that is eliminated is not identified.

mation. Supplementation of extracts of this mutant with purified SELD protein reconstituted the complete enzyme system. Preparations of the complementary enzyme purified from the mutant extracts utilized selenophosphate as substrate for replacement of sulfur in the 2-thiouridine of added thio-tRNAs with selenium (Glass et al., 1993; Veres and Stadtman, in press) (Fig. 2). Neither ATP nor the selenophosphate synthetase protein was required when selenophosphate was added as selenium donor. Chemically synthesized monoselenophosphate and the purified ^{75}Se-labeled selenophosphate synthesized enzymically were equally effective as the selenium donor for 2-selenouridine formation in this reaction. Furthermore, addition of selenide had no effect on selenophosphate utilization by the *Salmonella* enzyme in the absence of ATP and selenophosphate synthetase. From these results it appears that selenium substitution for sulfur occurs by a direct attack of selenophosphate on the carbon-sulfur bond of 2-thiouridine in the tRNA substrate. The precise mechanism of this interesting replacement reaction is the subject of continuing studies.

REFERENCES

Ehrenreich, A., K. Forchhammer, P. Tormay, B. Vepreck, and A. Böck. 1992. Selenoprotein synthesis in *E. coli.* Purification and characterization of the enzyme catalyzing selenium activation. *Eur. J. Biochem.* **206:**767–773.

Forchhammer, K., and A. Böck. 1991. Selenocysteine synthase from *Escherichia coli.* Analysis of the reaction sequence. *J. Biol. Chem.* **266:**6324–6328.

Glass, R. S., W. P. Singh, W. Jung, Z. Veres, T. D. Scholz, and T. C. Stadtman. 1993. Monoselenophosphate: synthesis, characterization and identity with the prokaryotic selenium donor, SePX. *Biochemistry* **32:**12555–12559.

Kim, I. Y., Z. Veres, and T. C. Stadtman. 1992. *Escherichia coli* mutant SELD enzymes. The cysteine 17 residue is essential for selenophosphate formation from ATP and selenide. *J. Biol. Chem.* **267:**19650–19654.

Kim, I. Y., Z. Veres, and T. C. Stadtman. 1993. Biochemical analysis of *Escherichia coli* selenophosphate synthetase mutants: lysine-20 is essential for catalytic activity and cysteine-17/19 for 8-azido-ATP derivatization. *J. Biol. Chem.* **268:**27020–27025.

Kramer, G. F., and B. Ames. 1988. Isolation and characterization of a selenium metabolism mutant of *Salmonella typhimurium. J. Bacteriol.* **170:**736–743.

Leinfelder, W., K. Forchhammer, B. Veprek, E. Zehelein, and A. Böck. 1990. *In vitro* synthesis of selenocysteinyl-tRNA^{UCA} from seryl-tRNA^{UCA}. Involvement and characterization of the selD gene product. *Proc. Natl. Acad. Sci. USA* **87:**543–547.

Leinfelder, W., K. Forchhammer, F. Zinoni, G. Sawers, M.-A. Mandrand-Berthelot, and A. Böck. 1988. *Escherichia coli* genes whose products are involved in selenium metabolism. *J. Bacteriol.* **170:**540–546.

Stadtman, T. C., J. N. Davis, E. Zehelein, and A. Böck. 1989. Biochemical and genetic analysis of *Salmonella typhimurium* and *Escherichia coli* mutants defective in specific incorporation of selenium into formate dehydrogenase and tRNAs. *BioFactors* **2:**35–44.

Veres, Z., I. Y. Kim, T. D. Scholy, and T. C. Stadtman. 1994. Selenophosphate synthetase: enzyme properties and catalytic reaction. *J. Biol. Chem.* **269:**10597–10603.

Veres, Z., and T. C. Stadtman. A purified selenophosphate-dependent enzyme from *Salmonella typhimurium* catalyzes the replacement of sulfur in 2-thio uridine residues in tRNAs with selenium. *Proc Natl. Acad. Sci. USA*, in press.

Veres, Z., L. Tsai, T. D. Scholz, M. Politino, R. S. Balaban, and T. C. Stadtman. 1992. Synthesis of 5-methylaminomethyl-2-selenouridine in tRNAs. ^{31}P NMR studies show the labile selenium donor synthesized by the selD gene product contains selenium bonded to phosphorus. *Proc. Natl. Acad. Sci. USA* **89:**2975–2979.

Wittwer, A. J., and T. C. Stadtman. 1986. Biosynthesis of 5-methylaminomethyl-2-selenouridine, a naturally occurring nucleoside in *E. coli* tRNA. *Arch. Biochem. Biophys.* **248:**540–550.

Chapter 19

Specific and General Responses of Cyanobacteria to Macronutrient Deprivation†

ARTHUR R. GROSSMAN,[1] DEVAKI BHAYA,[2] AND JACKIE L. COLLIER[1]

Department of Plant Biology, Carnegie Institution of Washington, 290 Panama Street, Stanford, California 94305,[1] and Centre for Biotechnology, Jawaharlal Nehru University, New Delhi 110067, India[2]

Cyanobacteria often have to cope with environments that are depleted in one or more essential nutrients and have developed a suite of responses to acclimate to nutrient-limited growth. Some of these responses are specifically triggered by the depletion of a particular nutrient. For example, most conditions of macronutrient limitation lead to the production of elevated levels of a high-affinity transport system that pumps the limiting nutrient into the cell. In some cases, an enzyme that is involved in converting a particular macronutrient from an inaccessible to an accessible form is produced. An example of such an enzyme is the periplasmic alkaline phosphatase that is specifically synthesized when cyanobacteria experience phosphate-limited growth. Cyanobacterial cells also exhibit general responses that are manifest in medium lacking any one of a number of different nutrients. These general responses include the diminution of oxygenic photosynthesis, the specific degradation of the phycobilisome (the major light-harvesting complex in the cell), the accumulation of intracellular inclusion bodies (Simon, 1987), and an alteration in the properties of the cell wall. In this chapter we will discuss our understanding of how sulfur and phosphorus deprivation cause the cells to alter their physiology and biochemistry and the ways in which these changes are regulated. The individual genes that have been isolated to date and are involved in the acclimation of cyanobacteria to sulfur and phosphorus deprivation are given in Table 1; these genes will be referred to throughout the text.

Specific Responses

The sulfur stress response. Sulfur is an essential macronutrient for the synthesis and decoration of a number of macromolecules in the cell, including proteins, carbohydrates, and lipids. Sulfur is incorporated into the essential amino acids cysteine and methionine, and sulfate can modulate protein function by bonding to tyrosine and serine residues. Sulfated polysaccharides produced by a number of algal species may be important in maintaining the integrity of the algal thallus, preventing desiccation, and regulating the cellular ionic balance (Kloareg and Quatrano, 1988). A sulfolipid is a prominent component of the photosynthetic membranes (Kleppinger-Sparace et al., 1990) and may play an important role in maintaining the structural integrity of the photosynthetic electron transport chain. A sulfated compound with a chitin-like backbone is produced by *Rhizobium* species and is necessary for the bacterium to establish a symbiotic association with leguminous plants (Lerouge et al., 1990). Other sulfated secondary metabolites might also play important roles as signaling compounds in the complex environment of the rhizosphere.

Much of the sulfur that is utilized by cyanobacterial cells is transported into the cell in the form of sulfate. This sulfate is reduced via the 3′-phosphoadenosine-5′-phosphosulfate (PAPS) (Schmidt and Christen, 1978) and adenosine-5′-phosphosulfate (APS) (Tsang and Schiff, 1975) sulfotransferase systems. However, other sources such as thiocyanate, cysteine, cystine, thiosulfate, and reduced glutathione can satisfy the requirement of the cells for sulfur (Lawry and Jensen, 1986; Schmidt et al., 1982). When cyanobacteria are deprived of sulfur, a number of activities accumulate that help the organism cope with the stressful conditions. One of the most apparent of these activities is the transport of sulfate into the cell. Kinetic measurements have demonstrated that the $K_{1/2}$ (the affinity of the sulfate transport system for its substrate) is essentially the same in cells that are grown under sulfur-replete and sulfur-deprived conditions. This suggests that the same transport system is used by cells grown in high- and low-sulfur medium. In contrast, the V_{max} for sulfate transport can increase 20-fold when the cells are starved for sulfur, suggesting that more of the transport system is being synthesized and in-

†CIW publication 1193.

TABLE 1. Genes involved in acclimation to sulfate and phosphorus stress

Gene	Function	Regulated by[a]	Organism	Reference
nblA	Phycobilisome degradation	S	*Synechococcus* sp. strain PCC 7942	Collier and Grossman, 1994
sbpA, cysA, cysT, cysW	Periplasmic sulfate transport system	S	*Synechococcus* sp. strain PCC 7942	Laudenbach and Grossman, 1991
rhdA	Not known	S	*Synechococcus* sp. strain PCC 7942	Laudenbach et al., 1991
cysU, cysV	Putative thiocyanate transport proteins	S	*Synechococcus* sp. strain PCC 7942	Laudenbach et al., 1991
cysP	Putative thiosulfate-binding protein	S	*Synechococcus* sp. strain PCC 7942	Laudenbach et al., 1991
Genes on plasmid pANL	Not known	S	*Synechococcus* sp. strain PCC 7942	Nicholson and Laudenbach, 1993
cysR	Transcriptional regulator	S	*Synechococcus* sp. strain PCC 7942	Laudenbach et al., 1991
cpcB3A3	Phycocyanin	S	*Calothrix* sp. strain PCC 7601	Mazel and Marliere, 1989
phoA	Alkaline phosphatase	P	*Synechococcus* sp. strain PCC 7942	Ray et al., 1991
pstS	Phosphate-binding protein	P	*Synechococcus* sp. strain WH 7803	Scanlan et al., 1993
sphS, sphR	Response sensor and regulator	P	*Synechococcus* sp. strain PCC 7942	Aiba et al., 1993

[a]S, sulfate; P, phosphorus.

corporated into the cytoplasmic membranes. Furthermore, it was genetically demonstrated that the two systems were the same by interposon mutagenesis of genes encoding polypeptide components of the sulfate transport system (discussed below).

To explore the regulated expression of genes that are activated when cyanobacteria are exposed to low-sulfur conditions, we used a heterologous probe to isolate genes encoding the polypeptide constituents of the sulfate transport system of *Synechococcus* sp. strain PCC 7942 (Green et al., 1989). The region of the genome that was isolated with this probe encoded the polypeptide constituents of the sulfate transport system but also encoded other polypeptides that are thought to be important for the acclimation of the organism to sulfur-limited growth (Grossman et al., 1992; Laudenbach et al., 1991). A physical map showing the positions of the different genes on the cloned cyanobacterial DNA fragment is presented in Fig. 1. Transcripts from all of these genes are present at low levels in sulfur-replete cells and accumulate to much higher levels when the cells are deprived of sulfur.

Both sequence analyses and insertional inactivation of each of the genes shown in Fig. 1 have helped in elucidating the function(s) of the individual genes. The genes that encode the four polypeptides of the sulfate transport system are clustered in this region of the genome. The gene designations are *cysA* for the gene encoding the nucleotide-binding protein, *sbpA* for the gene encoding the sulfate-binding protein, and *cysT* and *cysW* for genes encoding the hydrophobic proteins that form channels in the inner cytoplasmic membrane. The genes *sbpA*, *cysT*, and *cysW* are transcribed in one direction, whereas *cysA* is transcribed in the opposite direction. Other genes within this cluster have also tentatively been assigned functions in the acclimation process. For example, the gene designated *cysP*, which is located immediately downstream of *cysW*, is thought to be a thiosulfate-binding protein. In *Escherichia coli* the sulfate permease may function in the uptake of thiosulfate (a similar situation probably occurs in *Synechococcus* species). However, thiosulfate transport may be augmented by a specific periplasmic thiosulfate-binding protein (Hryniewicz et al., 1990) that is approximately 45% similar to the *Synechococcus cysP* sequence. The two polypeptides encoded by *cysU* and *cysV* may be part of a permease (they have considerable homology to *cysT* and *cysW*) specific for the transport of thiosulfate. Mutational analysis of this region is being performed to confirm the proposed gene functions.

Another gene depicted on the map, located in the middle of the sulfate permease operon, has been designated *cysR*. The *cysR* gene product resembles transcriptional regulators similar to those encoded by *fixK*, *crp*, and *fnr* (Laudenbach and Grossman, 1991). When the *cysR* gene is interrupted, the cyanobacterial cells can no longer grow on the sulfur source, thiocyanate. So far, thiocyanate is the only sulfur compound that is no

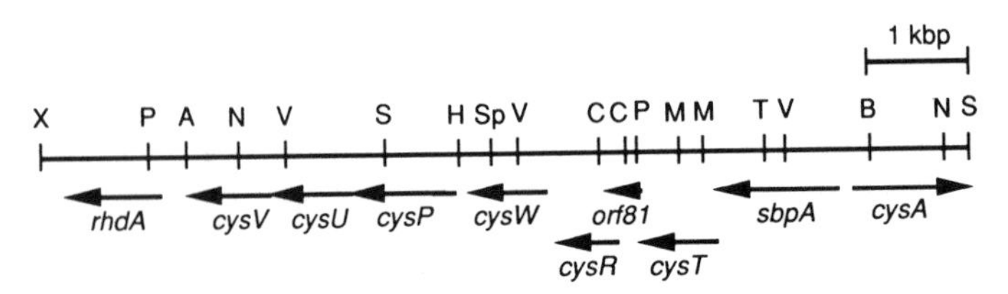

FIG. 1. Restriction map of a region of the *Synechococcus* sp. strain PCC 7942 genome that is regulated by sulfur availability. Restriction site abbreviations: X, *Xba*I; P, *Pst*I; A, *Sca*I; N, *Nae*I; V, *Eco*RV; S, *Sal*I; H, *Hin*dIII; Sp, *Sph*I; C, *Cla*I; M, *Sma*I; T, *Stu*I; B, *Bgl*II. The extent of each gene and the gene designation are given below the map (reprinted with permission from Kluwer Academic Publishers [Grossman et al., 1992]).

longer effective as the sole sulfur source in the *cysR*-defective strain. Recently, it has been shown that CysR controls genes localized to the large plasmid of *Synechococcus* sp. strain PCC 7942 (Nicholson and Laudenbach, 1993). Some of these genes may be involved in thiocyanate utilization. These results suggest that during sulfur-limited growth of *Synechococcus* sp. strain PCC 7942, systems involved in the acquisition of diverse sulfur-containing compounds are activated and aid the organism in weathering the deficiency (Table 1).

Positioned downstream of the putative thiosulfate transport genes is a gene designated *rhdA*, which encodes a protein with some similarity to the enzyme rhodanese (Laudenbach et al., 1991). Rhodanese is a thiosulfate-sulfur transferase that can cleave a sulfane bond and transfer the thiol group released to a thiophilic acceptor molecule (Westley, 1973). This protein is located in the periplasmic space and becomes very abundant in *Synechococcus* sp. strain PCC 7942 during sulfur-limited growth. It is still uncertain whether the cyanobacterial protein displays rhodanese activity or just functions in binding certain sulfur-containing molecules. Hence, the protein may have a catalytic role in the acclimation process or may simply deliver sulfur-containing compounds from the environment to the cytoplasmically localized permeases. This latter function would also help prevent leakage of sulfur-containing molecules from the cells.

The phosphate stress response. Phosphorus is most often assimilated by cells in the form of phosphate. Phosphate is required for the synthesis of nucleic acids and lipids, may covalently link to proteins and alter the activity of those proteins, and is essential for generating the high-energy currency of the cell (in the form of nucleoside triphosphates). It is also a nutrient which is often severely limiting in both the terrestrial and aqueous environments. When cyanobacteria are deprived of phosphorus, they synthesize periplasmic or extracellular phosphatases and exhibit an increased capacity to take up phosphate. Phosphate uptake in cyanobacteria may involve more than one transport system, and there may be as many as three transport systems in *Synechococcus* cells. Although phosphate-starved *Synechococcus* cells can exhibit a 50-fold increase in their capacity to take up phosphate, the $K_{1/2}$ for transport remains the same in starved and unstarved cells (Grillo and Gibson, 1979). Recently, a protein that is probably involved in the binding of phosphate and its delivery to the membrane-associated components of the phosphate transport system has been identified in the marine cyanobacterium *Synechococcus* sp. strain WH7803. This protein, which has a molecular mass of 32 kDa and is localized to the periplasmic space, has homology to the phosphate-binding protein of *Escherichia coli* (Scanlan et al., 1993). Furthermore, the abundance of the mRNA encoding this protein increases in cells that are depleted of phosphorus. It is not clear whether a similar binding protein is present in freshwater cyanobacteria.

Many cyanobacteria are thought to contain more than one phosphatase and synthesize phosphatases in response to phosphorus stress. Since most cyanobacteria are unable to take up organic phosphorus-containing molecules, the production of multiple periplasmic phosphatases under phosphorus-deficient conditions (Doonan and Jensen, 1980; Healey, 1982; Islam and Whitton, 1992; Marco and Orus, 1988a, 1988b; Whitton et al., 1990) allows the cells to scavenge phosphate by hydrolyzing it from a wide variety of phosphate sources. Some cyanobacteria are also able to use hydroxyapatite as a source of both calcium and phosphorus (Cameron and Julian, 1988). Genes for phosphatases from both *Nostoc* species (Xie et al., 1989) and *Synechococcus* sp. strain PCC 7942 (Ray et al., 1991) have been characterized. The alkaline phosphatase that accumulates in phosphorus-deficient *Synechococcus* sp. strain PCC 7942 is a periplasmic protein with a molecular mass of approximately 150 kDa (Block and Grossman, 1988). This alkaline phosphatase (gene designation *phoA*) appears to be transcriptionally controlled, and mRNA encoding PhoA accumulates only in phosphorus-deficient cells. The *Synechococcus* PhoA protein has some unique and interesting characteristics (Ray et al., 1991). The first one-third of the protein is homologous to the α subunit of the bacterial and chloroplast coupling factors. A region of about 50 amino acids shows strong homology to the UshA protein, a UDP sugar hydrolase in *E. coli*. The last half of the protein has sequences homologous to P-loops (motifs present in kinases) that are involved in binding nucleoside triphosphates via their terminal phosphate ester. The presence of multiple P-loops in the alkaline phosphatase might allow for the binding of phosphate groups attached to a variety of different organic compounds, thereby expanding the substrate specificity of the enzyme. This would be a highly desirable trait for an organism utilizing diverse phosphorylated compounds in its natural environment. Since phosphorus is frequently the limiting nutrient in freshwater ecosystems (Hecky and Kilham, 1988), the phosphatases of cyanobacteria probably play a vital role in natural cyanobacterial populations.

Progress has also been made in identifying elements that are important for controlling the responses of cyanobacteria to their environment and, in particular, to phosphorus-limited growth. Aiba et al. (1993) have shown that elements of a two-component regulatory system which involves both a polypeptide sensor (histidine kinase) and a regulator are involved in controlling the cyanobacterial response to phosphorus limitation. Complementation of an *E. coli* strain mutated in the histidine sensory kinase that controls the phosphate deprivation response with *Synechococcus* sp. strain PCC 7942 DNA allowed for the isolation of a cyanobacterial histidine kinase gene (*sphS*). Contiguous to this kinase gene was the gene coding for the regulator protein (*sphR*). Targeted inactivation of these genes in *Synechococcus* sp. strain PCC 7942 eliminated the ability of the cells to induce the alkaline phosphatase, as well as a 33-kDa protein of unknown function, during phosphorus-limited growth. Hence, two-component regulatory systems are key elements in the signal transduction pathway, and there may be a "pho regulon" in *Synechococcus* species analogous to that in *E. coli*. This would predict the presence of a "pho box"-like element in the promoters of phosphate-regulated genes. Sensor-regulator pairs are also probably involved in controlling chromatic adaptation (Chiang et al., 1992) and nitrogen fixation (Liang et al., 1992) in cyanobacteria.

General Responses

Over the last two decades, there has been sporadic examination of the general responses that accompany nutrient deprivation. Adverse nutrient conditions can cause dramatic alterations in the ultrastructure of the cyanobacterial cell and a marked decrease in the capacity of the cell for photosynthesis. *Synechococcus* cells starved of iron, nitrogen, or sulfur have less than half of the normal complement of thylakoid membranes (Sherman and Sherman, 1983; Wanner et al., 1986). The cells also accumulate large deposits of glycogen and polyphosphate granules and show a decrease in the number of carboxysomes and phycobilisomes. The alteration in the thylakoid membranes, the degradation of the light-harvesting phycobilisome, and the decrease in the capacity of the cells to perform oxygenic photosynthesis all suggest an intimate association between the availability and utilization of nutrients to drive anabolic processes and the ability of the cells to produce carbon skeletons via photosynthesis. Such a coupling is not surprising since the most energetically efficient utilization of resources would be achieved only if intracellular signals were generated that help the cell balance the production of fixed carbon with its utilization.

One of the most visually dramatic responses of cyanobacteria to nutrient-limited growth conditions is the reduction of pigment molecules in the cell (Grossman et al., in press). The first quantitation of pigment levels during nutrient deprivation was performed by Allen and Smith (1969). In that study it was demonstrated that in nitrogen-starved cells the phycobilisome was actively and rapidly degraded. This was noted spectrophotometrically as a loss in whole-cell A_{620}; this absorbance is due primarily to the presence of phycocyanin (PC), one of the major pigmented proteins present in the phycobilisome. Other studies have shown that sulfur-deprived cultures behave in a similar manner (Collier and Grossman, 1992). In contrast, in phosphate-deprived cells there is little or no PC degradation (Collier and Grossman, 1992). Decreased PC content in phosphorus-deprived cells is a consequence of dilution during continued cell division (cells deprived of phosphorus can continue to divide for four or five cycles as the rate of phycobilisome synthesis declines).

The loss of PC in both nitrogen- and sulfur-deprived cells is a consequence of the rapid and nearly complete degradation of the phycobilisome (Collier and Grossman, 1992; Yamanaka and Glazer, 1980). Degradation of this abundant complex could provide amino acids or carbon skeletons for the production of other cellular constituents required during nutrient deprivation. The use of phycobiliproteins as amino acid storage compounds may be especially important for marine cyanobacteria since nitrogen is frequently limiting in marine environments (Wyman et al., 1985). Phycobiliproteins are a poor source of sulfur amino acids, although phycobilisome degradation in sulfur-deprived cells suggests that the sulfur amino acids of the phycobiliproteins are recycled.

During sulfur-limited growth, some cyanobacteria specifically synthesize a set of PC subunits in which the only sulfur amino acids are the cysteine residues directly involved in chromophore attachment (Mazel and Marliere, 1989). Phycobilisomes contain no phosphorus at all, and cells limited for phosphorus do not degrade the phycobilisomes. Hence, the general losses of pigmentation in response to macronutrient limitation are not all the same.

The degradation of the phycobilisome during nitrogen and sulfur deprivation is an ordered process and is very similar in both cases (Collier and Grossman, 1992). The sequence of events that results in phycobilisome degradation is shown in Fig. 2. A brief summary of the phycobilisome structure may help the reader visualize the degradation process. The phycobilisome of *Synechococcus* sp. strain PCC 7942 is composed of a dicylindrical core and six sets of rods radiating from that core. The rods may each have three hexamers of PC, which are depicted as double disks in Fig. 2. Each of the hexamers is associated with a specific linker polypeptide: a 30-kDa linker polypeptide associates with the core-distal hexamer (most peripheral to the complex), a 33-kDa linker polypeptide associates with the middle hexamer, and a 27-kDa polypeptide associates with the core-proximal hexamer. These linker polypeptides are required for proper rod assembly. Elimination of nitrogen or sulfur from the growth medium provokes the rapid loss of the terminal PC hexamer of the phycobilisome rods and its associated 30-kDa linker polypeptide (Collier and Grossman, 1992; Yamanaka and Glazer, 1980). This is followed by some loss of the second PC hexamer and its associated 33-kDa linker polypeptide, along with a decrease in the number of rods per phycobilisome. The loss of these components results in a decrease in the phycobilisome sedimentation coefficient and a reduction in the ratio of PC to allophycocyanin (the major component of the phycobilisome core). The smaller phycobilisome can still function in harvesting light energy (Yamanaka and Glazer, 1980). Continued nutrient stress results in the complete degradation of the smaller phycobilisome, and the cells become bleached (the cells become yellow rather than the normal blue-green).

To learn more about the general phycobilisome degradation response, we mutated *Synechococcus* sp. strain PCC 7942 with nitrosoguanidine and screened for cells unable to degrade their phycobilisome. Colonies of such cells remained blue-green when grown on agar substantially free of sulfur, whereas wild-type colonies bleached. Some of the mutants obtained exhibited no phycobilisome degradation during either sulfur or nitrogen deprivation, although there was no net increase in phycobilisome levels during the stress treatment.

To define the lesion responsible for the nonbleaching phenotype, mutant cells were complemented back to the bleaching, wild-type phenotype and the DNA responsible for the complementation was characterized. Extensive genetic analyses demonstrated that the sequence on the genome that was required for complementation was approximately 250 nucleotides long and encoded a small protein of 59 amino acids. This gene, designated *nblA* (nonbleaching strain), was transcribed as a small mRNA and accumulated to high levels only in cells starved for nitrogen or sulfur. Low levels of the transcript were detected in both unstarved and phosphorus-starved cells.

A gene contiguous to *nblA*, and transcribed in the opposite orientation, has been named *txlA*

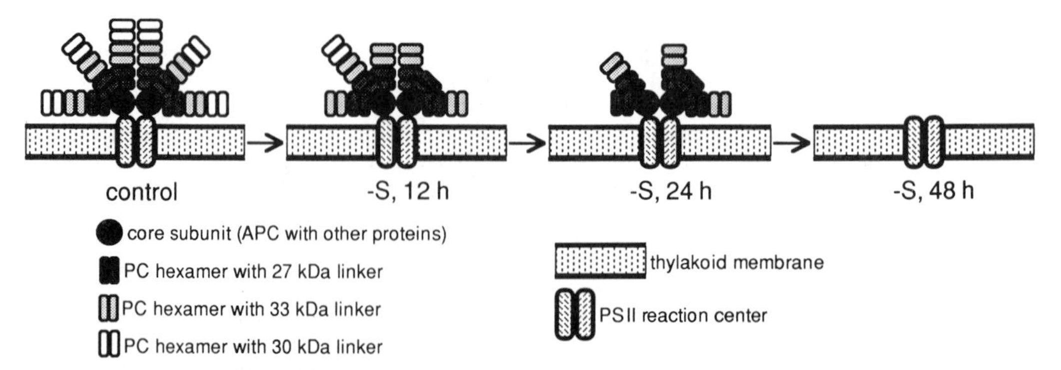

FIG. 2. Schematic representation of phycobilisome degradation following sulfur (or nitrogen) deprivation in *Synechococcus* sp. strain PCC 7942. The terminal phycocyanin (PC) hexamer and 30-kDa linker are lost first from the rods in a process which is complete by 18 h. After 12 h, some loss of the middle and proximal PC hexamers and 33- and 27-kDa linkers is observed. After 18 to 24 h, the remaining phycobilisome structure is degraded. The fate of the PS II reaction center complexes is not known; they may also be lost (reprinted with permission from Collier and Grossman, 1994).

(thioredoxin-like). This gene encodes a protein that has homology to both thioredoxin and protein disulfide isomerase. Preliminary evidence suggests that this gene may be regulated to some extent by the nutrient status of the environment and may play a role in the assembly and/or stability of the photosynthetic apparatus. More work is required to establish the precise function of the *txlA* gene product, and it will not be discussed further in this chapter.

The expression of *nblA* appears to be correlated with decreased cellular phycobilisome levels. When the wild-type *nblA* gene is placed on an autonomously replicating vector and transformed into *Synechococcus* sp. strain PCC 7942, the level of phycobilisomes on a per cell basis declines by 20 to 30% in nutrient-replete medium. This suggests that multiple copies of the gene may allow increased although still low-level expression of *nblA* and an elevated rate of phycobilisome degradation. Furthermore, the cells will degrade the phycobilisome during phosphate stress if *nblA* is placed under the control of the alkaline phosphatase promoter (only active during phosphate-limited growth) and introduced into the cell on the pPLAN vector (Collier and Grossman, 1994). No similarity was observed between the amino acid sequence encoded by *nblA* and any sequence in the GenBank data bases.

A number of interesting speculations can be made with respect to *nblA*. Since increased expression of this gene provokes the degradation of the phycobilisome, it is likely to be the primary gene whose activity must be elevated to elicit the degradation response during sulfur- or nitrogen-limited growth. Since the NblA polypeptide is very small and lacks similarity to any known protease, it is probably not a protease itself. However, it may function to activate a protease, such as the one studied by Wood and Haselkorn (1980), or may interact with the constituents of the phycobilisome and alter their susceptibility to proteolysis. This targeted degradation may involve covalent attachment, similar to the binding of ubiquitin to proteins in eukaryotes (Hershko, 1988), or the disruption of hydrophobic and/or ionic interactions among various constituents of the phycobilisome. Finally, *nblA* may activate certain genes that encode proteins directly involved in phycobilisome degradation. Further study of this system, with respect to both *nblA* and *txlA*, provides us with an opportunity to explore the role of posttranslational processes in the biosynthesis of the photosynthetic apparatus and the sensitivity of the photosynthetic apparatus to various environmental parameters. Elucidation of some of these processes may enable us to link changes in the metabolic state of the cells with coordinate changes in photosynthetic rates and the levels of the various components of the photosynthetic apparatus.

Our work discussed in this chapter was supported by National Science Foundation Grant DCB8916301.

We thank David Laudenbach for communicating some of his recent findings to us and Jane Edwards for technical help in the preparation of the manuscript.

REFERENCES

Aiba, H., M. Nagaya, and T. Mizuno. 1993. Sensor and regulator proteins from the cyanobacterium *Synechococcus* species PCC7942 that belong to the bacterial signal-transduction protein families: implication in the adaptive response to phosphate limitation. *Mol. Microbiol.* **8:**81–91.

Allen, M. M., and A. J. Smith. 1969. Nitrogen chlorosis in blue-green algae. *Arch. Microbiol.* **690:**111–120.

Block, M. A., and A. R. Grossman. 1988. Identification and purification of a derepressible alkaline phosphatase from *Anacystis nidulans* R2. *Plant Physiol.* **86:**1179–1184.

Cameron, H. J., and G. R. Julian. 1988. Utilization of hydroxyapatite by cyanobacteria as their sole source of phosphate and calcium. *Plant Soil* **109:**123–124.

Chiang, G. G., M. R. Schaefer, and A. R. Grossman. 1992. Complementation of a red-light indifferent cyanobacterial mutant. *Proc. Natl. Acad. Sci. USA* **89:**9415–9419.

Collier, J., and A. R. Grossman. 1994. A small polypeptide triggers complete degradation of light harvesting phycobilisomes in nutrient deprived cyanobacteria. *EMBO J.* **13:**1039–1047.

Collier, J. L., and A. R. Grossman. 1992. Chlorosis induced by nutrient deprivation in *Synechococcus* sp. strain PCC 7942: not all bleaching is the same. *J. Bacteriol.* **174:**4718–4726.

Doonan, B. B., and T. E. Jensen. 1980. Physiological aspects of alkaline phosphatase in selected cyanobacteria. *Microbios* **29:**185–207.

Green, L., D. Laudenbach, and A. R. Grossman. 1989. A region of a cyanobacterial genome required for sulfate transport. *Proc. Natl. Acad. Sci. USA* **86:**1949–1953.

Grillo, J. F., and J. Gibson. 1979. Regulation of phosphate accumulation in the cyanobacterium *Synechococcus. J. Bacteriol.* **140:**508–517.

Grossman, A. R., J. L. Collier, and D. E. Laudenbach. 1992. The response of *Synechococcus* sp. strain PCC 7942 to sulfur-limited growth, p. 3–10. *In* N. Murata (ed.), *Proc. IXth Int. Congr. Photosynth.*, vol. IV. Kluwer Academic Press, Dordrecht, The Netherlands.

Grossman, A. R., M. R. Schaefer, G. G. Chiang, and J. L. Collier. The responses of cyanobacteria to environmental conditions: light and nutrients. *In* D. Bryant (ed.), *The Molecular Biology of Cyanobacteria,* in press. Kluwer Academic Press, Dordrecht, The Netherlands.

Healey, F. P. 1982. Phosphate, p. 105–124. *In* N. G. Carr and B. A. Whitton (ed.), *The Biology of Cyanobacteria,* vol. 19. University of California Press, Berkeley.

Hecky, R. E., and P. Kilham. 1988. Nutrient limitation of phytoplankton in freshwater and marine environments: a review of recent evidence on the effects of enrichment. *Limnol. Oceanog.* **33:**796–822.

Hershko, A. 1988. Ubiquitin-mediated protein degradation. *J. Biol. Chem.* **263:**15237–15240.

Hryniewicz, M., A. Sirko, A. Palucha, A. Böck, and D. Hulanicka. 1990. Sulfate and thiosulfate transport in *Escherichia coli* K-12: identification of a gene encoding a novel protein involved in thiosulfate binding. *J. Bacteriol.* **172:**3358–3366.

Islam, M. R., and B. A. Whitton. 1992. Phosphorus content and phosphatase activity of the deepwater rice-field cyanobacterium (blue-green alga) *Calothrix* D764. *Microbios* **69:**7–16.

Kleppinger-Sparace, K. F., S. A. Sparace, and J. B. Mudd. 1990. Plant sulfolipids, p. 77–88. *In* H. Rennenberg, C. Brunold, L. J. Dekok, and I. Stulen (ed.), *Sulfur Nutrition and Sulfur Assimilation in Plants.* SPB Academic Press, The Hague, The Netherlands.

Kloareg, B., and R. S. Quatrano. 1988. Structure of the cell walls of marine algae and ecophysiological functions of the matrix polysaccharides. *Annu. Rev. Oceanogr. Mar. Biol.* **26**:259–315.

Laudenbach, D. E., D. Ehrhardt, L. Green, and A. R. Grossman. 1991. Isolation and characterization of a sulfur-regulated gene encoding a periplasmically localized protein with sequence similarity to rhodanese. *J. Bacteriol.* **173**:2751–2760.

Laudenbach, D. E., and A. R. Grossman. 1991. Characterization and mutagenesis of sulfur-regulated genes in a cyanobacterium: evidence for function in sulfate transport. *J. Bacteriol.* **173**:2739–2750.

Lawry, N. H., and T. E. Jensen. 1986. Condensed phosphate deposition, sulfur amino acid use, and unidirectional transsulfuration in *Synechococcus leopoliensis. Arch. Microbiol.* **144**:317–323.

Lerouge, P., P. Roche, C. Faucher, F. Maillet, G. Truchet, J. C. Promé, and J. Dénarie. 1990. Symbiotic host-specificity of *Rhizobium meliloti* is determined by sulphated and acylated glucosamine oligosaccaride signal. *Nature* (London) **344**:781–784.

Liang, J., L. Scappino, and R. Haselkorn. 1992. The *patA* gene product, which contains a region similar to CheY of *Escherichia coli*, controls heterocyst pattern formation in the cyanobacterium *Anabaena* 7120. *Proc. Natl. Acad. Sci. USA* **89**:5655–5659.

Marco, E., and M. I. Orus. 1988a. Alkaline phosphatase activity in two cyanobacteria. *Phyton* **48**:27–32.

Marco, E., and M. I. Orus. 1988b. Variation in growth and metabolism with phosphorus nutrition in two cyanobacteria. *J. Plant Physiol.* **132**:339–344.

Mazel, D., and P. Marliere. 1989. Adaptive eradication of methionine and cysteine from cyanobacterial light-harvesting proteins. *Nature* (London) **341**:245–248.

Nicholson, M. L., and D. E. Laudenbach. 1993. The involvement of a plasmid in the acclimation of cyanobacteria to sulfur stress. *In Proceedings of the Cyanobacterial Workshop*, Asilomar Conference Center, Pacific Grove, Calif.

Ray, J. M., D. Bhaya, M. A. Block, and A. R. Grossman. 1991. Isolation, transcription, and inactivation of the gene for an atypical alkaline phosphatase of *Synechococcus* sp. strain PCC 7942. *J. Bacteriol.* **173**:4297–4309.

Scanlan, D. J., N. H. Mann, and N. G. Carr. 1993. The response of the picoplanktonic marine cyanobacterium *Synechococcus* sp. WH7803 to phosphate starvation involves a protein homologous to the periplasmic phosphate-binding protein of *Escherichia coli. Mol. Microbiol.* **10**:181–191.

Schmidt, A., and U. Christen. 1978. A factor-dependent sulfotransferase specific for 3′-phosphoadenosine-5′-phosphosulfate (PAPS) in the cyanobacterium *Synechococcus* 6301. *Planta* **140**:239–244.

Schmidt, A., I. Erdle, and H.-P. Kost. 1982. Changes of c-phycocyanin in *Synechococcus* 6301 in relation to growth on various sulfur compounds. *Z. Naturforsch. Sect. C* **37**:870–876.

Sherman, D. M., and L. A. Sherman. 1983. Effect of iron deficiency and iron restoration on ultrastructure of *Anacystis nidulans. J. Bacteriol.* **156**:393–401.

Simon, R. D. 1987. Inclusion bodies in the cyanobacteria: cyanophycin, polyphosphate, polyhedral bodies, p. 199–225. *In* P. Fay and C. Van Baalen (ed.), *The Cyanobacteria.* Elsevier Science Publishers B.V., Amsterdam.

Tsang, M.L.-S., and J. A. Schiff. 1975. Studies of sulfate utilization by algae. Distribution of adenosine 5′-phosphosulfate (APS) and adenosine 3′-phosphate 5′-phosphosulfate (PAPS) sulfotransferase in assimilatory sulfate reducers. *Plant Sci. Lett.* **4**:301–307.

Wanner, G., G. Henkelmann, A. Schmidt, and H. Kost. 1986. Nitrogen and sulfur starvation of the cyanobacterium *Synechococcus* 6301. An ultrastructural, morphometrical, and biochemical comparison. *Z. Naturforsch. Sect. C* **41**:741–750.

Westley, J. 1973. Rhodanese. *Adv. Enzymol.* **39**:327–368.

Whitton, B. A., M. Potts, J. W. Simon, and S. L. J. Grainger. 1990. Phosphatase activity of the blue-green alga (cyanobacterium) *Nostoc commune* UTEX 584. *Phycologia* **29**:139–145.

Wood, N. B., and R. Haselkorn. 1980. Control of phycobiliprotein proteolysis and heterocyst differentiation in *Anabaena. J. Bacteriol.* **141**:1375–1385.

Wyman, M., R. P. F. Gregory, and N. G. Carr. 1985. Novel role for phycoerythrin in a marine cyanobacterium, *Synechococcus* strain DC2. *Science* **230**:818–820.

Xie, W.-Q., B. A. Whitton, J. W. Simon, K. Jager, D. Reed, and M. Potts. 1989. *Nostoc commune* UTEX 584 gene expressing indole phosphate hydrolase activity in *Escherichia coli. J. Bacteriol.* **171**:708–713.

Yamanaka, G., and A. N. Glazer. 1980. Dynamic aspects of phycobilisome structure. Phycobilisome turnover during nitrogen starvation in *Synechococcus* sp. *Arch. Microbiol.* **124**:39–47.

IV. PHOSPHATE REGULATION IN PATHOGENESIS AND SECONDARY METABOLISM

Chapter 20

Introduction: Phosphate and Survival of Bacteria

ARNOLD L. DEMAIN

Biology Department, Massachusetts Institute of Technology, Cambridge, Massachusetts 02139

In a perfect microbial world, microorganisms would happily grow to their fullest extent on an unending supply of nutrients. However, in the real world, they must survive under less than perfect conditions. Some do this by forming resistant structures (e.g., endospores) that can survive in soil for decades; others compete by producing chemical weapons (e.g., antibiotics) that can kill other forms of life; and some seek advantage by attacking higher forms of life (e.g., pathogenesis). In this section, the effects of phosphate and/or phosphorylation on these processes are examined.

The work of Keith Ireton and Alan D. Grossman (chapter 23) deals with differentiation in *Bacillus* species. In the complex development program of *Bacillus subtilis,* positive gene regulation depending on nutrient deprivation, high cell density, continued DNA synthesis, undamaged DNA, and an intact citric acid cycle lead to sporulation. Regulation is all mediated by the phosphorylative activation of a developmental transcription protein encoded by the gene *spo0A*. Another process, the development of the ability to be transformed by exogenous DNA (i.e., competence), is stimulated by high cell density, and this is mediated by phosphorylation of a different regulatory protein, ComA.

The biosynthesis of antibiotics is often negatively affected by levels of P_i which support maximum growth rates. This is the case with candicidin production by *Streptomyces griseus.* In chapter 24, Juan Martin, Paloma Liras, and coworkers show that the phosphate effect is mediated through the repression of a pathway enzyme, *p*-aminobenzoate synthase, carried out by interference in transcription of the *pabS* gene.

In *Salmonella typhimurium,* the PhoP regulon (composed of the PhoQ sensor kinase and the PhoP transcriptional regulator) is necessary for virulence to mice, survival in macrophages, and growth on succinate. The relationship of phosphorylation to these complex effects is discussed by Elizabeth Hohmann and Samuel Miller in chapter 21. Approximately 20 genes, widely located on the chromosome, require the regulon for transcriptional activation.

In chapter 22, Michael Vasil and coworkers investigate the role of phosphate on the pathogenesis of *Pseudomonas aeruginosa.* Expression of two phospholipase genes that are involved in virulence is derepressed by phosphate limitation. The complicated mechanisms underlying these effects and their relationship to osmoprotectants are just being revealed.

It is clear that phosphate and/or phosphorylation has major effects on the survival of soil bacteria and medically important bacteria. The basic mechanisms underlying such effects are just beginning to be understood. We expect to see much more work on these mechanisms in the future.

Chapter 21

The *Salmonella* PhoP Virulence Regulon

ELIZABETH L. HOHMANN AND SAMUEL I. MILLER

Infectious Disease Unit, Massachusetts General Hospital, Fruit Street, Boston, Massachusetts 02114

Salmonellae are motile gram-negative bacteria which are important enteric pathogens. These facultative intracellular parasites have significant host specificity. For example, *Salmonella typhimurium* infection results in a self-limiting gastroenteritis in humans, whereas infection of BALB/c mice results in a lethal typhoid fever-like illness. In contrast, *S. typhi* is a uniquely human pathogen and mice are completely resistant to infection. When *Salmonella* cells colonize and invade mammalian hosts, they must adapt to several different and rapidly changing hostile environments, including those found within the stomach (low pH) and intestine (bile salts, digestive enzymes, antimicrobial peptides, immunoglobulin A antibody, and competition with commensals). In addition, once the organisms invade host tissues, they are phagocytosed by eucaryotic cells and exposed to the oxidative and nonoxidative killing mechanisms of macrophages and polymorphonuclear leukocytes. The PhoP regulon, which includes a typical two-component regulatory system composed of a membrane sensor histidine kinase (PhoQ), a transcriptional activator (PhoP), and over 40 regulated gene products encoded by *phoP*-activated (*pag*) and -repressed (*prg*) genes, specifically responds to a macrophage intracellular environment to promote bacterial survival and disease.

Phosphatases of *S. typhimurium*

Kier et al. (1977) purified three separate periplasmic phosphatases of *S. typhimurium* LT2. These included a cyclic 2′,3′-nucleotide phosphodiesterase, an acid hexose phosphatase, and a nonspecific phosphatase. The 67-kDa cyclic nucleotide phosphodiesterase demonstrates optimal activity at pH 7.5 and requires divalent cations for activity. The hexose phosphatase functions as a dimer of 37-kDa subunits. The nonspecific acid phosphatase demonstrates optimal activity at pH 5.5 and functions as a dimer of 27-kDa subunits (Weppelman et al., 1977). Unlike *Escherichia coli*, *Salmonella* cells do not contain a 5′ nucleotidase or a nonspecific alkaline phosphatase (Kasahara et al., 1991). In *Salmonella* cells the acid hexose phosphatase and the cyclic phosphodiesterase are regulated by catabolite repression (cyclic AMP); however, the nonspecific acid phosphatase (NSAP) is part of the PhoP regulon (Kier et al., 1979). In contrast to the induction of *E. coli* alkaline phosphatase by phosphate limitation, *Salmonella* NSAP activity is nonspecifically induced by starvation regardless of whether the substrate limited is carbon, nitrogen, sulfur, or phosphate (Kier et al., 1979). As predicted by these data, expression of NSAP correlates inversely with growth rate and is maximal in stationary phase.

Two genetic loci, *phoN* at 95 min and *phoP* at 25 min, are required for acid phosphatase production. (Kier et al., 1979) Because temperature-sensitive mutants with mutations in *phoN* and constitutive expression mutants with mutations in *phoP* were isolated, Kier et al. (1977) postulated that *phoN* was the structural gene for acid phosphatase and that *phoP* was a regulatory locus; this hypothesis was subsequently proven (Kasahara et al., 1991; Miller et al., 1989).

The DNA sequence of *phoN* was recently published by two groups (Kasahara et al., 1991; Groisman et al., 1992c). The deduced amino acid sequence predicts that the acid phosphatase is a 250-amino-acid polypeptide with an amino-terminal leader peptide targeting the enzyme for secretion. NSAP activity was found in 98% of the *Salmonella* isolates tested, and isolates without such activity were found to have point mutations in *phoN* (Groisman et al., 1992c). The *phoN* locus has a G+C content of 43%, significantly lower than the overall content of the *Salmonella* genome (53%). In a screening of 14 other gram-negative bacteria for NSAP, only *Morganella morganii* and *Providentia stuartii*, organisms with low genomic G+C contents, had NSAP activity. These data, combined with the fact that DNA 5′ to the *phoN* gene is similar to the *oriT* of IncFII plasmids, led Groisman et al. (1992c) to postulate that *phoN* was acquired by lateral transmission in a plasmid-mediated event.

NSAP has been presumed to function as a scavenger of phosphate groups and phosphate esters which are otherwise not transportable across the cell membrane. The presence of such an enzyme could potentially provide a survival advantage under phosphate-limiting conditions. However, *phoN* mutants have been shown to grow more slowly only when *Salmonella* cells are forced to scavenge certain unusual phosphomonoesters (Kier et al., 1979). Perhaps as a result of the redundancy of phosphate-scavenging systems, *phoN* mutants have the virulence phenotype of wild-type organisms (Fields et al., 1989; Miller et al., 1989). The *phoN* locus requires the *phoP* regulatory locus for expression, and, similar to other *phoP*-activated genes, is probably induced within macrophages (see below). It could function to scavenge phosphate or to dephosphorylate periplasmic phosphate-containing substrates when it is induced specifically within the acidic macrophage environment.

Phenotypes of *phoP* Mutants

S. typhimurium infection of BALB/c mice has been extensively studied as a model of human typhoid fever (Carter and Collins, 1974). The *phoP* locus has been demonstrated by several groups of investigators to be essential to the full virulence of *S. typhimurium* for BALB/c mice (Fields et al., 1989; Miller et al., 1989; Galan and Curtiss, 1989). Mice can survive either oral or intraperitoneal innoculations with numbers of PhoP-null organisms representing 10,000 times the 50% lethal dose (LD_{50}) of wild-type strains (Galan and Curtiss, 1989; Miller et al., 1990a). It is likely that several different *phoP*-regulated virulence factors contribute to this impressive virulence defect. *phoP* mutants are defective in survival within cultured macrophages, and this phenotype has been correlated with animal virulence (Fields et al., 1986). *phoP* mutants are also remarkably sensitive to cationic peptides with antimicrobial activities (Fields et al., 1989; Miller et al., 1990b), including the intestinal defensins or cryptdins, peptides secreted into the mouse intestinal lumen from Paneth cells (Selsted et al., 1992) (Fig. 1). In addition, *phoP* mutants survive poorly at pH 3.3 or less, an environment encountered in the stomach (Foster and Hall, 1990). The extent of the contribution of acid and cationic protein sensitivity to the virulence defect of *phoP* mutants is unknown because no *pag* mutants with these phenotypes have been discovered. Cationic protein resistance is likely to be important to virulence, because several mutants sensitive to such compounds are attenuated for virulence (Groisman et al., 1992b).

The *phoP* Locus

The *phoP* locus contains two genes, *phoP* and *phoQ*, located in an operon (Fig. 2). Other enteric bacteria may contain homologs of these regulatory genes. A gene isolated from *E. coli* is predicted to encode a protein that is 93% similar to PhoP (Groisman et al., 1992; Kasahara et al., 1992). Southern hybridization experiments demonstrate that other gram-negative bacteria, as well as *Saccharomyces cerevisiae,* contain homologous DNA sequences, suggesting that regulators similar to PhoP could be present in a wide variety of microorganisms (Groisman et al., 1989). The deduced amino acid sequence of the gene products of *phoP* and *phoQ* are highly similar to other members of a family of two-component transcriptional regulators that respond to environmental stimuli by a phosphotransferase mechanism (Parkinson, 1993). In this family of protein pairs a protein kinase (PhoQ) responds to environmental signals by autophosphorylation of a histidine residue within its cytoplasmic domain. This sensor kinase phosphate (PhoQ-phosphate) then transfers phosphate to an aspartate(s) in the amino-terminal domain of a DNA-binding protein (PhoP). This phosphorylated DNA-binding protein (PhoP-phosphate) subsequently activates transcription of unlinked genes. The environmental signal is therefore transduced by phosphate transfer and results in transcriptional activation of unlinked genes.

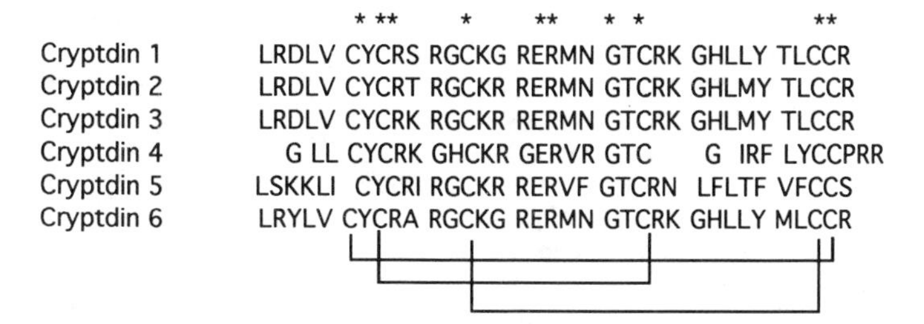

FIG. 1. Amino acid sequences of murine enteric defensins (cryptdins). Cryptdins 1 through 6 have been purified and sequenced at the peptide level. Solid bars under cryptdin 6 represent the characteristic intramolecular sulfide bonds of the defensin family of peptides, and asterisks above cryptdin 1 represent consensus residues in the defensin family.

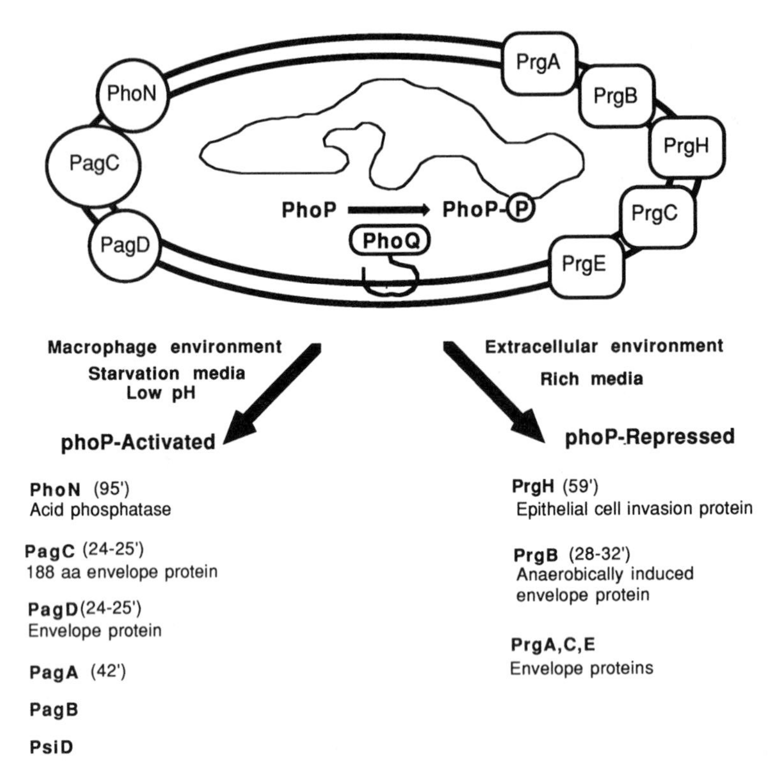

FIG. 2. Schematic representation of the PhoP regulon. PhoQ is a membrane-associated sensor kinase, and PhoP is a cytoplasmic transcriptional regulator. *phoP*-activated genes are induced by exposure to the macrophage intracellular environment, low pH, and starvation conditions. *phoP*-repressed genes are repressed by these environmental conditions and activated by extracellular environmental conditions and rich media.

The deduced amino acid sequence of the proteins encoded by the *phoP* locus predicts that the PhoP and PhoQ proteins are 224 and 487 amino acids long, respectively. Consistent with these data, the PhoP and PhoQ proteins are approximately 28 and 55 kDa, respectively, when sized by using Western hybridization analysis (Hohmann and Miller, unpublished). PhoP is of the OmpR subclass of the phosphorylated transcriptional activator family. Like the *ompR* locus, *phoP* transcription is neither autoregulated nor induced by environmental signals which activate *pag* (Hohmann and Miller, unpublished). Similar to EnvZ, PhoQ has two hydrophobic segments that are predicted to be transmembrane domains. The hydrophilic segment between these transmembrane domains is predicted to be located in the periplasm, and the kinase activity is predicted to be located in the cytoplasmic domain. Analysis of PhoQ topology with Tn*phoA*-generated gene fusions is consistent with this hypothesis (Miller et al., 1989; Hohmann and Miller, unpublished). (Active *phoA* fusions occur only when PhoA is fused to a domain localized in the periplasm, which allows reduction and dimerization of alkaline phosphatase [Manoil and Beckwith, 1985].) Similarity to known sensor histidine kinases is found only in the cytoplasmic domain of PhoQ at the predicted site of autophosphorylation and at the carboxy terminus. These findings are consistent with a transmembrane topology for PhoQ, permitting the extracellular environment to be sensed via the periplasmic domain. Consistent with this hypothesis, mutations in the periplasmic domain of PhoQ can alter *phoP*-activated gene expression in either a positive or negative fashion, i.e., either abolishing or activating *pag* gene expression (Hohmann and Miller, unpublished).

PhoP Constitutive Phenotype

Analysis of PhoP-regulated genes has been facilitated by the study of an allele of the *phoP* locus (*pho-24*) that results in a PhoP constitutive phenotype. Mutants with this allele were originally isolated on the basis of a high level of acid phosphatase expression in rich media (Kier et al., 1979). This phenotype, the result of a single-amino-acid change that alters a threonine in the periplasmic domain of PhoQ, may simulate the signal transduction that occurs within mammalian cells (Hohmann and Miller, unpublished). Two-dimensional polyacrylamide gel electrophoresis has been used to compare protein expression in wild-type and PhoP constitutive bacteria grown in rich medium (Miller and Mekalanos, 1990). Over 40 different proteins are altered in expression as a result of this single mutation, with approximately 20 increased and 20 markedly diminished in expression. Interestingly, PhoP constitutive bacteria are dramatically attenuated for virulence in mice with an LD_{50} similar to that of PhoP null mutants. Additionally, PhoP constitutive mutants are defective in survival within macrophages and after exposure to human serum (complement-mediated killing) (Behlau and Miller, unpublished). PhoP constitutive mutants are also defective in epithelial cell invasion (Behlau and Miller, 1993), whereas PhoP null mutants demonstrate wild-type invasion of such cells (Miller et al., 1992). This attenuated strain has normal, wild-type sensitivity to the rabbit defensin NP-1 (Miller et al., 1990b). These different virulence phenotypes of PhoP constitutive and PhoP null mutants together suggest that regulated expression of both *phoP*-activated and *phoP*-repressed genes is important for full virulence. Virulence is unlikely to be simply dependent on the expression of virulence factors but requires appropriate transcriptional regulation of these factors in response to environmental signals.

phoP-Activated Genes

To dissect the virulence phenotype regulated by the *phoP/phoQ* operon, a number of *phoP*-activated genes have been identified by using transposon mutagenesis (Gunn et al., submitted; Miller et al., 1989; Pulkkinen and Miller, 1991). These loci have been designated *pagA* through *pagP* and are widely spaced on the *Salmonella* chromosome, although several are clustered around 24 to 25 min. One gene at this location, *pagC*, encodes a single 188-amino-acid envelope protein that is similar to a family of outer membrane proteins of diverse function, including the *ail*-encoded eukaryotic invasion protein of *Yersinia enterocolitica* and the *lom*-encoded protein of bacteriophage lambda (Pulkkinen and Miller, 1991). *pagC*::Tn*phoA* mutants are deficient in survival within cultured murine macrophages and have a virulence defect in mice; this virulence defect is complemented by a plasmid containing a wild-type clone of *pagC*. Another gene, *pagD*, that encodes a single envelope protein is divergently transcribed from *pagC*, and Tn*phoA* insertions in this gene also attenuate *S. typhimurium*. Transposon insertions in the *pagA* and *pagB* loci do not as single mutations alter virulence in the murine model of typhoid fever (Miller et al., 1989).

phoP-Repressed Genes

phoP-repressed genes are defined as loci repressed as part of the PhoP constitutive phenotype. Five unlinked *phoP* repressed loci have been identified by Tn*phoA* mutagenesis and are widely dispersed on the *S. typhimurium* chromosome (Behlau and Miller, 1993). These genes are maximally expressed under conditions in which *phoP*-activated genes are not highly expressed (i.e., rich medium and log phase growth conditions). One *phoP*-repressed locus, *prgH*, is located at 59 min and is essential for *Salmonella*-induced epithelial cell endocytosis. *prgH*::Tn*phoA* mutants demonstrate wild-type survival within macrophages and have a mild virulence defect consistent with an inability to cross intestinal epithelia. *prgH* is highly linked to the *hil* (hyperinvasive) locus described by Lee et al. (1992). In PhoP constitutive bacteria, *prgH* transcription is markedly repressed. The fact that PrgH and PhoP constitutive mutants are defective in epithelial cell invasion supports the hypothesis that *phoP*-regulated genes may be important for at least two temporally separated phases of *Salmonella* pathogenesis; i.e., *phoP*-repressed gene expression is necessary for entry into epithelial cells, and *phoP*-activated gene expression is necessary for intracellular survival within macrophages.

Intracellular Activation of *pag* Transcription

Several hours after *S. typhimurium* organisms are phagocytosed by mouse macrophages, *phoP*-activated gene expression is dramatically induced, by approximately 100-fold (Alpuche Aranda et al., 1992). In contrast, after endocytosis by epithelial cells, no induction is observed. Data obtained by Abshire and Neidhardt (1993) have shown that exposing *Salmonella* cells to the intracellular milieu of the macrophage results in selective induction of a small subset of the many gene products induced by other stresses such as heat shock, peroxide, and DNA damage. *phoP*-activated gene products are among the most highly induced proteins after *Salmonella* cells enter macrophages (Abshire and Neidhardt, 1993). The specificity and magnitude of *phoP*-activated gene expression within macrophages suggest that a primary function of the PhoP-PhoQ regulon may be to modulate adaption to the intracellular environment of the macrophage rather than to perform metabolic activities such as phosphate scavenging.

phoP-activated gene induction requires acidification of the phagosome, because activation of *pag* transcription is blocked by the addition of weak bases that raise the pH of macrophage vacuoles (Alpuche Aranda et al., 1992). In addition, measurement of the pH from individual phagosomes indicates that the eventual acidification of phagosomes containing *Salmonella* cells to pH < 5.0 correlates with the period of maximal *phoP*-dependent gene expression (Alpuche Aranda et al., 1992). In vitro activation of *pag* by pH has also been demonstrated but to a much lesser degree than that achieved intracellularly (Alpuche Aranda et al., 1992), suggesting that additional, as yet undefined signals are also important for maximal PhoQ activation within macrophage phagosomes.

The PhoP Regulon and Spacious Phagosomes

In typical ingestion of microorganisms by macrophages, microorganisms are engulfed within a phagosome whose walls closely appose them (Silverstein et al., 1977). However, *S. typhimurium* enters mouse macrophages by induction of macropinocytosis and resides initially within large fluid-filled organelles termed spacious phagosomes (Alpuche Aranda et al., 1993) (Fig. 3). These organelles are morphologically similar to macropinosomes induced by growth factors and transforming agents, which are formed as a consequence of membrane ruffling (Racoosin and Swanson, 1992).

Internalization by macropinocytosis is observed after several different methods of bacterial opsonization, although after opsonization with immunoglobulin G, bacteria appear to enter small phagosomes that enlarge over time to form spacious phagosomes. Unlike the transient macropinosomes induced by colony-stimulating factor, the *Salmonella*-induced spacious phagosomes persist for a significant period as a result of fusion with each other or with fluid-filled macropinosomes. Therefore, *Salmonella* cells appear to persist within macrophages by slowing the shrinkage and acidification of spacious phagosomes. Spacious-phagosome formation may promote survival within macrophages by dilution of toxic lysosomal compounds or by allowing a gradual acclimatization to the harsh phagosome environment. The timing of spacious-phagosome shrinkage correlates with the induction of *phoP*-activated gene expression. PhoP constitutive bacteria are defective in spacious-phagosome formation and induction of macropinocytosis, whereas PhoP null mutants form spacious phagosomes normally, suggesting that two separate phases of *Salmonella* gene expression are important and are influenced by different macrophage microenvironments. Early in infection, *Salmonella* entry and spacious-phagosome formation are dependent on PhoP repressed gene expression; later, after acidification and shrinkage of the spacious phagosome, a switch to *pag* expression occurs.

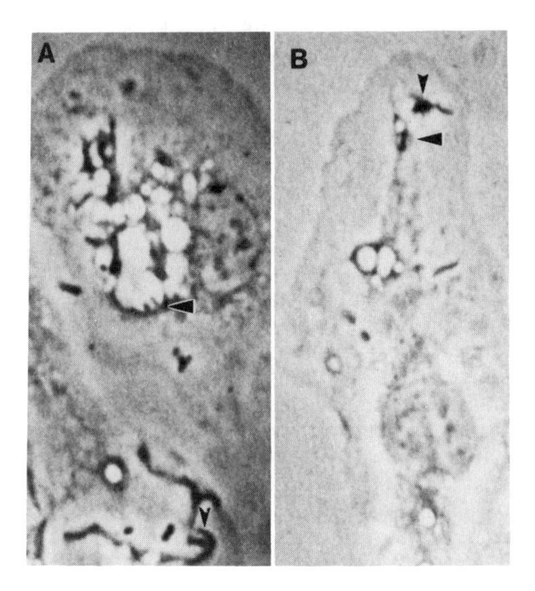

FIG. 3. Comparison of phagosomes formed by wild-type and PhoP constitutive *S. typhimurium* cells within BALB/c mouse bone marrow-derived macrophages. The images are selected frames from time-lapse phase-contrast videomicroscopy after phagocytosis of bacteria by macrophages. (A) Wild-type *S. typhimurium* cells within a spacious phagosome (larger horizontal arrowhead) and entering the macrophage within a membrane ruffle (smaller vertical arrowhead). Two bacteria adherent to the lower pole of the spacious phagosome are visible within the phase-bright phagosome. Near the vertical arrow, the prominent membrane ruffle is visible as a phase-dense line which is beginning to encircle another bacterium. (B) PhoP constitutive *S. typhimurium* cells enclosed within a conventional smaller phagosome with its membrane tightly adherent to the bacteria (larger horizontal arrowhead). A PhoP constitutive bacterium is shown entering the macrophage with minimal perturbation of the macrophage membrane (small vertical arrowhead).

The PhoP Regulon and Vaccine Development

Many investigators are currently working to develop better live bacterial vaccines for typhoid fever and other human and animal illnesses caused by nontyphoidal *Salmonella* species. *Salmonella* species are also attractive vectors that can be used to carry heterologous antigens, and they may be useful in the creation of live oral vaccines against multiple antigens. They are promising carriers for live multicomponent vaccines because of their relative safety in an immunosuppressed population and the ease with which they can be genetically manipulated. The addition of PhoP regulon mutations may confer added safety to such live vaccines. When used as a live vaccine, the PhoP constitutive mutant is extremely immunogenic, possibly because of its defect in spacious-phagosome formation. The use of *phoP*-regulated promoters to express heterologous antigens may be particularly useful in obtaining high-level expression of multiple heterologous antigens within macrophages. It is hoped that the rational development of multivalent vaccines will be advanced by the knowledge of the interactions between mammalian cells and *Salmonella* cells.

We thank the other members of our laboratory and our collaborators in other laboratories for scientific discussions and inspiration.

This work was supported by grants from the National Institutes of Health (AI30479 and AI00917) to S.I.M. E.L.H. is supported by an institutional grant from the National Institutes of Health (DK01410) to the Massachusetts General Hospital.

REFERENCES

Abshire, K. Z., and F. C. Neidhardt. 1993. Growth rate paradox of *Salmonella typhimurium* within host macrophages. *J. Bacteriol.* **175:**3744–3748.

Alpuche Aranda, C., J. A. Swanson, W. P. Loomis, and S. I. Miller. 1992. *Salmonella typhimurium* activates virulence gene transcription within acidified macrophage phagosomes. *Proc. Natl. Acad. Sci. USA* **89:**10079–10083.

Alpuche Aranda, C. M., E. L. Racoosin, J. A. Swanson, and S. I. Miller. 1993. *Salmonella* enter macrophages by macropinocytosis and survive within spacious phagosomes. *J. Exp. Med.* **179:**601–608.

Behlau, I., and S. I. Miller. 1993. A PhoP-repressed gene promotes *Salmonella typhimurium* invasion of epithelial cells. *J. Bacteriol.* **175:**4475–4484.

Behlau, I., and S. I. Miller. Unpublished data.

Carter, P. B., and F. M. Collins. 1974. The route of enteric infection in normal mice. *J. Exp. Med.* **139:**1189–1203.

Fields, P. I., E. A. Groisman, and F. Heffron. 1989. A *Salmonella* locus that controls resistance to microbicidal proteins from phagocytic cells. *Science* **243:**1059–1062.

Fields, P. I., R. V. Swanson, C. G. Haidaris, and F. Heffron. 1986. Mutants of *Salmonella typhimurium* that cannot survive within the macrophage are avirulent. *Proc. Natl. Acad. Sci. USA* **83:**5189–5193.

Foster, J. W., and H. K. Hall. 1990. Adaptive acidification tolerance response of *Salmonella typhimurium*. *J. Bacteriol.* **172:**771–778.

Galan, J. E., and R. Curtiss III. 1989. Virulence and vaccine potential of *phoP* mutants of *Salmonella typhimurium*. *Microb. Pathog.* **6:**422–443.

Gunn, J. S., C. M. Alpuche-Aranda, W. P. Loomis, W. J. Belden, and S. I. Miller. A virulence gene cluster required for *Salmonella typhimurium* survival within macrophage phagosomes. Submitted for publication.

Groisman, E. A., E. Chiao, C. J. Lipps, and F. Heffron. 1989. *Salmonella typhimurium phoP* virulence gene is a transcriptional regulator. *Proc. Natl. Acad. Sci. USA* **86:**7077–7081.

Groisman, E. A., F. Heffron, and F. Solomon. 1992a. Molecular genetic analysis of the *E. coli phoP* locus. *J. Bacteriol.* **174:**486–491.

Groisman, E. A., L. C. Para, M. Salcedo, C. J. Lipps, and F. Heffron. 1992b. Resistance to host antimicrobial peptides is necessary for *Salmonella* virulence. *Proc. Natl. Acad. Sci. USA* **89:**11939–11943.

Groisman, E. A., M. H. Saier, and H. Ochman. 1992c. Horizontal transfer of a phosphatase gene as evidence for mosaic structure of the *Salmonella* chromosome. *EMBO J.* **11:**1309–1316.

Hohmann, E. L., and S. I. Miller. Unpublished data.

Kasahara, M., A. Nakata, and H. Shinagawa. 1991. Molecular analysis of the *Salmonella typhimurium phoN* gene, which encodes nonspecific acid phosphatase. *J. Bacteriol.* **173:**6760–6765.

Kasahara, M., A. Nakata, and H. Shinagawa. 1992. Molecular analysis of the *E. coli phoP-phoQ* operon. *J. Bacteriol.* **174:**492–498.

Kier, L. D., R. Weppelman, and B. N. Ames. 1977. Resolution and purification of three periplasmic phosphatases of *Salmonella typhimurium*. *J. Bacteriol.* **130:**399–410.

Kier, L. D., R. M. Weppleman, and B. N. Ames. 1979. Regulation of nonspecific acid phosphatase in *Salmonella*: *phoN* and *phoP* genes. *J. Bacteriol.* **138:**155–161.

Lee, C. A., B. D. Jones, and S. Falkow. 1992. Identification of a *Salmonella typhimurium* invasion locus by selection for hyperinvasive mutants. *Proc. Natl. Acad. Sci. USA* **89:**1847–1851.

Manoil, C., and J. Beckwith. 1985. Tn*phoA*: a transposon probe for protein export signals. *Proc. Natl. Acad. Sci. USA* **82:**8129–8133.

Miller, S. I. 1990. Unpublished data.

Miller, S. I., A. M. Kukral, and J. J. Mekalanos. 1989. A two-component regulatory system (*phoP/phoQ*) controls *Salmonella typhimurium* virulence. *Proc. Natl. Acad. Sci. USA* **86:**5054–5058.

Miller, S. I., and J. J. Mekalanos. 1990. Constitutive expression of the PhoP regulon attenuates *Salmonella* virulence and survival within macrophages. *J. Bacteriol.* **172:**2485–2490.

Miller, S. I., J. J. Mekalanos, and W. S. Pulkkinen. 1990a. *Salmonella* vaccines with mutations in the *phoP* virulence regulon. *Res. Microbiol.* **141:**817–821.

Miller, S. I., W. S. Pulkkinen, M. E. Selsted, and J. J. Mekalanos. 1990b. Characterization of defensin resistance phenotypes associated with mutations in the *phoP* virulence regulon of *Salmonella typhimurium*. *Infect. Immun.* **58:**3706–3710.

Miller, V. L., K. B. Beer, W. P. Loomis, J. A. Olson, and S. I. Miller. 1992. An unusual *pagC*::Tn*phoA* mutation leads to an invasion- and virulence-defective phenotype in *Salmonella*. *Infect. Immun.* **60:**3763–3770.

Parkinson, J. S. 1993. Signal transduction schemes of bacteria. *Cell* **73:**857–871.

Pulkkinen, W. S., and S. I. Miller. 1991. A *Salmonella typhimurium* virulence protein is similar to a *Yersinia enterocolitica* and a bacteriophage lambda outer membrane protein. *J. Bacteriol.* **173:**86–93.

Racoosin, E. L., and J. A. Swanson. 1992. M-CSF-induced macropinocytosis increases solute endocytosis but receptor mediated endocytosis. *J. Cell Sci.* **102:**867–880.

Selsted, M. E., S. I. Miller, A. H. Henschen, and A. J. Oulette. 1992. Enteric defensins: antibiotic peptide components of intestinal host defense. *J. Cell Biol.* **118:**929–936.

Silverstein, S. C., R. M. Steinman, and Z. A. Cohn. 1977. Endocytosis. *Annu. Rev. Biochem.* **46:**669–677.

Weppelman, R., L. D. Kier, and B. N. Ames. 1977. Properties of two phosphates and a cyclic phosphodiesterase of *Salmonella typhimurium*. *J. Bacteriol.* **130:**411–419.

Chapter 22

Phosphate and Osmoprotectants in the Pathogenesis of *Pseudomonas aeruginosa*

MICHAEL L. VASIL,[1] ADRIANA I. VASIL,[1] AND VIRGINIA D. SHORTRIDGE[2]

Department of Microbiology, University of Colorado Health Sciences Center, Denver, Colorado 80262,[1] and Department of Laboratory Medicine, University of Washington Medical Center, Seattle, Washington 98195[2]

"Is nutritional immunity specific only for iron or might such other nutrilites as inorganic phosphate or zinc be involved? Human beings infected with gram negative (but not gram positive) bacterial pathogens reduce the quantity of their inorganic phosphate (P_i) in their plasma to a level that is suboptimal for growth." (Weinberg, 1974).

In the 20 years that have passed since Weinberg (1974) made this statement, a great deal has been learned about the role of iron deprivation in host-pathogen interactions. In contrast, during the same two decades little information has emerged about the possible roles of the limitation of P_i or other ions in the expression of bacterial virulence. Reasons for the relative paucity of data on this subject may be (i) that the expression of many known virulence factors (e.g., toxins) has frequently been associated with iron limitation but not with P_i limitation and (ii) the erroneous impression that microorganisms are probably not P_i restricted in mammalian hosts. Data from the early literature relating to the pathogenesis of *Pseudomonas aeruginosa* suggest that neither of these rationalizations is justified. Liu (1979), who originally described the plethora of extracellular virulence factors produced by this organism, also discovered that the expression of an extracellular heat-labile hemolysin of *P. aeruginosa* is induced by P_i limitation. This hemolysin, like the potent alpha-toxin of *Clostridium perfringens*, was subsequently found to have phospholipase C (PLC) activity (Fig. 1) (Esselmann and Liu, 1961). Because expression of an alkaline phosphatase as well as of the hemolytic PLC was induced by P_i limitation, Liu (1979) also proposed that these enzymes could be useful for scavenging P_i in the host. *P. aeruginosa* can also use the choline moiety of phosphatidylcholine as a sole carbon and nitrogen source (Fig. 1). In support of his hypothesis, Liu (1979) demonstrated that bronchial washings alone, which have plentiful amounts of phospholipids (e.g., phosphatidylcholine) but are limited in other nutrients including P_i, supported growth of *P. aeruginosa* as well as production of PLC. Further evidence of the role of P_i-limitation in the virulence of *P. aeruginosa* came during our earlier studies on the regulation of PLC production, when we (Gray et al., 1982; Gray and Vasil, 1981a; Vasil et al., 1985) isolated a mutant of *P. aeruginosa* that was not inducible for either PLC or alkaline phosphatase production under phosphate limitation (see below for a discussion of *plcA* mutation). This mutant, in comparison with the wild-type parental strain, was severely altered (~3-log-unit increase in 50% lethal dose) in its virulence in a mouse infection model (Vasil et al., 1985). More recently, Ostroff et al. (1989) demonstrated that the virulence of a wild-type strain of *P. aeruginosa* could be significantly modulated by altering the P_i concentration of the growth media used prior to infection in the mouse infection model. *P. aeruginosa* cells grown in low P_i (1 mM P_i) before their introduction into the mouse model were nearly 500-fold more virulent than were cells grown in 10 mM P_i prior to their inoculation into mice.

Because there is ample evidence that the production of a significant virulence determinant of *P. aeruginosa* (i.e., PLC) is regulated by P_i concentration and because this organism is eminently amenable to classical and molecular genetic analysis, this is an attractive system for use in addressing questions salient to the role of phosphate limitation on microbial virulence. This report describes our molecular and biochemical studies relating to the regulation of PLC from *P. aeruginosa*. When we initiated these efforts, we did not envision the complexity of this regulatory system, even though Liu (1979) prophetically warned that "production of phospholipase C by *P. aeruginosa* is a very elusive characteristic easily influenced by many environmental factors." Since that time, however, we have learned a great deal about the environmental influences that affect expression of this virulence determinant. It is also clear that there are still significant gaps to be filled.

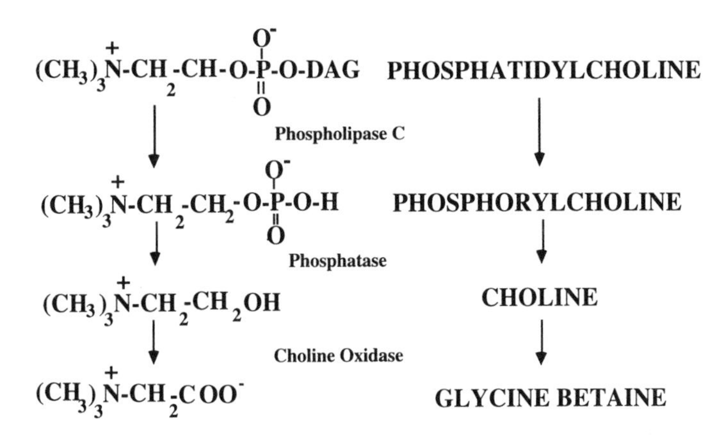

FIG. 1. Pathway of degradation of phospholipids (e.g., phosphatidylcholine) by *P. aeruginosa. P. aeruginosa* can also convert glycine betaine to glycine by successive demethylation and can then use glycine as a carbon and nitrogen source by metabolism of this amino acid. DAG, diacylglycerol.

Identification and Analysis of Two Genes of *P. aeruginosa* Encoding PLC Activity

A gene from *P. aeruginosa* encoding a hemolytic PLC activity was cloned and expressed in *Escherichia coli* (Vasil et al., 1982). Although the gene was not expressed from its own promoter in *E. coli,* it could be expressed by using the hybrid *E. coli tac* promoter. Sequencing and S1 nuclease analysis of the 5′ region of this gene revealed that it does not have a consensus *E. coli* promoter (Pritchard and Vasil, 1986). The structural gene encoding PLC is part of a three-gene operon. The other two genes, designated *plcR1* and *plcR2*, are located 3′ to the gene encoding PLC activity. The *plcR* gene products posttranslationally modify PLC, alter its conformation, and increase its hemolytic activity (Vasil and Vasil, unpublished). Pritchard and Vasil (1986) determined that transcription of this operon is enhanced in an environment low in P_i and repressed in an environment high in P_i (see Fig. 2).

To examine the role of PLC production in pathogenesis, gene replacement methods were used to delete the entire PLC operon (Ostroff et al., 1989) in a virulent strain of *P. aeruginosa.* This strain could then be compared with a wild-type strain for virulence in a mouse infection model. Surprisingly, however, it was found that this PLC deletion mutant still produced an extracellular PLC activity and that expression of this activity was regulated by P_i (Ostroff and Vasil, 1987; Ostroff et al., 1989). Because the mutant was constructed by gene replacement and because physical evidence indicated that the entire PLC operon had been deleted from *P. aeruginosa,* the only possibility was that this organism has two or more separate P_i-regulated genes encoding extracellular PLC activity. However, the extracellular PLC activity produced by the deletion mutant (ΔSR) is not hemolytic for human or sheep erythrocytes (Ostroff et al., 1989). Accordingly, the hemolytic PLC is called PLC-H while the nonhemolytic enzyme is designated PLC-N. The *plcN* gene was subsequently cloned, and it was found that the deduced amino acid sequences of PLC-H and PLC-N are 40% identical (Ostroff et al., 1990). Both genes seem to be well conserved in *P. aeruginosa* because both are associated with a constant-size restriction chromosomal fragment in all strains (>50) that have been examined thus far by Southern hybridization (Vasil and Vasil, unpublished). *plcH* and *plcN* are located at 34 and 67 min, respectively, on the ~75-min chromosome of *P. aeruginosa*; they are therefore not tandemly duplicated genes (Shortridge et al., 1992). Both enzymes are similar in size (78.3 kDa for PLC-H and 73.4 kDa for PLC-N) and have similar substrate specificities (phosphatidylcholine), and the expression of both is regulated by similar environmental signals in addition to P_i concentration (see below). It is therefore not possible to determine whether either or both enzymes were being examined in earlier studies (Ostroff et al., 1990; Ostroff and Vasil, 1987). Further analysis of the regulation of PLC expression and its role in the pathogenesis of *P. aeruginosa* must be carried out to investigate the production of both extracellular PLC activities by this organism.

Phosphate-Regulated Expression of PLC

We began the genetic analysis of PLC production by using chemical mutagenesis methods to isolate structural or regulatory mutants. Several mutants that are affected in P_i-regulated expres-

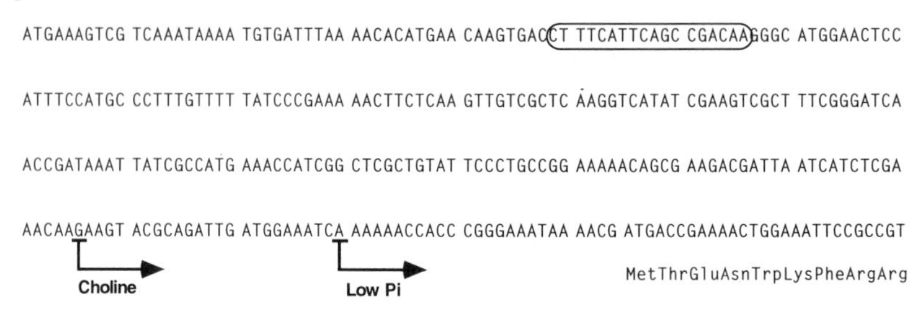

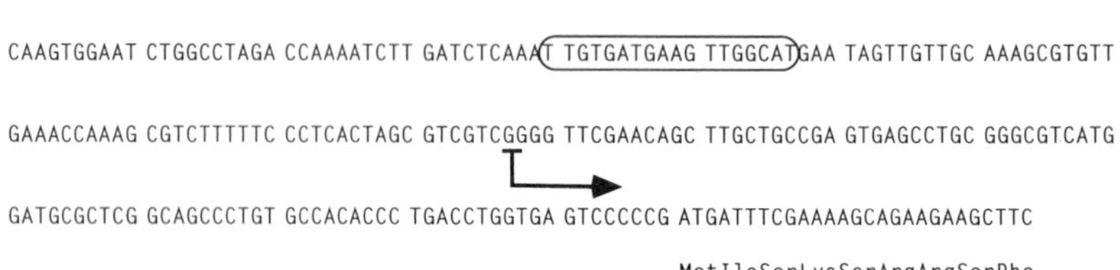

FIG. 2. The 5′ regulatory regions of *plcH* and *plcN* are shown (Ostroff et al., 1990; Pritchard and Vasil, 1986; Smith et al., 1988) (accession numbers are M13047 for *plcH* and M59304 for *plcN*). The circled 18 nucleotide bases 5′ to the transcriptional initiation sites represent putative *pho* boxs for each gene as described by Anba et al. (1990). The underlined bases with arrows indicate the experimentally determined (RNase protection of S1 nuclease) transcriptional initiation site for *plcH* or *plcN*. The arrow in the *plcH* regulatory region, designated "Low P_i," indicates the transcriptional initiation site during growth under low-P_i growth conditions (1 mM), whereas the arrow designated "Choline" indicates the transcription initiation site for *plcH* during growth under high-P_i conditions with 1 mM choline. The "Low P_i" initiation site determined by RNase protection is consistent with the site previously identified under low-P_i conditions by S1 analysis (Pritchard and Vasil, 1986). The transcription initiation site for *plcN* was determined by RNase protection (see text). This site is used regardless of whether choline is present. No transcription of *plcN* is detected under high-P_i conditions. The first nine deduced amino acid residues for PLC-H and PLC-N are shown.

sion of PLC were isolated (Gray, et al., 1981, 1982; Gray and Vasil, 1981b). Two mutants with different phenotypes were examined in more detail. The mutation (*plcA*) in one of these strains results in a phenotype in which PLC expression and alkaline phosphatase expression are not inducible even under low-P_i conditions (1 mM). This mutant is nonderepressible for four additional, normally, P_i-repressible, extracellular proteins. These data suggested that the *plcA* mutation was in a gene that had a positive regulatory effect on several P_i-regulated genes. The other mutant carries a mutation (*plcB*) that results in the constitutive production of PLC, alkaline phosphatase, and three additional extracellular proteins under high-P_i conditions (>10 mM). Both mutants also had decreased rates of P_i transport compared with the wild-type strain. Both mutations were mapped by classical mapping methods, and their respective loci were indistinguishable by transduction. The strain carrying the *plcB* mutation was not examined further. In contrast, it was possible to determine the nature of the *plcA* mutation at the molecular level. Anba et al. (1990) found that the *plcA* mutation could be complemented by a *P. aeruginosa* homolog of the *E. coli* Pho regulon transcriptional activator gene, *phoB*. The amino acid sequence of PhoB from *P. aeruginosa* is 59% identical and 76% similar to *E. coli* PhoB, and there is also a gene adjacent to *phoB* in *P. aeruginosa* that encodes a protein with significant homology to the *E. coli* Pho regulon sensor protein, PhoR. Thus, these and other data clearly indicate that *P. aeruginosa* has a P_i-inducible regulon like those of *E. coli* and *Bacillus subtilis*. In *P. aeruginosa*, however, there are at least two additional genes (*plc*), controlled by this regulon, which encode products not found in these other organisms. One of these gene products, PLC-H, has already been shown to probably have a significant role in bacterial virulence (Granstrom et al., 1984; Meyers and Berk, 1990; Ostroff et al., 1989; Vasil, 1988).

At present there are no data available to establish whether the PhoB protein of *P. aeruginosa* binds and directly activates either of the *plc* operons or whether there is an intermediate gene which is activated by PhoB, whose gene product in turn activates either or both of the *plc* genes. There is no obvious homology to a consensus PhoB-binding site (*pho* box) that could be found in the −35 region of the *plcH* gene, but there is a *pho* box-like sequence located 221 bp 5′ to the transcription initiation site for low P_i induction (Fig. 2). Likewise, there is no obvious *pho* box-like sequence in the promoter region of the *plcN* gene, but there is a sequence similar to the *E. coli* consensus sequence 77 bp 5′ to the transcription initiation codon (Fig. 2) (Anba et al., 1990). Because these sites are considerably further upstream than known *pho* boxes in *E. coli,* it will be necessary to do additional experiments, such as mutational analysis and DNA footprinting, to determine if PhoB directly regulates either *plc* gene. We have observed a similar situation with *P. aeruginosa* genes that are responsive to iron and are regulated by a *P. aeruginosa* homolog of the well-characterized *E. coli* iron repressor gene, *fur.* These iron-responsive, Fur-regulated genes do not have sequences in their 5′ regulatory regions that obviously resemble the consensus *fur* box which has been identified for iron-responsive genes of *E. coli* (Prince et al., 1993). It is possible that the PhoB and Fur homologs from *P. aeruginosa* regulate their respective P_i limitation and iron limitation responsive genes indirectly. Alternatively, the relatively higher G+C content of this organism (~67%) than that of *E. coli* (~50%) might mean that the sequence of the *pho* box or *fur* box cannot be identified by simply inspecting the 5′ region of genes that are P_i or iron regulated but that the sequence of their respective operators must be experimentally determined. It is interesting that there are quantitative and qualitative differences between the responses of each of the *plc* genes to P_i, as well as other environmental signals (see below). Accordingly, further analysis of the regulation of the *plc* genes may reveal some novel information on mechanisms of gene regulation by PhoB.

Regulation of PLC Expression by Osmoprotectants and the Pathogenic Potential of PLC in CF

Perhaps because it was noticed early that PLC expression is regulated by P_i, the influence of other environmental signals was not initially investigated. Recently it was found that the expression of an acid phosphatase, an acetylcholinesterase, and PLC activity could be detected when *P. aeruginosa* was grown even in media with high P_i (>10 mM) if 1 mM choline or glycine betaine were present (Lucchesi et al., 1989; Shortridge et al., 1992). These enzyme activities were detected at significantly lower levels (less than 1/10) when these compounds were omitted from the growth media. Shortridge et al. (1992) demonstrated that choline induced the expression of both PLCs at the transcriptional level and that phosphorylcholine as well as choline can induce the expression of PLC. There are some especially germane facts relating to the above observations and the role of PLC in virulence. First, choline and glycine betaine are derivatives of the substrate product of PLC, phosphorylcholine (Fig. 1). Since *P. aeruginosa* produces an alkaline phosphatase and an acid phosphatase that can convert phosphorylcholine (Filloux et al., 1988; Garrido et al., 1990) to choline and produces a choline oxidase, like that of *Rhizobium meliloti* (Smith et al., 1988), which can convert choline to glycine betaine (Fig. 1), these compounds will be available to further induce expression of PLC. Second, choline and glycine betaine are members of a class of compounds known as osmoprotectants (Csonka, 1989; Le Rudulier et al., 1984). This group of compounds can protect prokaryotic as well as eukaryotic organisms from the growth-inhibitory effects of high osmotic pressure. They are osmoprotective for *P. aeruginosa* as well (Vasil and Vasil, unpublished; D'Souza-Ault et al., 1993). In fact, D'Souza-Ault et al. (1993) have recently shown that even phosphorylcholine is osmoprotective for *P. aeruginosa.* It should be noted that only glycine betaine is internalized and used as an osmoprotectant by bacterial cells but phosphorylcholine and choline are osmoprotective for *P. aeruginosa* because they are converted to glycine betaine by this organism. Subsequently, we (Shortridge et al., 1992) reported that phosphorylcholine is as effective at inducing the expression of PLC as are choline and betaine. The generation of these compounds by *P. aeruginosa* is especially relevant to the pathogenic potential of this organism in cystic fibrosis (CF) patients. In these patients *P. aeruginosa* produces a chronic pulmonary infection that is largely localized to the bronchioles (Zach, 1990). At this anatomic site *P. aeruginosa* cells thrive in the thick dehydrated mucus secretions of these patients. As a result of the mutation in the cystic fibrosis transmembrane conductance regulator protein, chloride and water transport in the lungs of CF patients is altered and mucus secretions in CF patients are dehydrated in comparison with pulmonary mucus secretions in others. This dehydrated environment equates to one that is high in osmotic pressure. Choline phospholipids (i.e., sphingomyelin and phosphatidylcholine) are relatively abundant in these mucus secretions. Therefore, *P. aeruginosa* would have a relatively plentiful source of choline, as well as P_i, in this environment because it is able to generate these products by producing PLC. *P. aeruginosa* can use the P_i for critical reactions when it is P_i limited and can use choline and its derivative glycine betaine to protect itself from the high osmotic conditions in the lungs of CF patients. Recently several transgenic mouse models for CF have been described, and it may eventually be possible to examine PLC structural or regulatory mutants in these animals to test these hypotheses. Unfortunately, at present these animals do not survive long enough for us to establish a chronic pulmonary infection like those seen in CF patients (Whitsett, 1992; personal communication). We have, however (Granstrom et al., 1984), demonstrated that PLC-H is produced in CF patients by showing that 100% of the chronically colonized patients examined produced elevated titers of anti-PLC-H antibodies. These titers are not protective against CF because the organism is localized to the mucosal surfaces in the lung, but they generally reflect the response to antigens produced during infection. It has also recently been proposed that alterations in the phospholipid composition of the secretions from CF patients as compared with secretions from normal patients is a result of the production of PLC by *P. aeruginosa* (Girod et al., 1992). There is also a connection of PLC production to the development of the mucoid phenotype of *P. aeruginosa,* that is observed almost exclusively in strains isolated from CF patients (Krieg et al., 1988; Terry et al., 1991). These investigators found that phosphorylcholine stimulates the production of the mucoexpolysaccharide capsule when *P. aeruginosa* is grown under P_i-limiting conditions in a chemostat. Taken together, the observations described above suggest that production of PLC has a significant impact on the establishment of the chronic pulmonary infection with *P. aeruginosa* frequently seen in CF patients.

It is interesting that the responses of the genes encoding PLC-H or PLC-N to the concentration of P_i and to osmoprotectants are not identical (Shortridge et al., 1992). This was demonstrated by analyzing the expression of PLC by insertion and deletion (null) mutants, which could produce only PLC-H or PLC-N, as well as mutant that has the *phoB* (*plcA*) gene deleted. Not surprisingly, it was found that low-P_i-induced expression of either PLC-H or PLC-N was abolished in the *phoB* deletion mutant. However, choline induction of PLC-H expression under high-P_i conditions was not altered in the PhoB mutant, indicating that this response is independent of PhoB. In contrast, expression of PLC-N is always dependent on PhoB, regardless of whether choline is present.

Analysis of transcription by RNase protection assays of the *plc* genes under low-P_i growth conditions versus high-P_i growth conditions, with or without choline, revealed that there are two separate transcriptional initiation sites for the *plcH* gene (Fig. 2). There is a transcriptional initiation site when PLC-H expression is induced during growth under high-P_i conditions with choline, and there is a different transcriptional initiation site when PLC-H is expressed in an environment low in P_i without choline. In contrast, only one initiation site was detected when expression of PLC-N was induced during growth under low-P_i conditions whether choline was present or not (Fig. 2). No *plcN* transcript was detected under high-P_i growth conditions even if choline was present (data not shown).

Little is known regarding the mechanisms by which osmoprotectants could affect gene regulation. There are, however, some observations that offer some critical clues to understanding this question. It is known that increased osmotic pressure increases negative supercoiling of DNA, whereas osmoprotectants (e.g., glycine betaine) cause relaxation of supercoiling (Dorman, 1991; Drlica, 1990; Higgins et al., 1988). It has also been reported that anaerobiosis increases negative DNA supercoiling (Dorman, 1991; Drlica, 1990). In this regard it is worthwhile to note that transcription of the *bet* operon in *E. coli,* which encodes enzymes and transport proteins involved in the conversion of choline to glycine betaine, and the subsequent uptake of glycine betaine in this organism are inhibited under anaerobic conditions (Eshoo, 1988; Lamarck et al., 1991). We have found that PLC synthesis in *P. aeruginosa* is affected by conditions that are known to affect the superhelical density of DNA, including anaerobiosis (Vasil and Vasil, unpublished), osmotic pressure, osmoprotectants (Shortridge et al., 1992) and DNA gyrase inhibitors (Shortridge and Vasil, unpublished). It will therefore be of interest to further examine the impact of DNA supercoiling on the mechanisms of the regulation of PLC expression. It is possible that a better understanding of these processes will lead to improved therapeutic agents, particularly for CF patients with *P. aeruginosa* infections.

We have also initiated a more detailed analysis of the osmoprotectant-induced expression of PLC in *P. aeruginosa.* Several distinct classes of transposon-induced (Tn*5*) mutants which are inducible for PLC-H expression in a low-P_i growth environment but which are not induced by osmoprotectants have been isolated. These mutants could be altered in the transport of choline or glycine betaine, or they could be mutated in their ability to produce a transcriptional activator that directly or indirectly influences the transcription of the *plc* genes. Perhaps these activators could act by affecting the superhelical density of DNA near the respective promoters of the *plc* genes. Further molecular characterization of these mutants and other mutants should help elucidate the basic mechanisms by which osmoprotectants modulate gene expression.

Other Regulatory Influences

In addition to using choline or glycine betaine as an osmoprotectant, *P. aeruginosa* can utilize these compounds as the sole source of carbon and nitrogen (Smith et al., 1988). *Rhizobium* species also use choline as an osmoprotectant (Smith et al., 1988). However, they do so only in high-osmotic-pressure environments. Under normal osmotic pressure, *Rhizobium* species metabolize choline as a source of carbon and nitrogen. Thus, we examined whether PLC expression was induced when choline was the only carbon or nitrogen source available. In these experiments (Shortridge et al., 1992), it was found that the nitrogen source, whether it was from choline alone or from another compound (e.g., NH_4NO_3), did not influence the induction of PLC expression. PLC expression was induced by the presence of choline if it was the only carbon or nitrogen source available to *P. aeruginosa*. An insertion mutation in *rpoN*, encoding σ^{54}, a sigma factor required for optimal expression of nitrogen responsive genes and some virulence determinants of *P. aeruginosa* (Stock et al., 1989; Totten et al., 1990), abolishes the choline-induced expression of PLC. In contrast, induction of PLC expression at low P_i concentrations is not altered. Data presented in Table 1 demonstrate that expression of PLC activity in two or three *P. aeruginosa* strains carrying an insertion mutation in *rpoN* is not induced in the presence of choline. Curiously, in the third strain (PAO1 *rpoN*) the levels of PLC activity are highly variable. The reason for this phenomenon is unclear, but it is worthwhile noting that when several (50) individual colonies of the mutant were examined for production of PLC, there was a considerable variation in the ability of each of the colonies to produce PLC activity under the high-P_i conditions with choline. Most clones (42 of 50) produced significantly lower levels of PLC activity than did the wild type (e.g., similar to those seen in the other strains carrying the *rpoN* insertion mutation); however, the remaining clones (8 of 50) produced levels that were indistinguishable from the wild-type strain when they were grown under high-P_i conditions with choline (data not shown). These clones still had the original insertion mutation in the *rpoN* gene. Moreover, when individual colonies of wild-type clones were reexamined, a significant portion (20%) of them produced low levels of PLC activity under high-P_i conditions with choline, like those seen in the other mutant strains. The reverse was also true. Mutant (*rpoN*) clones that originally produced low levels of PLC activity in response to choline yielded colonies that could now produce wild-type levels of PLC under these conditions. Thus, although it is clear that *rpoN* has a significant effect on choline-induced expression, there appears to be an epigenetic phenomena in the PAO1 strain that can override the *rpoN* mutation. For some unknown reason, this anomaly is not observed with the other wild-type strains (PAKS and PA103). It is also probable that *rpoN* regulates the choline induction of PLC indirectly because neither *plcH* nor *plcN* has σ^{54}-like promoters (Fig. 2). It may be that σ^{54} is required for the expression of another gene encoding a transcriptional activator needed for the choline induction of PLC expression or that σ^{54} is involved in the expression of genes necessary for the uptake of choline or betaine.

TABLE 1. PLC-H production in *rpoN* (σ^{54}) mutants of three strains of *P. aeruginosa*[a]

Strain	Phospholipase C activity[b]	
	− Choline	+ Choline
PA103		
Wild type	3 ± 1	36 ± 3
rpoN	4 ± 1	4.5 ± 1
PAK		
Wild type	4 ± 1	24 ± 6
rpoN	3 ± 1.5	6 ± 6
PAO1		
Wild type	3 ± 1.5	22 ± 3.5
rpoN	2 ± 1.5	16.6 ± 14.5

[a]Insertion mutations in the *rpoN* genes of three *P. aeruginosa* strains were generated by methods described previously (Ostroff and Vasil, 1987; Ostroff et al., 1989; Shortridge et al., 1992).

[b]Units of PLC were assayed as previously described (Berka et al., 1981). In these assays most of the activity (>95%) detected is PLC-H, since the organisms were grown under high P_i (10 mM) conditions with or without choline and since only very small amounts (<0.1 U) of PLC-N are produced under these conditions.

Concluding Remarks

In the past few years, significant progress has been made toward beginning to understand the environmental factors that influence the expression the *plc* genes of *P. aeruginosa*. From the results presented here and in previous reports, it is evident that not only are the different environmental conditions complicated by themselves but that also there is significant overlapping levels of regulation. Because PLC production is a potentially significant virulence determinant of *P. aeruginosa* and the inducing conditions (e.g., P_i limitation, osmoprotectants) are clearly relevant to the pathogenesis of *P. aeruginosa* infections, it will be worthwhile to further dissect these responses at the molecular level. It is hoped that a coherent model can be developed in the next few years, illustrating how these critical factors direct the expression of both PLCs produced by this organism and how PLC expression affects the complex pathogenicity of *P. aeruginosa*.

Research described in this report from our laboratory was supported by a grant from the National Institute of Allergy and

Infectious Diseases (AI15940) and from the National Institute of Diabetes, Digestive, and Kidney Diseases (DK46440).

REFERENCES

Anba, J., M. Bidaud, M. L. Vasil, and A. Lazdunski. 1990. Nucleotide sequence of the *Pseudomonas aeruginosa phoB* gene for the phosphate regulon. *J. Bacteriol.* **172:**4685–4689.

Berka, R. M., G. L. Gray, and M. L. Vasil. 1981. Studies of phospholipase C (heat-labile hemolysin) in *Pseudomonas aeruginosa. Infect. Immun.* **34:**1071–1074.

Csonka, L. N. 1989. Physiological and genetic responses of bacteria to osmotic stress. *Microbiol. Rev.* **53:**121–147.

Dorman, C. J. 1991. DNA supercoiling and environment regulation of gene expression in pathogenic bacteria. *Infect. Immun.* **59:**745–748.

Drlica, K. 1990. Bacterial topoisomerases and the control of DNA supercoiling. *Trends Genet.* **6:**2136–2140.

D-Souza-Ault, M. R., L. T. Smith, and G. M. Smith. 1993. Roles of *N*-acetylglutaminylglutamine amide and glycine betaine in adaptation of *Pseudomonas aeruginosa* to osmotic stress. *Appl. Environ. Microbiol.* **59:**473–478.

Eshoo, M. W. 1988. *lac* fusion analysis of the *bet* genes of *Escherichia coli:* regulation by osmolarity, temperature, oxygen, choline and glycine betaine. *J. Bacteriol.* **170:**5208–5216.

Esselmann, M. T., and P. V. Liu. 1961. Lecithinase production by gram-negative bacteria. *J. Bacteriol.* **81:**939–945.

Filloux, A., M. Bally, C. Soscia, M. Murgier, and A. Lazdunski. 1988. Phosphate regulation in *Pseudomonas aeruginosa:* cloning of the alkaline phosphatase gene and identification of *phoB* and *phoR*-like genes. *Mol. Gen. Genet.* **212:**510–513.

Garrido, M. N., T. A. Lisa, S. Albelo, G. I. Lucchesi, and C. E. Domenech. 1990. Identification of the Pseudomonas aeruginosa acid phosphatase as a phosphorylcholone phosphatase activity. *Mol. Cell. Biochem.* **94:**89–95.

Girod, S., C. Galabert, A. Lecuire, J. M. Zahm, and E. Puchelle. 1992. Phospholipid composition and surface-active properties of tracheobronchial secretions from patients with cystic fibrosis and chronic obstruction pulmonary diseases. *Pediatr. Pulm.* **13:**22–27.

Granstrom, M., A. Ericsson, B. Strandvik, B. Wretlind, O. R. Pavlovskis, R. Berka, and M. L. Vasil. 1984. Relationship between antibody response to *Pseudomonas aeruginosa* exo-proteins and colonization/infection in patients with cystic fibrosis. *Acta Paediatr. Scand.* **73:**772–777.

Gray, G. L., R. M. Berka, and M. L. Vasil. 1981. A *Pseudomonas aeruginosa* mutant non-derepressible for orthophosphate-regulated proteins. *J. Bacteriol.* **147:**675–678.

Gray, G. L., R. M. Berka, and M. L. Vasil. 1982. Phospholipase C regulatory mutation of *Pseudomonas aeruginosa* that results in constitutive synthesis of several phosphate-repressible proteins. *J. Bacteriol.* **150:**1221–1226.

Gray, G. L., and M. L. Vasil. 1981a. Isolation and characterization of toxin-deficient mutants of *Pseudomonas aeruginosa. J. Bacteriol.* **147:**275–281.

Gray, G. L., and M. L. Vasil. 1981b. Mapping of a gene controlling the production of phospholipase C and alkaline phosphatase in *Pseudomonas aeruginosa. Mol. Gen. Genet.* **183:**403–405.

Higgins, C. F., C. J. Dorman, D. A. Stirling, L. Waddell, I. R. Booth, G. May, and G. Bremer. 1988. A physiological role for DNA supercoiling in the osmotic regulation of gene expression in *S. typhimurium* and *E. coli. Cell* **52:**569–576.

Krieg, D. P., J. P. Bass, and S. J. Mattingly. 1988. Phosphorylcholine stimulates capsule formation of phosphate-limited mucoid *Pseudomonas aeruginosa. Infect. Immun.* **56:**864–873.

Lamarck, T., I. Kaasen, M. W. Eshoo, P. Falkenberg, J. McDougall, and A. R. Strom. 1991. DNA sequence and analysis of the *bet* genes encoding the osmoregulatory choline-glycine betaine pathway of *Escherichia coli. Mol. Microbiol.* **5:**1049–1057.

Le Rudulier, D., A. R. Strom, A. M. Dandekar, L. T. Smith, and R. C. Valentine. 1984. Molecular biology of osmoregulation. *Science* **224:**1064–1068.

Liu, P. V. 1979. Toxins of *Pseudomonas aeruginosa*, p. 63–88. *In* R. G. Doggett (ed.), *Pseudomonas aeruginosa.* Academic Press, New York.

Lucchesi, G. I., T. A. Lisa, and C. E. Domenech. 1989. Choline and betaine as inducer agents of *Pseudomonas aeruginosa* phospholipase C activity in high phosphate medium. *FEMS Microbiol. Lett.* **57:**335–338.

Meyers, D. J., and R. S. Berk. 1990. Characterization of phospholipase C from *Pseudomonas aeruginosa* as a potent inflammatory agent. *Infect. Immun.* **58:**659–666.

Ostroff, R. M., A. I. Vasil, and M. L. Vasil. 1990. Molecular comparison of a nonhemolytic and a hemolytic phospholipase C from *Pseudomonas aeruginosa. J. Bacteriol.* **172:**5915–5923.

Ostroff, R. M., and M. L. Vasil. 1987. Identification of a new phospholipase C activity by analysis of an insertional mutation in the hemolytic phospholipase C structural gene of *Pseudomonas aeruginosa. J. Bacteriol.* **169:**4597–4601.

Ostroff, R. M., B. Wretlind, and M. L. Vasil. 1989. Mutations in the hemolytic-phospholipase C operon result in decreased virulence of *Pseudomonas aeruginosa* PAO1 grown under phosphate-limiting conditions. *Infect. Immun.* **57:**1369–1373.

Prince, R. W., C. D. Cox, and M. L. Vasil. 1993. Coordinate regulation of siderophore and exotoxin A production: molecular cloning and sequencing of the *Pseudomonas aeruginosa fur* gene. *J. Bacteriol.* **175:**2589–2598.

Pritchard, A. E., and M. L. Vasil. 1986. Nucleotide sequence and expression of a phosphate-regulated gene encoding a secreted hemolysin of *Pseudomonas aeruginosa. J. Bacteriol.* **167:**291–298.

Shortridge, V. D., A. Lazdunski, and Vasil. 1992. Osmoprotectants and phosphate regulate expression of phospholipase C in *Pseudomonas aeruginosa. Mol. Microbiol.* **5:**2823–2831.

Shortridge, V. D., and M. L. Vasil. Unpublished observations.

Smith, L. T., J.-L. Pocard, T. Bernard, and D. Le Rudulier. 1988. Osmotic control of glycine betaine biosynthesis and degradation in *Rhizobium meliloti. J. Bacteriol.* **170:**3142–3149.

Stock, J. B., A. J. Ninfa, and A. M. Stock. 1989. Protein phosphorylation and regulation of adaptive responses in bacteria. *Microbiol. Rev.* **53:**450–490.

Terry, J. M., S. E. Piña, and S. J. Mattingly. 1991. Environmental conditions which influence mucoid conversion in *Pseudomonas aeruginosa* PAO1. *Infect. Immun.* **59:**471–477.

Totten, P. A., J. Cano Lara, and S. Lory. 1990. The *rpoN* gene product of *Pseudomonas aeruginosa* is required for expression of diverse genes including the flagellin gene. *J. Bacteriol.* **172:**389–396.

Vasil, M. L. 1988. Phospholipase C of *Pseudomonas:* a redundant virulence factor? *Pediatr. Pulm. Suppl.* **2:**72–73.

Vasil, M. L., R. M. Berka, G. L. Gray, and H. Nakai. 1982. Cloning of a phosphate-regulated hemolysin gene (phospholipase C) from *Pseudomonas aeruginosa. J. Bacteriol.* **152:**431–440.

Vasil, M. L., R. M. Berka, G. L. Gray, and O. R. Pavlovskis. 1985. Biochemical and genetic studies of iron-regulated (exotoxin A) and phosphate regulated (hemolysin phospholipase C) virulence factors of *Pseudomonas aeruginosa. Antibiot. Chemother.* **36:**23–39.

Vasil, M. L., and A. I. Vasil. Unpublished data.

Weinberg, E. D. 1974. Iron and susceptibility to infectious disease. *Science* **134:**952–955.

Whitsett, J. A. 1992. Localization of human and murine CFTR mRNA in transgenic mice bearing hCFTR chimeric genes. *Pediatr. Pulm. Suppl.* **8:**89–90.

Whitsett, J. A. Personal communication.

Zach, M. S. 1990. Lung disease in cystic fibrosis—an updated concept. *Pediatr. Pulm. Suppl.* **8:**188–202.

Chapter 23

Integration of Multiple Developmental Signals by the Phospho-Transfer Pathway That Controls the Initiation of Sporulation in *Bacillus subtilis*

KEITH IRETON AND ALAN D. GROSSMAN

Department of Biology, Massachusetts Institute of Technology, Cambridge, Massachusetts 02139

Several environmental and physiological conditions are required for efficient spore production in *Bacillus subtilis*. These conditions include nutrient deprivation (Sonenshein, 1989), high cell density (Grossman and Losick, 1988; Ireton et al., 1993b), DNA replication (Ireton and Grossman, 1992, 1994), the absence of DNA damage (Ireton and Grossman, 1992), and the presence of an intact tricarboxylic acid (TCA) cycle (Sonenshein, 1989; Yousten and Hanson, 1972). In the absence of any one of these conditions, sporulation does not initiate efficiently. One of the primary events controlling the initiation of sporulation is phosphorylation of the developmental transcription factor encoded by *spo0A*. Phosphorylation of Spo0A is mediated by a multicomponent phosphorelay that transfers phosphate to Spo0A in a series of steps (Fig. 1) (Burbulys et al., 1991). The generation of a threshold level of Spo0A ~ P by this phospho-transfer pathway is a critical barrier that cells overcome to initiate sporulation (Chung et al., 1994). The function of the pathway is to couple the activation of Spo0A to the multiple environmental and physiological conditions required for the onset of sporulation. In this way, sporulation initiates only when all of these conditions are present.

Protein Phosphorylation and Spore Formation

Protein phosphorylation plays an important role in signal transduction in both eukaryotic and prokaryotic organisms. In many cases, the induction of a particular response or developmental event is controlled by phosphorylation of a key regulatory protein. Phosphorylation of a transcription factor (Spo0A) that is essential for development in the gram-positive bacterium *Bacillus subtilis* is one of the primary events controlling the initiation of the process of spore formation. Below, we summarize recent work on the mechanisms that control the initiation of sporulation in *B. subtilis*.

Spore formation in *B. subtilis* occurs in response to nutrient deprivation and results in the production of a dormant spore with several characteristic resistance properties. The first recognizable morphological event that occurs during spore development is the formation of a division septum located near one pole of the cell. This asymmetrically placed septum establishes two distinct cell types that have different developmental fates. The smaller compartment, the forespore, develops into the mature spore while enclosed inside the larger compartment, or mother cell. Proper development of the spore requires gene expression in both cell types (reviewed by Losick and Stragier [1992] and Piggot and Coote [1976]). It is not surprising that disruptions in DNA synthesis and DNA damage inhibit spore formation (Dunn et al., 1978; Ireton and Grossman, 1992, 1994; Mandelstam et al., 1971; Shibano et al., 1978), because proper gene expression requires an intact chromosome in each cell type.

Activation (Phosphorylation) of the Spo0A Transcription Factor and Initiation of Sporulation

One of the primary events controlling the initiation of sporulation is phosphorylation of the developmental transcription factor encoded by *spo0A*. Spo0A protein is essential for the initiation of sporulation and directly regulates (activates or represses, depending on the target gene) transcription of several genes early during development. Phosphorylation of Spo0A increases its ability to bind to DNA (Baldus et al., 1994; Bird et al., 1993; Strauch et al., 1990; Trach et al., 1991) and activate or repress transcription. For example, Spo0A ~ P is a direct transcriptional activator of *spoIIA*, *spoIIG*, and *spoIIE*; it has been shown to bind to 5′ regulatory regions of these loci and stimulate transcription from their promoters (Baldus et al., 1994; Bird et al., 1993; York et al., 1992). These *spoII* loci are essential for sporulation and for establishing cell-type-specific gene expression (reviewed by Losick and Stragier [1992]). In addition, Spo0A ~ P inhibits transcrip-

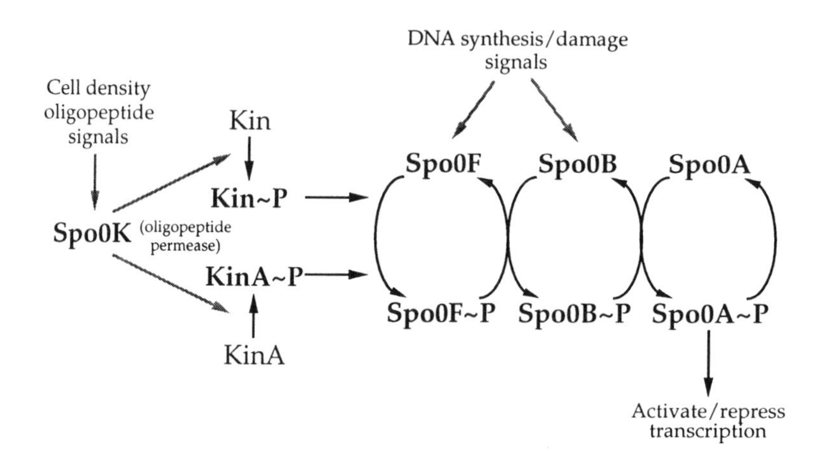

FIG. 1. The phosphorelay. Phosphorylation of Spo0A is achieved by the transfer of phosphate from histidine protein kinases to Spo0F, from Spo0F to Spo0B, and finally from Spo0B to Spo0A (Burbulys et al., 1991). The oligopeptide permease encoded by *spo0K* is probably involved in sensing cell density signals. Spo0B and/or Spo0F is the target for regulation by DNA synthesis and DNA damage signals (Ireton and Grossman, 1994). It is not known which steps are regulated by nutrient deprivation or the TCA cycle.

tion of *abrB* (Strauch et al., 1990), which encodes a repressor that inhibits expression of several genes during vegetative growth. By inhibiting expression of *abrB*, Spo0A ~ P indirectly stimulates transcription of genes under AbrB repression, including several genes that are essential for the initiation of spore development.

Spo0A is phosphorylated on an aspartate residue in its N-terminal domain (Fig. 2). The site of phosphorylation is in a region that is homologous to the common regulatory domain that defines bacterial response regulator proteins. Phosphorylation of most response regulator proteins occurs through a direct interaction between the regulator and a cognate histidine protein kinase. These kinases autophosphorylate on a histidine residue, and the phosphate from this histidine is transferred to an aspartate in the response regulator. Although the specific functions of the individual members of this extensive family of proteins are diverse, these proteins all seem to couple a particular adaptive or developmental response to changes in environmental or physiological conditions (Albright et al., 1989; Bourret et al., 1991; Stock et al., 1989).

In contrast to most response regulators, Spo0A does not readily accept phosphate directly from the major histidine protein kinase involved in sporulation, KinA. Rather, two additional proteins serve to transfer the phosphate from histidine protein kinases to the aspartate in the N-terminal domain of Spo0A. The Spo0F protein, also a response regulator, obtains phosphate from KinA ~ P and other histidine kinases (Trach and Hoch, 1993) and/or small molecule phosphate donors. The phosphate from Spo0F ~ P is then transferred to Spo0B, and then it is transferred from Spo0B ~ P finally to Spo0A. This multicomponent phospho-transfer pathway or phosphorelay (Fig. 1) has been characterized in vitro (Burbulys et al., 1991) and is consistent with the considerable amount of genetic and physiological information that has been accumulated (discussed by Burbulys et al. [1991] and Grossman [1991]).

The Regulatory N-Terminal Domain Inhibits the Ability of Spo0A To Function as a Transcription Factor

The activity of Spo0A seems to be controlled by an inhibitory mechanism in which the conserved N-terminal domain, containing the site of phosphorylation, "masks" the ability of the carboxyl terminus to function as a transcriptional activator. Various deletion mutations in the amino terminus of Spo0A create a constitutively active form of the transcription factor (Green et al., 1991; Ireton et al., 1993b), indicating that removal or disruption of the amino terminus unmasks the ability of the protein to activate transcription. In the wild-type protein, this unmasking is apparently achieved by phosphorylation of this inhibitory domain, which somehow prevents it from interfering with the activity of the carboxyl terminus (Fig. 3).

Masking of activity by the conserved N-terminal domain appears to be a relatively common mode of regulation for response regulator proteins. A form of the CheB methylesterase which lacks the N-terminal two-fifths of the protein has increased methylesterase activity in vitro (Simms et al., 1985). In addition, removal of the conserved N-terminal domain in the FixJ transcrip-

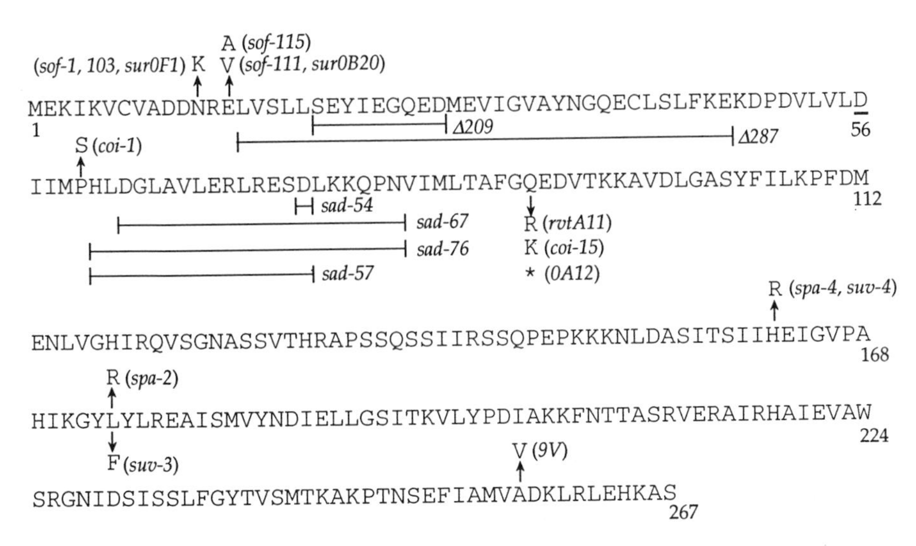

FIG. 2. Amino acid sequence of Spo0A. The sequence is indicated by the single-letter code. The site of phosphorylation (D-56) is underlined. Amino acid changes caused by various mutations are indicated, including several *sof* (Spiegelman et al., 1990) *surOF* and *surOB* (Shoji et al., 1988), *coi* (Olmedo et al., 1990), *pin* (Green et al., 1991), *spa* and *suv* (Grossman et al., 1992; Perego et al., 1991), and *sad* (Ireton et al., 1993b) mutations and *rvtA11* (Leighton, 1994).

tional activator creates a protein that is active in the absence of its cognate kinase (Kahn and Ditta, 1991).

The activity of other transcription factors is also often regulated by such a masking mechanism. In these cases, the activity of a DNA-binding or transcriptional activation domain is inhibited or masked by regulatory domain. For example, the steroid hormone-binding domain of the glucocorticoid receptor inhibits the activity of this protein as a transcriptional enhancer in the absence of hormone (Godowski et al., 1987; Hollenberg et al., 1987). The LuxR transcriptional activator contains amino-terminal regions that inhibit its ability to activate transcription in the absence of inducer (Choi and Greenberg, 1991). In addition, the major sigma factor of *E. coli* contains a region that masks the ability of free sigma-70 to bind DNA (Dombroski et al., 1992), and inhibition of DNA binding by this region is presumably relieved by interaction of sigma with core RNA polymerase. In *B. subtilis* two of the sporulation-specific sigma factors are synthesized as inactive precursors and become activated during development when the first several N-terminal amino acids are proteolytically removed (Losick and Stragier, 1992).

The Phosphorelay Integrates Multiple Environmental and Physiological Conditions

One function of the phosphorelay is to couple the activation of Spo0A and the onset of sporulation to the presence of the multiple developmental conditions that are required for spore formation. In this way, accumulation of Spo0A ~ P and initiation of sporulation occur only when all of the conditions are present.

Cells monitor both external and internal conditions to control the initiation of sporulation. External conditions include nutrient deprivation and high cell density, and these conditions provide the "motivation" to initiate spore formation. When facing imminent starvation, one of the ways that a cell of *B. subtilis* can survive until it encounters a new food source is by differentiating into a dormant spore that is resistant to a harsh environ-

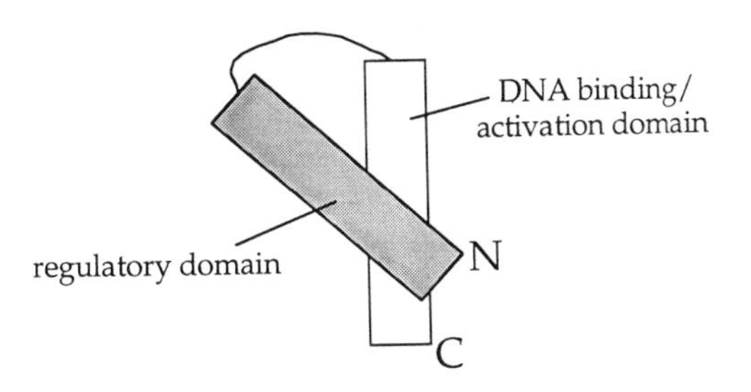

FIG. 3. Model for Spo0A function and the role of the N-terminal domain. The amino-terminal regulatory domain inhibits the function (DNA binding/transcriptional activation) of the carboxy-terminal domain. That inhibition is normally relieved by phosphorylation of an aspartate in the N-terminal regulatory domain. The *sad* deletion mutations abolish the inhibitory function of the amino-terminal domain (Ireton et al., 1993b).

ment. High cell density may also serve to indicate that sporulation is appropriate, since it may provide a signal for crowding and excessive competition for food sources. In its natural environment, the soil, *B. subtilis* cells will be at a low cell density when cells are dispersed, perhaps by water. Dispersal, even in the immediate absence of nutrients, might provide reasonable chances of finding another nutrient source.

In contrast to the external conditions, internal physiological conditions are needed to provide basic aspects of structure or metabolism that are essential for the successful generation of a spore. DNA replication and the absence of DNA damage are needed to ensure that there is a functional chromosome and proper gene expression in both the forespore and the mother cell (Losick and Stragier, 1992). A functional TCA cycle is needed to provide energy for the complex regulatory and morphological changes that occur during sporulation. Investigations whose results are summarized below indicate that both the internal and external conditions needed for the successful generation of the spore are monitored and coupled to the phosphorelay which controls the activation of Spo0A and the initiation of sporulation.

DNA Damage and the Activation of Spo0A

The activation of Spo0A is coupled to elongation of DNA synthesis and the absence of DNA damage early during development. DNA damage or inhibition of the elongation of DNA synthesis inhibit expression of genes controlled by Spo0A ~ P and prevent the initiation of sporulation. (Inhibiting elongation of DNA synthesis induces an SOS or DNA damage response.) These conditions inhibit activation of Spo0A by inhibiting one or more steps in the phosphorelay. Altered function mutations in *spo0A* (*sof-1*, *rvtA11*) that bypass the requirement for the phosphorelay in sporulation restore expression of Spo0A-controlled target genes (e.g., *spoIIA*, *spoIIG*) even under conditions of DNA damage. In addition, expression of these target genes is largely restored by a null mutation in *recA*, indicating that inhibition of elongation of DNA synthesis and DNA damage prevent the activation of Spo0A through induction of the SOS response (Ireton and Grossman, 1992). We postulate that at least one of the SOS-induced genes inhibits a step in the phosphorelay.

Coupling between Activation of Spo0A and Initiation of DNA Replication

The activation of Spo0A is also coupled to replication of chromosomal DNA early during development through a mechanism that is independent of SOS induction and *recA*. Like SOS induction, inhibition of the initiation of DNA replication prevents the onset of sporulation and expression of genes directly activated by Spo0A. The initiation of chromosomal replication is inhibited by shifting a *dnaB*(Ts) mutant to the nonpermissive temperature. *dnaB* is required for the initiation of replication and for attachment of the *oriC* region of the chromosome to the cell membrane (Hoshino et al., 1987; Winston and Sueoka, 1980). The defect in early developmental gene expression caused by blocking initiation of DNA replication is independent of *recA*, indicating that it occurs by a mechanism that is distinct from the DNA damage response. Experiments with altered function mutations in *spo0A* that make sporulation and expression of Spo0A-controlled genes independent of the phosphorelay indicate that this *recA*-independent response also regulates a step in the phosphorelay (Ireton and Grossman, 1994). Thus, both DNA damage (*recA*-dependent) and DNA replication (*recA*-independent) signals control the phosphorelay, perhaps by different mechanisms.

The mechanism by which signals generated by inhibition of DNA replication are sensed and transmitted to the phosphorelay has yet to be determined. Recent experiments indicate that the target of these signals in the phospho-transfer pathway, however, is likely to be Spo0F and/or Spo0B. DNA damage appears to affect the same target(s) (Ireton and Grossman, 1994).

Coupling between the "Metabolic Potential" of a Cell and Activation of Spo0A

Cells of *B. subtilis* are able to monitor the potential to produce energy necessary for spore development. This "metabolic potential," which is dependent on parts of the TCA cycle, is also coupled to the phosphorelay and the initiation of sporulation.

Mutations in some genes encoding enzymes of the TCA cycle, such as *citC* (encoding isocitrate dehydrogenase), have long been known to prevent the initiation of sporulation (Sonenshein, 1989; Yousten and Hanson, 1972). In addition, it has recently been demonstrated that a strain deleted for both of the genes encoding citrate synthase and also defective in *citC* is defective in expression of Spo0A-controlled genes (Jin and Sonenshein, 1993). The defect in expression of these genes is due to a defect in the activation of Spo0A, since the *rvtA11* and *sof-1* mutations in *spo0A* that bypass the requirement for the phosphorelay allow normal gene expression to occur in the absence of citrate synthase and isocitrate dehydrogenase (Ireton et al., 1993a).

How the TCA cycle affects the phosphorelay is not yet known. It seems most likely that disruption of the TCA cycle affects phosphorylation of

Spo0A indirectly, by altering the level of a metabolite that affects the phosphorelay. In principle, disruption of the TCA cycle could affect the activation of Spo0A by preventing the generation of a metabolite that stimulates the phosphorelay or by causing the accumulation of an inhibitory metabolite.

It is also possible that one of the TCA cycle enzymes itself (isocitrate dehydrogenase or aconitase?) regulates the activation of Spo0A. Although this might seem unlikely, there is a precedent for a TCA cycle enzyme playing a role in regulation of gene expression. In eukaryotes, the cytosolic form of aconitase is also an RNA-binding protein that, depending on the target transcript, regulates the stability or translation initiation of transcripts involved in iron uptake and storage (Cammack, 1993).

Nutrient Deprivation Regulates the Phosphorelay

The primary external signal required for the initiation of sporulation is nutrient depletion. Like DNA damage, DNA replication, and the TCA cycle, nutritional conditions also control the initiation of sporulation (in part) by regulating the phosphorelay. We have identified mutations in *spo0A* (*sad67*, *sad76*) that bypass the requirement for nutrient deprivation in the expression of Spo0A-controlled target genes. As expected, the altered forms of Spo0A encoded by these mutations are also active in the absence of the phosphorelay or of phosphorylation (Ireton et al., 1993b). These results indicate that starvation controls expression of a subset of early developmental genes by regulating the activation of Spo0A through the phosphorelay.

The nature of the nutritional signal(s) that triggers spore formation and the mechanism(s) by which these signals are sensed remains elusive. A change in the intracellular level of GTP is a candidate for the primary nutritional signal, since all well-characterized starvation conditions that induce spore formation result in a significant decrease in the intracellular GTP pool (Lopez et al., 1981; Sonenshein, 1989). In addition, treatments or mutations that decrease the level of GTP inside the cell cause efficient sporulation even in the presence of excess nutrients, indicating that reduction in the concentration of intracellular GTP is sufficient to induce sporulation (Freese et al., 1979; Lopez et al., 1979; Mitani et al., 1977).

If a change in the intracellular level of GTP is a key nutritional signal for sporulation, proteins that sense fluctuations in the GTP pool and regulate the phosphorelay in response to these fluctuations should exist. Mitchell and Vary (1989) demonstrated that the onset of sporulation produces changes in the levels or activities of several proteins that interact with GTP. In addition, a GTP-binding protein encoded by *obg* (*spo0B*-associated gene) is cotranscribed with the gene encoding the phosphorelay component, Spo0B. Although *obg* is essential for growth, it is not known whether it plays a role in sporulation (Trach and Hoch, 1989). It is tempting to speculate, however, that it plays a role in sensing nutritional (or other) signals and in coupling changes in the intracellular level of GTP to the flow of phosphate through the phosphorelay, possibly by affecting Spo0B.

It is important to note that nutrient deprivation also controls the expression of some genes during the initiation of sporulation independently of Spo0A. For example, transcription of genes encoding the enzymes aconitase (*citB*) and alanine dehydrogenase (*ald*) is induced early during sporulation, and this induction does not require the phosphorelay or *spo0A* (Dingman et al., 1987; Siranosian et al., 1993). Both of these genes are required for normal spore development. The regulatory factor or factors that control transcription of these genes in response to starvation have yet to be identified.

Cell Density Signals Regulate the Phosphorelay

Cell density signals also affect sporulation by regulating the phosphorelay. Efficient sporulation normally requires that a culture of *B. subtilis* be at a relatively high cell density when limited for nutrients (Grossman and Losick, 1988; Ireton et al., 1993b; Vasantha and Freese, 1979). This high-cell-density requirement is due to the presence of at least one extracellular factor, at least in part a peptide, that accumulates in the culture fluid at a high cell density (Grossman and Losick, 1988). The *sad67* mutation in *spo0A* bypasses the requirement for high cell density in spore formation, indicating that the sole role of cell density signals in sporulation is to regulate the phosphorelay (Ireton et al., 1993b).

The extracellular peptide factor(s) probably stimulates phosphorylation of Spo0A by acting through the oligopeptide permease encoded by the *spo0K* operon. *spo0K* is an operon of five genes that encode products homologous to those encoded by the oligopeptide permease operon from *Salmonella typhimurium* (Perego et al., 1991; Rudner et al., 1991). The requirement for *spo0K*, as well as for high cell density, in sporulation is bypassed by overexpression of the KinA histidine protein kinase (LeDeaux and Grossman, 1993; Rudner et al., 1991). These results may indicate that the transport system encoded by *spo0K* couples the accumulation of extracellular peptide factors to the activation of one or more histidine protein kinases that serve as sources of phosphate for Spo0F and Spo0B (Grossman, 1991; Ireton et al., 1993b; Rudner et al., 1991).

A Threshold Level of Spo0A ~ P Is Needed To Initiate Sporulation

In a culture of genetically identical cells, only a fraction of those cells produce spores. We have recently shown that the ability of some cells in a sporulating culture to go on to produce spores is associated with the ability of those cells to induce expression of early developmental genes activated by Spo0A ~ P (Chung et al., 1994).

Analysis of gene expression in single cells revealed that a sporulating culture is composed of two subpopulations, one that has initiated the developmental program and activated expression of genes controlled by Spo0A ~ P and another that remains uninduced for early developmental gene expression. Expression of developmental genes was determined by staining cells containing a *spo-lacZ* fusion with a dye that fluoresces upon hydrolysis by β-galactosidase and measuring fluorescence in individual cells by using a flow cytometer (Chung et al., 1994).

In a culture of sporulating cells, the size of the subpopulation expressing genes activated by Spo0A ~ P correlates well with the fraction of cells that eventually develop into spores. In addition, mutations in *kinA* or *spo0K* that diminish the amount of activated (phosphorylated) Spo0A cause a decrease in the size of the subpopulation expressing genes activated by Spo0A ~ P. Under these conditions, the fraction of cells expressing early developmental genes is very similar to the fraction of cells that produce spores (Chung et al., 1994). These results suggest that a threshold level of Spo0A ~ P must be attained to induce transcription of early developmental genes and that most of the cells that have induced transcription of these genes proceed to produce mature spores.

Activation of Spo0A Is a Developmental Checkpoint

Production of a dormant spore represents a last-ditch effort to survive in the absence of nutrients and requires a considerable investment of energy for inducing changes in gene expression and morphological development. The activation of Spo0A serves as a developmental checkpoint that ensures that sporulation does not initiate unless cells are starved, at a high cell density, and are able to provide energy and intact chromosomes for the two cell types generated later in development. In this way, inappropriate and unproductive attempts at sporulation are prevented.

Work done in our laboratory was supported by grants from NIH (GM41934) and from the Lucille P. Markey Charitable Trust.

REFERENCES

Albright, L. M., E. Huala, and F. M. Ausubel. 1989. Prokaryotic signal transduction mediated by sensor and regulator protein pairs. *Annu. Rev. Genet.* **23:**311–336.

Baldus, J. M., B. D. Green, P. Youngman, and C. P. Moran, Jr. 1994. Phosphorylation of *Bacillus subtilis* transcription factor Spo0A stimulates transcription from the *spoIIG* promoter by enhancing binding to weak 0A boxes. *J. Bacteriol.* **176:**296–306.

Bird, T. H., J. K. Grimsley, J. A. Hoch, and G. B. Spiegelman. 1993. Phosphorylation of Spo0A activates its stimulation of *in vitro* transcription from the *Bacillus subtilis spoIIG* operon. *Mol. Microbiol.* **9:**741–749.

Bourret, R. B., K. A. Borkovich, and M. I. Simon. 1991. Signal transduction pathways involving protein phosphorylation in prokaryotes. *Annu. Rev. Biochem.* **60:**401–441.

Burbulys, D., K. A. Trach, and J. A. Hoch. 1991. Initiation of sporulation in *B. subtilis* is controlled by a multicomponent phosphorelay. *Cell* **64:**545–552.

Cammack, R. 1993. A new use for an old enzyme. *Curr. Biol.* **3:**41–43.

Choi, S. H., and E. P. Greenberg. 1991. The C-terminal region of the *Vibrio fisheri* LuxR protein contains an inducer-independent *lux* gene activation domain. *Proc. Natl. Acad. Sci. USA* **88:**11115–11119.

Chung, J. D., G. Stephanopoulos, K. Ireton, and A. D. Grossman. 1994. Gene expression in single cells of *Bacillus subtilis:* evidence that a threshold mechanism controls the initiation of sporulation. *J. Bacteriol.* **176:**1977–1984.

Dingman, D. W., M. S. Rosenkrantz, and A. L. Sonenshein. 1987. Relationship between aconitase gene expression and sporulation in *Bacillus subtilis. J. Bacteriol.* **169:**3068–3075.

Dombroski, A. J., W. A. Walter, M. T. Record, Jr., D. A. Siegele, and C. A. Gross. 1992. Polypeptides containing highly conserved regions of transcription initiation factor sigma-70 exhibit specificity of binding to promoter DNA. *Cell* **70:**501–512.

Dunn, G., P. Jeffs, N. H. Mann, D. M. Torgersed, and M. Young. 1978. The relationship between DNA replication and the induction of sporulation in *Bacillus subtilis. J. Gen. Microbiol.* **108:**189–195.

Freese, E., J. E. Heinze, and E. M. Galliers. 1979. Partial purine deprivation causes sporulation of *Bacillus subtilis* in the presence of excess ammonia, glucose, and phosphate. *J. Gen. Microbiol.* **115:**193–205.

Godowski, P. J., S. Rusconi, R. Miesfeld, and K. R. Yamamoto. 1987. Glucocorticoid receptor mutants that are constitutive activators of transcriptional enhancement. *Nature* (London) **325:**365–368.

Green, B. D., M. G. Bramucci, and P. Youngman. 1991. Mutant forms of Spo0A that affect sporulation initiation: a general model for phosphorylation-mediated activation of bacterial signal transduction proteins. *Semin. Dev. Biol.* **2:**21–29.

Grossman, A. D. 1991. Integration of developmental signals and the initiation of sporulation in *B. subtilis. Cell* **65:**5–8.

Grossman, A. D., T. Lewis, N. Levin, and R. DeVivo. 1992. Suppressors of a *spo0A* missense mutation and their effects on sporulation in *Bacillus subtilis. Biochimie* **74:**679–688.

Grossman, A. D., and R. Losick. 1988. Extracellular control of spore formation in *Bacillus subtilis. Proc. Natl. Acad. Sci. USA* **85:**4369–4373.

Hollenberg, S. M., V. Giguere, P. Sequi, and R. M. Evans. 1987. Colocalization of DNA-binding and transcriptional activation functions in the human glucocorticoid receptor. *Cell* **49:**39–46.

Hoshino, T., T. McKenzie, S. Schmidt, T. Tanaka, and N. Sueoka. 1987. Nucleotide sequence of *Bacillus subtilis dnaB:* a gene essential for DNA replication initiation and membrane attachment. *Proc. Natl. Acad. Sci. USA* **84:**653–657.

Ireton, K., and A. D. Grossman. 1992. Coupling between gene expression and DNA synthesis early during development in *Bacillus subtilis. Proc. Natl. Acad. Sci. USA* **89:**8808–8812.

Ireton, K., and A. D. Grossman. 1994. A developmental checkpoint couples the initiation of sporulation to DNA replication in *Bacillus subtilis. EMBO J.* **13:**1566–1573.

Ireton, K., S. F. Jin, A. L. Sonenshein, and A. D. Grossman. 1993a. Unpublished results.

Ireton, K., D. Z. Rudner, K. J. Siranosian, and A. D. Grossman. 1993b. Integration of multiple developmental signals in *Bacillus subtilis* through the Spo0A transcription factor. *Genes Dev.* **7:**283–294.

Jin, S. F., and A. L. Sonenshein. 1993. Personal communication.

Kahn, D., and G. Ditta. 1991. Modular structure of FixJ: homology of the transcriptional activator domain with the −35 binding domain of sigma factors. *Mol. Microbiol.* **5:**987–997.

LeDeaux, J., and A. D. Grossman. 1993. Unpublished results.

Leighton, T. 1994. Personal communication.

Lopez, J. M., A. Dromerick, and E. Freese. 1981. Response of guanosine 5′-triphosphate concentration to nutritional changes and its significance for *Bacillus subtilis* sporulation. *J. Bacteriol.* **146:**605–613.

Lopez, J. M., C. L. Marks, and E. Freese. 1979. The decrease of guanine nucleotides initiates sporulation of *Bacillus subtilis. Biochim. Biophys. Acta* **587:**238–252.

Losick, R., and P. Stragier. 1992. Crisscross regulation of cell-type-specific gene expression during development in *B. subtilis. Nature* (London) **355:**601–604.

Mandelstam, J., J. M. Sterlini, and D. Kay. 1971. Sporulation in *Bacillus subtilis:* effect of medium on the form of chromosome replication and on initiation to sporulation in *Bacillus subtilis. Biochem. J.* **125:**635–641.

Mitani, T., J. F. Heinze, and E. Freese. 1977. Induction of sporulation in *Bacillus subtilis* by decoyinine or hadacidin. *Biochem. Biophys. Res. Commun.* **77:**1118–1125.

Mitchell, C., and J. C. Vary. 1989. Proteins that interact with GTP during sporulation of *Bacillus subtilis. J. Bacteriol.* **171:**2915–2918.

Olmedo, G., E. G. Ninfa, J. Stock, and P. Youngman. 1990. Novel mutations that alter the regulation of sporulation in *Bacillus subtilis:* evidence that phosphorylation of regulatory protein Spo0A controls the initiation of sporulation. *J. Mol. Biol.* **215:**359–372.

Perego, M., C. F. Higgins, S. R. Pearce, M. P. Gallagher, and J. A. Hoch. 1991. The oligopeptide transport system of *Bacillus subtilis* plays a role in the initiation of sporulation. *Mol. Microbiol.* **5:**173–185.

Perego, M., J. J. Wu, G. B. Spiegelman, and J. A. Hoch. 1991. Mutational dissociation of the positive and negative regulatory properties of the Spo0A sporulation transcription factor of *Bacillus subtilis. Gene* **100:**207–212.

Piggot, P. J., and J. G. Coote. 1976. Genetic aspects of bacterial endospore formation. *Bacteriol. Rev.* **40:**908–962.

Rudner, D. Z., J. R. LeDeaux, K. Ireton, and A. D. Grossman. 1991. The *spo0K* locus of *Bacillus subtilis* is homologous to the oligopeptide permease locus and is required for sporulation and competence. *J. Bacteriol.* **173:**1388–1398.

Shibano, Y., K. Tamura, M. Honjo, and T. Komano. 1978. Effect of 6-(para-hydroxyphenylazo)-uracil on sporulation in *Bacillus subtilis. Agric. Biol. Chem.* **42:**187–189.

Shoji, K., S. Hiratsuka, F. Kawamura, and Y. Kobayashi. 1988. New suppressor mutation *sur0B* of *spo0B* and *spo0F* mutations in *Bacillus subtilis. J. Gen. Microbiol.* **134:**3249–3257.

Simms, S. A., M. G. Keane, and J. Stock. 1985. Multiple forms of the CheB methylesterase in bacterial chemosensing. *J. Biol. Chem.* **260:**10161–10168.

Siranosian, K. J., K. Ireton, and A. D. Grossman. 1993. Alanine dehydrogenase (*ald*) is required for normal sporulation in *Bacillus subtilis. J. Bacteriol.* **175:**6789–6796.

Sonenshein, A. L. 1989. Metabolic regulation of sporulation and other stationary phase phenomena, p. 109–130. *In* I. Smith, R. Slepecky, and P. Setlow (ed.), *Regulation of Procaryotic Development.* American Society for Microbiology, Washington, D.C.

Spiegelman, G., B. Van Hoy, M. Perego, J. Day, K. Trach, and J. A. Hoch. 1990. Structural alterations in the *Bacillus subtilis* Spo0A regulatory protein which suppress mutations at several *spo0* loci. *J. Bacteriol.* **172:**5011–5019.

Stock, J. B., A. J. Ninfa, and A. M. Stock. 1989. Protein phosphorylation and regulation of adaptive responses in bacteria. *Microbiol. Rev.* **53:**450–490.

Strauch, M. A., V. Webb, G. Spiegelman, and J. A. Hoch. 1990. The Spo0A protein of *Bacillus subtilis* is a repressor of the *abrB* gene. *Proc. Natl. Acad. Sci. USA* **87:**1801–1805.

Trach, K., D. Burbulys, M. Strauch, J.-J. Wu, N. Dhillon, R. Jonas, C. Hanstein, C. Kallio, M. Perego, T. Bird, G. Spiegelman, C. Fogher, and J. A. Hoch. 1991. Control of the initiation of sporulation in *Bacillus subtilis* by a phosphorelay. *Res. Microbiol.* **142:**815–823.

Trach, K., and J. A. Hoch. 1989. The *Bacillus subtilis spo0B* stage 0 sporulation operon encodes an essential GTP-binding protein. *J. Bacteriol.* **171:**1362–1371.

Trach, K. A., and J. Hoch. 1993. Multisensory activation of the phosphorelay initiating sporulation in *Bacillus subtilis*: identification and sequence of the protein kinase of the alternate pathway. *Mol. Microbiol.* **8:**69–79.

Vasantha, N., and E. Freese. 1979. The role of manganese in growth and sporulation of *Bacillus subtilis. J. Gen. Microbiol.* **112:**329–336.

Winston, S., and N. Sueoka. 1980. DNA-membrane association is necessary for initiation of chromosomal and plasmid replication in *Bacillus subtilis. Proc. Natl. Acad. Sci. USA* **77:**2834–2838.

York, K., T. J. Kenney, S. Satola, C. P. Moran, Jr., H. Poth, and P. Youngman. 1992. Spo0A controls the sigma-A dependent activation of *Bacillus subtilis* sporulation-specific transcription unit *spoIIE. J. Bacteriol.* **174:**2648–2658.

Yousten, A. A., and R. S. Hanson. 1972. Sporulation of tricarboxylic acid cycle mutants of *Bacillus subtilis. J. Bacteriol.* **109:**886–894.

Chapter 24

Phosphate Control of Antibiotic Biosynthesis at the Transcriptional Level

JUAN F. MARTÍN, ANA T. MARCOS, ALICIA MARTÍN, JUAN A. ASTURIAS, AND PALOMA LIRAS

Area de Microbiología, Facultad de Biología, Universidad de León, 24071 León, Spain

It has been known for many years that the biosynthesis of antibiotics and other secondary metabolites is regulated negatively by easily utilizable phosphorus, carbon, and nitrogen sources. Biosynthesis of many antibiotics is repressed by phosphate concentrations above 1 mM. These metabolites are produced only when the producer cells reach phosphate starvation conditions. Little is known, however, about the molecular mechanism responsible for regulation of antibiotic biosynthesis at the transcriptional level. Recognition of particular promoters by different forms of RNA polymerase carrying specific sigma subunits may explain the changes in transcriptional activity of phosphate-, carbon-, or nitrogen-regulated promoters. The *dagA* gene of *Streptomyces coelicolor,* which encodes an extracellular agarase, is preceded by four promoters with distinct −10 and −35 regions. Transcription from these promoters is initiated 32 (*dagAp1*), 77 (*dagAp2*), 125 (*dagAp3*), and 220 (*dagAp4*) bases upstream of the *dagA* coding region. RNA polymerases that mediate in vitro transcription from *dagAp2* (RNA polymerase containing σ^{28}) and *dagAp3* (RNA polymerase containing σ^{49}) have been isolated (Buttner et al., 1988); the *dagAp4* and *dagAp1* transcribing activities remain uncharacterized (Buttner, 1989).

Production of Secondary Metabolites Occurs at Low Specific Growth Rates of the Producer Cells

In rich culture media, producer organisms form secondary metabolites (including antibiotics, alkaloids, mycotoxins, and pigments) at low specific growth rates, after most cell growth has already occurred (Martín and Demain, 1980). In chemically defined media, growth limitation and therefore production of secondary metabolites may occur from the beginning of the culture.

The delay in formation of antibiotics and other secondary metabolites until the growth rate has decreased below a certain threshold is genetically programmed (Martín and Liras, 1989). It seems that formation of secondary metabolites does not occur if abundant nutrients are available. In general, supplementation of a culture committed to antibiotic biosynthesis with an easily utilizable nutrient (e.g., glucose, phosphate, ammonium) results in suppression of secondary-metabolite formation and a burst of growth. Conversely, a nutritional shiftdown usually produces a rapid onset of the biosynthesis of these metabolites (Gil et al., 1985; Martín, 1989).

An intriguing question is how the cell controls the differential expression of genetic information required for growth and for antibiotic biosynthesis.

Regulation of Antibiotic Production

Several enzymes involved in antibiotic biosynthesis are either inhibited or repressed by easily assimilable phosphate, carbon, or nitrogen sources. P_i concentrations greater than 3 to 5 mM are frequently inhibitory to the production of plant, fungal, and bacterial secondary metabolites in liquid culture, although the growth of the producer cells is progressively stimulated by increasing P_i concentrations, up to 300 mM (Liras et al., 1977; Martín, 1989).

Many of the enzymes of the central pathways of primary metabolism (e.g., phosphofructokinase and glucose-6-phosphate dehydrogenase) are stimulated in phosphate-rich medium, thus providing greater amounts of intermediates required for macromolecular biosynthesis during the increased growth that occurs in the presence of elevated phosphate concentrations (Liras et al., 1977). Phosphate addition greatly stimulates formation of total RNA (mainly rRNA) in *Streptomyces griseus* (Asturias et al., 1990; Martín et al., 1988). In summary, it seems that phosphate stimulates expression of genes involved in the biosynthesis of macromolecules and of "housekeeping" genes (i.e., genes required for growth and proper operation of intermediary metabolism), whereas it frequently inhibits expression of genes encoding enzymes for the biosynthesis of secondary metabolites.

The biosynthesis of several groups of antibiotics, including aminoglycosides, tetracyclines, macrolides, polyenes, anthracyclines, ansamycins, and polyethers, is particularly sensitive to phosphate regulation. These compounds are synthesized by different biosynthetic pathways; they may play very different roles in the producer organisms, and their only common feature is that they are all typical secondary metabolites. An intriguing question is whether there is a common mechanism of phosphate control for all antibiotic biosynthetic pathways.

Not All Antibiotics and Other Secondary Metabolites Are Equally Sensitive to Phosphate Control

The biosynthesis of certain antibiotics is less sensitive to high concentrations of P_i than to carbon and nitrogen catabolite regulation. For example, biosynthesis of clavulanic acid by *Streptomyces clavuligerus* (a strain that synthesizes several different antibiotics) is very sensitive to phosphate regulation, whereas production of cephamycin by the same strain is not affected by 25 mM phosphate. It is therefore possible to dissociate cephamycin from clavulanic acid biosynthesis by adjusting the phosphate level in the medium (Romero et al., 1984). Sequential formation of thienamycin and cephamycin antibiotics in a culture of *Streptomyces cattleya* has been reported to be due to different degrees of nutrient limitation (Lilley et al., 1981), indicating that even though a general phenomenon of phosphate control of secondary metabolism gene expression exists, different genes are modulated to different extents. Phosphate control of the expression of a gene (*pabS*) involved in candicidin biosynthesis from *S. griseus* is also exerted when the gene is introduced into *Streptomyces lividans* (Rebollo et al., 1989), suggesting that the cellular components involved in phosphate repression exist in *Streptomyces* species different from a producer strain.

Phosphate Represses or Inhibits Antibiotic Synthases

Several antibiotic synthases which catalyze reactions in which P_i is neither a substrate nor a product of the reaction are repressed by phosphate (Martín, 1989). Some other enzymes, e.g., deacetoxycephalosporin C synthase in *S. clavuligerus* (Zhang et al., 1989), *Nocardia lactamdurans* (Cortés et al., 1987), and *Acremonium chrysogenum*, are inhibited by phosphate, indicating that phosphate control is exerted at at least two different levels (Table 1). There are some reports which suggest that the deacetoxycephalosporin C synthases of *S. clavuligerus* (Lebrihi et al., 1987; Lubbe et al., 1985) and *A. chrysogenum* (Zhang et al., 1988) are also repressed by phosphate, but no transcriptional studies of the *cefEF* gene (encoding the deacetoxycephalosporin C synthase/hydroxylase) in *A. chrysogenum* have been carried out to confirm this suggestion.

Phosphate Control of Candicidin Biosynthesis at the Transcriptional Level

The biosynthesis of the polyene macrolide antibiotic candicidin has been intensively studied as a model of phosphate control (Gil et al., 1985; Liras et al., 1977; Martín and Demain, 1976, 1977). The precursor of the *p*-aminoacetophenone moiety of candicidin, *p*-aminobenzoic acid (PABA), is formed from chorismic acid by a PABA synthase specific for secondary metabolism. Formation of PABA synthase is strongly repressed by phosphate (Gil et al., 1985). The gene encoding PABA synthase (*pabS*), together with its own promoter, was cloned in a 4.5-kb *Bam*HI fragment of DNA from *S. griseus* (Gil and Hopwood, 1983). The nucleotide sequence of the region in which the *pabS* gene is located was determined (Criado et al., 1993). Three open reading frames were found (Fig. 1). Open reading frame 2, of 2,171 nucleotides (nt), encoded a protein that showed extensive sequence identity with PABA synthases of gram-negative bacteria. In enterobacteria, PABA synthases A and B are separate proteins which act together to convert chorismic acid and glutamine to a diffusible intermediate (aminodeoxychorismic acid) that is later transformed into PABA by a third protein encoded by *pabC* (Green and Nichols, 1991). Alignment of the PABA synthase amino acid sequence with those for PABA synthases of enterobacteria indicated strong sequence identity between the NH_2 end of the *S. griseus* enzyme and PabA proteins from *Escherichia coli, Salmonella typhimurium, Klebsiella aerogenes,* and *Serratia marcescens.* Of 200 amino acid residues, 129 (64.5%) were similar in all four proteins. The COOH end of the deduced *S. griseus* amino acid sequence showed marked identity with PabB from *E. coli, S. typhimurium,* and *K. aerogenes.* Of 502 amino acids, 283 residues (56.5%) were identical. Thus, *pabS* (also named *pabAB*) from *S. griseus* encodes the PabA and PabB functions in a single gene. The PABA synthase sequence from *S. griseus* has 723 amino acids and a deduced M_r of 77,900.

Expression of the *pabS* gene is strongly repressed by phosphate when the gene is transcribed from its own promoter (Martín et al., 1988; Rebollo et al., 1989). More than 50% repression of

TABLE 1. Phosphate-regulated enzymes involved in antibiotic biosynthesis[a]

Antibiotic	Producing organism	Target enzyme	Mechanism of regulation[b]
Candicidin	*Streptomyces griseus*	*p*-Aminobenzoate synthase	R
Cephalosporin	*Acremonium chrysogenum*	Deacetoxycephalosporin C synthase[c]	D
Cephamycin	*Streptomyces clavuligerus*	α-Aminoadipyl-cysteinyl-valine synthetase	D
		Deacetoxycephalosporin C synthase[c]	I
		Isopenicillin N synthase[c]	I
	Nocardia lactamdurans	Deacetoxycephalosporin C synthase[c]	I
Gramicidin S	*Bacillus brevis*	Gramicidin S synthetase	D
Neomycin	*Streptomyces fradiae*	Neomycin phosphate phosphotransferase	R
Streptomycin	*Streptomyces griseus*	Streptomycin-6-phosphate phosphotransferase	R
Tetracycline	*Streptomyces aureofaciens*	Anhydrotetracycline oxygenase	R
Tylosin	*Streptomyces fradiae*	Valine dehydrogenase	D
		Methylmalonyl-CoA:pyruvate transcarboxylase[d]	D
		Propionyl-CoA carboxylase	D
		Protylonolide synthetase	D
	Streptomyces strain T59-235	dTDP-D-glucose-4,6-dehydratase	R
		dTDP-mycarose synthetase	R
		Macrocin *O*-methyltransferase	R

[a]Modified with permission from Liras et al. (1990).
[b]R, repression; I, inhibition; D, depression of the enzyme activity occurs, but the evidence does not prove unequivocally the existence of a repression mechanism.
[c]Enzymes involved in cephamycin and cephalosporin biosynthesis are less sensitive to phosphate control (concentrations of more than 25 mM phosphate are required to observe phosphate control) than are other antibiotic biosynthetic enzymes (usually sensitive to less than 5 mM phosphate).
[d]CoA, Coenzyme A.

PABA synthase formation from the cloned *pabS* gene was observed in medium supplemented with 0.1 mM phosphate. Phosphate control is exerted at the transcriptional level (Asturias et al., 1990; Martín et al., 1988). The specific mRNA for PABA synthase in *S. griseus* was quantified by using an internal fragment of the *pabS* gene as a probe. In batch cultures of *S. griseus*, synthesis of mRNA for PABA synthase was derepressed early in the fermentation and reached a peak at 12 h of incubation. A decrease of 95% in the formation of specific mRNA for PABA synthase was observed when the cultures were supplemented with 7.5 mM phosphate, even though formation of total RNA was greatly stimulated (Asturias et al., 1990).

Phosphate-Regulated Promoters in *Streptomyces* Species

Following the finding of the phosphate-regulated *pabS* gene, we investigated other phosphate-regulated promoters that may be involved in control of gene expression in *S. griseus*. We cloned a gene (*saf*) that increases the production of several extracellular enzymes in *S. griseus*, *S. lividans*, and *S. coelicolor* (Daza et al., 1991). When the *saf* gene was amplified in high-copy-number vectors, it retarded the formation of pigments and spores by *Streptomyces* species. The *saf* open reading frame encodes a small polypeptide containing 101 amino acids. The nature of the *saf* stimulatory effect is not known. We do not know whether the inhibitory effect of the *saf* gene product on pigment and spore production is due to the same mechanism as that of stimulation of enzyme production.

The promoter upstream from *saf* was identified by subcloning a DNA fragment in the promoter probe pIJ486. Using the *E. coli* promoter-probe shuttle vector, pULMJ51, we determined that the *saf* promoter region (P*saf*) is also active in *E. coli*. The transcription start points (*tsp*) of the

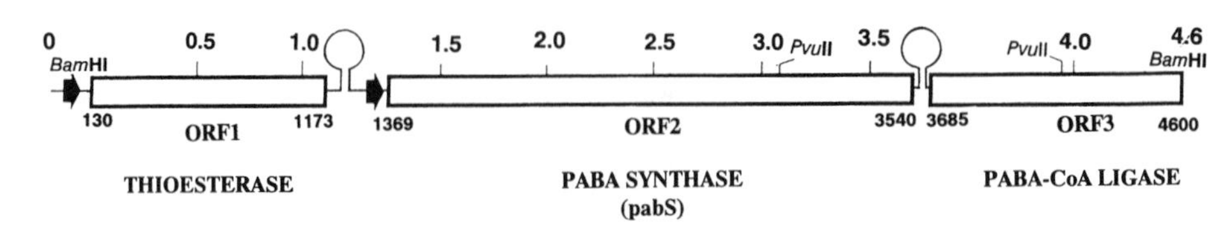

FIG. 1. Organization of the 4.6-kb region of the DNA of *S. griseus* IMRU 3570 carrying the phosphate-regulated *pabS* (also named *pabAB*) gene. Note the presence of strong terminator structures between the three open reading frames. Putative promoter sequences are indicated by thick arrows. ORF, open reading frame; CoA, coenzyme A.

saf promoter in *Streptomyces* species and *E. coli* have been determined by using high-resolution S1 mapping. The transcription start point is at the same position in both microorganisms. Expression from the *saf* promoter region was negatively regulated by phosphate in *Streptomyces* species but not in *E. coli*.

Analysis of the nucleotide sequence of the *saf* promoter region revealed the existence of multiple, short, direct, symmetrical and inverted repeat sequences (Fig. 2). Indeed, the transcription start point is contained in a 7-nt direct repeat, and is flanked by another direct repeat (Fig. 2). Similar structural features have been found in other *E. coli*-like *Streptomyces* promoters which show promoter activity in *E. coli* (Jaurin and Cohen, 1985). Similar short groups of repeat sequences were found in the promoter regions of several genes of different origin, including Simian virus 40 (Dhar et al., 1974), *trp* of *E. coli* (Bennett et al., 1976), and an *aph* gene of *Streptomyces fradiae* (Bibb et al., 1985) (Fig. 2). It is interesting that the neomycin phosphotransferase of *S. fradiae* encoded by the *aph* gene is known to be repressed by phosphate (Majumdar and Majumdar, 1970). Such regulation is probably exerted at the transcriptional level.

DNA-Binding Proteins Interact with Phosphate-Regulated Promoters

E. coli has a very complex regulatory mechanism for genes involved in uptake and metabolism of phosphate (Torriani and Ludtke, 1985). *phoA* encodes the periplasmic alkaline phosphatase, and *phoE* codes for an outer membrane pore protein that is induced under phosphate limitation. At least three genes, *phoB*, *phoR*, and *phoM*, are implicated in the regulation of the *pho* regulon. The *phoB* protein functions as an activator of transcription of the structural genes of the *pho* regulon, whereas *phoB* expression is in turn modulated by the products of *phoR* and *phoM*.

Comparison of the nucleotide sequences of the promoter regions of *phoA* (Kikuchi et al., 1981), *phoE* (Magota et al., 1984; Overbeeke et al., 1983) and *phoB* (Makino et al., 1986) revealed the presence of a consensus 18-bp sequence in the sense strand (i.e., a phosphate box) (Table 2). The *pho* box is followed 11 bp downstream by a potential Pribnow box, suggesting that the *pho* box functions as an aberrant −35 region in *E. coli* phosphate-controlled promoters (Tommansen et al., 1987). Similarly, a *pho* box has been identified in the promoter region of the *phoA* gene of *Zymononas mobilis* (Pond et al., 1989). *Pseudomonas aeruginosa* has a *pho* regulon that contains a phosphate starvation-inducible periplasmic phosphate-binding protein (that functions as an outer membrane porin) (*oprP*) and carries a *pho* box upstream of the *oprP* (Siehnel et al., 1988). A similar *pho* box was found in another phosphate-regulated gene of *P. aeruginosa* that encodes a heat-labile secreted hemolysin (phospholipase C) (Pritchard and Vasil, 1986). The two *Pseudomonas pho* boxes show 61% homology with the *E. coli* consensus sequence (Table 2). The gene of *P. aeruginosa* encoding the PhoB protein has been cloned (Filloux et al., 1988). Since *P. aeruginosa* and *E. coli* belong to different bacterial families

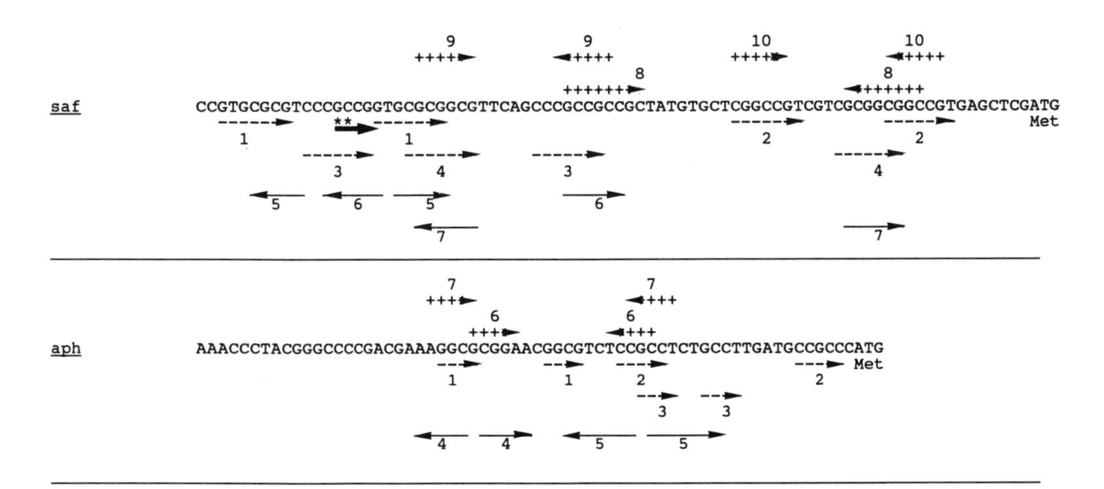

FIG. 2. Direct and inverted repeat sequences in the upstream region of the *saf* gene of *S. griseus* and the *aph* (aminoglycoside phosphotransferase) gene of *S. fradiae*, a neomycin producer. Thick arrow, transcription start point; thin arrows and continuous line, symmetrical repeats; dashed lines, direct repeats; plus signs, inverted repeats. Modified from Daza et al. (1991) with permission.

TABLE 2. Phosphate control sequences existing in phosphate-regulated promoters in different microorganisms[a]

Microorganism	Gene	Protein	Sequence (5′ → 3′)[b]	% Homology[b]
Escherichia coli	*phoB*	PhoB (regulator)	Consensus 5′ → 3′	95
	phoR	PhoR (regulator)	CTGTCATAAAACTGTCAT	95
	phoA	Alkaline phosphatase	T TT C	89
	phoS (*pstS*)	Phosphate-binding protein		95
	phoE	Outer-membrane protein		83
	ugpB	Glycerol-3-P binding protein		89
Enterobacter cloacae	*phoE*	Outer-membrane porin	TTGTCATAAAAGTTTCAT	83
Klebsiella pneumoniae	*phoE*	Outer-membrane porin	TTGTCATAAATATTTAAT	77
Zymomonas mobilis	*phoC*	Acid phosphatase	TTGTCTTATTATAGCCAC	66
Pseudomonas aeruginosa	*plC*	Phospholipase C	AGGTCATATCGAAGTCGC	61
	oprP	Periplasmic porin	TTGCAGTCTCGCTGTCAC	61
Streptomyces griseus	P_{114}	Unknown	CTGTCATGATACTGCGCG	66
	sph(=*aphD*)	Streptomycin phosphotransferase	CTGTGCCGACATCTGGCAT	72
Streptomyces lividans	ORFX[c]	Unknown product	CTTGCACCTCACGTCACG	66

[a]Modified from Liras et al. (1990).
[b]The homology is given as the percentage of nucleotides identical to the consensus *E. coli pho* box. All the *E. coli* sequences show small differences with respect to the *E. coli* consensus box (83 to 95% homology). The tandem repeat in the *E. coli pho* box is indicated by arrows. The underlined nucleotides are conserved with respect to the *E. coli* consensus *pho* box.
[c]ORFX is expressed from the P_{143} promoter region of the *tipA* gene but in the opposite orientation.

and have quite different G+C contents, the available evidence suggests a strong evolutionary conservation of the regulatory elements of the *pho* regulon.

Are There Phosphate Boxes in Phosphate-Regulated Promoters in *Streptomyces* Species?

Several genes involved in antibiotic biosynthesis have been cloned. Expression of some of the genes involved in streptomycin biosynthesis, e.g., the *sph* (= *aphD*) gene encoding streptomycin phosphotransferase, is known to be regulated by phosphate (Table 1). Computer analysis of the upstream region of the published sequence of the *sph* gene of *S. griseus* (Distler et al., 1987; Ohnuki et al., 1985) revealed a previously unreported putative phosphate control (PC) sequence, 324 nucleotides upstream of the ATG initiation triplet of the P2 promoter. A pseudo-PC sequence at 347 nucleotides upstream of the same promoter was also found, and both of them were located in the long transcribed leader region between the transcription initiation site and the ATG translation initiation triplet. The *sph* PC box contains 13 identical nucleotides of the 18 nucleotides of the *pho* box of *E. coli* (Table 2). A phosphate box was also located in a promoter, P_{114}, present in the DNA of most *Streptomyces* species, that was thought to be involved in control of candicidin biosynthesis. However, the relationship of this promoter sequence with the cluster of candicidin biosynthetic genes remains obscure, and the nature of the P_{114} gene is being elucidated (Marcos et al., unpublished [b]).

The promoter (P_{143}) of the *tipA* gene (which is induced by thiostrepton in *S. lividans*) shows promoter activity in both orientations. In the orientation opposite to *tipA*, expression of an unknown open reading frame, X, is regulated by phosphate (Murakami et al., 1989; Thompson, personal communication).

Since the promoter regions of *Streptomyces* genes have a high G+C content, the presence of A+T-rich PC boxes seems unlikely. Phosphate boxes in *Streptomyces* species might be more similar to the PHO5 activator sequences that confer phosphate control to the *CYC1* gene of *Saccharomyces cerevisiae* (Sengstag and Hinnen, 1988).

DNA-Binding Proteins Interact with Phosphate-Regulated Promoters in *Streptomyces* Species

To define the nucleotide sequence of putative phosphate boxes in *Streptomyces* species and to identify the activator (repressor) protein, DNA-binding proteins that interact with the *saf* promoter were sought in cell extracts of *S. griseus*. As shown in Fig. 3, a highly specific DNA-binding protein(s) that produces a shift in gel mobility of the P*saf* promoter was identified. Increasing concentrations of competing salmon sperm DNA did

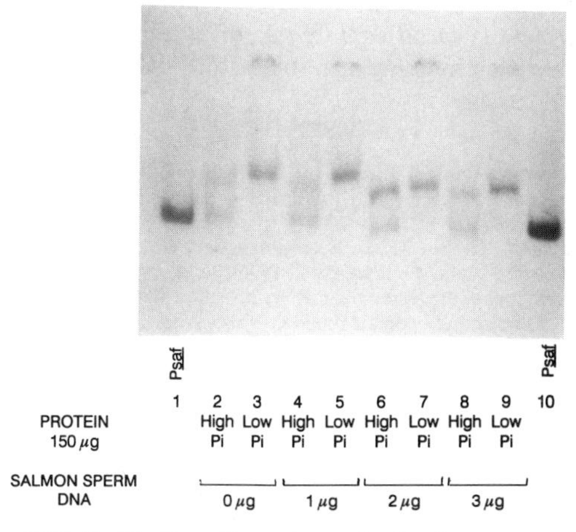

FIG. 3. Binding of proteins to the P*saf* promoter region as shown by gel shifting (retardation). Lanes: 1 and 10; P*saf* DNA without protein; 2, 4, 6, 8, retardation by cell extracts of *S. griseus* grown in phosphate-supplemented (10 mM) SPG medium; 3, 5, 7, 9, retardation of cell extracts of *S. griseus* grown in phosphate-limited medium.

not affect the formation of the complex. The gel retardation is eliminated by hydrolysis of the protein with proteinase K.

The protein interacting with the *saf* promoter may be modified by phosphorylation. Extracts of cells grown under phosphate starvation showed a high affinity for the *saf* promoter, whereas cells grown in phosphate-rich medium showed retardation of only a small fraction of the *saf* promoter DNA molecules (Fig. 3). These results suggest that the binding protein is modified (probably by phosphorylation) in phosphate-rich medium and interacts poorly with the *saf* promoter. Since it is known that phosphate represses expression of the *saf* gene, it is possible that a modified (phosphorylated) protein binds poorly to the promoter region and therefore does not effectively activate transcription of the *saf* gene; i.e., binding of the regulatory (unmodified) protein would be required for good expression of the gene. To confirm this hypothesis, the protein is now being purified to homogeneity. The regulatory protein may affect transcription by interaction with specific sequences in the DNA and also with a sigma factor of RNA polymerase as occurs in the *phoB* system of *E. coli* as described by Makino et al. (see chapter 2).

Sigma Factors and Phosphate-Regulated Promoters

Streptomyces species undergo complex differentiation phenomena (Chater, 1984) that are frequently associated with the production of antibiotics, pigments, and other secondary metabolites (Martín and Liras, 1989). Antibiotic biosynthetic genes may be recognized by specific sigma subunits of the RNA polymerase. Multiple forms of the RNA polymerase of *S. coelicolor* are known to exist (Buttner et al., 1988; Westpheling et al., 1985). Four sigma factors, σ^{49}, σ^{35}, σ^{28}, and σ^{whiG}, have been identified genetically or biochemically in *S. coelicolor* (Buttner, 1989; Buttner et al., 1990). Using an oligonucleotide probe designed from the conserved regions of the sigma factors of *E. coli* and *Bacillus subtilis,* Takahashi and coworkers identified four hybridizing regions (named *hrdA*, *hrdB*, *hrdC*, and *hrdD*) in the DNA of *S. coelicolor* A3(2) (Shiina et al., 1991; Tanaka et al., 1988, 1991). Independently, Buttner et al. (1990) cloned *hrdB*, *hrdC*, and *hrdD* from the same strain and showed by gene disruption that the *hrdB* gene product is essential for growth, whereas *hrdC* and *hrdD* mutants are viable and are apparently unaffected in differentiation. It is unclear whether *hrdA*, *hrdC*, and *hrdD* encode sigma factors with promoter recognition specificity different from that of the *hrdB*-encoded protein (Buttner, 1989).

To establish whether specific sigma factors interact with the phosphate-regulated promoters, we have cloned genes encoding sigma factors. *S. griseus* DNA was hybridized with either a probe internal to the *hrdB* gene of *S. coelicolor* or a probe based on the conserved *rpoD* box. Three hybridizing bands were clearly observed when the *hrdB* probe was used, whereas at least four bands were detected with the *rpoD* probe (the three more intense bands correspond to those observed by hybridization with the *hrdB* gene). The different hybridizing bands were cloned. After initial sequencing, one of the hybridizing bands in a 2.5-kb *Sma*I fragment was observed to correspond to the *hrdB* gene of *S. coelicolor* (Fig. 4). The deduced HrdB protein contained 510 amino acid residues and had a deduced molecular weight of 55,768. Comparison of the amino acid sequence with that of *S. coelicolor* HrdB showed that they were very similar (90.5% identical amino acids). It also showed homology to sigma factors from other gram-positive bacteria (Marcos et al., submitted). The cloned *S. griseus hrdB* gene was overexpressed in *E. coli* and is being used for reconstitution experiments with the *S. griseus* core RNA polymerase purified to near homogeneity (Marcos et al., unpublished [a]).

The *hrdB* gene is expressed in *S. griseus* as a 1.8-kb transcript. Expression was very efficient in phosphate-rich YED (yeast-extract dextrose) medium but not after nutritional shift-down to phosphate-limited SPG medium, which favours antibiotic (candicidin) biosynthesis (Martín and McDaniel, 1986), or to conditions that allow sporulation in liquid medium (Daza et al., 1989).

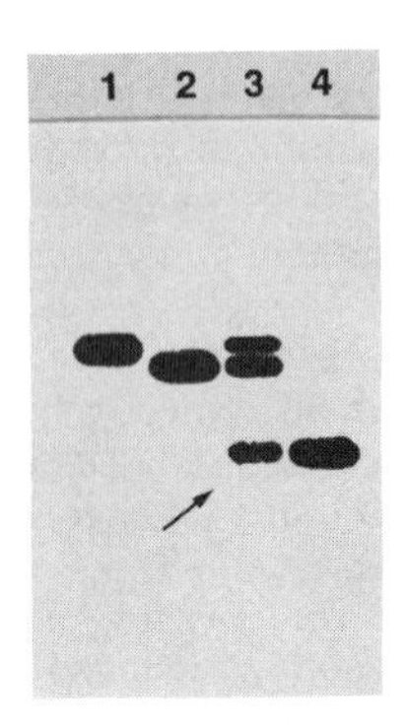

FIG. 4. *hrd* genes (homologous to *rpoD*) in the genome of *S. griseus* as shown by hybridization with an oligonucleotide based on the *rpoD* box of *S. coelicolor.* Lanes 1, cloned *hrdD*; 2, *hrdB*; 3, total DNA of *S. griseus* (a fourth band showed poor hybridization [arrow]; 4, *hrdA*.

Expression of *hrdB* was stimulated when the cells were again suspended in phosphate-rich media. These results indicate that the *hrdB*-encoded sigma factor is formed during active growth in phosphate-rich medium and is probably used to transcribe housekeeping genes essential for primary metabolism. The lack of expression of the *S. griseus hrdB* gene under sporulation or candicidin-producing conditions makes it unlikely that the *hrdB*-encoded sigma factor might be used to transcribe genes involved in candicidin biosynthesis.

Outlook

It is clear that many antibiotic biosynthetic enzymes are repressed by phosphate. This control takes place at the transcriptional level in the case of the *pabS* gene, which encodes the candicidin biosynthetic enzyme PABA synthase. Most probably, transcription of many other biosynthetic genes is also regulated by phosphate at the transcriptional level. Phosphate-regulated promoters occur in *Streptomyces* species and many other bacteria, although it is unknown if phosphate boxes similar to those recognized by the *phoB* protein of *E. coli* are present in these promoters. The finding of DNA-binding protein(s) that produces gel retardation of the *Psaf* promoter indicates that phosphate-regulated promoters in *Streptomyces* species may be activated by binding of a PhoB-like protein which may be inactivated by phosphorylation. Expression of phosphate-regulated promoters may require, in addition, specific sigma factors. Four DNA sequences homologous to the *rpoD* gene (*hrd*) of several gram-negative and gram-positive bacteria have been identified in the genome of *S. griseus.* It will be interesting to establish if any of the known *hrd* genes of *S. griseus* is specifically required for transcription of phosphate-regulated antibiotic biosynthetic genes.

REFERENCES

Asturias, J. A., P. Liras, and J. F. Martín. 1990. Phosphate control of *pabS* gene transcription during candicidin biosynthesis. *Gene* **93:**79–84.

Bennett, G. N., M. E. Schweingruber, K. D. Brown, C. Squires, and C. Yanofsky. 1976. Nucleotide sequence of the region preceding *trp* mRNA initiation site and its role in promoter and operator function. *Proc. Natl. Acad. Sci. USA* **73:**2351–2355.

Bibb, M. J., J. M. Ward, and S. N. Cohen. 1985. Nucleotide sequences encoding and promoting expression of three antibiotic resistance genes indigenous to *Streptomyces. Mol. Gen. Genet.* **199:**26–36.

Buttner, M. J. 1989. RNA polymerase heterogeneity in *Streptomyces coelicolor* A3(2). *Mol. Microbiol.* **3:**1653–1659.

Buttner, M. J., K. F. Chater, and M. J. Bibb. 1990. Cloning, disruption, and transcriptional analysis of three RNA polymerase sigma factor genes of *Streptomyces coelicolor* A3(2). *J. Bacteriol.* **172:**3367–3378.

Buttner, M. J., A. M. Smith, and M. J. Bibb. 1988. At least three different RNA polymerase holoenzymes direct transcription of the agarase gene (*dag*A) of *Streptomyces coelicolor* A3(2). *Cell* **52:**599–607.

Chater, K. F. 1984. Morphological and physiological differentiation in *Streptomyces,* p. 89–115. *In* R. Losick and L. Shapiro (ed.), *Microbial Development.* Cold Spring Harbor Laboratory, Cold Spring Harbor, N.Y.

Cortés, J., J. F. Martín, J. M. Castro, L. Láiz, and P. Liras. 1987. Purification and characterization of a 2-oxoglutarate-linked ATP-independent deacetoxycephalosporin C synthase of *Streptomyces lactamdurans. J. Gen. Microbiol.* **133:**3165–3174.

Criado, L. M., J. F. Martín, and J. A. Gil. 1993. The *pab* gene of *Streptomyces griseus,* encoding *p*-aminobenzoic acid synthase, is located between genes possibly involved in candicidin biosynthesis. *Gene* **126:**135–139.

Daza, A., J. F. Martín, A. Domínguez, and J. A. Gil. 1989. Sporulation of several species of *Streptomyces* in submerged cultures after nutritional downshift. *J. Gen. Microbiol.* **135:**2483–2491.

Daza, A., J. F. Martín, T. Vigal, and J. A. Gil. 1991. Analysis of the promoter region of *saf,* a *Streptomyces griseus* gene that increases production of extracellular enzymes. *Gene* **108:**63–71.

Dhar, R., S. M. Weissman, B. S. Zain, J. Pan, and A. M. Lewis. 1974. The nucleotide sequence preceding an RNA polymerase initiation site on SV40 DNA, II. The sequence of the early strand transcript. *Nucleic Acids Res.* **1:**595–613.

Distler, J., A. Ebert, K. Mansouri, K. Pissowotzki, M. Stockmann, and W. Piepesberg. 1987. Gene cluster for streptomycin biosynthesis in *Streptomyces griseus:* nucleotide sequence of three genes and analysis of transcriptional activity. *Nucleic Acids Res.* **15:**8041–8056.

Filloux, A., M. Bally, C. Soscia, M. Murgier, and A. Lazdunski. 1988. Phosphate regulation in *Pseudomonas aeruginosa:* cloning of the alkaline phosphatase gene and identification of *pho*B- and *pho*R-like genes. *Mol. Gen. Genet.* **212:**510–513.

Gil, J. A., and D. A. Hopwood. 1983. Cloning and expression of a p-aminobenzoic acid synthetase gene of the candicidin producer *Streptomyces griseus. Gene* **25:**119–132.

Gil, J. A., G. Naharro, J. R. Villanueva, and J. F. Martín. 1985. Characterization and regulation of p-aminobenzoic acid synthase from *Streptomyces griseus. J. Gen. Microbiol.* **131:**1279–1287.

Green, J. M., and B. P. Nichols. 1991. *p*-Aminobenzoate biosynthesis in *Escherichia coli.* Purification of aminodeoxychorismate lyase and cloning of *pab*C. *J. Biol. Chem.* **266:**12972–12975.

Jaurin, B., and S. N. Cohen. 1985. *Streptomyces* contains *Escherichia coli*-type A+T-rich promoters having novel structural features. *Gene* **39:**191–201.

Kikuchi, Y., K. Yoda, M. Yamasaki, and G. Tamura. 1981. The nucleotide sequence of the promoter and the amino-terminal region of alkaline phosphatase structural gene (*pho*A) of *Escherichia coli. Nucleic Acids Res.* **9:**5671–5678.

Lebrihi, A., P. Germain, and G. Lefebvre. 1987. Phosphate repression of cephamycin and clavulanic acid production by *Streptomyces clavuligerus. Appl. Microbiol. Biotechnol.* **26:**130–135.

Lilley, G., A. E. Clark, and G. C. Lawrence. 1981. Control of the production of cephamycin C and thienamycin by *Streptomyces cattleya* NRRL 8057. *J. Chem. Tech. Biotechnol.* **31:**127–134.

Liras, P., J. A. Asturias, and J. F. Martín. 1990. Phosphate control sequences involved in transcriptional regulation of antibiotic biosynthesis. *Trends Biotechnol.* **8:**184–189.

Liras, P., J. R. Villanueva, and J. F. Martín. 1977. Sequential expression of macromolecule biosynthesis and candicidin formation in *Streptomyces griseus. J. Gen. Microbiol.* **102:**269–277.

Lubbe, C., S. Wolfe, and A. L. Demain. 1985. Repression and inhibition of cephalosporin synthetases in *Streptomyces clavuligerus* by inorganic phosphate. *Arch. Microbiol.* **140:**317–320.

Magota, K., N. Otsuji, T. Miki, T. Horiuchi, S. Tsunasawa, J. Kondo, F. Sakiyama, M. Amemura, T. Morita, H. Shinagawa, and A. Nakata. 1984. Nucleotide sequence of the *phoS*, the structural gene for the phosphate-binding protein of *Escherichia coli. J. Bacteriol.* **157:**909–917.

Majumdar, M. K., and S. K. Majumdar. 1970. Isolation and characterization of three phosphoamido-neomycins and their conversion into neomycin by *Streptomyces fradiae. Biochem. J.* **120:**271–278.

Makino, K. H., H. Shinagawa, M. Amemura, and A. Nakata. 1986. Nucleotide sequence of the *pho*B gene, the positive regulatory gene for the phosphate regulon of *Escherichia coli* K-12. *J. Mol. Biol.* **190:**37–44.

Marcos, A. T., S. Gutiérrez, B. Díez, F. J. Fernández, V. Kumar, and J. F. Martín. Submitted for publication.

Marcos, A. T., P. Liras, and J. F. Martín. 1994. Unpublished results (a).

Marcos, A. T., R. Pérez-Redondo, and J. F. Martín. 1994. Unpublished results (b).

Martín, J. F. 1989. Molecular mechanism for the control by phosphate of the biosynthesis of antibiotic and secondary metabolites, p. 213–237. *In* S. Shapiro (ed.), *Regulation of Secondary Metabolism in Actinomycetes.* CRC Press, Inc., Boca Raton, Fla.

Martín, J. F., and A. L. Demain. 1976. Control by phosphate of candicidin biosynthesis. *Biochem. Biophys. Res. Commun.* **71:**1103–1109.

Martín, J. F., and A. L. Demain. 1977. Effect of exogenous nucleotides on the candicidin fermentation. *Can. J. Microbiol.* **23:**1334–1339.

Martín, J. F., and A. L. Demain. 1980. Control of antibiotic synthesis. *Microbiol. Rev.* **44:**230–251.

Martín, J. F., and L. E. McDaniel. 1986. Biosynthesis of candicidin by phosphate-limited resting cells of *Streptomyces griseus. Eur. J. Appl. Microbiol.* **3:**135–144.

Martín, J. F., and P. Liras. 1989. Organization and expression of genes involved in the biosynthesis of antibiotics and other secondary metabolites. *Annu. Rev. Microbiol.* **43:**173–206.

Martín, J. F., A. Daza, J. A. Asturias, J. A. Gil, and P. Liras. 1988. Transcriptional control of antibiotic biosynthesis at phosphate-regulated promoters and cloning of a gene involved in the control of the expression of multiple pathways in *Streptomyces,* p. 424–430. *In* Y. Okami, T. Beppu, and H. Ogawara (ed.), *Biology of Actinomycetes '88.* Japan Scientific Societies Press, Tokyo.

Murakami, T., T. G. Holt, and C. J. Thompson. 1989. Thiostrepton-induced gene expression in *Streptomyces lividans. J. Bacteriol.* **171:**1459–1466.

Ohnuki, T., T. Imanaka, and S. Aiba. 1985. Self-cloning in *Streptomyces griseus* of a *str* gene cluster for streptomycin biosynthesis and streptomycin resistance. *J. Bacteriol.* **164:**85–94.

Overbeeke, N., H. Bergmans, F. Van Mansfelt, and B. Lugtenberg. 1983. Complete nucleotide sequence of *pho*E, the structural gene for the phosphate limitation inducible outer membrane pore protein of *Escherichia coli* K12. *J. Mol. Biol.* **163:**513–532.

Pond, J. L., C. K. Eddy, K. F. Mackenzie, T. Conway, D. J. Borecky, and L. O. Ingram. 1989. Cloning, sequencing, and characterization of the principal acid phosphatase, and *phoC* product, from *Zymomonas mobilis. J. Bacteriol.* **171:**767–774.

Pritchard, A. E., and M. L. Vasil. 1986. Nucleotide sequence and expression of a phosphate-regulated gene encoding a secreted hemolysin of *Pseudomonas aeruginosa. J. Bacteriol.* **167:**291–298.

Rebollo, A., J. A. Gil, J. A. Asturias, P. Liras, and J. F. Martín. 1989. Cloning and characterization of a phosphate-regulated promoter involved in phosphate control of candicidin biosynthesis. *Gene* **79:**47–58.

Romero, J., P. Liras, and J. F. Martín. 1984. Dissociation of cephamycin and clavulanic acid biosynthesis in *Streptomyces clavuligerus. Appl. Microbiol. Biotechnol.* **20:**318–325.

Sengstag, C., and A. Hinnen. 1988. A 28-bp segment of the *Saccharomyces cerevisiae* PHO5 upstream activator sequence confers phosphate control to the CYC1-*lacZ* gene fusion. *Gene* **67:**223–228.

Shiina, T., K. Tanaka, and H. Takahashi. 1991. Sequence of *hrd*B, an essential gene encoding a sigma-like transcription factor of *Streptomyces coelicolor* A3(2): homology to principal sigma factors. *Gene* **107:**145–148.

Siehnel, R. J., E. A. Worobec, and R. E. W. Hancock. 1988. Regulation of components of the *Pseudomonas aeruginosa* phosphate-starvation-inducible regulon in *Escherichia coli. Mol. Microbiol.* **2:**347–352.

Tanaka, K., T. Shiina, and H. Takahashi. 1988. Multiple principal sigma factor homologs in eubacteria: identification of the *rpo*D box. *Science* **242:**1040–1042.

Tanaka, K., T. Shiina, and H. Takahashi. 1991. Nucleotide sequence of gene *hrd*A, *hrd*C, and *hrd*D from *Streptomyces coelicolor* A3(2) having similarity to *rpo*D genes. *Mol. Gen. Genet.* **229:**334–340.

Thompson, C. Personal communication.

Tommassen, J., M. Koster, and P. Overduin. 1987. Molecular analysis of the promoter region of the *Escherichia coli* K-12 *pho*E gene: identification of an element, upstream from the promoter, required for efficient expression of PhoE protein. *J. Mol. Biol.* **198:**633–641.

Torriani, A., and D. N. Ludtke. 1985. The *pho* regulon of *Escherichia coli,* p. 224–242. *In* M. Schaechter, F. C. Neidhart, J. Ingraham, and N. O. Kjeldgaard (ed.), *The Molecular Biology of Bacterial Growth.* Jones and Bartlett Publishers, Boston.

Westpheling, J., M. Raynes, and R. Losick. 1985. RNA polymerase heterogeneity in *S. coelicolor. Nature* (London) **313:**22–27.

Zhang, J., S. Wolfe, and A. L. Demain. 1988. Phosphate repressible and inhibitable β-lactam synthetases in *Cephalosporium acremonium* strain C-10. *Appl. Microbiol. Biotechnol.* **29:**242–247.

Zhang, J., S. Wolfe, and A. L. Demain. 1989. Phosphate regulation of ACV synthetase and cephalosporin biosynthesis in *Streptomyces clavuligerus. FEMS Microbiol. Lett.* **57:**145–150.

V. THE PHOSPHOTRANSFERASE SYSTEM

Chapter 25

Introduction: Bacterial Phosphotransferase Systems

HANS L. KORNBERG,[1] AND JOSEPH W. LENGELER[2]

Department of Biochemistry, University of Cambridge, Tennis Court Rd., Cambridge CB2 1QW, United Kingdom,[1] and Fachbereich Biologie/Chemie, Arbeitsgruppe Genetik, Universität Osnabrück, D-49069 Osnabrück, Federal Republic of Germany[2]

Exactly 30 years ago, Saul Roseman and his collaborators described a phosphoenol-pyruvate (PEP)-dependent carbohydrate:phosphotransferase system in *Escherichia coli* which became known as the PTS (Kundig et al., 1964). Originally considered only an unusually complex system to phosphorylate carbohydrates, it became gradually known as one of the central regulatory and signal transduction systems of many gram-positive and gram-negative eubacteria (for a recent review, see Postma et al. [1993]). As a central regulatory unit, the PTS is involved in transport and phosphorylation of a large number of carbohydrates, in signal transduction during the oriented movement (chemotaxis) toward PTS carbohydrates, and in the global regulation of many metabolic pathways. Since its discovery, more than 1,000 papers related to the PTS and its multiple functions have been published. We have learned many details about the properties and the function of the PTS proteins, in particular through cloning, sequencing, and mutation of the corresponding genes and also through purification and determination of the three-dimensional structure of several PTS proteins. In parallel, our appreciation of the complexity of this system has increased: at present we know of more than 20 different PTS proteins and of 10 non-PTS proteins that interact with the PTS. These communicate through "phosphoprotein talk" in a hierarchical way.

At the top of the hierarchy is a general protein kinase (catalytic subunit), designed enzyme I (EI). It autophosphorylates at a histidine residue by using PEP as the phosphodonor, and it transfers the phospho-group by means of several phosphocarrier proteins (or targeting subunits) to its many target loci. Of these proteins, the small histidine protein HPr is nearly universal with relatively unspecific recognition sites; the enzyme EIIA (EIIA) proteins (previously called enzymes III [EIII]), or domains, still react with a few proteins; and the EIIB proteins or domains normally react only with their corresponding partner EIIC through specific recognition sites. All EIIs analyzed thus far comprise the three domains EIIA, EIIB, and EIIC, either fused in various sequences or present in the form of free domains. EIICs form a substrate-specific and transmembrane facilitator channel through which the substrates enter the cell. Substrates are phosphorylated only subsequently through a phospho-group donated by EIIB. All EIIs form similar three-dimensional transporters, although they belong to different genetic families.

Signal transduction through the PTS is by protein phosphorylation, by protein-protein contact (often stoichiometric), and by protein diffusion, e.g., in chemotaxis or in catabolite repression and inducer exclusion. Catabolite repression and control of the generation of intracellular inducer are part of a global regulatory network. This control is superimposed over the transcription control of individual operons and regulons by specific activators and repressors. In enteric bacteria, $EIIA^{Glc}$ is central to this global control. In its phosphorylated form, $P \sim EIIA^{Glc}$ activates the enzyme adenylate cyclase (encoded by the gene *cya*), thus generating high cyclic AMP (cAMP) levels under starvation conditions. Feast conditions, by contrast, e.g., the presence of glucose in the medium, cause the dephosphorylation of $EIIA^{Glc}$, the deactivation of adenylate cyclase, and hence low cAMP levels. A large group of operons and regulons involved in peripheral catabolic metabolism require, for efficient transcription, the presence of a cAMP-binding protein (cAMP receptor protein

[CRP]), encoded by the gene *crp*. CRP needs cAMP for its binding to a specific consensus sequence of promoters. All operons and regulons controlled by CRP are considered to be members of a regulatory unit called the *crp* modulon. Their global control through cAMP, CRP, and $EIIA^{Glc}$ has been known traditionally as catabolite repression. This repression is reinforced by a second control mechanism, known as inducer exclusion. In its free form, $EIIA^{Glc}$ is not only unable to activate adenylate cyclase and the *crp* modulon but also binds to and inactivates in a reversible way many non-PTS transport systems and metabolic enzymes involved in the generation of intracellular inducer. Both control mechanisms together efficiently prevent induction and transcription of all members of the *crp* modulon. This includes the genes for the flagella (*flg*, *flh*, *fli*, and *mot* genes) and the central chemotactic unit (*che* genes). By sensing the presence of PTS carbohydrates in the medium and adjusting the phosphorylation state of the central PTS proteins, in particular $EIIA^{Glc}$, HPr, and EI, cells adapt rapidly to changing conditions in the environment, e.g., through positive chemotaxis towards PTS carbohydrates.

In this section, various aspects of different phosphotransferase systems and their EIIs are described as they are involved in transport, in phosphorylation, in signal transduction, and in global regulation of carbon and energy metabolism. The description includes various biochemical and genetic methods used to analyze such a complex system and cites results at the molecular and, for some complexes, even at the crystal structure level. It is remarkable that the PTS appears to be a unique and discrete major sensory and signal transduction system, surprisingly not related genetically to the widely distributed two-component sensory systems, discussed elsewhere in this book. Despite the major progress in our understanding at the qualitative level, progress at the quantitative level remains scarce. Old physiological problems such as diauxic growth and the glucose effect remain almost as much of a mystery as when they were first described (Epps and Gale, 1942; Monod, 1942; Magasanik, 1961; Kornberg, 1973) and remain to be explained.

REFERENCES

Epps, H. M. R., and E. F. Gale. 1942. The influence of the presence of glucose during growth on the enzymatic activities of *Escherichia coli:* comparison of the effect with that produced by fermentation acids. *Biochem. J.* **36:**619–623.

Kornberg, H. L. 1973. Fine control of sugar uptake by *Escherichia coli. Symp. Soc. Exp. Biol.* **27:**175–193.

Kundig, W., S. Gosh, and S. Roseman. 1964. Phosphate bound to histidine in a protein as an intermediate in a novel phosphotransferase system. *Proc. Natl. Acad. Sci. USA* **52:**1067–1074.

Magasanik, B. 1961. Catabolite repression. *Cold Spring Harbor Symp. Quant. Biol.* **26:**249–256.

Monod, J. 1942. *Recherches sur la Croissance des Cultures Bactériennes.* Hermann et Cie Paris.

Postma, P. W., J. W. Lengeler, and G. R. Jacobson. 1993. Phosphoenolpyruvate:carbohydrate phosphotransferase systems of bacteria. *Microbiol. Rev.* **57:**543–594.

Chapter 26

The Bacterial Phosphoenolpyruvate:Glycose Phosphotransferase System

SAUL ROSEMAN

Department of Biology and the McCollum-Pratt Institute, The Johns Hopkins University, Baltimore, Maryland 21218

This chapter is a general introduction to the phosphotransferase system (PTS). The last comprehensive review by Postma et al. (1993) comprises 46 pages of text and 551 references, although its stated purpose was only to update the review that they had published eight years earlier. Because the review by Postma et al. (1993) and earlier reviews (Meadow et al., 1990; Reizer et al., 1988; Saier, 1989) are so extensive, few primary references will be given in this article. A "minireview" dealing primarily with the glucose effect has also been published (Roseman and Meadow, 1990). Knowledge of the PTS has become so complex and has so many ramifications that it is difficult to grasp the take-home messages of even a well-written review.

This introduction is not for the PTS aficionado. It briefly discusses only a few points: the discovery of the PTS and the origin of its confusing nomenclature, some general properties and important functions of the system, and some problems and prospects for the future.

Why are so many laboratories interested in the PTS? One reason is its diverse physiological functions, several of which are outlined in Fig. 1. It translocates and simultaneously phosphorylates its sugar substrates (PTS sugars), is required for chemotaxis to the same sugars, regulates the ability of the cell to catabolize other compounds, and plays a critical role in maintaining the carbon and nitrogen cycles in marine (and probably terrestrial) environments and ecosystems.

Background and Nomenclature

The Introduction to the first paper on the PTS is shown in Fig. 2. The system was discovered while we were working on the metabolism of the sialic acids. *E. coli* K235 produces a polymer of sialic acid, called colominic acid, and extracts were assayed for an *N*-acetylmannosamine ATP-dependent kinase (the enzyme had been purified from rat liver, and we were studying its distribution). *N*-Acetylmannosamine could indeed be phosphorylated, but the phosphoryl donor in *E. coli* was phosphoenolpyruvate (PEP), and three protein fractions, called enzyme I (EI), HPr, and enzyme II (EII), were required for the reaction. EI catalyzed the transfer from PEP to a histidine residue in the phosphocarrier protein, HPr. The third protein fraction, EII, catalyzed the transfer from P-HPr to hexoses, hexosamines, and *N*-acylhexosamines of the D-gluco and manno configurations. The designation HPr was used to signify "heat-stable" protein, but we now know that the pure protein loses amide groups from two asparagine residues on heating to 100°C (Sharma et al., 1993), with concomitant loss in phosphocarrier activity.

Further work showed that the phosphotransfer sequence in the PTS can be viewed in its simplest terms as follows:

$$\mathrm{PEP} \Rightarrow \left[\frac{\text{General proteins}}{\text{Enzyme I or HPr}}\right] \Rightarrow \left[\frac{\text{Sugar-specific proteins}}{\text{Enzyme II complexes}}\right] \Rightarrow \text{Sugar}$$

Some EII complexes (such as $\mathrm{II^{Man}}$, the "mannose" complex described above) can be detected in cells grown on any carbon source, but most are not expressed unless the cell is grown on the appropriate sugar.

Following purification of the proteins, the phosphotransfer sequence was found to be that shown in Fig. 3, from PEP to EI to HPr to a sugar-specific protein such as III to the membrane protein to the sugar. During the last step, the sugar is simultaneously translocated across the membrane.

What makes the PTS literature so difficult are the many kinds of variations that have been reported. Some of these follow.

PTS sugars. Which sugars are PTS sugars? It depends on the organism. In obligate anaerobes, virtually all sugars, including lactose, are PTS sugars. In facultative anaerobes such as *E. coli,* however, some carbohydrates are PTS sugars while others (including lactose) are not, and in

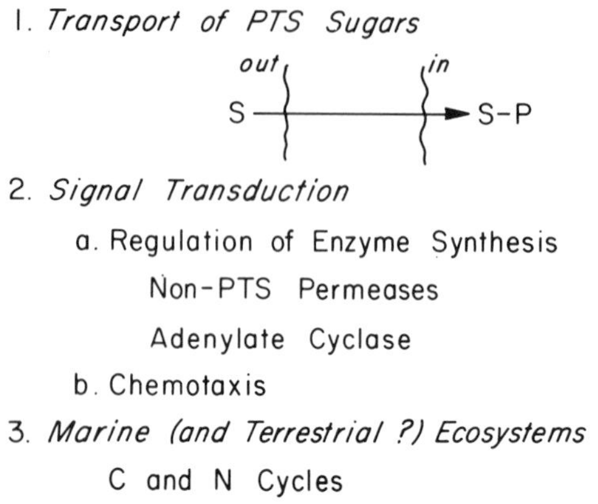

1. *Transport of PTS Sugars*

out / in

S ⟶ S-P

2. *Signal Transduction*

a. Regulation of Enzyme Synthesis

Non-PTS Permeases

Adenylate Cyclase

b. Chemotaxis

3. *Marine (and Terrestrial ?) Ecosystems*

C and N Cycles

FIG. 1. Some physiological functions of the PTS. 1. Sugar phosphate (S-P), the product of sugar uptake, is catabolized via the glycolytic pathway. 2. The PTS also regulates the activities of at least two enzymes, glycerol kinase and adenylate cyclase. 3. See the text for a description of the ecosystems.

obligate aerobes only a few carbohydrates are PTS sugars.

EII complexes in one organism. Figure 4 shows the variation in different EII complexes that are found in a single organism. For instance, in *E. coli,* II^{Mtl} (the mannitol permease) comprises a single polypeptide chain, the glucose-specific permease consists of two proteins (III^{Glc} and II^{Glc}), and the mannose permease, II^{Man}, is composed of three separately encoded proteins (IIA^{Man}, Pel, and IIB^{Man} [Pel is explained below]). Four, possibly even five separately encoded polypeptide chains per EII complex have been described.

Specificity of EII complexes. Some EII complexes are specific for a given sugar and its analogs, whereas others have a broader specificity. Two examples are given in Fig. 5. Glucose can enter *E. coli* and *Salmonella typhimurium* cells via the so-called specific glucose PTS permease, the III^{Glc}/II^{Glc} complex, or the II^{Man} complex, which functions with seven different sugars of the D-gluco and D-manno configurations.

EII complexes in different genera. The PTS proteins are not stringently conserved across bacterial genera, although amino acid sequence similarities are often observed at their active sites. Whereas II^{Man} from *E. coli* consists of three proteins, the corresponding complex from the closely related organism *Vibrio furnissii* is composed of four proteins (Bouma, 1991).

Functional domains. Variations on the basic theme extend to the active sites and functions of the EIIs. In general, it appears that there are two phosphate transfers between phospho-HPr and the sugar substrate, i.e., within the EII complex. In some cases, these involve transfer from P-HPr to a His residue and then again to another His in the same or a different polypeptide chain, whereas in others it appears to be from P-HPr to a His residue to a Cys thiol residue to the sugar (see chapter 27).

Chimeric proteins. Chimeric proteins have also been reported. These consist of single polypeptide chains that combine the functions of what are usually two distinct PTS proteins. For example, FPr, which is expressed when enteric bacteria are grown on fructose, is a combination of HPr and a III^{Fru} (also called IIA^{Fru}) and can substitute for HPr in mutants of the latter.

From these examples, it is clear that variations in the EII complexes are among the most perplexing aspects of the PTS. Early on, it was shown that II^{Man} could be separated into two membrane protein fractions, designated IIA^{Man} and IIB^{Man}. Subsequently, cloning and characterization of the genes and gene products showed that the *E. coli man* operon comprises three structural genes. In the older nomenclature these were designated IIA^{Man}, Pel, and IIB^{Man}; they are encoded by the *manX*, *manY*, and *manZ* genes, respectively. The Pel protein was given this designation before it was shown to be part of the II^{Man} complex. Pel mutants were resistant to penetration of the bacterial inner membrane by λ DNA after it is ejected from the phage, which is attached to the LamB protein in the outer membrane. IIA^{Man} accepts the phosphoryl group from P-HPr, whereas IIB^{Man} is the sugar receptor (Pel is also required for the complex to function). Simoni et al. (1968) first found that a cytoplasmic protein was required for the phosphorylation of some PTS sugars. Since we did not know the function of the new proteins, we called them factors III. We now know that III and IIA proteins serve identical functions, as phosphoryl acceptors from P-HPr. Both designations continue to be used, and for some of the chapters in this book, it is necessary to emphasize that

III = factor III = Enzyme III = II-A = IIA = Enzyme II-A

Nomenclature. Clarity and simplicity have decreased as more EII complexes are isolated. Two systems of nomenclature have recently been proposed for the EIIs. (i) The designation "class" is to be used (Meadow et al., 1990) to specify the number of separately encoded polypeptide chains in the complex, regardless of cellular location (integral membrane, peripheral membrane, cytoplasmic). Thus, class 1, 2, 3, or 4 specifies one to four polypeptide chains in the EII complexes, respectively. In this system, attempts were made to retain the designation II-B for the integral membrane sugar receptor when the number of polypep-

Reprinted from the PROCEEDINGS OF THE NATIONAL ACADEMY OF SCIENCES
Vol. 52, No. 4, pp. 1067–1074. October, 1964.

*PHOSPHATE BOUND TO HISTIDINE IN A PROTEIN AS AN INTERMEDIATE IN A NOVEL PHOSPHO-TRANSFERASE SYSTEM**

BY WERNER KUNDIG,† SUDHAMOY GHOSH,‡ AND SAUL ROSEMAN

RACKHAM ARTHRITIS RESEARCH UNIT AND DEPARTMENT OF BIOLOGICAL CHEMISTRY, UNIVERSITY OF MICHIGAN, ANN ARBOR

Communicated by Arthur Kornberg, August 10, 1964

Mammalian tissues contain a kinase involved in the intermediary metabolism of the sialic acids.[1, 2] This enzyme has been extensively purified,[3] studied in detail, and catalyzes the following reaction: *N*-Acyl-D-mannosamine + ATP $\xrightarrow{Mg^{++}}$ *N*-Acyl-D-mannosamine-6-P + ADP. To determine whether this kinase occurred in bacteria, such as *Aerobacter cloacae* and *Escherichia coli* K235,[4] that metabolize *N*-acetyl-D-mannosamine, extracts of these organisms were examined and found to contain a novel phospho-transferase system. The system obtained from *E. coli* K235 consisted of two enzymes, I and II, and a histidine-containing, heat-stable protein (HPr). The sequence of reactions is:

$$\text{Phosphoenolypyruvate (PEP)} + \text{HPr} \underset{Mg^{++}}{\overset{I}{\rightleftharpoons}} \text{Phospho-histidine-protein (P-HPr)} + \text{Pyruvate} \quad \text{(A)}$$

$$\text{P-HPr} + \text{Hexose} \xrightarrow[Mg^{++}]{II} \text{Hexose-6-P} + \text{HPr} \quad \text{(B)}$$

$$\text{PEP} + \text{Hexose} \xrightarrow[Mg^{++}]{I + II} \text{Hexose-6-P} + \text{Pyruvate} \quad \text{(A+B)}$$

The intermediate in the system, P-HPr, is protein-bound phosphohistidine.

FIG. 2. Introduction to the first paper on the PTS. Reprinted from Kundig et al. (1964) with permission.

tide chains exceeded one. III was used to designate the proteins that accept P from P-HPr in order to minimize confusion with the older literature. Unfortunately, neither acronym is particularly logical. (ii) The nomenclature suggested by Saier and Reizer (1992) has been adopted in the review by Postma et al. (1993) and is described in detail elsewhere in this volume (see chapter 31). The EIIs are grouped by sequence homology and by whether the two phosphoryl transfers within the EII complex involve only His residues, or one His and one Cys. In this system, functional domain is the key to classification, independent of whether the genes encoding the individual polypeptide chains have been split or fused during evolution. Thus, domains within a protein are designated identically to individual proteins if they serve the same function or, more frequently, if they are pre-

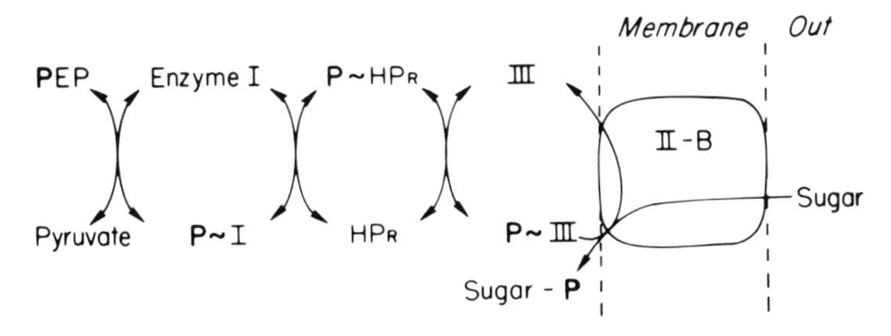

FIG. 3. Sequence of phosphotransfer reactions in the PTS. In this example, the EII complex comprises a pair of proteins, such as the III^{Glc}/II^{Glc} complex of *E. coli.* In this organism, EI monomer is 63.5 kDa and is phosphorylated at N-3 of the imidazole ring of a His residue, HPr is 9.1 kDa and is phosphorylated at N-1 of the His (it contains only one), III^{Glc} is 18.1 kDa and is phosphorylated at N-3 of a His, and II^{Glc} is 50.1 kDa and is phosphorylated on the −SH of a Cys.

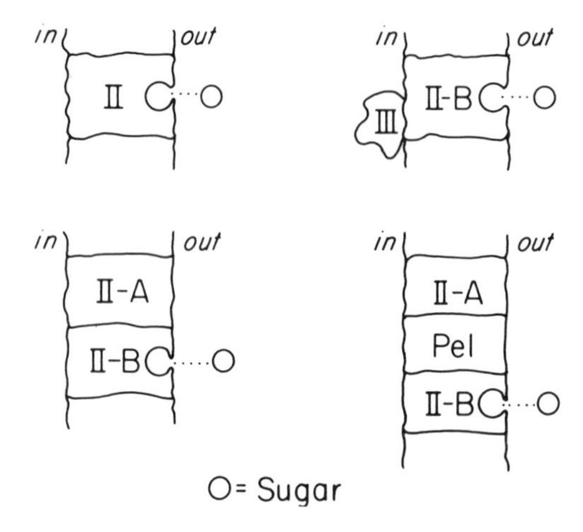

FIG. 4. Some examples of the variety of EII complexes. The II-A/Pel/II-B complex is the II^{Man} complex of *E. coli* and *S. typhimurium,* shown again in Fig. 5 with an alternate nomenclature (III^{Man}/II-P/II-B^{Man}). "Pel" is defined in the text.

sumed to serve the same function. Domains are called IIA, IIB, and IIC (and sometimes IID). IIA accepts the phosphoryl group from P-HPr, IIB accepts it from IIA, and IIC and IID replace II-B as the sugar receptor in the older nomenclature. Unfortunately, IIB designates different functions in the two proposed nomenclature systems.

Both nomenclature systems are less than ideal. The class designation does not define function. On the other hand, there are known exceptions to the functional-domain nomenclature, which may increase with time. A simplified new and improved nomenclature is required; hopefully, this will not require a 7-year GATT conference.

Phosphoryl Transfer and Phosphate Transfer Potential

As indicated in Fig. 3, the last step of phosphotransfer is to the sugar; during this reaction, the sugar is translocated across the cytoplasmic membrane. Thus, in a concerted step, by using the metabolic equivalent of one ATP, the sugar is translocated, trapped inside the cell, and placed directly into its catabolic pathway.

The phosphate transfer potential of PEP is greater than that of any other naturally occurring phosphate derivative (Fig. 6), about twice that of ATP. However, the first three proteins in the PTS sequence, phospho-EI, phospho-HPr, and phospho-III^{Glc}, have almost the same phosphotransfer potentials as PEP. Thus, there is an enormous change in free energy between phospho-III^{Glc} and internal Glc-6-P. Whether this change occurs in the P transfer to II^{Glc}, P transfer to the sugar, or translocation of the sugar and/or as heat remains to be determined, but -11 kcal (ca. -46 kJ) per mol of sugar phosphorylated is not yet accounted for.

For what follows, it is important to emphasize the reversibility of the phosphotransfer reactions, at least as far as the cytoplasmic proteins are concerned. Unlike the electron transport chain, which normally proceeds in one direction, the values in Fig. 6 predict that the reactions are freely reversible, starting from PEP through and including phospho-III^{Glc}, and this prediction has been confirmed in vitro. The physiological impact of this reversibility can be visualized in Fig. 5. If the proteins were fully phosphorylated in a cell growing on some non-PTS carbon source, such as glycerol, for example, and methyl α-glucoside was added to the medium, one would expect all of the PTS proteins to be quickly dephosphorylated, in-

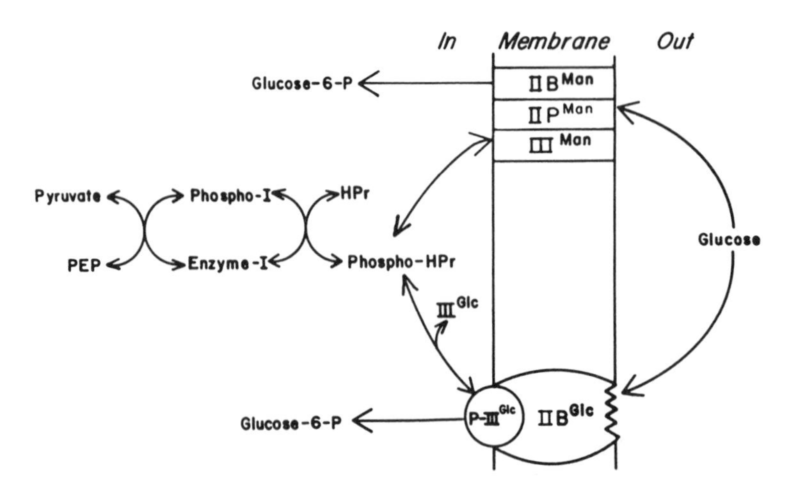

FIG. 5. Glucose transport systems in enteric bacteria.

	$\Delta G^{0'}$ (Kcal/mol)
PEP	-14.7
P ~ I	-14.5
P ~ HPr	-13.0
P ~ III	-14.4
Acyl ~ P	-10.1
ATP	-7.3
Glc-6-P	-3.3

FIG. 6. Apparent standard free energies of hydrolysis of phosphate derivatives. Reprinted from Weigel et al. (1982) with permission.

cluding phospho-III^{Glc} (discussed below). However, the same result could be achieved by adding mannose, mannitol, or any other PTS sugar. This effect should be greatest in leaky EI and HPr mutants, which cannot quickly restore the normal levels of the phosphorylated proteins, or under conditions where the cellular levels of PEP are low. The physiological consequences of rapid dephosphorylation on the activity of glycerol kinase are discussed in chapter 56.

Regulation of the PTS

As indicated above, PEP provides an enormous thermodynamic driving force for sugar uptake via the PTS. By using the apparent standard free energies of hydrolysis of PEP and Glc-6-P ($\Delta G^{0\prime}$), it is possible to calculate the K'_{eq} for the following reaction:

$$\text{PEP}_{\text{inside}} + \text{Sugar}_{\text{out}} = \text{Pyruvate}_{\text{inside}} + \text{Sugar-P}_{\text{inside}}$$

$$K'_{eq} = 10^8 = \frac{(\text{Pyruvate}_{\text{inside}})\ (\text{Sugar-P}_{\text{inside}})}{(\text{PEP}_{\text{inside}})\ (\text{Sugar}_{\text{out}})}$$

Where they have been measured, the internal concentrations of pyruvate and PEP are equal within an order of magnitude; therefore, at thermodynamic equilibrium, $(\text{Sugar-P}_{\text{inside}})/(\text{Sugar}_{\text{out}})$ = ca. 10^8; i.e., when $\text{Sugar}_{\text{out}}$ is 1 μM, $\text{Sugar-P}_{\text{inside}}$ would be 100 M! Therefore, there is an absolute requirement for stringent regulation of the PTS, and this in fact is the case. Figure 7 shows that glucose is rapidly taken up for about 1.5 min when it is added to starving cells but that uptake then virtually ceases, to be followed by a slow uptake which the cells can handle metabolically. It perhaps should be stressed that the uptake of sugar is the rate-limiting step in cell growth.

Regulation of the PTS is also evident in membrane vesicles, a system that is much simpler and subject to more experimental control than intact cells. Membrane vesicles of wild-type *E. coli* or *S. typhimurium* cells can take up glucose or its analog methyl α-glucoside in the presence of PEP. The kinetic studies show an initial high rate of uptake, which decreases in a few seconds. The decreased rate could not be attributed to end product inhibition (MeGlc-6-P and pyruvate), pyruvate catabolites, or the rate of uptake of PEP. It should be noted that the supply of PEP is in vast excess in these experiments.

Thus, the precise mechanism for PTS regulation of sugar uptake remains obscure. We have proposed that this regulation may be effected via EI. This interesting enzyme undergoes a monomer-dimer transition (monomer from *E. coli*, 63.5 kDa), and the dimer accepts the phosphoryl group from PEP, while all available evidence suggests that the monomer is the species that interacts with HPr. Chauvin et al. (1990) have recently reported the kinetics of association and dissociation of the EI monomer-dimer and find that they are surprisingly slow and affected by EI ligands. Thus, it is possible to speculate that a catabolite could bind to monomer or dimer and greatly influence this rate. If, for instance, the enzyme was maintained as a monomer, sugar uptake would stop.

Regulation by the PTS

One major area of research on the PTS is how it regulates other systems. It was one of the first

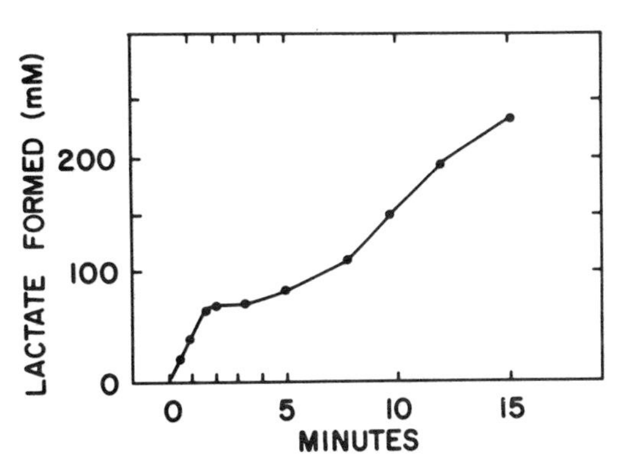

FIG. 7. Regulation of glucose uptake via the PTS. In these experiments, uniformly labeled glucose was added to a suspension of starving *S. lactis* cells, and at the indicated times aliquots were analyzed for lactate, glycolytic intermediates, etc. Almost all of the labeled carbon was found in the lactate, and the sum of all metabolites gave essentially the same plot as shown in the figure. The ordinate therefore represents glucose uptake. Reprinted from Mason et al. (1981) with permission.

signal-transducing systems discovered, although the molecular details on how it functions are still being clarified. Perhaps the most obvious regulatory role of the PTS is evident in the "glucose effect," a phenomenon first described in the late 1800s. It was shown that various bacterial proteins are not expressed or are very poorly expressed when the cells are grown on glucose. Jacques Monod (1942) was one of the first to quantitate the effect, and one of his many classic experiments is illustrated in Fig. 8.

E. coli (and other bacteria) was grown in a simple salts medium containing two sugars, and the rate of growth was observed. With certain pairs of sugars, such as maltose and xylose, a single exponential growth curve was obtained. With other pairs of sugars, such as glucose and xylose, a biphasic curve, which Monod called diauxie, was obtained. While not mentioned in Monod's thesis, it is clear from Fig. 8 that the cells can distinguish between external glucose and internal glucose and its catabolites generated from maltose. The explanation for the biphasic curve came much later and is shown in Fig. 9. In this case, the sugar pair is glucose and lactose. Glucose is consumed first, and the lactose operon is not expressed as long as glucose is present in the extracellular medium.

The diauxic effect in Fig. 9 is, in reality, a PTS effect. That is, virtually any PTS sugar can substitute for glucose and virtually any non-PTS sugar can substitute for lactose. How does the PTS regulate expression of the non-PTS operons? The model is shown in Fig. 10. Most (perhaps all) of the non-PTS operons under control of the PTS require cyclic AMP (cAMP) and inducer at certain minimal intracellular levels to initiate transcription. External glucose lowers both of these to the point at which transcription is suppressed.

The mechanism underlying this regulation was shown to involve the product of a single *pts* gene called *crr* (Saier and Roseman, 1976). Diauxie processes for a number of non-PTS systems were simultaneously relieved by a mutation of *crr*, and the gene was shown to encode the protein III^{Glc} in both *E. coli* (Meadow et al., 1982a, 1982b; Saffen et al., 1987) and *S. typhimurium* (Nelson et al., 1984). The simplest model (Fig. 11) consistent with all available data is that III^{Glc} binds to and inhibits various non-PTS permeases, leading to inhibition of uptake of the non-PTS inducing molecule ("inducer exclusion"). However, phospho-III^{Glc} does not inhibit the permeases. Regulation of adenylate cyclase by the PTS is less clear. Phospho-III^{Glc} is thought to be required for optimum activity of the cyclase, whereas III^{Glc} is inactive. However, other PTS proteins may also be involved. Finally, the mechanism by which glucose causes the expulsion of cAMP from the cell remains to be explained.

The basis of these models is that the "glucose" or PTS effect is mediated by the state of phosphorylation of at least III^{Glc}. In the presence of external glucose or the analog methyl α-glucoside, the phosphoryl groups are removed from the PTS proteins, leaving dephosphorylated III^{Glc}, which inhibits the permeases and at the same time reduces the level of adenylate cyclase activity. Again we note that any other PTS sugar would have the same effect because of the reversibility of the phosphotransfer reactions. As indicated in the next section, it is not clear why internally generated glucose or other PTS sugars do not have the same effect as external glucose.

III^{Glc} is a remarkable protein. The *E. coli* protein is only 18.1 kDa, but it recognizes and interacts with at least a dozen other proteins, some of which are shown in Fig. 12. Some of these interactions involve phosphate transfer, whereas others do not. The most interesting point is that there is no apparent sequence homology in the target pro-

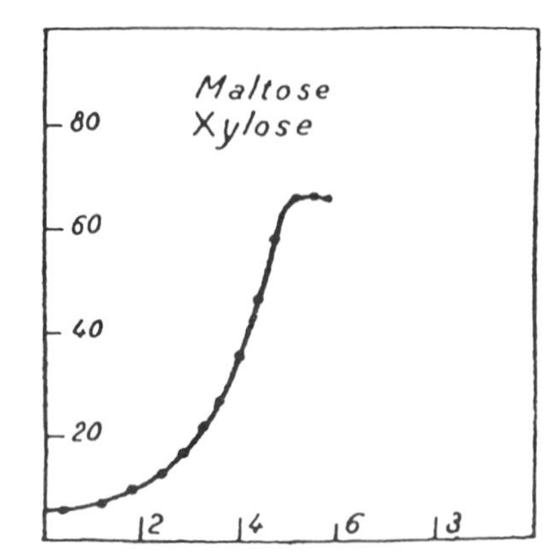

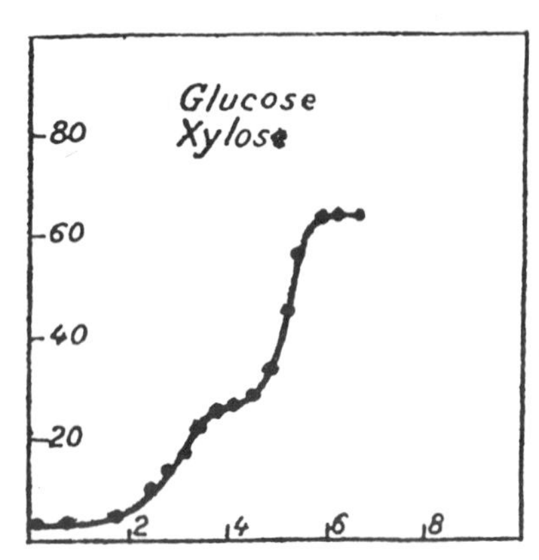

FIG. 8. Diauxic growth of *E. coli*. The ordinate represents cell number (determined by absorbance), and the abcissa shows time in hours. Reprinted from Monod (1942) with permission.

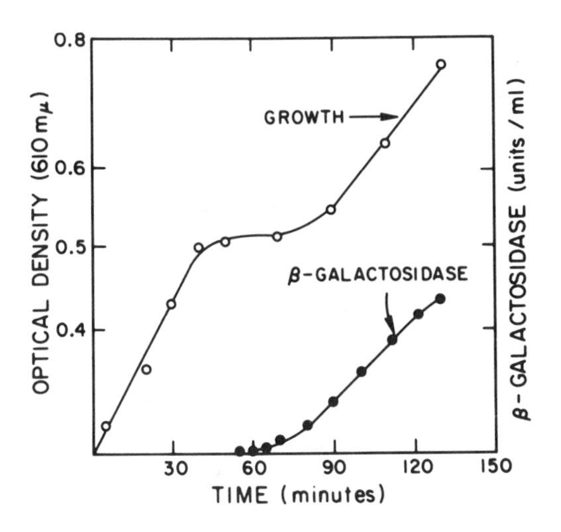

FIG. 9. Diauxic growth of *E. coli* on a mixture of glucose and lactose. Reprinted from Epstein et al. (1966) with permission.

teins. The molecular details of these interactions are beginning to be worked out. Both the crystal (X-ray diffraction [Worthylake et al., 1991]) and solution (nuclear magnetic resonance) structures of III^{Glc} have been established, as well as the nuclear magnetic resonance structure of phospho-III^{Glc} (Pelton et al., 1991, 1992). In addition, the crystal structure of the complex of III^{Glc} with one of its targets, glycerol kinase, has been resolved (Hurley et al., 1993) (see chapter 56).

Problems and Prospects

The problems that lie ahead are more interesting and far more numerous than those in the past. A few of these are outlined below.

Phosphate transfer reactions. Virtually every aspect of phosphate transfer reactions remains to be elucidated. In a four-component protein system, such as EI, HPr, and III^{Glc}/II^{Glc}, does phosphate transfer proceed via a sequence of bimolecular reactions as we normally write them, or are transient ternary or quarternary complexes formed? Are the latter formed during sugar transport and phosphorylation? Which amino acid residues in each of the species is involved in the binding and recognition phenomena? What are the binding constants of the interacting proteins? What are the rate constants of the different steps, and which is rate limiting? Does the EI monomer-dimer transition control the rate? Can these quantitative values account for the rates of sugar phosphorylation in vitro and in vivo? How is -11 kcal/mol of $\Delta G^{0\prime}$ expended between phospho-III^{Glc} and Glc-6-P?

PTS sugar transport. The molecular details of the translocation and phosphorylation steps are not known (see chapter 27). What is the mechanism for regulating PTS sugar uptake (Fig. 7)? Is it controlled by the monomer-dimer transition of EI? Precisely where and when during translocation is the sugar phosphorylated? What is the molecular configuration of the EII complex during this step? Why are there so many variants of the EIIs, even in a single organism? Is this related to different molecular mechanisms for the uptake of different PTS sugars, to regulation of each of these processes, or to regulation of other phenomena by these sugars?

The "glucose" effect. As indicated above, the PTS regulates transcription of non-PTS operons by controlling the levels of inducer and cAMP, and it does so by regulating the ratio of III^{Glc} to phospho-III^{Glc}. This simple model appears to be necessary, but is it sufficient? Are other PTS pro-

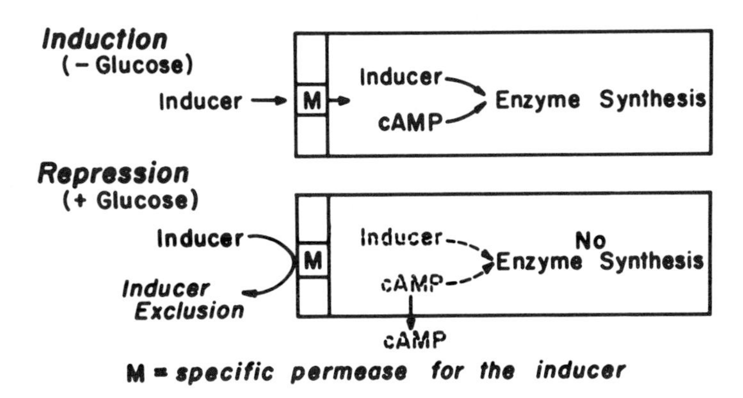

FIG. 10. The glucose effect in *E. coli*. Transcription of the appropriate operon requires minimal levels of cAMP and inducer, and one and/or the other is not attained in the presence of glucose or other PTS sugars or sugar analogs. At least in some strains of *E. coli*, adding glucose to the medium results in a rapid efflux of cAMP (Makman and Sutherland, 1965).

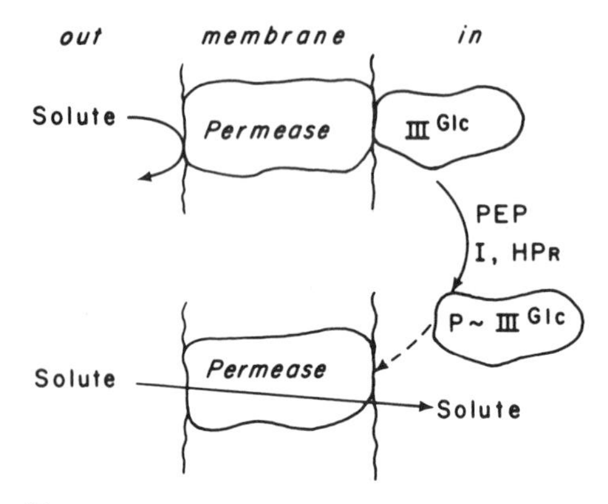

FIG. 11. PTS regulation of non-PTS permeases. III^{Glc} interacts stoichiometrically with the permease to inhibit transport, whereas phospho-III^{Glc} is inactive (or may be a positive effector). The inhibition is designated inducer exclusion.

teins (and possibly non-PTS proteins) directly involved? Do the PTS proteins, such as III^{Glc}, act as regulatory proteins directly in gene transcription? Is the structure of the III^{Glc}/glycerol kinase a model for how this protein will interact with the other targets (Fig. 12), i.e., by induced fit, possibly involving divalent cations (see chapter 56)?

The status of internally generated PTS sugars, such as glucose from lactose or maltose, is less than clear. The subject has gone through a cycle in which it was thought that internal PTS sugars are not phosphorylated by the PTS, to where they may be, to where some investigators believe that they are. However, if this is true, lactose, melibiose, and maltose (all of which yield intracellular glucose) should, for instance, give a diauxic effect with other non-PTS sugars such as xylose. As pointed out in the reviews, the fate of internally generated PTS sugars has not been resolved.

Regulation of expression of PTS proteins. In *E. coli,* the structural genes that encode HPr, EI, and III^{Glc} are contiguous (*ptsH*, *ptsI*, and *crr*, respectively). Growth of the cells on any PTS sugar increases expression of the proteins by a maximum of threefold (growth on glucose). Three promoters have been identified in this operon, two upstream from *ptsH* and one toward the distal end of *ptsI*, which regulates expression of *crr*. The operon also contains binding sites for the cAMP-cAMP receptor protein complex and contains other regulatory regions; it produces several different transcripts apparently depending on growth conditions. Genetic regulation of the genes encoding some EIIs have been studied, and these vary from the classical regulator-operator pairs to more complex phenomena involving control by phos-

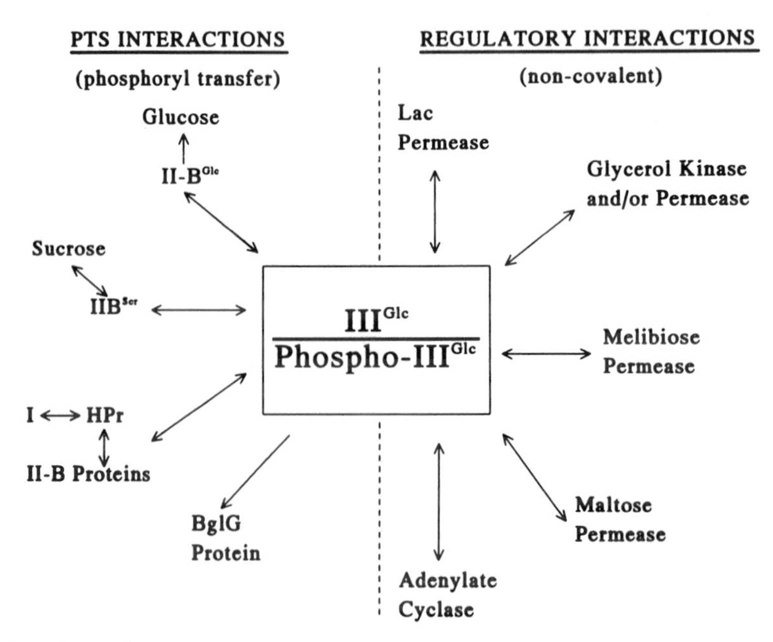

FIG. 12. Some interactions of III^{Glc} and Phospho-III^{Glc} with their target proteins. Phosphate transfer occurs between the proteins on the left, while those on the right form noncovalent complexes: III^{Glc} binds to and inhibits the permeases, whereas phospho-III^{Glc} is thought to activate adenylate cyclase.

phorylation of antiterminator proteins. These reports provide the foundation for what will undoubtedly prove to be a fruitful but complex area of gene regulation.

Chemotaxis. Chemotaxis to PTS sugars requires all of the PTS proteins. How is the PTS coupled to taxis? Do one or more of the PTS proteins interact directly with the flagellar motor or to one of the Che proteins, such as CheY, in order to control the direction of rotation of the flagella (see chapter 30)?

Phosphotransfer potentials: interactions with other systems. The PTS seems excessively complex. Sugar transport concomitant with phosphorylation "should" require only a membrane ATP-dependent kinase. Why the complexity? Why the very high energy levels of the phosphoproteins? Are they used to drive or to control other systems such as CheY mentioned above or some of the many two-component regulatory protein systems? It should be emphasized that the two-component systems, some of which are discussed in this volume, also involve phospho-His and acyl-P proteins. Thus far, only two interactions between the PTS and other metabolic or regulatory systems have been recorded. EI is reversibly phosphorylated at its active site by acetokinase (an acyl-P protein), and it has been suggested that this reaction links the PTS to the Krebs cycle (Fox et al., 1986). Recently, another ATP-dependent kinase which phosphorylates EI at its active-site histidine has been reported. The kinase requires NAD^+ and is inhibited by NADH, and it has been suggested to link the PTS to the electron transport chain (Dannelly and Roseman, 1992).

The PTS and evolution. Because of considerable conservation in the active sites of PTS proteins in gram-negative and gram-positive bacteria, it seems reasonable to conclude that the PTS was present in primordial prokaryotic cells and was at least retained in the *Eubacteria*. However, this leaves many questions. Did other transport systems evolve from the PTS? There is, in fact, a report of a lactose permease in *Streptococcus thermophilus* (Poolman et al., 1989, 1992) with a domain very similar to the one found in *E. coli* III^{Glc}. Is the PTS, or some portion of it, present in the *Archaebacteria* and/or eukaryotes? In fact, a recent paper describes an HPr-like protein in *Aspergillus fumigatus* (Barker et al., 1991).

Conclusion

The complexities of the PTS make the literature almost incomprehensible to the occasional reader. Perhaps this brief overview will help and has also succeeded in illustrating the many intriguing problems in this field and the diverse approaches that will be required to solve them.

REFERENCES

Barker, S., R. Mathews, W. Lee, A. Bostock, and J. Burnie. 1991. Identification of a gene encoding an HPr-like protein in *Aspergillus fumigatus*. *J. Med. Vet. Mycol.* **29**(6):381–386. (Abstract.)

Bouma, C. 1991. Ph.D. thesis. The Johns Hopkins University, Baltimore.

Chauvin, F., S. Roseman, and L. Brand. 1990. Monomer/dimer transition of Enzyme I of the *Escherichia coli* phosphotransferase system; kinetic studies. *Biophys. J.* **57**:429a. (Abstract.)

Dannelly, H. K., and S. Roseman. 1992. NAD^+ and NADH regulate an ATP-dependent kinase that phosphorylates Enzyme I of the *Escherichia coli* phosphotransferase system. *Proc. Natl. Acad. Sci. USA* **89**:11274–11276.

Epstein, W., S. Naono, and F. Gros. 1966. Synthesis of enzymes of the lactose operon during diauxic growth of *Escherichia coli*. *Biochem. Biophys. Res. Commun.* **24**:588. (Abstract.)

Fox, D. K., N. D. Meadow, and S. Roseman. 1986. Phosphate transfer between acetate kinase and Enzyme I of the bacterial phosphotransferase system. *J. Biol. Chem.* **261**:13498–13503.

Hurley, J. H., H. R. Faber, D. Worthylake, N. D. Meadow, S. Roseman, D. W. Pettigrew, and S. J. Remington. 1993. Structure of the regulatory complex of *Escherichia coli* III^{Glc} with glycerol kinase. *Science* **259**:673–677.

Kundig, W., S. Ghosh, and S. Roseman. 1964. Phosphate bound to histidine in a protein as an intermediate in a novel phosphotransferase system. *Proc. Natl. Acad. Sci. USA* **52**:1067–1074.

Makman, R. S., and E. W. Sutherland. 1965. Adenosine 3′,5′-phosphate in *Escherichia coli*. *J. Biol. Chem.* **240**:1309–1314. (Abstract.)

Mason, P. W., D. P. Carbone, R. A. Cushman, and A. S. Waggoner. 1981. The importance of inorganic phosphate in regulation of energy metabolism of *Streptococcus lactis*. *J. Biol. Chem.* **256**:1861–1866.

Meadow, N. D., D. K. Fox, and S. Roseman. 1990. The bacterial phosphoenolpyruvate:glucose phosphotransferase system. *Annu. Rev. Biochem.* **59**:497–542.

Meadow, N. D., J. M. Rosenberg, H. M. Pinkert, and S. Roseman. 1982a. Sugar transport by the bacterial phosphotransferase system. XX. Evidence that *crr* is the structural gene for the *Salmonella typhimurium* glucose-specific phosphocarrier protein III. *J. Biol. Chem.* **257**:14538–14542.

Meadow, N. D., D. W. Saffen, R. P. Dottin, and S. Roseman. 1982b. Molecular cloning of the *crr* gene, and evidence that it is the structural gene for III^{Glc}, a phosphocarrier protein of the bacterial phosphotransferase system. *Proc. Natl. Acad. Sci. USA* **79**:2528–2532.

Monod, J. 1942. *Recherches sur la Croissance des Cultures Bacteriennes*. Hermann et Cie, Paris.

Mopper, K., R. Dawson, G. Liebezeit, and V. Ittekkot. 1980. The monosaccharide spectra of natural waters. *Mar. Chem.* **10**:55–66.

Nelson, S. O., A. R. J. Schuitema, R. Benne, L. H. T. Van der Ploeg, J. S. Plijter, F. Aan, and P. W. Postma. 1984. Molecular cloning, sequencing, and expression of the *crr* gene: the structural gene for III^{Glc} of the bacterial PEP:glucose phosphotransferase system. *EMBO J.* **3**:1587–1593.

Pelton, J. G., D. A. Torchia, N. D. Meadow, and S. Roseman. 1992. Structural comparison of phosphorylated and unphosphorylated forms of III^{Glc}, a signal transducing protein from *Escherichia coli*, using three-dimensional NMR techniques. *Biochemistry* **31**:5215–5224.

Pelton, J. G., D. A. Torchia, N. D. Meadow, C.-Y. Wong, and S. Roseman. 1991a. Secondary structure of III^{Glc}, a signal transducing protein from *Escherichia coli*, determined by heteronuclear three-dimensional nuclear magnetic resonance spectroscopy. *Proc. Natl. Acad. Sci. USA* **88**:3479–3483.

Pelton, J. G., D. A. Torchia, N. D. Meadow, C.-Y. Wong, and S. Roseman. 1991b. ^{1}H, ^{15}N, ^{13}C NMR signal assignments of III^{Glc}, a signal-transducing protein of *Escherichia* *coli,* using three-dimensional triple-resonance techniques. *Biochemistry* **30:**10043–10057.

Poolman, B., R. Modderman, and J. Reizer. 1992. Lactose transport system in *Streptococcus thermophilus.* The role of histidine residues. *J. Biol. Chem.* **267:**9150–9157.

Poolman, B., T. J. Royer, S. E. Mainzer, and B. F. Schmidt. 1989. Lactose transport system of *Streptococcus thermophilus:* a hybrid protein with homology to the melibiose carrier and enzyme III of phosphoenolpyruvate-dependent phosphotransferase systems. *J. Bacteriol.* **171:**244–253.

Postma, P. W., J. W. Lengeler, and G. R. Jacobson. 1993. Phosphoenolpyruvate:carbohydrate phosphotransferase systems of bacteria. *Microbiol. Rev.* **57:**543–594.

Reizer, J., M. H. Saier, Jr., J. Deutscher, F. Grenier, J. Thompson, and W. Hengstenberg. 1988. The phosphoenolpyruvate:sugar phosphotransferase system in Gram-positive bacteria: properties, mechanism and regulation. *Crit. Rev. Microbiol.* **15:**297–338.

Roseman, S., and N. D. Meadow. 1990. Signal transduction by the bacterial phosphotransferase system: diauxie and the *crr* gene (J. Monod Revisited). Minireview. *J. Biol. Chem.* **265:**2993–2996.

Saffen, D. W., K. A. Presper, T. Doering, and S. Roseman. 1987. Sugar transport by the bacterial phosphotransferase system: molecular cloning and structural analysis of the *Escherichia coli ptsH, ptsI,* and *crr* genes. *J. Biol. Chem.* **262:**16241–16253.

Saier, M. H., Jr. 1989. Protein phosphorylation and allosteric control of inducer exclusion and catabolite repression by the bacterial phosphoenolpyruvate: sugar phosphotransferase system. *Microbiol. Rev.* **53:**109–120.

Saier, M. H., Jr., and J. Reizer. 1992. Proposed uniform nomenclature for the proteins and protein domains of the bacterial phosphoenolpyruvate:sugar phosphotransferase system. *J. Bacteriol.* **174:**1433–1438.

Saier, M. H., Jr., and S. Roseman. 1976. Sugar transport. IX. The *crr* mutation: its effect on repression of enzyme synthesis. *J. Biol. Chem.* **25:**6598–6605.

Sharma, S., P. K. Hammen, J. W. Anderson, A. Leung, F. Georges, W. Hengstenberg, R. Klevit, and E. B. Waygood. 1993. Deamidation of HPr, a phosphocarrier protein of the phosphoenolpyruvate:sugar phosphotransferase system, involves asparagine 38 (HPr-1) and asparagine 12 (Hpr-2) in isoaspartyl acid formation. *J. Biol. Chem.* **268:**17695–17704.

Simoni, R. D., M. Smith, and S. Roseman. 1968. Resolution of a staphylococcal phosphotransferase system into four protein components and its relation to sugar transport. *Biochem. Biophys. Res. Commun.* **31:**804–811.

Weigel, N., M. A. Kukuruzinska, A. Nakazawa, E. B. Waygood, and S. Roseman. 1982. Sugar transport by the bacterial phosphotransferase system. XIV. Phosphoryl transfer reactions catalyzed by Enzyme I of *Salmonella typhimurium. J. Biol. Chem.* **257:**14477–14491.

Worthylake, D., N. D. Meadow, S. Roseman, D.-I. Liao, O. Herzberg, and S. J. Remington. 1991. Three-dimensional structure of the *Escherichia coli* phosphocarrier protein III^{Glc}. *Proc. Natl. Acad. Sci. USA* **88:**10382–10386.

Chapter 27

Model for the Role of Domain Phosphorylation in the Mechanism of Carbohydrate Transport via Enzyme II of the Phosphoenolpyruvate-Dependent Mannitol Phosphotransferase System

G. T. ROBILLARD,[1] J. S. LOLKEMA,[2] AND H. BOER[1]

Department of Biochemistry and The Groningen Biomolecular Science and Biotechnology Institute, Nyenborgh 4, 9747 AG Groningen,[1] and Department of Microbiology, University of Groningen, Kerklaan 30, Haren,[2] The Netherlands

A considerable amount of kinetic data has been collected on two enzymes II (EII) of the phosphoenolpyruvate (PEP)-dependent carbohydrate transport system (PTS), *Escherichia coli* EII^{Mtl} and *Rhodopseudomonas sphaeroides* EII^{Frc}. The purpose of this short review is to bring together these data and try to develop a model for the role of the phosphoryl group in the transport and phosphorylation cycle. Elements which will be important for the model include the domain structure and association state of EII, their relation to the phosphoryl group transfer process, the orientation of the carbohydrate-binding sites, and the kinetics of binding and phosphorylation. Each element will be treated in the following paragraphs.

Domain Structure and Function of EII^{Mtl}

The function of EII is to catalyze the transport and phosphorylation of carbohydrates over the cytoplasmic membrane at the expense of PEP. This is accomplished by using the series of phosphorylated protein intermediates shown below, where EII is the actual transport protein.

$$\text{PEP} \rightarrow \text{P-EI} \rightarrow \text{P-HPr} \rightarrow \text{P-EII} \rightarrow \text{Carbohydrate-P}$$

The activity of EII can be measured in two ways: (i) as the rate of carbohydrate phosphorylation at the expense of PEP in the presence of the rest of the PTS components (the phosphorylation reaction) or (ii) as the rate of exchange of the phosphoryl group between unlabeled carbohydrate-P and radiolabeled carbohydrate (the exchange reaction). The latter reaction requires no PTS components other than EII.

Amino acid sequences have been derived from DNA sequences for a large number of EIIs from different sources. Comparison of these sequences has led to the general view that these proteins are constructed from domains, each with a separate enzymatic function. Depending on the EII in question, the domains can occur as covalently linked or separate entities (Saier et al., 1988). In the case of the mannitol-specific EII, EII^{Mtl}, a single polypeptide chain of 637 amino acids folds into three separate domains designated A, B, and C. The C-terminal 150 amino acids form the A domain, the next 140 amino acids form the B domain, and the final 340 residues at the N terminus constitute the C domain (see the central structure in Fig. 1) (Grisafi et al., 1989; Stephan and Jacobson, 1986; Stephan et al., 1989; Jacobson et al., 1983; White and Jacobson, 1990; van Weeghel et al., 1991a, 1991b, 1991c). The A and B domains are hydrophilic proteins which protrude into the cytoplasm, whereas the C domain is hydrophobic and is buried in the cytoplasmic membrane. In the *E. coli* EII^{Frc}, the B and C domains are covalently linked but the A domain is part of a different multienzyme complex (Geerse et al., 1989).

The A domain accepts a phosphoryl group from P-HPr and, in turn, phosphorylates the B domain. The P-B domain transfers its phosphoryl group to the carbohydrate which is bound to the C domain. The sites of phosphorylation have been determined by ^{32}P labeling and peptide isolation to be H-554 and C-384 on the A and B domains, respectively (Pas and Robillard, 1988). The function of the C domain is to bind and transport the carbohydrate over the cytoplasmic membrane (Grisafi et al., 1989; Lolkema et al., 1990). The role of these domains has been established from different lines of evidence, one of which has been to separate them at the gene level, express them separately, and study their functions individually. Such work on *E. coli* EII^{Mtl} has yielded the various domains and domain combinations shown in Fig. 1 (van

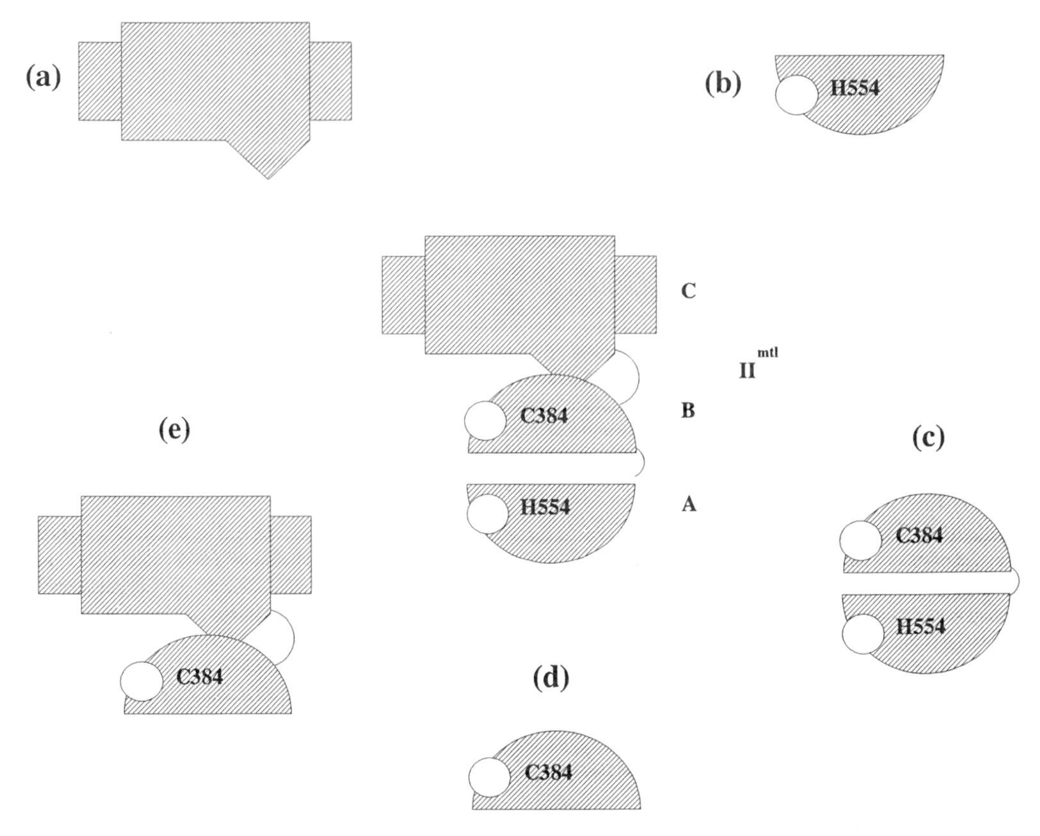

FIG. 1. Model of the domain structure of the *E. coli* EII^{Mtl} monomer (center) and the various domains and domain combinations which have been produced from the parent enzyme. (a) C domain, residues 1 to 346 (Boer et al., in press); (b) A domain, residues 490 to 637, with a Met in place of Ser-490 (van Weeghel et al., 1991c); (c) BA domains, residues 347 to 637 with Met, Asp, Pro in place of Ala, Ala, Thr (residues 347 to 349) (van Weeghel et al., 1991b); (d) B domain, residues 347 to 488, with the same changes in the three N-terminal amino acids as in the BA domains in panel c (Robillard et al., 1993); (e) The BC domains, residues 347 to 637 (Boer et al., unpublished).

Weeghel et al., 1991a, 1991b, 1991c; Robillard et al., 1993).

The activity of the A, B, and BA domains during this work has been monitored by a complementation assay (Fig. 2) which measures their ability to restore the PEP-dependent mannitol phosphorylation activity of site-directed mutants inactivated by virtue of an H554A or C384S mutation. In the case of the A, B, and BA domains, high levels of phosphorylation activity could be achieved, demonstrating that these domains retain their structure and phosphorylation function in the separated form. Transport activity could also be demonstrated with the separate domains by simultaneous in vivo expression of two plasmids, one encoding the C domain and one encoding the BA domain (Weng et al., 1992). Cells expressing protein from the two plasmids were able to grow on and ferment mannitol, whereas cells expressing either plasmid individually were not able to grow on mannitol (van Weeghel et al., 1991b).

Association State of EII^{Mtl} and the Individual Domains

The kinetics of mannitol phosphorylation and Mtl/Mtl-P exchange have been studied by using detergent solubilized membrane preparations containing EII^{Mtl}. Phosphorylation shows a nonlinear dependence on enzyme concentration; there is a basal activity which is stimulated fivefold by higher enzyme concentrations or the presence of Mg^{2+} (Lolkema and Robillard, 1990). Exchange also shows a nonlinear enzyme concentration dependence, but there is no activity at low protein concentrations. The nonlinear concentration dependence indicates that the enzyme associates, at least to a dimer, and that the associated form is more active in phosphorylation. In the case of Mtl/Mtl-P exchange, however, it is the only form of the enzyme which appears to be active (Leonard and Saier, 1983; Roossien et al., 1984; Lolkema and Robillard, 1990). This observation

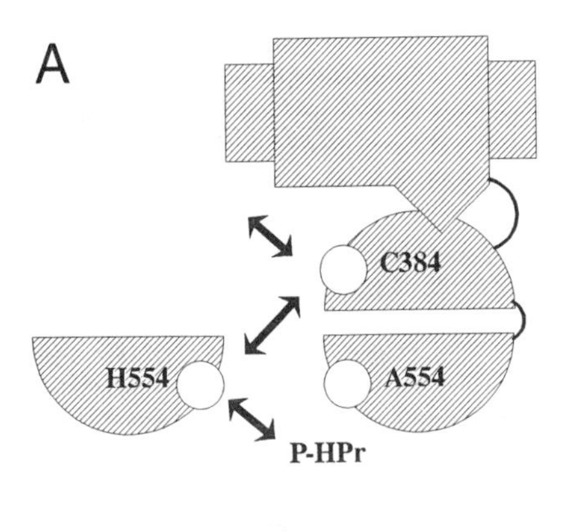

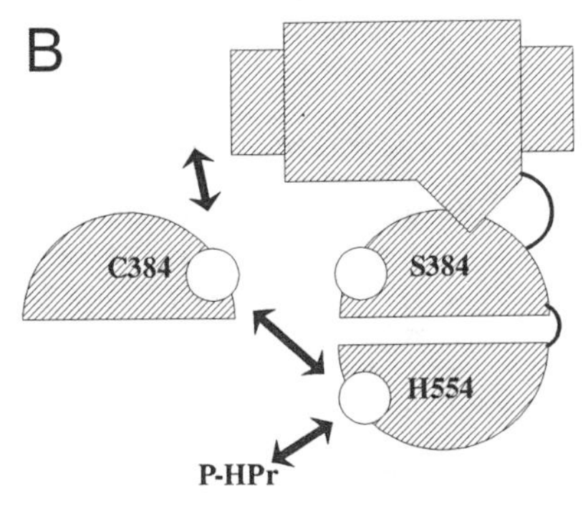

FIG. 2. Complementation schemes used to monitor the activity of EII^{Mtl} domains. (A) The activity of the A domain is measured by monitoring mannitol phosphorylation kinetics with membrane vesicles containing a H554A-EII^{Mtl} mutant which is inactive by virtue of the loss of its H554 phosphorylation site. (B) The activity of the B domain is measured in a similar way but with membranes containing a C384S-EII^{Mtl} mutant enzyme.

is crucial to understanding how the dimer functions in phosphoryl group transfer during vectorial phosphorylation and exchange. We will return to it below.

The size of the associated protein has been examined by gel permeation chromatography on TSK columns under conditions that favor the association of the enzyme (Lolkema et al., 1993a). To estimate the contribution of detergent to the total size of the complex, chromatography was carried out by using detergents with different micellar sizes. The size of EII^{Mtl}, expressed in terms of globular soluble proteins, varied from 315 kDa in decyl polyethylene glycol 300 (decyl-PEG), a detergent which forms large micelles, to 245 kDa in cholate, a detergent which forms small micelles. Since cholate contributes approximately 2 kDa to the complex, it appears that size of EII^{Mtl} in the associated state is equivalent of that of a 245-kDa globular protein. The size of the C domain was estimated to be 175 kDa in decyl-PEG. Correcting for the contribution of detergent yields the equivalent of a 105-kDa globular protein. A 245-kDa particle can accommodate three EII^{Mtl} molecules, and a 105-kDa particle can accommodate three C domains, indicating that the associated state is a trimer or a dimer with an asymmetric form. Considering the membrane-bound nature of these proteins and the domain structure of the intact protein, we are most probably dealing with a dimer with an asymmetric form. The association state of the soluble BA domain has also been examined; it was monomeric at all concentrations measured. Consequently, we conclude that EII^{Mtl} and IIC^{Mtl} are dimeric and that it is the C domain which is responsible for formation of the dimeric state in the intact enzyme.

Phosphoryl Group Transfer in the EII^{Mtl} Dimer

Phosphoryl group transfer in the dimer has been observed by monitoring the rates of mannitol phosphorylation and exchange in a heterodimeric system consisting of H554A-EII^{Mtl} and C384S-EII^{Mtl} (van Weeghel et al., 1991a). Each protein alone was inactive in mannitol phosphorylation by virtue of the lack of one of the two essential phosphorylation sites, but adding increasing amounts of the C384S-EII^{Mtl} to a fixed amount of H554A-EII^{Mtl} led to high levels of mannitol phosphorylation. This could be expected on the basis of the complementation shown in Fig. 2. What was surprising, however, were the results for a similar experiment measuring the Mtl/Mtl-P exchange rates. H554A-EII^{Mtl} is active in Mtl/Mtl-P exchange because only the C-384 phosphorylation site on the B domain is necessary for this reaction; C384S-EII^{Mtl}, however, is not active. Nevertheless, when increasing amounts of C384S-EII^{Mtl} were added to a reaction mixture containing a constant amount of H554A-EII^{Mtl}, the rates of exchange increased up to nearly a factor of 2. One explanation of these data is that the H554A-EII^{Mtl} homodimer works at half its maximum rate; i.e., only one subunit works at a time. In the heterodimers, each H554A-EII^{Mtl} unit is free to carry out the exchange reaction at the same rate as in the homodimer, but there are twice as many dimers containing H554A-EII^{Mtl} units.

The heterodimer data show that only one "active" B domain is necessary for exchange, while the kinetic observations show that the dimer is absolutely essential for exchange. Combining these observations yields two minimal models for the exchange reaction. The first is a model in which the route of phosphoryl group transfer between the B and C domains is across the dimer interface. The phosphoryl group originates from the Mtl-P bound to the C domain of the neighboring EII^{Mtl} subunit, hence the requirement for the dimer. The principle of microscopic reversibility requires that the back reaction occur in the same way. The second is a model in which the Mtl-P-binding site in the monomer would be facing the periplasmic face, thus preventing Mtl-P from phosphorylating the B domain. In the dimer, the orientation of the mannitol-binding sites could be coupled, with one inwardly facing site and one outwardly facing site. Mtl-P would then bind to the inwardly facing site and be in a position to pass its phosphoryl group back to the B domain; in a subsequent step, Mtl would bind and receive the phosphoryl group from the P-B domain. In this model, phosphoryl transfer would not necessarily have to occur across the dimer interface but could also occur between the B and C domain of the same subunit.

Orientation and Affinity of the Mannitol-Binding Sites

The orientation of the mannitol-binding sites has been monitored by flow dialysis measurements on inside-out (ISO) and right-side-out (RSO) membrane vesicles (Lolkema et al., 1990). High-affinity binding occurs in both vesicle preparations, indicating that high-affinity sites can be situated at either the periplasmic or the cytoplasmic surface. The rates of mannitol binding and the rates of exchange of bound and free mannitol are low enough, under certain circumstances, that the flow dialysis technique also gave kinetic information (Lolkema et al., 1992). Binding to ISO vesicles could be separated into a mannitol concentration-dependent and a concentration-independent step; these were interpreted, for a single subunit, in terms of the model in Fig. 3. In ISO vesicles, a high-affinity site situated at the cytoplasmic surface is in equilibrium with an occluded site. The mannitol concentration-independent step represents the recruitment of empty occluded sites to the cytoplasmic surface. The mannitol concentration-dependent step shows a rapid binding followed by a slower increase in the amount of mannitol bound. This was interpreted as binding to a cytoplasmic site followed by a conversion to a second bound form with higher affinity (a filled occluded site). Mannitol bound at the periplasmic site does not equilibrate rapidly with that at the cytoplasmic or occluded site, indicating that phosphorylation is required for a kinetically significant isomerization of these sites.

One could envisage that one role for phosphorylation of EII^{Mtl} would be to alter the affinity of the cytoplasmic or periplasmic sites for substrate. Lolkema et al. (1993b) have used perseitol, a mannitol substrate analog which does not become phosphorylated, to test the effect of phosphorylation on the binding affinity. The affinities of P-EII^{Mtl} for perseitol were deduced from the perseitol inhibition of mannitol binding and phosphorylation with intact or detergent-solubilized vesicles. The affinity of the phosphorylated enzyme was only a factor of 2 to 3 lower than that observed for the unphosphorylated enzyme. Thus, if these results may be extrapolated to mannitol, we can conclude that phosphorylation of the enzyme does not alter its affinity for mannitol.

The model in Fig. 3 makes no distinction between monomer and dimer because it was based on flow dialysis data which detected only high-affinity sites. Equilibrium dialysis-binding studies, however, indicate one high- and one low-affinity mannitol-binding site per EII^{Mtl} dimer. Therefore, the affinities of the binding sites in the EII^{Mtl} dimer may be coupled in a functional way, as suggested in the second model proposed above.

Steady-State Phosphorylation Kinetics

Fructose phosphorylation catalyzed by *R. sphaeroides* EII^{Frc} and mannitol phosphorylation catalyzed by *E. coli* EII^{Mtl} show striking similarities in the sugar concentration dependence of the steady-state rates. Since the observations and interpretation were made initially for the fructose system, we will treat the fructose data here and refer to the mannitol data only at one point at which the data sets differ. Double-reciprocal plots of the rate of sugar phosphorylation as a function of the sugar concentration are not linear when in-

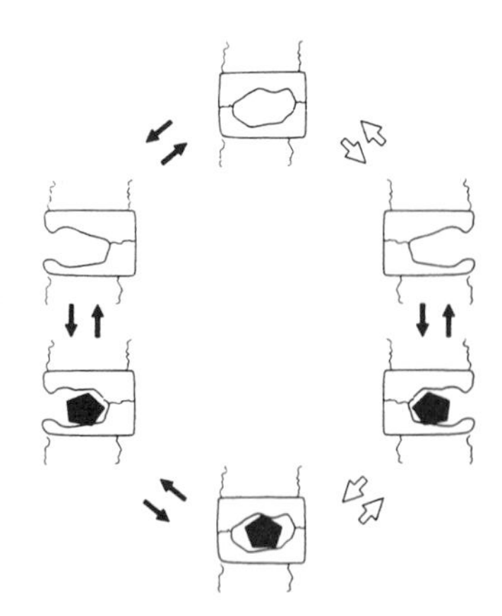

FIG. 3. Model for the orientation of mannitol-binding sites derived from the kinetics of binding (Lolkema et al., 1992). The occluded sites are in the middle of the drawing between the white and black arrows. The isomerization step represented by the white arrows is the step whose rate is enhanced by phosphorylation. Reprinted from Lolkema et al. (1992), copyright 1992, American Chemical Society.

tact ISO vesicles are used as the source of EII (Lolkema and Robillard, 1985). Rather, they show two regimes, one with a high affinity for substrate and one with an affinity 10 to 30 times lower. When the low-affinity rate data, calculated from the original data by subtracting the contribution from the high-affinity regimen, are again plotted in a double-reciprocal plot, the plots are not linear; the rate ceases to increase at high sugar concentrations. The low-affinity regime does not, therefore, reflect EII molecules with a lower affinity for sugar but, rather, a kinetic restriction in the phosphorylation reaction. The scheme which enabled us to simulate these kinetic data is reproduced in Fig. 4 (Robillard and Lolkema, 1988). Binding data show that a high-affinity binding site can face the cytoplasm or periplasm. In ISO vesicles the cytoplasm-facing site is on the outside in contact with solution. The normal phosphorylation and transport cycle would follow route 1 → 2 → 3 → 4 → 1, where phosphorylation would enable the reorientation of the sugar-binding site to the periplasmic face of the membrane, to be followed by sugar binding, transport, and phosphorylation. This would account for the high-affinity regime in the kinetics. In the case of properly sealed ISO vesicles, this binding site would face the inside of the vesicle, and mannitol, which does not diffuse passively across the membrane, could not bind to this site. Thus, the high-affinity regime must be due to leaky vesicles or membrane fragments where sugar could access the site readily. Route 1 → 5 → 3 → 4 → 1 would account for the low-affinity regime; it would be followed in properly sealed ISO vesicles in which sugar could bind only at the cytoplasm-facing site (S_i). Relative to the high-affinity route, it would be kinetically disfavored and would begin to contribute to the overall rate only at higher sugar concentrations, thus causing the apparent low affinity. The kinetics of EII^{Frc} and EII^{Mtl} differ in that, in

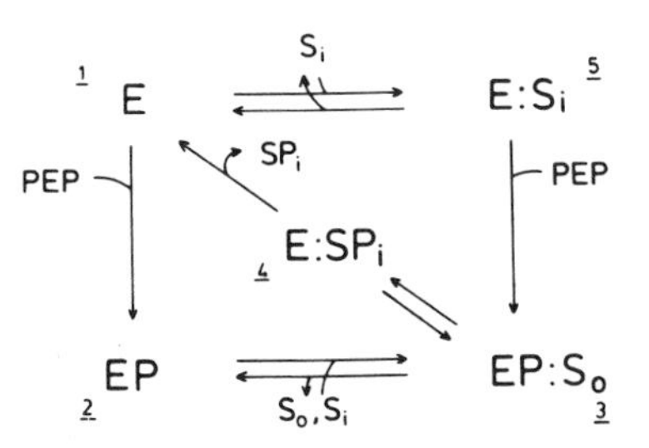

FIG. 4. Kinetic scheme for the *R. sphaeroides* EII^{Frc}-catalyzed fructose phosphorylation (Robillard and Lolkema, 1988). 'i' indicates the internal, cytoplasmic compartment; 'o' indicates the external, periplasmic compartment. Reprinted from Robillard and Lolkema (1988) with permission.

the latter, state 4 can be reached from state 5 directly rather than having to follow the route 5 → 3 → 4; i.e., sugar can bind at the cytoplasmic face and be phosphorylated and released as SP (Lolkema and Robillard, 1992). This means that phosphorylation is not absolutely coupled to the change in orientation of the binding site; rather, it increases the rate of isomerization between the two faces of the membrane (Lolkema et al., 1991).

Model for the Coupling of Phosphorylation and Transport

The above observations lead to a model for the alternating orientation of sugar-binding sites, coupled to the state of phosphorylation of the B domain. The main features are as follows.

1. EII^{Mtl} is a dimer by virtue of interactions between the membrane-bound C domains.
2. The subunits of the dimer function consecutively; while one is in the transport mode (binding site facing the periplasm), the other is in the phosphorylation mode (binding site facing the cytoplasm).
3. There is one sugar-binding site on each C domain of an EII monomer, which can face either the periplasm or the cytoplasm.
4. In the dimer, the affinity of the sites appears to be coupled such that there is one high-affinity and one low-affinity site.
5. Phosphorylation does not determine a specific orientation of the sugar-binding site (i.e., facing the cytoplasm versus the periplasm) or alter the affinity of these sites for the sugar. Instead, it alters the rate of isomerization of the sites between the cytoplasmic and periplasmic faces of the membrane.
6. The phosphoryl transfer step between C-384 on the B domain and mannitol on the C domain may occur across the dimer interface.

The proposed mechanism for transport and phosphorylation coupled to the alternating movements of binding sites in the dimer is shown in Fig. 5 for the case in which phosphorylation has to occur across the dimer interface. We use the Mtl/Mtl-P exchange reaction in intact cells to illustrate the mechanism because exchange requires a dimer and thus is a more restrictive case than the PEP-dependent phosphorylation reaction; furthermore, the model must be able to account for vectorial exchange, which has been measured in intact cells for both the mannitol and glucose PTS.

The exchange cycle begins when Mtl-P situated at the binding site facing the cytoplasm transfers its phosphoryl group to the B domain of the adjacent subunit (state 2 → 3). Phosphorylation of the B domain increases the rate of isomerization of the binding sites, bringing a radiolabeled mannitol from the periplasmic to the cytoplasmic side and the unlabeled mannitol from the cytoplasmic to the periplasmic side (state 3 → 4). Then radiolabeled mannitol phosphate is generated by phosphoryl group transfer again over the dimer interface; the labeled mannitol phosphate is released on the inside while unlabeled mannitol is released on the outside (state 4 → 5). This leaves one B domain dephosphorylated, thereby stopping the isomerization of the binding sites. The cycle is then repeated with the phosphorylation-dephosphorylation of the opposite subunit (states 5 → 8). The doubling of the exchange rate when heterodimers of H554A- and C384S-EII^{Mtl} are used supports the proposal that in the homodimer, only one subunit of the dimer works at a time.

The state of phosphorylation of the A domain has not been specified in the model in Fig. 5 since there is no information linking the phosphorylation state of this domain to the exchange activity of the C domain. The isomerization steps 3 → 4 and 7 → 8 are completely reversible, and many cycles can occur before a B domain dephosphorylates to produce Mtl-P. Vectorial phosphorylation driven by PEP, however, involves a different scenario. Here, the difference in free energy between PEP and the phosphoryl group on the A and B domains is large enough that both domains will be fully phosphorylated in the steady state. As soon as bound mannitol is converted to Mtl-P and released, the dephosphorylated b domain will receive another phosphoryl group from the P-A domain, enabling vectorial transport by the isomerization of an empty cytoplasmic site and a filled periplasmic site.

The model in Fig. 5 predicts the following.

1. A lag time would be observed in the exchange kinetics if one preincubated EII^{Mtl} with labeled mannitol and tried to start the reaction with Mtl-P; since the enzyme has a much higher affinity for mannitol than for Mtl-P, the presence of mannitol would prevent the enzyme from binding Mtl-P and slow the process of achieving a steady-state level of phosphorylation of the enzyme from Mtl-P.

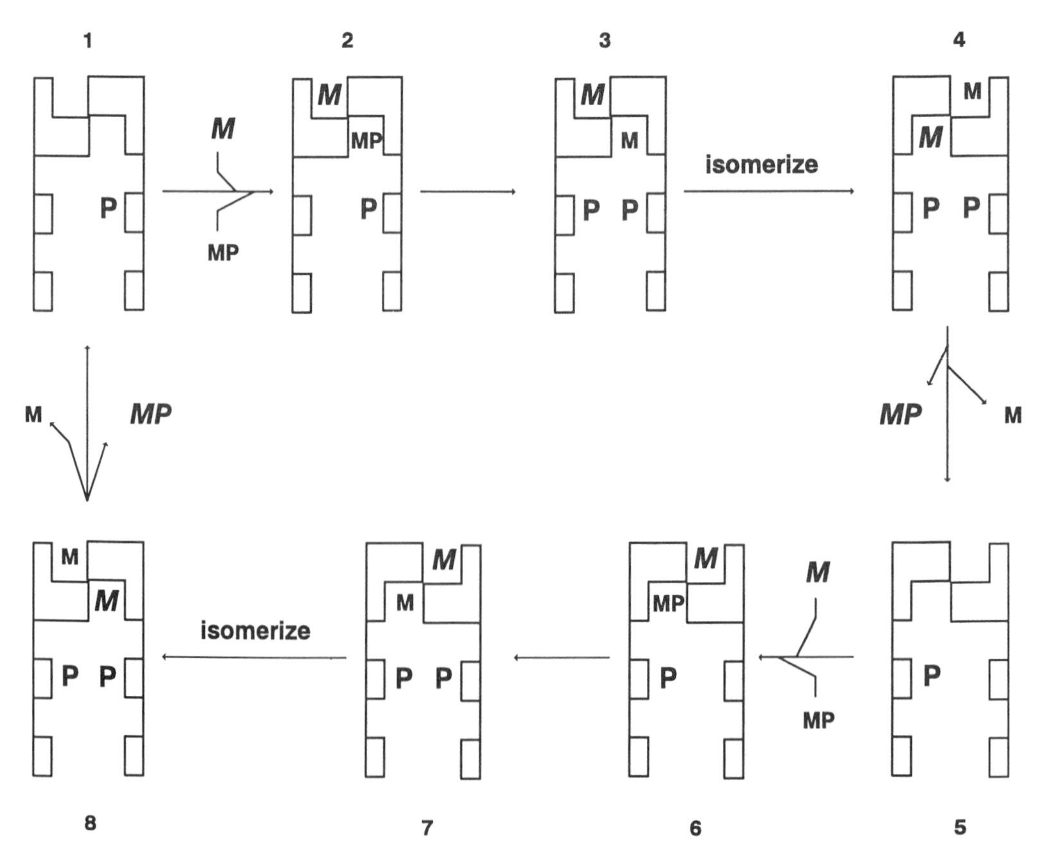

FIG. 5. Kinetic scheme for the coupled alternation of sugar-binding sites in the EII^{Mtl} dimer during vectorial Mtl/Mtl-P exchange. "P" indicates the state of phosphorylation of the B domain.

2. Exchange should occur in a system consisting of IIC^{Mtl} and wild-type EII^{Mtl}; adding increasing amounts of IIC^{Mtl} should lead to a doubling in the rate of exchange for reasons similar to the same observations in the H554A-EII^{Mtl}/C384S-EII^{Mtl} heterodimer exchange measurements. A doubling of the rate should also be seen in the PEP-dependent mannitol phosphorylation.
3. Exchange should occur in a system consisting of IIC^{Mtl} and a mutant protein incapable of binding mannitol but still possessing a functional B domain. The mutant would not bind mannitol, but its B domain could still accept the phosphoryl group from mannitol phosphate bound to the adjacent IIC^{Mtl} in the complex.
4. Heterodimers consisting of one wild-type EII^{Mtl} monomer and one mutant enzyme monomer incapable of binding mannitol and also incapable of being phosphorylated at the B domain should be inactive in Mtl/Mtl-P exchange because phosphoryl group transfer across the dimer interface would not be possible. If significant activity is observed, one would have to conclude that phosphoryl transfer between the B and C domains of the same subunit during an exchange reaction is possible.

Work is in progress to test these predictions.

REFERENCES

Boer, H., R. H. ten Hoeve-Duurkens, G. Schuurman-Wolters, and G. T. Robillard. Expression and kinetic characterization of the mannitol transport domain (IIC) of the phosphoenolpyruvate-dependent mannitol phosphotransferase system of *Escherichia coli*. *J. Biol. Chem.*, in press.

Boer, H., et al. Unpublished data.

Geerse, R. H., F. Izzo, and P. W. Postma. 1989. The PEP:fructose phosphotransferase system in *Salmonella typhimurium:* FPR combines Enzyme III^{fru} and pseudo-HPr activities. *Mol. Gen. Genet.* **216:**517–525.

Grisafi, P. L., A. Scholle, J. Sugiyama, C. Briggs, G. R. Jacobson, and J. W. Lengeler. 1989. Deletion mutants of the *Escherichia coli* K-12 mannitol permease: dissection of transport-phosphorylation, phospho-exchange and mannitol binding activities. *J. Bacteriol.* **171:**2719–2727.

Jacobson, G. R., D. M. Kelly, and D. R. Finlay. 1983. The intermembrane topology of the mannitol-specific enzyme of the *Escherichia coli* phosphotransferase system. *J. Biol. Chem.* **258:**2955–2959.

Leonard, J. E., and M. H. Saier, Jr. 1983. Mannitol-specific Enzyme II of the bacterial phosphotransferase system. *J. Biol. Chem.* **258:**10757–10760.

Lolkema, J. S., J. Kuiper, R. H. ten Hoeve-Duurkens, and G. T. Robillard. 1993a. Mannitol-specific Enzyme II of the phosphoenolpyruvate-dependent phosphotransferase system of *Escherichia coli:* physical size of Enzyme II^{mtl} and its domains IIBA and IIC in the active state. *Biochemistry* **32:**1396–1400.

Lolkema, J. S., and G. T. Robillard. 1985. Phosphoenolpyruvate-dependent fructose phosphotransferase system in *Rhodopseudomonas sphaeroides. Eur. J. Biochem.* **147:**69–75.

Lolkema, J. S., and G. T. Robillard. 1990. Subunit structure and activity of the mannitol-specific enzyme II of the *Escherichia coli* phosphoenolpyruvate-dependent phosphotransferase system solubilized in detergent. *Biochemistry* **29:**10120–10125.

Lolkema, J. S., and G. T. Robillard. 1992. The Enzymes II of the phosphoenolpyruvate-dependent carbohydrate transport systems, p. 135–167. *In* J. J. H. M. De Pont (ed.), *Molecular Aspects of Transport Proteins.* Elsevier Science Publishers B. V., Amsterdam.

Lolkema, J. S., D. Swaving-Dijkstra, and G. T. Robillard. 1992. Mechanics of solute translocation catalyzed by Enzyme II^{mtl} of the phosphoenolpyruvate-dependent phosphotransferase system in *Escherichia coli. Biochemistry* **31:**5514–5521.

Lolkema, J. S., D. Swaving-Dijkstra, R. H. ten Hoeve-Duurkens, and G. T. Robillard. 1990. The membrane-bound domain of the phosphotransferase Enzyme II^{mtl} of *Escherichia coli* constitutes a mannitol translocating unit. *Biochemistry* **29:**10659–10663.

Lolkema, J. S., R. H. ten Hoeve-Duurkens, D. Swaving-Dijkstra, and G. T. Robillard. 1991. Mechanistic coupling of transport and phosphorylation activity by Enzyme II^{mtl} of the *Escherichia coli* phosphoenolpyruvate-dependent phosphotransferase system. *Biochemistry* **30:**6716–6721.

Lolkema, J. S., E. S. Wartna, and G. T. Robillard. 1993b. Binding of the substrate analogue perseitol to phosphorylated and unphosphorylated Enzyme II^{mtl} of the phosphoenolpyruvate-dependent phosphotransferase system of *Escherichia coli. Biochemistry* **32:**5848–5854.

Pas, H. H., and G. T. Robillard. 1988. *S*-Phosphocysteine and phosphohistidine are intermediates in the phosphoenolpyruvate-dependent mannitol transport catalyzed by *Escherichia coli* EII^{mtl}. *Biochemistry* **27:**5835–5839.

Robillard, G. T., H. Boer, R. P. van Weeghel, G. Wolters, and A. Dijkstra. 1993. Expression and characterization of a structural and functional domain of the mannitol-specific transport protein involved in the coupling of mannitol transport and phosphorylation in the phosphoenolpyruvate-dependent phosphotransferase system of *Escherichia coli. Biochemistry* **32:**9553–9562.

Robillard, G. T., and J. S. Lolkema. 1988. Enzymes II of the phosphoenolpyruvate-dependent sugar transport systems: a review of their structure and mechanism of sugar transport. *Biochim. Biophys. Acta* **947:**493–519.

Roossien, F. F., M. Blaauw, and G. T. Robillard. 1984. Kinetics and subunit interactions of the mannitol-specific Enzyme II of the *Escherichia coli* phosphoenolpyruvate-dependent phosphotransferase system. *Biochemistry* **23:**4934–4939.

Saier, M. H., Jr., M. Yamada, J. Lengeler, B. Erni, K. Suda, P. Argos, K. Schnetz, B. Rak, C. A. Lee, G. C. Stewart, K. G. Peri, and E. B. Waygood. 1988. Sugar permeases of the bacterial phosphoenolpyruvate-dependent phosphotransferase system: sequence comparisons. *FASEB J.* **2:**199–208.

Stephan, M. M., and G. R. Jacobson. 1986. Membrane disposition of the *Escherichia coli* mannitol permease: identification of membrane-bound and cytoplasmic domains. *Biochemistry* **25:**8230–8234.

Stephan, M. M., S. S. Khandekar, and G. R. Jacobson. 1989. Hydrophilic C-terminal domain of the *Escherichia coli* mannitol permease: phosphorylation, functional independence and evidence for intersubunit phosphotransfer. *Biochemistry* **28:**7941–7946.

van Weeghel, R. P., Y. Y. van der Hoek, H. H. Pas, M. Elferink, W. Keck, and G. T. Robillard. 1991a. Details of mannitol transport in *Escherichia coli* elucidated by site-specific mutagenesis and complementation of phosphorylation site mutants of the phosphoenolpyruvate-dependent phosphotransferase system. *Biochemistry* **30:**1768–1773.

van Weeghel, R. P., G. Meyer, H. H. Pas, W. Keck, and G. T. Robillard. 1991b. Cytoplasmic phosphorylating domain of the mannitol-specific transport protein of the phosphoenolpyruvate-dependent phosphotransferase system in *Escherichia coli:* overexpression, purification and functional complementation with the mannitol binding domain. *Biochemistry* **30:**9478–9485.

van Weeghel, R. P., G. H. Meyer, W. Keck, and G. T. Robillard. 1991c. Phosphoenolpyruvate-dependent mannitol phosphotransferase system of *Escherichia coli:* overexpression, purification and characterization of the enzymatically active C-terminal domain of Enzyme II^{mtl} equivalent to Enzyme III^{mtl}. *Biochemistry* **30:**1774–1779.

Vogler, A. P., and J. Lengeler. 1988. Complementation of a truncated membrane-bound Enzyme II^{nag} from *Klebsiella pneumoniae* with a soluble Enzyme III of *Escherichia coli* K12. *Mol. Gen. Genet.* **213:**175–178.

Weng, Q.-P., J. Elder, and G. R. Jacobson. 1992. Site-specific mutagenesis of residues in the *Escherichia coli* mannitol permease that have been suggested to be important for its phosphorylation and chemoreception functions. *J. Biol. Chem.* **267:**19529–19535.

White, D. W., and G. R. Jacobson. 1990. Molecular cloning of the C-terminal domain of *Escherichia coli* D-mannitol permease: expression, phosphorylation and complementation with C-terminal permease deletion proteins. *J. Bacteriol.* **172:**1509–1515.

Chapter 28

Enzymes II of the Phosphotransferase System: Transport and Regulation

P. W. POSTMA, J. VAN DER VLAG, J. H. DE WAARD, W. M. G. J. YAP, K. VAN DAM, AND G. J. G. RUIJTER

E. C. Slater Institute, BioCentrum, University of Amsterdam, 1018 TV Amsterdam, The Netherlands

The phosphoenolpyruvate (PEP):carbohydrate phosphotransferase system (PTS) is a major system for the transport and phosphorylation of carbohydrates in enteric bacteria. In contrast to the situation with other carbohydrate transport systems, carbohydrates are accumulated in the cell as the phosphate esters which are directly channeled into the glycolytic pathway. Apart from its role in transport, the PTS plays a crucial role in chemotaxis toward PTS carbohydrates. Finally, components of the PTS are, depending on their phosphorylation state, major factors in the regulation of cellular metabolism.

In this chapter we will discuss the properties of some of the enzymes involved in PTS carbohydrate uptake, in particular the enzymes that transport and phosphorylate glucose to glucose-6-phosphate. We will also discuss the consequences for the cell when the PTS is actively transporting its substrates, i.e., how the PTS regulates the flux of carbon when cells such as *Escherichia coli* and *Salmonella typhimurium* are presented simultaneously with several carbohydrates.

Overview of the PTS

More than 15 different carbohydrates can be taken up and phosphorylated by the PTS in enteric bacteria (for a review, see Postma et al. [1993]). In all cases, the phospho group is derived from PEP and is transferred via two general PTS proteins, enzyme I (EI) and HPr, to the different carbohydrate-specific enzymes II (EIIs) (Fig. 1). Most PTS proteins are phosphoproteins; i.e., they are phosphorylated on a histidine residue or, in a few cases, on a cysteine residue. As shown schematically in Fig. 1, EIIs can be either a single polypeptide (for instance II^{Mtl}, specific for mannitol) or complexes consisting of two (e.g., II^{Glc}, specific for glucose), three (e.g., II^{Man}, specific for mannose), or four (e.g., II^{Sor}, specific for sorbose) different polypeptides. From recent biochemical and genetic studies it has become clear that all EIIs are made up of several different domains. In general, the IIA domain contains the first phosphorylation site, a histidine residue, whereas the IIB domain contains the second phosphorylation site, a cysteine or histidine residue. The IIC domain is involved in binding and translocating the carbohydrates. In a few cases (e.g., II^{Man}), the IIC domain is composed of two different subunits, IIC and IID, and in those systems both phosphorylation sites are histidine residues located on the IIA and IIB domains. EIIs can be grouped into five classes, based on sequence alignments (Postma et al., 1993; Saier and Reizer, 1992). Corresponding domains within one class share more than 25% identical amino acids.

Genetic, biochemical, and physiological studies have shown that IIA^{Glc} (formerly called III^{Glc}), involved in glucose transport and phosphorylation, is a central regulatory protein. Depending on its phosphorylation state, it is involved in activation of adenylate cyclase (phospho-IIA^{Glc}) and inhibition of several non-PTS uptake systems (IIA^{Glc}) (for a review, see Postma et al. [1993]). From the scheme in Fig. 1 it is clear that the phosphorylation state of IIA^{Glc} is determined not only by the rate of phosphorylation via phospho-HPr and dephosphorylation via $IICB^{Glc}$ but also by the activity of other EIIs. The PTS can be considered a signal transduction pathway in which each of the receptors, the EIIs, can signal the presence of its specific carbohydrate and can affect the phosphorylation state of all other PTS proteins, depending on the flux of phospho groups through the system.

Role of II^{Glc} in Glucose Transport and Phosphorylation

During glucose transport and phosphorylation, the phospho group from P-HPr is transferred first to His-90 of IIA^{Glc} and subsequently to Cys-421 of $IICB^{Glc}$. In the absence of EI and/or HPr, e.g., in *ptsI* or *ptsH* mutants, the flux is interrupted and phosphorylation of IIA^{Glc} and $IICB^{Glc}$ is not possible. Such mutants are unable to catalyze glucose transport via the nonphosphorylated $IICB^{Glc}$. Even if cells contain sufficient glucokinase to

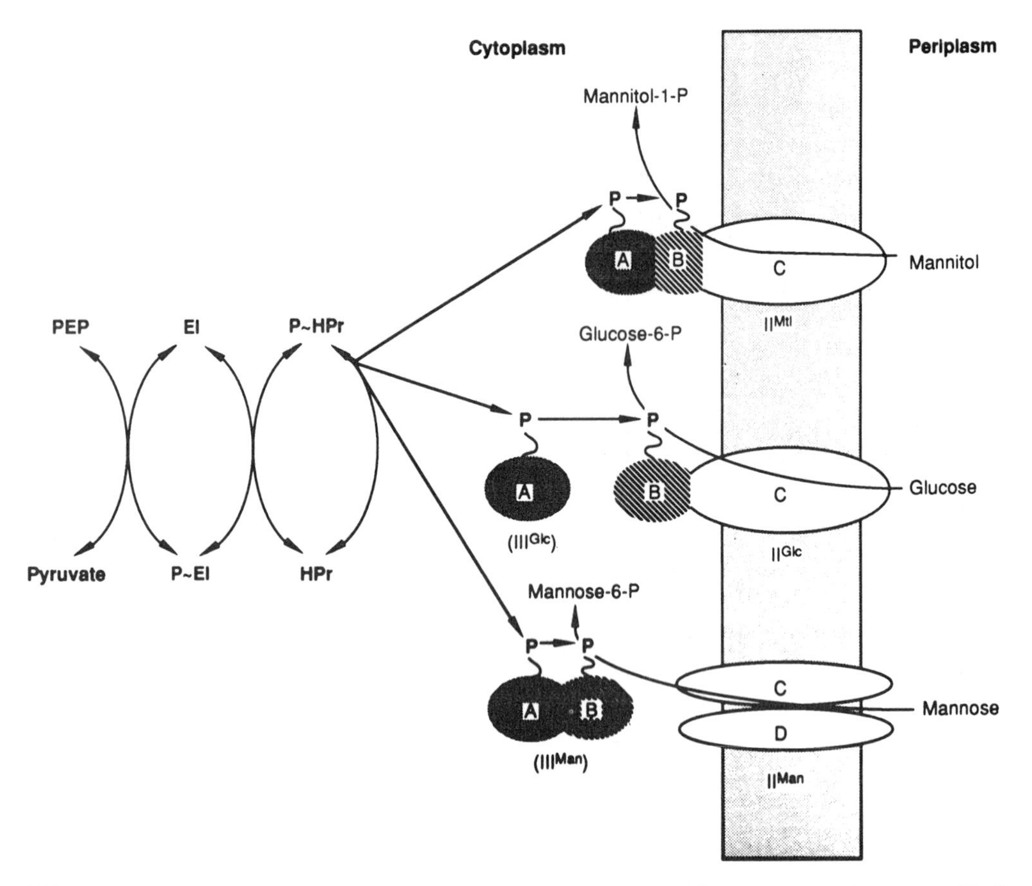

FIG. 1. PTS. EI and HPr are the general proteins of the PTS. Three different EIIs are shown, specific for mannitol (II^{Mtl}), glucose (II^{Glc}), and mannose (II^{Man}). Each contains two hydrophilic domains, IIA and IIB, with two phosphorylation sites (His in IIA and His or Cys in IIB) and a membrane-bound hydrophobic domain (IIC), sometimes split into two domains, IIC and IID. P ~ indicates the phosphorylated form of the various PTS proteins. This figure has been published previously (Postma et al., 1993).

phosphorylate the glucose intracellularly, the rate of transport in the absence of PTS-mediated phosphorylation is very low. It has been concluded that a nonphosphorylated EII is unable to catalyze transport in the absence of phosphorylation; i.e., facilitated diffusion does not occur (Postma and Stock, 1980). On the basis of kinetic data, it has been concluded from studies with mannitol and II^{Mtl} that the phosphorylated form of EII^{Mtl} can catalyze the inward movement of mannitol without concomitant phosphorylation; i.e., facilitated diffusion of the free carbohydrate is possible only via the phosphorylated form of an EII (Lolkema et al., 1991). Phosphorylation of II^{Mtl} increases the rate of mannitol translocation by several orders of magnitude.

One can ask whether it is possible to alter an EII such that it can catalyze transport of its carbohydrate without undergoing phosphorylation. Such "uncoupled" mutations have been isolated first in the chromosomal *S. typhimurium ptsG* gene, encoding $IICB^{Glc}$, and subsequently in the plasmid-localized *E. coli ptsG* gene by selection for growth of a *ptsHI* deletion mutant on glucose, allowing for glucose phosphorylation via glucokinase (Postma, 1981; Ruijter et al., 1992). In both cases mutant $IICB^{Glc}$ molecules that catalyzed the facilitated diffusion of glucose with an apparent affinity ranging from 10 to 0.6 mM were isolated. The parental II^{Glc} uptake system had an apparent K_m in the micromolar range. Although the maximal rate of glucose transport, catalyzed by these mutant proteins, was not greatly diminished, the apparent affinity for glucose decreased at least 100-fold. These results have been interpreted as follows: EII^{Glc} is a closed pore in the absence of phosphorylation and is opened by phosphorylation of the Cys-421 residue (in a wild-type strain), possibly during interaction with IIA^{Glc} (it has been shown in vitro that the transfer of the phospho group from purified $P\text{-}IICB^{Glc}$ to methyl α-glucoside [αMG] is enhanced by the presence of IIA^{Glc} [Erni,

1986]) or by uncoupling mutations in the *ptsG* gene like those described in this section.

The mutations in the *E. coli ptsG* gene that allow glucose transport without concomitant phosphorylation have been analyzed. Four different amino acid replacements have been found, all localized in or close to the (presumed) cytoplasmic loop between two membrane-spanning regions (Buhr and Erni, 1993). Although all of these *E. coli* $IICB^{Glc}$ mutant proteins can transport glucose without phosphorylation, some can still phosphorylate glucose (and αMG) when EI, HPr, and IIA^{Glc} are added. Thus, the uncoupling mutation has not eliminated PTS-mediated phosphorylation of $IICB^{Glc}$. In some cases, the apparent affinity, as measured by glucose oxidation, increased 50-fold in cells containing the mutant $IICB^{Glc}$ molecules in addition to EI, HPr, and IIA^{Glc} and approached wild-type values. However, in other mutants the affinity was not affected by phosphorylation. In all cases, phosphorylation was still dependent on the Cys-421 residue. Replacement by a serine residue eliminated glucose phosphorylation completely (Yap, unpublished). However, facilitated diffusion was not affected by replacement of the Cys-421 residue by a serine residue (Yap, unpublished).

Flux of Phospho Groups through the PTS

The PTS is a complex transport system which, depending on the particular carbohydrate, consists of three to six proteins: EI, HPr, and one to four carbohydrate-specific proteins. Since the flux of phospho groups from PEP to glucose is catalyzed by four different proteins, we have studied which, if any, of these proteins controls the rate of glucose uptake as well as oxidation of glucose and growth on glucose. Using metabolic control analysis (Kacser and Burns, 1973), one can determine to what extent an enzyme controls the flux through a pathway (equation 1) by determining the increase in activity of the overall flux (dJ/J) as a function of the increase in a particular enzyme (dE/E), keeping all other enzymes constant (equation 2).

$$S \xrightarrow{E_1} X_1 \xrightarrow{E_2} X_2 \;\ldots\ldots\; X_{N-1} \xrightarrow{E_N} P \quad (1)$$

$$C_E^J = \frac{dJ/J}{dE/E}, \qquad 0 \leq C_E^J \leq 1 \quad (2)$$

$$\Sigma C_E^J = 1 \quad (3)$$

where C_E^J is the control coefficient and J is the flux. In most cases, the flux control coefficients C_E^J have values between 0 (no control at all) and 1 (completely determining the flux). The summation theorem states that the sum of all control coefficients is 1 (equation 3). In general, in pathways that consist of a large number of enzymes such as those required for growth on glucose, control is divided among many different reactions; i.e., most control coefficients will be small. In contrast, in small pathways, such as the enzymes involved in glucose transport and phosphorylation, a single reaction may to a large extent control the flux. The flux control coefficients were determined for the glucose PTS for growth on glucose, glucose oxidation, and uptake and phosphorylation of glucose. It was found that in none of these pathways did EI, HPr, or IIA^{Glc} exert significant control. Thus, in each case C_E^J was zero for all three reaction pathways (van der Vlag, unpublished). The only enzyme found to control the flux of glucose phosphorylation was $IICB^{Glc}$: the flux control coefficient was 0.65 (Ruijter et al., 1991). Remaining control could be in the intracellular PEP and pyruvate concentrations, which, under these experimental conditions, could not be kept constant as required by control analysis. No data for these variables have yet been obtained.

Although it recently has been shown that the summation theorem leads to a more complicated form in cases such as the PTS which involve enzymes that transfer a group between themselves (van Dam et al., 1993), these results suggest that in the glucose PTS most control is exerted by $IICB^{Glc}$; i.e., the rate of dephosphorylation of P-IIA^{Glc} by the $IICB^{Glc}$-glucose complex is higher than the rate of phosphorylation of IIA^{Glc} by P-HPr. This may have important consequences for the phosphorylation state of IIA^{Glc} (see below).

Regulation by the Glucose PTS

Involvement of the PTS in the uptake and metabolism of non-PTS compounds, both in wild-type strains and in *ptsHI* mutants (defective in HPr and/or EI), has been studied for a long time. Mutants that contain *ptsHI* deletions are unable to grow on a number of non-PTS compounds including lactose, melibiose, maltose, and glycerol (class I compounds) and Krebs cycle intermediates, xylose, and rhamnose (class II compounds). Different processes underlie these phenomena (for a review, see Postma et al. [1993]).

For class II compounds the phenotype is due to a lowering of the intracellular cyclic AMP (cAMP) concentration. It has been shown that the phosphorylation state of IIA^{Glc} is important for adenylate cyclase activity. Cells with no IIA^{Glc} at all or cells in which IIA^{Glc} is in the nonphosphorylated state have a lowered cAMP level. It has been proposed that phospho-IIA^{Glc} is, directly or indirectly, an activator of adenylate cyclase.

Although class I compounds are also dependent on cAMP, another process is responsible for the observed phenotype. It has been shown that the various class I transport systems or metabolic enzymes are inhibited by the nonphosphorylated form of IIA^{Glc} (Fig. 2). This process is called inducer exclusion. It has been shown that IIA^{Glc} interacts specifically with and inhibits the lactose carrier (LacY), the melibiose carrier (MelB), a component of the maltose transport system (MalK), and glycerol kinase (GlpK) (for a review, see Postma et al. [1993]). In cells in which IIA^{Glc} is in the dephosphorylated state, either because EI and HPr are absent (a *ptsHI* mutant) or because the presence of a PTS carbohydrate such as glucose results in a net dephosphorylation of P-IIA^{Glc}, inducer exclusion can occur. Additional factors are important in determining the extent of inducer exclusion. IIA^{Glc} binds in a stoichiometric fashion to, e.g., the lactose carrier. In addition, binding of IIA^{Glc} to its target proteins occurs only if a substrate of the target protein is present. For instance, IIA^{Glc} binds to and inhibits the lactose carrier only when a β-galactoside is present. This serves to prevent IIA^{Glc} binding in a nonproductive manner. This is important since an *E. coli* or *S. typhimurium* cell contains a fixed number of IIA^{Glc} molecules (10,000 to 20,000). It follows that under conditions in which the number of target molecules exceeds the number of IIA^{Glc} molecules, some target proteins should be free and active: the cell can escape from inducer exclusion. This has been observed experimentally by inducing cells partially for one of the class I systems, by overproducing such a system, or by lowering the IIA^{Glc} concentration in a defined way.

By using plasmids that contain the *crr* gene under the control of a *tac* promoter, it can be shown that complete inhibition of glycerol kinase is

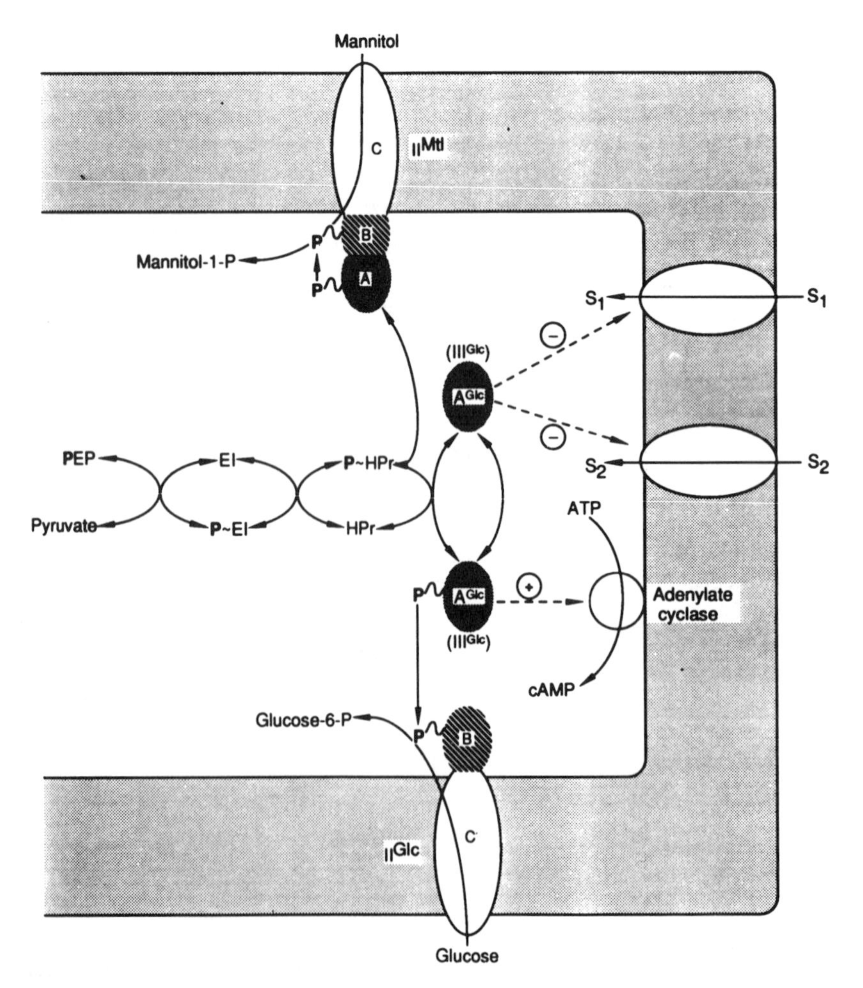

FIG. 2. Regulation by the PTS in enteric bacteria. Inhibition (−) of different non-PTS uptake systems (S_1 and S_2 represent lactose, melibiose, maltose, or glycerol) by IIA^{Glc} (III^{Glc}) and activation (+) of adenylate cyclase by P-IIA^{Glc} (P-III^{Glc}) are indicated. Other symbols are as in Fig. 1.

achieved when four molecules of IIA^{Glc} are present per glycerol kinase molecule (van der Vlag et al., 1994). The presence of less IIA^{Glc} leads to escape from inducer exclusion. These results support the crystallographic data obtained with a glycerol kinase-IIA^{Glc} complex. It was found that one IIA^{Glc} molecule is bound to each of the four monomers of glycerol kinase (Hurley et al., 1993).

IIA^{Glc} binds specifically to a number of target proteins which seem to share no obvious common traits. LacY and MelB are integral membrane proteins, MalK is a peripheral membrane protein, and GlpK is a cytoplasmic enzyme. In addition, no similarities have been found among those proteins. Mutations in LacY, MelB, and MalK that eliminate inducer exclusion in these mutant cells for a specific transport system have been isolated. Thus, in the *lacY* mutants the uptake of melibiose, maltose, or glycerol is still inhibited by PTS carbohydrates. A number of these mutations have been sequenced (Fig. 3) and seem to be scattered across large stretches of the proteins (Dean et al., 1990; Kühnau et al., 1991; Kuroda et al., 1992; Postma et al., 1988; Wilson et al., 1990). Some similarity among small domains of LacY, MalK, and GlpK has been noted, suggesting a possible binding site for IIA^{Glc}. However, the residues in GlpK, which are suggested to be in contact with IIA^{Glc} on the basis of the crystal structure, are completely different from those suggested on the basis of (limited) sequence similarity.

Since the crystal structure of the glycerol kinase-IIA^{Glc} complex has been determined, it is interesting to compare which residues are thought to interact with the GlpK monomers and which mutations have been isolated in IIA^{Glc} that prevent inducer exclusion. The residues of IIA^{Glc} that make contact with GlpK are amino acids 38 to 46, 71, 78, 88, 90, and 94 to 97. In *E. coli* three amino acid replacements in IIA^{Glc}, G47S, A76T, and S78F (Zeng et al., 1992), that eliminate inducer exclusion of class I compounds including glycerol, have been found. These results are in good agreement with the crystallographic data. Another *crr* mutation, formerly called *iex*, has also been sequenced (de Waard, unpublished). Interestingly, this G47C replacement results in a temperature-sensitive IIA^{Glc} (Nelson et al., 1984). It would be interesting to know the phenotype of the G47S mutant mentioned above. Similar mutations have been isolated in the *S. typhimurium crr* gene, resulting in G47S and E97K replacements (de Waard, unpublished). Although in no case has the binding of the mutant IIA^{Glc} to its target proteins been studied, the absence of inducer exclusion is most probably due to defective binding of IIA^{Glc}, since in both the *E. coli* and *S. typhimurium crr* mutants the amount of IIA^{Glc} or its activity in in vitro PEP-dependent phosphorylation of αMG was not changed very much.

Although emphasis has been put on the regulation by glucose, each PTS carbohydrate can exert the same effect since all EIIs are reversibly phosphorylated via P-HPr. This implies that any PTS carbohydrate can dephosphorylate P-HPr and, consequently, P-IIA^{Glc}. This is the basis of PTS-mediated repression of the transport and metabolism of all class I and class II compounds by any PTS carbohydrate. One can ask whether IIA^{Glc} is the sole mediator of this regulation in enteric bacteria. We have mentioned that *crr* mutations result in the absence of inducer exclusion and in lowered cAMP levels. It can be argued, however, that IIA^{Glc}-like molecules are present in the cell but at lower concentrations than those of IIA^{Glc}. This would allow inducer exclusion only at low concentrations of the target proteins. This is sup-

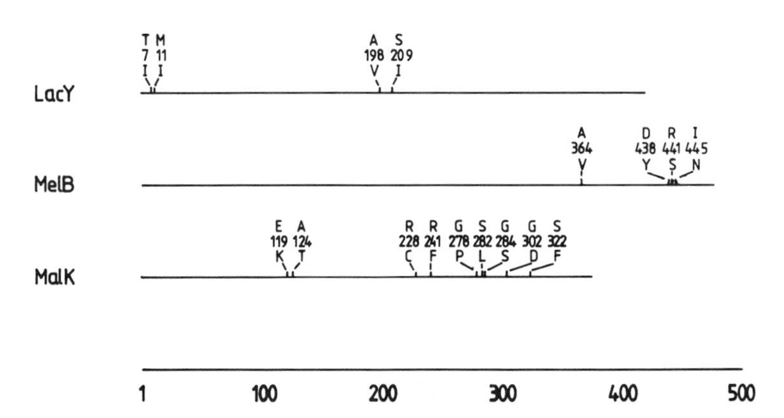

FIG. 3. Mutations resulting in absence of inducer exclusion. LacY, MelB, and MalK represent the lactose carrier, the melibiose carrier, and the MalK component of the maltose transport system, respectively. One-letter symbols for the amino acids are used. References to the original data are given in the text.

ported by the observation that in *S. typhimurium crr* deletion strains, inducer exclusion of glycerol and maltose can still be observed but only when their uptake systems are present at low levels (Nelson et al., 1982). From sequence analysis it has become clear that several EIIs contain IIA^{Glc}-like C-terminal domains, with similarities ranging from 30 to 40%. It has been shown that two of these EIIs, specific for *N*-acetylglucosamine (II^{Nag}) and β-glucosides (II^{Bgl}), can replace IIA^{Glc} in glucose and αMG transport (Vogler et al., 1988). Recently, we have found that II^{Nag} can replace IIA^{Glc} in inducer exclusion of glycerol and maltose (van der Vlag, unpublished). This implies that an integral membrane protein, II^{Nag}, can interact with and inhibit the same cytoplasmic enzyme glycerol kinase as the soluble IIA^{Glc}.

Conclusions

We have described in some detail the glucose PTS of *E. coli* and *S. typhimurium* which transports and phosphorylates glucose. In addition, the IIA^{Glc} component of the glucose PTS is a central regulatory molecule that can activate or inhibit non-PTS proteins, depending on its phosphorylation state. Most experiments support a model in which the presence of glucose or other PTS carbohydrates results in dephosphorylation of P-IIA^{Glc} and, as a consequence, causes inhibition of uptake of certain non-PTS compounds and lower adenylate cyclase activity. We have presented quantitative data on the control of the glucose flux via the glucose PTS which suggest that the major control is exerted by $IICB^{Glc}$. This is at present difficult to reconcile with the observation that in wild-type cells, addition of glucose or αMG results in an almost complete dephosphorylation of P-IIA^{Glc} (from $>90\%$ to $<10\%$).

REFERENCES

Buhr, A., and B. Erni. 1993. Membrane topology of the glucose transporter of *Escherichia coli. J. Biol. Chem.* **268:**11599–11603.

Dean, D. A., J. Reizer, H. Nikaido, and M. H. Saier. 1990. Regulation of the maltose transport system of *Escherichia coli* by the glucose-specific enzyme III of the phosphoenolpyruvate-sugar phosphotransferase system. Characterization of inducer exclusion-resistant mutants and reconstitution of inducer exclusion in proteoliposomes. *J. Biol. Chem.* **265:**21005–21010.

de Waard, J. H. Unpublished observations.

Erni, B. 1986. Glucose-specific permease of the bacterial phosphotransferase system: phosphorylation and oligomeric structure of the glucose-specific II^{Glc}-III^{Glc} complex of *Salmonella typhimurium. Biochemistry* **25:**305–312.

Hurley, J. H., D. Worthylake, H. R. Faber, N. D. Meadow, S. Roseman, D. W. Pettigrew, and S. J. Remington. 1993. Three-dimensional structure of the factor III^{Glc}-glycerol kinase regulatory complex. *Science* **259:**673–677.

Kacser, H., and J. A. Burns. 1973. The control of flux, p. 65–104. *In* D. D. Davies (ed.), *Rate Control of Biological Processes.* Cambridge University Press, Cambridge.

Kühnau, S., M. Reyes, A. Sievertsen, H. A. Shuman, and W. Boos. 1991. The activities of the *Escherichia coli* MalK protein in maltose transport, regulation and inducer exclusion can be separated by mutations. *J. Bacteriol.* **173:**2180–2186.

Kuroda, M., S. De Waard, K. Mizushima, M. Tsuda, P. Postma, and T. Tsuchiya. 1992. Resistance of the melibiose carrier to inhibition by the phosphotransferase system due to substitutions of amino acid residues in the carrier of *Salmonella typhimurium. J. Biol. Chem.* **267:**18336–18341.

Lolkema, J. S., R. H. Ten Hoeve-Duurkens, D. S. Kijkstra, and G. T. Robillard. 1991. Mechanistic coupling of transport and phosphorylation activity by enzyme II^{Mtl} of the *Escherichia coli* phosphoenolpyruvate-dependent phosphotransferase system. *Biochemistry* **30:**6716–6721.

Nelson, S. O., J. Lengeler, and P. W. Postma. 1984. Role of III^{Glc} of the phosphoenolpyruvate-glucose phosphotransferase system in inducer exclusion in *Escherichia coli. J. Bacteriol.* **160:**360–364.

Nelson, S. O., B. J. Scholte, and P. W. Postma. 1982. Phosphoenolpyruvate:sugar phosphotransferase system-mediated regulation of carbohydrate metabolism in *Salmonella typhimurium. J. Bacteriol.* **150:**604–615.

Postma, P. W. 1981. Defective enzyme II-B^{Glc} of the phosphoenolpyruvate:sugar phosphotransferase system leading to uncoupling of transport and phosphorylation in *Salmonella typhimurium. J. Bacteriol.* **147:**382–389.

Postma, P. W., C. P. Broekhuizen, A. R. J. Schuitema, A. P. Vogler, and J. W. Lengeler. 1988. Carbohydrate transport and metabolism in *Escherichia coli* and *Salmonella typhimurium:* regulation by the PEP:carbohydrate phosphotransferase system, p. 43–52. *In* F. Palmieri and E. Quagliariello (ed.), *Molecular Basis of Biomembrane Transport.* Elsevier Science Publishers, Amsterdam.

Postma, P. W., J. W. Lengeler, and G. R. Jacobson. 1993. Phosphoenolpyruvate:carbohydrate phosphotransferase systems of bacteria. *Microbiol. Rev.* **57:**543–594.

Postma, P. W., and J. B. Stock. 1980. Enzymes II of the phosphotransferase system do not catalyze sugar transport in the absence of phosphorylation. *J. Bacteriol.* **141:**476–484.

Ruijter, G. J. G., P. W. Postma, and K. van Dam. 1991. Control of glucose metabolism by enzyme II^{Glc} of the phosphoenolpyruvate-dependent phosphotransferase system in *Escherichia coli. J. Bacteriol.* **173:**6184–6191.

Ruijter, G. J. G., G. Van Meurs, M. A. Verwey, P. W. Postma, and K. Van Dam. 1992. Analysis of mutations that uncouple transport from phosphorylation in enzyme II^{Glc} of the *Escherichia coli* phosphoenolpyruvate-dependent phosphotransferase system. *J. Bacteriol.* **174:**2843–2850.

Saier, M. H., and J. Reizer. 1992. Proposed uniform nomenclature for the proteins and protein domains of the bacterial phosphoenolpyruvate:sugar phosphotransferase system. *J. Bacteriol.* **174:**1433–1438.

Van Dam, K., J. Van der Vlag, B. Kholodenko, and H. Westerhoff. 1993. The sum of the control coefficients of all enzymes on the flux through a group transfer pathway can be as high as two. *Eur. J. Biochem.* **212:**791–799.

van der Vlag, J. Unpublished data.

van der Vlag, J., K. van Dam, and P. W. Postma. 1994. Quantification of the regulation of glycerol and maltose metabolism by IIA^{Glc} of the phosphoenolpyruvate-dependent phosphotransferase system in *Salmonella typhimurium. J. Bacteriol.* **176:**3518–3526.

Vogler, A. P., C. P. Broekhuizen, A. Schuitema, J. W. Lengeler, and P. W. Postma. 1988. Suppression of III^{Glc}-defects by Enzymes II^{Nag} and II^{Bgl} of the PEP:carbohydrate phosphotransferase system. *Mol. Microbiol.* **2:**719–726.

Wilson, T. H., P. L. Yunker, and C. L. Hansen. 1990. Lactose transport mutants of *Escherichia coli* resistant to inhibition by the phosphotransferase system. *Biochim. Biophys. Acta* **1029:**113–116.

Yap, W. Unpublished observations.

Zeng, G. Q., H. de Reuse, and A. Danchin. 1992. Mutational analysis of the enzyme III^{Glc} of the phosphoenolpyruvate phosphotransferase system in *Escherichia coli. Res. Microbiol.* **143:**251–261.

Chapter 29

Modular Structure of the Enzymes II of Bacterial Phosphotransferase Systems

GARY R. JACOBSON

Department of Biology, Boston University, 5 Cummington Street, Boston, Massachusetts 02215

The bacterial phosphoenolpyruvate (PEP)-dependent carbohydrate phosphotransferase system (PTS) is responsible for the coupled transport and phosphorylation of many carbohydrates in a wide variety of bacteria (for recent reviews, see Meadow et al. [1990] and Postma et al. [1993]). Both of the general PTS phosphotransfer proteins, enzyme I (EI) and HPr, and the carbohydrate-specific enzyme II complexes (EIIs) participate in the transfer of the phospho group from PEP to the carbohydrate. The membrane-bound EIIs are also responsible for transport of the substrate as follows:

(1) PEP + EI $\rightleftharpoons$ P ~ (His)-EI + pyruvate
(2) P ~ EI + HPr $\rightleftharpoons$ P ~ (His)-HPr + EI
(3) P ~ HPr + EIIA $\rightleftharpoons$ P ~ (His)-EIIA + HPr
(4) P ~ EIIA + EIIB $\rightleftharpoons$ P ~ (Cys or His)-EIIB + EIIA
(5) $\text{P} \sim \text{EIIB} + \text{carbohydrate}_{out} \xrightarrow{\text{EIIC}} \text{EIIB} + \text{carbohydrate-P}_{in}$

EI and HPr are soluble proteins, and in most organisms they participate in PTS-mediated transport and phosphorylation of all PTS carbohydrates. The EII complexes, consisting of at least three different domains or proteins named EIIA, EIIB, and EIIC (in the nomenclature system of Saier and Reizer [1992]), are carbohydrate specific. EIIA and EIIB are hydrophilic domains or proteins, whereas the EIIC domain or protein is the integral-membrane component that possesses the substrate-binding and translocation site. The structures and roles of the EII domains or proteins in PTSs of different carbohydrate specificities suggest that the transport and phosphorylation mechanisms of all EIIs are very likely to be the same, and recent studies on the mannitol-specific EII (EII^{Mtl}) will be described as a model for these mechanisms.

In addition to their roles in carbohydrate transport and phosphorylation, the EIIs act as primary receptors for chemotaxis to their substrates, at least in enteric bacteria (Lengeler and Vogler, 1989). Moreover, the modular structure of the EIIs and their differing polypeptide organizations (from one to four proteins) probably reflect their different roles in nutrient acquisition and its regulation in bacterial cells. Thus, as described below, the separate EIIA protein of the glucose-specific EII (EII^{Glc}) regulates non-PTS permeases and adenylate cyclase, whereas recently discovered homologs of the EIIA domain of EII^{Mtl} may be regulatory links between the PTS and nitrogen assimilation in various bacteria (Reizer et al., 1992).

The EII Domains or Proteins Comprise Distinct Structural and Functional Modules

Complete or partial gene sequences are known for over 30 different EIIs from various bacteria. Sequence alignments of the gene products, biochemical studies, and various molecular genetic approaches have all provided evidence for the roles of each of the EII domains or proteins in phosphotransfer and transport as shown in the equations above (Postma et al., 1993). EIIA, EIIB, and EIIC may be domains of a single polypeptide, as in the *Escherichia coli* mannitol-specific $EIICBA^{Mtl}$ (Jacobson and Saraceni-Richards, 1993), in which case EIIA and EIIB are cytoplasmic domains attached to the integral-membrane EIIC domain. In other cases, these domains may be separated into two (e.g., $EIICB^{Glc}$ + $EIIA^{Glc}$ of *E. coli*) or even three (e.g., $EIIC^{Cel}$ + $EIIB^{Cel}$ + $EIIA^{Cel}$ of *E. coli*) distinct proteins, each encoded by a separate gene. In another class of EIIs, exemplified by the *E. coli* mannose-specific EII^{Man}, the EIIA and EIIB domains form a single protein and there are two membrane-bound proteins, EIIC and EIID (Figs. 1 and 2). Regardless of the domain organization, however, it is likely that the mechanisms of phosphotransfer and transport carried out by all EIIs are similar, as shown in the equations above and as illustrated schematically in Fig. 1.

Sequence comparisons of the various EIIs have also revealed that in EIIs in which more than one of these domains are part of a single polypeptide,

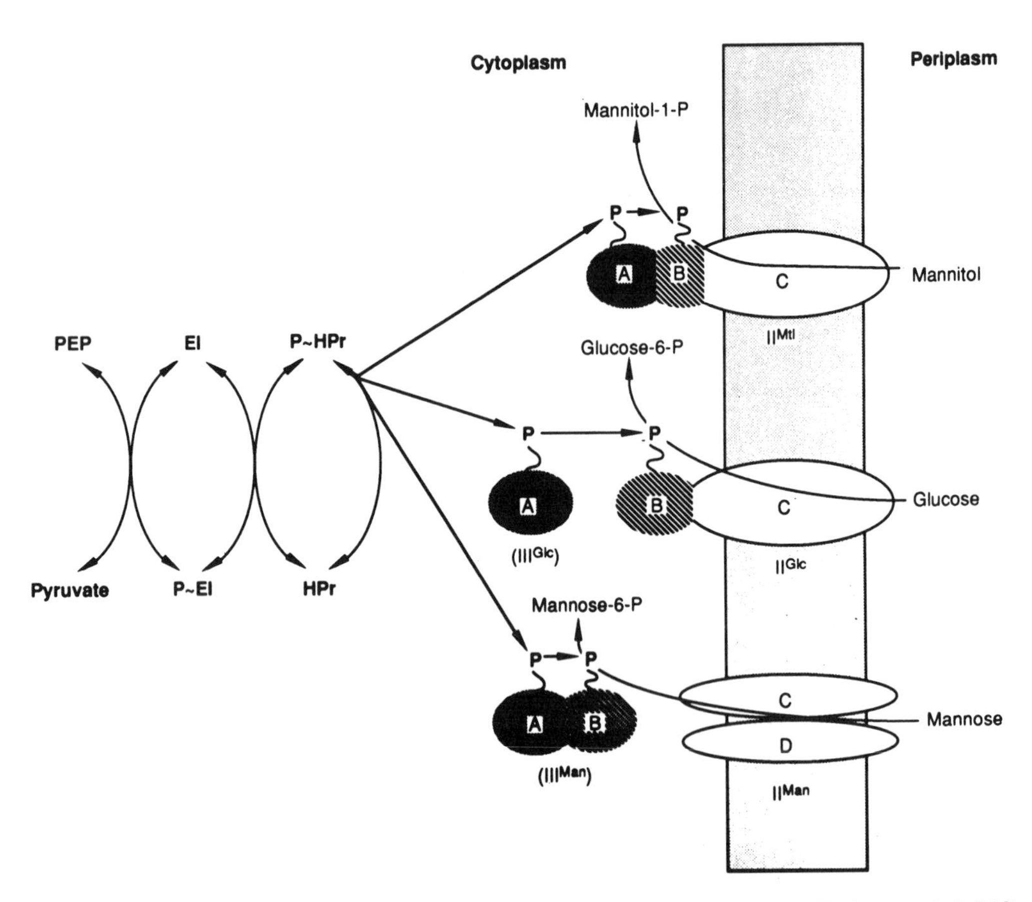

FIG. 1. Phosphotransfer pathways in the PTS and the domain organizations of the EIIs for mannitol (Mtl), glucose (Glc), and mannose (Man) in *E. coli.* A phospho group from PEP is transferred via the general proteins EI and HPr to an EIIA domain or protein at a histidine residue. This group is then transferred to an EIIB domain at a cysteine (for Mtl and Glc) or a histidine (for Man) residue prior to transfer to the incoming carbohydrate. The EIIC domain or protein is transmembrane and is responsible for binding and translocating the substrate. For EII^{Man}, a second transmembrane protein (EIID) is also important for mannose translocation. See the text for further details. Reprinted from Postma et al. (1993) with permission.

the positions of the domains in the primary sequence may be different, as shown in Fig. 2 (Postma et al., 1993; Saier and Reizer, 1992). For example, the order of the domains in EII^{Mtl} and EII^{Nag} of *E. coli* is C, B, A (from N to C terminus), whereas the order in EII^{Bgl} of *E. coli* is B, C, A. Moreover, both EII^{Glc} of *E. coli* and a plasmid-encoded, enteric EII^{Scr} use the same $EIIA^{Glc}$ for transport and phosphorylation of their substrates, but the domain order is C, B for the former and B, C for the latter (Fig. 2). Presumably, this "domain shuffling" can nonetheless be accommodated mechanistically, because in most cases the domains are connected by flexible linker regions (Lengeler, 1990; Postma et al., 1993). This is supported by the observations that the domains of EII^{Mtl} can be expressed as separate proteins in various combinations with retention of activity (White and Jacobson, 1990; van Weeghel et al., 1991a, 1991b; Robillard et al., 1993), that $EIIBCA^{Bgl}$ or $EIICBA^{Nag}$ can complement either $EIIBC^{Scr}$ or $EIICB^{Glc}$ in an *E. coli* strain lacking $EIIA^{Glc}$ (Schnetz et al., 1990; Vogler et al., 1988), and that EII^{Nag} from *Klebsiella pneumoniae* lacking its EIIA domain can be complemented by $EIIA^{Glc}$ in vivo (Vogler and Lengeler, 1988).

The enteric fructose-specific PTS and that from *Rhodobacter capsulatus* have additional complexities based on this theme of domain shuffling (Fig. 2). The membrane-bound components ($EIIB'BC^{Fru}$) in both cases show an apparent duplication (B′) of the B domain at the N terminus. Only the second of these contains a conserved cysteine residue, which is the probable second phosphorylation site (Prior and Kornberg, 1988; Wu and Saier, 1990), and the function of the B′ domain is unknown. Also, in both of these fructose PTSs,

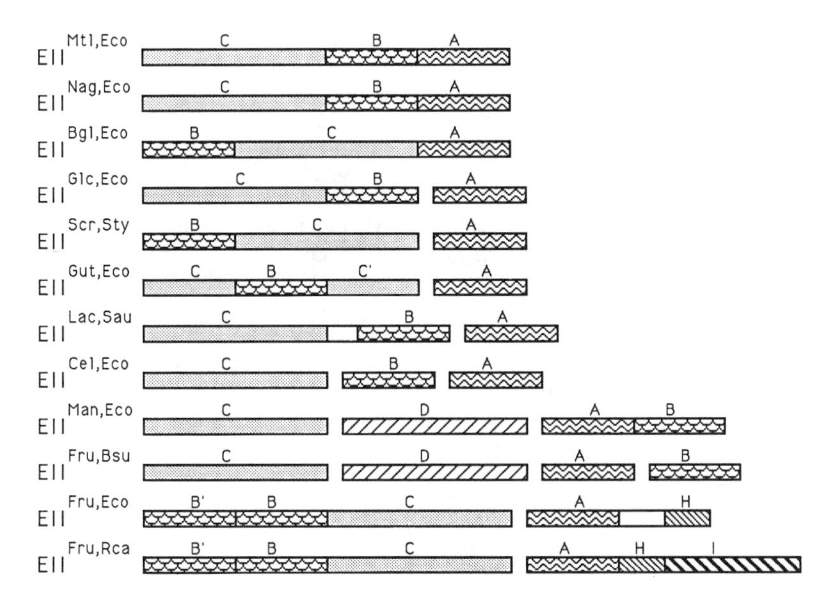

FIG. 2. Domain orders of various EIIs in different bacteria. The primary sequences are designated by the bars from the N termini on the left to the C termini on the right. Separate proteins are indicated by spaces between the bars. Domain abbreviations: A, EIIA; B, EIIB; B′, apparent duplication of EIIB (see text); C, EIIC; C′, a second predicted transmembrane region in the *E. coli* EII^{Gut}; D, a second transmembrane protein in EII^{Man} from *E. coli* and EII^{Fru} from *Bacillus subtilis;* H, HPr-like domain; I, EI-like domain. Carbohydrate abbreviations: Mtl, mannitol; Nag, *N*-acetylglucosamine; Bgl, β-glucosides; Glc, glucose; Scr, sucrose; Gut, glucitol; Lac, lactose; Cel, cellobiose; Man, mannose; Fru, fructose. Organism abbreviations: Eco, *E. coli;* Sty, *Salmonella typhimurium;* Sau, *Staphylococcus aureus;* Bsu, *Bacillus subtilis;* Rca, *Rhodobacter capsulatus.* Modified from Saier and Reizer (1992).

homologs of the general PTS protein EI and/or HPr are found fused to the $EIIA^{Fru}$ domain. In the enteric system, an HPr-like protein is fused to this domain, and this protein, called FPr, presumably replaces HPr in fructose phosphorylation (Geerse et al., 1989), whereas in *R. capsulatus* (which possesses only a fructose PTS), both HPr and EI (in that order) are fused to an N-terminal $EIIA^{Fru}$ domain (Wu et al., 1990).

E. coli EII^{Mtl} as a Model for EIIs of the PTS

Arguably, the best-understood EII of the PTS is EII^{Mtl} of *E. coli.* As discussed above, its three domains are all part of a single polypeptide, $EIICBA^{Mtl}$ (M_r 68,000). Biochemical and biophysical studies (Pas and Robillard, 1988; Pas et al., 1991), as well as a determination of the stereochemical course of mannitol phosphorylation by EII^{Mtl} (Mueller et al., 1990), are consistent with P ~ His-554 and P ~ Cys-384 in the EIIA and EIIB domains, respectively, as being bona fide catalytic intermediates in phosphotransfer to mannitol. In addition, the EIIC domain has been shown to be necessary and sufficient to bind and transport mannitol (Grisafi et al., 1989; Elferink et al., 1990; Lolkema et al., 1990). However, phosphorylation of the EIIA and EIIB domains greatly accelerates translocation by the EIIC domain, even though subsequent phosphorylation of mannitol by P ~ $EIIB^{Mtl}$ is not strictly coupled, mechanistically, to translocation (Lolkema et al., 1991b).

Recently, site-directed mutagenesis and other molecular genetic techniques have been used to probe the structure and function of EII^{Mtl}. The results of mutagenesis studies of His-554 and Cys-384 are consistent with their roles as phosphorylation sites 1 and 2, respectively (van Weeghel et al., 1991c; Weng et al., 1992). Moreover, it has been shown that inactive His-554 mutants can complement inactive Cys-384 mutants both in vitro (van Weeghel et al., 1991c) and in vivo (Weng et al., 1992). These results show that phosphotransfer from His-554 to Cys-384 can occur in an intermolecular fashion in an EII^{Mtl} oligomer (minimally a dimer) both in detergent solution and in the membrane. In addition, numerous studies have suggested that the physical, and possibly the most active, form of this protein is a dimer (reviewed by Jacobson [1992]), although the monomer apparently also possesses some activity in mannitol phosphorylation (Lolkema and Robillard, 1990).

Recently, studies in our laboratory have been directed toward understanding structure-function relationships in the integral-membrane EIIC domain of EII^Mtl^. Studies of the topology of EII^Mtl^ in the membrane by using the *phoA* fusion technique (Manoil and Beckwith, 1986) have suggested a structure for the EIIC domain consisting of at least six transmembrane regions, presumably α-helices (Sugiyama et al., 1991). As shown in Fig. 3, this model predicts that these transmembrane regions are connected by three relatively short periplasmic loops (numbered 2, 4, and 6 in Fig. 3) and two larger cytoplasmic regions (numbered 3 and 5 in Fig. 3). Very recently, a model for EII^Glc^ based on *phoA* and *lacZ* fusion analyses has been proposed (Buhr and Erni, 1993). This model suggests that there are eight transmembrane helices, but in many other respects it is similar to the model for EII^Mtl^ shown in Fig. 3, including the presence of a large cytoplasmic loop corresponding, approximately, to loop 5 in EII^Mtl^.

Also indicated in Fig. 3 are residues in the EIIC^Mtl^ domain that have been studied by mutagenesis. Interestingly, residues that have been identified so far as important for EII^Mtl^ activity are all located in an 85-residue segment (putative cytoplasmic loop 5), and different mutations in this region affect virtually all of the activities of this protein (reviewed by Jacobson and Saraceni-Rich-

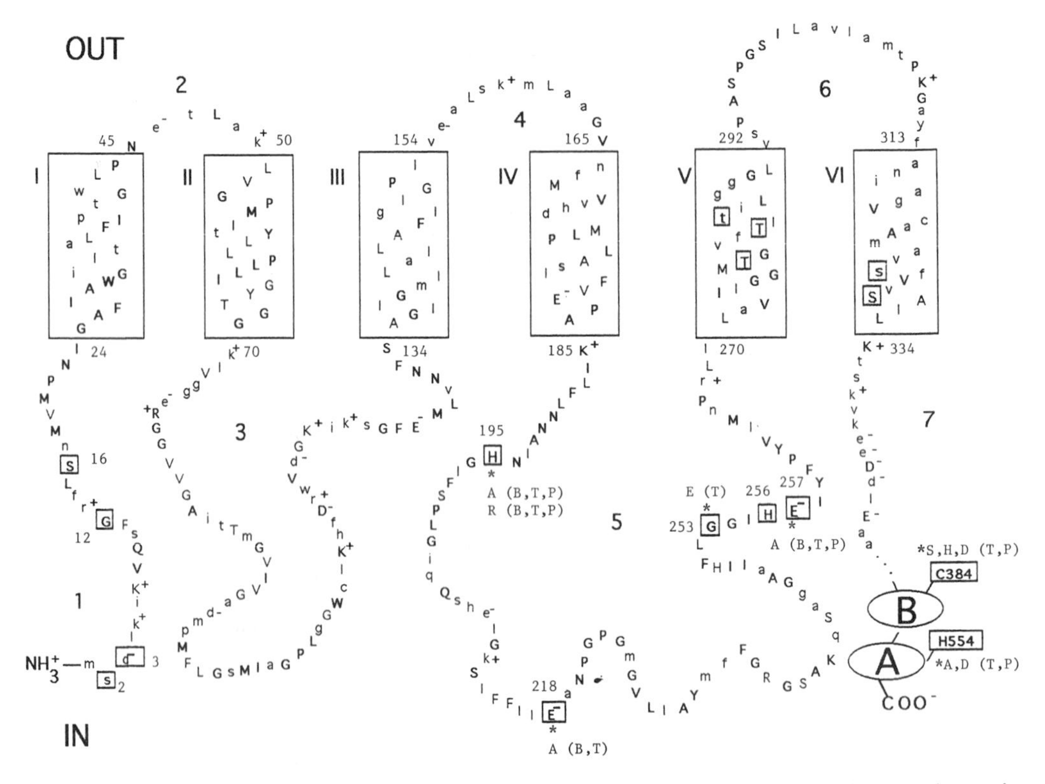

FIG. 3. Topology model for the EIIC^Mtl^ domain showing residues studied by mutagenesis. There are six putative transmembrane domains (I to VI) connected by three periplasmic "loops" (regions 2, 4, and 6) and two cytoplasmic "loops" (regions 3 and 5). The localization of the N-terminal region (region 1) is uncertain. Region 7, beginning with Lys-334, extends to the C terminus (residue 637) and is composed of the EIIA (A) and EIIB (B) domains. The boxed residues have been studied by various mutagenesis techniques. An asterisk (*) denotes a residue that, when replaced by certain residues (indicated after the asterisk), results in significant alteration of one or more activities of EII^Mtl^. The activities affected are indicated in parentheses after each replacement residue (B, mannitol binding; T, mannitol transport; P, mannitol phosphorylation). His-554 in the EIIA domain and Cys-384 in the EIIB domain (domains not shown to scale) are the obligatory phosphorylation sites during phosphotransfer from P-HPr to mannitol to form mannitol 1-phosphate. Capital letters denote residues that are identical or highly conserved in the sequences of the EIIC^Mtl^ domains from two gram-positive organisms, *Staphylococcus carnosus* (Fischer and Hengstenberg, 1992) and *Streptococcus mutans* (A. Honeyman, personal communication). Lowercase letters denote residues that are not highly conserved among these three sequences. See the text for further details. Modified from Jacobson and Saraceni-Richards (1993) with permission.

ards [1993]). Thus, His-195 appears to have a role in both mannitol binding and phosphorylation (Weng and Jacobson, 1993), Glu-218 is important for coupling of mannitol transport to phosphorylation (Lengeler and Heuel, personal communication), Gly-253 plays a role in mannitol transport (Manayan et al., 1988), and Glu-257, which is conserved in most EIIs of the PTS, appears to be necessary for mannitol binding (Jacobson and Saraceni-Richards, 1993).

If loop 5 is indeed not a transmembrane region of EII^{Mtl}, it has been suggested that it could either form a type of "cap" or "gate" for a channel formed by the transmembrane helices (Postma et al., 1993) or project up into such a channel to itself form at least part of the mannitol-binding and translocation site (Lengeler, 1990). In either case, it would be reasonable to postulate that phosphorylation of Cys-384 in the EIIB domain could produce a conformational change in the EIIC domain, perhaps involving loop 5, that would promote mannitol translocation into the cell (Jacobson and Saraceni-Richards, 1993). Consistent with this hypothesis, it has been shown that certain mutations in Cys-384 can affect the K_D for mannitol binding to the EIIC domain (Lolkema et al., 1991a; Weng and Jacobson, 1993) and that His-195 in loop 5 of the EIIC domain may be involved in phosphotransfer from Cys-384 to mannitol (Weng and Jacobson, 1993). On the other hand, phosphorylation of EII^{Mtl} does not greatly affect its affinity for D-mannoheptitol (perseitol), a mannitol analog that binds to EII^{Mtl} but is not transported or phosphorylated (Jacobson et al., 1983; Lolkema et al., 1993). Thus, if phosphorylation of EII^{Mtl} produces a conformational change in the EIIC domain, this change may affect mannitol translocation more than it affects mannitol binding.

Involvement of EIIAs and Their Homologs in Regulation of Non-PTS Nutrient Assimilatory Systems

As discussed elsewhere in this volume, the PTS not only is a major carbohydrate assimilatory system in many heterotrophic bacteria but also plays a central role in regulation of other nutrient assimilatory systems in many of these organisms (reviewed by Postma et al. [1993] and Saier [1989]). Thus, $EIIA^{Glc}$, depending on its phosphorylation state, regulates other carbohydrate assimilatory systems (such as *lac*) through the process called inducer exclusion. It also regulates the activity of adenylate cyclase, which results in PTS-mediated regulation of genes whose maximal transcription depends on the cyclic AMP-catabolite activator protein complex. This regulatory role of $EIIA^{Glc}$ may be one reason for its being expressed as a separate protein rather than as a domain of the EII^{Glc} complex. As a separate, soluble protein, $EIIA^{Glc}$ can freely diffuse within the cell to alternatively interact with P ~ HPr, $EIICB^{Glc}$, adenylate cyclase, or non-PTS permeases such as the lactose permease (Postma and Lengeler, 1985).

Recently, we have also discovered that $EIIA^{Mtl}$ of *E. coli* and related EIIA domains or proteins (e.g., the $EIIA^{Fru}$ domain of FPr) have sequence similarity to a class of proteins that may be involved in regulating other nutrient assimilatory systems in various bacteria (Reizer et al., 1992). These latter proteins are encoded by genes located downstream of the gene (*rpoN*) encoding a minor sigma factor, σ^{54}. σ^{54} is known to be required for the transcription of a number of genes, including many which are necessary for nitrogen assimilation under N-limiting conditions (Kustu et al., 1989). The predicted protein products of these genes that are downstream of *rpoN* (called ORF162 in *E. coli* and *K. pneumoniae*) have been suggested to be negative regulators of σ^{54}-dependent transcription (Merrick and Coppard, 1989), and the most striking sequence similarity between these proteins and the EIIAs is found around and including the known phosphorylation site in $EIIA^{Mtl}$ (Reizer et al., 1992).

The foregoing observations have led to the hypothesis that ORF162 and its homologs could be phosphorylated by P ~ HPr of the PTS and that this could be a regulatory link between carbon assimilation via the PTS and N assimilation controlled by σ^{54}-dependent promoters if the putative regulatory activity of ORF162 were controlled by its phosphorylation state (Reizer et al., 1992). As a first step in testing this hypothesis, we recently overexpressed ORF162 from *K. pneumoniae* in *E. coli* and demonstrated its PTS-dependent phosphorylation (Begley and Jacobson, in press). Furthermore, overexpression of ORF162 affects the growth rate of cells in a manner depending on the growth medium and on the presence of an intact σ^{54} protein. These observations are consistent with the hypothesis that ORF162 may be a regulatory link between carbon and N assimilation, but clearly much more work remains to confirm this hypothesis and to determine the exact role of ORF162 in the cell. In any event, it is possible that the covalent linkages of $EIIA^{Mtl}$ to the other domains of EII^{Mtl} and of $EIIA^{Fru}$ to an HPr-like protein may help to avoid their interference with the putative regulatory role of ORF162.

Conclusions

Despite the different modular organizations of the EIIs of the PTS, their mechanisms of carbohydrate transport and phosphorylation are likely to be similar. Current studies on these mechanisms are focused on determining regions and residues in the EIIs that are important in function, but these studies must eventually be complemented by direct structural determinations. The different modular structures of the EIIs may, in part, be due to the roles of the PTS in regulating other nutrient assimilatory systems. Some of these roles have yet to be completely elucidated and perhaps even to be discovered.

Research in my laboratory was supported by Public Health Service grant GM28226 from the National Institute of General Medical Sciences.

REFERENCES

Begley, G. S., and G. R. Jacobson. Overexpression, phosphorylation and growth effects of ORF162, a *Klebsiella pneumoniae* protein that is encoded by a gene linked to *rpoN*, the gene encoding σ^{54}. *FEMS Microbiol. Lett.*, in press.

Buhr, A., and B. Erni. 1993. Membrane topology of the glucose transporter of *Escherichia coli. J. Biol. Chem.* **268:**11599–11603.

Elferink, M. G. L., A. J. M. Driessen, and G. T. Robillard. 1990. Functional reconstitution of the purified phosphoenolpyruvate-dependent mannitol-specific transport system of *Escherichia coli* in phospholipid vesicles: coupling between transport and phosphorylation. *J. Bacteriol.* **172:**7119–7125.

Fischer, R., and W. Hengstenberg. 1992. Mannitol-specific enzyme II of the phosphoenolpyruvate-dependent phosphotransferase system of *Staphylococcus carnosus.* Sequence and expression in *Escherichia coli* and structural comparison with the enzyme $II^{Mannitol}$ of *Escherichia coli. Eur. J. Biochem.* **204:**963–969.

Geerse, R. H., F. Izzo, and P. W. Postma. 1989. The PEP:fructose phosphotransferase system in *Salmonella typhimurium:* FPr combines Enzyme III^{Fru} and pseudo-HPr activities. *Mol. Gen. Genet.* **216:**517–525.

Grisafi, P. L., A. Scholle, J. Sugiyama, C. Briggs, G. R. Jacobson, and J. W. Lengeler. 1989. Deletion mutants of the *Escherichia coli* K-12 mannitol permease: dissection of transport-phosphorylation, phospho-exchange, and mannitol-binding activities. *J. Bacteriol.* **171:**2719–2727.

Honeyman, A. Personal communication.

Jacobson, G. R. 1992. Interrelationships between protein phosphorylation and oligomerization in transport and chemotaxis via the *Escherichia coli* mannitol phosphotransferase system. *Res. Microbiol.* **143:**113–116.

Jacobson, G. R., and C. Saraceni-Richards. 1993. The *Escherichia coli* mannitol permease as a model for transport via the bacterial phosphotransferase system. *J. Bioenerget. Biomembr.* **25:**621–626.

Jacobson, G. R., L. Tanney, D. M. Kelly, K. B. Palman, and S. Corn. 1983. Substrate and phospholipid specificity of the purified mannitol permease of *Escherichia coli. J. Cell. Biochem.* **23:**231–240.

Kustu, S., E. Santero, J. Keener, D. Popham, and D. Weiss. 1989. Expression of σ^{54} (*ntrA*)-dependent genes is probably united by a common mechanism. *Microbiol. Rev.* **53:**367–376.

Lengeler, J. W. 1990. Molecular analysis of the Enzyme II complexes of the bacterial phosphotransferase system (PTS) as carbohydrate transport systems. *Biochim. Biophys. Acta* **1018:**155–159.

Lengeler, J. W., and H. Heuel. Personal communication.

Lengeler, J. W., and A. P. Vogler. 1989. Molecular mechanisms of bacterial chemotaxis towards PTS-carbohydrates. *FEMS Microbiol. Rev.* **63:**81–92.

Lolkema, J. S., D. S. Dijkstra, R. H. Ten Hoeve-Duurkens, and G. T. Robillard. 1990. The membrane-bound domain of the phosphotransferase enzyme II^{Mtl} of *Escherichia coli* constitutes a mannitol translocating unit. *Biochemistry* **29:**10659–10663.

Lolkema, J. S., D. S. Dijkstra, R. H. Ten Hoeve-Duurkens, and G. T. Robillard. 1991a. Interaction between the cytoplasmic and membrane-bound domains of enzyme II^{Mtl} of the *Escherichia coli* phosphoenolpyruvate-dependent phosphotransferase system. *Biochemistry* **30:**6721–6726.

Lolkema, J. S., and G. T. Robillard. 1990. Subunit structure and activity of the mannitol-specific Enzyme II of the *Escherichia coli* phosphoenolpyruvate-dependent phosphotransferase system solubilized in detergent. *Biochemistry* **29:**10120–10125.

Lolkema, J. S., R. H. Ten Hoeve-Duurkens, D. S. Dijkstra, and G. T. Robillard. 1991b. Mechanistic coupling of transport and phosphorylation activity by enzyme II^{Mtl} of the *Escherichia coli* phosphoenolpyruvate-dependent phosphotransferase system. *Biochemistry* **30:**6716–6721.

Lolkema, J. S., E. S. Wartna, and G. T. Robillard. 1993. Binding of the substrate analogue perseitol to phosphorylated and unphosphorylated enzyme II^{Mtl} of the phosphoenolpyruvate-dependent phosphotransferase system of *Escherichia coli. Biochemistry* **32:**5848–5854.

Manayan, R., G. Tenn, H. B. Yee, J. D. Desai, M. Yamada, and M. H. Saier, Jr. 1988. Genetic analyses of the mannitol permease of *Escherichia coli:* isolation and characterization of a transport-deficient mutant which retains phosphorylating activity. *J. Bacteriol.* **170:**1290–1296.

Manoil, C., and J. Beckwith. 1986. A genetic approach to analyzing membrane protein topology. *Science* **233:**1403–1408.

Meadow, N. D., D. K. Fox, and S. Roseman. 1990. The bacterial phosphoenolpyruvate:glycose phosphotransferase system. *Annu. Rev. Biochem.* **59:**497–542.

Merrick, M. J., and J. R. Coppard. 1989. Mutations in genes downstream of the *rpoN* gene (encoding σ^{54}) of *Klebsiella pneumoniae* affect expression from σ^{54}-dependent promotors. *Mol. Microbiol.* **3:**1765–1775.

Mueller, E. G., S. S. Khandekar, J. R. Knowles, and G. R. Jacobson. 1990. Stereochemical course of the reactions catalyzed by the bacterial phosphoenolpyruvate:mannitol phosphotransferase system. *Biochemistry* **29:**6892–6896.

Pas, H. H., G. H. Meyer, W. H. Kruizinga, K. S. Tamminga, R. P. van Weeghel, and G. T. Robillard. 1991. ^{31}Phospho-NMR demonstration of phosphocysteine as a catalytic intermediate on the *Escherichia coli* phosphotransferase system EII^{Mtl}. *J. Biol. Chem.* **266:**6690–6692.

Pas, H. H., and G. T. Robillard. 1988. S-phosphocysteine and phosphohistidine are intermediates in the phosphoenolpyruvate-dependent mannitol transport catalyzed by *Escherichia coli* EII^{Mtl}. *Biochemistry* **27:**5835–5839.

Postma, P. W., and J. W. Lengeler. 1985. Phosphoenolpyruvate:carbohydrate phosphotransferase system of bacteria. *Microbiol. Rev.* **49:**232–269.

Postma, P. W., J. W. Lengeler, and G. R. Jacobson. 1993. Phosphoenolpyruvate:carbohydrate phosphotransferase systems of bacteria. *Microbiol. Rev.* **57:**543–594.

Prior, T. I., and H. L. Kornberg. 1988. Nucleotide sequence of *fruA*, the gene specifying enzyme II^{Fru} of the phosphoenolpyruvate-dependent sugar phosphotransferase system in *Escherichia coli* K12. *J. Gen. Microbiol.* **134:**2757–2768.

Reizer, J., A. Reizer, M. H. Saier, and G. R. Jacobson. 1992. A proposed link between nitrogen and carbon metabolism involving protein phosphorylation in bacteria. *Protein Sci.* **1:**722–726.

Robillard, G. T., H. Boer, R. P. van Weeghel, G. Wolters, and A. Dijkstra. 1993. Expression and characterization of a structural and functional domain of the mannitol-specific transport protein involved in the coupling of mannitol transport and phosphorylation in the phosphoenolpyruvate-dependent phosphotransferase system of *Escherichia coli. Biochemistry* **32:**9553–9562.

Saier, M. H., Jr. 1989. Protein phosphorylation and allosteric control of inducer exclusion and catabolite repression by the bacterial phosphoenolpyruvate:sugar phosphotransferase system. *Microbiol. Rev.* **53:**109–120.

Saier, M. H., Jr., and J. Reizer. 1992. Proposed uniform nomenclature for the proteins and protein domains of the bacterial phosphoenolpyruvate:sugar phosphotransferase system. *J. Bacteriol.* **174:**1433–1438.

Schnetz, K., S. L. Sutrina, M. H. Saier, Jr., and B. Rak. 1990. Identification of catalytic residues in the β-glucoside permease of *Escherichia coli* by site-specific mutagenesis and demonstration of interdomain cross-reactivity between the β-glucoside and glucose systems. *J. Biol. Chem.* **265:**13464–13471.

Sugiyama, J. E., S. Mahmoodian, and G. R. Jacobson. 1991. Membrane topology analysis of the *Escherichia coli* mannitol permease by using a nested-deletion method to create *mtlA-phoA* fusions. *Proc. Natl. Acad. Sci. USA* **88:**9603–9607.

Van Weeghel, R. P., G. H. Meyer, W. Keck, and G. T. Robillard. 1991a. Phosphoenolpyruvate-dependent mannitol phosphotransferase system of *Escherichia coli:* overexpression, purification, and characterization of the enzymatically active C-terminal domain of Enzyme II^{Mtl} equivalent to Enzyme III^{Mtl}. *Biochemistry* **30:**1774–1779.

Van Weeghel, R. P., G. H. Meyer, H. H. Pas, W. Keck, and G. T. Robillard. 1991b. Cytoplasmic phosphorylating domain of the mannitol-specific transport protein of the phosphoenolpyruvate-dependent phosphotransferase system in *Escherichia coli*: overexpression, purification and functional complementation with the mannitol-binding domain. *Biochemistry* **30:**9478–9485.

Van Weeghel, R. P., Y. Y. van der Hoek, H. H. Pas, M. Elferink, W. Keck, and G. T. Robillard. 1991c. Details of mannitol transport in *Escherichia coli* elucidated by site-specific mutagenesis and complementation of phosphorylation site mutants of the phosphoenolpyruvate-dependent mannitol-specific phosphotransferase system. *Biochemistry* **30:**1768–1773.

Vogler, A. P., C. P. Broekhuizen, A. Schuitema, J. W. Lengeler, and P. W. Postma. 1988. Suppression of III^{Glc}-defects by Enzymes II^{Nag} and II^{Bgl} of the PEP:carbohydrate phosphotransferase system. *Mol. Microbiol.* **2:**719–726.

Vogler, A. P., and J. W. Lengeler. 1988. Complementation of a truncated membrane-bound Enzyme II^{Nag} from *Klebsiella pneumoniae* with a soluble Enzyme III in *Escherichia coli* K12. *Mol. Gen. Genet.* **213:**175–178.

Weng, Q.-P., J. Elder, and G. R. Jacobson. 1992. Site-specific mutagenesis of residues in the *Escherichia coli* mannitol permease that have been suggested to be important for its phosphorylation and chemoreception functions. *J. Biol. Chem.* **267:**19529–19535.

Weng, Q.-P. and G. R. Jacobson. 1993. Role of a conserved histidine residue, His-195, in the activities of the *Escherichia coli* mannitol permease. *Biochemistry* **32:**11211–11216.

White, D. W., and G. R. Jacobson. 1990. Molecular cloning of the C-terminal domain of *Escherichia coli* D-mannitol permease: expression, phosphorylation, and complementation with C-terminal permease deletion proteins. *J. Bacteriol.* **172:**1509–1515.

Wu, L.-F., and M. H. Saier, Jr. 1990. Nucleotide sequence of the *fruA* gene, encoding the fructose permease of the *Rhodobacter capsulatus* phosphotransferase system, and analyses of the deduced protein sequence. *J. Bacteriol.* **172:**7167–7178.

Wu, L.-F., J. M. Tomich, and M. H. Saier, Jr. 1990. Structure and evolution of a multidomain multiphosphoryl transfer protein. Nucleotide sequence of the *fruB(HI)* gene in *Rhodobacter capsulatus* and comparisons with homologous genes from other organisms. *J. Mol. Biol.* **213:**687–703.

Chapter 30

Signal Transduction through Phosphotransferase Systems

JOSEPH W. LENGELER, KATJA BETTENBROCK, AND RENATE LUX

Fachbereich Biologie/Chemie, Universität Osnabrück, D-49069 Osnabrück, Federal Republic of Germany

Bacteria, like other living cells, contain sensors to monitor their surroundings. Many stimuli are sensed directly on the outside of the cell and are signaled through the membrane, before they trigger a cellular response. This response can involve changes in the synthesis of proteins, modulation of enzyme activities, changes in behavior (e.g., motility), or changes in cell differentiation. As a general rule, the response is transient and is maintained only as long as the environmental stimulus persists. Often, the organism adapts; i.e., it ends the response although the stimulus is still active. The entire bacterial cell thus may be considered the equivalent of a transiently responding sensory system.

Signal Transduction Systems and Global Regulatory Networks Are Linked through Protein Phosphorylation

Many bacterial sensory systems have been analyzed, and all appear to share a common theme: signal transduction through protein-protein interactions. Such interactions are modulated by means of a biochemical modification of the proteins in response to changes in the stimuli. Protein modification frequently involves phosphorylation of histidines, aspartates, and serines (for recent reviews, see Parkinson [1993] and Stock [1993]). The most popular bacterial signaling pathways (simpler and more complex variants exist) are the "two-component chemosensor systems," which comprise four functional domains or modules (for recent reviews, see Bourret et al. [1991], Parkinson and Kofoid [1992], and Stock et al. [1989]). The first component (or sensor) consists of a receptor (input) domain coupled (or fused) to a transmitter domain. Sensors for external stimuli bridge the membrane, whereas sensors for changes in metabolic pools are intracellular. The intracellular sensors sense changes in the environment indirectly through changes in transport system activities. The transmitter autophosphorylates at a conserved histidine residue in an ATP-dependent reaction. This autophosphorylation is modulated by binding and dissociation of a stimulating molecule at the receptor domain. Activation of a particular transmitter enables it to phosphorylate the receiver domain of a second protein at an aspartyl residue. This third module is coupled (or fused) to the fourth (output) domain, and together they form the response regulator of a two-component system. In response to its phosphorylation by the transmitter (kinase) domain, the receiver modulates the activity of the output domain, which then regulates either enzyme activities or gene expression. Gene regulation often involves alternative sigma factors or other transcription factors for groups of genes organized in a modulon (see below). Except for protein phosphorylation, any biochemical modification which allows a rapid, reversible, and specific modulation of protein activities could serve the same purpose. The transmitter and receiver in particular must be well tuned to minimize unwanted cross talk between different regulatory systems (Wanner, 1992). During evolution, various signaling systems have diverged into different genetic families, probably also to minimize cross talk.

Complex bacterial activities require coordination of blocs or networks of metabolic pathways, e.g., the genes and enzymes involved in peripheral carbohydrate transport and metabolism, in control of nutrient and energy supply, or in cell movement, to name a few of the 40-odd systems described thus far in enteric bacteria. The term "modulon" has been proposed (Iuchi and Lin, 1988) for such networks of operons and regulons in which each member is not only regulated by individual repressors and activators but also controlled by global regulation (Chuang et al., 1993; Guest, 1992). This general and pleiotropic regulatory system is invariably epistatic, i.e., dominant over the individual systems. Most global regulatory systems are connected to two-component chemosensors and control their modulon in response to drastic, global changes in the environment such as sudden food starvation, anoxya, and a series of stress situations which generate a state of general alarm in a cell (or population). The alarm may also be indicated to the cell as changes in specific "alarmones," e.g., cyclic AMP (cAMP) or ppGpp and other nucleotide derivatives, which function as intracellular indicator molecules (Neidhardt et al., 1987). In this case, intracellular sensors become active. These often monitor entire metabolic networks and require modules in addition to the four modules of normal two-component systems. For the corresponding modulons, e.g., those involved in carbon supply, nitrogen supply, cellular respiration, and cell movement, a controlled cross talk seems advantageous (Guest, 1992).

The Bacterial PTS Is a Sensory and Global Regulatory Unit

The phosphoenolpyruvate (PEP)-dependent carbohydrate:phosphotransferase system (PTS) is a regulatory unit such as those described above (Meadow et al., 1990). The complexity of the PTS reflects its multiple roles in transport and phosphorylation of a large number of carbohydrates, in movement toward these carbon sources (chemotaxis), and as a global regulatory system for peripheral catabolic and glycolytic pathways (for a recent review, see Postma et al. [1993]). It can be viewed as a series of carbohydrate transport systems present in a single bacterium, e.g., the 16 PTSs described thus far in *Escherichia coli* (Meadow et al., 1990; Postma et al., 1993; Robillard and Lolkema, 1988). Each of these systems includes a substrate-specific part (or enzyme II [EII]) which transports and phosphorylates the different substrates. All 16 EIIs of a cell are phosphorylated by way of a single protein kinase enzyme I (EI). Phosphorylation of the EIIs, however, is not direct but occurs through ancillary phospho-carrier proteins. These are equivalent to the targeting subunits of other targeted protein kinases, which direct a catalytic subunit, here EI, to its many target loci, here the 16 EIIs (Fig. 1) (Hubbard and Cohen, 1993).

On the other hand, the PTS can be viewed as a global regulatory unit composed of at least 20 to 30 proteins (domains or modules) that transfer phosphoryl groups between themselves (Postma et al., 1993; Roseman and Meadow, 1990; Saier, 1989). Together they monitor the environment through their transport activities. This information is used to control the expression of individual operons through mRNA antitermination, of the *crp* modulon through catabolite repression and inducer exclusion, and of a group of glycolytic genes and enzymes through gene or protein activation. The information is also signaled to the flagellar motors, where it is used to control the chemotactic behavior of the cells (Taylor and Lengeler, 1990). All of these mechanisms, except for the chemotaxis to PTS carbohydrates, are described in other chapters. Therefore, we will concentrate on the general strategies used by PTS proteins to "talk" to each other through protein phosphorylation, in particular the communication of the catalytic domain EI with its many target loci through the use of targeting subunits. We will also include what little information is available on the coupling of the PTS to the chemotactic system in enteric bacteria as derived from our own results.

Protein Phosphorylation during Carbohydrate Transport and Phosphorylation through PTSs

The different steps and molecules involved in transport and phosphorylation of a substrate through any PTS (see Meadow et al. [1990],

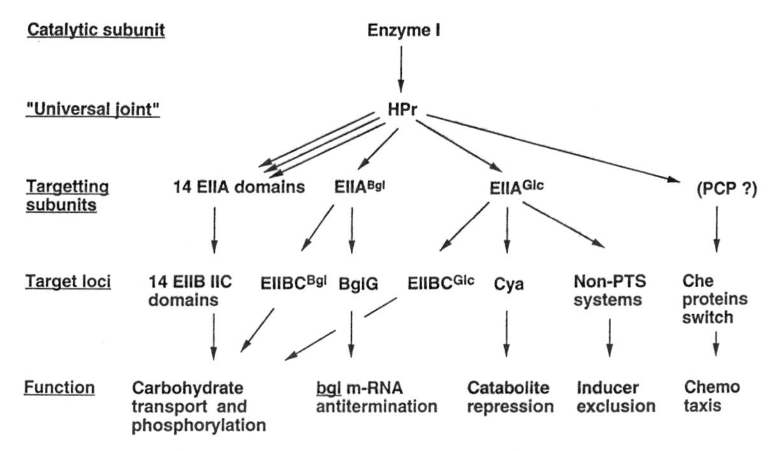

FIG. 1. The various functions of the PTS as a regulatory unit. The general protein kinase EI as the catalytic subunit is linked through the "universal joint" HPr and several EIIA proteins or domains ("targeting subunits") to the various target loci and functions indicated. All of these are discussed in the text. Not indicated is the indirect linkage of the fructose PTS through FruR to a series of glycolytic genes and enzymes (Postma et al., 1993; Saier, 1989).

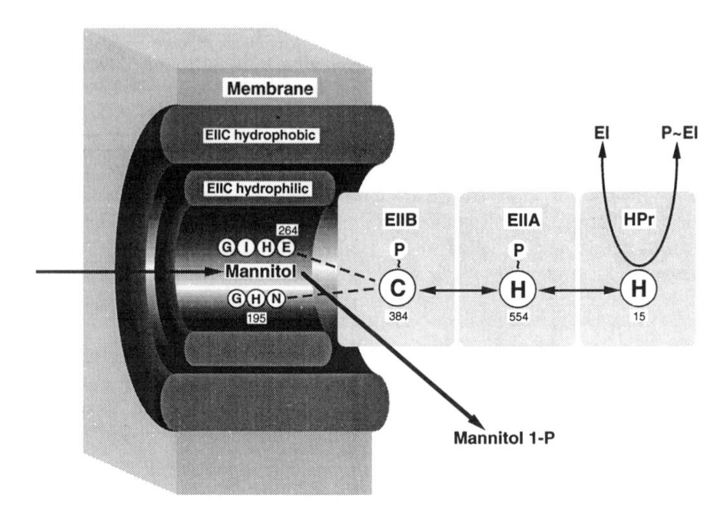

FIG. 2. Essential structural elements of the mannitol-specific PTS from *E. coli.* In addition to the catalytic kinase EI, the histidine protein HPr (phosphorylated at His-15), domains EIIA (with His-554), and EIIB (with Cys-384) are essential in substrate phosphorylation. The membrane-bound domain EIIC contains six (to eight) transmembrane helices, which form a hydrophobic tube, and hydrophilic loops involved in substrate binding and translocation. For further details, see the text.

Postma et al. [1993], and Robillard and Lolkema [1988] and references therein) are shown schematically in Fig. 2, as exemplified for the mannitol-specific EII of *E. coli.* Each EII comprises a membrane-bound channel domain (EIIC), which corresponds to an autonomous and substrate-specific facilitator. Six (to eight) transmembrane helices form, according to the model, a hydrophobic tube into which a large hydrophilic loop appears to be folded back (Lengeler, 1990). This loop contains a series of essential residues, e.g., several for the mannitol-specific PTS or EII^{Mtl} of *E. coli* indicated in Fig. 2, which seem to be part of the substrate-binding site. EIIC alone is able to bind all its substrates with normal affinity and specificity, but for high-affinity substrates it corresponds to a closed channel. Such substrates seem to convert the interior of the channel into a "locked" state, which prevents efficient carbohydrate flux. Low-affinity substrates, by contrast, seem unable to lock the channel and are transported by facilitated diffusion through an isolated EIIC channel (our unpublished results).

Under normal conditions, the membrane-bound EIIC domain forms a complex with a cytoplasmic domain EIIB, to which it is fused as a general rule. EIIB can be phosphorylated at a cysteine residue by way of the protein kinase EI, but this occurs only with the help of two ancillary proteins (targeting subunits). EI autophosphorylates at a histidine residue at the expense of PEP. It donates the phosphoryl group to a small histidine protein or HPr, from where the group is transferred to a second phospho-carrier protein, EIIA, in a fully reversible reaction. Similar to EI, HPr is a general protein, whereas EIIAs are normally sugar specific and are often fused to their preferred target EIIB. EIIAs, phosphorylated again at a histidine residue, finally phosphorylate the different EIIB domains. After crossing the membrane, substrates bound to the catalytic center of EIIC accept the phospho group from phospho-EIIB and are released into the cytoplasm as carbohydrate phosphates. It is not known whether phospho-EIIB unlocks EIIC, thus allowing substrate flux and consequent phosphorylation, or whether the phosphorylation of the substrate unlocks the channel. Permanently unlocked EII^{Glc} and EII^{Mtl} mutants have been isolated and invariably have drastically increased K_m values for their substrates (see Postma et al. [1993] and references therein). This result seems to favor the second possibility, i.e., that substrate phosphorylation unlocks the channel.

The more than 20 EIIs sequenced thus far can be grouped into at least four genetic families (Lengeler, 1990; Postma et al., 1993). For members of a family, e.g., the glucose family, well-conserved consensus sequences and a high heterologous suppression are characteristic, whereas members of different families, e.g., of the glucose and mannitol families, show only local similarities. Three-dimensional structures as determined by X-ray and two-dimensional nuclear magnetic resonance spectroscopy have become available for HPr, $EIIA^{Glc}$, and $EIIA^{Mtl}$ (see Kroon et al. [1993], Postma et al. [1993], and Stock [1993] and references therein). Although both EIIA domains interact with HPr and its essential His-15, it is difficult to find a common theme for the details of their active sites. In both EIIAs, however, the amino acids recognized before as members of the family specific consensus sequence (Lengeler et al., 1990) form a shallow hydrophobic depression on the surface of the molecule which surrounds the two histidine residues involved in phosphorylation. This surface is complementary to a corresponding structure of HPr containing the essential His-15 and Arg-17 residues, such that the phospho group of P ~ HPr can be buried at the interface and transferred from His-15 of HPr to, e.g., His-90 of $EIIA^{Glc}$. It is tempting to speculate that such a relatively nonspecific contact site constitutes a universal joint which allows the general targeting subunit HPr to communicate with dozens of structurally unrelated EIIA domains and perhaps of other proteins so far unidentified. Compared with the consensus sequences of HPr and EIIAs, which comprise basically structural and widely scattered residues, the consensus sequences of EIIB and EIIC are clearly more restricted (Lengeler et al., 1990), as expected for specific subunit interactions.

PTS-Mediated Control of Carbohydrate Metabolism and Protein Interactions

The complex role of the PTS in carbohydrate metabolism is also mediated by interactions of specific PTS proteins with various target proteins (Fig. 1), with the state of phosphorylation of the PTS targeting subunits controlling the regulatory output (see Postma et al. [1993], Roseman and Meadow [1990], Saier [1989], and Stock [1993], and references therein). Thus, the unphosphorylated $EIIA^{Glc}$ of enteric bacteria binds to and inhibits non-PTS permeases and enzymes involved in metabolism of non-PTS carbohydrates, whereas the phosphorylated form has no effect. For EIIA and glycerol kinase, a cocrystal of the two proteins has been isolated (see chapter 56). The results of this analysis show that phosphorylation of His-90 from $EIIA^{Glc}$ prevents formation of an inhibitory complex by introducing a bulky negatively charged group into a hydrophobic surface that binds glycerol kinase. $EIIA^{Glc}$ also modulates the activity of the enzyme adenylate cyclase, which synthesizes cAMP. The concerted action of "inducer exclusion" and of "catabolite repression" then causes cells to utilize one carbon source, e.g., glucose, in preference to others (diauxie) (Roseman and Meadow, 1990). A third mechanism controls induction of PTS-dependent metabolic pathways and the corresponding genes by transcription antitermination, e.g., in the β-glucoside PTS (*bgl* genes) in *E. coli* (Amster-Choder et al., 1992; Débarbouillé et al., 1991; Schnetz and Rak, 1990). The three domains EIIA, EIIB, and EIIC of EII^{Bgl} are fused into one large protein which forms, at the inner face of the membrane, a complex with an antiterminator protein BglG. In the presence of substrate, EII^{Bgl} phosphorylates the incoming β-glucoside molecules, leaving the antiterminator in a free, active form (induction). In the absence of substrate, EII^{Bgl} converts BglG to a phosphorylated, inactive form which is unable to prevent the constitutive mRNA termination of the *bgl* genes (repression).

Two new mechanisms essential in the communication of proteins become apparent in these systems. (i) $EIIA^{Glc}$ and its target proteins form a stoichiometric complex such that the number of proteins, their dissociation constants, and their biochemical modifications are important (see Postma et al. [1993], and Roseman and Meadow [1990] and references therein). (ii) Targeting subunits may diffuse within the cell and may thus transport information by bringing catalytic subunits to various target loci, sequestering them from others, etc. (Segall et al., 1985). For example, there is increasing evidence that EI through HPr and EIIA forms a complex with the membrane-bound EIIBC domains (Wu et al., 1990) or that promoter DNAs form a complex through attached regulator proteins with the corresponding membrane-bound receptors and transporter molecules to allow fast and specific regulation in response to extracellular stimuli (Island and Kadner, 1993). The presence of EIIs in which the different functional domains are fused compared with others in which the domains are autonomous proteins thus probably reflects sophisticated intersubunit interactions.

Signal Transduction during PTS-Dependent Chemotaxis

Most of the PTS functional domains (proteins) and all of the mechanisms mentioned thus far seem relevant for our understanding of the PTS as a chemosensor system in chemotaxis (see Postma et al. [1993], and Taylor and Lengeler [1990] and references therein). Many bacteria respond to the presence of PTS-carbohydrates by positive chemotaxis. Stimulation requires uptake and phosphorylation of a substrate through an EII. The intracellular accumulation of carbohydrate-phosphate alone is unable to trigger chemotactic stimulation, nor are extensive metabolism of these compounds and the generation of energy necessary for chemotaxis. Mutants lacking EI and HPr do not show a chemotactic response to any PTS-carbohydrate, even if the corresponding EIIs are present. Such unphosphorylated EIIs and mutant forms which cannot be phosphorylated are still able to bind their substrate with a normal efficiency but fail to signal. Finally, despite intensive searching, no mutations in EIIs which eliminate chemotaxis but keep transport and phosphorylation intact were found. This analysis included intergenic hybrids in which, for example, $EIICB^{Glc}$ from *E. coli* was combined with $EIIA^{Nag}$ from the nonmotile *Klebsiella pneumoniae* or with $EIIBC^{Scr}$ and $EIIA^{Nag}$, both from *K. pneumoniae*. All heterologous and hybrid EIIs active in transport were also active in chemotaxis, invariably showing the specificity of the EIIC domain. From these and other data reviewed recently (Taylor and Lengeler, 1990), it was concluded that the signal for PTS-dependent chemotaxis is a change in the phosphorylation level of EI and HPr as a consequence of substrate transport through an EII and, more precisely, the dephosphorylation of EI or of EI and of HPr. This signal consequently must be communicated to the central processing unit of the chemotaxis machinery through protein diffusion, complexing, and (de)phosphorylation.

Another major chemotaxis pathway (Fig. 3) in enteric bacteria involves methyl-accepting chemotaxis proteins (MCPs) as the membrane-bound sensors (see Bourret et al. [1991], Parkinson [1993], and Parkinson and Kofoid [1992] and references therein). MCPs bind but do not transport attractants and repellents. Upon stimulation, they transmit a signal through the cytoplasmic membrane to a cytoplasmic transmitter domain which communicates with the central processing unit of the chemotaxis machinery. Four general chemotaxis proteins, called CheA, CheW, CheY, and CheZ, together with the switch of the flagellar motor (proteins FliG, FliM, and FliN), form the processing unit. Its central elements are the two-component system modules CheA as transmitter and CheY as receiver. CheA is an ATP-dependent protein kinase which, in unstimulated cells, autophosphorylates slowly at a histidine residue. Positive MCP-dependent stimuli inhibit autophosphorylation of CheA, thus causing prolonged runs of the cells and a positive chemotaxis. Negative stimuli, however, cause binding of CheA to the transmitter domain of an MCP with the help of the targeting subunit CheW. This increases the phosphorylation of CheA, which in the next step transfers its phosphoryl group to an aspartate of CheY. Phospho-CheY in turn diffuses to the flagellar motor switch and locks it in the clockwise

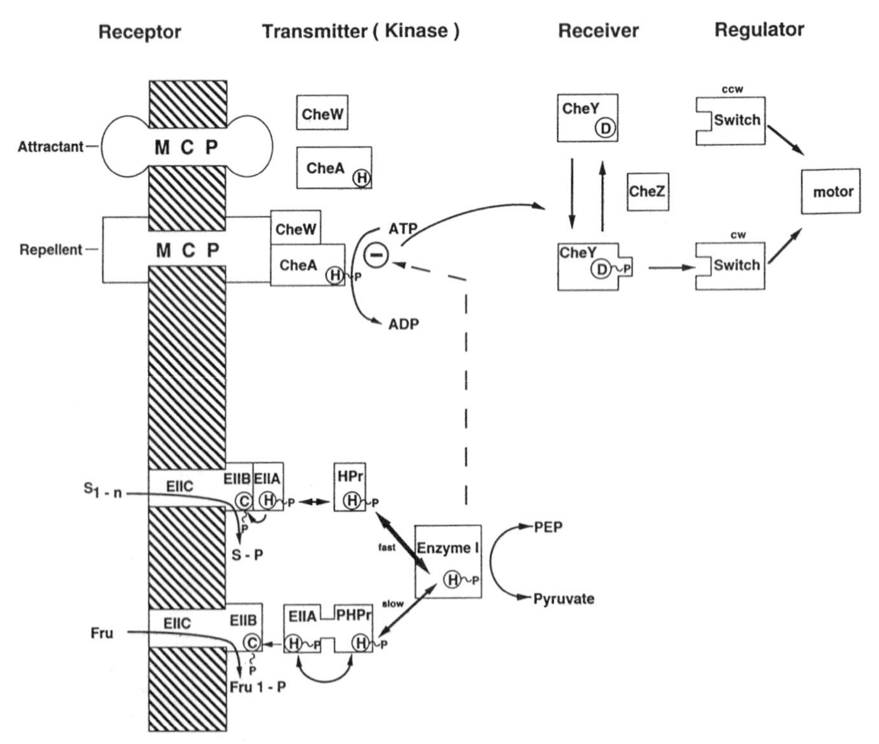

FIG. 3. Signal transduction pathways involved in the chemotaxis of enteric bacteria. The MCP-dependent and PTS-dependent signaling systems are presented schematically together with the central processing unit (formed by proteins CheA, CheW, CheY, and CheZ) and the flagellar motor switch. Phosphorylated amino acids involved in the process (for details, see the text) are indicated. Clockwise (cw) and counterclockwise (ccw) rotation of the flagellar motor, the former induced by phospho-CheY bound to the switch (FliG, FliM, FliN), are indicated. Abbreviations: S, substrate; S-P, substrate phosphate; Fru, D-fructose; Fru 1-P, D-fructose-1-phosphate. The dashed line indicates the postulated inhibition of CheA as the consequence of a rapid dephosphorylation of EI by HPr during PTS-dependent transport and phosphorylation.

or tumble position, thus causing negative chemotaxis. Dephosphorylation of phospho-CheY, most probably through its antagonist CheZ, unlocks the switch. Any decrease in the pool of phospho-CheY, through either inhibition of CheA phosphorylation or stimulation of CheZ, thus corresponds to a positive signal.

Mutants lacking all MCPs retain normal PTS-dependent chemotaxis. In a strain deleted for all the genes encoding MCPs, the four Che proteins of the processing unit, and two additional proteins (CheB and CheR) involved in adaptation to MCP-dependent stimuli, reintroduction of CheA, CheY, and CheW, but not of CheZ, is sufficient to restore PTS-dependent chemotaxis (Taylor and Lengeler, 1990). The former three proteins are most probably needed only to restore tumbling and a change in the direction of swimming. CheZ, however, might be dispensable because positive PTS-generated stimuli modulate phospho-CheY pools in a CheZ-independent way. During uptake of PTS-carbohydrates in an HPr-dependent process, enteric bacteria show a positive chemotaxis. The reaction does not require either EIIA^{Glc} (except for chemotaxis to glucose and sucrose through EIICB^{Glc} and EIIBC^{Scr}, respectively), adenylate cyclase, the cAMP receptor protein, cAMP, or cGMP. Mutants deleted for the corresponding genes retain positive chemotaxis provided the flagella and the EIIs are synthesized (Grübl et al., 1990). Acetyl phosphate is also not required for the PTS-dependent chemotaxis (Dailey and Berg, 1993). This phospho-group donor has been shown to phosphorylate CheY in vitro and might phosphorylate EI in vivo, but according to the results of Dailey and Berg (1993), a role of acetyl phosphate in PTS-dependent chemotaxis seems to be ruled out.

Positive chemotaxis does apparently require fast dephosphorylation of EI. Mutants which lack HPr can be suppressed to a Pts$^+$ phenotype through constitutive overexpression of *fruF*, encoded in the *fru* operon (Fig. 3). This gene codes for a complex protein, FPr (376 amino acids), in which EIIA^{Fru} has been fused to an HPr-like domain (or pseudo-HPr). FPr proteins from *E. coli* and *Salmonella typhimurium* resemble each other closely (90% identical residues [our unpublished results]), as do HPr and pseudo-HPr (33% identical residues). Suppression of *ptsH* mutants by FPr (cloned on high-copy-number vectors) for growth on and transport or phosphorylation of PTS-carbohydrates yields activities above normal (170%) (Fig. 4). The suppressed strains do not show PTS-dependent chemotaxis; i.e., transport/phosphorylation and chemotactic response have been uncoupled. Even D-fructose (Fru), the natural substrate, does not cause chemotaxis during uptake through the fructose PTS. This behavior is reminiscent of mutants with a Pro-11→Glu exchange in HPr. The Pro-11→Glu mutants showed a small decrease (about twofold) in transport activity but no chemotactic response (Grübl et al., 1990). The defect decreased primarily the rephosphorylation of the mutated HPr by EI. Under normal conditions, autophosphorylation of EI at the expense of PEP is the rate-limiting step during uptake of a substrate through a PTS, because wild-type HPr is rephosphorylated rapidly by phospho-EI. This rephosphorylation, as measured by using purified proteins, is reduced in the Pro-11→Glu mutant (Grübl et al., 1990) and in FPr (our unpublished results) to a level still sufficient for (nearly) normal transport and phosphorylation but not sufficient to trigger a chemotactic reaction. Overexpression of the mutant HPr restored chemotaxis, whereas overexpression of FPr (from the isolated *fruF* gene cloned behind a *lacZp* promoter on a multicopy vector) did not. This difference could reflect the possibility that the increased mass of FPr (376 residues compared with 85 for HPr) prevents its efficient diffusion from EIIs to the flagellar motor switch. In agreement with this hypoth-

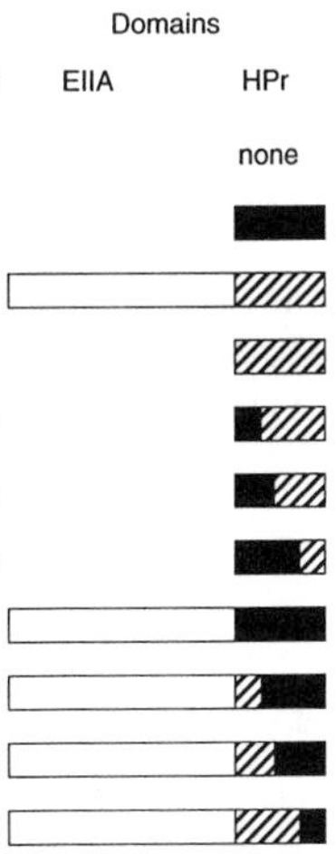

Plasmids	Domains: EIIA	Domains: HPr	Che	Growth	Transport
none		none	-	-	0
pHPr			3 +	3 +	18
pFPr			-	3 +	31
pPHPr			2 +	3 +	27
pHFPr17			-	2 +	8
pHFPr33			-	-	0
pHFPr54			-	-	0
pFHPr1			(-)	2 +	12
pFHPr17			-	-	0
pFHPr33			-	-	0
pFHPr54			-	-	0

FIG. 4. Activity of various HPr and FPr derivatives in PTS-dependent chemotaxis. JLV92, a *ptsH5* mutant of *E. coli* lacking HPr, was transformed with the plasmids indicated and tested for PTS-dependent activities. The high-copy-number plasmid pUC18/19 was used to overexpress from a *lacZp* promoter the various domains indicated. These are *ptsH* and HPr from *E. coli*, *fruF* and FPr from *S. typhimurium*, the HPr-like domain or pseudo-HPr (PHPr) from FPr, and HPr-FPr hybrids fused at residue 17, 33, or 54 or FPr-HPr hybrids as indicated. Chemotaxis toward mannitol and *N*-acetylglucosamine was tested on swarm plates and in capillary tube tests (−, negative, to 3+, strong reaction); growth for the same carbon sources was tested on McConkey indicator plates (1%) and in minimal media (0.2%) (−, no growth, to 3+, 55 min generation time); and transport for [^{14}C]mannitol (5 μM) is given in nanomoles per minute per milligram of protein.

esis, it could be shown that overexpression of pseudo-HPr, after deletion of the IIA^{Fru} domain, restored PTS chemotaxis to near normal levels (Fig. 4), whereas an FPr hybrid, in which a complete HPr replaces pseudo-HPr, showed a drastically reduced chemotactic response. On the other hand, it is also possible that the EIIA^{Fru} domain of FPr masks (by steric hindrance) a hypothetical site on HPr (and on pseudo-HPr) that does specifically interact with the next protein in the signal transduction pathway necessary for PTS taxis. The result corroborates previous conclusions that the phosphorylation and dephosphorylation levels of EI and HPr required for efficient transport are lower than those required for efficient chemotaxis. It has yet to be determined if and how positive stimulation through the PTS modulates the level of phospho-CheY and/or the flagellar motor switch. Preliminary results from direct in vitro tests involving purified Che and PTS proteins indicate an inhibition of the CheA protein kinase activity by nonphosphorylated EI (our unpublished results). According to these results and the results discussed above, any conditions which cause a net and transient dephosphorylation of EI are equivalent to a positive signal in PTS taxis, which acts by modulating the CheA kinase activity. To understand the system, we must describe in quantitative terms the stoichiometric interactions which are at the basis of the communication between two complex signal transduction systems, the PTS and the two-component chemotaxis system.

We thank R. Eisermann (Bochum) and P. W. Postma (Amsterdam) for the generous gift of plasmids, C.-A. Alpert and H. Heuel for help with drawings, and E. Placke for help in editing the manuscript.

We thank the Deutsche Forschungsgemeinschaft for financial support through Sonderforschungsbereich 171, Teilprojekt C3.

REFERENCES

Amster-Choder, O., F. Houman, and A. Wright. 1992. Modulation of the dimerization of a transcriptional antiterminator protein by phosphorylation. *Science* **257:**1395–1398.

Bourret, R. B., K. A. Borkovich, and M. I. Simon. 1991. Signal transduction pathways involving protein phosphorylation in prokaryotes. *Annu. Rev. Biochem.* **60:**401–441.

Chuang, S., D. L. Daniels, and F. R. Blattner. 1993. Global regulation of gene expression in *Escherichia coli*. *J. Bacteriol.* **175:**2026–2036.

Dailey, F. E., and H. C. Berg. 1993. Change in direction of flagellar rotation in *Escherichia coli* mediated by acetate kinase. *J. Bacteriol.* **175:**3236–3239.

Débarbouillé, M., I. Martin-Verstraete, M. Arnaud, A. Klier, and G. Rapoport. 1991. Positive and negative regulation controlling expression of the *sac* genes in *Bacillus subtilis*. *Res. Microbiol.* **142:**757–764.

Grübl, G., A. P. Vogler, and J. W. Lengeler. 1990. Involvement of the histidine protein (HPr) of the phosphotransferase system in chemotactic signaling of *Escherichia coli* K-12. *J. Bacteriol.* **172:**5871–5876.

Guest, J. R. 1992. Oxygen-regulated gene expression in *Escherichia coli*. *J. Gen. Microbiol.* **138:**2253–2263.

Hubbard, M. J., and P. Cohen. 1993. On target with a new mechanism for the regulation of protein phosphorylation. *Trends Biochem. Sci.* **18:**172–177.

Island, M. D., and R. J. Kadner. 1993. Interplay between the membrane-associated UhpB and UhpC regulatory proteins. *J. Bacteriol.* **175:**5028–5034.

Iuchi, S., and E. C. C. Lin. 1988. *arcA* (*dye*), a global regulatory gene in *Escherichia coli* mediating repression of enzymes in aerobic pathways. *Proc. Natl. Acad. Sci. USA* **85:**1888–1892.

Kroon, G. J. A., J. Grötzinger, K. Dijkstra, R. M. Scheek, and G. T. Robillard. 1993. Backbone assignments and secondary structure of the *Escherichia coli* enzyme-II mannitol A domain determined by heteronuclear three-dimensional NMR spectroscopy. *Protein Sci.* **2:**1331–1341.

Lengeler, J. W. 1990. Molecular analysis of the Enzyme II-complexes of the bacterial phosphotransferase system (PTS) as carbohydrate transport systems. *Biochim. Biophys. Acta* **1018:**155–159.

Lengeler, J. W., F. Titgemeyer, A. P. Vogler, and B. M. Wöhrl. 1990. Structure and homologies of carbohydrate:phosphotransferase system (PTS) proteins. *Philos. Trans. R. Soc. London Ser. B* **326:**489–504.

Meadow, N. D., D. K. Fox, and S. Roseman. 1990. The bacterial phosphoenolpyruvate:glycose phosphotransferase system. *Annu. Rev. Biochem.* **59:**497–542.

Neidhardt, F. C., J. L. Ingraham, and M. Schaechter. 1987. *Physiology of the Bacterial Cell.* Sinauer Associates, Inc. Sunderland, Mass.

Parkinson, J. S. 1993. Signal transduction schemes of bacteria. *Cell* **73:**857–871.

Parkinson, J. S., and E. C. Kofoid. 1992. Communication modules in bacterial signaling proteins. *Annu. Rev. Genet.* **26:**71–112.

Postma, P. W., J. W. Lengeler, and G. R. Jacobson. 1993. Phosphoenolpyruvate:carbohydrate phosphotransferase systems of bacteria. *Microbiol. Rev.* **57:**543–594.

Robillard, G. T., and J. S. Lolkema. 1988. Enzymes II of the phosphoenolpyruvate-dependent sugar transport systems: a review of their structure and mechanism of sugar transport. *Biochim. Biophys. Acta* **947:**493–519.

Roseman, S., and N. D. Meadow. 1990. Signal transduction by the bacterial phosphotransferase system. *J. Biol. Chem.* **265:**2993–2996.

Saier, M. H., Jr. 1989. Protein phosphorylation and allosteric control of inducer exclusion and catabolite repression by the bacterial phosphoenolpyruvate:sugar phosphotransferase system. *Microbiol. Rev.* **53:**109–120.

Schnetz, K., and B. Rak. 1990. β-Glucoside permease represses the *bgl* operon of *Escherichia coli* by phosphorylation of the antiterminator protein and also interacts with glucose-specific enzyme III^{Glc}, the key element in catabolite control. *Proc. Natl. Acad. Sci. USA* **87:**5074–5078.

Segall, J. E., A. Ishihara, and H. C. Berg. 1985. Chemotactic signalling in filamentous cells of *Escherichia coli*. *J. Bacteriol.* **161:**51–59.

Stock, J. 1993. Phosphoprotein talk. *Curr. Biol.* **3:**303–305.

Stock, J. B., A. J. Ninfa, and A. M. Stock. 1989. Protein phosphorylation and regulation of adaptive responses in bacteria. *Microbiol. Rev.* **53:**450–490.

Taylor, B. L., and J. W. Lengeler. 1990. Transductive coupling by methylated tranducing proteins and permeases of the phosphotransferase system in bacterial chemotaxis, p. 69–90. *In* R. C. Aloia, C. C. Curtain, and L. M. Gordon (ed.), *Membrane Transport and Information Storage.* Wiley Liss, New York.

Wanner, B. L. 1992. Is cross regulation by phosphorylation of two-component response regulator proteins important in bacteria? *J. Bacteriol.* **174:**2053–2058.

Wu, L.-F., J. M. Tomich, and M. H. Saier, Jr. 1990. Structure and evolution of a multidomain multiphosphoryl transfer protein. Nucleotide sequence of the *fruB(HI)* gene in *Rhodobacter capsulatus* and comparisons with homologous genes from other organisms. *J. Mol. Biol.* **213:**687–703.

Chapter 31

The Bacterial Phosphotransferase System: a Multifaceted Regulatory System Controlling Carbon and Energy Metabolism

MILTON H. SAIER, JR., JING JING YE, TOM M. RAMSEIER, FRIEDRICH TITGEMEYER, AND JONATHAN REIZER

Department of Biology, University of California at San Diego, La Jolla, California 92093-0116

Gram-negative and gram-positive bacteria exhibit major differences in the preferred metabolic pathways used for carbon catabolism and in the regulatory mechanisms used for controlling these pathways (Reizer and Peterkofsky, 1987; Steinmetz, 1993). However, representatives of both groups of organisms possess phosphoenolpyruvate (PEP)-dependent sugar phosphotransferase systems (PTS), which, though differing in detail, are similar in general properties and mechanism of action (Reizer et al., 1988b). Moreover, within each of these two major classifications of bacteria there exist major differences among the species. For example, most low-G+C gram-positive bacteria possess a complete PTS including enzyme I that phosphorylates HPr on histidine 15 and a set of sugar-specific enzymes II, as well as an HPr kinase and HPr(Ser-P) phosphatase that are lacking in gram-negative bacteria. *Lactobacillus brevis,* on the other hand, has only HPr and the kinase and phosphatase that reversibly phosphorylate it on serine 46 (Reizer et al., 1988a, 1988b), whereas *Acholeplasma laidlawii* has these three proteins as well as enzyme I (Hoischen et al., 1993). Both *L. brevis* and *A. laidlawii* lack detectable enzyme II activities, and thus, the primary function of the PTS in these organisms appears to be regulation of non-PTS functions and not sugar transport.

Among the most pronounced metabolic and transcriptional differences observed for different bacterial genera are those dealing with the mechanisms by which carbon and energy metabolism is regulated. Thus, the heat-stable glucose-specific enzyme IIA (enzyme IIA^{Glc}) in *Escherichia coli* serves as the central regulatory protein to control non-PTS carbohydrate utilization both at the transcriptional level (by controlling cytoplasmic cyclic AMP and inducer levels) and at the metabolic level (by allosterically controlling the activities of various non-PTS transport systems and catabolic enzymes) (Saier, 1989, 1993). Another heat-stable protein, HPr(Ser-P), serves this function in *L. brevis* (Ye et al., 1994a). In *E. coli,* the cyclic AMP receptor protein (CRP) controls catabolite repression of many target genes, particularly those initiating sugar catabolism (Botsfield and Harman, 1992), whereas the fructose repressor, FruR (sometimes together with CRP, sometimes independently), appears to determine sensitivity to catabolite repression of key enzymes involved in central pathways of carbon metabolism (Ramseier et al., 1993; Ramseier, unpublished). A FruR homolog, CcpA, is believed to play an analogous role in *Bacillus subtilis* [sometimes together with, but sometimes independently of, HPr(Ser-P)] (Hueck et al., in press; Deutscher et al., 1994; Saier et al., 1992).

Many observations suggest that our current knowledge only begins to scratch the surface of the problem concerning the involvement of the PTS in regulation. Thus, for example, the PTS appears to phosphorylate proteins that are homologous to the IIA^{Fru} and IIA^{Mtl} proteins (Reizer et al., 1992), and these homologous proteins apparently function in conjunction with σ^{54} holo-RNA polymerase to regulate transcription of nitrogen utilization operons. Moreover, novel proteins of the PTS and novel transcriptional regulatory proteins controlling PTS gene expression are being discovered at a bewildering rate.

Recent progress in understanding the molecular details of how the PTS, a multifaceted regulatory system, controls the activities and syntheses of (i) its own protein components, (ii) transport systems and catabolic enzymes initiating non-PTS carbon source catabolism, and (iii) central pathways of carbon metabolism is summarized in this brief update of the topic.

Novel Proteins of the PTS

Many years ago we reported the presence in *Salmonella typhimurium* of a fructose-inducible HPr protein, FPr, that could substitute for HPr (Saier et al., 1970, 1976; Sutrina et al., 1988; Geerse et al., 1989). More recently we identified a cryptic enzyme I (I* or I^{Fru}), which, when expressed as a result of a mutation in a gene designated *ptsJ*, could substitute for enzyme I (Chin et al., 1987b). This *ptsJ* gene was mapped by transduction to a region between the *ptsHIcrr* operon and the *cysPTWAM* operon at min 49 on the *Salmonella* chromosome (Chin et al., 1987b). Recently we have subcloned and sequenced this region and have identified an open reading frame (ORF) that codes for a putative transcriptional regulatory protein of the GntR family (Buck and Guest, 1989; Reizer et al., 1991). PtsJ exhibits most striking sequence similarity (35% amino acid identity) to an ORF in *Rhodobacter sphaeroides* (Neidle and Kaplan, 1992), which probably regulates transcription of photosynthetic genes (Titgemeyer, unpublished). Moreover, sequencing downstream of the previously sequenced portion of the mannitol (*mtl*) operon of *E. coli* (Figge et al., 1994) led to the discovery of a gene homologous to a single ORF (Urf2A) that maps upstream of the *gapB* gene in a cluster of glycolytic genes in *E. coli* (Alefounder and Perham, 1989). When the former ORF is inactivated by insertional mutagenesis, the *mtl* operon is expressed constitutively (Figge et al., 1994). We therefore tentatively designate this gene *mtlR*. Upstream of both the *mtlA* and *gapB* operons is a DNA consensus sequence for FruR binding, suggesting that FruR, in conjunction with each of these two homologous putative transcriptional regulators (MtlR and Urf2A, respectively), may function cooperatively in repression of their respective target operons. We believe that MtlR and Urf2A represent the matriarch and patriarch of a novel family of transcriptional regulators.

Recently, we specifically inactivated the *S. typhimurium* trifunctional gene *fruB(MH)*, which encodes the protein bearing a fructose-specific enzyme IIA domain (B), a putative modulator domain (M), and an HPr domain (H). This mutation prevents utilization of fructose at low sugar concentrations. Secondary mutations in this genetic background give rise to pseudorevertants with restored fructose utilization (Vartak and Michotey, unpublished). Activation of a cryptic PTS encoded by genes that are normally silent might be responsible for the restored fructose utilization activity. In this connection, and as a result of the *E. coli* genome sequencing project, three novel PTS operons, two coding for enzyme II complexes similar in sequence to the fructose-mannitol systems and the third similar in sequence to the β-glucoside–sucrose systems, have been identified (Reizer et al., in press). The putative enzyme II proteins encoded by these ORFs exhibit unique characteristics that distinguish them from homologous proteins. Their functions remain to be established, but their discovery suggests the presence of additional structural and regulatory proteins of the PTS which will eventually allow us to expand upon the currently known repertoire of PTS functions.

Regulation of Non-PTS Permeases and Enzymes by Enzyme IIA^{Glc} in Enteric Bacteria

In an early report, Saier et al. (1978) described mutants of *S. typhimurium* and *E. coli* which specifically rendered the maltose, melibiose, glycerol, and lactose uptake systems resistant to PTS-mediated regulation, and they mapped the relevant mutations. Recently these and other mutant genes have been sequenced, identifying the regions in target maltose, melibiose, and lactose permeases that render them resistant to PTS-mediated regulation (Dean et al., 1990; Wilson et al., 1990; Kühnau et al., 1991; Kuroda et al., 1992). The mutations in LacY were found in its central cytoplasmic loop (Wilson et al., 1990). We have extended the studies on the lactose permease in collaboration with H. R. Kaback's group by demonstrating that several insertional and point mutations in the central loop abolish both PTS regulation and binding of Enzyme IIA^{Glc} to the LacY protein (Hoischen et al., unpublished). Simple assays for the direct binding of enzyme IIA^{Glc} to wild-type and mutant LacY proteins that do not depend on the transport activity of LacY have recently been developed (Titgemeyer and Saier, unpublished).

LacY, a member of the major substrate/proton cotransport facilitator superfamily (Marger and Saier, 1993), shows greatest sequence similarity to the plasmid-encoded raffinose and sucrose permeases of *E. coli*. We have demonstrated that the raffinose permease (RafB) is regulated by enzyme IIA^{Glc} (Titgemeyer et al., 1994). Sequence similarity between the putative binding regions of LacY, RafB, and MalK allowed us to postulate a consensus sequence for enzyme IIA^{Glc} binding. This sequence is lacking in the sucrose permease (CscB) as a result of two microdeletions in the loop region of this permease (Titgemeyer et al., 1994). The consensus sequence was identified in arabinose isomerase of *E. coli*, and inhibition of this enzyme by purified enzyme IIA^{Glc} was demonstrated (Hoischen, unpublished).

Regulation of TMG:H^+ Symport by HPr(Ser-P) via the Lactose Permease in *L. brevis*

In *L. brevis*, lactose is taken up by proton symport (Romano et al., 1987). Indirect in vivo and in vitro evidence has suggested that the permease might be regulated by the fructose-1,6-diphosphate (FDP) and gluconate-6-phosphate (Gnt-6-P)-activated HPr kinase-dependent phosphorylation of serine 46 in HPr [HPr(Ser-P)] (Romano et al., 1987; Reizer et al., 1988a). Recently, Ye et al. (1994a) developed a vesicle system derived from *L. brevis* cells for studying thiomethyl β-galactoside (TMG) uptake and its regulation by the PTS, using PTS proteins from *B. subtilis*, purified by X. Cui and J. Reizer. The results establish that HPr(Ser-P) and its S46D mutant analog, but not free HPr or the S46A mutant HPr, inhibit [^{14}C]TMG accumulation via this permease. Representative results are presented in Table 1.

Vesicles provided with an energy source (arginine) accumulated [^{14}C]TMG against a large concentration gradient, and, in contrast to intact cells, the presence of glucose or FDP was not inhibitory. The presence of HPr outside the vesicles had no effect, regardless of whether glucose or FDP was present. However, when wild-type HPr was electroporated into the vesicles, TMG uptake was strongly inhibited by the presence of glucose. FDP also inhibited uptake, provided that it was electroporated into the vesicles together with HPr. When the S46A mutant HPr was electroporated into the vesicles, no inhibition by glucose or FDP was seen, and the S46D mutant HPr was inhibitory regardless of the presence or absence of glucose or FDP (Table 1). Control experiments revealed that only in the presence of glucose or intravesicular FDP was wild-type (but not the S46A mutant) HPr phosphorylated on serine 46. These results clearly indicate that intravesicular HPr(Ser-P) or the S46D mutant protein, but not free HPr or HPr(S46A), inhibits TMG uptake. We propose an allosteric regulatory mechanism analogous to that by which enzyme IIA^{Glc} inhibits the lactose permease in *E. coli*.

Regulation of Phosphorylation-Coupled TMG Uptake and TMG-P Phosphatase Activities by HPr(Ser-P) in *Lactococcus lactis*

TMG uptake-coupled phosphorylation via the PTS and cytoplasmic TMG-P hydrolysis followed by TMG explusion are thought to be regulated by a mechanism involving HPr(Ser) phosphorylation (Reizer and Panos, 1980; Thompson and Saier, 1981; Reizer et al., 1983). In experiments analogous to those described above for *L. brevis*, vesicles of *L. lactis* were prepared, and the PEP-dependent uptake of TMG was studied (Ye et al., 1994b). Results were similar to those reported in Table 1 for the *L. brevis* system, except that, as expected, wild-type HPr or the S46A mutant protein substantially stimulated TMG uptake. Glucose strongly inhibited TMG accumulation, as did intravesicular (but not extravesicular) FDP. The inhibitory effects of glucose and FDP were abolished when HPr(S46A) was used instead of wild-type HPr, and HPr(S46D) inhibited accumulation in both the presence and absence of glucose or FDP. A process involving inhibition

TABLE 1. Inhibition of TMG/H^+ symport in vesicles prepared from *L. brevis* cells: involvement of HPr(Ser) phosphorylation

HPr (100 μM)	Arginine (20 mM)[a]	Glucose (5 mM)	FDP (5 mM)[b]	TMG uptake (nmol/mg of protein after 15 min)
None	−	−	−	1.2
	+	−	−	13.2
	+	+	−	9.6
	+	−	+ (inside)	10.9
w.t. HPr[c] (outside)	−	−	−	1.3
	+	−	−	10.2
	+	+	−	9.4
	+	−	+ (inside)	10.1
w.t. HPr (inside)	−	−	−	1.1
	+	−	−	11.7
	+	+	−	2.8
	+	−	+ (outside)	11.2
	+	−	+ (inside)	1.4
S46D HPr (inside)	−	−	−	1.0
	+	−	−	4.3
	+	+	−	2.4
S46A HPr (inside)	−	−	−	1.4
	+	−	−	10.8
	+	+	−	9.8

[a]L-Arginine provided the source of energy for TMG uptake in right-side-out vesicles.

[b]FDP is an allosteric activator of the HPr(Ser) kinase that phosphorylates wild-type HPr on serine 46.

[c]w.t., wild type.

of uptake as well as stimulation of efflux was suggested. The results showed that TMG accumulation via this PTS-dependent system is controlled by HPr(Ser) phosphorylation. Taken together with the results of the previous section, it therefore appears that phosphorylation of serine 46 in HPr by the metabolite-activated, ATP-dependent protein kinase regulates accumulation of β-galactosides involving both PTS and non-PTS mechanisms.

Transcriptional Regulation of Genes Encoding Central Enzymes of Carbon Metabolism by FruR, the Fructose Repressor in Enteric Bacteria

Previous studies in our laboratory have suggested that the fructose repressor of enteric bacteria functions in catabolite repression of central enzymes of carbohydrate metabolism (Chin et al., 1987a, 1989; Saier and Chin, 1990). Collaboratively with A. Cozzone's group in France, T. Ramseier has conducted experiments leading to the conclusion that the fructose repressor, FruR, binds directly to the upstream regulatory regions of several operons concerned with carbon and energy metabolism and that it thereby activates or represses these genes (Ramseier et al., 1993; Cortay et al., in press). Among the operons to which FruR bound were the *fru*, *pps*, *ace*, *icd*, and *pts* operons of *E. coli* and/or *S. typhimurium* (Ramseier et al., 1993). A consensus sequence for FruR binding was derived, and this sequence was found upstream of several operons concerned with carbon metabolism. These operons included the *zwf*, *edd-eda*, *pykF*, and *gapB* operons (Ramseier and Bledig, unpublished). In vivo experiments suggested that a number of genes not previously known to be regulated by FruR are in fact targets of this protein. All such genes appear to be involved in carbon and energy metabolism. It appears that glycolysis, gluconeogenesis, the Krebs cycle, the glyoxalate shunt, the pentose phosphate pathway, the Entner-Doudoroff pathway and electron transfer may all be regulated by FruR. Moreover, mutants of *S. typhimurium* lacking FruR have been found to be virtually avirulent in mice (Saier and Chin, 1990). The results clearly show that FruR is a principal pleiotropic regulatory protein controlling the direction and pathway of carbon flux in enteric bacteria.

Transcriptional Regulation of Genes Encoding Carbohydrate Catabolic Enzymes by CcpA and HPr(Ser-P) in *B. subtilis*

In *B. subtilis*, a gram-positive bacterium lacking cyclic AMP, a protein homologous to FruR, the CcpA protein (Henkin et al., 1991), is believed to mediate catabolite repression (see Saier [1991] for a review). Although the direct binding of CcpA to its target DNA sequences has not been demonstrated, several target catabolite repression element (CRE) sequences, when altered, have been shown to render specific *Bacillus* operons resistant to catabolite repression (Henkin et al., 1991; Hueck et al., in press). Additionally, replacement of wild-type chromosomally encoded HPr with HPr(S46A) results in the loss of catabolite repression of some operons (*gnt*, *mtl*, and *gut*) but not others (*xyl* and *glp*) (Deutscher et al., 1994; see Saier et al., 1992). It appears that these two proteins, sometimes together, sometimes independently, determine sensitivity to catabolite repression, but the precise molecular mechanisms are unknown.

Conclusion

The results summarized above indicate the existence of mechanisms whereby PTS proteins and auxiliary PTS proteins regulate carbon and energy metabolism by directly regulating gene transcription and the activities of various transport systems and enzymes. The recent discovery of novel PTS proteins as well as recent biochemical and physiological experiments indicate that a plethora of PTS-mediated regulatory mechanisms may be operative in bacteria as dissimilar as *E. coli, B. subtilis, L. brevis, S. pyogenes, L. lactis,* and *A. laidlawii.*

We acknowledge the valuable contributions of several of our laboratory co-workers whose advances have been mentioned in the text of this review: Stefan Bledig, Mike Chin, Xuewen Cui, Josef Deutscher, Thien Dinh, Christian Hoischen, Christoph Hueck, Gary Jacobson, Romy Mason, Valerie Michotey, Pete Pitaknarongphorn, Aiala Reizer, Tony Romano, Sarah Sutrina, and Narendra Vartak. We thank Mary Beth Hiller for invaluable assistance in the preparation of the manuscript.

Work in our laboratory was supported by Public Health Service grants 5RO1AI 21702 and 2RO1AI 14176 from the National Institute of Allergy and Infectious Diseases.

REFERENCES

Alefounder, P. R., and R. N. Perham. 1989. Identification, molecular cloning and sequence analysis of a gene cluster encoding the class II fructose 1,6-bisphosphate aldolase, 3-phosphoglycerate kinase and a putative second glyceraldehyde 3-phosphate dehydrogenase of *Escherichia coli*. *Mol. Microbiol.* **3:**723–732.

Botsford, J. L., and J. G. Harman. 1992. Cyclic AMP in prokaryotes. *Microbiol. Rev.* **56:**100–122.

Buck, D., and J. R. Guest. 1989. Overexpression and site-directed mutagenesis of the succinyl-CoA synthetase of *Escherichia coli* and nucleotide sequence of a gene (*g30*) that is adjacent to the *suc* operon. *Biochem. J.* **260:**737–747.

Chin, A. M., D. A. Feldheim, and M. H. Saier, Jr. 1989. Altered transcriptional patterns affecting several metabolic pathways in strains of *Salmonella typhimurium* which overexpress the fructose regulon. *J. Bacteriol.* **171:**2424–2434.

Chin, A. M., B. U. Feucht, and M. H. Saier, Jr. 1987a. Evidence for the regulation of gluconeogenesis by the fructose phosphotransferase system in *Salmonella typhimurium*. *J. Bacteriol.* **169:**897–899.

Chin, A. M., S. Sutrina, D. A. Feldheim, and M. H. Saier, Jr. 1987b. Genetic expression of enzyme I* activity of the phosphoenolpyruvate:sugar phosphotransferase system in *ptsHI* deletion strains of *Salmonella typhimurium*. *J. Bacteriol.* **169:**894–896.

Cortay, J.-C., D. Nègre, M. Scarabel, T. M. Ramseier, N. B. Vartak, J. Reizer, M. H. Saier, Jr., and A. J. Cozzone. Regulation of the acetate operon: overproduction, purification and characterization of the FruR protein. *J. Biol. Chem.*, in press.

Dean, D. A., J. Reizer, H. Nikaido, and M. H. Saier, Jr. 1990. Regulation of the maltose transport system of *Escherichia coli* by the glucose-specific Enzyme III of the PTS: characterization of inducer exclusion-resistant mutants and reconstitution of inducer exclusion in proteoliposomes. *J. Biol. Chem.* **265:**21005–21010.

Deutscher, J., J. Reizer, C. Fischer, A. Galinier, M. H. Saier, Jr., and M. Steinmetz. 1994. Loss of protein kinase catalyzed phosphorylation of HPr, a phosphocarrier protein of the phosphotransferase system, by mutation of the *ptsH* gene confers catabolite repression resistance to several catabolic genes of *Bacillus subtilis*. *J. Bacteriol.* **176:**3336–3344.

Figge, R. M., T. M. Ramseier, and M. H. Saier, Jr. 1994. The mannitol repressor (MtlR) of *Escherichia coli*. *J. Bacteriol.* **176:**840–847.

Geerse, R. H., F. Izzo, and P. W. Postma. 1989. The PEP:fructose phosphotransferase system in *Salmonella typhimurium:* FPr combines Enzyme III^{fru} and pseudo-HPr activities. *Mol. Gen. Genet.* **216:**517–525.

Henkin, T. M., F. J. Grundy, W. L. Nicholson, and G. H. Chambliss. 1991. Catabolite repression of α-amylase gene expression in *Bacillus subtilis* involves a *trans*-acting gene product homologous to the *Escherichia coli lacI* and *galR* repressors. *Mol. Microbiol.* **5:**575–584.

Hoischen, C. Unpublished results.

Hoischen, C., J. Reizer, A. Dijkstra, S. Rottem, and M. H. Saier, Jr. 1993. Presence of protein constituents of the gram-positive bacterial phosphotransferase regulatory system in *Acholeplasma laidlawii*. *J. Bacteriol.* **175:**6599–6604.

Hoischen, C., F. Titgemeyer, T. Dinh, and S. Pitaknarongphorn. Unpublished results.

Hueck, C., W. Hillen, and M. H. Saier, Jr. Analysis of a *cis*-active sequence mediating catabolite repression in gram-positive bacteria. *Res. Microbiol.*, in press.

Kühnau, S., M. Reyes, A. Sievertsen, H. A. Shuman, and W. Boos. 1991. The activities of the *Escherichia coli* MalK protein in maltose transport, regulation, and inducer exclusion can be separated by mutations. *J. Bacteriol.* **173:**2180–2186.

Kuroda, M., S. de Waard, K. Mizushima, M. Tsuda, P. Postma, and T. Tsuchiya. 1992. Resistance of the melibiose carrier to inhibition by the phosphotransferase system due to substitutions of amino acid residues in the carrier of *Salmonella typhimurium*. *J. Biol. Chem.* **267:**18336–18341.

Marger, M. D., and M. H. Saier, Jr. 1993. A major superfamily of transmembrane facilitators catalyzing uniport, symport and antiport. *Trends Biochem. Sci.* **18:**13–20.

Neidle, E. L., and S. Kaplan. 1992. *Rhodobacter sphaeroides rdsA*, a homolog of *Rhizobium meliloti fixG*, encodes a membrane protein which may bind cytoplasmic [4Fe-4S] clusters. *J. Bacteriol.* **174:**6444–6454.

Ramseier, T. M. Unpublished results.

Ramseier, T. M., and S. Bledig. Unpublished results.

Ramseier, T. M., D. Nègre, J.-C. Cortay, M. Scarabel, A. J. Cozzone, and M. H. Saier, Jr. 1993. *In vitro* binding of the pleiotropic transcriptional regulatory protein, FruR, to the *fru, pps, pts, icd* and *ace* operons of *Escherichia coli* and *Salmonella typhimurium*. *J. Mol. Biol.* **234:**28–44.

Reizer, A., J. Deutscher, M. H. Saier, Jr., and J. Reizer. 1991. Analysis of the gluconate (*gnt*) operon of *Bacillus subtilis*. *Mol. Microbiol.* **5:**1081–1089.

Reizer, J., V. Michotey, A. Reizer, and M. H. Saier, Jr. Novel phosphotransferase system genes revealed by bacterial genome analysis: unique, putative fructose- and glucoside-specific systems. *Protein Sci.*, in press.

Reizer, J., M. J. Novotny, C. Panos, and M. H. Saier, Jr. 1983. The mechanism of inducer expulsion in *Streptococcus pyogenes:* a two-step process activated by ATP. *J. Bacteriol.* **156:**354–361.

Reizer, J., and C. Panos. 1980. Regulation of β-galactoside phosphate accumulation in *Streptococcus pyogenes* by an expulsion mechanism. *Proc. Natl. Acad. Sci. USA* **77:**5497–5501.

Reizer, J., and A. Peterkofsky (ed.). 1987. *Sugar Transport and Metabolism in Gram-Positive Bacteria.* Ellis Horwood, Chichester, England.

Reizer, J., A. Peterkofsky, and A. H. Romano. 1988a. Evidence for the presence of heat-stable protein (HPr) and ATP-dependent HPr kinase in heterofermentative lactobacilli lacking phospho*enol*pyruvate:glycose phosphotransferase activity. *Proc. Natl. Acad. Sci. USA* **85:**2041–2045.

Reizer, J., A. Reizer, M. H. Saier, Jr., and G. R. Jacobson. 1992. A proposed link between nitrogen and carbon metabolism involving protein phosphorylation in bacteria. *Protein Sci.* **1:**722–726.

Reizer, J., M. H. Saier, Jr., J. Deutscher, F. Grenier, J. Thompson, and W. Hengstenberg. 1988b. The phosphoenolpyruvate:sugar phosphotransferase system in Gram-positive bacteria: Properties, mechanism and regulation. *Crit. Rev. Microbiol.* **15:**297–338.

Romano, A. H., G. Brino, A. Peterkofsky, and J. Reizer. 1987. Regulation of β-galactoside transport and accumulation in heterofermentative lactic acid bacteria. *J. Bacteriol.* **169:**5589–5596.

Saier, M. H., Jr. 1989. Protein phosphorylation and allosteric control of inducer exclusion and catabolite repression by the bacterial phosphoenolpyruvate:sugar phosphotransferase system. *Microbiol. Rev.* **53:**109–120.

Saier, M. H., Jr. 1991. A multiplicity of potential carbon catabolite repression mechanisms in prokaryotic and eukaryotic microorganisms. *New Biol.* **3:**1137–1147.

Saier, M. H., Jr. 1993. Regulatory interactions involving the proteins of the phosphotransferase system in enteric bacteria. *J. Cell. Biochem.* **51:**62–68.

Saier, M. H., Jr., and M. Chin. 1990. Energetics of the bacterial phosphotransferase system in sugar transport and the regulation of carbon metabolism, p. 273–299. *In* T. A. Krulwich (ed.), *The Bacteria: A Treatise on Structure and Function. Vol. XII. Bacterial Energetics.* Academic Press, New York.

Saier, M. H., Jr., J. Reizer, and J. Deutscher. 1992. Protein phosphorylation and the regulation of sugar transport in Gram-negative and Gram-positive bacteria, p. 181–190. *In* S. Papa, A. Azzi, and J. M. Tager (ed.), *Proceedings of the International Symposium on Adenine Nucleotides in Cellular Energy Transfer and Signal Transduction.* Birkhäuser Verlag, Basel, Switzerland.

Saier, M. H., Jr., R. D. Simoni, and S. Roseman. 1970. The physiological behavior of Enzyme I and heat-stable protein mutants of a bacterial phosphotransferase system. *J. Biol. Chem.* **245:**5870–5873.

Saier, M. H., Jr., R. D. Simoni, and S. Roseman. 1976. Sugar transport. Properties of mutant bacteria defective in proteins of the phosphoenolpyruvate:sugar phosphotransferase system. *J. Biol. Chem.* **251:**6584–6597.

Saier, M. H., Jr., H. Straud, L. S. Massman, J. J. Judice, M. J. Newman, and B. U. Feucht. 1978. Permease-specific mutations in *Salmonella typhimurium* and *Escherichia coli* that release the glycerol, maltose, melibiose, and lactose transport systems from regulation by the phosphoenolpyruvate:sugar phosphotransferase system. *J. Bacteriol.* **133:**1358–1367.

Steinmetz, M. 1993. Carbohydrate catabolism: pathways, enzymes, genetic regulation, and evolution, p. 157–170. *In* A. L. Sonenshein, J. A. Hoch, and R. Losick (ed.), *Bacillus subtilis and Other Gram-Positive Bacteria: Biochemistry, Physiology, and Molecular Genetics.* American Society for Microbiology, Washington, D.C.

Sutrina, S. L., A. M. Chin, F. Esch, and M. H. Saier, Jr. 1988. Purification and characterization of the fructose-inducible HPr-like protein, FPr, and the fructose-specific Enzyme III of the phosphoenolpyruvate:sugar phosphotransferase system of *Salmonella typhimurium*. *J. Biol. Chem.* **263:**5061–5069.

Thompson, J., and M. H. Saier, Jr. 1981. Regulation of methyl-β-D-thiogalactopyranoside-6-phosphate accumulation in *Streptococcus lactis* by exclusion and expulsion mechanisms. *J. Bacteriol.* **146:**885-894.

Titgemeyer, F., R. E. Mason, and M. H. Saier, Jr. 1994. Regulation of the raffinose permease of *Escherichia coli* by the glucose-specific enzyme IIA of the phosphoenolpyruvate:sugar phosphotransferase system. *J. Bacteriol.* **176:**543–546.

Titgemeyer, F., and M. H. Saier, Jr. Unpublished results.

Vartak, N. B., and V. Michotey. Unpublished results.

Wilson, T. H., P. L. Yunker, and C. L. Hansen. 1990. Lactose transport of *Escherichia coli* resistant to inhibition by the phosphotransferase system. *Biochim. Biophys. Acta* **1029:**113–116.

Ye, J. J., J. Reizer, X. Cui, and M. H. Saier, Jr. 1994a. ATP-dependent phosphorylation of serine in HPr regulates lactose:H^+ symport in *Lactobacillus brevis*. *Proc. Natl. Acad. Sci. USA* **91:**3102–3106.

Ye, J. J., J. Reizer, X. Cui, and M. H. Saier, Jr. 1994b. Inhibition of the phosphoenolpyruvate:lactose phosphotransferase system and activation of a cytoplasmic sugar-phosphate phosphatase in *Lactococcus lactis* by ATP-dependent metabolite-activated phosphorylation of serine-46 in the phosphocarrier protein, HPr. *J. Biol. Chem.* **269:**11837–11844.

VI. POLYPHOSPHATES AND PHOSPHATE RESERVES

Chapter 32

Introduction: Polyphosphates and Phosphate Reserves

IGOR S. KULAEV

Institute of Biochemistry and Physiology of Microorganisms, Russian Academy of Sciences, Pushchino, Moscow Region, 142292 Russia

In recent years, a number of investigations on the structure, localization, metabolism, and the physiological roles of inorganic polyphosphates have been published. An important stimulus in the studies of their biochemistry was the symposium organized by Torriani-Gorini in Concarneau (Torriani-Gorini et al., 1987). In that forum, polyphosphate metabolism in bacteria and also in eukaryotes was considered in detail. It was a remarkable symposium. The polyphosphate and PP_i sections were organized by the well-known biochemist, Harland G. Wood. Unfortunately, premature death (in his mid-80s) stopped his brilliant research into the enzymes of PP_i and polyphosphate metabolism (Wood, 1985). Quite recently, J. P. Ebel also died. His classic work on the biochemistry of polyphosphates is the basis of all subsequent studies in this field (Ebel, 1951). The work of Ebel et al. (1958), along with investigations by Wiame and Lefebre (1946), Spiegelman and Kamen (1947), and Belozersky (1958), was the first to indicate the closely related functioning of polyphosphates and nucleic acids.

Such findings are supported by recent research. In particular, Reusch and Sadoff (1988) showed the possible participation of polyphosphates in the uptake of DNA into the cell during genetic transformation. In my mind, the studies by Reusch and Sadoff (1988) induced Arthur Kornberg—the pioneer of polyphosphate studies—to return to the enzymes of polyphosphate metabolism in *Escherichia coli* (Akiyama et al., 1993). Kornberg's new fundamental data are presented in chapter 34. Polyphosphate metabolism enzymes are investigated now not only in Arthur Kornberg's laboratory but also elsewhere. A new enzyme involved in the biosynthesis of polyphosphates and mannoproteins in yeasts has been isolated and partially purified in our laboratory (Kulaev, 1990). In chapter 35, the relation between the structure and function of polyphosphate metabolism in bacteria is discussed.

These studies are important to our understanding of the physiological roles of polyphosphates, which are still not clear. We can only say that polyphosphates are multifunctional compounds (Kulaev, 1979). Their functions most probably changed significantly in the course of evolution. In prokaryotes, they largely play a role in bioenergetics (Kulaev, 1990), but they also operate as osmotically inert reserves of phosphorus and energy. In lower eukaryotes, this second function of polyphosphates becomes dominant. In yeasts, major bioenergetic processes involve not polyphosphates but ATP. Interestingly, different yeast organelles contain specific pools of inorganic polyphosphates. These polyphosphate fractions have a different metabolism and play their own roles in the functioning of the organelles. The physiological role of polyphosphates in the nuclei of eukaryotic cells is completely unclear now. In our laboratory (Kulaev 1979), as in some others, polyphosphates were isolated from a nonhistone fraction of nuclear proteins from some eukaryotes, including the tissues of higher animals. Possibly, polyphosphates play a significant role in the regulation of gene expression.

It is with great interest that we turn to chapter 33, in which Goldstein discusses the solubilization of exogenous mineral phosphates by gram-negative bacteria in the environment.

In conclusion, I should say that important theoretical and applied aspects of phosphonate utilization are also discussed in this volume (chapter 36). It is proper to recall that Harry Rosenberg—a well-known investigator of phosphorus metabolism in bacteria—pioneered studies of microbial degradation of phosphonates (La Nauze et al., 1970).

REFERENCES

Akiyama, M. E., E. Crooke, and A. Kornberg. 1993. An expolyphosphatase of *Escherichia coli.* The enzyme and its *ppx* gene in a polyphosphate operon. *J. Biol. Chem.* **268:**633–639.

Belozersky, A. N. 1958. The formation and functions of polyphosphates in the developmental processes of some lower organisms. *Communications and Reports of the Fourth International Biochemistry Congress,* Vienna.

Ebel, J. P. 1951. Recherches chimiques et biologiques sur les poly-et metaphosphates. Thèse Doct. Sci. Phys. Strasbourg, France.

Ebel, J. P., G. Dirheimer, and M. Yacoub. 1958. Sur l'existence d'une combinaison acide ribonucleique-polyphosphates dans la levure. *Bull. Soc. Chim. Biol.* **40:**738.

Kulaev, I. S. 1979. *The Biochemistry of Inorganic Polyphosphates.* John Wiley & Sons, Inc., New York.

Kulaev, I. S. 1990. The physiological role of inorganic polyphosphates in microorganisms: some evolutionary aspects, p. 223–233. *In* E. A. Dawes (ed.), *Novel Biodegradable Microbial Polymers.* Kluwer Academic Publishers, Dordrecht, The Netherlands.

La Nauze, J. M., H. Rosenberg, and D. S. Shaw. 1970. The enzymic cleavage of the carbon-phosphorus bond: purification and properties of phosphonatase. *Biochim. Biophys. Acta* **212:**332–350.

Reusch, R. N., and H. L. Sadoff. 1988. Putative structure and functions of a poly-β-hydroxybutyrate/calcium polyphosphate channel in bacterial plasma membranes. *Proc. Natl. Acad. Sci. USA* **85:**4176–4180.

Spiegelman, S., and M. J. Kamen. 1947. Some basic problems in the relation of nucleic acid turnover to protein synthesis. *Cold Spring Harbor Symp. Quant. Biol.* **12:**211–220.

Torriani-Gorini, A., F. G. Rothman, S. Silver, A. Wright, and E. Yagil (ed.). 1987. *Phosphate Metabolism and Cellular Regulation in Microorganisms.* American Society for Microbiology, Washington, D.C.

Wiame, J. M., and P. H. Lefebre. 1946. Conditions de formation does la levure d'compose nucleique polyphosphate. *C. R. Soc. Biol.* **140:**921–923.

Wood, H. G. 1985. Inorganic pyrophosphate and polyphosphates as sources of energy. *Curr. Top. Cell. Regul.* **26:**355–369.

Chapter 33

Involvement of the Quinoprotein Glucose Dehydrogenase in the Solubilization of Exogenous Mineral Phosphates by Gram-Negative Bacteria

ALAN H. GOLDSTEIN

Department of Biology, California State University, Los Angeles, California 90032

Mineral Phosphates as a Source of Soluble Phosphorus for Cell Growth

Phosphorus is an essential mineral macronutrient for biological growth and development. All phosphorus currently cycling through the biosphere was originally derived from minerals that composed the parent material of soils. For the purpose of this discussion, I will limit myself to the availability of P in one component of the biosphere, namely soils. At any given time, a substantial component of P in any soil is in the form of poorly soluble mineral phosphates. As discussed below, these mineral phosphates are, in general, not directly bioavailable. A second major component of soil P is in organic matter, much of which is high-molecular-weight material (e.g., RNA). Most soil microorganisms obtain P from the environment via membrane transport and subsequent assimilation. In general, these microorganisms cannot access (transport) these two major sources of soil P directly but must first bioconvert them to either soluble ionic phosphate (P_i, $H_2PO_4^-$ or HPO_4^{2-}) or low-molecular-weight organic phosphate. These processes may occur sequentially, e.g., the transformation of RNA to nucleoside monophosphates via RNase followed by release by P_i via acid or alkaline phosphatase.

The genetic basis for these bioconversion systems is, of course, the result of evolutionary adaptation to the physical chemistry and biochemistry of P in the soil system. For a number of reasons, the molecular genetics and enzymology of bacterial biotransformations of organic phosphates have received a great deal of attention. As a result, our understanding of these systems is advanced. By contrast, we have only recently begun to elucidate the molecular genetic and biochemical basis of bacterial transformations of poorly soluble mineral phosphates. I will describe some of the variables of state with respect to the effective dissolution of a major family of poorly soluble mineral phosphates, the calcium phosphates. This group of related minerals makes up the major fraction of unavailable mineral P in many soils. I will then discuss the molecular genetics and biochemistry of a metabolic pathway recently shown to form the basis for efficacious biotransformation of poorly soluble calcium phosphates by gram-negative bacteria. Finally, I will report on preliminary studies that demonstrate the presence of this pathway in gram-negative bacteria living in the root zone (rhizobacteria) of plants growing in calcareous desert soils expected to be deficient in P_i but high in poorly soluble calcium phosphates.

Mineral Phosphates in Soils

The ability of some microorganisms to dissolve poorly soluble mineral phosphates, specifically ground bone, was one of the first phenotypes to be described by agricultural microbiologists (reviewed by Goldstein [1986]). Interest in this phenotype was a natural consequence of accumulated information about the role of phosphorus in plant growth, especially with respect to agricultural crop production. According to Brentnall (1991), the first mention of the beneficial effect of bones as a fertilizer can be found in 17th century European publications. In 1769, the Swedish scientist J. G. Gahn discovered that calcium phosphate is the main component of bones, and about 30 years later, the conclusion was reached that the enhancing (fertilizer) effect of ground bones on plant growth was due mainly to calcium phosphate. This enhancement results from the fact that the concentration of soluble P_i in natural soils tends to be extremely low. Soluble P_i is normally present in soil solution at concentrations of less than 1 ppm (10 μM $H_2PO_4^-$). This limited solubility is the result of the ability of P_i to form a wide range of stable insoluble compounds. Under alkaline to neutral pH conditions (the situation for most arid and semiarid soils), these compounds are mainly insoluble calcium phosphates.

The potential importance of poorly soluble mineral phosphates in nutrient cycling and growth of microbial populations can be appreciated when one recognizes that throughout most of the United States, the surface 30 cm of soil contains an average of 0.05% P (Fried and Broeshart, 1967). These soils usually have P_i concentrations in solution of less than 10 μM. This concentration can drop well below 1 μM in arid and semiarid regions. Because of the potential of P_i for the enhancement of crop growth and/or the development of a renewable low-input fertilizer technology, microbial biotransformation of poorly soluble mineral phosphates to P_i has been studied intensively by agricultural microbiologists. However, until recently, little was known about the genetic and biochemical bases for the mineral phosphate-solubilizing (Mps) phenotype. The scope of this discussion will be limited to dissolution of these poorly soluble calcium phosphates by gram-negative bacteria, for which, during the past few years, significant progress in our understanding of Mps metabolism has been achieved.

Gluconic Acid-Mediated Dissolution of Calcium Phosphates by Gram-Negative Bacteria

Many types of calcium phosphate compounds can be found in soils. These compounds have a wide range of solubilities, which, in general, follow an inverse relationship with the Ca/P ratios. For example, monocalcium phosphate [$Ca(H_2PO_4)_2$, Ca/P = 0.50] has a water solubility of 150,000 ppm at pH 7, whereas fluoroapatite [$Ca_{10}(PO_4)_6F_2$, Ca/P = 1.66] has a water solubility of 0.003 ppm. Poorly soluble mineral phosphates such as fluorapatite or hydroxyapatite can be effectively dissolved in aqueous solution only under acidic conditions. This dissolution is the result of acid-mediated proton substitution for calcium as shown in equation 1 for fluorapatite and a generic acid HX that dissociates to form $H^+ + X^-$:

$$Ca_{10}(PO_4)_6F_2^{(s)} + 20HX^{(aq)} \rightarrow 10CaX_2^{(aq)} + 6H_3PO_4^{(aq)} + 2HF^{(aq)} \quad (1)$$

This form of complete dissolution will, of course, occur only under extremely acidic conditions. Depending on soil pH and the presence of other ionic components, pseudo-steady-state conditions will be achieved, resulting in the release of some P_i along with the formation of a number of crystalline calcium phosphate compounds (e.g., tricalcium phosphate, octacalcium phosphate) and amorphous calcium phosphate oxides and hydrous oxides.

The bacterial Mps phenotype has historically been associated with the production of low-molecular-weight organic acids (reviewed by Goldstein [1986] and Goldstein et al. [1993]). Recently, Goldstein and coworkers have identified the metabolic and genetic bases for high-efficiency solubilization of poorly soluble calcium phosphates by *Erwinia herbicola* (Liu et al., 1992) and *Pseudomonas cepacia* (Goldstein et al., 1993). We have shown that solubilization is the result of acidification of the periplasmic space (and ultimately the external medium) by the direct oxidation of glucose (nonphosphorylating oxidation) or other aldose sugars by the quinoprotein glucose dehydrogenase (GDH) (Fig. 1). We have proposed that the direct oxidation pathway forms the basis for the Mps phenotype in gram-negative bacteria.

Depending on the species, glucose-derived gluconic acid may be further oxidized in the periplasm to 2-ketogluconic or 2,5-diketogluconic acid (Anderson et al., 1985). As shown in Fig. 2, the enzymes of the direct oxidation pathway are oriented on the outer face of the cytoplasmic membrane so that their oxidized substrates are released to the periplasmic space (Ameyama et al., 1981; Duine, 1991; Lessie and Phibbs, 1984). The products of direct periplasmic oxidation may be taken up by specific transport systems and further catabolized. GDH oxidizes a broad range of aldose sugars.

Little is known about the regulation of the genes coding for quinoproteins or how quinoprotein-mediated oxidative metabolism is regulated. Nonphosphorylating oxidation is one of the four major metabolic pathways for glucose (aldose) utilization by bacteria (Lessie and Phibbs, 1984; Gottschalk, 1986). Most species have at least two of these pathways. The quinoprotein GDH controls the unique step in direct oxidation (Divine et al., 1979). It transfers electrons from aldose sugars to the electron transport chain via two-electron, two-proton oxidations mediated by the cofactor pyrroloquinoline-quinone (PQQ) (Duine, 1991). The redox states of PQQ are shown in Fig. 1B.

The biochemical complexity of aldose utilization is, of course, the result of the interactive and dynamic nature of cell growth. Metabolic capabilities must be tuned for environmental conditions including organic and mineral nutrient availability. It is now known that periplasmic oxidation of aldose sugars can contribute electrons directly to the respiratory electron transport pathway. In addition, protons generated from these oxidations can contribute directly to the transmembrane proton motive force (van Schie et al., 1985; Duine, 1991). Evidence exists to suggest that GDH can play a regulatory and bioenergetic role in this aspect of energy metabolism (van Schie et al., 1985; van Schie, 1987). For several bacterial species that use the direct oxidation pathway, it has further been shown that uptake of solutes such as alanine, lactose, and proline can be directly controlled by electron transfer activity in the respiratory chain.

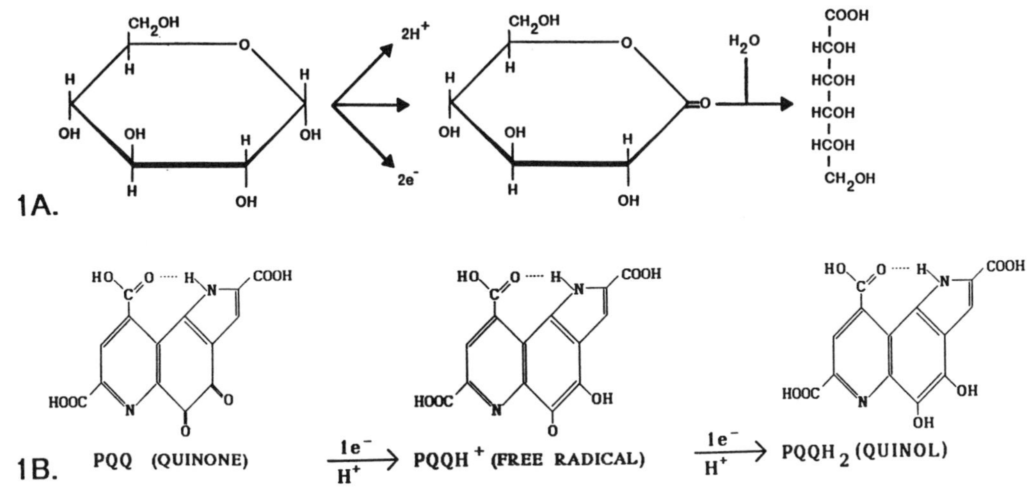

FIG. 1. (A) The quinoprotein GDH-catalyzed glucose oxidation to gluconic acid proceeds via the gluconolactone intermediate. (B) The prosthetic group PQQ is a two-electron, two-proton redox carrier. The redox reaction proceeds via a free radical intermediate. The midpoint redox potential of the PQQ/$PQQH_2$ couple is +90 mV at pH 7 (Anthony, 1988).

The molecular mechanism(s) whereby periplasmic oxidation is coupled to respiratory electron transport varies among genera and species. The biochemical or genetic regulatory mechanisms by which a given species switches between the direct phosphorylative and periplasmic oxidative modes remain unknown.

Although GDH may have a bioenergetic raison d'être in some species under certain conditions, a generalized physiological role of this pathway has not been shown. In fact, Lessie et al. (1979) showed that GDH-deficient strains of *P. cepacia* grew normally with glucose as a carbon source. Likewise, it has been difficult to assess the utility of this pathway in terms of microbial ecology (see van Schie [1987] and references therein). The gluconic acid phenotype is widely distributed among the gram-negative genera, but it has not been possible to identify a major bioenergetic or ecological advantage for this trait. For example, many pseudomonads express the direct oxidation pathway in the presence of glucose but oxidize less than 1% of the glucose present. The bioenergetic purpose of such a "dissimilatory bypass" is obscure. Conversely, some *Acinetobacter* species (e.g., *Acinetobacter calcoaceticus* LMD 79.41) can stoichiometrically convert glucose to gluconic acid at concentrations of 1 mol/liter or higher but are incapable of uptake of glucose or gluconic acid for energy metabolism (van Schie, 1987). As discussed below, a metabolic rationale for at least some of these phenotypes can be found in the role of periplasmic (and ultimately extracellular) acidification in the dissolution of poorly soluble mineral phosphates. There may be other ecological roles for extracellular acidification as well.

The complexity of the direct oxidation phenotype is reflected in the molecular genetics of GDH and related quinoprotein dehydrogenases. There is no information on the genetic or biochemical mechanisms that regulate the synthesis or assembly of the apoglucose dehydrogenase/PQQ holoenzyme. Data show that there are significant differences in regulation of this system among bacterial species. Virtually every combinatorial form of expression is observed, i.e., constitutive apoglucose dehydrogenase expression coupled with inducible PQQ biosynthesis versus constitutive PQQ biosynthesis coupled with inducible biosynthesis of the apoenzyme. In *P. aeruginosa*, holoenzyme activity is inducible by glucose, gluconate, mannitol, or glycerol, whereas in *A. calcoaceticus*, the enzyme is synthesized constitutively (van Schie, 1987). *Acinetobacter lwoffii* and *Escherichia coli* do not show acid production in the presence of glucose without the addition of exogenous PQQ (Goosen et al., 1989). Cell extracts of these two bacteria also show glucose oxidation upon addition of PQQ, indicating that GDH apoenzyme was produced constitutively. *A. lwoffii* does not metabolize glucose at all but nevertheless synthesizes apoenzyme constitutively. It has been postulated that, for organisms such as *E. coli* and *A. lwoffii*, PQQ present in the environment may be considered nutritionally to be a vitamin. The location of the GDH apoenzyme on the outer face of the cytoplasmic membrane facilitates binding of exog-

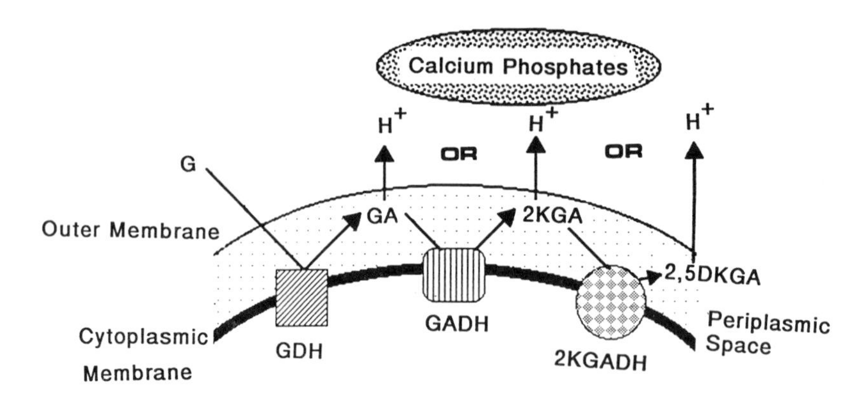

FIG. 2. The direct oxidation pathway functions on the outer face of the cytoplasmic membrane so that both acidic protons and oxidized substrate are initially produced in the periplasmic space. The ultimate fate of the oxidized glucose varies from species to species. Each oxidation is a two-electron, two-proton process. It is assumed that, for reasons of electroneutrality, the redox protons are constrained to bioenergetic or membrane transport processes. However, one acidic proton may be released into the extracellular space for every glucose molecule catabolized. This one proton may dissociate from any one of the three acidic species produced by the pathway depending on the physicochemical conditions in the periplasmic space and the extracellular environment. Therefore, three protons are shown representing these three possible acidic dissociations. Insoluble mineral phosphates are, by definition, not diffusible. Therefore, this model postulates that these materials do not enter the periplasmic space. Conversely, bacteria with the direct oxidation pathway are commonly observed to lower the pH of the medium so that movement of acidic protons or undissociated organic acids out of the periplasmic space is assumed. P_i is released from the mineral phosphate by proton substitution for Ca^{2+}, as discussed in the text. Abbreviations: G, glucose; GA, gluconic acid; 2KGA, 2-ketogluconic acid; 2,5DKGA, 2,5-diketogluconic acid; GADH, gluconate dehydrogenase; 2KGADH, 2-ketogluconate dehydrogenase. Adapted from Goldstein et al. (1993) and Anderson et al. (1985).

enous PQQ to form the holoenzyme. Conversely, in *P. stutzeri,* PQQ can be detected only when ethanol is present in the culture medium as an inducing agent. The biosynthetic pathway for PQQ has not been elucidated.

Although the GDH pathway may make a contribution to the energy status of some cells by generating a transmembrane proton motive force, it seems improbable that this role can justify the stoichiometric conversion of glucose to gluconic acid at concentrations such as those discussed above. However, in P_i-limited, high-calcium-phosphate soil ecosystems, the GDH-mediated dissimilatory bypass system can play a vital role by increasing the availability of P_i. In fact, we have recently shown that high-efficiency solubilization of rock phosphate ore by *P. cepacia* and *E. herbicola* is the result of gluconic acid produced in the periplasmic space by the GDH-catalyzed direct oxidation of glucose. Initially, the GDH pathway was identified via cloning of functional mineral phosphate-solubilizing (*mps*) genes (Goldstein and Liu, 1987; Liu et al., 1992). These studies ultimately demonstrated that activation of the GDH pathway resulted in both gluconic acid production and mineral phosphate solubilization. Recent studies have confirmed that gluconic acid is, in fact, capable of mediating dissolution of rock phosphates are via direct acidification, as shown in equation 1. Representative data from these studies are shown in Fig. 3.

From these data it may be seen that, even with relatively low levels of gluconic acid, a significant amount of P_i is released. For example, 60 mM gluconic acid (equivalent to approximately 1% gluconic acid) resulted in the release of approximately 0.1 mM P_i. As discussed above, this type of local enhancement of P_i availability within the soil system could have a significant impact on cell growth. The concentration of gluconic acid and the amount of poorly soluble mineral phosphate present in any particular region of the soil system would, of course, be highly variable. Furthermore, effective concentrations, acidity, ion activities, etc., would be dramatically different in the soil where the bioelectrochemistry and physical electrochemistry are dominated by interfacial phenomena. However, many workers have noted an enhanced level of gluconic acid-producing bacteria in the region of the soil adjacent to plant roots where the availability of aldose sugars would be higher than in the bulk soil (reviewed by Goldstein [1986]).

Nucleic Acid Probe for the Mps Trait

To explore the possible role of the direct oxidation pathway as the physiological genetic basis of

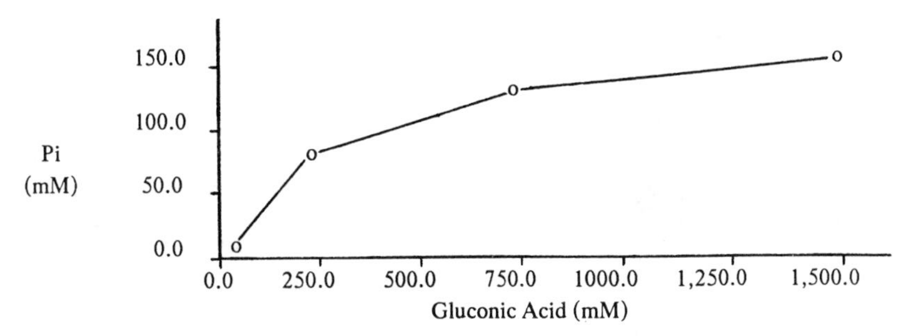

FIG. 3. Release of soluble P_i from fluorapatite rock phosphate ore. We stirred 30% rock phosphate ore (<200 mesh) at room temperature with the indicated concentration of gluconic acid. After 5 days, residual rock phosphate ore was removed by centrifugation and the soluble P_i concentration was determined by ion chromatography. Values shown are the averages from three separate experiments. Standard deviations were less than 10% for all points. The first point corresponds to 60 mM gluconic acid, resulting in the release of 1 mM P_i, as discussed in the text.

the Mps phenotype, we have used DNA sequence homology between GDHs to construct probes to assay for the presence of candidate GDH genes in rhizobacteria isolated from P_i-limited environments.

Probe construction was based on sequence data for cloned GDH genes from *A. calcoaceticus, E. coli,* and *Gluconobacter oxydans* (Cleton-Jansen, 1991; Cleton-Jansen et al., 1990). These workers analyzed regions of deduced amino acid homology in the sequenced cloned GDH genes and other PQQ-dependent dehydrogenase genes (e.g., methanol dehydrogenase and alcohol dehydrogenase) and delineated common features that may provide important clues to elucidating the molecular architecture of the PQQ-binding side and the mechanism of electron transfer to the respiratory chain. There is striking overall homology in certain regions among all quinoproteins, although the substrate specificities of methanol dehydrogenase and alcohol dehydrogenase differ completely from that of GDH.

One highly conserved region in all eight proteins occurs at the C terminus from residues 736 to 782 of the *A. calcoaceticus* GDH. This region is of special significance since it is also homologous to GDH B of *A. calcoaceticus,* the only known nonmembrane-bound quinoprotein glucose dehydrogenase. It has been postulated, therefore, that this region is involved in PQQ binding. This homology is even more striking when one compares only the three membrane-bound quinoprotein GDHs cloned and sequenced to date (Fig. 4).

Using data obtained from the literature, we focused our efforts on the region of GDH from amino acids 756 to 769 (numbering for the *A. calcoaceticus* protein, [Cleton-Jansen et al., 1990]). The deduced sequence translates as PAGGQATPMTYEI. We have synthesized a 39-mer oligonucleotide probe by using the codon biases identified from sequencing an *E. herbicola pqq* gene (Liu et al., 1992). This probe was used in Southern blot DNA-DNA hybridizations to identify bands from restriction nuclease digests of genomic DNA from two bacterial strains previously shown to produce gluconic acid and to be highly efficient in solubilization of rock phosphate ore, *E. herbicola* (Eho 10) and *P. cepacia* (E-37). The probe was then used to screen DNA from unidentified gram-negative Mps^+ and Mps^- rhizobacteria isolated from soil from the Anzo Barrego desert region of southern California. The results of the screening (Fig. 5) showed that the three Mps^+ bacterial isolates have DNA sequences with homology to the probe whereas the two Mps^- isolates have no apparent homology under our hybridization conditions. This research is only preliminary, but it suggests the presence of the direct oxidation pathway in the Mps^+ isolates.

Conclusion

In conclusion, it has long been known that acid dissolution of insoluble calcium phosphates occurs via direct proton attack at the crystal surface, with subsequent ionic substitution. It has also been known historically that the bacterial Mps^+ phenotype was often associated with production of gluconic or 2-ketogluconic acid. We now know

1. 740 RAFNVTNGKKLWEARLPAGGQATPMTYEINGKQYVVIMAGG 780
2. 734 RAYNMSNGEKLWQGRLPAGGQATPMTYEVNGKQYVVISAGG 774
3. 740 RAYLNTTGKVLWQDRLPAGAQATPITYAINGKQYIVTYAGG 780

FIG. 4. Region of extremely high amino acid homology shared by the three cloned membrane-bound quinoprotein GDHs. 1, *A. calcoaceticus* (GDH-A); 2, *E. coli* (GDH); 3, *G. oxydans* (GDH). Underlined amino acids are identical. Data from Cleton-Jansen (1991) and Cleton-Jansen et al. (1990).

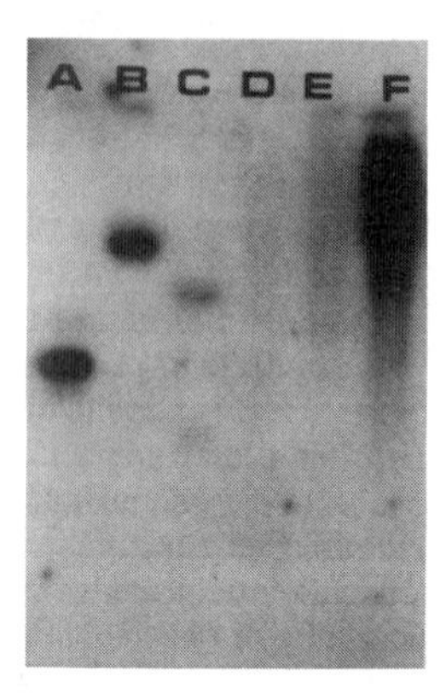

FIG. 5. Three-day exposure of a Southern blot DNA-DNA hybridization of *Hin*dIII-digested DNA (2 μg per lane) from three Mps+ and two Mps− gram-negative rhizobacteria isolated from soils collected from the Anzo Barrego desert region of southern California. Lanes A, B, and C are Mps+, whereas lanes D and E are Mps−. Lane F is the *E. herbicola* Eho10 Mps+ control. Strains in lanes A, B, and D were from the root surface of creosote bush (*Larrea tridentada*), whereas strains in lanes C and E were from an *Atriplex* plant (species not determined). Washed roots (10 cm depth) were shaken overnight at room temperature in HAP medium supplemented with 1 mM sodium phosphate (Goldstein and Liu, 1987). Isolates were initially subcultured via dilution plating onto Luria-Bertani medium. The Mps phenotype was determined by replica plating onto HAP medium as previously described (Goldstein and Liu, 1987). Isolation of DNA, restriction digests, Southern blots, and DNA-DNA hybridizations were carried out by standard methods (Sambrook et al., 1989). The hybridization was at 42°C in 4× SSPE buffer (1× SSPE in 0.18 M NaCl, 10 mM NaH_2PO_4, and 1 mM EDTA [ph 7.7]) with 10% formamide and 0.08% sodium dodecyl sulfate. The wash was at 30°C in the same buffer.

that during high-efficiency dissolution of poorly soluble mineral phosphates by gram-negative Mps+ bacteria, these protons are supplied by the direct periplasmic oxidation of aldose sugars by a metabolic pathway that includes a quinoprotein dehydrogenase embedded in the outer face of the cytoplasmic membrane. A preliminary model for this system is shown in Fig. 2. A number of questions remain to be answered in terms of both the physiological ecology and the molecular genetics of the Mps phenotype. In the near future, we hope to begin to look at the distribution of the direct oxidation pathway in bacterial populations from desert soils.

Finally, it will be of interest to look into the effect (if any) of P_i availability on expression of the GDH pathway in bacterial isolates obtained from P_i-limited environments. Early attempts to study the Mps phenotype directly were hampered by the changes in the buffering capacity of the medium itself upon release of P_i, which, in turn, modified the rate of dissolution of insoluble calcium phosphate. The availability of nucleic acid and antibody probes for GDH should allow us to examine the possible relationship between the nutritional status of the cell with respect to P and expression of the direct oxidation pathway.

REFERENCES

Ameyama, M., K. Matsushita, Y. Ohno, E. Shinagawa, and O. Adachi. 1981. D-Glucose dehydrogenase of *Gluconobacter suboxydans:* solubilization, purification and characterization. *Agric. Biol. Chem.* **45:**851–861.

Anderson, S., C. B. Marks, R. Lazarus, J. Miller, K. Stafford, J. Seymour, D. Light, W. Rastetter, and D. Estell. 1985. Production of 2-keto-L-gluconate, an intermediate in L-ascorbate synthesis, by a genetically modified *Erwinia herbicola. Science* **230:**144–149.

Anthony, C. 1988. Quinoproteins and energy transduction, p. 293–315. *In* C. Anthony (ed.), *Bacterial Energy Transduction.* Academic Press, Inc., New York.

Brentnall, B. A. 1991. Phosphate fertilizers: economic aspects, p. 425. *In Ullman's Encyclopedia of Industrial Chemistry,* vol. A19. VCH Verlagsgesellschaft, Weinheim, Germany.

Cleton-Jansen, A.-M. 1991. A molecular genetic analysis of the substrate specificity of quinoprotein glucose dehydrogenase, p. 78–80. Ph.D. Dissertation. University of Leiden, Leiden. The Netherlands.

Cleton-Jansen, A.-M., N. Goosen, O. Fayet, and P. van de Putte. 1990. Cloning, mapping, and sequencing of the gene encoding *Escherichia coli* quinoprotein glucose dehydrogenase. *J. Bacteriol.* **172:**6308–6315.

Duine, J. A. 1991. Quinoproteins: enzymes containing the quinoid cofactor pyrroloquinoline quinone, topaquinone or tryptophan-tryptophan quinone. *Eur. J. Biochem.* **200:**271–284.

Duine, J. A., J. Frank, and J. K. Van Zeeland. 1979. Glucose dehydrogenase from *Acinetobacter calcoaceticus:* a 'quinoprotein'. *FEBS Lett.* **108:**443–446.

Fried, M., and H. Broeshart. 1967. *The Soil-Plant System in Relation to Inorganic Nutrition,* p. 6–54. Academic Press, Inc., New York.

Goldstein, A. H. 1986. Bacterial mineral phosphate solubilization: historical perspectives and future prospects. *Am. J. Alternative Agric.* **1:**57–65.

Goldstein, A. H., and S. T. Liu. 1987. Molecular cloning and regulation of a mineral phosphate solubilizing gene from *Erwinia herbicola. Bio/Technology* **5:**72–74.

Goldstein, A. H., R. D. Rogers, and G. Mead. 1993. Mining by microbe. *Bio/Technology* **11:**1250–1254.

Goosen, N., H. P. Horsman, R. G. Huinen, and P. van de Putte. 1989. *Acinetobacter calcoaceticus* genes involved in biosynthesis of the coenzyme pyrrolo-quinoline-quinone: nucleotide sequence and expression in *Escherichia coli* K-12. *J. Bacteriol.* **171:**447–455.

Gottschalk, G. 1986. *Bacterial Metabolism,* 2nd ed. Springer-Verlag KG, Berlin.

Lessie, T. G., T. Berka, and S. Zamanigian. 1979. *Pseudomonas cepacia* mutants blocked in the direct oxidative pathway of glucose degradation. *J. Bacteriol.* **139:**323–325.

Lessie, T. G., and P. V. Phibbs, Jr. 1984. Alternative pathways of carbohydrate utilization in pseudomonads. *Annu. Rev. Microbiol.* **38:**359–387.

Liu, S. T., L.-Y. Lee, C.-Y. Tai, C.-H. Horng, and Y.-S. Chang, J. H. Wolfram, R. Rogers, and A. H. Goldstein. 1992. Cloning of an *Erwinia herbicola* gene necessary for gluconic acid production and enhanced mineral phosphate solubilization in *Escherichia coli* HB101: nucleotide sequence and probable involvement in biosynthesis of the coenzyme pyrroloquinoline quinone. *J. Bacteriol.* **174:**5814–5819.

Sambrook, J., E. F. Fritsch, and T. Maniatis. 1989. *Molecular Cloning: A Laboratory Manual,* 2nd ed. Cold Spring Harbor Laboratory, Cold Spring Harbor, N.Y.

van Schie, B. J. 1987. The physiological function of gluconic acid production in *Acinetobacter* species and other gram negative bacteria, p. 37. Ph.D. Dissertation. Delft Technical University, Delft, The Netherlands.

van Schie, B. J., K. J. Hellingwerf, J. P. van Dijken, M. G. L. Elferink, J. M. van Dijl, J. G. Kuenen, and W. N. Konigns. 1985. Energy transduction by electron transfer via a pyrrolo-quinoline quinone-dependent glucose dehydrogenase in *Escherichia coli, Pseudomonas aeruginosa,* and *Acinetobacter calcoaceticus* (var. *lwoffii*). *J. Bacteriol.* **163:**493–499.

Chapter 34

Inorganic Polyphosphate: a Molecular Fossil Come to Life

ARTHUR KORNBERG

Department of Biochemistry, Stanford University School of Medicine, Stanford, California 94305-5307

Inorganic polyphosphate (polyP) is a linear polymer of tens or many hundreds of P_i residues linked by high-energy phosphoanhydride bonds (Fig. 1). Although it is widely distributed in bacteria, fungi, protozoa, plants, and mammals (Kulaev, 1979; Kulaev and Vagabov, 1983; Kulaev et al., 1987; Wood and Clark, 1988), little is known about its cellular functions. Dismissed and ignored as a "molecular fossil," polyP should be viewed as a "molecule for many reasons." PolyP may act as a (i) source of energy, (ii) phosphate reservoir, (iii) donor for sugar and adenylate kinases, (iv) chelator for divalent cations, (v) buffer against alkaline stress, (vi) regulator of transcription, and (vii) structural element in competence for DNA entry and transformation.

Our approach to clarifying the functions of this ubiquitous, abundant, but largely forgotten polymer has been to isolate the enzymes which make and use polyP so that we might (i) identify the genes that encode the enzymes, (ii) knock out the genes or overexpress them to observe the physiological consequences, (iii) determine the location of the enzymes in the wild-type and overproducer strains, (iv) use these enzymes as reagents of high specificity for the synthesis and analysis of polyP chains of distinctive lengths and states of complexation, (v) explore the enzymatic mechanisms responsible for the metabolism of these remarkable polymers, and (vi) apply the biochemical and genetic knowledge to problems as theoretical as prebiotic evolution and as practical as the removal of the phosphate that contaminates waterways worldwide.

In this chapter I describe three polyP enzymes from *Escherichia coli* and two from *Saccharomyces cerevisiae*. The *E. coli* enzymes include a polyP kinase (PPK), which catalyzes the readily reversible synthesis of polyP from the terminal phosphate of ATP, and two exopolyphosphatases (exopolyPases), one of which is identical to the guanosine pentaphosphate hydrolase that generates guanosine tetraphosphate, the crucial effector that regulates gene expression in the stringent response to amino acid deprivation. Phenotypes of *E. coli* mutants lacking PPK suggest important roles for polyP in response to stresses such as heat and oxidants and to the deprivations that threaten survival in the stationary phase of the growth cycle. The yeast enzymes are polyPases, one an exopolyPase and the other an endopolyPase. The purified enzymes have served as valuable reagents to prepare labeled, well-defined substrates (with PPK) and to determine (with PPK and the polyPases) the features and abundance of polyP in bacterial, yeast, and animal cells.

E. coli PPK

The enzyme. For a description of PPK, see Ahn and Kornberg (1990), Kornberg et al. (1956), and Kornberg (1957). Synthesis of long polyP chains (ca. 1,000 residues) is catalyzed by PPK. The terminal phosphate of ATP is polymerized in a freely reversible reaction:

$$n\text{ATP} \leftrightarrow n\text{ADP} + \text{polyP}_n$$

PPK is attached as a peripheral membrane protein in cell lysates and must be detached for purification; purified PPK reassociates with the membrane fraction. Tryptophan fluorescence of PPK is enhanced by interaction with liposomes prepared from phospholipids with acidic head groups but without regard for the saturation of the fatty acyl groups.

The PPK reaction is highly processive; oligophosphate (oligoP) intermediates are not detected either in the synthesis of polyP from ATP or in the reverse reaction. Autophosphorylation of a histidine residue in PPK in the 3-position appears to be an intermediate stage in the reaction inasmuch as a [^{32}P]PPK reacts readily with ADP to form ATP and can donate the phosphate for incorporation into polyP in the course of synthesis from unlabeled ATP. Paradoxically, the enzyme is not labeled by [^{32}P]polyP while being converted to ATP. Also puzzling is that chains of intermediate lengths (e.g., 15 to 100 residues) can serve as substrates for conversion to ATP but are not used as primers for elongation. Thus, the mechanisms whereby the long chains are formed and depolymerized and the basis for discrimination between long and short chains are still obscure.

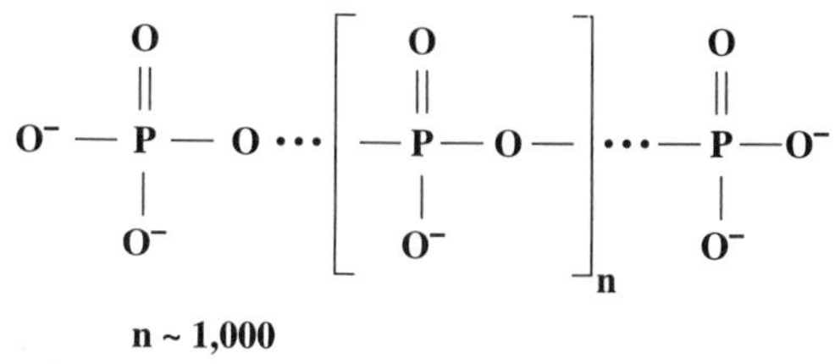

$$O^- - \overset{\overset{O}{\|}}{\underset{\underset{O^-}{|}}{P}} - O \cdots \left[- \overset{\overset{O}{\|}}{\underset{\underset{O^-}{|}}{P}} - O - \right]_n \cdots - \overset{\overset{O}{\|}}{\underset{\underset{O^-}{|}}{P}} - O^-$$

$n \sim 1{,}000$

FIG. 1. Structure of inorganic polyP.

The gene, *ppk*. The gene encoding PPK has been cloned, sequenced, knocked out, and overexpressed (Akiyama et al., 1992). It is located at 53.4 min on the *E. coli* linkage map. The open reading frame encodes a sequence of 687 amino acids (mass of 80,278 Da). The enzyme is a homotetramer. It has proven to be a stable and reliable reagent for determining polyP abundance in cells by measuring ATP produced from either [^{32}P]polyP (with ADP) or unlabeled polyP (with [^{14}C]ADP); the labeled ATP product is separated and measured by thin-layer chromatography.

Mutants. For a description of mutants, see Crooke et al. (1994). When the *ppk* gene was disrupted by insertion of a kanamycin resistance gene, polyP levels were reduced from near 2.0 μg/10^{11} cells to 0.16 μg. Mutant cells showed a growth lag following dilution of a stationary-phase culture but then grew at a similar rate and to a similar extent to those of the wild type. The most striking phenotypes were the responses of the *ppk* mutant to stress and deprivation. Heating at 55°C for 2 min was lethal to 98% of stationary-phase mutant cells but produced only a barely detectable decrease in the viability of the wild type. Exposure to an oxidant (42 mM H_2O_2 for 60 min) killed 90% of the mutant cells but had less than a 1% effect on the wild type. Catalase levels of mutant cells, measured by removal of H_2O_2, were reduced to about 50% of those in the wild type.

Survival of mutant cells in stationary phase in a minimal medium was sharply reduced. Only 2% of the mutant cells survived after 2 days compared with no detectable loss of the wild type. In the mutant population, the large-colony type was replaced by a small-colony variant, a feature which persisted upon subculture. The basis for selection of this variant requires further study.

E. coli ExopolyPases

ExopolyPase. The exopolyPase (PPX) (Akiyama et al., 1993) is highly processive in catalyzing the hydrolysis of terminal residues of long-chain polyP (~500 to 1,000 residues) to P_i. The gene (*ppx*) that encodes a polypeptide of 513 amino acids (58,133 Da) is located 7 bp downstream of *ppk*. Transcription of *ppx* depends on the *ppk* promoter, indicative of a polyP operon of *ppk* and *ppx* (Fig. 2). PPX, purified from overproducing cells, is judged to be a homodimer. Optimal activity is supported by Mg^{2+} (1 mM) and K^+ (175 mM). The effectiveness of intermediate-size polyP chains as substrates is far less than that of long chains.

Guanosine pentaphosphate phosphohydrolase. An exopolyPase activity purified from mutants lacking PPX (Keasling et al., 1993) has surprisingly proved to be guanosine pentaphosphate phosphohydrolase (GPP) (Hara and Sy, 1983). Evidence for identity is based on (i) sequences of five tryptic digestion fragments of the homogeneous protein being found in the translated gene for GPP (i.e., *gppA*), (ii) the size of the protein (100 kDa), and (iii) the constant ratio of expolyPase activity to GPP activity through the last steps of a 300-fold purification leading to homogeneity.

GPP (Keasling et al., 1993) liberates P_i by processive hydrolysis of polyP chains (500 to 1,000 residues) or by hydrolysis of the 5′-γ-phosphate of guanosine 5′-triphosphate 3′-diphosphate (pppGpp) to the corresponding ppGpp. The K_m for long-chain polyP as substrate (~0.5 nM, expressed as polymer concentration) is 5 to 6 orders of magnitude lower than that for pppGpp (0.13 mM); the k_{cat} for the polyPase activity is 1.1 s^{-1}, whereas that for pppGpp hydrolase is 0.023 s^{-1}. Thus, the overall potency of the enzyme as an exopolyPase, expressed as k_{cat}/K_m, would appear to be 10^7 times greater than as a pppGpp hydrolase. However, this discrepancy is more apparent than real in view of two considerations. First, the cellular abundance of pppGpp in the stringent response reaches concentrations which satisfy the K_m of the hydrolase activity. Second, the higher k_{cat} value for the exopolyPase is calculated on the basis of P_i residues released rather than the conversion of the entire polyP to P_i; this high k_{cat} derives from the processivity of the enzyme on the polymeric substrate, an action in which the time expended in dissociation and reassociation of the substrate is saved.

Assuming the active site for hydrolysis to be the same for both activities, one must account for the enormous difference in the affinities for polyP and pppGpp as substrates. The processive hydrolysis of the 1,000-residue chain of polyP continues until the length is reduced to about 40 residues. These intermediate-size chains accumulate until the long chains are nearly all removed, at which point these too are degraded to P_i and end up as chains of four residues. It may be significant that the end products of both polyP and pppGpp hydrolysis are tetraphosphates.

The recent revelation that the amino-terminal domains of PPX and GPP share structural homol-

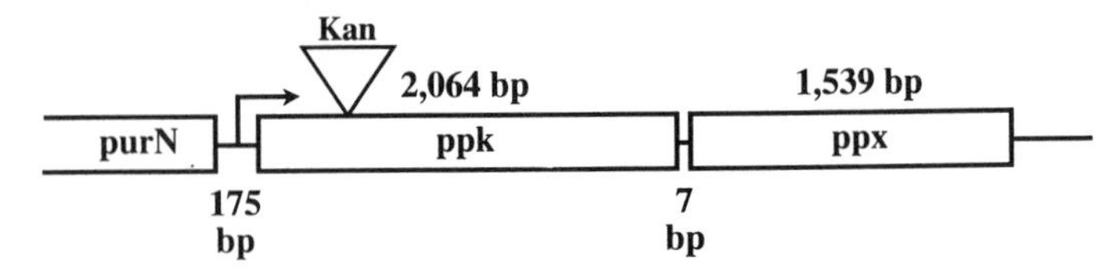

FIG. 2. The *ppk-ppx* operon in *E. coli.*

ogies with the ATP-binding clefts of actin, hexokinase, and ATPases (F. Bazan, personal communication) directs attention to the carboxyl-terminal region, which probably serves as recognition domains of these enzymes. Further examination of the sequences of PPX and GPP may provide insights into how GPP distinguishes long-chain from short-chain polyP and the polyP chains from pppGpp.

PolyP as a Structural Element in Cellular Competence for DNA Entry and Transformation

A membrane domain containing a complex of polyhydroxybutyrate, Ca^{2+}, and polyP in a residue ratio of 2:1:2 was identified in the competent state of a variety of bacteria, including *E. coli* (Reusch and Sadoff, 1988). The physical nature of this structure and the mechanism by which it provides a channel for DNA entry are still uncertain. The capacity of ppk^- mutants to develop reduced but adequate competence despite the lack of long polyP chains has led to the discovery of a polyP species of about 60 residues synthesized by a pathway distinct from the PPK pathway (Castuma and Kornberg, unpublished).

Roles of PolyP in Regulation

The finding that a key enzyme in the stringent response for regulation of protein synthesis is a long-chain exopolyPase suggests an involvement of polyP in cellular adjustments to environmental stresses. Preliminary studies of *E. coli* exposed to serine hydroxamate, a molecule that induces the stringent response, confirm earlier qualitative observations of an increase in polyP (Cashel, personal communication). PolyP has also been tied into the intricate pathways of phosphate regulation. An *E. coli* strain with a mutation in *phoU*, a negative regulator of the *pho* regulon, achieves high levels of polyP under certain conditions (Rao et al., 1985). The presence of two putative *phoB* boxes upstream of the *ppk* operon and diminished expression of *ppk* in *phoB* mutants (Rao and Kornberg, unpublished) are further indications of polyP involvements in phosphate metabolism and deserve further study.

Still other observations point to the likely importance of polyP in regulation of gene expression. In the stationary phase in *E. coli,* a marked decrease in transcription and a selection of stress-related promoters over those of biosynthesis genes have been observed. These features have been attributed to binding of RNA polymerase by polyP (Ishihama, personal communication). The holoenzyme, isolated from stationary-phase cells when treated with PPK and ADP to remove polyP, displayed the features of the exponential-phase holoenzyme. Related to these observations may be the poor survival and deficient responses to stress already indicated for *ppk* mutants.

The likely capacity of a strong polyanion such as polyP to bind basic proteins (e.g., histones) and the domains of proteins, such as those of the polymerases that bind polyanionic templates (i.e., RNA and DNA), suggests a variety of regulatory functions in cells. More extensive studies of polyP in a wide array of cells and tissues should provide clearer views of the ways in which polyP may be used to effect metabolic changes and determine developmental patterns.

Yeast PolyPases

Soluble exopolyPase of *S. cerevisiae* (scPPX1). The soluble exopolyPase of *S. cerevisiae,* purified near 7,000-fold to apparent homogeneity, is monomeric with a molecular mass of 40 kDa (Wurst and Kornberg, 1994). It acts as an exoenzyme in a processive mode, releasing P_i residues from long polyP chains until PP_i is reached (Fig. 3).

$$\text{PolyP}_n + (n - 2)\text{H}_2\text{O} \rightarrow (n - 2)\text{P}_i + \text{PP}_i$$

PolyP of all lengths examined are used as substrates, with a preference for those of about 250 residues (Table 1). These are degraded with a k_{cat}/K_m ratio near the limit for diffusion-controlled reactions. At 37°C, the enzyme releases about 500 P_i molecules per second. It does not act on PP_i, ATP, or the cyclic form of tripolyP. For optimal activity the enzyme requires magnesium, manganese, or cobalt.

TABLE 1. Influence of chain length on kinetics of the yeast scPPX1 reaction[a]

Mean chain length (range)[b]	K_m (μM)	k_{cat}/K_m (10^{-6} M^{-1} s^{-1})	k_{cat} (s^{-1})	k_{cat} (P_i release) (s^{-1})
3	140	1.3	180	550
10 (7–13)	3.9	16	63	630
25 (20–30)	0.16	91	15	360
50 (32–65)	0.060	140	8.4	420
100 (70–130)	0.024	140	3.3	330
250 (100–400)	0.004	290	1.2	290
500 (300–1,000)	0.050	23	1.0	520

[a]$PolyP_{500}$ was partially hydrolyzed with HCl and fractionated by column chromatography, with selected fractions serving as substrates. The values are given in terms of polymer concentrations, except for the column on the right.

[b]The values for the chain lengths are averages with a size range that includes at least 90% of the total substrate shown in parentheses.

The enzyme is a powerful catalyst, increasing the uncatalyzed hydrolytic rate of a phosphoanhydride bond by 10^{11}-fold. The high k_{cat}/K_m value, especially for $polyP_{250}$, suggests that its action is limited more by diffusion than by substrate binding and the catalytic steps required for the complete degradation of a polyP chain. The lower k_{cat}/K_m values for the short chains are more likely to be caused by substrate binding than by the hydrolysis itself. However, the preference of yeast scPPX1 for longer chains in terms of K_m values cannot be attributed simply to the relatively higher concentration of total P_i residues at a given concentration of long chains compared with the same concentration of shorter chains. Even when concentrations are expressed in terms of P_i residues, the K_m values for longer chains are considerably lower than those for shorter ones. Furthermore, $polyP_{500}$, the longest chain examined, displayed a higher K_m than did a chain half that size. These results suggest that the enzyme can discriminate at the substrate-binding step. After formation of the enzyme-substrate complex, P_i residues are released at a rate independent of substrate length (Table 1).

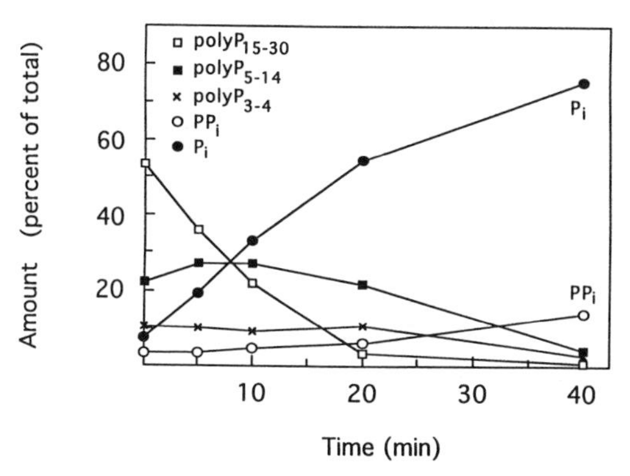

FIG. 3. Products of scPPX1 hydrolysis of short substrates. ^{32}P-labeled $polyP_{500}$ (16 nM) was partially hydrolyzed for 26 min at 96°C near pH 2. A mixture of chain lengths (1 to 45 residues; 200 μM in P_i residues) was incubated with the purified enzyme (7.3 ng/ml) for different intervals. Phospho-imager analyses of polyacrylamide gel electrophoresis (20% polyacrylamide) patterns were divided into five classes (P_i, PP_i, $polyP_{3-4}$, $polyP_{5-14}$, and $polyP_{15-45}$).

The specific activity of scPPX1 is 36-fold higher than that of *E. coli* PPX and 3.6-fold higher than that of another polyPase from *S. cerevisiae* (Andreeva and Okorokov, 1993). The latter is also distinguishable from scPPX1 by its association with the cell membrane, dependence on detergents, a higher K_m for long chains, relative indifference to salt, and a more pronounced activation by Co^{2+}.

Yeast scPPX1 is probably localized in the cytosol; the gene has been located on chromosome VIII. Mutants prepared by insertion of the *LEU3* gene grow at a normal rate, but more refined studies of polyP metabolism may reveal the physiological consequences of this defect. The 200-fold overexpression of the gene in *E. coli* has permitted the ready preparation of the enzyme, at present the most potent reagent for analysis of polyP chains of a wide range of sizes.

EndopolyPase of *S. cerevisiae* (scPPN). In mutants deficient in scPPX1, another polyPase has been identified, with the following distinctive characteristics: (i) it is probably localized in the vacuole; (ii) it is dependent on protein processing, as indicated by its relative absence in extracts of strains deficient in vacuolar proteases; and (iii) it attacks polyP chains to yield shorter chains without release of P_i.

PolyP in Animal Cells and Tissues

Although the presence of polyP in animal cells and tissues had been widely observed (Gabel and Thomas, 1971), the analytic methods were not sufficiently definitive and reliable to identify and quantitate polyP. We have used two novel assays for the identity and abundance of polyP: (i) converting it to ATP with PPK and (ii) converting it to P_i with scPPX1. Results of the two assays have agreed to within 5%.

Various cell lines (NIH 3T3, Jurkat, mouse L, 293, and Vero) incorporated $^{32}P_i$ into polyP (in a 60-min interval) in amounts from 2.5 to 11.4 pmol (expressed as P_i) per 10^6 cells; the chain lengths ranged from 30 to 100 residues. From mouse and rat tissues (brain, kidney, liver, lung, and heart), the levels of polyP (50 to 800 residues long) varied between 0.27 and 1.1 nmol/mg of protein, an abundance roughly 1% that of DNA, and with some enrichment in the nucleus.

Summary

Inorganic polyP, previously regarded as a molecular fossil from prebiotic times, should now be viewed as a dynamic molecule in which its energetically active phosphoanhydride bonds can be attached at the terminal linkages by enzymes that transfer water, glucose, AMP, or ADP or by hydrolytic attack on internal linkages. Widely distributed throughout plant and animal kingdoms, polyP may be used in a variety of ways depending on the species, cell, subcellular localization, and the physiologic adjustments to growth, differentiation, stress, and deprivation. Adaptation to "life in the slow lane" seems especially worthy of attention in view of the current focus in biological studies on exponential or rapid growth. The longevity of an individual and survival of a species depend on being able to manage in the stationary phase.

In these studies, emphasis has been placed on the isolation of enzymes from *E. coli* and *S. cerevisiae* to obtain definitive reagents for analysis of polyP in extracts of cells and tissues and to provide, through "reverse genetics," insights into the physiological roles of polyP in cellular growth, metabolism, and development.

I acknowledge the talented colleagues who came to work with me on DNA replication and let me persuade them to spend their precious time on an arcane subject. I owe them—Kyunghe Ahn, Masahiro Akiyama, Leroy Bertsch, Celina Castuma, Elliott Crooke, Jay Keasling, Anand Kumble, Pradeep Ramulu, Narayana Rao, Rohini Vij, and Helmut Wurst—my gratitude for their courage in joining me in an effort to make a largely forgotten polymer unforgettable. I also want to acknowledge Akiro Ishihama for information regarding the *sur* phenotype of the *ppk* mutant. To Annamaria Torriani-Gorini, whose vision, initiative, and skills made this symposium possible, I am greatly indebted.

I am grateful to the NIH and Human Frontier Science Program for support of my research.

REFERENCES

Ahn, K., and A. Kornberg. 1990. Polyphosphate kinase from *Escherichia coli. J. Biol. Chem.* **265:**11734–11739.

Akiyama, M., E. Crooke, and A. Kornberg. 1992. The polyphosphate kinase gene of *Escherichia coli.* Isolation and sequence of the *ppk* gene and membrane location of the protein. *J. Biol. Chem.* **267:**22556–22561.

Akiyama, M., E. Crooke, and A. Kornberg. 1993. An exopolyphosphatase of *Escherichia coli.* The enzyme and its *ppx* gene in a polyphosphate operon. *J. Biol. Chem.* **268:**633–639.

Andreeva, N. A., and L. A. Okorokov. 1993. Purification and characterization of highly active and stable polyphosphatase from *Saccharomyces cerevisiae* cell envelope. *Yeast* **9:**127–139.

Bazan, F. Personal communication.

Cashel, M. Personal communication.

Castuma, C., and A. Kornberg. Unpublished results.

Crooke, E., M. Akiyama, N. N. Rao, and A. Kornberg. 1994. Genetically altered levels of inorganic polyphosphate in *Escherichia coli. J. Biol. Chem.* **269:**6290–6295.

Gabel, N. W., and V. Thomas. 1971. Evidence for the occurrence and distribution of inorganic polyphosphate in vertebrate tissues. *J. Neurochem.* **18:**1229–1242.

Hara, A., and J. Sy. 1983. Guanosine 5′-triphosphate, 3′-diphosphate, 5′-phosphohydrolase. *J. Biol. Chem.* **258:**1678–1683.

Ishihama, A. Personal communication.

Keasling, J. D., L. Bertsch, and A. Kornberg. 1993. Guanosine pentaphosphate phosphohydrolase of *Escherichia coli* is a long-chain exopolyphosphatase. *Proc. Natl. Acad. Sci. USA* **90:**7029–7033.

Kornberg, A., S. R. Kornberg, and E. S. Simms. 1956. Metaphosphate synthesis by an enzyme from *Escherichia coli. Biochim. Biophys. Acta* **20:**215–227.

Kornberg, S. R. 1957. Adenosine triphosphate synthesis from polyphosphate by an enzyme from *Escherichia coli. Biochim. Biophys. Acta* **26:**294–300.

Kulaev, I. S. 1979. *The Biochemistry of Inorganic Polyphosphates.* John Wiley & Sons, Inc., New York.

Kulaev, I. S., and V. M. Vagabov. 1983. Polyphosphate metabolism in microorganisms. *Adv. Microb. Physiol.* **24:**83–171.

Kulaev, I. S., V. M. Vagabov, and Y. A. Shabalin. 1987. New data on biosynthesis of polyphosphates in yeast, p. 233–238. *In* A. Torriani-Gorini, F. G. Rothman, S. Silver, A. Wright, and E. Yagil (ed.), *Phosphate Metabolism and Cellular Regulation in Microorganisms.* American Society for Microbiology, Washington, D.C.

Rao, N. N., and A. Kornberg. Unpublished observations.

Rao, N. N., M. F. Roberts, and A. Torriani. 1985. Amount and chain length of polyphosphates in *Escherichia coli* depend on cell growth conditions. *J. Bacteriol.* **162:**242–247.

Reusch, R. N., and H. L. Sadoff. 1988. Putative structure and functions of a poly-β-hydroxybutyrate/calcium polyphosphate channel in bacterial plasma membranes. *Proc. Natl. Acad. Sci. USA* **85:**4176–4180.

Wood, H. G., and J. E. Clark. 1988. Biological aspects of inorganic polyphosphates. *Annu. Rev. Biochem.* **57:**235–260.

Wurst, H., and A. Kornberg. 1994. A soluble exopolyphosphatase of *Saccharomyces cerevisiae. J. Biol. Chem.* **269:**10996–11001.

Chapter 35

Molecular Genetics of Polyphosphate Accumulation in *Escherichia coli*

HARDOYO, KATSUFUMI YAMADA, AYAKO MURAMATSU, YUKI ANBE, JUNICHI KATO, AND HISAO OHTAKE

Department of Fermentation Technology, Hiroshima University, Higashi-Hiroshima, Hiroshima 724, Japan

The outbreak of algal blooms is one of the environmental concerns resulting from eutrophication in lakes and other surface waters (Codd and Bell, 1985). Algal blooms degrade water quality by producing an offensive odor and taste. The nuisance growth of algae renders boating and fishing difficult and discourages swimming. Excessive growth of algae consumes dissolved oxygen when the algae are decomposed by aerobic bacteria, causing mass mortality of fish and other aquatic organisms. Algal toxin production is also a serious problem in drinking water supplies (Wicks and Thiel, 1990).

Phosphorus compounds are essential constituents in organisms. Since P_i is often found to be the limiting factor for algal growth in nature, the removal of P_i from wastewaters is believed to be important in the control of eutrophication of surface waters (Levin et al., 1975; Deinema et al., 1980). Activated-sludge processes, which are commonly used for treating wastewaters, are very effective in removing organic pollutants, but they remove P_i relatively poorly. Under ordinary operating conditions, activated sludges are capable of removing an average of only 20 to 40% of the P_i concentrations normally found in municipal wastewaters (Carberry and Tenney, 1973). Municipal wastewaters are relatively low in sources of carbon, and this limits the removal of P_i by sludge microorganisms. Therefore, to make activated sludges more effective in removing P_i, it appears essential to enable sludge microorganisms to take up and store P_i in excess of their requirements for growth.

Many microorganisms accumulate excess P_i in the form of polyphosphate (polyP) under unfavorable growth conditions such as low pH, anaerobiosis, or sulfur starvation (Harold, 1966). Some bacterial species have also been found to take up P_i far in excess of their requirements for growth and accumulate polyP after being subjected to P_i starvation (Harold, 1963; Ohtake et al., 1985). In such microorganisms, the key enzyme for polyP synthesis is generally polyP kinase (PPK). The PPK-encoding gene (*ppk*) was first cloned from *Escherichia coli* by Akiyama et al. (1992). More recently, we cloned and sequenced the PPK-encoding gene from *Klebsiella aerogenes* (formerly named *Aerobacter aerogenes*) (Kato et al., 1993b). With the cloning of the gene encoding PPK, it became possible to genetically engineer polyP accumulation in bacteria. We report here the use of genetic engineering to improve the ability of bacteria to accumulate polyP by using *E. coli* as a test organism.

Manipulation of Genes Involved in PolyP Accumulation

In *E. coli*, P_i is transported through the cytoplasmic membrane via either the high-affinity phosphate-specific transport (Pst) system or the low-affinity phosphate inorganic transport (Pit) system. The Pst system of *E. coli* is a multicomponent system consisting of several membrane proteins and a soluble periplasmic binding protein. The entire *pst* region of the *E. coli* chromosome has been cloned and sequenced, revealing the existence of five genes: *pstS*, *pstC*, *pstA*, *pstB*, and *phoU*, transcribed anticlockwise in the *E. coli* chromosome (Amemura et al., 1985; Surin et al., 1985). The products of the first four genes are required for the transport of P_i, and, together with the fifth gene, they are involved in the regulation of the *pho* regulon. We constructed two recombinant plasmids, pEP02.2 and pEP02.5, both of which carried the *pst* genes from *E. coli* MV1184 (Fig. 1). The *pst* genes were expressed at high levels from the *tac* promoter in pEP02.2 and from the *tet* promoter in pEP02.5 under conditions of P_i excess.

The enzymes involved in polyP synthesis are polyP kinase (ATP:polyP phosphotransferase) and 1,3-diphosphoglycerate:polyP phosphotransferase (Wood and Clark, 1988). In *E. coli*, the *ppk* gene encoding PPK possesses an open reading frame which translates to a protein of 687 amino acids and a calculated mass of 80,278 Da (Akiyama et

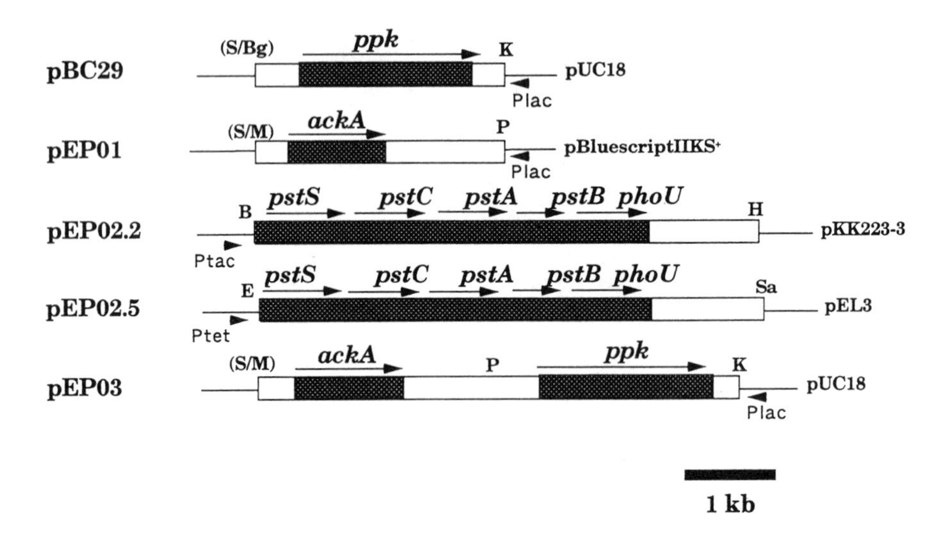

FIG. 1. Recombinant plasmids used in the present study (Kato et al., 1993a). Open boxes (with shaded regions) indicate the cloned *E. coli* DNAs, and thin lines indicate vectors. Arrows show the open reading frames and their transcriptional directions. The sites and directions of *lac*, *tac*, and *tet* promoters are indicated by arrowheads. Restriction sites: B, *Bam*HI; Bg, *Bgl*I; E, *Eco*RI; H, *Hin*dIII; K, *Kpn*I; M, *Mlu*I; P, *Pst*I; S, *Sma*I; Sa, *Sal*I. The sites in parentheses were destroyed during construction.

al., 1992). The nucleotide sequence analysis of *K. aerogenes ppk* showed that the PPK protein shared 93% identical amino acid residues with *E. coli* PPK protein (Kato et al., 1993b). Possible *pho* box sequences were found in the promoter regions of both the *ppk* genes, indicating that their expression is regulated by the *phoB* product (Kato et al., 1993b). In this study we used pBC29 (Akiyama et al., 1993), which contained *E. coli ppk* together with its own promoter, to increase the gene dosage of *ppk* in *E. coli* MV1184 (Fig. 1).

PPK polymerizes the terminal phosphate of ATP into polyP in a freely reversible reaction (nATP $\rightleftharpoons$ nADP+ polyP$_n$). We used acetate kinase (ACK) as an ATP regeneration system for polyP synthesis. ACK, which catalyzes the formation of ATP and acetate in the presence of acetyl phosphate and ADP, is known to be involved in ATP synthesis (Thauer et al., 1977). Overproduction of ACK was therefore expected to elevate the intracellular ATP concentration in *E. coli*. Plasmid pEP01 contains *E. coli ackA*, encoding ACK, and its own promoter (Lee et al., 1990). Although the transcriptional direction of the *ackA* gene is the reverse of that of the *lac* promoter, the levels of ACK activity in *E. coli* MV1184(pEP01) were more than 100-fold higher than the parental level. Plasmid pEP03 contained both *ppk* and *ackA*.

The utilization and degradation of polyP is catalyzed by polyPases and by specific kinases including polyPglucokinase and polyPfructokinase (Wood and Clark, 1988). The *ppx* gene encoding an exopolyphosphatase (PPX) has been found downstream of the *ppk* gene in *E. coli* (Akiyama et al., 1993), as well as in *K. aerogenes* (Kato et al., 1993b), constituting a polyP operon. Insertion of *kan* into *E. coli ppk* on the chromosome of *E. coli* NM522 (*recA*$^+$) disrupted not only PPK but also PPX activities.

Rate-Limiting Step for P$_i$ Accumulation

P$_i$ uptake experiments were performed with *E. coli* MV1184(pUC18), a control strain, and its recombinant derivatives (Fig. 2). Cells were grown in Luria broth (L broth) and used for P$_i$ uptake experiments in T medium (Harold, 1963) without being subjected to P$_i$ limitation. The growth of these recombinant strains, except for MV1184(pBC29 plus pEP02.2), was almost equivalent to that of the control strain. The control strain removed about 20% of the P$_i$ from the medium during the first 3 h. However, no P$_i$ uptake occurred after growth had stopped. Similar results were also obtained with *E. coli* EJ500, HB101, and W3110 (data not shown). These *E. coli* strains, including MV1184, did not accumulate high levels of polyP, even after being subjected to P$_i$ limitation. Strain MV1184(pBC29) removed twice as much P$_i$ from the medium as did the control strain. Thus, increasing the dosage of *ppk* alone doubled the phosphorus content of *E. coli* MV1184 (Fig. 3). The rate of P$_i$ removal greatly increased when pEP03 (*ppk ackA*) was introduced into MV1184, indicating that ACK functioned as an effective system for ATP regeneration. Over-

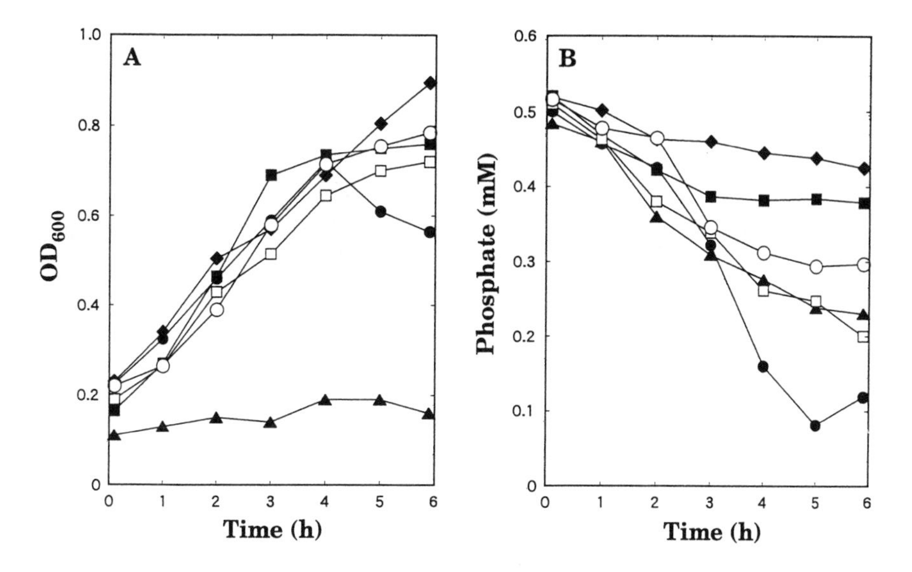

FIG. 2. Growth of (A) and P_i removal by (B) *E. coli* MV1184 containing pUC18 (■), pBC29 (○), pEP01 (♦), pEP02.2 (□), pEP03 (●), or pEP02.2 plus pBC29 (▲) in T medium containing 0.5 mM P_i. Cells were grown in 2× YT medium (Sambrook et al., 1989) at 37°C with shaking for 7 h, inoculated into L broth (a 1% inoculum), and incubated for 9 h under the same conditions. For cultures of recombinant strains containing pEP02.2, 0.5 mM IPTG was added to L broth, and the mixture was incubated for 9 h to induce the expression of the *pst* genes. Cells were transferred to T medium containing 0.5 mM potassium phosphate until the optical density at 600 nm (OD_{600}) reached about 0.2. The cell suspension was incubated at 37°C with constant stirring, and samples were taken at intervals for determination of growth and P_i concentration. P_i was assayed as described previously (Ohtake et al., 1985). Data from Kato et al. (1993a).

production of ACK did not cause derepression of the *pho* genes in MV1184. No increase in alkaline phosphatase activity was observed during P_i uptake experiments with MV1184(pEP03) (data not shown). This recombinant strain removed approximately 90% of the P_i from the medium within 4 h. However, it later released P_i to the medium, probably because of cell lysis (Fig. 2).

Interestingly, strain MV1184(pEP02.2) (*pst* genes) removed more P_i from the medium than did MV1184(pBC29). The phosphorus content of MV1184(pEP02.2) was approximately twice that of MV1184(pBC29) (Fig. 3). This finding may indicate that P_i transport across the cell membrane is a rate-limiting step for P_i accumulation in *E. coli* MV1184 under conditions of P_i excess. Drastic changes in growth and P_i uptake were observed when pBC29 (*ppk*) and pEP02.2 (*pst* genes) were simultaneously introduced into MV1184. Growth of this recombinant was limited in T medium, even though it grew well in 2× YT medium (Sambrook et al., 1989) and L broth. Nevertheless, this recombinant removed approximately threefold more P_i than did the control strain. Consequently, the phosphorus content of this recombinant reached 12.3% on dry weight basis, or approximately 10-fold more than that of the control. Such a high level of P_i accumulation in bacteria has never been reported before. The fractionation of cellular phosphorus revealed that acid-soluble and acid-insoluble polyPs accounted for approximately 65% of the total cellular phosphorus of MV1184(pBC29, pEP02.2) (data not shown). No detectable release of P_i was observed with this recombinant. In pEP02.2, the *pst* genes were placed under the control of the *tac* promoter, and isopropyl-β-D-thiogalactopyranoside (IPTG) was required for inducing their expression. Plasmid pEP02.5 was constructed to solve this problem by placing the *pst* genes under the (constitutive) control of the *tet* promoter. MV1184(pEP02.5) removed P_i at a rate similar to that seen with MV1184(pEP02.2) without addition of IPTG (data not shown).

Strain MV1184 containing both pEP02.2 (*pst* genes) and pEP03 (*ppk ackA*) was also constructed. However, introduction of both plasmids appeared to be detrimental to MV1184. The cell viability of this recombinant decreased soon after the cells entered the stationary phase of growth in L broth. As a result, subsequent P_i uptake experiments in T medium could not be performed with this recombinant. The ability of *E. coli* NM522(pBC29) to accumulate P_i was not improved by removing PPX activity by disrupting the polyP operon on its chromosome (data not shown). This is probably because (i) *ppk* was carried on a high-copy-number plasmid, pUC18, and

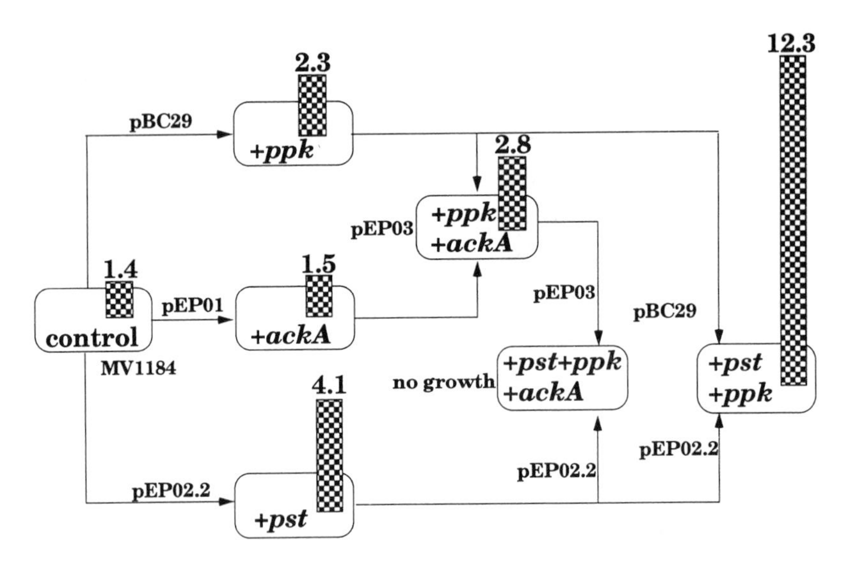

FIG. 3. Phosphorus content of MV1184(pUC18) and the five recombinant derivatives. The number above each bar represents the content of cellular phosphorus expressed as a percentage of dry cell weight. Arrows indicate each step of genetic improvement of *E. coli* MV1184 for enhanced P_i accumulation. Total cellular phosphorus was determined by ammonium persulfate digestion followed by the P_i determination (Ohtake et al., 1985). Data from Kato et al. (1993a).

(ii) several enzymes, other than PPX, were also responsible for the degradation of polyP (Wood and Clark, 1988).

PolyP Accumulation Capacity

E. coli cells were found to release polyP into the medium when they accumulated excessive levels of polyP. This seems to be the mechanism by which a further increase in cellular polyP is limited. The release of polyP was first observed during P_i uptake experiments with MV1184(pBC29, pEP02.2) (Kato et al., 1993a). In these experiments, the total phosphorus content in the culture supernatant was found to increase at around 4 h after the start of incubation, even though the P_i concentration continued to decrease. As a consequence, the phosphorus content of this recombinant, which reached a maximum of 16% (49% as P_i) at 4 h, decreased to 12% by 6 h. Most of the phosphorus compounds, released to the medium, were acid labile (1 N HCl for 7 min at 100°C) and were hydrolyzed by alkaline phosphatase, liberating P_i. When the phosphorus compounds, precipitated with ethanol and subjected to polyacrylamide gel electrophoresis, were stained with toluidine blue, a metachromatic shift similar to that observed with purified polyP (Griffin et al., 1965) was detected. The rate of polyP release, estimated from the increase in the concentration of acid-labile phosphorus in the culture supernatant, was dependent on that of P_i uptake. No polyP release was observed after cells completely removed P_i from the medium, but it resumed soon after P_i was added to the medium. Once *E. coli* cells accumulated excessive levels of polyP, the rate of polyP release became essentially equivalent to that of P_i uptake (Fig. 4). Nevertheless, these cells did not release a detectable amount of polyP unless P_i was added to the culture. When P_i uptake was inhibited by 0.1 mM carbonyl cyanide *m*-chlorophenylhydrazone (CCCP), no polyP release occurred. In addition, neither P_i uptake nor polyP release was observed at 4°C.

The mechanism for polyP release is unclear. We detected a pool of polyP in MV1184(pBC29 pEP02.2) by means of in vivo high-resolution ^{31}P nuclear magnetic resonance spectroscopy (Kato and Ohtake, unpublished). The intensity of the polyP signal significantly increased when the cell pellets were treated with EDTA, a membrane-impermeable chelator. In contrast, the signal was broadened by the addition of a shift reagent, praseodymium (Rao et al., 1985), so that it could not be seen above the background. These results suggest the presence of surface polyP that can be readily released to the medium. Akiyama et al. (1992) reported that PPK, purified from overexpressing cells, preferentially associated with the outer membrane, even though ADP and ATP are substrates for the enzyme. They speculated that the enzyme may be located in Bayer patches (Bayer, 1968), described as fusions of inner and outer membranes communicating directly between

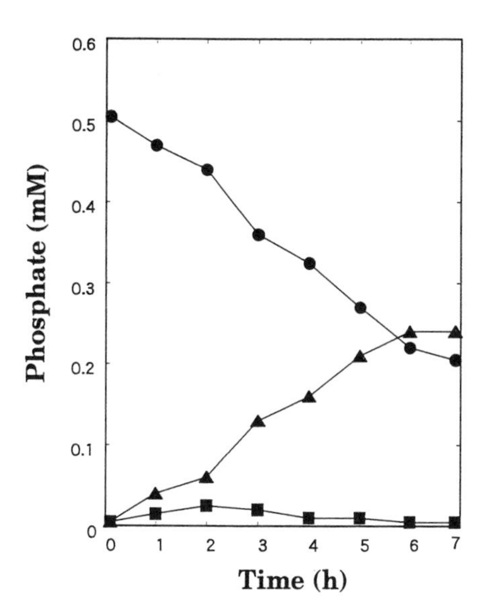

FIG. 4. Concentration changes of P_i (●), acid-labile phosphorus (▲), and acid-stable phosphorus (■) in the culture supernatant during the P_i uptake experiment with MV1184(pBC29, pEP02.2). Cells were grown as described in the legend to Fig. 2 and harvested by centrifugation. The cell pellets were resuspended in T medium containing 0.5 mM P_i until an optical density at 600 nm of about 0.12 was reached. The culture was incubated at 37°C with constant stirring. Concentrations of acid-labile and acid-stable phosphorus were determined as previously described (Kato et al., 1993a).

the cell exterior and interior compartments. If so, the orientation of PPK may be related to the release of polyP. Understanding of this mechanism is essential to prevent *E. coli* recombinants from releasing polyP. In parallel with P_i accumulation, *E. coli* cells were found to take up Mg^{2+} from the medium. Increasing the concentration of Mg^{2+} in the medium improved the growth of and P_i removal by *E. coli* cells. However, the addition of Mg^{2+} did not prevent the cells from releasing polyP.

Concluding Remarks

The ability of *E. coli* to take up and store excess P_i was improved by modifying the genetic regulation and increasing the dosage of the genes involved in the key steps of P_i transport and polyP metabolism. P_i transport across the inner membrane is a possible rate-limiting step for P_i accumulation in MV1184 under conditions of P_i excess. Strain MV1184(pBC29, pEP02.2) accumulated a maximum of 16% of its dry weight as phosphorus (49% as P_i), of which approximately 65% was stored in the form of polyP. However, *E. coli* cells, which accumulated excessive levels of polyP, released it into the medium. This seems to be the mechanism by which a further increase in cellular polyP is limited.

We gratefully acknowledge M. Akiyama and A. Kornberg for providing pBC29 and for open exchanges of information. We also thank A. Nakata for providing pMK800.

This work was supported in part by the Ebara Research Co. Ltd.

REFERENCES

Akiyama, M., E. Crooke, and A. Kornberg. 1992. The polyphosphate kinase gene of *Escherichia coli. J. Biol. Chem.* **267:**22556–22561.

Akiyama, M., E. Crooke, and A. Kornberg. 1993. An exopolyphosphatase of *Escherichia coli:* the enzyme and its *ppx* gene in a polyphosphate operon. *J. Biol. Chem.* **268:**633–639.

Amemura, M., K. Makino, H. Shinagawa, A. Kobayashi, and A. Nakata. 1985. Nucleotide sequence of the genes involved in phosphate transport and regulation of the phosphate regulon in *Escherichia coli. J. Mol. Biol.* **184:**241–250.

Bayer, M. E. 1968. Areas of adhesion between wall and membrane of *Escherichia coli. J. Gen. Microbiol.* **53:**395–404.

Carberry, J. B., and M. W. Tenney. 1973. Luxury uptake of phosphate by activated sludge. *J. Water Pollut. Control Fed.* **45:**2444–2462.

Codd, G. A., and S. G. Bell. 1985. Eutrophication in freshwaters. *Water Pollut. Control* **84:**225–232.

Deinema, M. H., L. H. A. Habets, J. Scholten, E. Turkstra, and H. A. A. M. Webers. 1980. The accumulation of polyphosphate in *Acinetobacter* spp. *FEMS Microbiol. Lett.* **9:**275–279.

Griffin, J. B., N. M. Davidian, and R. Penniall. 1965. Studies of phosphorus metabolism by isolated nuclei. VII. Identification of polyphosphate as a product. *J. Biol. Chem.* **240:**4427–4434.

Harold, F. M. 1963. Accumulation of inorganic polyphosphate in *Aerobacter aerogenes.* I. Relationship to growth and nucleic acid synthesis. *J. Bacteriol.* **86:**216–221.

Harold, F. M. 1966. Inorganic polyphosphate in biology: structure, metabolism and function. *Bacteriol. Rev.* **30:**772–794.

Kato, J., and H. Ohtake. Unpublished data.

Kato, J., K. Yamada, A. Muramatsu, Hardoyo, and H. Ohtake. 1993a. Genetic improvement of *Escherichia coli* for the enhanced biological removal of phosphate. *Appl. Environ. Microbiol.* **59:**3744–3749.

Kato, J., T. Yamamoto, K. Yamada, and H. Ohtake. 1993b. Cloning, sequence and characterization of the polyphosphate kinase-encoding gene (*ppk*) of *Klebsiella aerogenes. Gene* **137:**237–242.

Lee, T.-Y., K. Makino, H. Shinagawa, and A. Nakata. 1990. Overproduction of acetate kinase activates phosphate regulon in the absence of the *phoR* and *phoM* functions in *Escherichia coli. J. Bacteriol.* **172:**2245–2249.

Levin, G. V., G. J. Topol, and A. G. Tarnay. 1975. Operation of full-scale biological phosphorus removal plant. *J. Water Pollut. Control Fed.* **47:**577–590.

Ohtake, H., K. Takahashi, Y. Tsuzuki, and K. Toda. 1985. Uptake and release of phosphate by a pure culture of *Acinetobacter calcoaceticus. Water Res.* **19:**1587–1594.

Rao, N. N., M. F. Roberts, and A. Torriani. 1985. Amount and chain length of polyphosphates in *Escherichia coli* dependent on cell growth conditions. *J. Bacteriol.* **162:**205–211.

Sambrook, J., E. F. Fritsch, and T. Maniatis. 1989. *Molecular Cloning: a Laboratory Manual,* 2nd ed. Cold Spring Harbor Laboratory, Cold Spring Harbor, N.Y.

Surin, B. P., H. Rosenberg, and G. B. Cox. 1985. Phosphate-specific transport system of *Escherichia coli:* nucleotide sequence and gene-polypeptide relationships. *J. Bacteriol.* **161:**189–198.

Thauer, R. K., K. Jungermann, and K. Decker. 1977. Energy conservation in chemotrophic anaerobic bacteria. *Bacteriol. Rev.* **41:**100–180.

Wicks, J. R., and P. G. Thiel. 1990. Environmental factors affecting the production of peptide toxins in floating scums of the cyanobacterium *Microcystis aeruginosa* in a hypertrophic african reservoir. *Environ. Sci. Technol.* **24:**1413–1418.

Wood, H. G., and J. E. Clark. 1988. Biological aspects of inorganic polyphosphates. *Annu. Rev. Biochem.* **57:**235–260.

Chapter 36

Phosphate-Regulated Genes for the Utilization of Phosphonates in Members of the Family *Enterobacteriaceae*

BARRY L. WANNER

Department of Biological Sciences, Purdue University, West Lafayette, Indiana 47907

Bacteria such as *Escherichia coli* and *Enterobacter aerogenes* use three kinds of phosphorus sources for growth: inorganic phosphate (P_i), organophosphates, and phosphonates (Pn) (Wanner, 1990). When the preferred P source (P_i) is in excess, P_i is taken up by a low-affinity P_i transporter (such as Pit), and genes for high-affinity P_i uptake and for use of alternative P sources are repressed. The latter are coregulated as members of the Pho regulon and correspond to a subset of phosphate-starvation-inducible (*psi*) genes (Wanner, 1993). Although we now know much about both P_i and organophosphate metabolism, we know relatively little about Pn metabolism. Pn are a large class of organophosphorus molecules that have direct carbon-phosphorus (C-P) bonds in place of the more familiar carbon-oxygen-phosphorus ester bond. Therefore, bacteria that use Pn as a P source must be able to break the C-P bond. Over 7 years ago (shortly before the first PHO Meeting in Concarneau, France), the *E. coli psiD* locus was shown to be required for Pn utilization (Wackett et al., 1987b). Subsequent studies showed that the *psiD* locus is itself a 14-gene cluster that appears to encode both a binding protein-dependent Pn transporter and a C-P lyase (Chen et al., 1990; Metcalf and Wanner, 1991, 1993a, 1993c; Makino et al., 1991; Wanner and Metcalf, 1992). In addition, *Enterobacter aerogenes* was shown to have genes for two separate Pn degradative pathways: a C-P lyase pathway and a phosphonatase pathway (Lee et al., 1992). The existence of two pathways for Pn use (both of which are under Pho regulon control) implies that Pn are an important alternative phosphorus source in nature.

The *E. coli* Pho regulon is composed of more than 30 *psi* genes. These genes encode Psi proteins that mostly appear to have a role in the process of P assimilation, because their genes are induced when P_i is limiting for growth (Wanner, 1993). Such a role has been demonstrated for all of the many Psi proteins (Wanner, 1987a, 1987b) whose role is now known (Table 1). These include the PstSCAB proteins for high-affinity P_i uptake (Webb et al., 1992), the UgpBAEC proteins for uptake of the transportable organophosphate ester *sn*-glycerol-3-phosphate and its associated UgpQ protein for hydrolysis of diesters taken up by this transporter (Overduin et al., 1988; Brzoska et al., 1994), bacterial alkaline phosphatase (Bap) for extracellular (periplasmic) degradation of nontransportable organophosphates (Chang et al., 1985), the PhoE porin for entry into the periplasm of large polyanions (where they may be degraded [Tommassen et al., 1992]), the PhnC to PhnP proteins for uptake and intracellular breakdown of phosphonates (Metcalf and Wanner, 1993c), and others. Also, it has been proposed that the PhoU protein, which is encoded together with the Pst transporter in the *pstSCAB-phoU* operon, may have a role in intracellular P_i metabolism (Steed and Wanner, 1993). It is therefore reasonable to propose that other *psi* genes under Pho regulon control (such as *psiE*, *psiF*, and *psiH* [*phoH*]), whose functions are unknown (Wanner, 1993), also play a role in P assimilation. In addition, several non-Psi proteins may provide alternative means of P assimilation, although a role(s) for these proteins in P assimilation may be secondary (Wanner, 1990). Such non-Psi proteins include Pit for the low-affinity P_i transporter (Rosenberg, 1987), which may function only in certain laboratory *E. coli* K-12 strains (our unpublished results), GlpT and UhpT for uptake systems for glycerol-3-phosphate and hexose phosphates, respectively (whose synthesis is carbon regulated for use of these compounds as a carbon source [Larson et al., 1992; Island et al., 1992]), and the periplasmic degradative enzymes AppA and Agp (whose syntheses are also not P_i regulated [Dassa et al., 1991; Pradel et al., 1990]).

The greatest number of Psi proteins with a known function are involved in Pn utilization. The *E. coli psiD* (*phn*) locus encodes 14 Psi proteins (named PhnC to PhnP) for Pn uptake and breakdown by a C-P lyase pathway (Wackett et al., 1987; Wanner and Boline, 1990; Chen et al., 1990; Metcalf and Wanner, 1991, 1993a, 1993c; Makino et al., 1991). *E. aerogenes* contains a homologous set of genes, which are apparently

TABLE 1. Phosphorus acquisition in *E. coli* and related bacteria[a]

Environmental P source	Assimilation mechanisms (uptake or metabolism)	
	Psi proteins	Non-Psi proteins
P_i	PstSCAB, PhoU	Pit
Phosphate esters		
Transportable	UgpBAEC, UgpQ	GlpT, UhpT
Nontransportable	Bap, PhoE	AppA, Agp
Pn	PhnC to PhnP	None

[a]A variety of environmental P sources may be used as the sole P source for growth. Assimilation mechanisms involve uptake systems as well as a variety of intracellular or extracellular enzymes, depending on the P source and growth conditions. Psi, phosphate starvation inducible; PstSCAB, P_i-specific (high affinity) transporter; PhoU, regulatory protein associated with transporter, which may also have an unknown role in P_i metabolism; Pit, low-affinity P_i transporter; UgpBAEC, uptake system for *sn*-glycerol-3-phosphate; UgpQ, phosphodiesterase; GlpT, carbon-regulated glycerol-3-phosphate uptake system; UhpT, transporter for hexose phosphates; Bap, bacterial alkaline phosphatase; PhoE, anion porin; AppA, acid phosphatase; Agp, glucose-1-phosphatase: PhnC-to-PhnP, Pn uptake and C-P lyase components. See the text for citations.

arranged in two (or more) unlinked loci (Lee et al., 1992). In addition, *E. aerogenes* has a set of genes for an alternative Pn degradative pathway, the phosphonatase pathway, which is absent in *E. coli.* The latter but not the former genes are also present in *Salmonella typhimurium* (Wanner and Boline, 1990). Altogether, there now appear to exist 20 or more different *psi* genes for Pn utilization among these bacteria (Metcalf et al., unpublished).

In this chapter, I will briefly summarize our current understanding about Pn utilization by the *E. coli* C-P lyase pathway. Our own work began shortly before the 1986 PHO Meeting with the discovery that the *psiD* (*phn*) locus was required for Pn utilization (Wackett et al., 1987b). Some recent studies on Pn degradation have been summarized elsewhere (Wanner, 1992, in press; Wanner and Metcalf, 1992). In addition, two treatises on Pn metabolism have been published (Hilderbrand, 1982; Hori et al., 1984). The reader is referred to these works for earlier studies on Pn. There has also been recent progress on Pn biosynthesis (see, e.g., Seidel et al. [1992] and Wohlleben et al. [1992]), which is not covered in this chapter.

Early Studies on Pn Degradation

The ability of bacteria to metabolize Pn (as well as phosphite, also called phosphonic acid) has been known for a long time (Casida, 1960; Harkness, 1966; Rosenberg and La Nauze, 1967; Zeleznick et al., 1963; Cook et al., 1978, 1979; Daughton et al., 1979). These early studies established the existence of two pathways for Pn breakdown, a phosphonatase pathway (LaNauze et al., 1970; Dumora et al., 1983, 1989; Olsen et al., 1988) and a C-P lyase pathway (Shames et al., 1987; Wackett et al., 1987a; Cordeiro et al., 1986). These pathways differ in their catalytic mechanism of labilizing the C-P bond and, consequently, in their substrate specificity. The phosphonatase pathway involves a two-step process in which the substrate α-aminoethylphosphonate is first converted to an aldehyde by an α-aminoethylphosphonate transaminase. The formation of an aldehyde at the 2-carbon leads to destabilization of the C-P bond and allows for its hydrolytic cleavage by an enzyme (phosphonoacetaldehyde hydrolase), which has the trivial name phosphonatase. Both the transaminase and phosphonatase have been purified. Because C-P bond cleavage by the phosphonatase requires a keto group on the 2-carbon, the phosphonatase pathway acts only on compounds with a substitution on the 2-carbon that can be converted to an aldehyde. The natural substrate of this pathway is α-aminoethylphosphonate, which is quite common in nature. In contrast, the C-P lyase pathway leads to a direct dephosphonation. It acts on a wide variety of substituted and unsubstituted alkylphosphonates as well as arylphosphonates. Those Pn shown to be broken down by the *E. coli* C-P lyase (on the basis of mutant studies) include the 2-carbon-substituted Pn α-aminoethylphosphonate, the unsubstituted Pn ethylphosphonate and methylphosphonate, phosphonoacetate, and phosphite (Fig. 1) (Metcalf and Wanner, 1991). Early studies also demonstrated that phosphite can be metabolized by an oxidative pathway (Malacinski and Konetzka, 1966, 1967), which is distinct from the C-P lyase pathway.

The direct dephosphonation of methylphosphonate, ethylphosphonate, or benzylphosphonate leads to formation of the corresponding hydrocarbon methane, ethane, and benzene, respectively. Because the C-P bonds in compounds such as these are resistant to strong-acid or strong-base catalysis (Fig. 2) (Shames et al., 1987), C-P bond fission by a lyase cannot involve a hydrolytic mechanism. Both radical and redox mechanisms

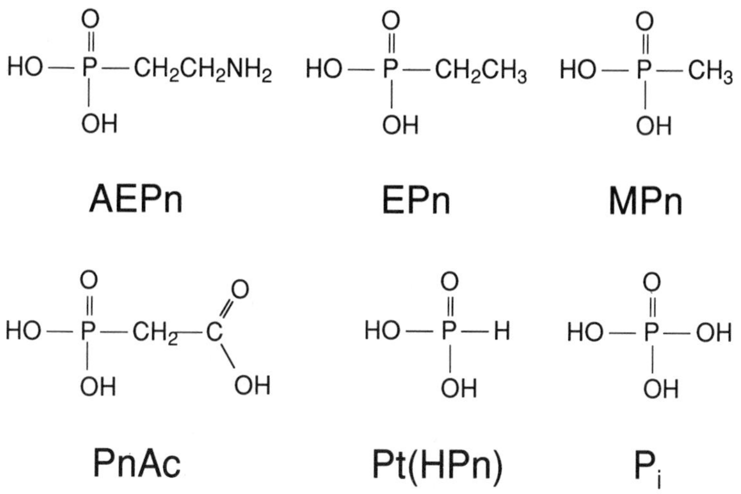

$HO-P(=O)(OH)-CH_2CH_2NH_2$ AEPn

$HO-P(=O)(OH)-CH_2CH_3$ EPn

$HO-P(=O)(OH)-CH_3$ MPn

$HO-P(=O)(OH)-CH_2-C(=O)OH$ PnAc

$HO-P(=O)(OH)-H$ Pt(HPn)

$HO-P(=O)(OH)-OH$ P_i

FIG. 1. Structures of some representative Pn and phosphate. AEPn, α-aminoethylphosphonate; EPn, ethylphosphonate; MPn, methylphosphonate; PnAc, phosphonoacetate; Pt (HPn), phosphite (also called phosphonic acid). Modified from Fig. 6 of Metcalf and Wanner (1991) with permission.

have been proposed on the basis of in vivo studies of product formation (Shames et al., 1987; Avila et al., 1987, 1991; Avila and Frost, 1989). Studies on the catalytic mechanism have been hampered by the inability to detect a C-P lyase in a cell-free system (see the review by Wanner and Metcalf [1992]). Recently, a third pathway for C-P bond cleavage that is specific for phosphonoacetate has also been identified (McMullan and Quinn, 1994); like the phosphonatase pathway, it appears to involve a hydrolase.

Involvement of the Pho Regulon in Pn Degradation

Because of the presence of the C-P lyase pathway, *E. coli* produces methane as a by-product of methylphosphonate utilization as a P source. Since *E. coli* does not ordinarily make methane, methane formation is a sensitive assay for a functional C-P lyase. The finding that the amount of methane produced by *E. coli* increased during P_i limitation provided the first clue that the synthesis

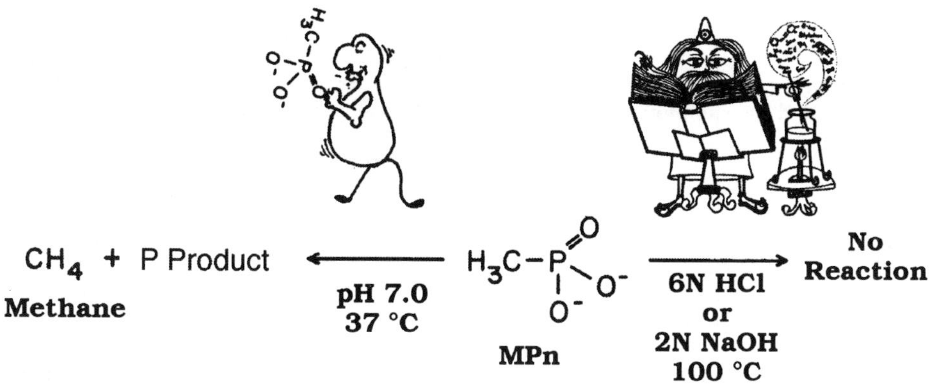

CH_4 + P Product (Methane) ← (pH 7.0, 37 °C) — $H_3C-P(=O)(O^-)O^-$ (MPn) — (6N HCl or 2N NaOH, 100 °C) → No Reaction

FIG. 2. Cleavage of the C-P bond of an alkylphosphonate. The C-P bond of an alkylphosphonate such as methylphosphonate (MPn) is resistant to acid-base catalysis. Indeed, one method for determination of Pn phosphorus involves measuring the amount of organophosphorus remaining after strong-acid hydrolysis (6 N HCl at 120°C for 4 h), since the C-P bond resists such treatment (Snyder and Law, 1970). However, the C-P bond is broken by a direct dephosphonation by bacteria containing a C-P lyase. C-P bond cleavage by a lyase leads to release of a P product (whose identity has yet to be established [Wanner and Metcalf, 1992; Wanner, in press]), which can be used for growth, and the corresponding hydrocarbon. This figure has been adapted from an unpublished one created by L. P. Wackett and C. T. Walsh.

of the *E. coli* C-P lyase might be under Pho regulon control. This was substantiated by testing the effect of Pho regulon regulatory mutations on methane production (Wackett et al., 1987b). In brief, methane production was abolished in a *phoB* mutant, which lacks the transcriptional activator of the Pho regulon, and was rendered constitutive in mutants that expressed the Pho regulon constitutively. Also, both the *phoA* gene (for bacterial alkaline phosphatase) and the *pst* genes (for high-affinity P_i uptake) were shown not to be required.

In an earlier study, a large collection of mutants in which the *lacZ* gene had been genetically fused to a number of different *psi* promoters were isolated (Wanner et al., 1981; Wanner and McSharry, 1982). Because these mutants were made by using a transposon, each had a particular *psi* gene interrupted by the transposon insertion. By testing these *psi* mutants for Pn use, it was discovered that mutants carrying an insertion in the *psiD* locus were Pn negative. Such mutants were unable to use any of a variety of Pn as a P source, and they were also blocked in methane production from methylphosphonate (Wackett et al., 1987b).

It was also observed that wild-type *E. coli* K-12 used Pn only after a pronounced lag (Wackett et al., 1987b). This lag period was necessary because it allowed for the accumulation of mutants, which subsequently used Pn without an extended lag. In other words, Pn utilization was cryptic in *E. coli* K-12 because it required an activating mutation. In contrast, Pn utilization was functional in many other strains (in particular *E. coli* B), because they used Pn without a (pronounced) lag. Genetic mapping experiments showed that the activating mutation(s) was linked to the *psiD* locus and that *E. coli* K-12 and *E. coli* B differed at this locus (Wanner and Boline, 1990).

Molecular Biology of C-P Bond Cleavage by C-P Lyase and Phosphonatase Pathways

Genes for Pn use were cloned by complementation of the *psiD* mutants for growth on methylphosphonate as a sole P source. To clone genes for a functional C-P system, an *E. coli* B gene library was used to complement a mutant with the *psiD* locus deleted. Unexpectedly, all plasmids that allowed for Pn use had very large DNA inserts. Because mapping and cloning experiments showed that at least 10 kbp was necessary for complementation (Wanner and Boline, 1990), a 15.6-kbp insert of one such plasmid was sequenced in its entirety (Chen et al., 1990). This revealed a potential operon of 17 genes, named in alphabetical order *phnA* through *phnQ* (Fig. 3). It was also shown that most of these open reading frames were translated in a T7 promoter expression system. Subsequent experiments showed that an 11.4-kbp region carrying genes *phnC* through *phnP* was sufficient for complementation of a *phn* (*psiD*) deletion mutant (Metcalf and Wanner, 1993a). A *psi* promoter immediately preceding the *phnC* gene was identified (Metcalf and Wanner, 1991, 1993a; Makino et al., 1991). Computational analyses of the predicted *phn* gene products indicated that some were likely to form a binding protein-dependent transporter (Chen et al., 1990). This was substantiated by mutational studies (Metcalf and Wanner, 1991, 1993c). The cryptic *E. coli* K-12 *phn* locus has been cloned by two laboratories (Loo et al., 1987; Makino et al., 1991). DNA sequence analysis of the *phnC* to *phnP* genes from *E. coli* K-12 revealed the presence of a frameshift mutation within the *E. coli* K-12 *phnE* gene (Makino et al., 1991). The functional *E. coli* B *phn* locus has a tandem 8-bp repeat within the *phnE* gene, and the cryptic *E. coli* K-12 *phn* locus has three copies of the 8-bp repeat (which causes a frameshift mutation). An activating mutation occurs by deletion of one 8-bp repeat.

Mutational Studies on a 14-Gene Cluster for Pn Degradation

All 14 genes, *phnC* to *phnP*, were mutated to determine the role of their gene products (Fig. 4). Because these genes are likely to be arranged in an operon, mutational effects had to be distinguished from effects due to polarity on a downstream gene. Polarity effects were assessed by measuring the expression of a reporter gene (the *uidA* gene for β-glucuronidase [Metcalf and Wanner, 1993b]) that had been introduced immediately downstream of the *phnC*-to-*phnP* gene cluster. Transposon-induced mutations were isolated on a plasmid, the sites were determined by restriction mapping and DNA sequencing, and these mutations were then recombined onto the chromosome (Metcalf and Wanner, 1993c). Many of these mutations were made by using Tn*phoA* and Tn*phoA'* elements, which allow for determination of transcription, translation, and cell surface localization determinants at the same site in a gene (Wilmes-Riesenberg and Wanner, 1992; Wanner, 1994). Results indicated that the *phnC*, *phnD*, and *phnE* gene products are required for transport; that the *phnG*, *phnH*, *phnI*, *phnJ*, *phnK*, *phnL*, and *phnM* gene products are required for catalysis; that the *phnN* and *phnP* gene products are conditionally required for catalysis (and therefore may have an accessory role in catalysis); and that the *phnF* and *phnO* gene products are unnecessary for transport or catalysis (Metcalf and Wanner, 1993c) and therefore may have a regulatory role because they have motifs in common with other regulatory proteins (Chen et al., 1990; Haydon and Guest, 1991). Accordingly, PhnC may be the ATP-de-

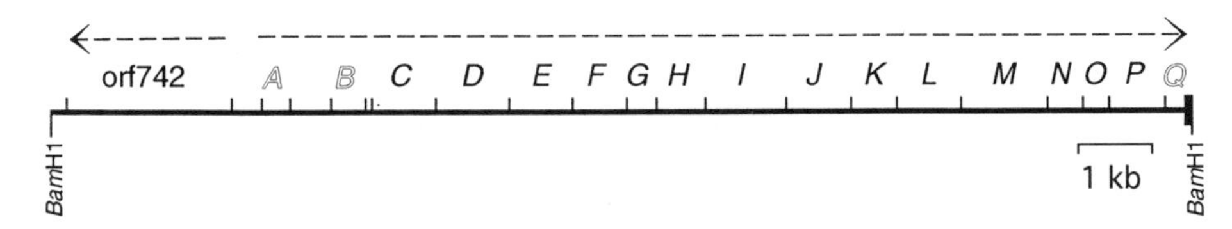

FIG. 3. Organization of genes in a Pn-positive complementing plasmid. The complete DNA sequence of a 15.6-kbp fragment has been determined (Chen et al., 1990). Arrows show the orientation of transcription inferred from the DNA sequence. A large open reading frame in the leftward orientation (orf742) was shown not to be required for Pn utilization. Genes named *phnA* to *phnQ* (in alphabetical order) were originally thought to be required. All three *psiD*::Mu d1 insertions were shown (by DNA sequence analysis) to lie within an approximately 100-bp segment of the *phnD* gene (Metcalf et al., 1990). Those genes in hollow letters were subsequently shown not to be required (Metcalf and Wanner, 1993a).

pendent permease component, PhnD may be the periplasmic binding protein, and PhnE may be the integral membrane component of a binding protein-dependent Pn transport system (Metcalf and Wanner, 1991). Also, PhnG, PhnH, PhnI, PhnJ, PhnK, PhnL, and PhnM may form a C-P lyase enzyme complex (Metcalf and Wanner, 1993c), and this complex may be membrane associated because PhnM is probably an integral membrane protein on the basis of its sequence (Chen et al., 1990). It should be mentioned that our evidence for a Pn transport system is indirect. It is based on the finding (Metcalf and Wanner, 1991) that mutations in the *phnC*, *phnD*, and *phnE* genes (but not in the other eleven *phn* genes [Metcalf and Wanner, 1993c]) abolish the use of an organophosphate (phosphoserine) for growth under conditions where its use is dependent upon a cytoplasmic phosphatase (the *serB* gene product). Transport assays were not done because of the unavailability

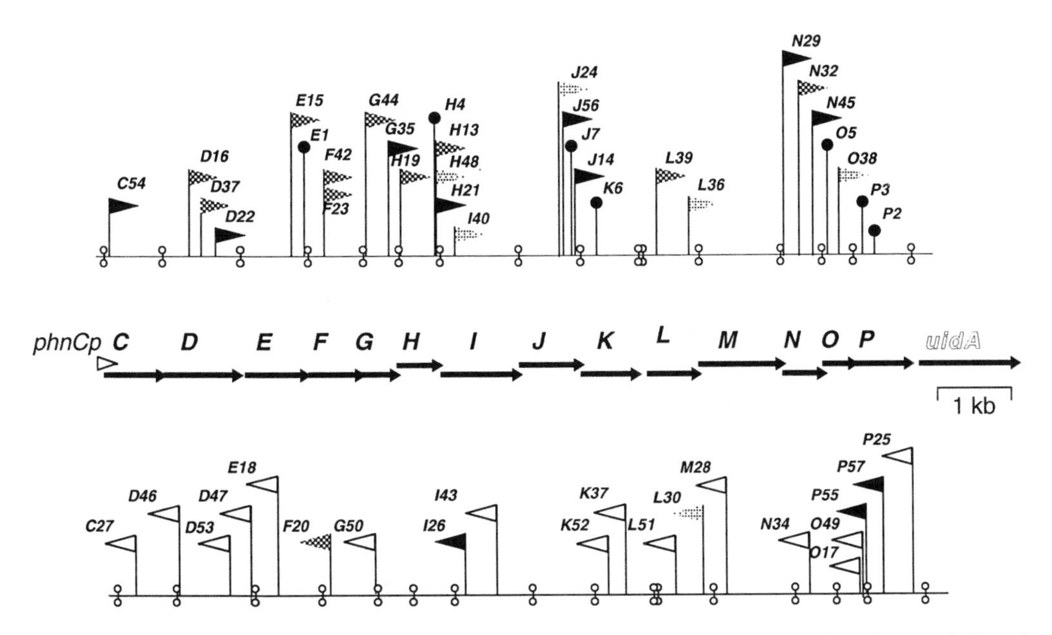

FIG. 4. Mutational analysis of the *phnC*-to-*phnP* gene cluster. Sites of 50 transposon insertions are indicated. Solid circles show insertion sites of Tn*5* elements (Metcalf and Wanner, 1991). Flags show insertion sites of Tn*phoA'* elements (Metcalf and Wanner, 1993c). The top line shows all Tn*5* insertions and the Tn*phoA'* insertions in line with rightward transcription. The bottom line shows Tn*phoA'* insertions in the opposite orientation. Dumbbells mark the starts and stops of the *phn* gene. The shading of flags indicates the reading frame at the insertion site. Solid flags indicate those that would form in-frame gene fusions when switched to Tn*phoA* and Tn*phoA'* elements (Wilmes-Riesenberg and Wanner, 1992; Wanner, 1994). Reprinted from Metcalf and Wanner (1993c) with permission.

of a suitable radiolabeled Pn from commercial sources. Preliminary data obtained by using radiolabeled α-aminoethylphosphonate purified from membranes of ^{32}-P-labeled *Tetrahymena* cells are consistent with the conclusion that the *phnCDE* genes encode a Pn transporter (Yakovleva and Wanner, unpublished). New studies are now directed toward testing these conclusions and further extending this work.

Prospectus

New studies on Pn degradation are likely to be greatly facilitated by recent molecular biology and genetic studies on the genes for both the C-P lyase and phosphonatase pathways. Immediate goals will involve assignment of gene functions by further mutational studies of the genes and biochemical characterization of the gene products. In the case of the C-P lyase pathway, this will involve development of a cell-free system. It is now clear that this may require the isolation of an enzyme complex. In the case of the phosphonatase pathway, new studies will involve determination of the catalytic mechanism and the residues involved in the active site. This should be aided by new knowledge of the primary sequences of the gene products for Pn uptake and breakdown by the phosphonatase pathway (Metcalf et al., unpublished [b]).

My laboratory is supported by NIH grant GM35392 and NSF grant DMB9108005.

I thank Larry Wackett, Spencer Shames, and Chris Walsh for innumerable helpful discussions concerning Pn degradation over the past several years. I am also especially grateful to Deepak Agrawal, Weihong Jiang, Ki-Sung Lee, Bill Metcalf, and Galina Yakovleva who contributed to studies on Pn degradation in my laboratory.

REFERENCES

Avila, L. Z., K. M. Draths, and J. W. Frost. 1991. Metabolites associated with organophosphonate C-P bond cleavage: chemical synthesis and microbial degradation of [^{32}P]-ethylphosphonic acid. *Bioorg. Med. Chem. Lett.* **1:**51–54.

Avila, L. Z., and J. W. Frost. 1989. Phosphonium ion fragmentations relevant to organophosphonate biodegradation. *J. Am. Chem. Soc.* **111:**8969–8970.

Avila, L. Z., S. H. Loo, and J. W. Frost. 1987. Chemical and mutagenic analysis of aminomethylphosphonate biodegradation. *J. Am. Chem. Soc.* **109:**6758–6764.

Brzoska, P., M. Rimmele, K. Brzostek, and W. Boos. 1994. The *pho* regulon-dependent Ugp uptake system for glycerol-3-phosphate of *Escherichia coli* is *trans* inhibited by P_i. *J. Bacteriol.* **176:**15–20.

Casida, L. E. 1960. Microbial oxidation and utilization of orthophosphite during growth. *J. Bacteriol.* **80:**237–241.

Chang, C. N., W.-J. Kuang, and E. Y. Chen. 1985. Nucleotide sequence of the alkaline phosphatase gene of *Escherichia coli. Gene* **44:**121–125.

Chen, C.-M., Q. Ye, Z. Zhu, B. L. Wanner, and C. T. Walsh. 1990. Molecular biology of carbon-phosphorus bond cleavage: cloning and sequencing of the *phn* (*psiD*) genes involved in alkylophosphonate uptake and C-P lyase activity in *Escherichia coli* B. *J. Biol. Chem.* **265:**4461–4471.

Cook, A. M., C. G. Daughton, and M. Alexander. 1978. Phosphonate utilization by bacteria. *J. Bacteriol.* **133:**85–90.

Cook, A. M., C. G. Daughton, and M. Alexander. 1979. Benzene from bacterial cleavage of the carbon-phosphorus bond of phenylphosphonates. *J. Biochem.* **184:**453–455.

Cordeiro, M. L., D. L. Pompliano, and J. W. Frost. 1986. Degradation and detoxification of organophosphonates: cleavage of the carbon to phosphorus bond. *J. Am. Chem. Soc.* **108:**332–334.

Dassa, J., H. Fishi, C. Marck, M. Dion, M. Kieffer-Bontemps, and P. L. Boquet. 1991. A new oxygen-regulated operon in *Escherichia coli* comprises the genes for a putative third cytochrome oxidase and for pH 2.5 acid phosphatase (*appA*). *Mol. Gen. Genet.* **229:**341–352.

Daughton, C. G., A. M. Cook, and M. Alexander. 1979. Biodegradation of phosphonate toxicants yields methane or ethane on cleavage of the C-P bond. *FEMS Microbiol. Lett.* **5:**91–93.

Dumora, C., A.-M. Lacoste, and A. Cassaigne. 1983. Purification and properties of 2-aminoethylphosphonate:pyruvate aminotransferase from *Pseudomonas aeruginosa. Eur. J. Biochem.* **133:**119–125.

Dumora, C., A.-M. Lacoste, and A. Cassigne. 1989. Phosphonoacetaldehyde hydrolase from *Pseudomonas aeruginosa:* purification properties and comparison with *Bacillus cereus* enzyme. *Biochim. Biophys. Acta* **997:**193–198.

Harkness, D. R. 1966. Bacterial growth on aminoalkylphosphonic acids. *J. Bacteriol.* **92:**623–627.

Haydon, D. J., and J. R. Guest. 1991. A new family of bacterial regulatory proteins. *FEMS Microbiol. Lett.* **79:**291–296.

Hilderbrand, R. L. 1982. *The Role of Phosphonates in Living Systems.* CRC Press, Inc., Boca Raton, Fla.

Hori, T., M. Horiguchi, and A. Hayashi. 1984. *Biochemistry of Natural C-P Compounds.* Maruzen, Ltd., Kyoto, Japan.

Island, M. D., B.-Y. Wei, and R. J. Kadner. 1992. Structure and function of the *uhp* genes for the sugar phosphate transport system in *Escherichia coli* and *Salmonella typhimurium. J. Bacteriol.* **174:**2754–2762.

LaNauze, J. M., H. Rosenberg, and D. C. Shaw. 1970. The enzymic cleavage of the carbon-phosphorus bond: purification and properties of phosphonatase. *Biochim. Biophys. Acta* **212:**332–350.

Larson, T. J., J. S. Cantwell, and A. T. Van Loo-Bhattacharya. 1992. Interaction at a distance between multiple operators controls the adjacent, divergently transcribed *glpTQ-glpACB* operons of *Escherichia coli* K-12. *J. Biol. Chem.* **267:**6114–6121.

Lee, K.-S., W. W. Metcalf, and B. L. Wanner. 1992. Evidence for two phosphonate degradative pathways in *Enterobacter aerogenes. J. Bacteriol.* **174:**2501–2510.

Loo, S. H., N. K. Peters, and J. W. Frost. 1987. Genetic characterization of an *Escherichia coli* mutant deficient in organophosphonate degradation. *Biochem. Biophys. Res. Commun.* **148:**148–152.

Makino, K., S.-K. Kim, H. Shinagawa, M. Amemura, and A. Nakata. 1991. Molecular analysis of the cryptic and functional *phn* operons for phosphonate use in *Escherichia coli* K-12. *J. Bacteriol.* **173:**2665–2672.

Malacinski, G., and W. A. Konetzka. 1966. Bacterial oxidation of orthophosphite. *J. Bacteriol.* **91:**578–582.

Malacinski, G. M., and W. A. Konetzka. 1967. Orthophosphite-nicotinamide adenine dinucleotide oxidoreductase from *Pseudomonas fluorescens. J. Bacteriol.* **93:**1906–1910.

McMullan, G., and J. P. Quinn. 1994. In vitro characterization of a phosphate starvation-independent carbon-phosphorus bond cleavage activity in *Pseudomonas fluorescens* 23F. *J. Bacteriol.* **176:**320–324.

Metcalf, W. W., W. Jiang, K.-S. Lee, and B. L. Wanner. Unpublished results (a).

Metcalf, W. W., W. Jiang, and B. L. Wanner. Unpublished results (b).

Metcalf, W. W., P. M. Steed, and B. L. Wanner. 1990. Identification of phosphate-starvation-inducible genes in *Escherichia coli* K-12 by DNA sequence analysis of *psi::lacZ*(Mud1) transcriptional fusions. *J. Bacteriol.* **172**:3191–3200.

Metcalf, W. W., and B. L. Wanner. 1991. Involvement of the *Escherichia coli phn* (*psiD*) gene cluster in assimilation of phosphorus in the form of phosphonates, phosphite, P_i esters, and P_i. *J. Bacteriol.* **173**:587–600.

Metcalf, W. W., and B. L. Wanner. 1993a. Evidence for a fourteen-gene, *phnC* to *phnP*, locus for phosphonate metabolism in *Escherichia coli. Gene* **129**:27–32.

Metcalf, W. W., and B. L. Wanner. 1993b. Construction of new β-glucuronidase cassettes for making transcriptional fusions and their use with new methods for allele replacement. *Gene* **129**:17–25.

Metcalf, W. W., and B. L. Wanner. 1993c. Mutational analysis of an *Escherichia coli* fourteen-gene operon for phosphonate degradation using Tn*phoA*′ elements. *J. Bacteriol.* **175**:3430–3442.

Olsen, D. B., T. W. Hepburn, M. Moose, P. S. Mariano, and D. Dunaway-Mariano. 1988. Investigation of the *Bacillus cereus* phosphonoacetaldehyde hydrolase. Evidence for a Schiff base mechanism and sequence analysis of an active-site peptide containing the catalytic lysine residue. *Biochemistry* **27**:2229–2234.

Overduin, P., W. Boos, and J. Tommassen. 1988. Nucleotide sequence of the *ugp* genes of *Escherichia coli* K-12: homology to the maltose system. *Mol. Microbiol.* **2**:767–775.

Pradel, E., C. Marck, and P. L. Boquet. 1990. Nucleotide sequence and transcriptional analysis of the *Escherichia coli agp* gene encoding periplasmic acid glucose-1-phosphatase. *J. Bacteriol.* **172**:802–807.

Rosenberg, H. 1987. Phosphate transport in prokaryotes, p. 205–248. *In* B. P. Rosen and S. Silver (ed.), *Ion Transport in Prokaryotes.* Academic Press, Inc., San Diego, Calif.

Rosenberg, H., and J. M. La Nauze. 1967. The metabolism of phosphonates by microorganisms. The transport of aminoethylphosphonic acid in *Bacillus cereus. Biochim. Biophys. Acta* **141**:79–90.

Seidel, H. M., D. L. Pompliano, and J. R. Knowles. 1992. Phosphonate biosynthesis: molecular cloning of the gene for phosphoenolpyruvate mutase from *Tetrahymena pyriformis* and overexpression of the gene product in *Escherichia coli. Biochemistry* **31**:2598–2608.

Shames, S. L., L. P. Wackett, M. S. LaBarge, R. L. Kuczkowski, and C. T. Walsh. 1987. Fragmentative and stereochemical isomerization probes for homolytic carbon to phosphorus bond scission catalyzed by bacterial carbon-phosphorus lyase. *Bioorg. Chem.* **15**:366–373.

Snyder, W. R., and J. H. Law. 1970. A quantitative determination of phosphonate phosphorus in naturally occurring aminophosphonates. *Lipids* **5**:800–802.

Steed, P. M., and B. L. Wanner. 1993. Use of the *rep* technique for allele replacement to construct mutants with deletions of the *pstSCAB-phoU* operon: evidence of a new role for the PhoU protein in the phosphate regulon. *J. Bacteriol.* **175**:6797–6809.

Tommassen, M. J., Struyvé, and H. de Cock. 1992. Export and assembly of bacterial outer membrane proteins. *Antonie Leeuwenhoek* **61**:81–85.

Wackett, L. P., S. L. Shames, C. P. Venditti, and C. T. Walsh. 1987a. Bacterial carbon-phosphorus lyase: products, rates, and regulation of phosphonic and phosphinic acid metabolism. *J. Bacteriol.* **169**:710–717.

Wackett, L. P., B. L. Wanner, C. P. Venditti, and C. T. Walsh. 1987b. Involvement of the phosphate regulon and the *psiD* locus in the carbon-phosphorus lyase activity of *Escherichia coli* K-12. *J. Bacteriol.* **169**:1753–1756.

Wanner, B. L. 1987a. Phosphate regulation of gene expression in *Escherichia coli*, p. 1326–1333. *In* F. C. Neidhardt, J. L. Ingraham, K. B. Low, B. Magasanik, M. Schaechter, and H. E. Umbarger (ed.), *Escherichia coli and Salmonella typhimurium: Cellular and Molecular Biology*, vol. 2. American Society for Microbiology, Washington, D.C.

Wanner, B. L. 1987b. Bacterial alkaline phosphatase gene regulation and the phosphate response in *Escherichia coli*, p. 12–19. *In* A. Torriani-Gorini, F. G. Rothman, S. Silver. A. Wright, and E. Yagil (ed.), *Phosphate Metabolism in Microorganisms.* American Society for Microbiology, Washington, D.C.

Wanner, B. L. 1990. Phosphorus assimilation and its control of gene expression in *Escherichia coli*, p. 152–163. *In* G. Hauska and R. Thauer (ed.), *The Molecular Basis of Bacterial Metabolism.* Springer-Verlag KG, Heidelberg, Germany.

Wanner, B. L. 1992. Genes for phosphonate biodegradation in *Escherichia coli*, p. 1–6. *In* S. K. Ballal (ed.), *Southern Association of Agricultural Scientists Bulletin of Biochemistry and Biotechnology*, vol. 5. Southern Association of Agricultural Scientists, Cookeville, Tenn.

Wanner, B. L. 1993. Gene regulation by phosphate in enteric bacteria. *J. Cell. Biochem.* **51**:47–54.

Wanner, B. L. 1994. Gene expression in bacteria using Tn*phoA* and Tn*phoA*′ elements to make and switch *phoA* gene, *lacZ* (op), and *lacZ* (pr) fusions, p. 291–310. *In* K. W. Adolph (ed.), *Methods in Molecular Genetics*, vol. 3. Academic Press, Inc., Orlando, Fla.

Wanner, B. L. An overview, molecular genetics of carbon-phosphorus bond cleavage in bacteria. *Biodegradation*, in press.

Wanner, B. L., and J. A. Boline. 1990. Mapping and molecular cloning of the *phn* (*psiD*) locus for phosphonate utilization in *Escherichia coli. J. Bacteriol.* **172**:1186–1196.

Wanner, B. L., and R. McSharry. 1982. Phosphate-controlled gene expression in *Escherichia coli* using Mu*dl*-directed *lacZ* fusions. *J. Mol. Biol.* **158**:347–363.

Wanner, B. L., and W. W. Metcalf. 1992. Molecular genetic studies of a 10.9-kbp operon in *Escherichia coli* for phosphonate uptake and biodegradation. *FEMS Microbiol. Lett.* **100**:133–140.

Wanner, B. L., S. Wieder, and R. McSharry. 1981. Use of bacteriophage transposon Mu*dl* to determine the orientation for three *proC*-linked phosphate-starvation-inducible (*psi*) genes in *Escherichia coli* K-12. *J. Bacteriol.* **146**:93–101.

Webb, D. C., H. Rosenberg, and G. B. Cox. 1992. Mutational analysis of the *Escherichia coli* phosphate-specific transport system, a member of the traffic ATPase (or ABC) family of membrane transporters: a role for proline residues in transmembrane helices. *J. Biol. Chem.* **267**:24661–24668.

Wilmes-Riesenberg, M. R., and B. L. Wanner. 1992. Tn*phoA* and Tn*phoA*′ elements for making and switching fusions for study of transcription, translation, and cell surface localization. *J. Bacteriol.* **174**:4558–4575.

Wohlleben, W., R. Alijah, J. Dorendorf, D. Hillemann, B. Nussbaumer, and S. Pelzer. 1992. Identification and characterization of phosphinothricin-tripeptide biosynthetic genes in *Streptomyces viridochromogenes. Gene* **115**:127–132.

Yakovleva, G., and B. L. Wanner. Unpublished results.

Zeleznick, L. D., T. C. Myers, and E. B. Titchener. 1963. Growth of *Escherichia coli* on methyl- and ethylphosphonic acids. *Biochim. Biophys. Acta* **78**:546–547.

VII. PHOSPHOLIPIDS

Chapter 37

Introduction: Lipid Involvement in Protein Translocation in the Prokaryotic Secretion Pathway

EEFJAN BREUKINK AND BEN DE KRUIJFF

Department of Biochemistry of Membranes, Centre for Biomembranes and Lipid Enzymology, University of Utrecht, Padualaan 8, 3584 CH Utrecht, The Netherlands

Newly synthesized precursor proteins are exported out of the cytosol of *Escherichia coli* via secretion mechanisms that generally involve the signal sequence of the precursor, the Sec machinery with proteinaceous components, ATP, and a proton motive force. By using a genetic approach in which the synthesis of acidic phospholipids was blocked by disruption of the *pgsA* gene, it was shown that acidic phospholipids were involved in protein translocation (De Vrije et al., 1988). These authors showed that, under conditions of low acidic phospholipid content in the inner membrane of *E. coli,* the translocation of preproteins was severely hampered both in vivo and in vitro. With the use of an *E. coli* strain (HDL 11) in which the *pgsA* gene was placed under control of the *lac* operon, it was shown that the increase of acidic phospholipid content in the inner membrane was directly proportional to the increase in the translocation efficiency (Kusters et al., 1991). It was also shown, by using a lipid transfer protein-based method, that the translocation activity of inner membranes depleted of acidic phospholipids could be fully restored to wild-type levels by reintroduction of phosphatidylglycerol (PG) or other acidic phospholipids into the membrane (Kusters et al., 1991). In this chapter we will try to dissect the role of acidic phospholipids in the prokaryotic secretion pathways by using two approaches. The first approach will focus on the interaction of SecA with acidic phospholipids and their implications for protein translocation. The second approach will focus on the role of signal sequence-lipid interactions in protein translocation.

SecA-Lipid Interactions

The SecA protein is a cytosolic and peripherally membrane-associated, negatively charged ATPase and plays a central role in the translocation of precursor proteins across the inner membrane of *E. coli.* It interacts with the chaperonin SecB (Hartl et al., 1990) and the integral membrane protein SecY (Lill et al., 1989) as well as with the precursor. It probably uses ATP to initiate membrane translocation of precursors and translocate other domains (Schiebel et al., 1991). Several observations suggest that acidic phospholipids play a role in the translocation activity of SecA. The first observation was made by Lill et al. (1990), who showed that the ATPase activity of SecA was specifically stimulated by acidic phospholipids. This was followed by the observation of Hendrick and Wickner (1991) that the acidic phospholipids together with SecY and SecE provide a high-affinity binding site for SecA at the membrane. An interesting observation concerning the subcellular localization of SecA was made by Oliver et al. (1990). They showed that a portion of SecA which was localized at the membrane could not be extracted by methods which would normally extract peripherally associated membrane proteins. The possibility that SecA could insert into the lipid phase of the membrane during the translocation of a precursor protein was suggested to explain the nonextractable fraction of SecA in the inner membrane (Oliver et al., 1990). We investigated this possibility by using purified components in the monolayer system and showed that SecA could efficiently penetrate a lipid monolayer only in the

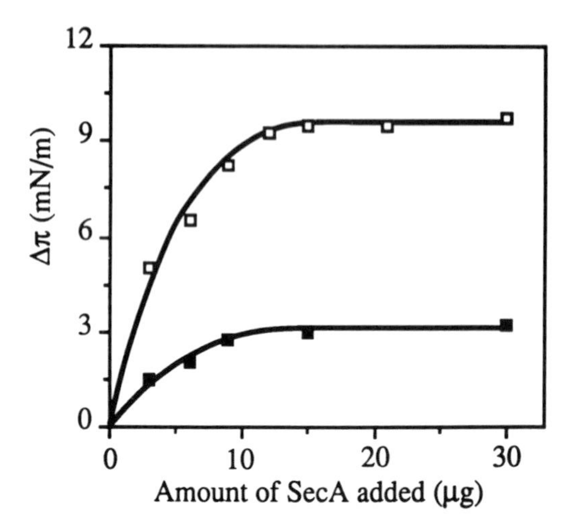

FIG. 1. Increase in surface pressure ($\Delta\pi$) of monolayers of dioleoyl-PC (■) and dioleoyl-PG (□) mediated by SecA injection in the subphase (5 ml of 100 mM NaCl, 50 mM Tris-HCl [pH 7.6]; temperature, 31°C). The initial pressure was 20 mN/m. (See Breukink et al. [1992] for details.)

presence of acidic phospholipids (Breukink et al., 1992). As shown in Fig. 1, the increase in the surface pressure from the lipid monolayer as a result of SecA insertion is much higher for a PG monolayer than for a monolayer of zwitterionic phosphatidylcholine (PC). This leads to the suggestion that the translocation defect in HDL 11 cells with a low anionic lipid content could be explained partly by the decreased membrane insertion of SecA. This suggestion is supported by the observation that efficient penetration of SecA occurred only for inner membrane lipid extracts of wild-type cells and not for lipid extracts isolated from HDL 11 cells (Breukink et al., 1992). Interestingly, nucleotides had large and specific effects on SecA insertion into PG monolayers. It was shown that the presence of ATP inhibited the SecA monolayer insertion, whereas the presence of nonhydrolyzable ATP analogs did not influence the SecA monolayer insertion. These findings, together with the observations of Schiebel et al. (1991), who showed that ATP binding could initiate the translocation of preOmpA, led to the hypothesis of a SecA binding, insertion, and deinsertion cycle (Breukink et al., 1992), in which SecA could facilitate the translocation of a precursor across the inner membrane. Van der Wolk et al. (1993) showed that *Bacillus subtilis* wild-type SecA could also penetrate a PG monolayer and that this penetration was, like that of *E. coli* SecA, modulated by nucleotides. This suggests that the lipid insertion property of SecA is conserved among the prokaryotes. Furthermore, these authors observed that wild-type *B. subtilis* SecA, but not a mutant SecA which was deficient in its translocation ATPase, could be recycled from isolated inner membrane vesicles upon addition of ATP. This observation supported the above proposed working model for SecA. More support for the insertion of SecA in the lipid phase of the membrane came from Ulbrandt et al. (1992), who showed, using vesicle systems instead of monolayer systems, that SecA inserted deeply in the lipid phase of the membrane and that this was accompanied by a partial unfolding of the protein. This unfolding event was also seen by Shinkai et al. (1991), who showed that acidic phospholipids could trigger a conformational change in SecA which resulted in enhanced susceptibility toward the V8 protease. Upon studying the interaction of SecA with lipid vesicles containing acidic phospholipids, we observed an increase in the turbidity of the solution (Breukink et al., 1993). This increase in turbidity was due to vesicle aggregation caused by SecA. The aggregation effect was highly dependent on the ionic strength of the solution (Breukink et al., 1993), whereas the insertion of SecA in a lipid monolayer was not (Breukink et al., 1991). Thus, the aggregation effect was attributed to the existence of two different lipid-binding sites present in the SecA monomer; one binding site was more hydrophobic, leading to insertion of SecA in the lipid phase of the membrane, and the other was more electrostatic, leading to a more superficial interaction of SecA with the membrane lipids. Support for this proposal came from a monolayer experiment, as shown in Fig. 2, in which SecA previously inserted into a PG monolayer (as shown by a surface pressure increase) was able to bind radioactively labeled PG vesicles (as shown by an increase in the surface radioactivity). Moreover, the increase in the ionic strength of the subphase (from 100 to 500 mM NaCl) had an effect only on the SecA-mediated vesicle binding, not on the SecA insertion.

Other studies on the SecA-lipid interaction also supported the existence of at least two sites of interaction with phospholipids on the SecA protein. Cabelli et al. (1991) showed that the membrane-binding determinant probably lies in the amino terminal one-third of the protein, while Ulbrandt et al. (1992) showed that the region after the middle one-third of the SecA primary structure is probably involved in the interaction with the membrane lipids. These findings and the observation that the SecA molecule probably consists of more than one domain (Weaver et al., 1992) make the existence of a second lipid-binding site more likely. The function of a second lipid-binding site on the SecA molecule may lie in the targeting of the SecA-SecB-precursor complex via two-dimensional diffusion to the integral SecY/E complex to form translocase (see Fig. 5). An interesting observation was made by Kusters et al. (1992): in the

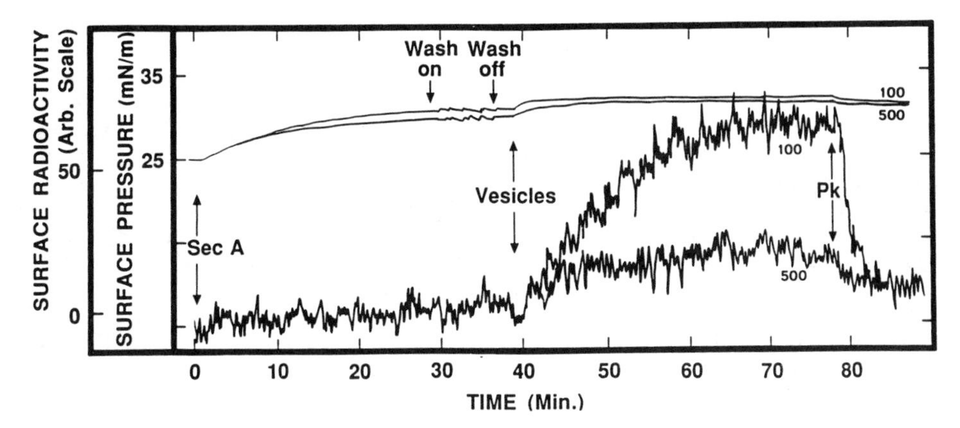

FIG. 2. SecA causes close contact between a dioleoyl-PG monolayer and dioleoyl-PG vesicles which is dependent on the ionic strength of the subphase. The experiments done in the presence of 100 and 500 mM NaCl are labeled 100 and 500, respectively. The upper two tracings are those of the surface pressure, and the lower two tracings are those of the surface radioactivity. Arrows mark the different events of SecA addition (SecA), washing the subphase (Wash on, Wash off), addition of radiolabeled vesicles (Vesicles), and addition of proteinase K (Pk). (See Breukink et al. [1993] for details.)

event of decreased acidic phospholipid levels in the inner membrane of the HDL 11 strain, this strain tries to compensate for the decreased translocation by increasing the SecA level in the cytosol. This phenomenon may be explained by the loss of an important target of the SecA molecule, the acidic phospholipids, which causes the targeting to the integral SecY/E complex via the membrane to be less effective, which may be compensated for by increasing the cytosolic SecA content.

Signal Sequence-Lipid Interactions

The signal sequence is absolutely essential for plasma membrane translocation of periplasmic

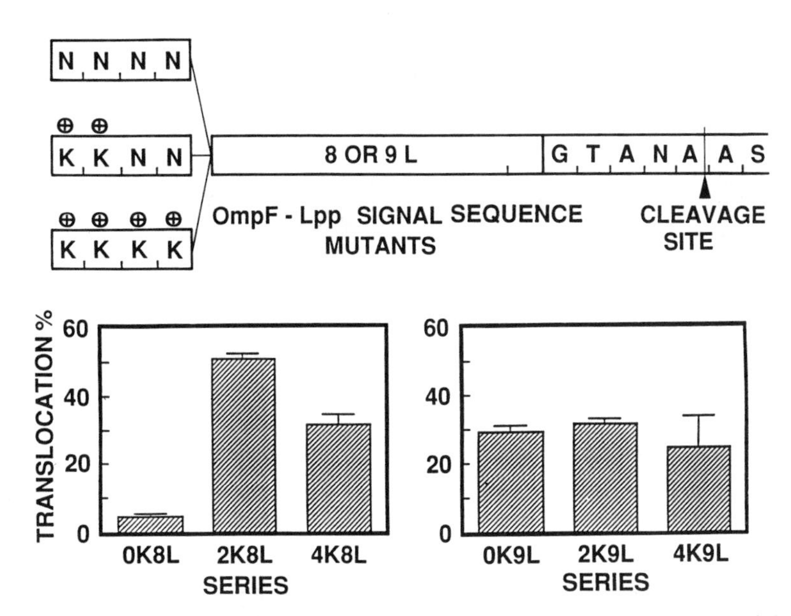

FIG. 3. Graphic representation of the OmpF-Lpp signal sequence mutants with different ratios of charge (lysines) to hydrophobicity (leucines), and their translocation across *E. coli* wild-type inverted membrane vesicles. (See Phoenix et al. [1993b] for details.)

and outer membrane proteins and is cleaved from the protein during or shortly after translocation. Research on chemically synthesized peptides corresponding to wild-type and mutant signal sequences has revealed important characteristics of a signal sequence, one of which is its strong and preferential interaction with acidic phospholipids in model membrane systems. This is consistent with the overall structure of the signal sequence, consisting of a positively charged N terminus followed by a hydrophobic core and a more polar C-terminal region flanking the cleavage site. Studies showed that a threshold value of acid phospholipids is needed for efficient insertion of the PhoE signal peptide into the lipid phase. This threshold value is in the physiological range of the anionic lipid concentration in the inner membrane (Batenburg et al., 1988; Demel et al., 1990). This insertion is accompanied by alpha-helix formation (Keller et al., 1992), which is probably an essential characteristic of signal sequences since a striking correlation between alpha-helix formation and functionality in the translocation process has been established (Briggs et al., 1985; McKnight et al., 1989). Monolayer studies and computer prediction methods on the PhoE signal peptide lead to the proposal that the signal peptide follows a two-phase insertion route. It inserts first in the lipid phase in a looped conformation, which later stretches to span the bilayer (Batenburg et al., 1988).

The availability of the HDL 11 strain and chimeric OmpF-Lpp precursor proteins with various ratios of charge (lysines) to hydrophobicity (leucines) in the signal sequence (Hikita and Mizushima, 1992) has permitted us to test whether signal sequence-lipid interactions are important within the functional translocation pathway. At wild-type acidic phospholipid levels the translocation of the precursors containing eight leucines in the signal sequence is dependent on the number of positive charges at the N terminus (Fig. 3), suggesting that for this precursor series an electrostatic interaction plays a role in translocation. However, the precursors containing nine leucines in the signal sequence translocated independently of the number of positive charges at the N terminus (Fig. 3), suggesting that in some way the need for an electrostatic interaction has been overcome by an increase in the length and hydrophobicity of the signal sequence. Except for the OK8L precursor, all preproteins were more or less dependent on SecA for their translocation activity (Hikita and Mizushima, 1992). As shown in Fig. 4, these precursor proteins with different ratios of charge to hydrophobicity in the signal sequence showed differences in their dependence on acidic phospholipids in their translocation. The precursors con-

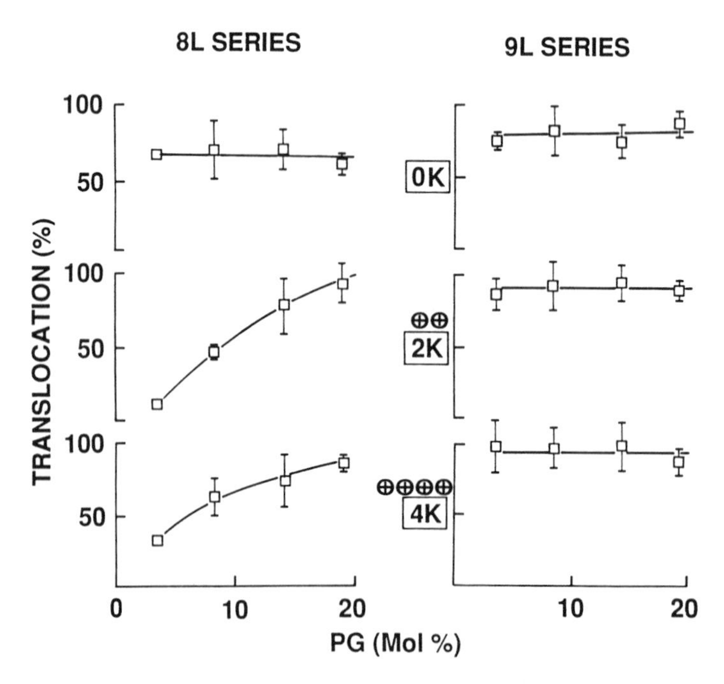

FIG. 4. PG dependence of the translocation of the mutants presented in Fig. 3. Translocation across PG-depleted membranes was related to translocation across wild-type vesicles (Fig. 3), which was assumed to be 100% efficient for this system. The average results from three data sets are shown, along with their standard deviations. (See Phoenix et al. [1993b] for details.)

taining eight leucines and two or four lysines in the signal sequence showed a linear relationship between the translocation and the number of acidic phospholipids in the membrane, whereas the precursor containing eight leucines but no lysines showed no dependency (Fig. 4). These data suggest that direct interaction between the signal sequence and the acidic phospholipids takes place during the translocation of precursor proteins. Figure 4 furthermore shows that the translocation of the precursors containing zero, two, or four lysines and nine leucines in the signal sequence was independent of the number of acidic lipids present in the inner membrane. Additional evidence for the initial electrostatic interaction of the signal sequence with the acidic phospholipids came from studies with positively charged compounds which have the potential to interact with anionic lipid head groups (Phoenix et al., 1993a). These authors showed that the positively charged compounds (doxorubicin and polylysine) could inhibit the translocation of the precursors containing eight leucines and two or four lysines but not the translocation of the precursors containing nine leucines.

Apparently the minimal number of acidic phospholipids which is still present in strain HDL 11 (9% [Kusters et al., 1991]) is sufficient for SecA function in the case of the precursors containing nine leucines. The explanation for this loss of translocation dependency on acidic phospholipids for the nine-leucine precursors may lie in the increased hydrophobicity (and length), which allows a more direct interaction with the membrane interior, thus bypassing the requirement for an initial electrostatic interaction with the acidic phospholipids in the membrane.

Using the above results, it is hard to dissect the SecA-lipid interaction from the signal sequence-lipid interaction during translocation. This calls for a precursor which is independent of SecA for its translocation to dissect the role of acidic lipids. Such a precursor is the precursor of the M13 coat protein, procoat, which inserts in the membrane in a signal sequence-dependent manner (Gallusser and Kuhn, 1990) but independently of SecA (Wolfe et al., 1985). It was shown that the translocation of this precursor was also dependent on acidic phospholipids (Kusters et al., 1994). Apparently the signal sequence of a precursor protein which is independent of SecA for its translocation directly interacts with the acidic phospholipids in the membrane.

How do we visualize the interaction of the signal sequence with acidic lipids during the translocation process? It has been shown that SecA interacts with the precursor recognizing both portions of its mature part and the signal sequence (Lill et al., 1990). In the model shown in Fig. 5, we propose that SecA plays an important role in

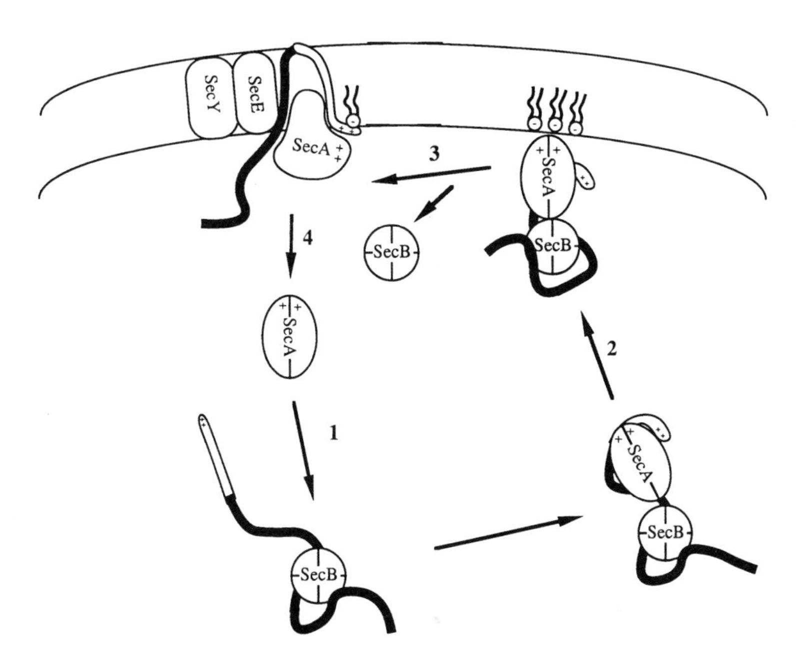

FIG. 5. Model for the initial stages in the translocation of SecA- and SecB-dependent precursor proteins. SecA is used to direct the precursors to SecY/E via two-dimensional diffusion at the membrane. See the text for more details.

the initial phases of the translocation. In Fig. 5, SecA binds to the SecB-precursor complex (step 1), directing it to the membrane (step 2) by means of its binding site, which leads to a superficial membrane interaction. This step depends on the presence of acidic phospholipids for the SecA-membrane interaction. Alternatively, SecA is already bound to the membrane lipids and the precursor-SecB complex may either bind first to the lipids and then interact with SecA or directly dock on the membrane-bound SecA. The next step (step 3) involves targeting of the complex by SecA via two-dimensional diffusion at the membrane surface to the SecY/E complex (which forms, together with the acidic phospholipids, the high-affinity binding site for SecA at the membrane). During this third step, the signal sequence inserts into the membrane, probably in a looped conformation, in an acidic lipid-dependent way. The ATP binding by SecA and the insertion of SecA into the lipid phase of the membrane initiate the translocation of the precursor, while during this process the signal sequence unloops. Upon ATP-hydrolysis, SecA is released from the membrane and the precursor, to allow for further translocation (step 4) in which both SecA and the proton motive force play an important role. It should be emphasized that besides the lipid-protein interactions described in this model, additional functions of anionic lipids in translocation are possible.

We thank R. Kusters, D. Phoenix, S. Mizushima, A. Kuhn, C. Hikita, and R. Keller for their contributions to this work.

Support for this work came from the Netherlands Foundation for Chemical Research (SON) with financial aid from the Netherlands Organization for Scientific Research (NWO).

REFERENCES

Batenburg, A. M., R. A. Demel, A. J. Verkleij, and B. De Kruijff. 1988. Penetration of the signal sequence of *Escherichia coli* PhoE protein into phospholipid model membranes leads to lipid specific changes in signal peptide structure and alterations in lipid organisation. *Biochemistry* **27:**5687–5685.

Breukink, E., R. A. Demel, G. De Korte-Kool, and B. De Kruijff. 1992. SecA insertion into phospholipids is stimulated by negatively charged lipids and inhibited by ATP: a monolayer study. *Biochemistry* **31:**1119–1124.

Breukink, E., R. C. A. Keller, and B. De Kruijff. 1993. Nucleotide and negatively charged lipid dependent vesicle aggregation caused by SecA. Evidence that SecA contains two lipid binding sites. *FEBS Lett.* **331:**19–24.

Briggs, M. S., L. M. Gierasch, A. Zlotnick, J. D. Lear, and W. F. Degrado. 1985. In vivo function and membrane binding properties are correlated for *Escherichia coli* LamB signal sequences. *Science* **228:**1096–1099.

Cabelli, R. J., K. M. Dolan, L. Qian, and D. B. Oliver. 1991. Characterisation of membrane-associated and soluble states of SecA protein from wild-type and *SecA51(TS)* mutant strains of *Escherichia coli. J. Biol. Chem.* **266:**24420–24427.

Demel, R. A., E. Goormaghtigh, and B. De Kruijff. 1990. Lipid and peptide specificities in signal peptide-lipid interactions in model membranes. *Biochim. Biophys. Acta* **1027:**155–162.

De Vrije, T., R. L. De Swart, W. Dowhan, J. Tommassen, and B. De Kruijff. 1988. Phosphatidylglycerol is involved in protein translocation across *Escherichia coli* inner membranes. *Nature* (London) **334:**173–175.

Gallusser, A., and A. Kuhn. 1990. Initial steps in protein membrane insertion. Bacteriophage M13 procoat protein binds to the membrane surface by electrostatic interaction. *EMBO J.* **9:**2723–2729.

Hartl, F.-U., S. Lecker, E. Schiebel, J. P. Hendrick, and W. Wickner. 1990. The binding cascade of SecB to SecA to SecY/E mediates preprotein targeting to the *E. coli* plasma membrane. *Cell* **63:**269–279.

Hendrick, J. P., and W. Wickner. 1991. SecA protein needs both acidic phospholipids and SecY/E protein for functional high-affinity binding to the *Escherichia coli* plasma membrane. *J. Biol. Chem.* **266:**24596–24600.

Hikita, C., and S. Mizushima. 1992. The requirement of a positive charge at the aminoterminus can be compensated for by a longer central hydrophobic stretch in the functioning of signal peptides. *J. Biol. Chem.* **267:**12375–12379.

Keller, R. C. A., J. A. Killian, and B. De Kruijff. 1992. Anionic phospholipids are essential for alpha-helix formation of the signal peptide of prePhoE upon interaction with phospholipid vesicles. *Biochemistry* **31:**1672–1677.

Kusters, R., E. Breukink, A. Gallusser, A. Kuhn, and B. De Kruijff. 1994. A dual role for phosphatidylglycerol in protein translocation across the *Escherichia coli* inner membrane. *J. Biol. Chem.* **269:**1560–1563.

Kusters, R., W. Dowhan, and B. De Kruijff. 1991. Negatively charged phospholipids restore prePhoE translocation across phosphatidylglycerol-depleted *Escherichia coli* inner membranes. *J. Biol. Chem.* **266:**8659–8662.

Kusters, R., R. Huijbregts, and B. De Kruijff. 1992. Elevated cytosolic concentrations of SecA compensate for a protein translocation defect in *Escherichia coli* cells with reduced levels of negatively charged phospholipids. *FEBS Lett.* **308:**97–100.

Lill, R., K. Cuningham, L. A. Brundage, K. Ito, D. B. Oliver, and W. Wickner. 1989. SecA protein hydrolysis ATP and is an essential component of protein translocation ATPase of *Escherichia coli. EMBO J.* **8:**961–966.

Lill, R., W. Dowhan, and W. Wickner. 1990. The ATPase activity of SecA is regulated by acidic phospholipids, SecY, and the leader and mature domains of precursor proteins. *Cell* **60:**271–280.

McKnight, C. J., M. S. Briggs, and L. M. Gierasch. 1989. Functional and nonfunctional LamB signal sequences can be distinguished by their biophysical properties. *J. Biol. Chem.* **264:**17293–17297.

Oliver, D. B., R. J. Cabelli, and G. P. Jarosik. 1990. SecA protein: autoregulated initiator of secretory precursor protein translocation across the *E. coli* plasma membrane. *J. Bioenerg. Biomembr.* **22:**311–336.

Phoenix, D. A., F. A. De Wolf, R. W. H. M. Staffhorst, C. Hikita, S. Mizushima, and B. De Kruijff. 1993a. Phosphatidylglycerol dependent protein translocation across the *Escherichia coli* inner membrane is inhibited by the anti cancer drug doxorubicin. *FEBS Lett.* **324:**113–116.

Phoenix, D. A., R. Kusters, C. Hikita, S. Mizushima, and B. De Kruijff. 1993b. OmpF-Lpp signal sequence mutants with varying charge hydrophobicity ratios provide evidence for a phosphatidylglycerol-signal sequence interaction during protein translocation across the *Escherichia coli* inner membrane. *J. Biol. Chem.* **268:**17069–17073.

Schiebel, E., A. J. M. Driessen, F.-U. Hartl, and W. Wickner. 1991. $\Delta\mu_{H+}$ and ATP function at different steps of the catalytic cycle of preprotein translocase. *Cell* **64:**927–939.

Shinkai, A., L. H. Mei, H. Tokuda, and S. Mizushima. 1991. The conformation of SecA, as revealed by its protease sensitivity is altered upon interaction with ATP, presecretory proteins, everted membrane vesicles, and phospholipids. *J. Biol. Chem.* **266:**5827–5833.

Ulbrandt, N. D., E. London, and D. B. Oliver. 1992. Deep penetration of a portion of *Escherichia coli* SecA protein into model membranes is promoted by anionic phospholipids and by partial unfolding. *J. Biol. Chem.* **267:**15184–15192.

Van der Wolk, J., M. Klose, E. Breukink, R. A. Demel, B. De Kruijff, R. Freundl, and A. J. M. Driessen. 1993.

Characterisation of a *Bacillus subtilis* SecA mutant protein deficient in translocation ATPase and release from the membrane. *Mol. Microbiol.* **8**:31–42.
Weaver, A. J., A. W. McDowall, D. B. Oliver. and J. Deisenhofer. 1992. Electron microscopy of thin-sectioned three-dimensional crystals of SecA protein from *Escherichia coli:* structure in projection at 40 Å resolution. *J. Struct. Biol.* **109**:87–96.
Wolfe, P. B., M. Rice, and W. Wickner. 1985. Effects of two sec genes on protein assembly into the plasma membrane of *Escherichia coli. J. Biol. Chem.* **260**:1836–1841.

Chapter 38

Roles of Phospholipids in *Escherichia coli*

WILLIAM DOWHAN

Department of Biochemistry and Molecular Biology, University of Texas Medical School, P.O. Box 20208, Houston, Texas 77225

A major role of phospholipids in all cells is to form the membrane bilayer which acts as both a permeability barrier and the matrix for membrane-associated functions. In addition there are an increasing number of examples in which phospholipids have been directly shown to be precursors to other macromolecules, substrates for posttranslational modifications, and either regulatory molecules or direct precursors to regulatory molecules. Since phospholipids have no inherent catalytic activity and the involvement of phospholipids can be demonstrated only by studying changes in a cellular process or enzyme activity, the uncovering of functions for phospholipids has usually come incidental to in vitro studies not initially aimed at understanding phospholipid functions. Systematic investigation in vivo of the roles of phospholipids has been limited by the difficulty in changing cellular phospholipid composition and the potential pleiotrophic effect of making such changes. When changes in phospholipid metabolism have been made, the major obstacle has been to relate changes in phospholipid composition to major cellular processes independent of the general effects of loss of membrane barrier function or cell viability.

Recent molecular genetic studies of *Escherichia coli* have shown that major systematic changes in phospholipid composition can be brought about while maintaining cell viability (Dowhan, 1992d). More important has been the demonstration that despite clear pleiotrophic effects brought about by major changes in phospholipid metabolism and composition, meaningful correlations can be made in vivo which can be verified in vitro to establish clear evidence for specific roles of various phospholipid species in *E. coli*. This chapter will briefly describe the types of strains currently available and the use of these strains to document specific functions of glycerophosphate-based lipids (hereafter referred to as phospholipids).

Zwitterionic Phospholipids

Synthesis of PE. Phosphatidylethanolamine (PE) accounts for 70 to 80% of the total phospholipids of *E. coli* and is enriched in the outer membrane relative to the inner membrane of the cell envelope (Raetz and Dowhan, 1990). The biosynthesis of PE is shown in Fig. 1 and is dependent on the *pssA* (Dowhan, 1992c) and *psd* (Dowhan and Li, 1992) gene products. Both genes and their respective gene products have been extensively studied (Raetz and Dowhan, 1990). Construction of a null allele of the *pssA* gene reduces the PE content of cells to less than 0.007% of the total phospholipids and, rather than resulting in cell death or growth arrest, results in a strain dependent on divalent metal ions ($Ca^{2+} > Mg^{2+} > Sr^{2+}$) for viability and growth (DeChavigny et al., 1991). Optimal growth occurs above 30 mM for the first two cations and at 15 mM for the third; millimolar concentrations of Ba^{2+} or spermidine and monovalent cations up to 1 M do not support growth. The phospholipid-to-protein ratio is unchanged in the mutant in which the zwitterionic phospholipid is replaced by the two major acidic phospholipids, phosphatidylglycerol (PG) and cardiolipin (CL). This mutant does have a reduced growth rate in rich medium and is a tryptophan auxotroph. Although both phenotypes are a result of the lack of a functional *pssA* gene, the molecular basis for these cell properties is not known. In *E. coli* the intracellular Mg^{2+} concentration approaches 100 mM (Chang et al., 1986) and Ca^{2+} is actively excluded from the cytoplasm (Gangola and Rosen, 1987), which suggests that the requirement for divalent metal ions is for structures or functions which occur either on the outer surface of the inner membrane, in the periplasmic space, or on the outer membrane. The outer membrane is a likely site for this growth requirement because this membrane is normally enriched in PE relative to the inner membrane (Raetz et al., 1979) and is known to require divalent metal ions for maintenance of structural integrity (Nikaido and Vaara, 1987). Since this mutant is viable but is defective or compromised in several cellular functions, it has been a useful strain in which to investigate the roles of PE in cell processes.

Phospholipid polymorphism. The primary structural organization of membrane phospho-

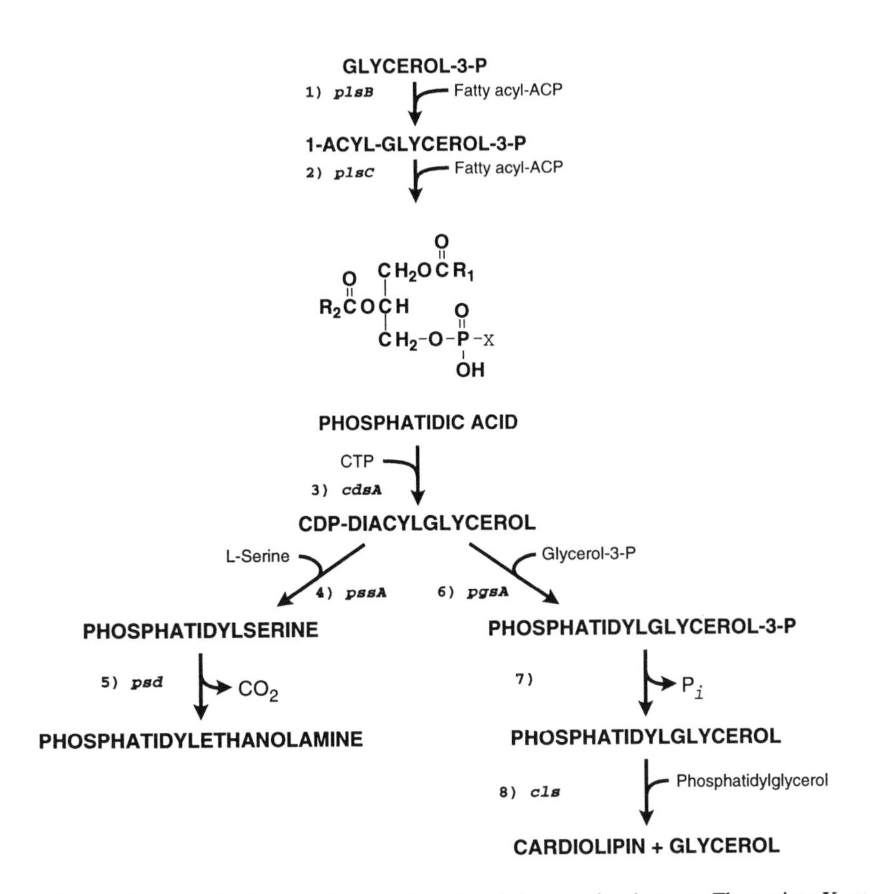

FIG. 1. Pathway of phospholipid biosynthesis in *E. coli* and the associated genes. The moiety X attached to phosphatidic acid (step 2) is defined for the product of each enzyme, and the charge of the head group at physiological pH is indicated for the relevant phospholipids. The name of each gene is listed with the respective step catalyzed by the following enzymatic activities: 1, *sn*-glycerol-3-phosphate acyltransferase; 2, 1-acyl-*sn*-glycerol-3-phosphate acyltransferase (X = OH); 3, CDP-diacylglycerol synthase (X = CMP); 4, phosphatidylserine synthase (X = serine; one positive charge and two negative charges); 5, phosphatidylserine decarboxylase (X = ethanolamine; one positive and one negative charge); 6, phosphatidylglycerophosphate synthase (X = *sn*-glycerol-3-phosphate); 7, phosphatidylglycerophosphate phosphatase (X = glycerol; one negative charge); 8, cardiolipin synthase (X = PG; two negative charges).

lipids is in a bilayer structure to maintain a permeability barrier to water-soluble metabolites and large macromolecules. Many model membrane studies and some in vivo studies have shown that under certain conditions some phospholipids can form various nonbilayer configurations, which, if extensive, would compromise the permeability barrier. However, the potential for such discontinuity in the bilayer might be essential for cell membrane growth and fusion, movement of macromolecules through the bilayer, or membrane-related signaling and sensing mechanisms (Cullis and de Kruijff, 1979). PE has long been known to undergo a temperature-dependent phase transition from liquid-crystalline bilayer (L_α) to inverted hexagonal (H_{II}) nonbilayer (Fig. 2) depending on the degree of saturation of the fatty acid chains; CL also undergoes such a transition in the presence of millimolar concentrations of the above three divalent metal ions in the same order of effectiveness (Lindblom and Rilfors, 1989) as they support growth in the *pssA93* null allele-containing strains. Interestingly, this null mutation is incompatible, even in the presence of divalent cations, with a null allele of the *cls* gene (CL synthase), which is normally associated with no other phenotype. These results suggested that phospholipid polymorphism may have some in vivo relevance and is dependent on the non-bilayer-forming phospholipids, PE and CL.

To further establish a correlation between non-bilayer phospholipid content and cell viability, Rietveld et al. (1993) analyzed the physical state by performing ^{31}P nuclear magnetic resonance spectroscopy of phospholipids isolated from the *pssA93* null allele. Mutant cells grown in Ca^{2+}

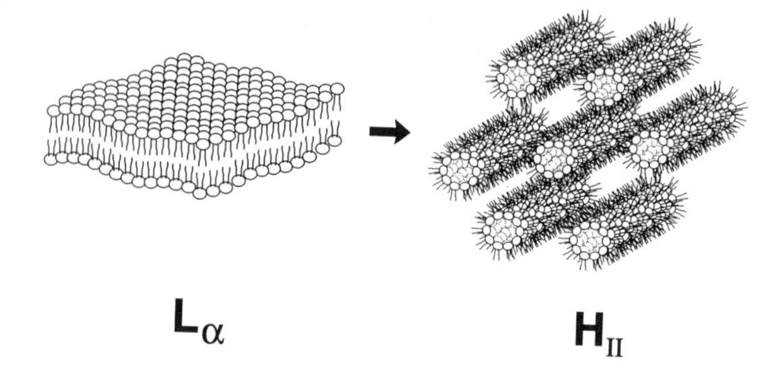

FIG. 2. Phospholipids such as PE and CL can exist in either the liquid-crystalline state (L_α) or the inverted hexagonal state (H_{II}) depending on temperature and solvent conditions.

have a significantly reduced content of CL relative to cells grown in Mg^{2+} or Sr^{2+}, consistent with the higher effectiveness of Ca^{2+} cation in inducing the H_{II} phase for CL. When physical properties of phospholipids extracted from cells grown in one of these three cations were analyzed in vitro by using the same cation as the cells were grown in, the phospholipids underwent a transition from the bilayer to the H_{II} phase with a midpoint 10°C above the growth temperature, as was found with phospholipids from wild-type cells in the absence of cations. However, if phospholipids from cells grown in Ca^{2+} were analyzed in the presence of Mg^{2+}, the midpoint of the transition was 30°C above the growth temperature. For phospholipids from cells grown in the presence of Mg^{2+} and analyzed in the presence of Ca^{2+}, the transition midpoint was 10°C below the growth temperature. Sr^{2+} induced a transition (10°C above the growth temperature) in the extracted phospholipids only in the concentration range of 5 to 15 mM, at which it also supported growth, but not at 20 mM, at which it could not support growth of the mutant; as expected, Ba^{2+} was ineffective in inducing a transition. The combined results from model membrane studies, the growth properties and phospholipid compositions of this mutant, and the physical and chemical properties of phospholipids extracted from this mutant strongly support an important role for phospholipid polymorphism in cell function. As might be expected, prevention of a high proportion of nonbilayer phospholipid structure at the growth temperature would be expected, in order to maintain membrane barrier function (reduced CL when grown in Ca^{2+}), but the requirement for some non-bilayer-forming potential near the growth temperature (a midpoint for the transition 10°C above the growth temperature) suggests that cells regulate their phospholipid compositions to make this potential available. Therefore, at least part of the growth requirement for divalent metal ions of strains lacking PE may be to allow CL to substitute for PE in providing phospholipid polymorphism (Rietveld et al., 1993). Clearly the overall physical state of the membrane must be bilayer to maintain membrane integrity, but areas of nonbilayer structure or the potential to form nonbilayers might be necessary for membrane growth, protein insertion, translocation of macromolecules across the bilayer, and maintenance of sealed interfaces between proteins and phospholipids.

Energy transduction. Energy transduction in biological systems is dependent on a membrane of low ion permeability and on membrane-associated components which carry out the interconversion of various forms of energy. Therefore, it is not surprising that this process is dependent on the nature of the membrane phospholipid matrix. In mitochondrial function, PE has been implicated as specifically required for reconstitution of the coenzyme Q-containing cytochrome bc_1 complex III (Schagger et al., 1990) and CL appears to be required for the cytochrome *c* oxidase complex IV (Robinson, 1993). Few studies involving the specific role of phospholipids in *E. coli* oxidative phosphorylation have been carried out. However, Jaworowski et al. (1981) noted that the purified NADH dehydrogenase I of the inner membrane of *E. coli* copurified with a significant amount of phospholipid and its lipid cofactor coenzyme Q, suggesting a possible strong association of the enzyme with phospholipid.

In light of the above observations, the properties of the components of the electron transport chain in strains of *E. coli* lacking PE were investigated. The primary alteration is a fivefold reduction in the rate of whole-chain electron transfer from NADH to oxygen in isolated inverted membrane vesicles, which may partly explain the slower growth of these mutants (Mileykovskaya and Dowhan, 1992, 1993); the rate of whole-chain oxidation of succinate or lactate was unaffected in these isolated membrane vesicles. All three of these metabolites utilize coenzyme Q as an intermediate electron transfer cofactor, but the level of this cofactor was found to be normal in the mutant. On the basis of the specific activity of the individual succinate, lactate, and NADH dehydrogenases, as determined by transfer of electrons to artificial acceptors, the levels of these three enzymes were normal in the mutant. Finally, when inverted, closed membrane vesicles were used, all three electron donors plus the hydrolysis of ATP resulted in a similar membrane potential (positive inward) to that generated by wild-type vesicles, substantiating the impermeability of the mutant membranes to protons. These results taken together point to a specific requirement for PE in the transfer of electrons from NADH to coenzyme Q catalyzed by NADH dehydrogenase I, the step affected by the absence of PE. Further in vitro studies focusing on the requirement for PE in reconstitution of the activity of this enzyme now have relevance to the functioning of the enzyme in vivo.

Solute transport. Specific permeability of the membrane bilayer is catalyzed by carrier proteins, of which the lactose permease (*lacY* gene product) is the most widely studied. The uptake of lactose or its analogs against a concentration gradient is accomplished by the permease coupling uphill substrate uptake with downhill proton influx (Kaback et al., 1990). Earlier work had shown that PE was absolutely required for reconstitution of energy-dependent uptake of lactose into closed membrane vesicles (Chen and Wilson, 1984); acidic phospholipids and other zwitterionic phospholipids such as monomethyl- and dimethyl-PE and phosphatidylcholine (PC) could not substitute for PE. The physiological significance of this observation was shown by the 10- to 20-fold reduction in the rate of lactose permease activity in cells lacking PE (Bogdanov and Dowhan, 1993). The reduction in rate was not due to a decrease in the amount of the permease or to the inability of either whole cells or isolated membrane vesicles to generate the necessary electrochemical potential or ΔpH across the membrane. The permease in mutants was still able to effectively catalyze facilitated exchange of substrate across the membrane but was not able to couple facilitated transport with the energy gradient necessary for accumulation. By using LacY-PhoA protein fusions (Calamia and Manoil, 1990), preliminary results suggest that the carrier is misassembled in membranes lacking PE and may have its topology reversed with respect to the plane of the bilayer (Bogdanov and Dowhan, unpublished). This interesting observation indicates that the phospholipid composition of the membrane plays an important role in the insertion of the permease into the membrane. Glucose, mannitol, and proline transport is also affected in cells lacking PE (Bogdanov and Dowhan, unpublished). Clearly, such misassembly cannot occur with all membrane proteins and may affect only a class of nonessential proteins such as some of the transport proteins. A broader survey of the solute transport systems may uncover common structural motifs which require interaction with PE for proper membrane insertion.

Motility and chemotaxis. The expression of the flagellum chemotaxis system of *E. coli* is dependent on over 40 genes organized into 13 operons (Jones and Aizawa, 1991; Macnab, 1987a, 1987b) which are under a regulatory hierarchy with the *flhD* master operon at the top. The next tier of operons governs the formation of basal body and hook assembly, and these are followed by the operons which encode filaments, motor activity, and chemotaxis. A number of stress-related conditions shut off flagellation by repressing transcription of the *flhD* master operon. The absence of PE can now be added to the list of conditions which suppress the synthesis of the flagellation machinery (Shi et al., 1993). Strains temperature sensitive in either the phosphatidylserine (PS) decarboxylase or the PS synthase are nonmotile and do not display chemotaxis when grown at temperatures approaching the restrictive temperature. A null mutation in the *pssA* gene also renders cells incapable of motility and chemotaxis at all temperatures, even in the presence of divalent metal ions which suppress the growth phenotype of this mutant. The lack of chemotaxis is correlated with the reduction or absence of flagellar synthesis, as evidenced by both the lack of gene products and a lack of transcription of representative genes fused to the *lacZ* gene as a reporter. The lack of expression of these related genes is due to the repression of transcription of the *flhD* master operon. Therefore, the lack of PE in the membrane (which cannot be substituted for by its immediate precursor, PS) is sensed by the master regulon, and this leads to repression of the whole flagellum chemotaxis system. The degree of expression of this system appears to be roughly related to the amount of PE in the membrane, since intermediate expression occurs under conditions intermediate between fully permissive and fully nonpermissive for the temperature-sensitive mutations in the *pssA* and *psd* genes. The mechanism by which cells transmit these perturbations in membrane phospholipid composition to the master regulon is unknown; however, this effect could be a result of lack of feedback inhibition due to poor assembly of the flagellar components in membranes lacking PE, part of the general repressive response to stress, or a secondary response to the many cellular changes elicited by changes in membrane phospholipid composition (Shi et al., 1993). Interestingly, mutants defective in the respiratory chain do not undergo flagellation (Bar et al., 1977).

Acidic Phospholipids

Synthesis of PG and CL. The committed step to the synthesis of the major acidic phospholipids (Fig. 1) of *E. coli* is catalyzed by the *pgsA* gene product (Dowhan, 1992b). After removal of the phosphate from the intermediate, a portion of the resulting PG (which makes up about 20 to 25% of the total phospholipids), is converted to CL (5 to 10% of the total phospholipids) by CL synthase (Hirschberg and Kennedy, 1972); these two lipids make up the majority of the acidic or anionic glycerol-based phospholipids of *E. coli.* Unlike PE, for which the ionic head group turns over less than 5% per generation (Kanfer and Kennedy, 1963), the ionic head group of PG turns over about 30 to 40% per generation (Schulman and Kennedy, 1977). This turnover is due largely to the formation of the membrane-derived oligosaccharide (in inverse proportion to the osmolarity of the growth medium [Jackson et al., 1986]), the formation of CL, and the posttranslational transfer of the unacylated glycerol moiety to several outer membrane lipoproteins including the major outer membrane lipoprotein (*lpp* gene product) covalently attached to the peptidoglycan (Wu et al., 1983). Although each of these turnover events is important to *E. coli,* none is absolutely essential to cell viability (Nishijima et al., 1988; Weissborn et al., 1992; Wu et al., 1983). However, PG itself is essential, as demonstrated by the arrest of growth in cells which have lost a functional copy of the *pgsA* gene (Heacock and Dowhan, 1987). As noted previously, CL appears not to be essential as long as PE is present (DeChavigny et al., 1991; Nishijima et al., 1988).

Although PG is essential for cell growth, its normal level is not required to maintain cell membrane integrity, and arrest in cell growth at limiting PG levels does not result in cell death. There also appears to be a hierarchy of requirements or uses for PG which, if suppressed or bypassed, allows cell growth to continue as long as some newly synthesized PG is made available to the cell. These conclusions are based on results obtained with strains of *E. coli* in which both the rate of synthesis and the content of the PG can be regulated. In such strains the normal chromosomal copy of the *pgsA* gene was inactivated by gene disruption and the only functional copy of the gene was expressed by a Φ(*lacOP-pgsA*) fusion between the chromosomal copy of the *lac* promoter and the *pgsA* gene (Heacock and Dowhan, 1989). In these strains the level of gene product and of PG can be controlled by regulating the amount of the gratuitous inducer of the *lac* operon, isopropyl-β-D-thiogalactopyranoside (IPTG), in the growth medium. Such a gene arrangement in an otherwise wild-type genetic background renders the growth rate of cells dependent on IPTG. Removal of IPTG from the growth medium results in arrest of growth after the steady-state PG level is reduced about 10-fold to 2 to 3% of the total phospholipids. Addition of IPTG to an arrested culture (even after 6 to 8 h) results in reinitiation of growth after a short lag period, indicating that cells simply arrest but remain viable at low PG contents without undergoing lysis or death.

There is some indication that the steady-state level of PG is less important than the level of newly synthesized PG available for whatever functions become limiting in cells with a *pgsA* mutation. A mutant with a point mutation in the *pgsA* gene with a barely detectable level of enzymatic activity was found to grow normally, which originally led to the proposition that acidic phospholipids may not be absolutely essential (Miyazaki et al., 1985). However, strains with this allele were not completely lacking in phosphatidylglycerophosphate synthase activity, and the viability of these strains was dependent on the genetic background. The growth arrest phenotype of this *pgsA3* allele could be suppressed by a mutation in the *lpp* gene (Asai et al., 1989), which does not result in a significant change in the level of either enzymatic activity or PG but does result in a reduced rate of turnover of PG for the posttranslational modification of the lipoprotein. A similar suppression of the dependence on IPTG for growth of the Φ(*lacOP-pgsA*) fusion has also been seen (Dowhan, 1992a; Kusters et al., 1991); however, the *lpp* mutation does not suppress the phenotype of the null allele of the *pgsA* gene (Dowhan, 1992a; Heacock and Dowhan, 1987). The introduction of a mutation in the synthesis of CL results in an increase in the steady-state level of PG, but the Φ(*lacOP-pgsA*) fusion is still dependent on IPTG for growth in this genetic background (Dowhan, 1992a). These results suggest that the newly synthesized PG pool remaining after modification of the newly synthesized lipoprotein may be limiting for growth. Still unknown is which functions are ultimately limiting in cells carrying the null allele of the *pgsA* gene.

Which functions other than modification of the lipoprotein might require PG? Further molecular genetic and biochemical studies with the above IPTG-dependent Φ(*lacOP-pgsA*) fusion have revealed several processes which are dependent on acidic phospholipids. Among these are the translocation of newly synthesized outer membrane proteins across the cytoplasmic membrane (Breukink et al., 1992; de Vrije et al., 1988; Hendrick and Wickner, 1991; Kusters et al., 1991; Lill et al., 1990), the assembly of the pore-forming bacterial toxin colicin A (van der Goot et al., 1993), and the initiation of DNA replication (see below); the details of phospholipid involvement in the first two functions are discussed in chapters 37 and 39.

Initiation of DNA replication. Considerable in vitro evidence suggests that initiation of DNA replication involves either acidic phospholipids directly or a membrane surface containing acidic phospholipids (see Kornberg and Baker [1992] for a historical summary of the evidence). The cytoplasmic DnaA protein (encoded by the *dnaA* gene) initiates replication by binding to specific DNA sequences at the *oriC* (origin of chromosomal replication) locus of the *E. coli* chromosome, bringing about strand separation (Sekimizu et al., 1988) to allow synthesis of the RNA primer required for initiation of DNA synthesis. This event is followed by organization of other cytoplasmic proteins into the replication complex. The active form of the DnaA protein contains tightly bound ATP, which slowly hydrolyzes to tightly bound ADP; the ADP form of the protein is inactive but still has high affinity for the *oriC* locus. A potential mechanism for activation of inactive DnaA protein for successive rounds of initiation or for the initial round of replication for newly synthesized but inactive DnaA protein may involve acidic phospholipids. Acidic phospholipids such as PG and CL were found to facilitate the exchange of ATP and ADP with *oriC*-bound DnaA protein to activate the inactive protein in an in vitro reconstituted system (Sekimizu and Kornberg, 1988; Yung and Kornberg, 1988). Association of the DnaA protein with *oriC* was essential for activation by acidic phospholipids (Crooke et al., 1992) whereas activation of free DnaA protein associated with phospholipids required either DnaK protein (a heat shock/chaperone protein) or phospholipase treatment (Hwang et al., 1990). Possible involvement of phospholipids in initiation is further suggested by a correlation between cessation of phospholipid synthesis in a *plsB* mutant (first acyltransferase of phospholipid biosynthesis) and all of the macromolecular synthesis including specifically initiation of DNA replication (McIntyre et al., 1977; Pierucci and Rickert, 1985). In addition, cessation of initiation occurs in response to a reduction in the fluidity of the membrane bilayer (Fralick and Lark, 1973), which is consistent with the requirement for unsaturated acidic phospholipids in the activation of DnaA protein (Sekimizu and Kornberg, 1988). Although much of the in vitro evidence is convincing, there is only minimal in vivo evidence for the involvement of acidic phospholipids in initiation.

Recent experiments with cells carrying the Φ(*lacOP-pgsA*) fusion (Heacock and Dowhan, 1989), which makes cell growth dependent on IPTG, provide in vivo genetic evidence for the involvement of acidic phospholipids in the initiation of DNA replication (Xia and Dowhan, unpublished). DnaA protein-dependent initiation (Fig. 3A) can be bypassed by mutations which allow initiation at the four alternate *oriK* replication sites (constitutive stable DNA replication). In *rnh* (encodes RNase H) mutants the RNA primers necessary for initiation of DNA replication, which are normally degraded at all sites on the chromosome except the DnaA protein-protected *oriC* site, are not degraded; this allows constitutive stable DNA replication at the *oriK* sites (Fig. 3B) independent of DnaA protein and *oriC* (Horiuchi et al., 1984; Ogawa et al., 1984; von Meyenburg et al., 1987). Constitutive stable DNA replication is normally dependent on RecA protein, which serves to open the DNA duplex (Kogoma et al., 1985) much as the DnaA protein does at *oriC*. Therefore, if the reduction of acidic phospholipid synthesis in the absence of IPTG is limiting for DnaA protein function in initiation, an *rnh* mutation should make the growth of the phospholipid mutants independent of IPTG in a RecA protein-dependent manner. Introduction of a null allele of the *rnh* gene into the above IPTG-dependent strain conferred IPTG-independent growth which was dependent on a functional RecA protein (Fig. 3B). The dependence on RecA protein was bypassed by introduction of a *lexA* mutation, which has previously been shown to make initiation of DNA replication at *oriK* independent of RecA protein (Torrey and Kogoma, 1987). These results are consistent with functional DnaA protein being the limiting factor in the absence of acidic phospholipid synthesis, since suppression of the growth phenotype could occur through induction of non-DnaA protein-dependent initiation of DNA synthesis. Since DnaA protein does have affinity for acidic phospholipids and membranes (Sekimizu and Kornberg, 1988; Yung and Kornberg, 1988), an association of the protein-DNA replicon in vivo with membranes containing acidic phospholipids is implicit in the model shown in Fig. 3A.

Evidence for acidic phospholipid involvement was also obtained without relying on the complex genetic manipulation necessary to induce constitutive stable DNA replication. In cells carrying the Φ(*lacOP-pgsA*) fusion, stable maintenance of plasmids dependent on an *oriC* locus for replication, and therefore on DnaA protein, was found to be dependent on acidic phospholipid synthesis (i.e., on IPTG), whereas stable maintenance of plasmids with the ColE1 origin, known to be independent of DnaA protein, was independent of IPTG.

The above in vivo molecular genetic evidence, coupled with the detailed in vitro biochemical dissection of the requirements for DnaA protein-dependent initiation of DNA replication, strongly supports the involvement of acidic phospholipids in this process. However, initiation of replication is not the only essential process dependent on acidic phospholipids since an *rnh* mutation cannot

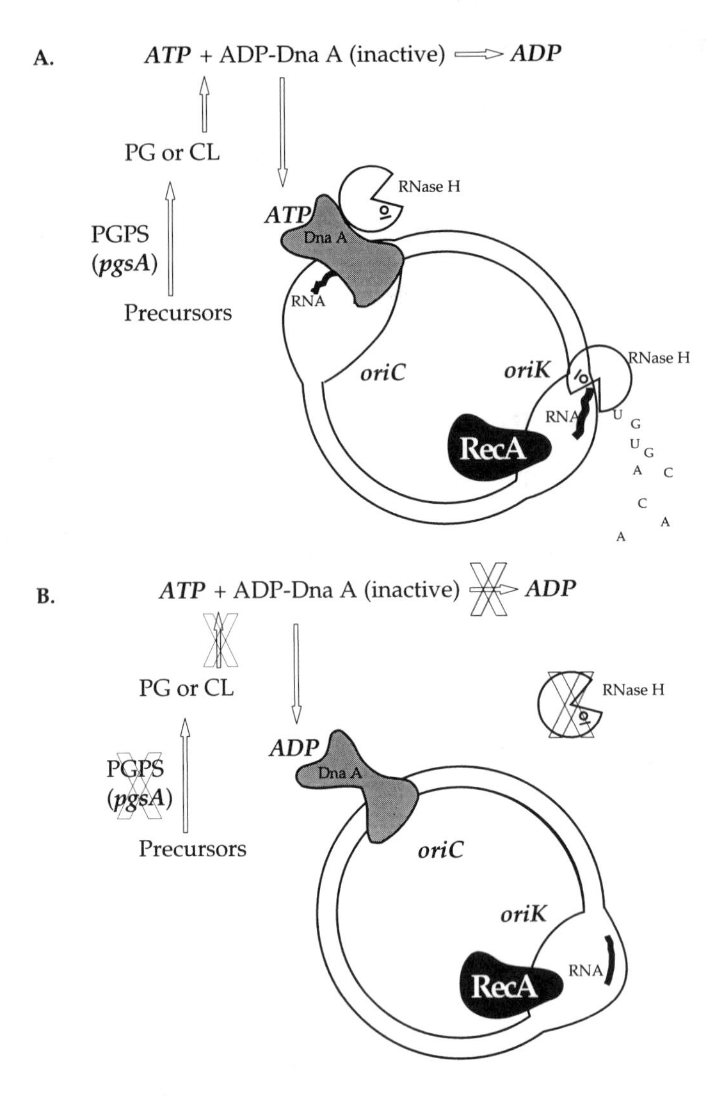

FIG. 3. (A) DnaA protein-dependent initiation at *oriC* in the presence of RNase H with the postulated requirement for acidic phospholipids and the *pgsA* gene product. (B) DnaA protein-independent, RecA-dependent initiation of DNA replication at an *oriK* site in the absence of both RNase H and the *pgsA* gene product.

suppress growth arrest of cells carrying the null allele of the *pgsA* gene.

Summary

In recent years, molecular genetic approaches have led to the construction of mutants with specific mutations in phospholipid metabolism, and this has made possible a more systematic and direct dissection of the role or involvement of individual phospholipids in important cellular processes (Dowhan, 1992d). This approach has uncovered hitherto unsuspected roles for phospholipids, such as participation in the complex macromolecular assembly responsible for translocation of proteins across the inner membrane of *E. coli*. These observations have led to a number of in vitro studies, which allow a more detailed understanding of the mechanism of protein translocation. In other cases, such as the initiation of DNA replication, the function of the *lac* permease, and the membrane insertion of colicin A, this approach has been very useful in obtaining in vivo verification of the functions of phospholipids for which there was only in vitro evidence. Further systematic studies of the type outlined above may lead to the uncovering of other suppressors or bypass mutations, which will undoubtedly demonstrate additional previously unknown functions for phospholipids. Extension of this approach to eukaryotic cells such as *Saccharomyces ceresiae* should be possible, since many of the genes of phospholipid metabolism from this organism have been cloned (Carman and Henry, 1989).

The work cited from my laboratory was supported in part by grant GM 20487 from the National Institutes of Health.

REFERENCES

Asai, Y., Y. Katayose, C. Hikita, A. Ohta, and I. Shibuya. 1989. Suppression of the lethal effect of acidic-phospholipid deficiency by defective formation of the major outer membrane lipoprotein in *Escherichia coli. J. Bacteriol.* **171:**6867–6869.

Bar, T. J., B. J. Howlett, and D. E. Koshland, Jr. 1977. Flagellar formation in *Escherichia coli* electron transport mutants. *J. Bacteriol.* **130:**787–792.

Bogdanov, M., and W. Dowhan. 1993. Phosphatidylethanolamine (PE) is required for *in vivo* function of the lactose permease of *Escherichia coli. FASEB J.* **7:**abstr. 518.

Bogdanov, M., and W. Dowhan. Unpublished data.

Breukink, E., R. A. Demel, G. de Korte-Kool, and B. de Kruijff. 1992. SecA insertion into phospholipids is stimulated by negatively charged lipids and inhibited by ATP: A monolayer study. *Biochemistry* **31:**1119–1124.

Calamia, J., and C. Manoil. 1990. *lac* permease of *Escherichia coli:* topology and sequence elements promoting membrane insertion. *Proc. Natl. Acad. Sci. USA* **87:**4937–4941.

Carman, G. M., and S. A. Henry. 1989. Phospholipid biosynthesis in yeast. *Annu. Rev. Biochem.* **58:**635–669.

Chang, C.-F., H. Shuman, and A. P. Somlyo. 1986. Electron probe analysis, X-ray mapping and electron energy-loss spectroscopy of calcium, magnesium, and monovalent ions in log-phase and in dividing *Escherichia coli* B cells. *J. Bacteriol.* **167:**935–939.

Chen, C.-C., and T. H. Wilson. 1984. The phospholipid requirement for activity of the lactose carrier of *Escherichia coli. J. Biol. Chem.* **259:**10150–10158.

Crooke, E., C. E. Castuma, and A. Kornberg. 1992. The chromosome origin of *Escherichia coli* stabilizes DnaA protein during rejuvenation by phospholipids. *J. Biol. Chem.* **267:**16779–16782.

Cullis, P. R., and B. de Kruijff. 1979. Lipid polymorphism and the functional roles of lipids in biological membranes. *Biochim. Biophys. Acta* **559:**399–420.

DeChavigny, A., P. N. Heacock, and W. Dowhan. 1991. Phosphatidylethanolamine may not be essential for the viability of *Escherichia coli. J. Biol. Chem.* **266:**5323–5332.

de Vrije, T., R. L. de Swart, W. Dowhan, J. Tommassen, and B. de Kruijff. 1988. Phosphatidylglycerol is involved in protein translocation across *Escherichia coli* inner membranes. *Nature* (London) **334:**173–175.

Dowhan, W. 1992a. Role of phospholipids in cell function. *NATO ASI Ser.* **H63:**11–32.

Dowhan, W. 1992b. Phosphatidylglycerophosphate synthase from *Escherichia coli. Methods Enzymol.* **209:**313–321.

Dowhan, W. 1992c. Phosphatidylerine synthase from *Escherichia coli. Methods Enzymol.* **209:**287–298.

Dowhan, W. 1992d. Strategies for generating and utilizing phospholipid synthesis mutants in *Escherichia coli. Methods Enzymol.* **209:**7–20.

Dowhan, W., and Q. X. Li. 1992. Phosphatidylserine decarboxylase from *Escherichia coli. Methods Enzymol.* **209:**348–359.

Fralick, J. A., and K. G. Lark. 1973. Evidence for the involvement of unsaturated fatty acids in initiating chromosome replication in *Escherichia coli. J. Mol. Biol.* **80:**459–475.

Gangola, P., and B. P. Rosen. 1987. Maintenance of intracellular calcium levels in *Escherichia coli. J. Biol. Chem.* **262:**12570–12574.

Heacock, P. N., and W. Dowhan. 1987. Construction of a lethal mutation in the synthesis of the major acidic phospholipids of *Escherichia coli. J. Biol. Chem.* **262:**13044–13049.

Heacock, P. N., and W. Dowhan. 1989. Alterations of the phospholipid composition of *Escherichia coli* through genetic manipulation. *J. Biol. Chem.* **264:**14972–14977.

Hendrick, J. P., and W. Wickner. 1991. SecA protein needs both acidic phospholipids and SecY/E protein for functional high-affinity binding to the *Escherichia coli* plasma membrane. *J. Biol. Chem.* **266:**24596–24600.

Hirschberg, C. B., and E. P. Kennedy. 1972. Mechanism of the enzymatic synthesis of cardiolipin in *Escherichia coli. Proc. Natl. Acad. Sci. USA* **69:**648–651.

Horiuchi, T., H. Maki, and M. Sekiguchi. 1984. RNase H-defective mutants of *Escherichia coli:* a possible discriminatory role of RNase H in initiation of DNA replication. *Mol. Gen. Genet.* **195:**17–22.

Hwang, D. S., E. Crooke, and A. Kornberg. 1990. Aggregated DnaA protein is dissociated and activated for DNA replication by phospholipase or DnaK protein. *J. Biol. Chem.* **265:**19244–19248.

Jackson, B. J., J. M. Gennity, and E. P. Kennedy. 1986. Regulation of the balanced synthesis of membrane phospholipids: experimental test of models for regulation in *Escherichia coli. J. Biol. Chem.* **261:**13464–13468.

Jaworowski, A., G. Mayo, D. C. Shaw, H. D. Campbell, and I. G. Young. 1981. Characterization of the respiratory NADH dehydrogenase of *Escherichia coli* and reconstitution of NADH oxidase in *ndh* mutant membrane vesicles. *Biochemistry* **20:**3621–3628.

Jones, C. J., and S. Aizawa. 1991. The bacterial flagellum and flagellar motor: structure, assembly and function. *Adv. Microb. Physiol.* **32:**109–172.

Kaback, H. R., E. Bibi, and P. D. Roepe. 1990. β-Galactoside transport in *Escherichia coli:* a functional dissection of the *lac* permease. *Trends Biochem. Sci.* **15:**309–314.

Kanfer, J., and E. P. Kennedy. 1963. Metabolism and function of bacterial lipids. *J. Biol. Chem.* **238:**2919–2922.

Kogoma, T., K. Skarstad, E. Boye, K. von Meyenburg, and H. B. Steen. 1985. RecA protein acts at the initiation of stable DNA replication in *rnh* mutants of *Escherichia coli* K-12. *J. Bacteriol.* **163:**439–444.

Kornberg, A., and T. A. Baker. 1992. *DNA Replication,* p. 749–752. W. H. Freeman & Co., New York.

Kusters, R., W. Dowhan, and B. de Kruijff. 1991. Negatively charged phospholipids restore prePhoE translocation across phosphatidylglycerol depleted *Escherichia coli* membranes. *J. Biol. Chem.* **266:**8659–8662.

Lill, R., W. Dowhan, and W. Wickner. 1990. The ATPase of SecA is regulated by acidic phospholipids, SecY, and the leader and mature domains of precursor proteins. *Cell* **60:**271–280.

Lindblom, G., and L. Rilfors. 1989. Cubic phases and isotropic structures formed by membrane lipids—possible biological relevance. *Biochim. Biophys. Acta* **988:**221–256.

Macnab, R. M. 1987a. Flagella, p. 70–83. *In* F. C. Neidhardt, J. L. Ingraham, K. B. Low, B. Magasanik, M. Schaechter, and H. E. Umbarger (ed.), *Escherichia coli and Salmonella typhimurium: Cellular and Molecular Biology,* vol. 1. American Society for Microbiology, Washington, D.C.

Macnab, R. M. 1987b. Motility and chemotaxis, p. 732–759. *In* F. C. Neidhardt, J. L. Ingraham, K. B. Low, B. Magasanik, M. Schaechter, and H. E. Umbarger (ed.), *Escherichia coli and Salmonella typhimurium: Cellular and Molecular Biology,* vol. 2. American Society for Microbiology, Washington, D.C.

McIntyre, T. M., B. K. Chamberlain, R. E. Webster, and R. M. Bell. 1977. Mutants of *Escherichia coli* defective in membrane phospholipid synthesis. Effects of cessation and reinitiation of phospholipid synthesis on macromolecular synthesis and phospholipid turnover. *J. Biol. Chem.* **255:**4487–4493.

Mileykovskaya, E. I., and W. Dowhan. 1992. Alterations in the electron transfer chain in mutant strains of *Escherichia coli* lacking phosphatidylethanolamine. *FASEB J.* **6:**abstr. 1125.

Mileykovskaya, E. I., and W. Dowhan. 1993. Alterations in the electron transfer chain in mutant strains of *Escherichia coli* lacking phosphatidylethanolamine. *J. Biol. Chem.* **268:**24824–24831.

Miyazaki, C., M. Kuroda, A. Ohta, and I. Shibuya. 1985. Genetic manipulation of membrane phospholipid composition in *Escherichia coli: pgsA* mutants defective in phosphatidylglycerol synthesis. *Proc. Natl. Acad. Sci. USA* **82:**7530–7534.

Nikaido, H., and M. Vaara. 1987. Outer membrane, p. 7–22. *In* F. C. Neidhardt, J. L. Ingraham, K. B. Low, B. Magasanik, M. Schaechter, and H. E. Umbarger (ed.), *Escherichia coli and Salmonella typhimurium: Cellular and Molecular Biology,* vol. 1. American Society for Microbiology, Washington, D.C.

Nishijima, S., Y. Asami, N. Uetake, S. Yamogoe, A. Ohta, and I. Shibuya. 1988. Disruption of the *Escherichia coli cls* gene responsible for cardiolipin synthesis. *J. Bacteriol.* **170:**775–780.

Ogawa, T., G. G. Pickett, T. Kogoma, and A. Kornberg. 1984. RNase H confers specificity in the *dnaA*-dependent initiation of replication at the unique origin of the *Escherichia coli* chromosome *in vivo* and *in vitro. Proc. Natl. Acad. Sci. USA* **81:**1040–1044.

Pierucci, O., and M. Rickert. 1985. Duplication of *Escherichia coli* during inhibition of net phospholipid synthesis. *J. Bacteriol.* **162:**374–382.

Raetz, C. R. H., and W. Dowhan. 1990. Biosynthesis and function of phospholipids in *Escherichia coli. J. Biol. Chem.* **265:**1235–1238.

Raetz, C. R. H., G. D. Kantor, M. Nishijima, and K. F. Newman. 1979. Cardiolipin accumulation in the inner and outer membranes of *Escherichia coli* mutants defective in phosphatidylserine synthetase. *J. Bacteriol.* **139:**544–551.

Rietveld, A. G., J. A. Killian, W. Dowhan, and B. de Kruijff. 1993. Polymorphic regulation of membrane phospholipid composition in *Escherichia coli. J. Biol. Chem.* **268:**12427–12433.

Robinson, N. 1993. Functional binding of cardiolipin to cytochrome *c* oxidase. *Bioenerg. Biomembr.* **25:**153–165.

Schagger, H., T. Hagen, B. Roth, U. Brandt, T. Link, and G. von Jagow. 1990. Phospholipid specificity of bovine heart bc_1 complex. *Eur. J. Biochem.* **190:**123–130.

Schulman, H., and E. P. Kennedy. 1977. Relation of turnover of membrane phospholipids to synthesis of membrane-derived oligosaccharides of *Escherichia coli. J. Biol. Chem.* **252:**4250–4255.

Sekimizu, K., D. Bramhill, and A. Kornberg. 1988. Sequential early stages in the *in vitro* initiation of replication at the origin of the *Escherichia coli* chromosome. *J. Biol. Chem.* **263:**7124–7130.

Sekimizu, K., and A. Kornberg. 1988. Cardiolipin activation of DnaA protein, the initiation protein of replication in *Escherichia coli. J. Biol. Chem.* **263:**7131–7135.

Shi, W., M. Bogdanov, W. Dowhan, and D. R. Zusman. 1993. The *pss* and *psd* genes are required for motility and chemotaxis in *Escherichia coli. J. Bacteriol.* **175:**7711–7714.

Torrey, T., and T. Kogoma. 1987. Genetic analysis of constitutive stable DNA replication in *rnh* mutants of *Escherichia coli. Mol. Gen. Genet.* **208:**420–427.

van der Goot, F. G., N. Didat, F. Pattus, W. Dowhan, and L. Letellier. 1993. Role of acidic lipids in the translocation and channel activity of colicins A and N in *Escherichia coli* cells. *Eur. J. Biochem.* **213:**217–221.

von Meyenburg, K., E. Boye, K. Skarstad, L. Koppes, and T. Kogoma. 1987. Mode of initiation of constitutive stable DNA replication in RNase H-defective mutants of *Escherichia coli* K-12. *J. Bacteriol.* **169:**2650–2658.

Weissborn, A. C., M. K. Rumley, and E. P. Kennedy. 1992. Isolation and characterization of *Escherichia coli* mutants blocked in production of membrane-derived oligosaccharides. *J. Bacteriol.* **174:**4856–4859.

Wu, H. C., M. Tokunaga, H. Tokunaga, S. Hayashi, and C.-Z. Giam. 1983. Posttranslational modification and processing of membrane lipoproteins in bacteria. *J. Cell. Biochem.* **22:**161–171.

Xia, W., and W. Dowhan. Unpublished data.

Yung, B. Y.-M., and A. Kornberg. 1988. Membrane attachment activates DnaA protein, the initiation protein of chromosome replication in *Escherichia coli. Proc. Natl. Acad. Sci. USA* **85:**7202–7205.

Chapter 39

Role of Negatively Charged Phospholipids in the Mode of Action of Pore-Forming Colicins: an Attempt To Relate In Vitro and In Vivo Studies

LUCIENNE LETELLIER,[1] F. GISOU VAN DER GOOT,[2] JUAN M. GONZÁLEZ-MAÑAS,[2] JEREMY H. LAKEY,[2] WILLIAM DOWHAN,[3] AND FRANC PATTUS[4]

Laboratoire des Biomembranes, UA 1116, Centre National de la Recherche Scientifique, Université Paris Sud, Bât 433, 91405 Orsay Cedex,[1] and Département Récepteurs et Protéines Membranaires, ESBS, Pôle Universitaire d'Illkirch, F-67400 Illkirch,[4] France; European Molecular Biology Laboratory, Meyerhofstrasse 1, 69242 Heidelberg, Germany[2]; and Department of Biochemistry and Molecular Biology, University of Texas Medical School, Houston, Texas 77225[3]

Several reports suggest that acidic lipids are involved in the translocation of proteins through the *Escherichia coli* envelope or may stimulate the activity of proteins (i.e., ATPase activity of SecA) required for translocation (de Vrije et al., 1988, 1990; Lill et al., 1990; Kusters et al., 1991; Ulbrandt et al., 1992). These processes involve vectorial translocation from the cytoplasm to the periplasmic space or the outer membrane.

Bacterial toxins are an interesting example of a "reverse translocation" from the external medium to the inner membrane. One example, colicin A, is secreted by and active against *E. coli* cells and closely related species. To exert its lethal activity, it first binds to receptor proteins (BtuB/OmpF) located in the outer membrane. Translocation through the envelope follows, and this translocation requires the participation of proteins encoded by the *tolQRAB* gene cluster (Lazdunski et al., 1988). Finally it reaches the inner membrane, where it forms a voltage-gated ionic channel whose opening leads to a massive efflux of cytoplasmic potassium (Bourdineaud et al., 1990). Colicin A and its translocation machinery have been shown to form a complex located in contact sites between the outer and inner membrane (Guihard et al., 1994) (Fig. 1). Colicin A and related colicins (E1, B, K, N, la, and lb) form voltage-gated channels in planar lipid bilayers (Schein et al., 1978; Bullock et al., 1983, Slatin, 1988; Pattus et al., 1990), and there is a general agreement that the in vitro pore-forming activity of these molecules is dependent on the presence of acidic lipids (Pattus et al., 1983; Massotte et al., 1989).

It is therefore of interest to know whether the translocation and the in vivo activity of pore-forming colicins also require acidic lipids. In this chapter we review the recent knowledge of the effect of negatively charged lipids on the insertion of colicin A into lipid vesicles and on the efflux of K^+ ions from *E. coli* cells in which the content of diphosphatidylglycerol can be controlled. For a more general review on colicin function, the reader is referred to Lakey et al. (1994).

Role of Negatively Charged Lipids in the Insertion of Colicin A into Liposomes

The intrinsic fluorescence of the colicin A C-terminal pore-forming fragment does not change after insertion into normal phospholipid vesicles and is thus an unsuitable probe for monitoring the membrane insertion process. Membrane insertion was therefore monitored by measuring the quenching of this fluorescence by brominated dioleoylphosphatidylglycerol (Br-DOPG) vesicles (González-Mañas et al., 1992). Bromine atoms located at the midpoint of the phospholipid acyl chain quench the tryptophan fluorescence, indicating contact between fluorophores of the membrane inserted protein and the hydrophobic core of the bilayer. Addition of Br-DOPG vesicles to a protein solution quenches the tryptophan fluorescence in a time-dependent manner. This quenching can be fitted to a single-exponential function and thus can be interpreted as a one-step process. This allows calculation of an apparent rate constant of protein insertion into the membrane. Detailed analysis of the effect of protein and lipid concentration on the kinetic insertion constant led to the following conclusion.

The insertion process can be described as follows:

$$[\text{colicin} + \text{vesicles}] \xrightarrow{\text{very fast}} [\text{associated colicin}] \xrightarrow{\text{slow}} [\text{inserted colicin}]$$

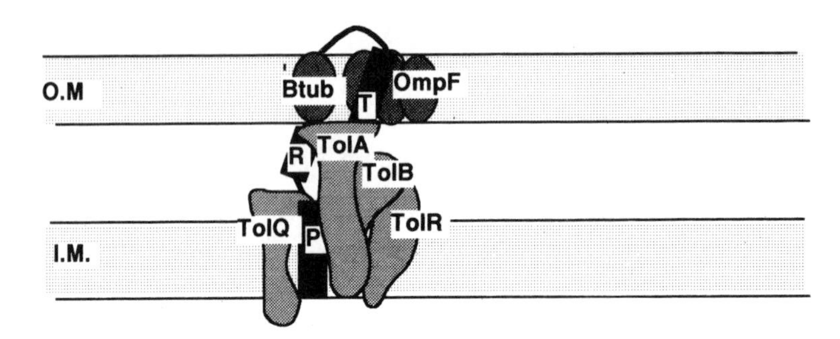

FIG. 1. Mode of action of colicin A in vivo. Colicin A is a three-domain protein with a C-terminal pore-forming domain (P), a central receptor binding domain (R), and a N-terminal "translocation domain" (T). The entry of colicin occurs in three steps. (i) Colicin A first binds to a receptor on the outer membrane (O.M.) of the target cell. (ii) Translocation through the cell envelope requires *tolQRAB* gene products. (iii) The pore-forming domain inserts into the inner membrane (I.M.) and forms a voltage-gated channel. During translocation. colicin A is in an unfolded state. The insertion-competent state of the pore-forming domain, the molten globule, may form within the periplasm upon refolding after translocation.

The first step, which is a bimolecular reaction, is not seen by the quenching assay but appears to be completed within the mixing time of the solution (6 to 10 s). Once the protein has "landed" on the surface of the vesicle, the transition to an insertion-competent state may take place, and it is either this process or the insertion itself which is rate limiting and hence measured by the quenching which takes place during the insertion. This second step depends on the surface concentration of lipids and protein. Preliminary fluorescence energy transfer experiments, with lipids labeled on the polar head group, indicate that association of colicin A to the vesicles is indeed a fast process.

Negatively charged lipids and acidic pH are required for colicin A insertion into membranes in vitro. For example, diluting the negative charges of Br-DOPG by brominated dioleoylphosphatidylcholine (Br-DOPC) dramatically reduces the insertion kinetic constant. The first role of the negatively charged lipids has been postulated to be promotion of electrostatic interaction with positively charged residues on the colicin A pore-forming domain. However, a second role may be the catalysis of the formation of the insertion intermediate. Recently we have shown the colicin A pore-forming domain undergoes a transition to the "molten-globule" state at acidic pH in solution (van der Goot et al., 1991); the main characteristics of this state are conservation of secondary structure, absence of tertiary structure, and water inaccessibility to the core of the molecule (the tryptophan emission wavelength remains unchanged) (Dolgikh et al., 1981). The molten globule has been observed as an early intermediate in protein folding and was suggested as a possible model for the membrane insertion-competent state of proteins (Bychkova et al., 1988).

The kinetic scheme shown above indicates that an intermediate accumulates at the membrane surface. The key role of pH and negatively charged lipids on the kinetics suggests that this intermediate is related to the molten-globule state. Negatively charged lipids are known to lower the surface pH with respect to the bulk pH. This ΔpH is due to the presence of the electrostatic potential which is formed at the surface of the vesicle. When this ΔpH is taken into account, and only then, an excellent correlation can be found between the increase of the rate constant, k, and the proportion of colicin A adopting the characteristics of the molten-globule state in solution (van der Goot et al., 1991). The tryptophans remain inaccessible within this insertion-competent state and therefore should not be quenched by the bromine atoms within the core of the bilayer. This state can be identified with the colicin E1 insertion-competent intermediate described by Merrill et al. (1990).

In conclusion, the roles of negatively charged lipids in vitro can be separated into two aspects, a direct role in the electrostatic interaction between the protein and the membrane surface and an indirect role by affecting the local pH at the membrane surface.

Figure 2 summarizes our present knowledge of the mechanism of membrane insertion of the colicin pore-forming fragment into lipid vesicles in vitro on the basis of its three-dimensional structure (Parker et al., 1992) and of more than 10 years on intensive research. The mechanism by which membrane potential induces the conducting state of the channel is still unknown.

Role of Negatively Charged Phospholipids In Vivo

In vivo pore-forming activity in whole cells can be analyzed by monitoring the efflux of cytoplasmic potassium induced by the colicin. The

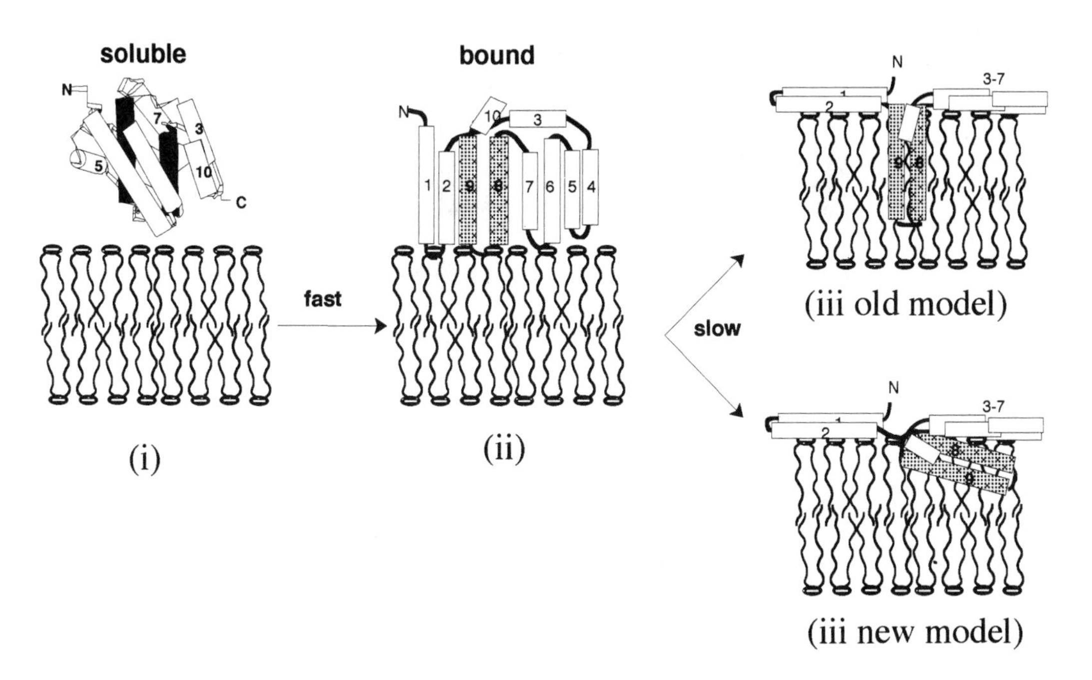

FIG. 2. Model for colicin A pore-forming domain insertion into lipid vesicles based on the X-ray structure and in vitro studies (adapted from Parker et al. [1992] and Lakey et al., [1994]). (i) This domain is an "inside-out" membrane protein consisting of 10 alpha-helices. Two of them (8 and 9) form a hydrophobic hairpin initially buried within the protein in solution. (ii) In its bound state at the vesicle surface, the pore-forming domain is thought to undergo a transition to the molten-globule state. Negatively charged lipids play a key role in steps i through iii.

translocation step can be characterized by the lag time between addition of colicin and appearance of this efflux (Bourdineaud et al., 1990; Bénédetti et al., 1992).

The major phospholipids found in *E. coli* cells are phosphatidylethanolamine (PE) (75 to 80%), PG (20%), and di-PG (1 to 5%) (Raetz, 1978). Therefore the overall negative charge is mainly due to PG and di-PG. Strains in which the PG content can be controlled and varied have been constructed (Kusters et al., 1991; Heacock and Dowhan, 1989). We have used strains HDL1001 and HDL11. In strain HDL1001 the production of PG-phosphate synthetase, the enzyme which is implicated in the synthesis of PG and di-PG, is under the control of the *lac* promoter. HDL11 is a derivative of HDL1001 which carries an additional mutation in *lpp,* thus preventing the synthesis of the major outer membrane lipoproteins (Kusters et al., 1991; Heacock and Dowhan, 1989).

The in vivo activity of colicin A has been shown to depend on the energetic state of the bacteria (membrane potential, respiratory activity, internal ATP) (Cramer et al., 1983; Letellier, 1992). To study the role of acidic lipids in in vivo cell functions by using mutant strains, it is crucial that the change in lipid composition is not accompanied by an alteration of energetic parameters. Strain HDL1001 did not fulfill these requirements. It is obvious that the mutations allowing the control of the PG level in this strain have dramatic side effects on several membrane functions (Fig. 3) (van der Goot et al., 1993). The internal level of ATP and the degree of respiration are far lower at low levels than high levels of PG. Furthermore, this strain is unable to actively transport potassium. In contrast, the energetic state of HDL11 was similar to that of a wild-type strain. This strain was therefore more suitable for the present work.

By measuring the potassium efflux induced by colicin A and the lag time preceding this efflux, it was possible to gain information on the effect of PG on the channel activity and translocation of the toxin (Fig. 4). Reducing the PG content led to an increase of the lag time from 30 to 65 s when the percentage of PG was decreased from 13.5 to 1.2% of total phospholipids and to a decrease in the rate of K^+ efflux by a factor of 3. Surprisingly, no dependency on PG was observed with the related pore-forming colicin N.

A

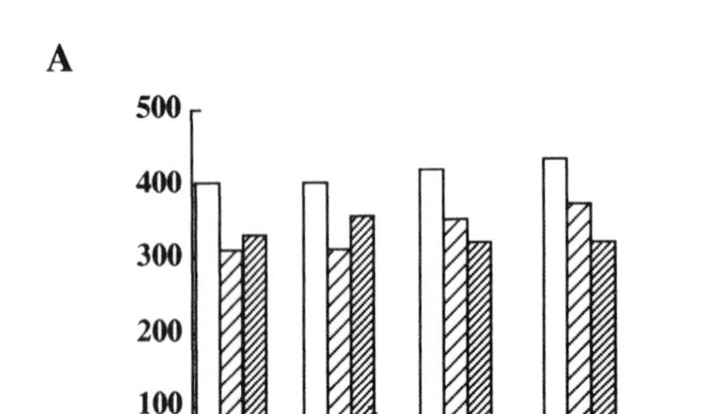

B

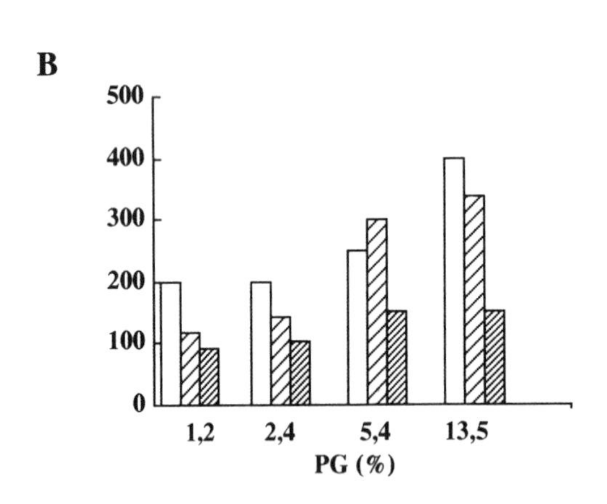

FIG. 3. Energetic parameters of strains HDL11 (A) and HDL1001 (B) as a function of PG concentration (percentage of total lipids). The intracellular potassium concentration (K_{in}), the oxygen consumption (V_{O_2}), and the intracellular ATP concentration were measured at 37°C. Symbols: □, K_{in} (nanomoles per milligram); ▨, V_{O_2} × 3 (nanomoles per minute per milligram); ▨, ATP × 60 (nanomoles per milligram).

A

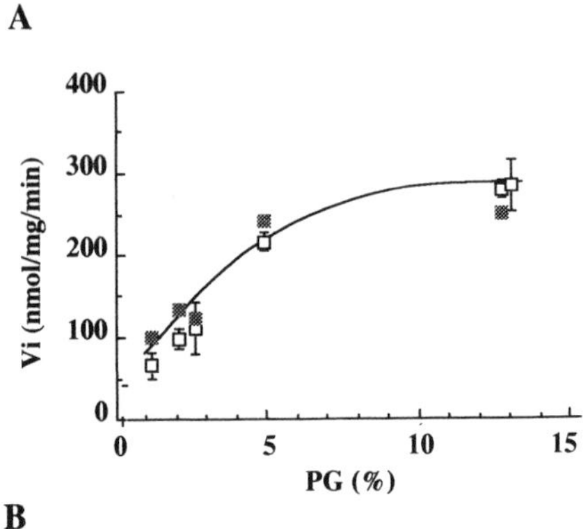

B

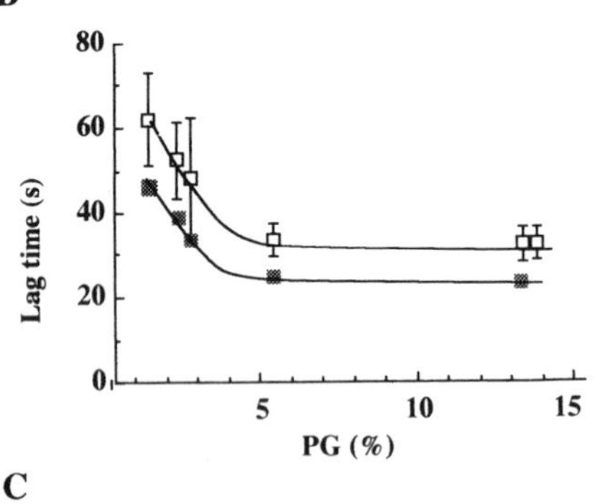

C

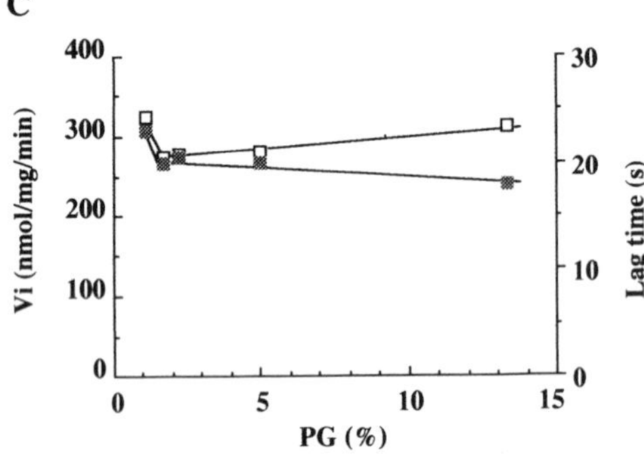

FIG. 4. (A and B) Effect of PG concentration in HDL11 cells on the initial rate of colicin-A potassium efflux (A) and the lag time preceding this efflux (B). Both the potassium efflux and the lag time were measured in a buffer containing either 110 mM (□) or 10 mM sodium phosphate (by-pass of the btuB receptor) (■). The lag time is the time that elapsed between colicin addition and the potassium efflux. The colicin A multiplicity was 200. (C) Effect of the PG content on the initial rate of colicin N potassium efflux (□) and on the lag time preceding this efflux (■) in a buffer containing 110 mM sodium phosphate. Panels A and B reprinted from van der Goot et al. (1993) with permission.

Since the lag time reflects the time needed for binding to BtuB/OmpF, translocation through the Tol machinery, and insertion into the inner membrane (Bénédetti et al., 1992), this means that at least one of these three steps was affected by the decrease in the level of PG. Bypassing the BtuB receptor by decreasing the ionic strength of the incubation buffer led to a decrease in the lag time but did not abolish the PG dependence. Therefore, binding to the receptor is not affected by the PG concentration. Interestingly, the lag time preceding the opening of the colicin N channel was the same regardless of the PG level. Because colicin N uses OmpF as a receptor and TolA and TolQ for its translocation, this suggests that these proteins are not affected by the lack of PG. Therefore, the PG dependence observed for the lag time preceding colicin A-induced potassium efflux might be related to a direct effect on the colicin itself. The increase in lag time and the decrease in the rate of potassium efflux induced by colicin A observed at low PG levels could both be explained by the fact that PG is required for insertion into the inner membrane: at low PG levels, insertion could be slower and less channels could insert.

The in vivo activity of colicin N as a function of the PG level was very different from that of colicin A. We have observed that the rate of potassium efflux induced by the colicin N channel was the same regardless of the level of acidic lipids. The pore-forming fragment of colicin N is very similar to that of colicin A: their amino acid sequences are highly homologous, they show the same channel properties in lipid bilayers (Wilmsen et al., 1990) and the threshold of membrane potential for opening of the channel in vivo is the same as for colicin A. However, the calculated pI value of the C-terminal pore-forming fragment of colicin N (pI = 10.25) is far higher than that of colicin A (pI = 5.82). This is mainly due to the lack of eight aspartic acidic residues in colicin N. One could therefore imagine that the number of negatively charged phospholipids required for the insertion of colicin N would be smaller than for that of colicin A. In strain HDL11 containing a low level of PG, the number of total negatively charged phospholipids would still be sufficient for a proper insertion of colicin N but too low for that of colicin A.

In conclusion, although an effect of the PG content on colicin A activity is also observed in vivo, the dependence appears to be less drastic than in in vitro experiments. Because of experimental problems to achieve even lower levels of PG than those obtained with strain HDL11 (HDL1001 is not suitable for this study), it is not yet clear whether low levels of PG are absolutely essential for the colicin activity.

Relating In Vivo and In Vitro Activity

There is growing evidence that the channel formed by colicin A in artificial membranes in the absence of other proteins is the same as the channel which promotes K^+ efflux from sensitive *E. coli* cells after incubation with colicin A. Both are nonselective and permeable to K^+ and phosphate ions (Guihard et al., 1993); they display similar voltage gating, and none of the colicin A mutants obtained so far displayed properties that contradict with this assumption. The greater importance of negatively charged lipids in vitro than in vivo may be explained by their dual role in vitro. Negatively charged lipids catalyze the formation of the molten-globule state in vitro by decreasing the local pH at the vesicle surface. The molten-globule state can be induced by several means, such as temperature, or as a kinetic intermediate of protein renaturation (see van der Goot et al. [1991] and Dolgikh et al. [1981] and references therein).

Recent circular dichroism and differential scanning calorimetry experiments have shown that the colicin A C-terminal fragment also undergoes a transition to the molten-globule state at high temperature and neutral pH (Muga et al., 1993). The temperature of the transition decreases with pH. The recent evidence that colicin A unfolds during translocation through the outer membrane of the target cell (Bénédetti et al., 1992) suggests that the molten globule may also occur in vivo without the need for an acidic compartment in the periplasmic space and without the need for a negatively charged membrane surface. Decreasing the pH in vitro is equivalent to the unfolding by the translocation machinery in vivo. As soon as the unfolded pore-forming domain of colicin A finds renaturating conditions in the periplasm of the target cell (Fig. 1), it may refold into the molten-globule state in a few milliseconds. Thus, the acidic lipids may affect only the electrostatic interaction between the pore-forming domain and the membrane.

REFERENCES

Bénédetti, H., R. Lloubès, C. Lazdunski, and L. Letellier. 1992. Colicin A unfolds during its translocation in *Escherichia coli* cells and spans the whole cell envelope when its pore has formed. *EMBO J.* **11:**441–447.

Bourdineaud, J. P., P. Boulanger, C. Lazdunski, and L. Letellier. 1990. In vivo properties of colicin A: channel activity is voltage dependent but translocation may be voltage independent. *Proc. Natl. Acad. Sci. USA* **87:**1037–1041.

Bullock, J. O., F. S. Cohen, J. R. Dankert, and W. A. Cramer. 1983. Comparison of the macroscopic and single channel conductance properties of colicin E1 and its COOH-terminal tryptic peptide. *J. Biol. Chem.* **258:**9908–9912.

Bychkova, V., R. Pain, and O. Ptitsyn. 1988. The 'molten globule' state is involved in the translocation of proteins across membranes? *FEBS Lett.* **238:**231–234.

Collarini, M., G. Amblard, C. Lazdunski, and F. Pattus. 1987. Gating processes of channels induced by colicin A, its C-terminal fragment and colicin E1 in planar lipid bilayers. *Eur. Biophys. J.* **14:**147–153.

Cramer, W. A., J. R. Dankert, and Y. Uratani. 1983. The membrane channel-forming bacteriocidal protein, colicin E1. *Biochim. Biophys. Acta* **737:**173–193.

de Vrije, T., A. M. Battenburg, J. A. Killian, and B. de Kruijff. 1990. Lipid involvement in protein translocation in *Escherichia coli. Mol. Micribiol.* **4:**143–150.

de Vrije, T., R. L. de Swart, W. Dowhan, J. Tommassen, and B. de Kruijff. 1988. Phosphatidylglycerol is involved in protein translocation across *Escherichia coli* inner membrane. *Nature* (London) **334:**173–175.

Dolgikh, D., R. Gilmanshin, E. Brazhnikov, V. Bychkova, G. Semisotnov, S. Venyaminov, and O. Ptitsyn. 1981. Alpha-lactalbumin: compact state with fluctuating tertiary structure? *FEBS Lett.* **136:**311–315.

González-Mañas, J. M., J. H. Lakey, and F. Pattus. 1992. Brominated phospholipids as a tool for monitoring the membrane insertion of colicin A. *Biochemistry* **31:**7294–7300.

Guihard, G., H. Benedetti, M. Besnard, and L. Letellier. 1993. Phosphate efflux through the channels formed by colicins and phage T5 in *E. coli* cells is responsible for the fall in the cytoplasmic ATP. *J. Biol. Chem.* **268:**17775–17780.

Guihard, G., P. Boulanger, H. Benedetti, R. Lloubès, M. Besnard, and L. Letellier. 1994. Colicin A and the tol proteins involved in its translocation are preferentially located in contact sites between the inner and outer membranes of *Escherichia coli* cells. *J. Biol. Chem.* **269:**5874–5880.

Heacock, P. N., and W. Dowhan. 1989. Alteration of the phospholipid composition of *Escherichia coli* through genetic manipulation. *J. Biol. Chem.* **264:**14972–14977.

Kusters, R., W. Dowhan, and B. de Kruijff. 1991. Negatively charged phospholipids restore prePhoE translocation across phosphatidylglycerol-depleted *Escherichia coli* inner membranes. *J. Biol. Chem.* **266:**8659–8662.

Lakey, J. H., F. G. Van der Goot, and F. Pattus. 1994. All in the family: the toxic activity of pore-forming colicins. *Toxicology* **18:**391–395.

Lazdunski, C., D. Baty, V. Geli, D. Cavard, J. Morlon, R. H. P. Lloubès, M. Knibiehler, M. Chartier, S. Varenne, M. Frenette, J. L. Dasseux, and F. Pattus. 1988. The membrane channel-forming colicin A: synthesis, secretion, structure, action and immunity. *Biochim. Biophys. Acta* **947:**445–464.

Letellier, L. 1992. In vivo properties of colicin A: channel activity and translocation across the envelope of *Escherichia coli*. *Biochim. Biophys. Acta* **1101:**218–220.

Lill, R., W. Dowhan, and W. Wickner. 1990. The ATPase activity of SecA is regulated by acidic phospholipids, SecY, and the leader and mature domains of precursor proteins. *Cell* **60:**271–280.

Massotte, D., J.-L. Dasseux, P. Sauve, M. Cyrklaff, K. Leonard, and F. Pattus. 1989. Interaction of the pore-forming domain of colicin A with phospholipid vesicles. *Biochemistry* **28:**7713–7719.

Merrill, A. R., F. S. Cohen, and W. A. Cramer. 1990. On the nature of the structural change of the colicin E1 channel peptide necessary for its translocation-competent state. *Biochemistry* **29:**5829–5836.

Muga, A., J. M. Gonzalez-Mañas, J. H. Lakey, F. Pattus, and W. K. Surewicz. 1993. pH dependent stability and membrane insertion of the pore-forming domain of colicin A. *J. Biol. Chem.* **268:**1553–1557.

Parker, M. W., J. P. Postma, F. Pattus, A. D. Tucker, and D. Tsernoglou. 1992. Refined structure of the pore-forming domain of colicin A at 2.4 Å resolution. *J. Mol. Biol.* **224:**639–657.

Pattus, F., M. C. Martinez, B. Dargent, D. Cavard, R. Verger, and C. Lazdunski. 1983. Interaction of colicin A with phospholipid monolayers and liposomes. *Biochemistry* **22:**5698–5707.

Pattus, F., D. Massotte, H. U. Wilmsen, J. Lakey, D. Tsernoglou, A. Tucker, and M. W. Parker. 1990. Colicins: prokarytic killer-pores. *Experientia* **46:**180–192.

Raetz, C. R. H. 1978. Enzymology, genetics, and regulation of membrane phospholipid synthesis in *Escherichia coli*. *Microbiol. Rev.* **42:**614–659.

Slatin, S. L. 1988. Colicin E1 in planar lipid bilayers. *Int. J. Biochem.* **20:**737–744.

Schein, J. S., B. L. Kagan, and A. Finkelstein. 1978. Colicin K acts by forming voltage-dependent channels in phospholipid bilayer membranes. *Nature* (London) **276:**159–163.

Ulbrandt, N. D., E. London, and D. Oliver. 1992. Deep penetration of a proportion of *Escherichia coli* SecA protein into model membranes is promoted by anionic phospholipids and by partial unfolding. *J. Biol. Chem.* **267:**15184–15192.

van der Goot, F. G., N. Didat, F. Pattus, W. Dowhan, and L. Letellier. 1993. Role of acidic lipids in the translocation and channel activity of colicins A and N in *Escherichia coli* cells. *Eur. J. Biochem.* **213:**217–221.

van der Goot, F. G., J. M. González-Mañas, J. H. Lakey, and F. Pattus. 1991. A 'molten-globule' membrane-insertion intermediate of the pore-forming domain of colicin A. *Nature* (London) **354:**408–410.

Wilmsen, H. U., A. P. Pugsley, and P. Pattus. 1990. Colicin N forms voltage- and pH-dependent channels in planar lipid bilayer membranes. *Eur. Biophys. J.* **18:**149–158.

Winiski, A. P., A. C. McLaughlin, R. V. McDaniel, M. Eisenberg, and S. McLaughlin. 1986. An experimental test of the discreteness of charge effects in positive and negative lipid bilayers. *Biochemistry* **25:**8206–8214.

VIII. PROTEIN EXPORT AND FOLDING

Chapter 40

Introduction: Protein Export and Folding

JAN TOMMASSEN

Department of Molecular Cell Biology and Institute of Biomembranes, Utrecht University, Padualaan 8, 3584 CH Utrecht, The Netherlands

Understanding protein folding is an important aspect of studies on protein traffic. Exported proteins are synthesized in the cytoplasm in a precursor form. It has been demonstrated that the folding of a precursor in the cytoplasm into its stable tertiary structure will inhibit its translocation (Randall and Hardy, 1986). Consequently, the premature folding of a precursor has to be prevented. The cotranslational translocation of a precursor may be an important means of preventing premature folding. Furthermore, the attached signal sequence (Park et al., 1988) and the interaction with molecular chaperones (Collier et al., 1988) may retard folding of the precursor and prolong its translocation-competent state. This translocation-competent state is not necessarily a completely unfolded structure. It has been shown that translocation-competent proOmpA contains a considerable degree of secondary and tertiary structure (Lecker et al., 1990). Consequently, active unfolding may be required to allow translocation of the precursor in an expanded linear form. The unfolding reaction could be mediated by SecA at the expense of ATP (Driessen, 1992) or by the translocation reaction itself (Arkowitz et al., 1993). However, it should also be noted that translocation does not necessarily occur in a fully expanded form, since the translocation of a precursor with an internal disulfide bridge has been demonstrated in vitro (Tani et al., 1990). The degree of tertiary structure that can be adopted by a precursor before and during translocation remains to be determined.

Folding of (domains of) the nascent chains on the extracytoplasmic side of the membrane may contribute to the energy requirements of the translocation reaction (von Heijne and Blomberg, 1979) and may prevent reverse translocation. Many denatured proteins refold spontaneously in vitro into their native structure. However, this refolding is, in general, slow, and therefore enzymes and chaperones may be involved in protein folding in vivo. Enzymes that catalyze the formation of disulfide bridges (Bardwell et al., 1991; Kamitani et al., 1992) or the *cis-trans* isomerization of peptidylprolyl bonds in proteins (Hayano et al., 1991) in the periplasm have been described. In addition, chaperones such as the *lipB* gene product of *Pseudomonas glumae* (Frenken et al., 1993) may be required to guide the folding of specific proteins by preventing inappropriate interactions.

Protein traffic does not stop in the periplasm. Outer membrane proteins will have to insert into the appropriate membrane. Insertion is not determined (only) by primary sequence elements: long stretches of hydrophobic residues are lacking in outer membrane proteins. Consequently, folding of these proteins into a native protein-like structure and exposure of hydrophobic residues to the exterior are essential to enable the insertion into the membrane by hydrophobic interactions (Bosch et al., 1986). Therefore, the recent resolution of the crystal structures of two *Escherichia coli* outer membrane proteins, PhoE and OmpF (Cowan et al., 1992), will be very beneficial in studies on the insertion process.

Finally, in contrast to *E. coli*, other gram-negative bacteria secrete a variety of proteins into the extracellular medium. Many proteins are secreted via the periplasm in a two-step pathway that is conserved in different bacteria (Pugsley, 1993). These proteins fold in the periplasm into their tertiary structure, and are translocated across the outer membrane as folded proteins. Inhibition of folding in the periplasm interfered with their translocation across the outer membrane (Pugsley, 1992; Frenken et al., 1993). Consequently, it will be essential to study the structure and the folding of these exoproteins to understand the secretion mechanism and the interaction with components of the secretion apparatus.

REFERENCES

Arkowitz, R. A., J. C. Joly, and W. Wickner. 1993. Translocation can drive the unfolding of a preprotein domain. *EMBO J.* **12:**243–253.

Bardwell, J. C. A., K. McGovern, and J. Beckwith. 1991. Identification of a protein required for disulfide bond formation *in vivo*. *Cell* **67:**581–589.

Bosch, D., J. Leunissen, J. Verbakel, M. de Jong, H. van Erp, and J. Tommassen. 1986. Periplasmic accumulation of truncated forms of outer-membrane PhoE protein of *Escherichia coli* K-12. *J. Mol. Biol.* **189:**449–455.

Collier, D. N., V. A. Bankaitis, J. B. Weiss, and P. J. Bassford, Jr. 1988. The antifolding activity of SecB promotes the export of the *E. coli* maltose-binding protein. *Cell* **53:**273–283.

Cowan, S. W., T. Schirmer, G. Rummel, M. Steiert, R. Ghosh, R. A. Pauptit, J. N. Jansonius, and J. P. Rosenbusch. 1992. Crystal structures explain functional properties of two *E. coli* porins. *Nature* (London) **358:**727–733.

Driessen, A. J. M. 1992. Precursor protein translocation by the *Escherichia coli* translocase is directed by the protonmotive force. *EMBO J.* **11:**847–853.

Frenken, L. G. J., A. de Groot, J. Tommassen, and C. T. Verrips. 1993. Role of the *lipB* gene product in the folding of the secreted lipase of *Pseudomonas glumae*. *Mol. Microbiol.* **9:**591–599.

Hayano, T., N. Takahashi, S. Kato, N. Maki, and M. Suziki. 1991. Two distinct forms of peptidylprolyl-*cis-trans*-isomerase are expressed separately in periplasmic and cytoplasmic compartments of *Escherichia coli* cells. *Biochemistry* **30:**3041–3048.

Kamitani, S., Y. Akiyama, and K. Ito. 1992. Identification and characterization of an *Escherichia coli* gene required for the formation of correctly folded alkaline phosphatase, a periplasmic enzyme. *EMBO J.* **11:**57–62.

Lecker, S. H., A. J. M. Driessen, and W. Wickner. 1990. ProOmpA contains secondary and tertiary structure prior to translocation and is shielded from aggregation by association with SecB protein. *EMBO J.* **9:**2309–2314.

Park, S., G. Liu, T. B. Topping, W. H. Cover, and L. L. Randall. 1988. Modulation of folding pathways of exported proteins by the leader sequence. *Science* **239:**1033–1035.

Pugsley, A. P. 1992. Translocation of a folded protein across the outer membrane in *Escherichia coli*. *Proc. Natl. Acad. Sci. USA* **89:**12058–12062.

Pugsley, A. P. 1993. The complete general secretory pathway in gram-negative bacteria. *Microbiol. Rev.* **57:**50–108.

Randall, L. L., and S. J. S. Hardy. 1986. Correlation of competence for export with lack of tertiary structure of the mature species: a study *in vivo* of maltose-binding protein in *E. coli*. *Cell* **46:**921–928.

Tani, K., H. Tokuda, and S. Mizushima. 1990. Translocation of proOmpA possessing an intramolecular disulfide bridge into membrane vesicles of *Escherichia coli*. Effect of membrane energization. *J. Biol. Chem.* **265:**17341–17347.

von Heijne, G., and C. Blomberg. 1979. Trans-membrane translocation of proteins. The direct transfer model. *Eur. J. Biochem.* **97:**175–181.

Chapter 41

Membrane Protein Assembly: Can Protein-Lipid Interactions Explain the "Positive Inside" Rule?

GUNNAR VON HEIJNE

Karolinska Institute Center for Structural Biochemistry, NOVUM, S-141 57 Huddinge, Sweden

The question asked in the title of this paper—"Can protein-lipid interactions explain the 'positive inside' rule?"—has the simplest of answers—"We don't know"—yet indicates that the rules relating amino acid sequence to membrane protein topology cannot so far be given a clear explanation in mechanistic terms. In this chapter, I will try to review what is and is not known about the mechanisms of membrane protein assembly in *Escherichia coli* and how one might imagine that positively charged amino acids could act to control the membrane insertion process and hence the topology of inner membrane proteins.

The "Positive Inside" Rule

It has long been known that the cytoplasmic or periplasmic location of loops and tails in bacterial inner membrane proteins correlates with their content of basic amino acids: the "positive inside" rule (von Heijne, 1986). In fact, the distribution of arginines and lysines between the two sides of the membrane is so skewed that it can be used very effectively together with hydrophobicity plots to predict the correct transmembrane topology directly from the amino acid sequence (von Heijne, 1992).

The positive inside rule is now known to be more than a simple statistical correlation, and can in fact be used as a basis for manipulating the topology of membrane proteins (Boyd and Beckwith, 1990; von Heijne and Manoil, 1990). The best-studied protein in this regard is leader peptidase (Lep) from the inner membrane of *Escherichia coli,* which has served as a model system to demonstrate many aspects of how charged amino acids can control membrane protein topology.

Lep has two hydrophobic transmembrane segments (H1 and H2); a short, highly charged cytoplasmic loop (P1); and a large C-terminal periplasmic domain (P2) (Fig. 1). The orientation is thus $N_{out}-C_{out}$, with both the N and C termini in the periplasm. As suggested by the positive inside rule, this orientation can be inverted by redesigning the molecule such that the N-terminal tail carries a larger number of positively charged residues than the P1 domain (von Heijne, 1989; Nilsson and von Heijne, 1990). The charge can be carried by either lysines or arginines or even by histidines if the cytoplasmic pH is sufficiently low (Andersson et al., 1992), whereas negatively charged residues have only slight effects (Andersson et al., 1992) unless the protein is finely poised on the threshold between the wild-type and inverted topologies (Andersson and von Heijne, 1993a).

Thus, work on Lep as well as on other inner membrane proteins has clearly demonstrated that positively charged residues are the main topological determinants, and similar results have been obtained for eukaryotic plasma membrane proteins (von Heijne and Gavel, 1988; Beltzer et al., 1991; Parks and Lamb, 1991, 1993; Sipos and von Heijne, 1993). Statistical studies suggest that this may hold also for proteins of the thylakoid membrane (Gavel et al., 1991) and for mitochondrially encoded mitochondrial inner membrane proteins (Gavel and von Heijne, 1992).

Mechanistic Aspects: the Sec Machinery

According to the positive inside rule, periplasmically exposed domains in inner membrane proteins should have a low content of positively charged residues. This very simple rule breaks down, however, for domains that are longer than 60 to 70 residues, for which one often finds a more balanced content of charged residues (von Heijne and Gavel, 1988) similar to what is found in globular periplasmic proteins.

What could be the reason for this apparent distinction between short and long periplasmic domains? Studies on the membrane assembly of the phage M13 procoat protein (Wolfe et al., 1985) and inverted Lep constructs (von Heijne, 1989) suggested an answer: whereas periplasmic proteins depend on the *sec* machinery (Schatz and Beckwith, 1990) for translocation across the inner membrane, the short periplasmic segments in procoat and inverted Lep can be readily translocated

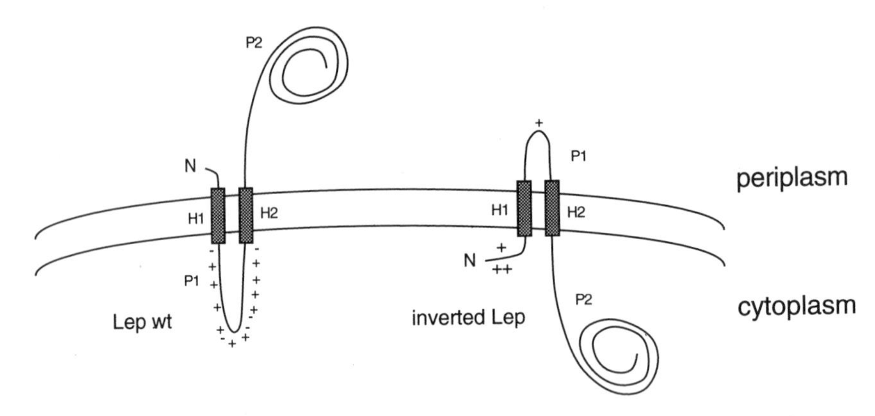

FIG. 1. Orientation of wild-type (wt) (left) and inverted (right) Lep in the inner membrane.

even under conditions where the *sec* machinery is compromised. Accordingly, such *sec*-independent translocation would thus be possible only when the translocated region was short and contained few positively charged residues, whereas the more forgiving *sec* machinery could act efficiently only on longer segments.

In support of this idea, when the periplasmic domain of procoat is lengthened from ~20 to ~100 residues by introduction of a segment from the outer membrane protein OmpA, its translocation becomes *sec* dependent (Kuhn, 1988), whereas extensions of the cytoplasmic C-terminal domain have no effect on *sec* dependence (Kuhn et al., 1987). In the case of inverted Lep, *sec* dependence increases almost linearly with increasing length of the translocated loop from ~25 to ~60 residues and then plateaus (Andersson and von Heijne, 1993b). A positive counterexample is provided by a large periplasmic loop (~180 residues) found in the MalF protein that appears to translocate quite efficiently across the membrane in cells with an impaired *sec* machinery (McGovern and Beckwith, 1991). We have recently made fusions between an N-terminal part of MalF including this loop and the P2 domain from Lep (serving as a convenient topological reporter), and we found a small but significant effect on translocation of the large MalF loop when the *sec* machinery was blocked (Andersson and von Heijne, unpublished). The question of the possible relation between *sec* dependence and chain length is thus still not fully settled, and further work must be done.

Mechanistic Aspects: the Proton Motive Force

The proton motive force (PMF) across the inner membrane (positive and acidic outside) is known to play a critical role during *sec*-dependent translocation of periplasmic and outer membrane proteins (Bakker and Randall, 1984; Yamada et al.,1989; Shiozuka et al., 1990; Schiebel et al., 1991), but its importance for the assembly of inner membrane proteins has not been much studied. The *sec*-independent assembly of the M13 procoat protein is sensitive to the protonophore carbonyl cyanide *m*-chlorophenylhydrazone (CCCP) (Date et al., 1980), although mutations in the periplasmic domain can make the protein rather insensitive to the drug (Zimmermann et al., 1982).

We have recently analyzed the effect of CCCP on the translocation of a series of *sec*-independent inverted Lep constructs that differ only by the number of negatively charged residues in their periplasmic loop. Translocation is completely unaffected by CCCP when there are fewer than three negatively charged residues but becomes increasingly sensitive as this number is increased to three or four (Andersson and von Heijne, 1994) (Fig. 2). It thus appears that the PMF lowers the barrier for translocation of negatively charged residues and increases it for positively charged ones, suggesting a possible reason for the positive inside rule. Indeed, on the basis of this idea, we have been able to design a protein with an N-terminal negatively charged tail and one positive charge in the P1 loop; this protein inserts into the inner membrane with the wildtype orientation in the presence of a PMF; when the PMF is dissipated, however, ~50% of the molecules insert in the inverted orientation.

Interestingly, a similar effect on the initial insertion of the signal sequence into the translocation machinery during *sec*-dependent translocation was recently reported (Geller et al., 1993). In this case, negatively charged residues placed immediately downstream of the signal sequence were found to promote insertion in the presence but not in the absence of a membrane potential, while positively charged residues in the same location had the opposite effect and blocked insertion in the presence of a potential. In this respect, *sec*-independent translocation of short loops in inner

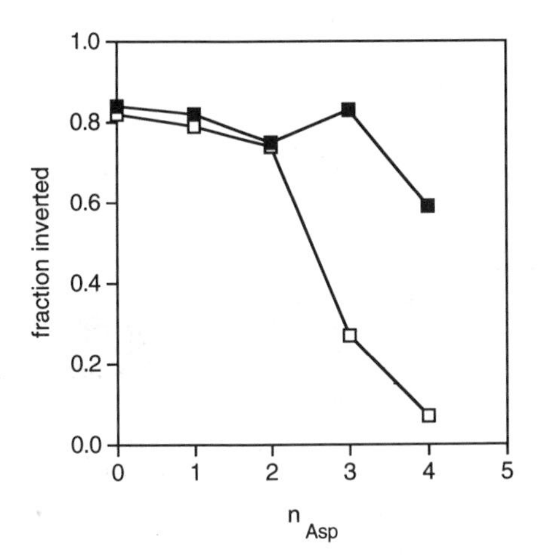

FIG. 2. CCCP dependence of the translocation of the P1 domain in a series of inverted Lep constructs with three lysines in the N-terminal tail and one lysine plus *n* aspartic acids in the P1 loop as a function of *n*. Symbols: ■, CCCP absent; □, CCCP present.

membrane proteins may be mechanistically akin to the initial insertion of the signal sequence during *sec*-dependent export.

Mechanistic Aspects: Phospholipids

Acidic phospholipids are required for the proper functioning of the SecA protein (Lill et al., 1990; Kusters et al., 1991; Ulbrandt et al., 1992) and are thus critically important for *sec*-dependent protein translocation. Whether they are also needed for *sec*-independent assembly is not known, although an effect on the assembly of M13 procoat into inverted inner membrane vesicles in vitro has been found (Kusters, 1992).

We have analyzed the membrane assembly of a series of Lep-derived constructs in strain HDL11 (Kusters et al., 1991) in which the levels of the acidic phospholipid phosphatidylglycerol can be controlled by varying the level of the inducer isopropyl-β-D-thiogalactopyranoside in the growth medium. A small effect on the kinetics of translocation of the P2 domain in constructs with the wild-type orientation was apparent when induced and noninduced cells were compared; this is as expected, since this step is *sec* dependent. So far, however, we have not seen any major effects on the topology of various constructs under conditions where phosphatidylglycerol levels are low, suggesting that acidic phospholipids are not directly involved in processes determining topology.

Conclusions

The importance of positively charged residues as topological determinants during membrane protein assembly in bacteria is by now well established. The mechanistic basis for the positive inside rule is far from clear, however. Recent data suggest that the PMF may play a critical role by creating an asymmetry between positively and negatively charged residues and, further, that the length of a translocated domain may be related to whether it will be translocated by a *sec*-dependent or a *sec*-independent mechanism (note that the term "*sec*-independent" is only operationally defined—no apparent effect on translocation in vivo under conditions of an impaired *sec* machinery—but should not be taken to necessarily mean that no component[s] of the *sec* machinery may be involved in some capacity [Andersson and von Heijne, 1993b]). If and how phospholipids affect membrane protein assembly is still largely unknown.

This work was supported by grants from the Swedish Natural Sciences Research Council (NFR), the Swedish Technical Research Council (TFR), and the Swedish National Board for Industrial and Technical Development (NUTEK).

REFERENCES

Andersson, H., E. Bakker, and G. von Heijne. 1992. Different positively charged amino acids have similar effects on the topology of a polytopic transmembrane protein in *Escherichia coli. J. Biol. Chem.* **267:**1491–1495.

Andersson, H., and G. von Heijne. 1993a. Position-specific Asp-Lys pairing can affect signal sequence function and membrane protein topology. *J. Biol. Chem.* **268:**21389–21393.

Andersson, H., and G. von Heijne. 1993b. *sec*-Dependent and *sec*-independent assembly of *E. coli* inner membrane proteins—the topological rules depend on chain length. *EMBO J.* **12:**683–691.

Andersson, H., and G. von Heijne. Unpublished results.

Andersson, H., and G. von Heijne. 1994. Membrane protein topology: effects of $\Delta\mu H^+$ on the translocation of charged residues explain the 'positive inside' rule. *EMBO J.* **13:**2267–2272.

Bakker, E. P., and L. L. Randall. 1984. The requirement for energy during export of beta-lactamase in Escherichia coli is fulfilled by the total proton-motive force. *EMBO J.* **3:**895–900.

Beltzer, J. P., K. Fiedler, C. Fuhrer, I. Geffen, C. Handschin, H. P. Wessels, and M. Spiess. 1991. Charged residues are major determinants of the transmembrane orientation of a signal-anchor sequence. *J. Biol. Chem.* **266:**973–978.

Boyd, D., and J. Beckwith. 1990. The role of charged amino acids in the localization of secreted and membrane proteins. *Cell* **62:**1031–1033.

Date, T., C. Zwizinski, S. Ludmerer, and W. Wickner. 1980. Mechanisms of membrane assembly: effects of energy poisons on the conversion of soluble M13 coliphage procoat to membrane-bound coat protein. *J. Biol. Chem.* **77:**827–831.

Gavel, Y., J. Steppuhn, R. Herrmann, and G. von Heijne. 1991. The positive-inside rule applies to thylakoid membrane proteins. *FEBS Lett.* **282:**41–46.

Gavel, Y., and G. von Heijne. 1992. The distribution of charged amino acids in mitochondrial inner membrane proteins suggests different modes of membrane integration for nuclearly and mitochondrially encoded proteins. *Eur. J. Biochem.* **205:**1207–1215.

Geller, B., H. Y. Zhu, S. Y. Cheng, A. Kuhn, and R. E. Dalbey. 1993. Charged residues render pro-OmpA potential dependent for initiation of membrane translocation. *J. Biol. Chem.* **268:**9442–9447.

Kuhn, A. 1988. Alterations in the extracellular domain of M13 procoat protein make its membrane insertion dependent on *secA* and *secY*. *Eur. J. Biochem.* **177:**267–271.

Kuhn, A., G. Kreil, and W. Wickner. 1987. Recombinant forms of M13 procoat with an OmpA leader sequence or a large carboxy-terminal extension retain their independence of secY function. *EMBO J.* **6:**501–505.

Kusters, R. 1992. Studies on the role of phospholipids in protein translocation across the *Escherichia coli* inner membrane. Thesis. University of Utrecht, Utrecht, The Netherlands.

Kusters, R., W. Dowhan, and B. de Kruijff. 1991. Negatively charged phospholipids restore PrePhoe translocation across phosphatidylglycerol-depleted *Escherichia coli* inner membranes. *J. Biol. Chem.* **266:**8659–8662.

Lill, R., W. Dowhan, and W. Wickner. 1990. The ATPase activity of SecA is regulated by acidic phospholipids, SecY, and the leader and mature domains of precursor proteins. *Cell* **60:**271–280.

McGovern, K., and J. Beckwith. 1991. Membrane insertion of the *Escherichia coli* MalF protein in cells with impaired secretion machinery. *J. Biol. Chem.* **266:**20870–20876.

Nilsson, I. M., and G. von Heijne. 1990. Fine-tuning the topology of a polytopic membrane protein. Role of positively and negatively charged residues. *Cell* **62:**1135–1141.

Parks, G. D., and R. A. Lamb. 1991. Topology of eukaryotic type-II membrane proteins—importance of N-terminal positively charged residues flanking the hydrophobic domain. *Cell* **64:**777–787.

Parks, G. D., and R. A. Lamb. 1993. Role of NH2-terminal positively charged residues in establishing membrane protein topology. *J. Biol. Chem.* **268:**19101–19109.

Schatz, P. J., and J. Beckwith. 1990. Genetic analysis of protein export in *Escherichia coli*. *Annu. Rev. Genet.* **24:**215–248.

Schiebel, E., A. J. M. Driessen, F.-U. Hartl, and W. Wickner. 1991. $\Delta\mu H^+$ and ATP function at different steps of the catalytic cycle of preprotein translocase. *Cell* **64:**927–939.

Shiozuka, K., K. Tani, S. Mizushima, and H. Tokuda. 1990. The proton motive force lowers the level of ATP required for the in vitro translocation of a secretory protein in Escherichia coli. *J. Biol. Chem.* **265:**18843–18847.

Sipos, L., and G. von Heijne. 1993. Predicting the topology of eukaryotic membrane proteins. *Eur. J. Biochem.* **213:**1333–1340.

Ulbrandt, N., E. London, and D. Oliver. 1992. Deep penetration of a portion of *Escherichia coli* SecA protein into model membranes is promoted by anionic phospholipids and by partial unfolding. *J. Biol. Chem.* **267:**15184–15192.

von Heijne, G. 1986. The distribution of positively charged residues in bacterial inner membrane proteins correlates with the transmembrane topology. *EMBO J.* **5:**3021–3027.

von Heijne, G. 1989. Control of topology and mode of assembly of a polytopic membrane protein by positively charged residues. *Nature* (London) **341:**456–458.

von Heijne, G. 1992. Membrane protein structure prediction—hydrophobicity analysis and the positive-inside rule. *J. Mol. Biol.* **225:**487–494.

von Heijne, G., and Y. Gavel. 1988. Topogenic signals in integral membrane proteins. *Eur. J. Biochem.* **174:**671–678.

von Heijne, G., and C. Manoil. 1990. Membrane proteins—from sequence to structure. *Protein Eng.* **4:**109–112.

Wolfe, P. B., M. Rice, and W. Wickner. 1985. Effects of two sec genes on protein assembly into the plasma membrane of *Escherichia coli*. *J. Biol. Chem.* **260:**1836–1841.

Yamada, H., H. Tokuda, and S. Mizushima. 1989. Proton motive force dependent and independent protein translocation revealed by an efficient in vitro assay system of *Escherichia coli*. *J. Biol. Chem.* **264:**1723–1728.

Zimmermann, R., C. Watts, and W. Wickner. 1982. The biosynthesis of membrane-bound M13 coat protein. Energetics and assembly intermediates. *J. Biol. Chem.* **257:** 6529–6536.

Chapter 42

Recognition of Ligands as Nonnative by SecB, a Molecular Chaperone Involved in Export of Protein in *Escherichia coli*

LINDA L. RANDALL[1] AND SIMON J. S. HARDY[2]

Department of Biochemistry and Biophysics, Washington State University, Pullman, Washington 99164-4660,[1] and Department of Biology, University of York, York, YO1 5DD, United Kingdom[2]

Protein Export in *Escherichia coli*

Selected proteins synthesized in the cytosol of *E. coli* are exported through the cytoplasmic membrane to the periplasmic space or to the outer membrane by a process that is similar to the transfer of proteins from the cytosol of eukaryotic cells into the lumen of the endoplasmic reticulum. In both systems proteins destined to leave the cytosolic compartments are synthesized as precursors containing amino-terminal extensions, the signal or leader sequences, and these precursors interact with chaperones, whose function is to maintain the polypeptides in a state that is competent for translocation. The best-studied chaperone in the export pathway of *E. coli* is SecB (for a comprehensive review, see Collier [1993]). SecB can interact with precursor polypeptides posttranslationally as well as when they are still nascent polypeptides on the ribosome (Kumamoto, 1989; Kumamoto and Francetic 1993). The complex of precursor and SecB binds to SecA, which in turn interacts with the membrane-associated translocation apparatus comprising SecY, SecE, and Band 1 (Hartl et al., 1990). Coupled to the hydrolysis of ATP, SecA undergoes a cyclic binding and release of the precursor to mediate translocation of the polypeptide through the membrane (Schiebel et al., 1991). Efficient translocation also requires proton motive force (Yamada et al., 1989; Schiebel et al., 1991). Leader peptidase removes the leader peptide during export, and the membrane proteins SecD and SecF play a role at a late step (Schatz and Beckwith, 1990), likely to be folding of the protein or release of the protein into the periplasm (Fig. 1) (for a review, see Wickner et al. [1991]).

SecB Is a Molecular Chaperone

SecB is not essential for all protein export but is involved in the efficient export of a subset of proteins including the outer membrane proteins (OmpA, PhoE, and LamB and the periplasmic maltose-binding (Kumamoto and Beckwith, 1985) and oligopeptide-binding (Collier et al., 1988) proteins. Proteins efficiently exported independently of SecB function include the outer membrane lipoprotein and periplasmic ribose-binding protein (Hayashi and Wu, 1985; Kumamoto and Beckwith, 1985). SecB is a negatively charged, cytosolic protein of monomeric molecular weight 16,400 (Weiss et al., 1988), which functions as a homotetramer (Watanabe and Blobel, 1989a). It has been shown to bind to precursors of several exported proteins in vitro (Lecker et al., 1989) and in vivo (Kumamoto, 1989) and to maintain the precursor of maltose-binding protein in an unfolded protease-sensitive state both in vivo (Kumamoto and Gannon, 1988) and in a cell-free system (Collier et al., 1988; Weiss et al., 1988). Furthermore, when maltose-binding protein is diluted out of denaturant into a solution containing excess purified SecB, its folding, as monitored by changes in the intrinsic fluorescence of tryptophan, is completely blocked (Liu et al., 1989). Taken together, all these data compellingly indicate that when SecB is acting as a chaperone for soluble proteins, the illicit or nonproductive interactions that it prevents are not primarily those of its ligand with other proteins but, rather, are normal folding interactions that would render the ligand incapable of being exported. In contrast, when it is chaperoning outer membrane proteins, which are probably intrinsically insoluble, its main function may be to prevent aggregation.

SecB differs from other molecular chaperones in that ATP seems to have no effect on binding or release of ligand. However, it should be remembered that SecB interacts with SecA, which does hydrolyze ATP when it binds precursor at the membrane translocation apparatus (Hartl et al., 1990). Whether either binding or hydrolysis of ATP is necessary for the transfer of precursor from SecB to SecA is not known.

The distinguishing characteristic of all polypeptides that interact with SecB during export is the possession of a leader sequence. However, SecB

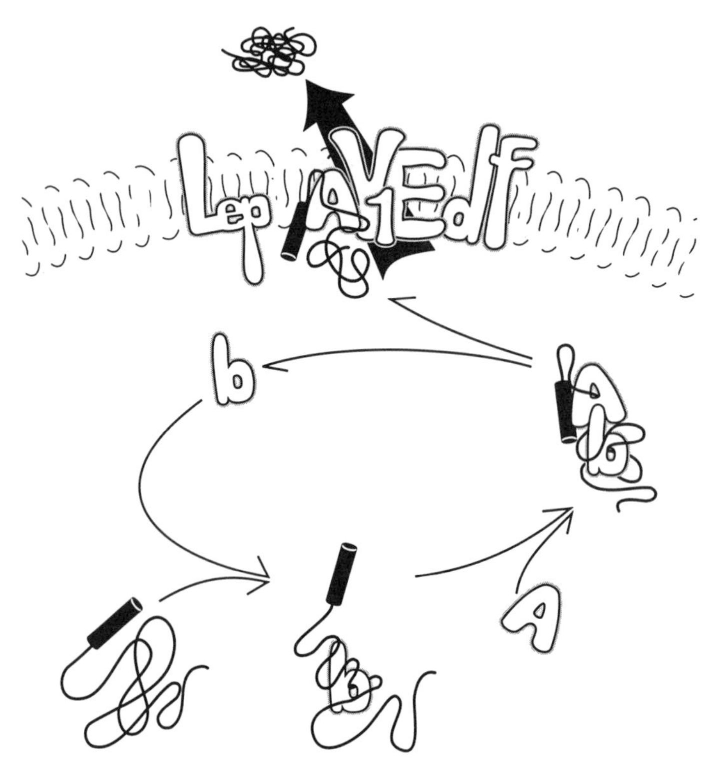

FIG. 1. Working model for protein export. The protein components of the export apparatus are represented by letters as follows: A, SecA; b, SecB; Y, SecY; E, SecE; d, SecD; f, SecF; Lep, leader peptidase. Band 1 is represented by 1, and the precursor is the elongated curved line with the leader portion shown as a cylinder.

does not recognize the leader directly. Rather, the leader plays an indirect, albeit important, role by retarding the folding of the polypeptide to allow SecB to bind before the polypeptide can fold into its native structure, rendering it incompatible with translocation through the membrane (Liu et al., 1989). Tryptophan fluorescence studies of the rate of protein folding in vitro demonstrated that the leader sequence of maltose-binding protein retarded the folding relative to the mature form (Park et al., 1988). This retardation was shown to be important to the binding of SecB in vivo by making use of the fact that proteins defective in export sequester available SecB and thus interfere with the export of other, normal proteins. This interference phenomenon was used as an in vivo assay for the ability of various species of maltose-binding protein to bind SecB. The expression of maltose-binding protein that lacked the leader sequence but was otherwise wild type and should therefore fold rapidly in the cytoplasm did not interfere with export. The interpretation is that without a leader sequence, the polypeptide could not bind SecB. However, the function of the leader in mediating SecB binding could be replaced by a single amino acyl substitution in the mature region of the protein, which drastically reduced the rate of folding. The expression of such a protein that completely lacked the leader but that folded slowly interfered with the export of a SecB-dependent protein but not with that of a SecB-independent protein (Liu et al., 1989). Thus, the role of the leader in binding SecB is indirect. It retards folding to allow SecB to bind. In addition to mediating entry into the export pathway by allowing binding of SecB, the leader sequence has other functions at the membrane, as shown by the large number of studies of proteins with mutated leaders that still bind to SecB in vivo and are delivered to the membrane but not exported (see Bassford [1990] for a review). The conclusion that SecB does not bind specifically with the leader sequence itself but interacts within the body of the unfolded precursor is supported by three other types of investigation. First, when leader sequences were exchanged between a protein whose export is dependent on the function of SecB and one whose export is independent of it, the allegiance of SecB did not follow the leader (Gannon et al., 1989; Collier et al., 1990). Second, the interference phenomenon described above (in which proteins that cannot be exported because of defects in their leader sequence nevertheless interfere with the export of wild-type proteins [Bankaitis and Bassford, 1984; Collier et al., 1988]), was independent of the leader sequence on the interfering species but dependent on the presence of sequences within the body of the protein. Third, and most direct, stable complexes can be formed in vitro between SecB and either unfolded mature maltose-binding protein or unfolded precursor maltose-binding protein. When both mature and precursor maltose-binding protein were simultaneously presented for binding to SecB under conditions of limiting SecB, there was no preferential formation of complex with precursor maltose-binding protein, indicating that the two forms of maltose-binding protein had similar affinities for SecB and therefore that the leader sequence did not contribute significantly to the binding (Randall et al., 1990). All these data indicate clearly that the binding sites for SecB on maltose-binding protein do not lie within the leader sequence. It should be noted, however, that there is one report that binding of SecB to maltose-binding protein is dependent on the presence of a leader sequence (Watanabe and Blobel, 1989b). For LamB there is evidence that the leader, in addition to other regions of the protein, is involved in binding to SecB (Altman et al., 1990).

Selective Binding of SecB to Nonnative Proteins

If the leader sequence itself is not the binding site for SecB, how does this molecular chaperone interact exclusively with polypeptides that contain leader sequences? We believe that the answer is that the possession of a leader sequence slows the folding of a precursor such that it remains in a state that allows interaction with SecB for much longer than other proteins which do not contain leaders, and this enables it to bind to the chaperone. In other words, on completion of a polypeptide that can initially interact with SecB, and there are clearly many of these (Hardy and Randall, 1991), there is a kinetic partitioning between folding and interacting with SecB. Proteins lacking leader sequences fold very rapidly. The rate of folding is much higher than the rate of association with the chaperone, and consequently only a minority of these molecules interact with SecB. Proteins possessing leader sequences fold slowly, such that the rate of binding to SecB is higher than the rate of folding. Thus, the great majority of molecules bind to the chaperone and enter the protein export pathway.

SecB interacts with unfolded proteins but has no detectable affinity for native proteins (Hardy and Randall, 1991). This raises the following interesting question: how does the chaperone recognize a ligand as nonnative? To determine what is recognized, Randall (1992) established an assay based on the sensitivity to proteolysis of a complex comprising SecB bound to the ligand carboxamidomethylated bovine pancreatic trypsin inhibitor (R·BPTI) compared with the sensitivity to proteolysis of the free components. When uncomplexed with ligand, SecB is quantitatively cleaved by low concentrations of proteinase K to a form that is lacking about 50 amino acyl residues from the carboxyl terminus. In contrast, the free ligand is resistant to proteolysis under the same conditions. The formation of a complex between SecB and R·BPTI concomitantly renders R·BPTI sensitive and SecB resistant to proteolysis. The presence in the reaction mixture of native BPTI, previously shown not to bind to SecB (Hardy and Randall, 1991), does not protect SecB from proteolysis. Because the concentration of SecB in the assay (0.6 μM) is much greater than the estimated dissociation constant (5 nM) for R·BPTI (Hardy and Randall, 1991), one can use protection from proteolysis to estimate the stoichiometry of the complex as approximately 1 mol of ligand bound to 1 mol of monomeric SecB. Thus, a tetrameric unit of SecB has multiple peptide binding sites.

This assay was used to survey a large number of peptides of known sequence. Peptides that conferred at least 50% protection of 0.6 μM SecB at a concentration of 15 μM or less were considered to be ligands. Those that did not meet this criterion did not confer any protection even at the highest concentrations tested, which in some cases were much higher than 15 μM. The ligands as well as the peptides that showed no binding in the experiments are given in Table 1, together with their sequences. The only obvious features that distinguish the ligands from the peptides that do not bind appear to be a net positive charge and a length greater than 11 residues. In addition, flexibility appears to be important, since the ligands confer better protection if they are not constrained by disulfide bonds or, in the case of the zinc finger tested, by the presence of zinc ions.

The observed protection from proteolysis has at least two possible explanations. Interaction of a peptide with SecB might directly exclude proteinase K from the sensitive site through steric hindrance. Alternatively, binding of the ligand might induce a conformational change in SecB that results in protection of the sensitive site. The notion that a conformational change is involved is supported by the observation that SecB, which is sensitive to proteolysis at low ionic strength, can be rendered completely resistant to proteolysis even in the absence of ligand if the ionic strength is increased to between 100 and 150 mM with either NaCl or KCl or if 10 mM magnesium acetate or 1 mM $CaCl_2$ is present. SecB has an isoelectric point of 4, so it is likely that at low ionic strength, negative charges on SecB keep regions of the polypeptide in an open, protease-sensitive conformation. High ionic strength or neutraliza-

TABLE 1. Peptides assayed for interaction with SecB[a]

Peptide	Sequence	Charge[b]	Interaction[c]
P_β	RYFNAKAGLCQTF	+2	+
S1	VIEVVQGAYRAIRHIPRRIR	+4	+
S1b	DRVIEVVQGAYRAIRHIPRRIRQG	+4	+
S4	NNNTRKSIRIQRGPGRAFVTIGKIG	+6	+
Melittin	GIGAVLKVLTTGLPALISWIKRKRQQ-NH_2	+6	+
Zinc finger (no Zn^{2+})	RSFVCEVCTRAFARQEHLKRHYRSHTNEK	+4	+
Defensin HNP-1	ACYCRIPACIAGERRYGTCIYQGRLWAFCC	+3	+
Defensin NP-1	VVCACRRALCLPRERRAGFCRIRGRIHPLCCRR	+9	+
Defensin NP-5	VFCTCRGFLCGSGERASGSCTINGVRHTLCCRR	+4	+
Somatostatin	AGCKNFFWKTFTSC	+2	+
Mastoparan	INLKALAALAKKIL-NH_2	+4	+
Bradykinin	RPPGFSPFR	+2	−
M-K bradykinin	MKRPPGFSPFR	+3	−
S026-B	Ac-SLNAAKSELDKAIG-NH_2	0	−
P_α	NNFKSAEDCMRTAGGA	0	−
Glucagon	HSEGTFTSDYSKYLDSRRAQDFVQWLMNT	−1	−
GCN4-p1	Ac-RMKQLEDKVEELLSKNYHLENEVARLKKLVGER	0	−
Fos-p1	Ac-CGGLTDTLQAETDQLEDKKSAL-QTEIANLLKEKEKLEFILAAY	−5	−

[a] P_α and P_β are from BPTI; S1, S1b, and S4 are from human immunodeficiency virus gp41; and the zinc finger is from ADR1. S026-B is from T4 lysozyme; GCN4-p1 and Fos-p1 are leucine zippers; Ac, acetyl (Randall, 1992).

[b] Approximate net charge at pH 7.6.

[c] +, at a concentration of 15 μM or less the peptide conferred at least 50% protection of 0.6 μM SecB from proteolysis by proteinase K; −, no protection from proteolysis.

tion of charges on SecB by the binding of ligands or divalent cations would allow the formation of a tighter structure. Circular dichroic spectra of SecB either at high ionic strength or with ligand bound show more structure than do corresponding spectra at low ionic strength or with the binding site unoccupied (Fasman and Randall, unpublished).

If the peptide-binding sites are negatively charged, with little if any preference for particular side chains, polymers of basic amino acids might be readily accommodated. Indeed, polylysine and polyarginine efficiently protect SecB from degradation by proteinase K, while polymers of glutamic acid and of prolyl glycyl proline do not. The minimum length of a lysine polymer that allows binding is eight residues. The peptide-binding sites detected by this assay are likely to be part of the physiological binding site since poly-Lys competes for binding with R·BPTI, which in turn has been shown to compete for binding with the natural ligand, nonnative maltose-binding protein.

These studies led to the conclusion that SecB has multiple binding sites for flexible stretches of polypeptide. Since during export SecB binds polypeptides that are not tightly folded, it is reasonable to expect that SecB also has hydrophobic binding sites. The fluorescent probe 1-anilino-naphthalene-8-sulfonate (ANS), which binds to hydrophobic patches and in so doing both increases its fluorescence intensity and shifts its emission maximum from 520 to 472 nm (Ptitsyn et al., 1990; Semisotnov et al., 1991), was used to probe for a hydrophobic site on SecB. Neither SecB nor any of the peptide ligands studied binds ANS. However, when a ligand was added to SecB a hydrophobic site was exposed and ANS was bound. It is interesting that the increase in ANS fluorescence is not coincident with the increase in protease resistance, which is a measure of peptide binding, but is shifted to higher concentrations of ligand. This observation is consistent with the interpretation that simultaneous occupation of multiple peptide-binding sites is required to induce a conformational change that in turn exposes an ANS binding site on SecB.

Although it is possible that the hydrophobic area on the SecB tetramer functions in the interaction of SecA with the complex between SecB and its ligand, it seems more likely that this area serves as an additional binding site between nonnative polypeptides and SecB. If this idea is correct, we would expect that the hydrophobic patch would be unavailable for ANS binding when SecB is bound to a physiological ligand because it would already be filled by that ligand. This expectation is fulfilled for maltose-binding protein, since the ANS fluorescence of the complex between maltose-binding protein and SecB at physiological ionic strength is less, not greater, than the sum of the ANS fluorescence observed with the individual components, namely, SecB and folding intermediates of maltose-binding protein (Randall, 1992).

Characterization of a complex between SecB and maltose-binding protein, a natural ligand, showed that SecB interacts at multiple sites on the ligand. The entire binding region covers approximately half of the primary sequence of maltose-binding protein and comprises contiguous sites positioned around the center of the sequence (Topping and Randall, 1994).

A Model

Our model for binding between SecB and its ligands is shown in Fig. 2. At low ionic strength the SecB tetramer is in an open conformation and can be cleaved by protease. Binding of flexible positively charged lengths of peptide or an increase in ionic strength progressively changes the conformation so that protease sensitivity decreases and a hydrophobic area is partially exposed. At physiological ionic strength SecB is in a conformation that is resistant to proteolysis and the hydrophobic patch is partially exposed even in the absence of ligand. Once several lengths of peptide are bound at either low or high ionic strength, the tetramer undergoes a change so that the hydrophobic area is fully exposed, providing an interactive surface for hydrophobic regions of nonnative polypeptide ligands. Exposure of the hydrophobic site only after initial interaction with a ligand would be advantageous because if SecB always displayed a hydrophobic patch, it would tend to aggregate. The existence of two sites with different binding requirements would allow the chaperone to have high selectivity for nonnative proteins, even though each site demonstrates broad specificity. SecB might transiently interact with flexible loops of native proteins, but recognition of a polypeptide as nonnative would require subsequent occupation of the hydrophobic site on SecB by regions of the polypeptide ligand that would not be accessible in the native state. The same mechanism can be invoked to explain why direct recognition of the leader cannot account for the high selectivity in vivo for binding of precursors to SecB during protein export (Randall et al., 1990). Even though the leader, which is positively charged, might fill one of the peptide-binding sites, the interaction would not be stable unless a second site were filled. The leader slows the folding of the precursor, thereby ensuring that the necessary regions of the polypeptide are available to fill the sites.

We have only fragmentary knowledge concerning the binding of ligands to any of the chaperones. Chaperones, which interact with polypeptides that are in a nonnative state, can be divided into two broad classes. One group are those with amino acyl sequences related to that of GroEL (the

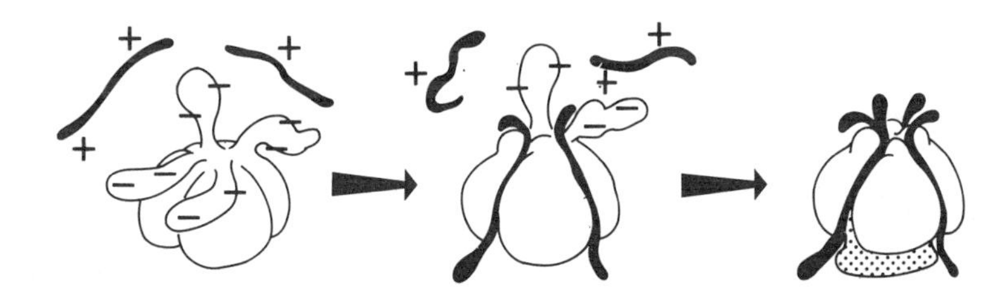

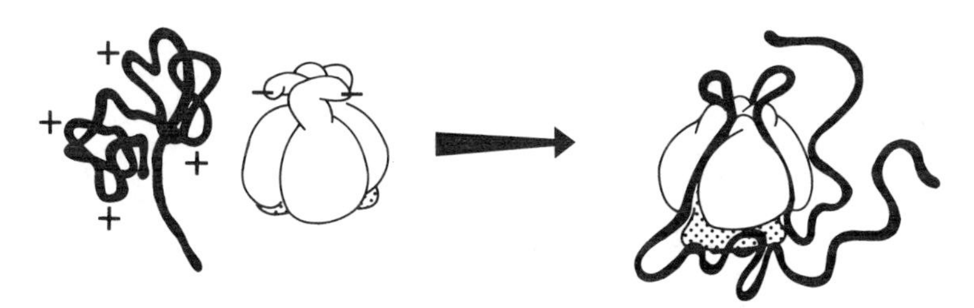

FIG. 2. Working model for the binding of ligands to SecB. At low ionic strength, binding of peptide ligands that carry net negative charge occurs independently at multiple sites on the SecB tetramer. Saturation of the peptide binding sites induces a conformation change to expose a hydrophobic site (represented by the stippled area). At physiological ionic strength the conformation of SecB is such that the hydrophobic site is partially exposed, but simultaneous interaction with stretches of the nonnative polypeptide at the peptide-binding sites induces complete exposure of the hydrophobic site, which then interacts with hydrophobic regions on the ligand.

hsp60 family). These chaperones are involved primarily in the facilitation of folding and are large complexes comprising two rings of seven identical subunits. Landry et al. (1992), using nuclear magnetic resonance spectroscopy, showed that GroEL bound a peptide in an α-helical conformation. Image analysis of electron micrographs of GroEL suggest that the ligand binds within a central cavity of the GroEL cylinder (Langer et al., 1992). The large chaperonins, with sequence homology to GroEL (the hsp60 family) may be fundamentally different in their mechanism of interaction with ligands from the second broad class, which includes the smaller chaperones of the hsp70 families and SecB. Many members of this class are involved in protein localization as well as in protein folding. The same peptide ligand that bound to GroEL as an α-helix was shown to bind in an extended conformation to DnaK (Landry et al., 1992), a representative member of the hsp70 family in *E. coli*. The site on DnaK might be similar to the hydrophilic peptide sites on SecB that seem to select flexible stretches for binding. A study of the eukaryotic hsp70 chaperone, BiP (Flynn et al., 1991), identified a binding site that selected for aliphatic residues. This site might correspond functionally to the hydrophobic site on SecB. It may be that the smaller chaperones use a mechanism similar to that of SecB during selective interaction with their nonnative ligands. It will be interesting to see how many different ways there are to selectively bind nonnative polypeptides.

REFERENCES

Altman, E., S. Emr, and C. Kumamoto. 1990. The presence of both the signal sequence and a region of mature LamB protein is required for the interaction of LamB with the export factor SecB. *J. Biol. Chem.* **265:**18154–18160.

Bankaitis, V. A., and P. J. Bassford, Jr. 1984. The synthesis of export-defective proteins can interfere with normal protein export in *E. coli*. *J. Biol. Chem.* **259:**12193–12200.

Bassford, P. J., Jr. 1990. Export of the periplasmic maltose-binding protein of *Escherichia coli*. *J. Bioenerg. Biomembr.* **22:**401–439.

Collier, D. N. 1993. SecB: a molecular chaperone of *Escherichia coli* protein secretion pathway. *Adv. Protein Chem.* **44:**151–193.

Collier, D. N., V. A. Bankaitis, J. B. Weiss, and P. J. Bassford, Jr. 1988. The antifolding activity of SecB promotes the export of the *E. coli* maltose-binding protein. *Cell* **53:**273–283.

Collier, D. N., S. M. Strobel, and P. J. Bassford, Jr. 1990. SecB-independent export of *Escherichia coli* ribose-binding protein (RBP): some comparisons with export of maltose-binding protein (MBP) and studies with RBP-MBP hybrids. *J. Bacteriol.* **172:**6875–6884.

Fasman, G. D., and L. L. Randall. Unpublished results.

Flynn, G. C., J. Pohl, M. T. Flocco, and J. E. Rothman. 1991. Peptide-binding specificity of the molecular chaperone BiP. *Nature* (London) **353:**726–730.

Gannon, P. M., P. Li, and C. A. Kumamoto. 1989. The mature portion of *Escherichia coli* maltose-binding protein (MBP) determines the dependence of MBP on SecB for export. *J. Bacteriol.* **169:**1286–1290.

Hardy, S. J. S., and L. L. Randall. 1991. A kinetic partitioning model of selective binding of nonnative proteins by the bacterial chaperone SecB. *Science* **251:**439–443.

Hartl, F.-U., S. Lecker, E. Schiebel, J. P. Hendrick, and W. Wickner. 1990. The binding cascade of SecB to SecA to SecY/E mediates preprotein targeting to the *E. coli* plasma membrane. *Cell* **63:**269–279.

Hayashi, S., and H. Wu. 1985. Accumulation of prolipoprotein in *Escherichia coli* mutants defective in protein secretion. *J. Bacteriol.* **161:**949–954.

Kumamoto, C. A. 1989. *Escherichia coli* SecB protein associates with exported protein precursors *in vivo*. *Proc. Natl. Acad. Sci. USA* **86:**5320–5324.

Kumamoto, C. A., and J. Beckwith. 1985. Evidence for specificity at an early step in protein export in *Escherichia coli*. *J. Bacteriol.* **163:**267–274.

Kumamoto, C. A., and O. Francetic. 1993. Highly selective binding of nascent polypeptides by an *Escherichia coli* chaperone protein in vivo. *J. Bacteriol.* **175:**2184–2188.

Kumamoto, C. A., and P. M. Gannon. 1988. Effects of *Escherichia coli secB* mutations on pre-maltose-binding protein conformation and export kinetics. *J. Biol. Chem.* **263:**11554–11558.

Landry, S. J., R. Jordan, R. McMacken, and L. M. Gierasch. 1992. Different conformations for the same polypeptide bound to chaperones DnaK and GroEL. *Nature* (London) **355:**455–457.

Langer, T., G. Pfeifer, J. Martin, W. Baumeister, and F.-U. Hartl. 1992. Chaperonin-mediated protein folding: GroES binds to one end of the GroEL cylinder, which accommodates the protein substrate within its central cavity. *EMBO J.* **11:**4757–4765.

Lecker, S., R. Lill, T. Ziegelhoffer, C. Georgopoulos, P. J. Bassford, Jr., C. A. Kumamoto, and W. Wickner. 1989. Three pure chaperone proteins of *Escherichia coli*, SecB, trigger factor, and GroEL, form soluble complexes with precursor proteins *in vitro*. *EMBO J.* **9:**2703–2709.

Liu, G., T. B. Topping, and L. L. Randall. 1989. Physiological role during export for the retardation of folding by the leader peptide of maltose-binding protein. *Proc. Natl. Acad. Sci. USA* **86:**9213–9217.

Park, S., G. Liu, T. B. Topping, W. H. Cover, and L. L. Randall. 1988. Modulation of folding pathways of exported proteins by the leader sequence. *Science* **239:**1033–1035.

Ptitsyn, O. B., R. H. Pain, G. C. Semisotnov, E. Zerovnik, and O. I. Razgulyaev. 1990. Evidence for a molten globule state as a general intermediate in protein folding. *FEBS Lett.* **262:**20–24.

Randall, L. L. 1992. Peptide binding by chaperone SecB: implications for recognition of nonnative structure. *Science* **257:**241–245.

Randall, L. L., T. B. Topping, and S. J. S. Hardy. 1990. No specific recognition of leader peptide by SecB, a chaperone involved in protein export. *Science* **248:**860–863.

Schatz, P. J., and J. Beckwith. 1990. Genetic analysis of protein export in *Escherichia coli*. *Annu. Rev. Genet.* **24:**215–248.

Schiebel, E., A. J. M. Driessen, F.-U. Hartl, and W. Wickner. 1991. $\Delta\mu H^+$ and ATP function at different steps of the catalytic cycle of preprotein translocase. *Cell* **64:**927–939.

Semisotnov, G. V., N. A. Rodionova, O. I. Razgulyaev, V. N. Uversky, A. K. Gripasj, and R. I. Gilmanschin. 1991. A study of the 'molten globule' intermediate state in protein folding by a hydrophobic fluorescent probe. *Biopolymers* **31:**119–129.

Topping, T. B., and L. L. Randall. 1994. Determination of the binding frame within a physiological ligand for the chaperone SecB. *Protein Sci.* **3:**730–736.

Watanabe, M., and G. Blobel. 1989a. Cytosolic factor purified from *E. coli* is necessary and sufficient for export of a preprotein and is a homotetramer of SecB. *Proc. Natl. Acad. Sci. USA* **86:**2728–2732.

Watanabe, M., and G. Blobel. 1989b. SecB functions as a cytosolic signal recognition factor for protein export in *E. coli*. *Cell* **58:**695–705.

Weiss, J. B., P. H. Ray, and P. J. Bassford, Jr. 1988. Purified SecB protein of *E. coli* retards folding and promotes membrane translocation of maltose-binding protein *in vitro*. *Proc. Natl. Acad. Sci. USA* **85:**8978–8982.

Wickner, W., A. Driessen, and F.-U. Hartl. 1991. The enzymology of protein translocation across the *Escherichia coli* plasma membrane. *Annu. Rev. Biochem.* **60:**101–124.

Yamada, H., H. Tokuda, and S. Mizushima. 1989. Proton motive force-dependent and -independent protein translocation revealed by an efficient *in vitro* assay system of *Escherichia coli*. *J. Biol. Chem.* **264:**1723–1728.

Chapter 43

Molecular Mechanism of Protein Translocation across the Cytoplasmic Membrane of *Escherichia coli*

SHOJI MIZUSHIMA

Tokyo College of Pharmacy, 1432-1 Horinouchi, Hachioji, Tokyo 192-03, Japan

Extensive genetic studies have revealed that at least six gene products (Sec A, SecB, SecD, SecE, SecF, and SecY proteins) are involved in general protein translocation across the cytoplasmic membrane in *Escherichia coli* (Bieker et al., 1990; Bieker-Brady and Silhavy, 1992). To perform extensive biochemical studies on the process of protein translocation, we have constructed *sec* gene-carrying plasmids, which allow the overproduction of individual Sec proteins. The overproduction in turn led us to purify these Sec proteins and to reconstitute protein translocation activity. The success of the reconstitution further led to the discovery of a novel membrane protein, p12, involved in protein translocation. In this chapter I will summarize our recent work on the roles of individual Sec proteins and p12 in the protein translocation reaction. The roles of phospholipids in the translocation reaction will also be discussed.

Overproduction and Purification of Sec Proteins

SecB has been overproduced by Weiss et al. (1988). We have overproduced all other Sec proteins (Kawasaki et al., 1989; Matsuyama et al., 1990, 1992). Individual *sec* genes were placed under the control of the *tac* promoter on a plasmid, which also possessed the *lac*I^q gene, and overproduction was induced with isopropyl-β-D-thiogalactopyranoside (IPTG). Overproduction ranging from 20- to 3,000-fold was achieved. SecA, SecE, and SecF could be overproduced alone, whereas the overproduction of SecD or SecY alone resulted in their rapid breakdown (Matsuyama et al., 1990, 1992). The overproduction of SecD and SecY was achieved with the simultaneous overproduction of SecF and SecE, respectively, suggesting the occurrence of interaction between SecD and SecF, and SecY and SecE (Matsuyama et al., 1990; Sagara et al., 1994).

All the Sec proteins thus overproduced have been purified (Akimaru et al., 1991; Akita et al., 1990; Matsuyama et al., 1992; Tokuda et al., 1991). (Fig. 1). SecA was purified from the cytosolic fraction, which contained most of this protein in the overproducing cells. SecD, SecE, SecF, and SecY were purified from the membrane fraction solubilized with octylglucoside or Sarkosyl.

Reconstitution of Protein Translocation Activity

Reconstitution of protein translocation activity was attempted by using these purified proteins and *E. coli* phospholipids. Translocationally active proteoliposomes were reconstituted from SecE, SecY, and phospholipids (Akimaru et al., 1991). The reconstituted activity was SecA and ATP dependent. The omission of one of these five factors resulted in the complete loss of translocation activity. SecD and SecF did not enhance the translocation activity (Matsuyama et al., 1992) (Fig. 2).

With a fixed amount of SecE (140 pmol), the reconstituted activity exhibited saturation at about 30 pmol of Sec Y, the SecE/SecY molar ratio being 4.8. On the other hand, with a fixed amount of SecY (10 pmol), more than 20-fold as many SecE molecules were required to exhibit saturation (Akimaru et al., 1991; Hanada et al., unpublished). Although the real reason for this apparent discrepancy is unclear, these results suggest that SecE is present in excess over SecY in functional stoichiometry and that the two proteins may not form a stable complex in liposomes.

SecE possesses three transmembrane segments. A reconstitution study on genetically engineered SecEs revealed that a truncated SecE fragment containing only the carboxyl-terminal segment was 50% as active as intact SecE (Nishiyama et al., 1992), which was consistent with a genetic study (Schatz et al., 1991). In the case of SecY, which possesses 10 membrane-spanning segments, all of the segments are seemingly required for its function (Nishiyama et al., unpublished). These facts suggest that SecY is the principal component of the translocation machinery.

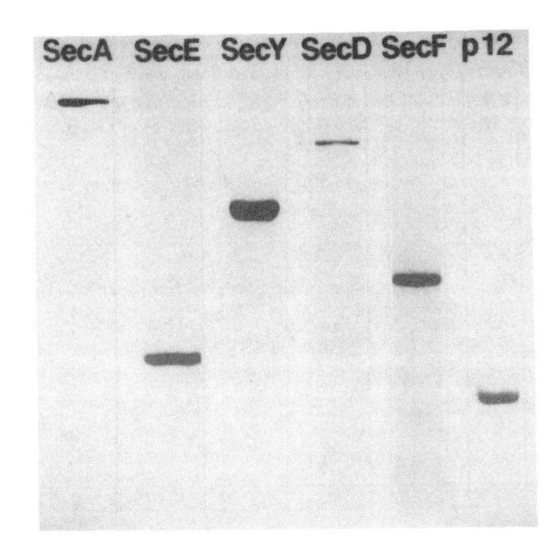

FIG. 1. Sodium dodecyl sulfate (SDS)-polyacrylamide gel electrophoretic profiles of purified Sec proteins and p12. After electrophoresis, gels were stained with Coomassie brilliant blue.

A Novel Membrane Protein, p12, Is a Component of the Translocation Machinery

Reconstitution studies led to the discovery of another membrane protein, p12, involved in protein translocation (Nishiyama et al., 1993). p12 is a 12-kDa cytoplasmic membrane protein present in the trichloroacetic acid-soluble fraction of the octylglucoside-solubilized membrane. When purified p12 was added together with SecY and SecE, the translocation activity of the reconstituted proteoliposomes was enhanced more than 20-fold. The enhanced translocation was ATP and SecA dependent. Furthermore, anti-p12 antibody raised against a peptide representing the C-terminal portion inhibited the translocation of proteins into everted membrane vesicles derived from the cytoplasmic membrane. With fixed amounts of SecE (81 pmol) and SecY (18 pmol), the reconstitution activity did not exhibit saturation even at 500 pmol of p12. Thus, equimolar stoichiometric relationships among the three proteins were not observed (Fig. 3). Furthermore, p12 did not change the stoichiometric relationship between SecE and SecY. These facts suggest that p12 may not form a stable complex with other components in the lipid membrane, similarly to SecY and SecE, as discussed in the preceding section on SecY and SecE.

The gene encoding p12 has been identified and sequenced. The deduced amino acid sequence suggests the presence of two or three possible membrane-spanning domains. With the cloned gene, the overproduction of p12 was attempted. The overproduction supported the simultaneous overproduction of SecY, suggesting the occurrence of interaction between p12 and SecY in the membrane. The SecE-p12 interaction was also suggested from an in vivo experiment with the *secE*(Cs) mutant (Nishiyama et al., 1993).

SecD Is Involved in the Release of Translocated Proteins from the Cytoplasmic Membrane

The effect of anti-SecD immunoglobulin G (IgG) on protein translocation across the cytoplasmic membrane was studied by using spheroplasts (Matsuyama et al., 1993). Inhibition of the translocation of OmpA and maltose-binding protein (MBP) into the medium by the IgG was observed, with concomitant accumulation of their precursor and mature forms in the spheroplasts (Fig. 4). This effect was specific to anti-SecD IgG. Anti-SecE and anti-SecY IgGs, which were raised against epitopes located in the periplasmic domains, did not interfere with the translocation. The mature form of MBP thus accumulated was sensitive to trypsin, which was externally added to the spheroplasts. In contrast, MBP released into the medium was resistant to trypsin, as is the native MBP. The precursor form accumulated in the spheroplasts was resistant to externally added trypsin. These results indicate that SecD is most probably involved in the release of translocated

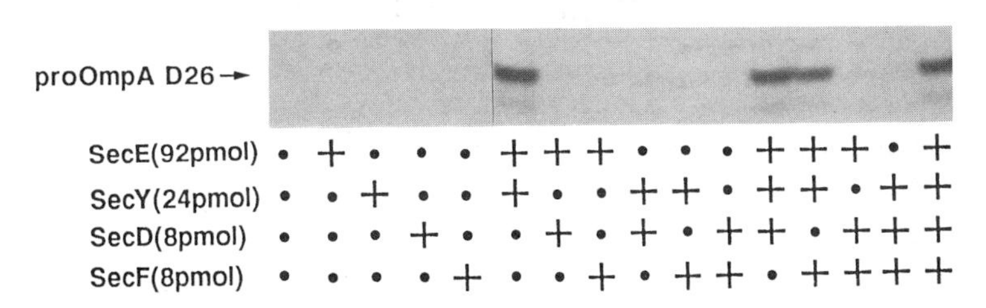

FIG. 2. Effects of SecE, SecY, SecD, and SecF on reconstitution of translocation active proteoliposomes. Reconstitution of proteoliposomes was carried out with these Sec proteins and *E. coli* phospholipids, and protein translocation activity was assayed by using proOmpA D26 as the substrate in the presence of SecA and ATP. The photograph shows proOmpA D26 that became proteinase K resistant as a result of translocation into proteoliposomes. Reprinted from Matsuyama et al. (1992) with permission.

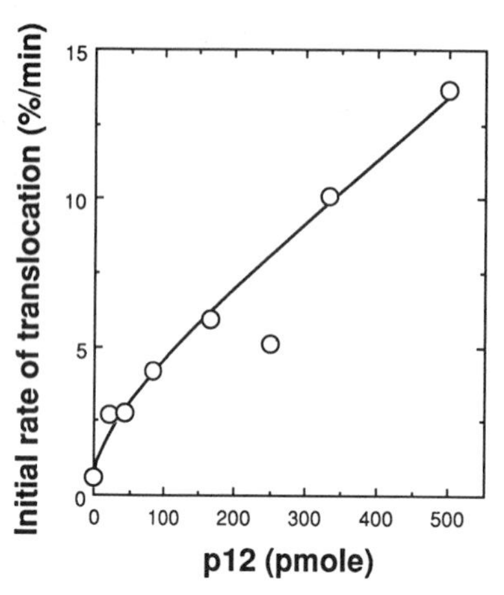

FIG. 3. Effects of the amounts of p12 on the translocation activity of reconstituted proteoliposomes. The indicated amounts of purified p12 were mixed with 81 pmol of SecE and 18 pmol of SecY; this was followed by reconstitution. Initial rates of translocation of proOmpA D26 into the reconstituted proteoliposomes were plotted against the amount of p12 used in the reconstitution. Reprinted from Nishiyama et al. (1993) with permission.

proteins from the cytoplasmic membrane. However, the possibility that SecD is involved in the very final stage of the translocation, i.e., the export of the C terminus from the membrane, cannot be excluded.

We failed to elucidate the function of SecD in the reconstitution study (Matsuyama et al., 1992). We also failed to demonstrate the inhibition of in vitro protein translocation by anti-SecD IgG after entrapment of the IgG in everted cytoplasmic membrane vesicles (Fujita et al., unpublished). We do not think that these results contradict the present ones. First, the function of SecD can be directly demonstrated only as the release of translocated proteins from the membrane, which cannot be detected by the conventional assay procedure, which utilizes proteinase K resistance as a translocation criterion. Second, the accumulation of a precursor protein is probably the result of the stacking of the nonreleased mature protein. The accumulation may be observed, therefore, only when the translocation machinery is functioning catalytically. It is probable, however, that the translocation exhibited by everted membrane vesicles does not represent a catalytic process, since the molar amounts of translocation machinery and of a preprotein in a conventional reaction mixture were estimated to be roughly the same (Matsuyama et al., 1992; Schiebel et al., 1991).

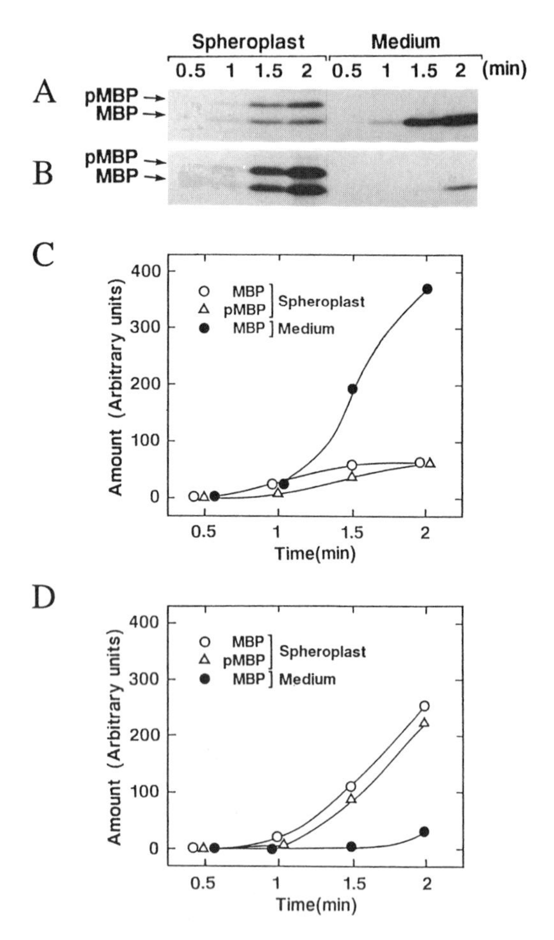

FIG. 4. Effect of anti-SecD IgG on the release of MBP into the medium. Spheroplasts were treated with nonimmune (A and C) or anti-SecD (B and D) IgG on ice for 3 min. The minimal medium containing Tran^{35}S label was added at zero time to initiate labeling at 30°C. Aliquots were removed at the indicated times, chilled in the presence of cold methionine and cysteine, and then fractionated into spheroplasts and the medium. Individual fractions were subjected to immunoprecipitation with anti-MBP antiserum, and then analyzed by SDS-polyacrylamide gel electrophoresis and fluorography (A and B). The amounts of the precursor and mature forms of MBP on fluorograms were densitometrically determined and plotted (C and D). The number of methionine residues in proMBP (nine residues) and MBP (six residues) was taken into consideration. Reprinted from Matsuyama et al. (1993) with permission.

SecF Stabilized SecD and SecY

Very recently, we observed that the overproduction of SecF resulted in the simultaneous overproduction of SecD encoded by the *tac-secD* gene (Sagara et al., 1994). A pulse-chase experiment revealed that the overproduction was due to stabilization of SecD by SecF. The SecF overproduction resulted in the overproduction of SecY as well. This SecF effect was not due to its effect on SecD or SecE, since SecF overproduction did not affect expression of the *secD* and *secE* genes.

All things considered, it is likely that SecF interacts with both SecD and SecY, which are involved in the early and late stages, respectively, of the translocation reaction. The number of molecules of SecF was estimated to be smaller than those of SecD and SecY. It is unclear how SecF molecules, present in smaller numbers, stabilize the other Sec proteins.

Roles of Phospholipids in Protein Translocation

Phospholipids participate in the translocation reaction not only as structural components of the membrane but also as functional components of the translocation machinery. Phosphatidylglycerol and cardiolipin, which are acidic phospholipids, are required for protein translocation (de Vrije et al., 1988; Kusters et al., 1991). They also stimulate SecA translocation ATPase activity (Lill et al., 1990) and modulate the conformation of SecA (Shinkai et al., 1991). Furthermore, the insertion of SecA into a phospholipid monolayer was enhanced when the layer contained these acidic phospholipids (Breukink et al., 1992). It is clear, therefore, that acidic phospholipids are involved in the functioning of SecA.

Recently, in a collaboration with the de Kruijff group, we found that acidic phospholipids are involved in the recognition of signal peptides in the translocation process. Protein translocation usually requires positively charged amino acid residues at the N terminus of the signal peptide. We found, however, that this requirement can be compensated for when signal peptides possess a longer hydrophobic stretch (12 alternate alanine and leucine residues [see Fig. 5]), although the translocation reaction is still SecA dependent (Hikita and Mizushima, 1992). Interestingly, the translocation led by such signal peptides no longer required acidic phospholipids (Phoenix et al., 1993). In other words, it required neither positively charged amino acid residues nor negatively charged phospholipids. In addition, SecA interacts with preproteins by recognizing the positive charge of the signal peptide (Akita et al., 1990), whereas no interaction was observed between SecA and preproteins that have no positively charged residue but have a longer hydrophobic stretch. This is the case even though the translocation of such preproteins is SecA-dependent (Mori et al., unpublished).

Although it is still difficult to explain how these phenomena are related to the initial event of protein translocation, one possible explanation is as follows (Fig. 5). For the translocation of most of the preproteins possessing a positively charged signal peptide, the initial event may be the recognition of preproteins by the cytosolic SecA, as discussed previously (Mizushima et al., 1992). Of course, this does not exclude the involvement of chaperone proteins, such as SecB, before this event. The SecA-preprotein complex then moves on to the membrane to initiate transmembrane translocation. Acidic phospholipids in the membrane are involved in SecA recognition. They may also be involved in the recognition of the positive charge to facilitate the transmembrane spanning of the signal peptide domain. Preproteins possessing no positive charge but possessing a longer hydrophobic stretch in the signal peptide domain, on the other hand, may not be able to form a tight complex with the cytosolic SecA but interact hydrophobically and directly with the membrane. The longer hydrophobic stretch may facilitate the transmembrane spanning of the signal peptide. Since the translocation of such preproteins also requires SecA, a SecA molecule which preexists in or simultaneously binds to the membrane plays a role in the following stage of the translocation reaction. It should be noted, however, that all of the signal peptides so far reported for natural prokaryotic preproteins possess amino-terminal positively charged residues.

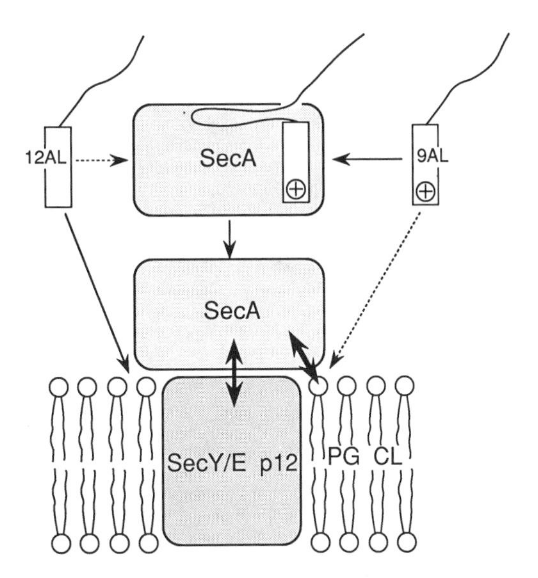

FIG. 5. Functional interaction between preproteins, SecA, and phospholipids in the initial stage of protein translocation. 9AL and 12AL represent signal peptides possessing 9 and 12 alternate alanine and leucine residues as the hydrophobic stretch, respectively (Hikita and Mizushima, 1992). PG, phosphatidylglycerol; CL, cardiolipin. For details, see the text.

Model for Protein Translocation

Our current model for the translocation of preproteins across the cytoplasmic membrane of *E. coli* is shown in Fig. 6. The process up to the SecA-mediated transfer of preproteins to the trans-

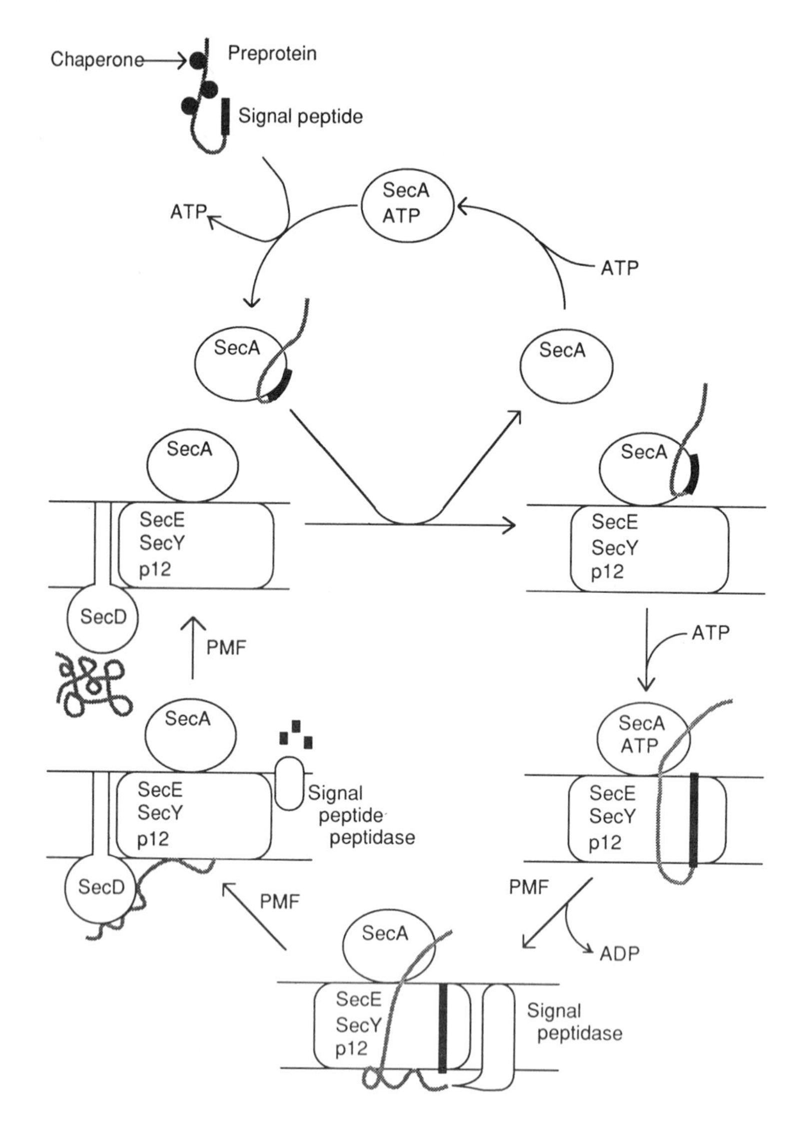

FIG. 6. Model for protein translocation. Although SecD, SecF, signal peptidase, and signal peptide peptidase are depicted only at a certain stage of translocation, they are assumed to be in the membrane throughout the translocation reaction. PMF, proton motive force. For details, see the text.

location machinery in the membrane was discussed in the preceding section. Two possible ways by which preproteins are transferred from the cytosolic SecA to the membrane are considered. One is the transfer of a preprotein alone from the cytosolic to the membrane-bound SecA. Alternatively, the preprotein-SecA complex may replace SecA preexisting on the membrane. In any event, a translocation complex comprising a preprotein, SecA, SecE, SecY, p12, SecD, and probably SecF can be formed. It is questionable whether these proteins always exist as a stable complex in the membrane. Rather, the results of stoichiometric studies on SecY, SecE, and p12, described in the preceding sections, suggest that these proteins may exist separately in the membrane and gather to form a complex only when a preprotein or a preprotein-SecA complex reaches them. The assembly of a translocation machinery on demand has been proposed on the basis of a genetic analysis (Bieker-Brady and Silhavy, 1992). This mechanism prevents leakage of cellular solutes through the pores for proteins to be translocated. This idea was discussed when the signal hypothesis was proposed (Blobel and Dobberstein, 1975a, 1975b).

The translocation of proteins across the membrane requires two types of energy, ATP and a proton motive force. They are involved in different stages of the translocation reaction (Schiebel et al., 1991; Tani et al., 1989). Their roles have been discussed (Mizushima and Tokuda, 1990; Wickner et al., 1991) and thus will be not discussed here.

Finally, translocated proteins are released from the membrane with the aid of SecD (Matsuyama et al., 1993) and probably the proton motive force (Geller, 1990), and the release seemingly facilitates folding of proteins in the periplasmic space. Although the in vivo stabilization of SecD and SecY by SecF suggests that SecF also plays a role in the translocation event, no biochemical evidence of this function has been presented.

I thank Chikako Motoyama for secretarial support.

The work from my laboratories in the Institute of Applied Microbiology, the University of Tokyo, and Tokyo College of Pharmacy has been supported by grants from the Ministry of Education, Science and Culture of Japan; Human Frontier Science Program Organization; and Naito Memorial Foundation.

REFERENCES

Akimaru, J., S. Matsuyama, H. Tokuda, and S. Mizushima. 1991. Reconstitution of a protein translocation system containing purified SecY, SecE and SecA from *Escherichia coli*. *Proc. Natl. Acad. Sci. USA* **88:**6545–6549.

Akita, M., S. Sasaki, S. Matsuyama, and S. Mizushima. 1990. SecA interacts with secretory proteins by recognizing the positive charge at the amino terminus of the signal peptide in *Escherichia coli*. *J. Biol. Chem.* **265:**8164–8169.

Bieker, K. L., G. J. Phillips, and T. J. Silhavy. 1990. The *sec* and *prl* genes of *Escherichia coli*. *J. Bioenerg. Biomembr.* **22:**291–310.

Bieker-Brady, K., and T. J. Silhavy. 1992. Suppressor analysis suggests a multistep, cyclic mechanism for protein secretion in *Escherichia coli*. *EMBO J.* **11:**3165–3174.

Blobel, G., and B. Dobberstein. 1975a. Transfer of proteins across membranes. I. Presence of proteolytically processed and unprocessed nascent immunoglobulin light chains on membrane-bound ribosomes of murine myeloma. *J. Cell. Biol.* **67:**835–851.

Blobel, G., and B. Dobberstein. 1975b. Transfer of proteins across membranes. II. Reconstitution of functional rough microsomes heterologous components. *J. Cell. Biol.* **67:**852–862.

Breukink, E., R. A. Demel, G. de Korte-Kool, and B. de Kruijff. 1992. SecA insertion into phospholipids is stimulated by negatively charged lipids and inhibited by ATP: a monolayer study. *Biochemistry* **31:**1119–1124.

de Vrije, T., R. L. de Swat, W. Dowhan, J. Tommassen, and B. de Kruijff. 1988. Phosphatidylglycerol is involved in protein translocation across *Escherichia coli* inner membranes. *Nature* (London) **334:**173–175.

Fujita, Y., S. Matsuyama, and S. Mizushima. Unpublished observations.

Geller, B. L. 1990. Electrochemical potential releases a membrane-bound secretion intermediate of maltose-binding protein in *Escherichia coli*. *J. Bacteriol.* **172:**4870–4876.

Hanada, M., S. Mizushima, and H. Tokuda. Unpublished observations.

Hikita, C., and S. Mizushima. 1992. The requirements of a positive charge at the amino terminus can be compensated for by a longer central hydrophobic stretch in the functioning of signal peptides. *J. Biol. Chem.* **267:**12375–12379.

Kawasaki, H., S. Matsuyama, S. Sasaki, M. Akita, and S. Mizushima. 1989. SecA protein is directly involved in protein secretion in *Escherichia coli*. *FEBS Lett.* **242:**431–434.

Kusters, R., W. Dowhan, and B. de Kruijff. 1991. Negatively charged phospholipids restore prePhoE translocation across phosphatidylglycerol-depleted *Escherichia coli* inner membrane vesicles. *J. Biol. Chem.* **266:**8659–8662.

Lill, R., W. Dowhan, and W. Wickner. 1990. The ATPase activity of SecA is regulated by acidic phospholipids, SecY, and the leader and mature domains of precursor proteins. *Cell* **60:**271–280.

Matsuyama, S., J. Akimaru, and S. Mizushima. 1990. SecE-dependent overproduction of SecY in *Escherichia coli*. Evidence for interaction between two components of the secretory machinery. *FEBS Lett.* **269:**69–100.

Matsuyama, S., Y. Fujita, and S. Mizushima. 1993. SecD is involved in the release of translocated secretory proteins from the cytoplasmic membrane of *Escherichia coli*. *EMBO J.* **12:**265–270.

Matsuyama, S., Y. Fujita, K. Sagara, and S. Mizushima. 1992. Overproduction, purification and characterization of SecD and SecF, integral membrane components of the protein translocation machinery of *Escherichia coli*. *Biochim. Biophys. Acta* **1122:**77–84.

Mizushima, S., and H. Tokuda. 1990. *In vitro* translocation of bacterial secretory proteins and energy requirements. *J. Bioenerg. Biomembr.* **22:**389–399.

Mizushima, S., H. Tokuda, and S. Matsuyama. 1992. Molecular characterization of Sec proteins comprising the protein secretory machinery of *Escherichia coli*, p. 21–32. *In* W. Neupert and R. Lill (ed.), *Membrane Biogenesis and Protein Targeting*. Elsevier Science Publishers BV, Amsterdam.

Mori, H., M. Araki, C. Hikita, S. Matsuyama, and S. Mizushima. Unpublished observations.

Nishiyama, K., S. Mizushima, and H. Tokuda. 1992. The carboxyl-terminal region of SecE interacts with SecY and is functional in the reconstitution of protein translocation activity in *Escherichia coli*. *J. Biol. Chem.* **267:**7170–7177.

Nishiyama, K., S. Mizushima, and H. Tokuda. 1993. A novel membrane protein involved in protein translocation across the cytoplasmic membrane of *Escherichia coli*. *EMBO J.* **12:**3409–3415.

Nishiyama, K., S. Mizushima, and H. Tokuda. Unpublished observations.

Phoenix, D. A., R. Kusters, C. Hikita, S. Mizushima, and B. de Kruijff. 1993. OmpF-Lpp signal sequence mutants with varying charge hydrophobicity ratios provide evidence for a phosphatidylglycerol-signal sequence interaction during protein translocation across the *Escherichia coli* inner membrane. *J. Biol. Chem.* **268:**17069–17073.

Sagara, K., S. Matsuyama, and S. Mizushima. 1994. SecF stabilizes SecF and SecY, components of the protein translocation machinery of the *Escherichia coli* cytoplasmic membrane. *J. Bacteriol.* **176:**4111–4116.

Schatz, P. J., K. L. Bieker, K. M. Ottemann, T. J. Silhavy, and J. Beckwith. 1991. One of the three transmembrane stretches is sufficient for the functioning of the SecE protein, a membrane component of the *E. coli* secretion machinery. *EMBO J.* **10:**1749–1757.

Schiebel, E., A. J. M. Driessen, F.-U. Hartl, and W. Wickner. 1991. $\Delta\mu H^+$ and ATP function at different steps of the catalytic cycle of preprotein translocase. *Cell* **64:**927–939.

Shinkai, A., H.-M. Lu, H. Tokuda, and S. Mizushima. 1991. The conformation of SecA, as revealed by its protease sensitivity, is altered upon interaction with ATP, presecretory proteins, everted membrane vesicles, and phospholipids. *J. Biol. Chem.* **266:**5827–5833.

Tani, K., K. Shiozuka, H. Tokuda, and S. Mizushima. 1989. *In vitro* analysis of the process of translocation of OmpA across the *Escherichia coli* cytoplasmic membrane. A translocation intermediate accumulates transiently in the absence of the proton motive force. *J. Biol. Chem.* **264:**18582–18588.

Tokuda, H., J. Akimaru, S. Matsuyama, K. Nishiyama, and S. Mizushima. 1991. Purification of SecE and reconstitution of SecE-dependent protein translocation activity. *FEBS Lett.* **279:**233–236.

Weiss, J. B., P. H. Ray, and P. J. Bassford Jr. 1988. Purified SecB protein of *Escherichia coli* retards folding and promotes membrane translocation of the maltose-binding protein *in vitro*. *Proc. Natl. Acad. Sci. USA* **85:**8978–8982.

Wickner, W., A. J. M. Driessen, and F.-U. Hartl. 1991. The enzymology of protein translocation across the *Escherichia coli* plasma membrane. *Annu. Rev. Biochem.* **60:**101–124.

Chapter 44

Escherichia coli Alkaline Phosphatase Biogenesis: Influence of Overproduction and Amino Acid Substitutions

MARINA NESMEYANOVA, ANDREW KARAMYSHEV, ANDREW KALININ, IRINA TSFASMAN, ALLA BADYAKINA, MICHAEL KHMELNITSKY, MICHAEL SHLYAPNIKOV, AND VLADIMIR KSENZENKO

Institute of Biochemistry and Physiology of Microorganisms, Russian Academy of Sciences, 142292 Pushchino, Russia

Alkaline phosphatase (AP) (nonspecific phosphomonoesterase [EC 3.1.3.1]) is one of the most important enzymes of *Escherichia coli* phosphate metabolism, supplying the cell with P_i from exogenous sources (Torriani, 1960). It is the product of a single structural *phoA* gene constituent of the Pho regulon, synthesized under phosphate starvation and secreted into the periplasm.

The secretion-coupled biogenesis of AP is a multistep process (Fig. 1). It involves (i) synthesis of AP as a precursor (prePhoA) containing an amino-terminal signal sequence that targets the protein to the cytoplasmic membrane and initiates its transmembrane movement (Inouye et al., 1982); (ii) translocation of the precursor across the cytoplasmic membrane; (iii) precursor processing—cleavage of signal peptide by membrane-bound leader peptidase (Lp) on the outer surface of the membrane after translocation of the precursor is over (Inouye and Beckwith, 1977)—as a result, the mature AP subunits are formed and released into the periplasm; (iv) proteolytic modification of the subunits by protease Iap splitting off the N-terminal arginine (Nakata et al., 1987); and (v) their simultaneous dimerization with formation of intrachain disulfide bonds (Schlesinger and Barrett, 1965). AP is active as a dimer, and its multimers are active as well (Reynolds and Schlesinger, 1969). During the last steps of AP assembly in the periplasm, multiple forms of AP are formed (Fig. 1). Isozyme I is a dimer of identical subunits containing N-terminal arginine. Isozyme II is a dimer of heterologous subunits, one with arginine and the other without it. Isozyme III is a dimer with both subunits lacking arginine. This secreted protein is a suitable model for a study of exoprotein secretion and assembly.

Under the conditions optimal for AP expression, *E. coli* cells accumulate for the most part a completely processed isozyme III located in the periplasm (Nakata et al., 1987; Nesmeyanova et al., 1981). Because of the complexity of the *E. coli* cell wall and the presence of the outer membrane, this enzyme, just like most of the other secreted proteins of this bacterium, is not secreted into the medium. Other forms of the enzyme, isozymes I and II, are present in small quantities, and AP precursor (prePhoA) is essentially absent. PrePhoA, like other preproteins, is found only under conditions of specific inhibition of processing or secretion in the presence of protonophore or anesthetics (Halegoua and Inouye, 1979; Lazdunski et al., 1979).

We observed, however, that AP biogenesis is significantly altered when *phoA* is expressed as a constituent of multicopy plasmids, resulting in an increased enzyme synthesis (Nesmeyanova et al., 1991b; Tsfasman et al., 1993). Different stages of protein biogenesis and secretion also depend to a great extent on the primary structure of AP-specific domains, in particular, the processing site and N-terminal domain of the mature polypeptide chain.

This paper presents the data on the impact of overproduction and amino acid substitutions in AP on different stages of its biogenesis.

Biogenesis of Overproduced AP

Previously it was shown that AP encoded by the *phoA* gene of plasmids is synthesized by *E. coli* cells in an increased quantity and acquires the ability to be secreted not only into the periplasm but also into the culture medium (Lazzaroni et al., 1985). We studied this phenomenon in detail, since overproduction provokes significant changes in the secreted protein biogenesis and its peculiarities would allow a better understanding of the individual steps of the mechanism.

Immunoblot analysis of the cells overproducing AP revealed an additional enzyme form differing in its electrophoretic mobility from the mature AP and corresponding to its precursor. It was present in various *E. coli* strains in large quantities and sedimented with the total membranes under cell fractionation.

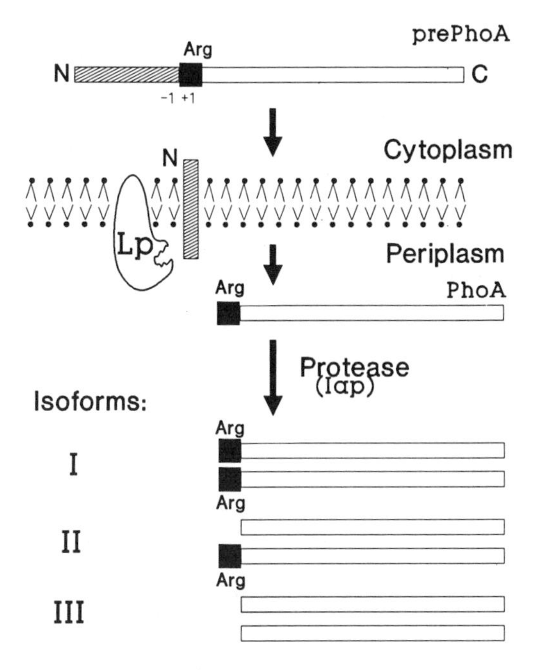

FIG. 1. Schematic illustration of *E. coli* AP biogenesis. Lp, leader (signal) peptidase; Iap, protease.

The solubilization of membrane proteins from a total membrane fraction as described by Schnaitman (1971) showed no significant prePhoA quantity among either the cytoplasmic or outer membrane proteins. However, prePhoA was dissolved from the residual cell particles in buffer containing 8 M urea. This indicates that the predominant localization of prePhoA is in the cytoplasmic aggregates cosedimented with the membranes.

To obtain evidence that the prePhoA aggregates contain the native precursor with the signal peptide, purified precursor and the aggregates in the membrane fraction were treated with leader peptidase (Zwizinski and Wickner, 1980). These precursors in the presence of urea were converted into mature protein (Tsfasman et al., 1993).

Accumulation of AP precursor as insoluble aggregates in the cytoplasm is probably due to a limited quantity of secretion sites (Ito et al., 1981) and to imbalance between synthesis of the enzyme polypeptide chains and their translocation across cytoplasmic membrane and processing. Formation of aggregates might be considered an alternative pathway of removing from the cytoplasm the excess protein intended for secretion and therefore foreign to the cytoplasm. The aggregates found are obviously similar to inclusion bodies formed during both overproduction of homologous proteins and intracellular expression of foreign proteins (Marston, 1986).

Other fundamental questions are how the overproduction affects the subsequent steps of the enzyme assembly, where they are completed, and whether the extracellular enzyme is identical to the periplasmic one or undergoes some further modification.

We studied the protein composition and AP isozyme spectrum of the periplasm and the culture medium of *E. coli* K802 carrying the plasmid with the *phoA* gene (pHI-7 [Inouye et al., 1981]) as well as in the periplasm of the original strain. The AP overproduction resulted in a selective secretion of the enzyme into the medium (Fig. 2A) and increased levels of isozyme I and II (Fig. 2B) The isozyme spectrum of AP in the culture medium showed no significant differences from the periplasmic one.

Accumulation of incompletely modified isozymes during AP overproduction shows that there is an imbalance between protein synthesis and secretion into the periplasm and the posttranslocational proteolytic modification in this compartment. This imbalance might result from an insufficient quantity of periplasmic protease Iap required to split off the N-terminal arginine to complete the proteolytic modification of the increased quantity of mature AP. Besides, we cannot exclude the possibility of incomplete processing as a result of inhibition (probably by a feedback mechanism) by an excess of the end product (isozyme III). The formation of AP multimers may be a mechanism to eliminate such inhibition.

A significant feature of overproduced AP biogenesis is the location in the culture medium,

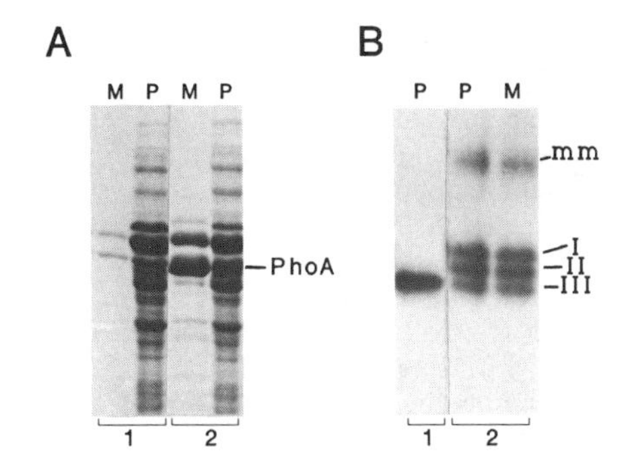

FIG. 2. Protein and isozyme spectra of periplasm (P) and culture medium (M) from *E. coli* K802 (Sambrook et al., 1989) (lanes 1) and *E. coli* K802 transformed by plasmid pHI-7 carrying the *phoA* gene (Inouye et al., 1981) (lanes 2). mm, PhoA multimers. Roman numerals indicate the isoforms of AP. (A) Electrophoregram of proteins in sodium dodecyl sulfate-polyacrylamide gel electrophoresis; (B) Pho AP activity in the gel with α-naphthylphosphate as the substrate. Electrophoresis was carried out under nondenatured conditions (7.5% polyacrylamide).

where it is excreted with preservation of intact envelope and in the absence of cell lysis.

The character of PhoA excretion into the medium depends on the nature of the strains and correlates with the peculiarities of chemical composition and ultrastructure organization of the cell envelope specific for each strain.

We observed different mechanisms of PhoA excretion into the medium. In *E. coli* C90 we observed a nonspecific excretion of the same type as that of colicins (Pugsley and Cole, 1987) and of *lky* mutants (Lazzaroni and Portalier, 1982); a higher content of lysophospholipids and 2-keto-3-deoxy-D-manno-octonate (lipopolysaccharide [LPS] fragment), as well as formation of small bubbles on the cell surface, which is indicative of membrane fragmentation (Bogdanov et al., 1989).

The changes of the envelope, primarily the increase in lysophospholipid content, are probably caused by the activation of phospholipase A2 and provoke an increase in the envelope permeability for all periplasmic proteins. This is confirmed by the specific activities of PhoA and the protein composition of the periplasm and the culture medium (Nesmeyanova et al., 1990).

Other strains, *E. coli* K802 and *E. coli* DH1, transformed by plasmid pHI-7, selectively excreted 70 to 90% of the AP into the medium, compared with other periplasmic proteins (Nesmeyanova et al., 1991b). Thus, AP was a predominant protein in the culture medium, and its specific activity was several times higher than in the periplasm. The mechanisms of selective excretion were different in the two strains. The outer membranes of *E. coli* K802/pHI-7 had lower lipoprotein and higher LPS contents than did the original K802 strain, and the cells were characterized by a greater periplasmic space and large spherical exocellular vesicles, both filled with AP (Nesmeyanova et al., 1991a).

Neither membrane fragmentation nor formation of the outer membrane vesicles was detected in *E. coli* DH1/pHI-7. Its cell envelope contained a higher ratio of cardiolipin to phosphatidylglycerol than in the original strain. AP secretion into the periplasm is usually accompanied by decrease of this ratio; its increase, by contrast, is characteristic of secretion directly across the two membranes without release into the periplasm, as we have observed in the case of hemolysin (Manuvakhova et al., 1990).

The above mechanisms are known to be used by *E. coli* cells for normal protein secretion (Hirst and Welch, 1988; Poirier and Holt, 1983) and are probably used for secretion of the overproduced periplasmic proteins. Thus, AP biogenesis at its overproduction in various *E. coli* strains has common features including accumulation of an intermediate form of the enzyme and its alternative localization. This is the result of a limited number of translocation sites in the cytoplasmic membranes (Ito et al., 1981) and probably of the absence in bacteria of a translational block that occurs in eucaryotic organisms (Pugsley, 1989).

Effect of Amino Acid Substitutions in AP on Its Biogenesis

The major factor determining biogenesis and secretion of the protein across the membranes is the structure of its specific domains, such as the N terminus and the processing site. Therefore, we investigated the effect of amino acid substitutions in these two domains of AP. By oligonucleotide-directed mutagenesis, we obtained mutations in *phoA* gene resulting in the following substitutions in these regions of the protein: (i) Ala(+1) for Arg and Gln(+4) for Glu (PhoA76); (ii) Gln(+4) for Glu (PhoA54); and (iii) Val(−1) for Ala (PhoA46) (Fig. 3A).

The N-terminal amino acid of AP is of particular interest: it is a positively charged arginine, rather rarely found in this position of other secreted proteins. During biogenesis, the cleavage of N-terminal arginine and production of AP isoforms take place (Fig. 1). To study the role of the N-terminal amino acid, we obtained various amino acid substitutions of Arg(+1) by introduction of an amber mutation into the corresponding position of the *phoA* gene and subsequent expression of the mutated gene in strains containing different amber suppressor tRNA (Karamyshev et al., 1994) (Fig. 3B).

Most of the amino acid substitutions exerted an effect on different stages of AP biogenesis. Some of them affected the protein processing in vivo as judged by conversion of AP precursor into mature protein in pulse-chase experiments (Michaelis et al., 1986). These experiments showed that amino acid substitutions in the processing site, Val(−1) for Ala (PhoA46) and Pro(+1) for Arg [PhoAPro(+1)], resulted in complete inhibition of signal peptide cleavage. The precursors of these mutated proteins did not turn into mature forms even 60 min after label incorporation was chased. However, inhibition of the processing did not prevent precursor translocation across the cytoplasmic membrane. The translocation of the precursor into the periplasm was confirmed by immunoblot analysis of subcellular fractions. It is known that *E. coli* AP acquires enzyme activity only in the periplasm, where necessary conditions for the formation of intrachain disulfide bonds and the active enzyme assembly are met, whereas nontranslocated AP precursor has no enzyme activity (Boyd et al., 1987). Although the PhoA46 and PhoAPro(+1) mutant proteins were found in the cells as precursors, the cells have high enzymatic activity.

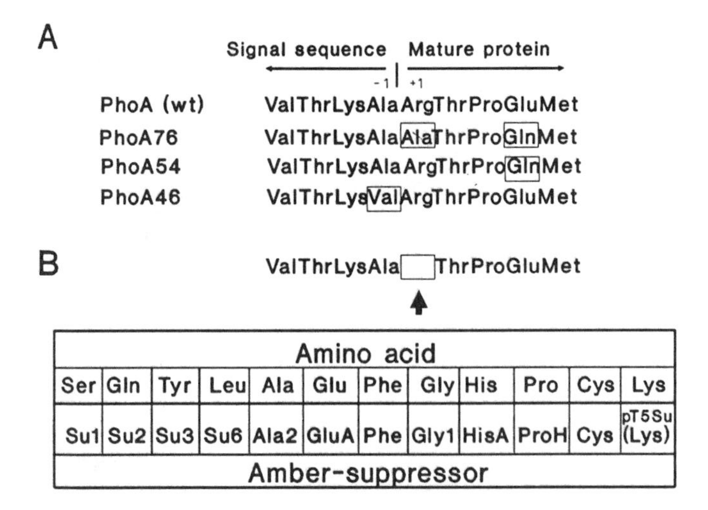

FIG. 3. Amino acid substitutions in the cleavage site and N terminus of mature AP. (A) Substitutions (boxed) resulted from mutations introduced into the corresponding sites of the *phoA* gene by oligonucleotide-directed mutagenesis. (B) Substitutions by suppression of *phoA*(Am) expression in *E. coli* MK251 (Karamyshev et al., 1994) carrying the pPHOA7(Am) plasmid and the indicated amber suppressors. *E. coli* amber suppressors are described by Kleina et al. (1990) and Normanly et al. (1990), and the lysine insertion amber suppressor pT5Su(Lys) of phage T5 was constructed by Ksenzenko et al. (unpublished).

It is known that the sites of signal peptide cleavage by leader peptidases are rather regular and described by the rule "−3, −1": only small nonbranched and noncharged amino acids are found in this region (von Heijne, 1984). This means that only Ala, Ser, Gly, Thr, or Gln is present in position −1, and aromatic (Phe, Tyr, and Trp), charged (Asp, Glu, Lys, Arg, and His), and extremely polar (Asn and Gln) amino acid residues are never found in position −3. Such structural regularity in the −3, −1 region is considered to be necessary for recognition by leader peptidase of the signal peptide cleavage site but insignificant for protein translocation across the cytoplasmic membrane (Fikes and Bassford, 1987; Pollitt et al., 1986; Shen et al., 1991). Our results entirely conform with these data. In contrast to the sites −3 and −1, position +1 is rather irregular. However, Gln, Ile, Leu, Met, and Pro were not found in this position in procaryotic secreted proteins (von Heijne, 1986). Our results show that introduction of proline into the +1 position leads to complete inhibition of processing. It is known that proline has a great effect on the secondary structure of the polypeptide chain (Shulz and Schirmer, 1979). Its action on processing probably results from the change in the processing-site structure, which makes it impervious to attack by signal peptidase. Similar results have been recently obtained for the phage M13 procoat protein (Shen et al., 1991) and maltose-binding protein (Barkocy-Gallagher and Bassford, 1992).

The rate of in vivo processing of other mutant proteins was so high that differences between them could not be revealed. However, the effect of some amino acid substitutions on the rate of processing was observed in vitro by using isolated precursors of mutant proteins PhoA54 and PhoA76 and wild-type prePhoA. The substitution of Gln(+4) for Glu (PhoA54) has no marked effect on the rate of signal peptide cleavage, whereas simultaneous substitution of Ala(+1) for Arg and Gln(+4) for Glu (PhoA76) significantly increases it, suggesting that the interaction of leader peptidase with signal peptide cleavage site is more efficient in the presence of alanine in the +1 position. It should be noted that it is precisely this amino acid residue which is present in the +1 position in many procaryotic secreted proteins (von Heijne, 1984).

All investigated amino acid substitutions in AP affected its isozymic spectrum. This was shown with the use of electophoresis under nondenaturing conditions with subsequent manifestation of AP activity directly in the gel (Fig. 4). The wild-type PhoA is present in the periplasm mainly as three isoforms, while the isozymic spectra of mutant proteins differ significantly.

PhoA54 protein was found mainly as three isoforms having a lower electrophoretic mobility than isoforms of wild-type protein, which is due to the increase in the total positive charge of the protein as a consequence of the Gln(+4) substitution for Glu. This substitution does not affect the

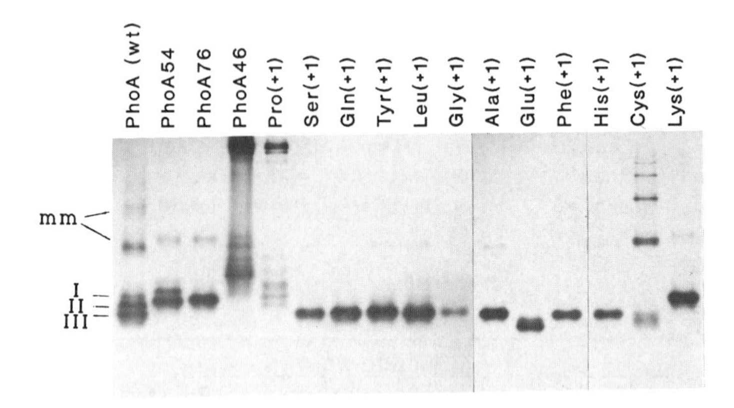

FIG. 4. Isozyme spectra of wild-type and mutant APs. Roman numerals, isoforms of wild-type AP; mm, multimers. All samples were analyzed by electrophoresis under nondenaturing conditions followed by the treatment with the AP substrate α-naphthylphosphate and Fast Red Dye RR.

N-terminal Arg cleavage during posttranslocational modification—otherwise only one form would be found. The mutant protein PhoA76, with simultaneous substitutions of Ala(+1) for Arg and Gln(+4) for Glu, was found only as one band with the electrophoretic mobility of wild-type PhoA isozyme I. This result could be expected, since the above amino acid substitutions preserve the total charge of the protein identical to the charge of wild-type PhoA isoform I. However, it does not allow one to elucidate whether the N-terminal amino acid was cleaved.

Protein PhoAGlu(+1) was found mainly as an isoform having a higher electrophoretic mobility than that of wild-type protein isoform III, and protein PhoALys(+1) as an isoform corresponding, by its electrophoretic mobility, to isoform I of wild-type AP. The results indicate that the alteration of N-terminal Lys and Glu prevents the Iap protease cleavage. If cleaved, other isoforms would be observed. Earlier it was shown that addition of arginine to the culture medium inhibits the transformation of AP isoform I into isoforms II and III, which formed the basis for the suggestion about Iap protease specificity (Piggot et al., 1972; Schlesinger et al., 1975). However, there is still no direct evidence for Iap protease specificity toward N-terminal arginine. Our results show that Iap protease is not specific, at least toward glutamic acid and lysine, which supports the suggestion that it is specific toward arginine.

Isozyme spectra of proteins containing Ser, Gln, Tyr, Leu, Gly, Ala, or Phe in the +1 position have peculiar features expected for Arg(+1) substitution by an uncharged amino acid. All proteins mentioned were found as one isoform corresponding, by its electrophoretic mobility, to isoform III of the wild-type protein. The same isozyme spectrum was revealed for PhoAHis(+1) as well. Since the pK′ of the histidine R-group is equal to 6.0, under electrophoresis conditions at pH 8.3 it is noncharged and consequently has no effect on the electrophoretic mobility of the protein.

Mutant proteins with modified processing [PhoA46 and PhoAPro(+1)] and the protein with Cys in +1 position were found as several high-molecular-weight forms. Such results for proteins PhoA46 and PhoAPro(+1) might be explained by a nonspecific aggregation of proteins with each other or by formation of their complexes with other cell components as a result of the presence of noncleaved signal peptide. In the case of protein PhoACys(+1), there are probably additional sulfide bonds between several subunits, with formation of multimers having enzyme activities. It is most likely that in this case the N-terminal cysteine is not cleaved by protease.

The data obtained in this work show that single amino acid substitutions both in the signal peptide cleavage site and in the N-terminal domain of the mature polypeptide chain of AP have an impact on individual stages of its biogenesis, confirming a significant role of these domains in the above process.

REFERENCES

Barkocy-Gallagher, G. A., and P. J. Bassford, Jr. 1992. Synthesis of precursor maltose-binding protein with proline in the +1 position of the cleavage site interferes with the activity of *Escherichia coli* signal peptidase I *in vivo*. *J. Biol. Chem.* **267:**1231–1238.

Bogdanov, M. V., Y. S. Tarakhovski, M. S. Manuvakhova, G. M. Gongadze, and M. A. Nesmeyanova. 1989. Compositional and structural peculiarities of envelope of *Escherichia coli* cells secreting alkaline phosphatase into medium. *Biol. Membr.* **6:**301–308. (In Russian.)

Boyd, D., C.-D. Guan, S. Willard, W. Wright, K. Strauch, and J. Beckwith. 1987. Enzymatic activity of alkaline phosphatase precursor depends on its cellular location, p. 89–93. *In* A. Torriani-Gorini, F. G. Rothman, S. Silver, A. Wright, and E. Yagil (ed.), *Phosphate Metabolism and Cellular Reg-*

ulation in Microorganisms. American Society for Microbiology, Washington, D.C.

Fikes, J. D., and P. J. Bassford, Jr. 1987. Export of unprocessed maltose-binding protein to the periplasm of *Escherichia coli* cells. *J. Bacteriol.* **169:**2352–2359.

Halegoua, S., and M. Inouye. 1979. Translocation and assembly of outer membrane proteins of *Escherichia coli:* selective accumulation of precursors and novel assembly intermediates caused by phenethyl alcohol. *J. Mol. Biol.* **130:**39–61.

Hirst, T. R., and R. A. Welch. 1988. Mechanisms for secretion of extracellular proteins by gram-negative bacteria. *Trends Biochem. Sci.* **13:**265–268.

Inouye, H., W. Barnes, and J. Beckwith. 1982. The signal sequence of alkaline phosphatase of *Escherichia coli. J. Bacteriol.* **149:**434–439.

Inouye, H., and J. Beckwith. 1977. Synthesis and processing of alkaline phosphatase precursor *in vitro. Proc. Natl. Acad. Sci. USA* **74:**1440–1444.

Inouye, H., S. Michaelis, A. Wright, and J. Beckwith. 1981. Cloning and restriction mapping of alkaline phosphatase structural gene (*phoA*) of *Escherichia coli* and generation of deletion mutant in vitro. *J. Bacteriol.* **146:**668–675.

Ito, K., P. J. Bassford, Jr., and J. Beckwith. 1981. Protein localization in *E. coli:* is there a common step in the secretion of periplasmic and other membrane proteins? *Cell* **24:**707–717.

Karamyshev, A. L., M. G. Shlyapnikov, M. I. Khmelnitsky, M. A. Nesmeyanova, and V. N. Ksenzenko. 1994. The study of biogenesis and secretion of alkaline phosphatase and its mutant forms in *E. coli.* I. Introduction of mutations into alkaline phosphatase gene. *Mol. Biol.* **28:**150–157. (In Russian.)

Kleina, L. G., J.-M. Masson, J. Normanly, J. Abelson, and J. H. Miller. 1990. Construction of *Escherichia coli* amber suppressor tRNA genes. II. Synthesis of additional tRNA genes and improvement of suppressor efficiency. *J. Mol. Biol.* **213:**705–717.

Ksenzenko, V. N., N. V. Pan'kova, A. L. Karamyshev, and M. G. Shlyapnikov. Unpublished data.

Lazdunski, C., D. Baty, and I. Pages. 1979. Procain, a local anesthetic interacting with the cell membrane, inhibits the processing of precursor forms of periplasmic proteins in *Escherichia coli. Eur. J. Biochem.* **96:**49–57.

Lazzaroni, J. C., D. Atlan, and R. C. Portalier. 1985. Excretion of alkaline phosphatase by *Escherichia coli* K12 pho constitutive mutants transformed with plasmids carrying the alkaline phosphatase structural gene. *J. Bacteriol.* **164:**1376–1380.

Lazzaroni, J. C., and R. C. Portalier. 1982. Production of extracellular alkaline phosphatase by *Escherichia coli* K12 periplasmic-leaky mutants carrying PhoA plasmids. *Eur. J. Appl. Microbiol. Biotechnol.* **16:**146–150.

Manuvakhova, M. S., M. V. Bogdanov, and M. A. Nesmeyanova. 1990. A study of the role of the *Escherichia coli* envelope in hemolysin secretion. *Biol. Membr.* **7:**390–398. (In Russian.)

Marston, F. A. O. 1986. The purification of eukaryotic polypeptides synthesized in *Escherichia coli. Biochem. J.* **240:**1–12.

Michaelis, S., J. F. Hunt, and J. Beckwith. 1986. Effects of signal sequence mutations on the kinetics of alkaline phosphatase export to the periplasm in *Escherichia coli. J. Bacteriol.* **167:**160–167.

Nakata, A., H. Shinagawa, and F. C. Rothman. 1987. Molecular mechanism of isozyme formation of alkaline phosphatase in *Escherichia coli,* p. 139–141. *In* A. Torriani-Gorini, F. G. Rothman, S. Silver, A. Wright, and E. Yagil (ed.), *Phosphate Metabolism and Cellular Regulation in Microorganisms.* American Society for Microbiology, Washington, D.C.

Nesmeyanova, M. A., O. B. Motlokh, M. N. Kolot, and I. S. Kulaev. 1981. Multiple forms of alkaline phosphatase from *Escherichia coli* cells with repressed and derepressed biosynthesis of the enzyme. *J. Bacteriol.* **146:**453–459.

Nesmeyanova, M. A., N. E. Suzina, I. M. Tsfasman, and A. O. Badyakina. 1991a. Secretion of periplasmic PhoA protein under its overproduction into the medium by outer membrane vesicles. Peculiarities of chemical content of vesicles and membranes in *Escherichia coli* secreting cells. *Mol. Biol.* **25:**974–988. (In Russian.)

Nesmeyanova, M. A., I. M. Tsfasman, A. L. Karamyshev, and N. E. Suzina. 1991b. Secretion of the overproduced periplasmic PhoA protein into the medium and accumulation of its precursor in *phoA*-transformed *Escherichia coli* strains: involvement of outer membrane vesicles. *World J. Microbiol. Biotechnol.* **7:**394–406.

Nesmeyanova, M. A., M. V. Zakharova, E. T. Fedotikova, and O. N. Lisina. 1990. Excretion of periplasmic alkaline phosphatase into the medium in *E. coli* cells transformed with plasmids carrying the gene of this enzyme. *Biotechnologia* **6:**20–23. (In Russian.)

Normanly, J., L. G. Kleina, J.-M. Masson, J. Abelson, and J. H. Miller. 1990. Construction of *Escherichia coli* amber suppressor tRNA genes. III. Determination of tRNA specificity. *J. Mol. Biol.* **213:**719–726.

Piggot, P. J., M. D. Sklar, and L. Gorini. 1972. Ribosomal alterations controlling alkaline phosphatase isozymes in *Escherichia coli. J. Bacteriol.* **110:**291–299.

Poirier, P. T., and S. C. Holt. 1983. Acid and alkaline phosphatases of *Capnocytophaga* species. I. Production and cytological localization of the enzymes. *Can. J. Microbiol.* **29:**1350–1360.

Pollitt, S., S. Inouye, and M. Inouye. 1986. Effect of amino acid substitutions at the signal peptide cleavage site of the *Escherichia coli* major outer membrane lipoprotein. *J. Biol. Chem.* **261:**1835–1837.

Pugsley, A. P. 1989. *Protein Targeting.* Academic Press, Inc., San Diego, Calif.

Pugsley, A. P., and S. T. Cole. 1987. An unmodified form of the ColE2-lysis protein, an envelope lipoprotein, regains reduced ability to promote colicin E2 release and lysis of producing cells. *J. Gen. Microbiol.* **133:**2411–2420.

Reynolds, I. A., and M. J. Schlesinger. 1969. Alterations in the structure and function of *E. coli* alkaline phosphatase due to Zn^{++} binding. *Biochemistry* **8:**588–593.

Sambrook, J., E. F. Fritsch, and T. Maniatis. 1989. *Molecular Cloning: a Laboratory Manual,* vol. 3, 2nd ed. Cold Spring Harbor Laboratory, Cold Spring Harbor, N.Y.

Schlesinger, M. J., and K. Barrett. 1965. The reversible dissociation of the alkaline phosphatase of *Escherichia coli.* I. Formation and reactivation of subunits. *J. Biol. Chem.* **24:**4284–4292.

Schlesinger, M. J., W. Bloch, and P. M. Kelley. 1975. Differences in the structure, function, and formation of two isozymes of *Escherichia coli* alkaline phosphatase, p. 333–342. *In* C. L. Markert (ed.), *Isozymes. I. Molecular Structure.* Academic Press, Inc., New York.

Schnaitman, C. A. 1971. Solubilization of the cytoplasmic membrane of *Escherichia coli* by Triton X-100. *J. Bacteriol.* **108:**545–552.

Shen, L. M., J.-I. Lee, S. Cheng, H. Jutte, A. Kuhn, and R. E. Dalbey. 1991. Use of site-directed mutagenesis to define the limits of sequence variation tolerated for processing of the M13 procoat protein by the *Escherichia coli* leader peptidase. *Biochemistry* **30:**11775–11781.

Shulz, G. E., and R. H. Schirmer. 1979. *Principles of Protein Structure.* Springer-Verlag, Inc., New York.

Torriani, A. 1960. Influence of inorganic phosphate in the formation of phosphatase by *Escherichia coli. Biochim. Biophys. Acta* **38:**460–466.

Tsfasman, I. M., A. O. Badyakina, N. E. Nagornaya, and M. A. Nesmeyanova. 1993. Peculiarities of alkaline phosphatase biogenesis in *E. coli* at its oversynthesis. *Mol. Biol.* **27:**805–816. (In Russian.)

von Heijne, G. 1984. How signal sequences maintain cleavage specificity. *J. Mol. Biol.* **173:**243–251.

von Heijne, G. 1986. A new method for predicting signal sequence cleavage sites. *Nucleic Acids Res.* **14:**4683–4690.

Zwizinski, C., and W. Wickner. 1980. Purification and characterization of leader (signal) peptidase from *Escherichia coli. J. Biol. Chem.* **255:**7973–7977.

Chapter 45

Pathways of Disulfide Bond Formation in Proteins In Vivo

JAMES BARDWELL,[1] ALAN DERMAN,[2] DOMINIQUE BELIN,[3] GEORG JANDER,[4] WILL PRINZ,[4] NANCY MARTIN,[4] AND JON BECKWITH[4]

Institut für Biophysik und Physikalische Biochemie, Universität Regensburg, Regensburg, D-93040 Germany[1]; Department of Biochemistry and Biophysics, University of California, San Francisco, California 94143-0448[2]; Département de Pathologie, Centre Médicale Universitaire, Geneva, Switzerland[3]; and Department of Microbiology and Molecular Genetics, Harvard Medical School, Boston, Massachusetts 02115[4]

The formation of disulfide bonds appears to be limited mainly to proteins outside the cytoplasmic compartment (Schulz and Schirmer, 1979). Secreted proteins such as insulin, immunoglobulins, serine proteases, and many bacteria toxins contain essential disulfide bonds. Membrane-bound proteins, including membrane receptors, often contain multiple disulfide bonds. In contrast, the rare cytoplasmic protein exhibiting disulfide bonds is usually an oxidoreductase which utilizes the cysteine pair as part of its active site. Furthermore, if one causes the normally secreted proteins with disulfide bonds, β-lactamase (Pollitt and Zalkin, 1983) and alkaline phosphatase (Derman and Beckwith, 1991), to be retained in the bacterial cytoplasm, the cystine linkages do not form. Conversely, β-galactosidase, which has only free cysteines in its normal cytoplasmic location, is thought to form disulfide linkages when its export to the bacterial periplasm is promoted (Bardwell et al., 1991).

What explains the difference in the ability to form disulfide bonds between the different cellular compartments? Until recently, it had been thought by many that disulfide bond formation in proteins in vivo was a spontaneous process, taking place in an uncatalyzed fashion in the appropriate oxidizing environment. According to this explanation, the bacterial periplasm and the eukaryotic lumen of the endoplasmic reticulum exhibit the proper redox potential for disulfide bond formation, whereas the cytoplasm is too reducing for the process to take place. Recent evidence from our laboratory and others has forced a reevaluation of this picture.

Disulfide Bond Formation Is Catalyzed in Exported and Membrane Proteins

We have developed a genetic selection in *Escherichia coli* that has revealed the existence of proteins that are required for the formation of cystine linkages in proteins. We came upon these mutants as a by-product of our studies on membrane protein assembly. A MalF–β-galactosidase hybrid protein (protein 102) exhibits no β-galactosidase enzymatic activity because of the presumed structure depicted in Fig. 1. The proposal for this structure is based on the knowledge that at least a portion of the β-galactosidase is protruding into the periplasm (McGovern and Beckwith, 1991) and that the cell cannot translocate the entire molecule across the cytoplasmic membrane (Lee et al., 1989). We believe that this structure is stabilized by "unwanted" cystine linkages that form in that portion of the β-galactosidase molecule that is translocated. We have shown that externally added dithiothreitol, which will reduce disulfide bonds, restores β-galactosidase activity to the hybrid protein (unpublished results). Thus, mutants that failed to catalyze disulfide bond formation might also be expected to have the same effect, allowing cells to grow on lactose as the sole carbon source.

We selected spontaneous Lac$^+$ derivatives of a strain carrying *malF-lacZ* fusion 102. The mutations fell into two genes which we have named *dsbA* and *dsbB* (Bardwell et al., 1991, 1993). The two classes of mutations have very similar phenotypes; they caused a severe reduction in disulfide bond formation in a number of proteins tested. These include alkaline phosphatase, β-lactamase, OmpA, and the cloned mouse serine proteases, urokinase, and tissue plasminogen activator. Other laboratories have presented evidence for similar defects in disulfide bond formation in *dsb* mutants for acid phosphatase (Bocquet, personal communication) and pullulunase (Pugsley, 1992) in *E. coli*, a *Vibrio cholerae* pilus (Peek and Taylor, 1992) and a cholera toxin subunit (Yu et al., 1992), and a component of *E. coli* flagella (Dailey and Berg, 1993). Kinetic studies with OmpA in particular show that the rate of disulfide bond formation is reduced by approximately 2 orders of magnitude in a *dsbA* mutant. The residual oxidizing activity seen in these backgrounds may

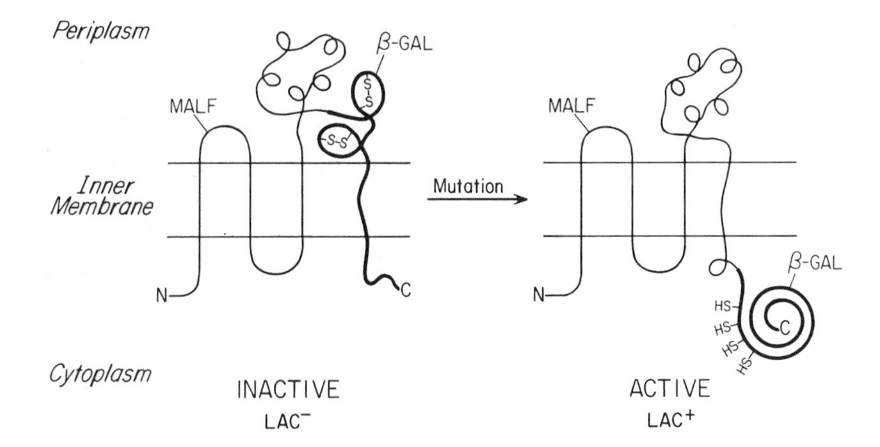

FIG. 1. Properties of the MalF–β-galactosidase hybrid protein. The thick lines represent β-galactosidase, and the thin lines represent MalF. When β-galactosidase (β-GAL) is active, we have assumed that the enzyme is entirely in the cytoplasm with a portion of the formerly periplasmic domain of MalF pulled into the membrane. We have no direct evidence for these details of the structure. The mutation that leads to a Lac⁺ phenotype would be in a gene unlinked to the gene fusion (e.g., *dsbA*).

be due to spontaneous oxidation of cysteines or to a low-activity backup system for disulfide bond formation. The *dsb* mutants are viable. The *dsbA* and *dsbB* genes or their homologs have also been found in *E. coli* and other gram-negative organisms in several other laboratories (Kamitani et al., 1992; Peek and Taylor, 1992; Yu et al., 1992; Dailey and Berg, 1993; Tomb, 1992; Missiakas et al., 1993).

We have mapped the *dsbA* and *dsbB* genes on the *E. coli* chromosome (Fig. 2), cloned them, and determined their DNA sequences. The DsbA protein is an abundant periplasmic protein with a motif containing a cysteine-X-Y-cysteine sequence found in a number of disulfide bond oxidoreductases. The protein has been crystallized, and its structure has been determined to a 2-Å resolution (Martin et al., 1993). A portion of it closely resembles thioredoxin. A second helical domain not present in thioredoxin is found in DsbA. Features of the structure suggest that DsbA may contain a peptide-bonding cleft absent in thioredoxin. In vitro, DsbA can directly oxidize cysteines in proteins such as alkaline phosphatase (Akiyama et al., 1992) and hirudin, a thrombin inhibitor (Wunderlich et al., 1993), without the presence of any other oxidant such as glutathione. DsbA, like eukaryotic protein disulfide bond isomerase (PDI), may also catalyze rearrangement of disulfide bonds in proteins already oxidized.

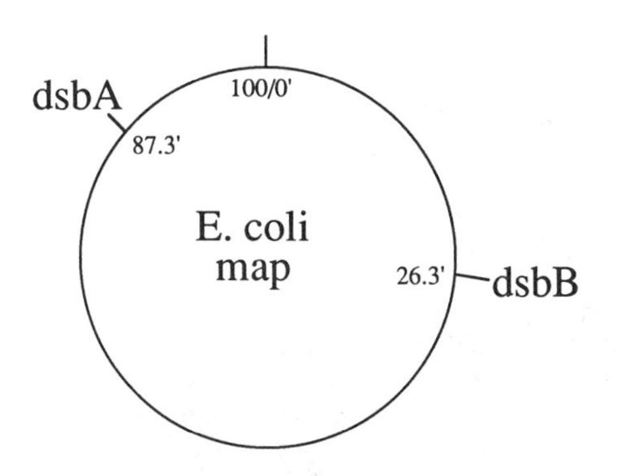

FIG. 2. Chromosomal locations of *dsbA* and *dsbB*.

DsbA contains a very reactive disulfide bond, consistent with its ability to transfer disulfides to its substrates (Zapun et al., 1993; Wunderlich et al., 1993). Directed-mutagenesis studies have shown, with the DsbA homolog of *V. cholerae*, that the cysteines in the motif region are essential for the function of the protein (Yu et al., 1993). Further, selection for mutants that make full-length DsbA in our system has yielded mainly *dsbA* mutations, eliminating either of the two cysteines.

DsbB is a cytoplasmic membrane protein with four membrane-spanning segments. It has a pair of cysteines in each of its periplasmic domains, one of them being in a cysteine-X-Y-cysteine arrangement. Use of in vivo obtained mutations and in vitro directed mutations shows that all four cysteines are important for DsbB activity (Jander et al., unpublished; Raina et al., personal communication). Several lines of evidence suggest that DsbB is involved in the reoxidation of DsbA (Bardwell et al., 1993) (Fig. 3).

The important conclusion from these studies is that disulfide bond formation in proteins in *E. coli* is a catalyzed rather than a spontaneous process.

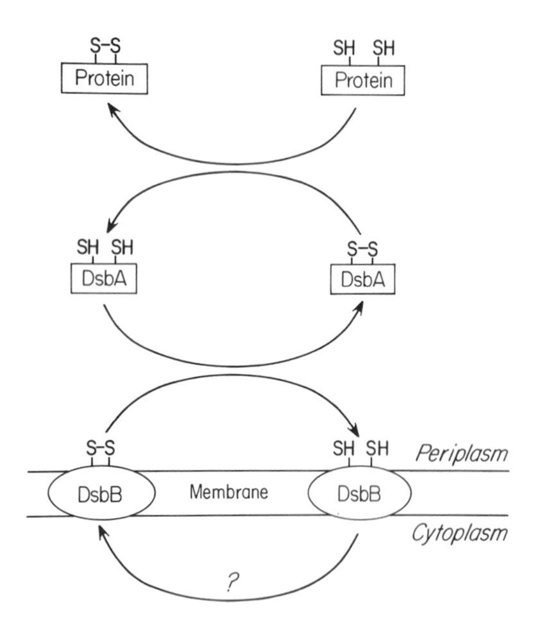

FIG. 3. A model for the function of the Dsb proteins. The DsbA protein transfers its disulfide linkage to proteins, such as alkaline phosphatase, that are exported to the cell envelope. The reduced DsbA is then reoxidized by the membrane protein DsbB. Genetic evidence suggests that this occurs via the cysteines of DsbB. Other cytoplasmic membrane components may be necessary for the reoxidation of DsbB.

In the absence of DsbA, disulfide bond formation in all periplasmic and outer membrane proteins tested is dramatically reduced. We believe that the requirement for a catalyst will be found to be generally true of disulfide bond formation. Recent support for this assumption with regard to eukaryotes comes from studies on mutants in the analogous cysteine-X-Y-cysteine motifs of *S. Cerevisiae* PDI (LaMantia and Lennarz, 1993). These mutants are viable but are dramatically reduced in disulfide bond formation in carboxypeptidase Y. It seems likely, therefore, that PDI, which is ubiquitous in eukaryotes, provides the same function as DsbA. Although it was known that PDI was capable of catalyzing disulfide bond formation, the name isomerase was used because its main function was thought to be the reshuffling of disulfide bonds during protein folding. This conception of its function was, in turn, based partly on the long-held assumption that the *formation* of disulfide bonds was spontaneous.

Mechanisms

Two major areas of research arise from the discovery of the Dsb system. The first involves the question of how DsbA promotes disulfide bond formation in proteins. The following questions should be pursued. Does DsbA act on a protein as it is being secreted into the periplasm, after it is fully secreted, after partial folding, or after it has assumed a fully folded structure? Interaction with a fully folded structure may be difficult if the cysteines are buried. Does DsbA promote disulfide bond formation only between pairs of cysteines that are correctly bonded in the final folded structure, or do disulfide bonds form at random or in an ordered fashion, only to be reshuffled (by DsbA) to achieve the final structure? Are there periplasmic chaperone proteins involved in the folding process that are required to maintain proteins in the proper state to be oxidized? One finding to be kept in mind in answering some of these questions is that β-galactosidase, a protein ordinarily lacking disulfide bonds, is apparently oxidized when translocated into the periplasm. In this case, the cysteines that are oxidized may not be optimally arranged for disulfide bond formation in the normal cytoplasmic folded structure.

The second area of research involves the issue of the source of the oxidizing potential of DsbA. In addition to *dsbB,* two other genes have been found that may be involved in the pathway that leads to reoxidation of DsbA after it has performed its function (Raina et al., personal communication). Does the Dsb pathway ultimately derive its oxidizing potential from components of electron transport, or is there a process for activating this system involving atmospheric oxygen?

One other question often asked is why the *dsb* mutants are viable. First, it is not clear that there are any disulfide-bond-containing proteins that are essential for cell viability. Second, there is sufficient nonenzymatic background oxidation of proteins that if there are any essential disulfide-bonded proteins, enough of the active protein may be formed to allow cell growth. If it is possible to eliminate the background disulfide bond formation by mutation or by varying growth conditions, this question may be answered.

Finally, it would be interesting to determine whether yeast PDI can substitute for DsbA in *E. coli.*

Forcing Disulfide Bond Formation in the Cytoplasm

Two possible reasons can be advanced for the paucity of disulfide bonds in cytoplasmic proteins. The absence of a catalytic system such as the Dsb system may be sufficient explanation. Alternatively, there may be a system that actively keeps cysteines reduced.

We have taken a genetic approach to determining the basis of the absence of disulfide bond formation in the cytoplasm. To this end, we have sought mutants of *E. coli* that allow disulfide bonds to form in cytoplasmic proteins, which ordinarily do not contain such bonds. The protein we have chosen for these studies is the *E. coli* alkaline phosphatase totally lacking its amino-terminal signal sequence (Derman et al., 1993b). Without this signal, 99% of the alkaline phosphatase is localized to the cytoplasm, where its disulfide bonds do not form; as a result, it is enzymatically inactive (Derman and Beckwith, 1991). We have devised genetic selections that require activation of the cytoplasmic alkaline phosphatase (Fig. 4). These selections demand, in effect, the formation of disulfide bonds to activate alkaline phosphatase in the cytoplasm.

E. coli alkaline phosphatase is a nonspecific phosphomonoesterase. We constructed strains in which a cytoplasmic-nonspecific phosphoesterase active would be required for growth. An *E. coli* mutant missing the fructose-1,6-bisphosphatase gene (*fbp*) is unable to grow on glycerol as the sole carbon source, since it is unable to achieve gluconeogenesis. We selected at 37°C for glycerol-utilizing mutants of an *fbp* strain carrying a plasmid coding for an alkaline phosphatase missing its signal sequence (Derman et al., 1993a). The *phoA* gene is under the control of the *lac* repressor and the *tac* promoter. A second selection which yielded the same class of mutants utilized a strain missing the enzyme serine-phosphate phosphatase (Fig. 4).

Ten mutants were characterized, and all were found to map to the *trxB* gene on the *E. coli* chromosome. The *trxB* gene codes for the enzyme thioredoxin reductase, which is responsible for the reduction of thioredoxin. In this mutant background, the cytoplasmic alkaline phosphatase now exhibited about 25% of the enzymatic activity expected for a fully active protein, compared with no activity in the *trxB*$^+$ background. Furthermore, the protein contained disulfide bonds. Fractionation studies showed that both the enzymatic activity and the disulfide-bonded protein were localized to the cytoplasm. Kinetic studies on the acquisition of disulfide bonds by alkaline phosphatase showed a half time for formation of about 5 min.

In addition to alkaline phosphatase, we have found that when a version of the mouse urokinase gene missing its signal sequence is cloned into the *trxB* background, substantial levels of activity are found, whereas in the wild-type background no activity is seen. The truncated version of urokinase used has six disulfide bonds. In contrast to the results with alkaline phosphatase missing its signal sequence, urokinase does not acquire an active conformation efficiently in growing *trxB* cells. However, incubation of mutant cells on ice is sufficient to allow the accumulation of active enzyme. This process, which is blocked by the sulfhydryl-blocking agent iodoacetaminde, probably reflects the formation and/or reshuffling of disulfide bonds in the mutant cytoplasm.

One explanation for these phenotypes of the *trxB* mutants is that thioredoxin is responsible for maintaining in the cytoplasm a reducing environment that prevents disulfide bonds from forming. Then, in the *trxB* mutants, thioredoxin would no longer be reduced, allowing disulfide bonds to form. If this were the case, a strain carrying a mutation in the *trxA* gene, which codes for thiore-

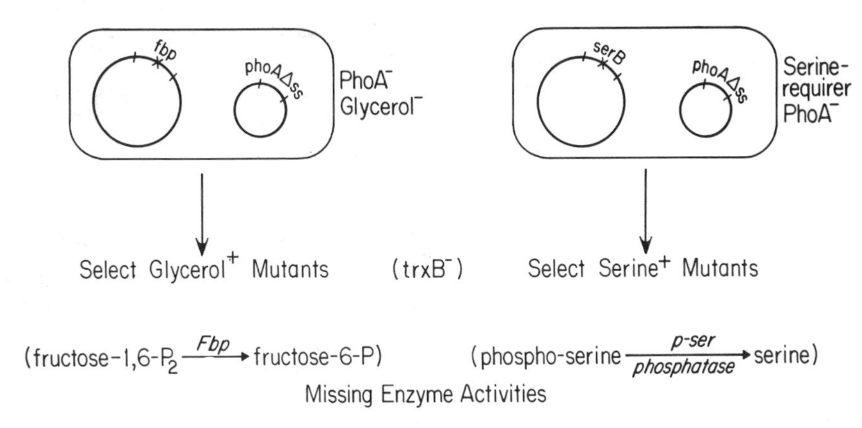

FIG. 4. Genetic selections for cytoplasmically active alkaline phosphatase. In each case, the strains contain an altered alkaline phosphatase gene (*phoA*) on a plasmid. This *phoA* gene is deleted for the DNA coding for amino acids 2 to 22, thus eliminating the entire signal sequence from the protein. The chromosome of each strain carries a mutation eliminating a cellular phosphatase, fructose-1,6-bis-phosphatase (*fbp*) in one case and serine-phosphate phosphatase (*serB*) in the other. The former strain is unable to utilize glycerol as the sole carbon source, and the latter strain cannot grow on medium that lacks the amino acid serine.

doxin, should have the same phenotype as the *trxB* mutants. However, we have found that a *trxA* mutation restores only a small amount of alkaline phosphatase activity (1%) compared with *trxB* mutants (20 to 25%). Another explanation is that, in the *trxB* mutant, oxidized thioredoxin accumulates and acts to promote disulfide bond formation in proteins. Again, a test of this hypothesis suggests that it is not correct. A double mutant, *trxA trxB*, exhibits the same levels of alkaline phosphatase activity (20%) as does the single-mutant *trxB* strain. On the basis of these findings, we postulate that thioredoxin reductase may have another substrate which it can reduce. Either this other substrate is itself responsible for maintaining the appropriate redox environment, or, in the *trxB* mutants, the oxidized form of this protein promotes disulfide bond formation.

Questions

A number of interesting questions arise from these studies. First, how can the cell survive a different redox environment in which disulfide bonds can form in the cytoplasm? Would disulfide bonds not form in other cytoplasmic proteins, thus severely affecting cell growth? A possible answer comes from a consideration of the folding properties of cytoplasmic versus exported proteins. Proteins that are destined to be exported are designed to fold slowly in the cytoplasm either because of intrinsic features of the protein or because of their interaction with chaperones such as SecB. Cytoplasmic proteins, on the other hand, presumably fold much faster. The kinetics of disulfide bond formation in the *trxB* mutants are relatively slow, exhibiting a half time of about 5 min for alkaline phosphatase. Particularly for cysteines that become buried within the three-dimensional structure of a protein, the rapid folding of cytoplasmic proteins might prevent disulfide bond formation. Further, proteins such as alkaline phosphatase may have their cysteines positioned for the formation of disulfide bonds, and the formation of those bonds may be much more rapid than for cysteines in cytoplasmic proteins that are not so favorably positioned. Further studies on cytoplasmic proteins are necessary to determine whether they are in any way affected in the *trxB* background.

Second, how can the cell survive the activation in the cytoplasm of a nonspecific phosphomonoesterase? Should it not hydrolyze a variety of important phosphorylated metabolic intermediates and important molecules such as ATP? One argument is that the K_m of alkaline phosphatase is so high that it will not do much harm in the cell. In addition, the concentrations of P_i in the cytoplasm will inhibit the activity of the enzyme to some extent. The reason that the selection works with the *serB* and *fbp* mutants is that the substrates have accumulated to such high concentrations that the high K_m and inhibition of the enzyme are not a problem.

One might also worry that disulfide bond formation in the cytoplasm would interfere with the secretion of proteins which will ultimately contain disulfide bonds. However, the rate of disulfide bond formation in the cytoplasm in *trxB* mutants is much lower than the rate of protein secretion. We have been able to study the effect of disulfide bond formation in the cytoplasm on a slowly secreted protein, by using a signal sequence mutant of alkaline phosphatase (*phoA68*), which exports a fraction of the protein (15%) slowly. In a *phoA68 trxB* background, only about one-third (5%) of the alkaline phosphatase is exported as in a *trxB*$^+$ cell. The remaining two-thirds of the alkaline phosphatase (10%) is trapped in the cytoplasm in active form. There appears to be a competition between disulfide bond formation and secretion. It may be that any molecules of alkaline phosphatase that form their disulfide bonds in the cytoplasm are retained there, because the at least partially folded conformation may prevent translocation across the membrane.

This work was supported by research grants from the American Cancer Society and National Institutes of Health and an American Cancer Society Research Professorship to J.B., by a grant from the Swiss National Foundation to D.B., and by fellowships from the Helen Hay Whitney Foundation and the Medical Research Council of Canada to J.C.A.B.

We thank Dan Fraenkel for his suggestions of the genetic selections for active alkaline phosphatase in the cytoplasm.

REFERENCES

Akiyama, Y., S. Kamitani, N. Kusukawa, and K. Ito. 1992. *In vitro* catalysis of oxidative folding of disulfide-bonded proteins by the *Escherichia coli dsbA (ppfA)* gene product. *J. Biol. Chem.* **267:**22440–22445.

Bardwell, J. C. A., J.-O. Lee, G. Jander, N. Martin, D. Belin, and J. Beckwith. 1993. A pathway for disulfide bond formation *in vivo. Proc. Natl. Acad. Sci. USA* **90:**1038–1042.

Bardwell, J. C. A., K. McGovern, and J. Beckwith. 1991. Identification of a protein required for disulfide bond formation *in vivo. Cell* **67:**581–589.

Bocquet, P. Personal communication.

Dailey, F. E., and H. C. Berg. 1993. Mutants in disulfide bond formation that disrupt flagellar assembly in *Escherichia coli. Proc. Natl. Acad. Sci. USA* **90:**1043–1047.

Derman, A. I., and J. Beckwith. 1991. *Escherichia coli* alkaline phosphatase fails to acquire disulfide bonds when retained in the cytoplasm. *J. Bacteriol.* **173:**7719–7722.

Derman, A. I., W. A. Prinz, D. Belin, and J. Beckwith. 1993a. Mutations that allow disulfide bond formation in the cytoplasm of *Escherichia coli. Science* **262:**1744–1747.

Derman, A. I., J. W. Puziss, P. J. Bassford, Jr., and J. Beckwith. 1993. A signal sequence is not required for protein export in *prlA* mutants of *Escherichia coli. EMBO J.* **12:**879–888.

Jander, G., N. Martin, and J. Beckwith. Unpublished results.

Kamitani, S., Y. Akiyama, and K. Ito. 1992. Identification and characterization of an *Escherichia coli* gene required for the formation of correctly folded alkaline phosphatase, a periplasmic enzyme. *EMBO J.* **11:**57–62.

LaMantia, M., and W. J. Lennarz. 1993. The essential function of yeast protein disulfide isomerase does not reside in its isomerase activity. *Cell* **74:**899–908.

Lee, C., P. Li, H. Inouye, and J. Beckwith. 1989. Genetic studies on the inability of β-galactosidase to be translocated across the *Escherichia coli* cytoplasmic membrane. *J. Bacteriol.* **171:**4609–4616.

Martin, J. L., J. C. A. Bardwell, and J. Kuriyan. 1993. Crystal structure of DsbA protein required for disulphide bond formation in vivo. *Nature* (London) **365:**464–468.

McGovern, K., and J. Beckwith. 1991. Membrane insertion of the *Escherichia coli* MalF protein in cells with impaired secretion machinery. *J. Biol. Chem.* **266:**20870–20876.

Missiakas, D., C. Georgopoulos, and S. Raina. 1993. Identification and characterization of the *Escherichia coli* gene *dsbB,* whose product is involved in the formation of disulfide bonds *in vivo. Proc. Natl. Acad. Sci. USA* **90:**7084–7088.

Peek, J. A., and R. K. Taylor. 1992. Characterization of a periplasmic thiol:disulfide interchange protein required for the functional maturation of secreted virulence factors of *Vibrio cholerae. Proc. Natl. Acad. Sci. USA* **89:**6210–6214.

Pollitt, S., and H. Zalkin. 1983. Role of primary structure and disulfide bond formation in β-lactamase secretion. *J. Bacteriol.* **153:**27–32.

Pugsley, A. P. 1992. Translocation of a folded protein across the outer membrane in *Escherichia coli. Proc. Natl. Acad. Sci. USA* **89:**12058–12062.

Raina, S., D. Missiakis, and C. Georgopouleos. Personal communication.

Schulz, G. E., and R. H. Schirmer. 1979. *Principles of Protein Structure.* Springer-Verlag, New York.

Tomb, J.-F. 1992. A periplasmic protein disulfide oxidoreductase is required for transformation of *Haemophilus influenzae* Rd. *Proc. Natl. Acad. Sci. USA* **89:**10252–10256.

Wunderlich, M., A. Otto, R. Seckler, and R. Glockshuber. 1993. Bacterial protein disulfide isomerase: efficient catalysis of oxidative protein folding at acidic pH. *Biochemistry* **32:**12251–12256.

Yu, J., S. McLaughlin, R. B. Freedman, and T. R. Hirst. 1993. Cloning and active site mutagensis of *Vibrio cholerae* DsbA, a periplasmic enzyme that catalyzes disulfide bond formation. *J. Biol. Chem.* **268:**4326–4330.

Yu, J., H. Webb, and T. R Hirst. 1992. A homologue of the *Escherichia coli* DsbA protein involved in disulphide bond formation is required for enterotoxin biogenesis in *Vibrio cholerae. Mol. Microbiol.* **6:**1949–1958.

Zapun, A., J. C. A. Bardwell, and T. E. Creighton. 1993. The reactive and destabilizing disulfide bond of DsbA, a protein required for protein disulfide bond formation *in vivo. Biochemistry* **32:**5083–5092.

Chapter 46

Biogenesis of Outer Membrane Porin PhoE of *Escherichia coli*

JAN TOMMASSEN, MARLIES STRUYVÉ, PATRICK VAN GELDER, AND HANS DE COCK

Department of Molecular Cell Biology, Institute of Biomembranes, Utrecht University, Padualaan 8, 3584 CH Utrecht, The Netherlands

Integral cytoplasmic membrane proteins generally contain one or several hydrophobic α-helices that span the membrane. Such stretches of hydrophobic residues are lacking in outer membrane proteins, probably because they would prevent the export of these proteins across the inner membrane. In fact, inspection of their primary sequences suggests that outer membrane proteins are very hydrophilic overall. Therefore, the three-dimensional folding of these proteins, thereby exposing the hydrophobic residues to the exterior, seems to be a prerequisite for insertion into the outer membrane. Consequently, the biogenesis of outer membrane proteins is thought to involve multiple steps (Fig. 1). After their synthesis as precursors in the cytoplasm or on membrane-bound polysomes, the proteins are transported in an unfolded state across the inner membrane. This process is mediated by the export machinery, consisting of several Sec proteins (reviewed by Schatz and Beckwith [1990]). In the periplasm, the processed proteins fold into a native-like tertiary structure, which will enable them to insert into the outer membrane by hydrophobic interactions. Trimerization of trimeric outer membrane proteins probably occurs after their insertion.

To gain insight in the molecular details of the individual events that occur during the biogenesis of outer membrane proteins, we are studying the PhoE protein of *Escherichia coli* K-12 as a model system. PhoE protein is a trimeric, pore-forming outer membrane protein. Its synthesis is induced under phosphate limitation (Overbeeke and Lugtenberg, 1980), and, by forming anion-selective pores (Benz et al., 1985), it functions to facilitate the uptake of phosphate and phosphorylated compounds. In this chapter we will describe our recent in vitro studies on the folding, trimerization, and outer membrane insertion of PhoE. Furthermore, we will discuss the intragenic requirements within PhoE that are essential for the correct assembly of the protein into the outer membrane.

In Vitro Folding and Assembly of PhoE Protein

To study the folding and assembly of PhoE, the protein was synthesized in a radioactive form in vitro in an extract of *E. coli* cells (de Cock et al., 1990a, 1990b). The folding of the protein was probed in immunoprecipitations with monoclonal antibodies (MAbs) that recognize conformational epitopes in the native PhoE protein. To assess a possible effect of the signal sequence on the folding of the protein (Park et al., 1988), both the wild-type PhoE precursor and mutant forms with defective signal sequences were studied. The PhoE proteins could be precipitated with the conformation-dependent MAbs (Fig. 2), showing that the in vitro synthesized proteins fold into a native-like structure. However, the efficiency of the immunoprecipitations depended on the PhoE species under study and varied between 1 and 25% of the total amount of protein synthesized (de Cock et al., 1990a). The wild-type PhoE precursor was the least efficiently precipitated, whereas the mutant species with the most defective signal sequence was the most efficiently precipitated. These results are consistent with the notion that an important function of the signal sequence is to maintain a precursor of an exported protein in a translocation-competent, i.e., nonnative, state (Park et al., 1988).

In vivo assembled PhoE trimers are highly resistant to denaturation. They withstand harsh conditions, like 60°C in 2% sodium dodecyl sulfate (SDS). When the in vitro synthesized PhoE proteins that could be precipitated with the conformation-dependent MAbs were analyzed by SDS-polyacrylamide gel electrophoresis (SDS-PAGE) without prior heating of the samples, no trimers could be detected on the gels. However, a PhoE form with a higher electrophoretic mobility than that of the denatured PhoE was observed (de Cock et al., 1990a, 1990b) (Fig. 3, m*PhoE). This faster-migrating PhoE form was converted to the denatured form by heating the samples before electrophoresis. This electrophoretic behavior is reminiscent of the well-described heat modifiability of monomeric outer membrane proteins such as OmpA (Heller, 1978): correctly folded OmpA is known to migrate faster during SDS-PAGE than the denatured form does. We assume that the faster-migrating PhoE form corresponds to a folded monomer containing a native-like

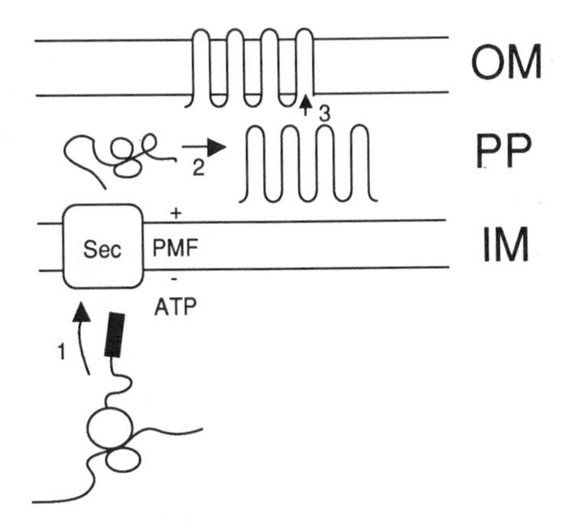

FIG. 1. Model for the biogenesis of outer membrane proteins. 1, Precursor proteins are co- or posttranslationally targeted to the translocation apparatus (Sec), located in the inner membrane (IM). Translocation of the linear precursors requires ATP and the proton motive force (PMF). The signal sequence is removed during or after translocation. 2, After the release of the mature protein in the periplasm (PP), it will fold into a native-like, insertion-competent configuration. 3, Insertion requires hydrophobic interactions between the folded protein and outer membrane components. After insertion into the outer membrane (OM), proteins fold into their final structure. Porins will assemble into their trimeric configuration.

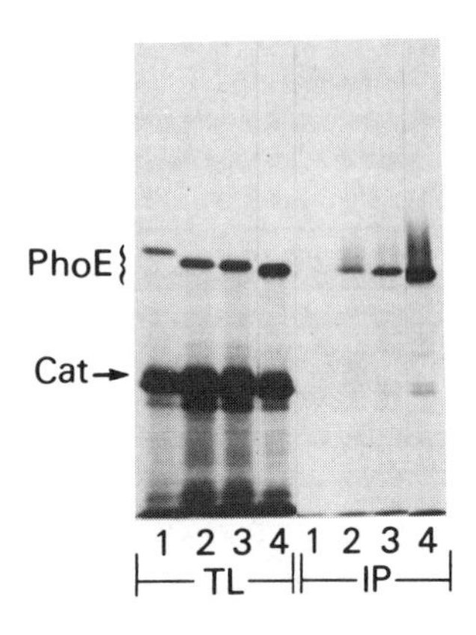

FIG. 2. Folding of in vitro synthesized PhoE, probed in immunoprecipitations with conformation-dependent MAbs. Plasmids pJP29, pJP367, pJP366, and pJP370 (de Cock et al., 1990a) were used to direct the in vitro synthesis of the wild-type PhoE precursor (lane 1) or mutant precursors in which only eight (lane 2), six (lane 3), or two (lane 4) amino acid residues, respectively, of the signal sequence are present. The lanes marked TL show the translation products. Cat represents chloramphenicol transacetylase, which is also encoded by the plasmids. The lanes marked IP show the results of immunoprecipitations with the conformation-dependent, PhoE-specific MAb PP1-5 (de Cock et al., 1990a).

structure. It might represent an assembly intermediate of PhoE, a supposition which is underscored by the recent detection of this form in vivo during pulse-chase experiments (van Gelder and Tommassen, unpublished).

Addition of isolated outer membranes was observed to induce trimerization of the in vitro synthesized PhoE proteins (Fig. 3) (de Cock et al., 1990b). The outer membrane component that is reponsible for this effect has not yet been identified. Purified lipopolysaccharides (LPS) or mixed LPS-phospholipid vesicles were inactive. Nevertheless, LPS is either directly or indirectly involved in the trimerization process, since outer membranes of mutants with defects in the core region of the LPS were less efficient in inducing trimerization (de Cock et al., submitted). The latter result is consistent with the observation that such mutants are defective in the assembly of several outer membrane proteins in vivo (Ames et al., 1974; Koplow and Goldfine, 1974; Tommassen and Lugtenberg, 1981). Like in vivo assembled PhoE trimers, the trimers obtained in vitro were highly resistant to denaturation (Fig. 3). However, in contrast to the in vivo assembled trimers, they were sensitive to proteases (de Cock et al., 1990a). Furthermore, they were not inserted into the added membranes, since they remained in the supernatant after pelleting of the membranes by centrifugation. Insertion of the in vitro synthesized PhoE proteins in a trypsin-resistant trimeric configuration was observed when small amounts of Triton X-100 (optimal concentration, approximately 0.06%) were added, together with the membranes (de Cock et al., submitted). It is not exactly known how the detergent stimulates the correct assembly of PhoE. The in vitro assembly of OmpF, secreted by *E. coli* spheroplasts, was also reported to depend on the presence of small amounts of Triton X-100 (Sen and Nikaido, 1990). Similarly, small amounts of a detergent permitted the refolding and insertion of OmpA into large vesicles of dimyristoyl phosphatidylcholine (Surrey and Jähnig, 1992). However, in the case of OmpA, no detergent was required when small vesicles, in which the lipid packing is not optimal, were used. Therefore, it was suggested that the detergent is required to introduce defects in the lipid organization in the case of large vesicles (Surrey and Jähnig, 1992). In conclusion, the assembly of PhoE was reconstituted in vitro. Three distinct steps, i.e., folding of the monomer, trimerization, and insertion, can be distinguished, and the requirements for each individual step can be studied.

Intragenic Requirement for PhoE Assembly

Structure of PhoE protein. Recently, the crystal structure of PhoE was resolved by X-ray dif-

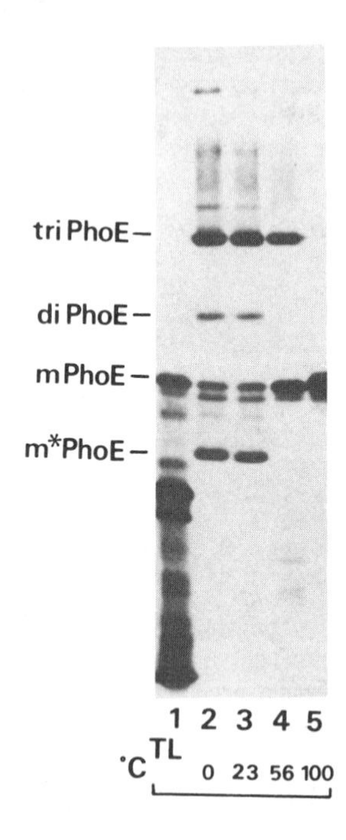

FIG. 3. Trimerization of in vitro synthesized PhoE protein. Plasmid pJP370 (de Cock et al., 1990a) was used to direct the in vitro synthesis of a quasi-mature form of PhoE in which only two amino acid residues of the signal sequence are present (lane 1, TL). Subsequently, outer membranes were added to induce trimerization (de Cock et al., 1990b), and folded PhoE forms were immunoprecipitated with a conformation-dependent, PhoE-specific MAb. Samples were treated at the indicated temperatures (0 to 100°C) in sample buffer before SDS-PAGE. The gel was run at a low amperage (20 mA) in a temperature-controlled room at 4°C to avoid denaturation of some PhoE forms. triPhoE, diPhoE, mPhoE, and m*PhoE represent trimers, dimers, denatured monomers, and folded monomers, respectively.

fraction analysis (Cowan et al., 1992). The structure is consistent in many details with the topology model that was previously proposed (van der Ley and Tommassen, 1987; Tommassen, 1988). Each monomer forms a β-barrel with 16 antiparallel amphipathic β-strands. The hydrophobic side of each β-strand is exposed to the lipids or to the subunit interface (Fig. 4). Two rings of aromatic residues, mainly phenylalanine and tyrosine, surround the trimeric complex at the borders of the hydrophobic membrane domain and the periplasmic and surface rims. The periplasmic turns are very short, whereas the exposed loops are larger. Loop L3 is not exposed at the cell surface but runs inside the β-barrel and forms a constriction within the pore; this constriction is responsible for many of the pore properties. The N terminus folds back to form a salt bridge with the C terminus at the subunit interface (Cowan et al., 1992).

In polytopic inner membrane proteins, each membrane-spanning segment is hydrophobic and could therefore potentially insert into the membrane, independent of the other membrane-spanning segments. In contrast, outer membrane proteins have only a hydrophobic exterior after folding into the β-barrel structure. Mutations that prevent the folding into this tertiary structure would therefore be expected to interfere with outer membrane insertion. Indeed, large deletions in PhoE (removing over 30 amino acid residues) prevented outer membrane incorporation and resulted in the periplasmic accumulation of the mutant proteins (Bosch et al., 1986). Similar results have been reported for other outer membrane proteins (see, e.g., Klose et al., 1988). Although apparently the correct tertiary structure is required for outer membrane insertion, some parts of the protein might be of special importance, e.g., by constituting a sorting signal for selective insertion in the appropriate membrane. Other parts might be of special importance for the correct folding of the protein. To identify such putative important parts, mutagenesis was carried out on the different domains of PhoE.

External loops. Comparison of the primary structures of the three related *E. coli* porins PhoE, OmpC, and OmpF (Mizuno et al., 1983) shows that the cell surface-exposed loops are hypervariable whereas the membrane-spanning segments are well conserved. This observation suggests that the cell surface-exposed loops do not contain important information for folding or outer membrane insertion. Indeed, the insertion of stretches of additional amino acid residues in the exposed loops (see, e.g., Agterberg et al., 1987) or loop L3 (Struyvé et al., 1993) generally did not disturb the assembly of the protein into the outer membrane. Also, mutant proteins with deletions removing almost the complete loop L4 and L5 (i.e., Δ156–165 and Δ195–206, respectively [Fig. 4]) were still correctly assembled into the outer membrane as evidenced by the binding of PhoE-specific MAbs to the intact cells or by the sensitivity of the cells to the PhoE-specific phage TC45 (Agterberg et al., 1989).

Membrane-spanning β-strands. Since the membrane-spanning segments of the porins are well conserved, they were expected to be less tolerant to mutagenesis than the exposed loops. Nevertheless, a mutant protein with an N-terminal deletion removing a part of the first membrane-spanning segment (i.e., Δ3–11) was still normally assembled in the outer membrane (Bosch et al., 1988). Also, mutant proteins lacking the complete first membrane-spanning segment were analyzed. Part of the total amount of the mutant PhoE protein Δ3–30 was still correctly assembled in the

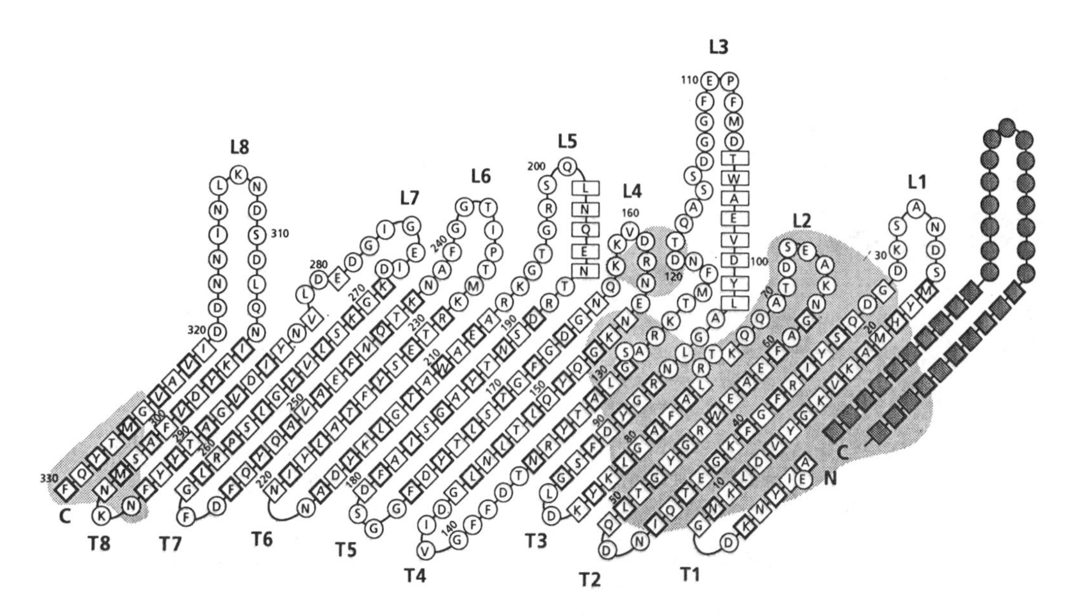

FIG. 4. Topology of PhoE in the outer membrane (Cowan et al., 1992). The amino acid sequence is shown in the single-letter code. The positions of the amino (N) and carboxy (C) termini are indicated. The secondary-structure assignments are diamonds for β-strands, rectangles for α-helices, and circles for turns and loops. Bold diamonds in the β-strands indicate that the amino acyl side chains are external, i.e., directed toward the lipids or the subunit interface. The interface region is shaded. The loops at the cell surface are denoted L1 to L8, and the turns at the periplasmic side are denoted T1 to T8. Loop L3 folds inside the β-barrel. The last two β-strands are repeated on the right-hand side (shaded) to emphasize the cyclic nature of the structure.

outer membrane, as evidenced by its protease resistance and by binding of PhoE-specific MAbs and phage TC45 to the cells. However, no trimers were detected by SDS-PAGE, indicating that this mutant protein inserts in the outer membrane as a monomer or as a very labile trimer (Bosch et al., 1988). This observation is consistent with the fact that the first membrane-spanning segment does not face the lipids but is located in the trimers at the subunit interface (Fig. 4). Immunoelectron microscopy revealed that the mutant protein Δ3–30 accumulated partially in the periplasm, indicating that the efficiency of outer membrane insertion is reduced. In contrast, deletion of the last membrane-spanning segment (Δ316–330) completely abolished outer membrane insertion (Bosch et al., 1989).

The hydrophobic sides of the amphipathic β-strands of PhoE are facing the lipids or the subunit interface (Fig. 4). Substitution of hydrophilic residues for hydrophobic ones in these β-strands would therefore be expected to interfere with stable insertion of the protein within the membrane or with trimerization, respectively. To assess this hypothesis, site-directed mutagenesis was performed (Struyvé et al., submitted [b]). Single mutations, e.g., A79D (denoting substitution of aspartic acid for alanine at position 79), L81S, Y151D, Y174K, G185E, L223Q, and T225N, were usually tolerated. However, double mutations resulting in the replacement of two hydrophobic residues in a single β-strand by hydrophilic ones (e.g., A79D/L81S and L223Q/T225N) had a strong adverse effect on the assembly of the protein in vivo. The nature of this assembly defect was studied in further detail in vitro (Struyvé et al., submitted [b]). After its in vitro synthesis, the mutant protein A79D/L81S could be precipitated with conformation-dependent MAbs, although this took place somewhat less efficiently than the wild-type protein. However, the immunoprecipitated mutant protein did not migrate during SDS-PAGE in the position of the folded monomer (Fig. 3) but in the position of the denatured PhoE protein. This result indicates that the mutant protein folds in vitro into a native-like structure, but the folded monomer is unstable and denatures during sample preparation. The mutant protein L223Q/T225N was only very poorly precipitated with the conformation-dependent MAbs, indicating that it is severely disturbed in folding. These results suggest that the acquisition of a hydrophobic exterior, as a result of the formation of the β-barrel, requires interaction with lipid components to stabilize this conformation. This possibility is reinforced by the recent observation that the in vitro folding of PhoE into the folded monomer requires the presence of Triton X-100 in the buffers (unpublished observation). Therefore, the folding of PhoE might occur at the periplasmic surface of either one of the membranes, rather than free in the periplasm as depicted in Fig. 1, unless this free intermediate is stabilized by lipid components or chaperones.

Homologies in unrelated outer membrane proteins. If outer membrane proteins contain a specific sorting signal, required for the selective insertion into the outer membrane, such a signal could be conserved in different unrelated proteins. However, if these proteins insert into the outer membrane in a folded configuration (as depicted in Fig. 1), such a putative signal might be a conformational one and might therefore be difficult to detect by comparison of the primary structures of the proteins. Nevertheless, a few short stretches of amino acid residues with vague homology were reported to be present in the porins, OmpA, and LamB (Nikaido and Wu, 1984). Most of these regions appeared to be insignificant in the sorting process, since their deletion did not affect the sorting of OmpA or PhoE or since these homologies were not observed when additional outer membrane protein sequences became available. For the most pronounced similarity region, sequence comparisons have recently been updated (Nikaido and Reid, 1990). It appears that only a glycine residue, corresponding to Gly-144 in PhoE, is absolutely conserved in this segment. To assess a possible role of Gly-144 in the biogenesis of PhoE, it was changed by site-directed mutagenesis into Leu (de Cock et al., 1991). Pulse-chase experiments revealed that the mutant protein was less efficiently assembled in the outer membrane in vivo than was the wild-type protein. In vitro, the G144L mutant protein was inefficiently precipitated with the conformation-dependent PhoE-specific MAbs. This result indicates that folding rather than sorting is affected by the mutation. This conclusion was underscored by the inspection of the crystal structure of PhoE (Cowan et al., 1992), which showed that the torsion angle at the position of Gly-144 could be accommodated by leucine only with difficulty.

A probably more relevant region of homology was recently detected at the C termini of different outer membrane proteins (Struyvé et al., 1991). The consensus sequence consists of a phenylalanine (Trp in some proteins) as the C-terminal residue and hydrophobic residues at positions 3 (preferentially Tyr), 5, 7, and 9 from the C terminus. This consensus sequence was found at the C termini of outer membrane proteins with widely different functions, including porins, receptors, and enzymes, and of diverse origin, including *E. coli* and *Pseudomonas, Neisseria,* and *Chlamydia* species. OmpA protein consists of two domains, an N-terminal domain embedded in the membrane and a C-terminal domain extending into the periplasm (Schweizer et al., 1978; Vogel and Jähnig, 1986). The consensus sequence is not found in the periplasmic extension but at the C terminus of the membrane-embedded domain. Deletion of a segment containing the consensus sequence in PhoE, i.e., Δ316–330 (Bosch et al., 1989) completely prevented outer membrane incorporation. Moreover, whereas large deletions in the membrane-embedded part of OmpA prevented the correct assembly of the protein into the outer membrane, immunoelectron microscopy revealed that the mutant proteins were still associated with the outer membrane unless the deletion covered a segment containing the consensus sequence (Klose et al., 1988). In the latter case, the mutant proteins accumulated in the periplasm. These results are at least consistent with the hypothesis that the C-terminal consensus sequence represents a targeting signal.

Since the C-terminal phenylalanine is the most pronounced feature of the consensus sequence, its function was further analyzed by site-directed mutagenesis (Struyvé et al., 1991). Substitution or deletion of this residue resulted in mutant PhoE proteins, whose expression was lethal to the cells. All mutant proteins were less efficiently incorporated in the outer membrane than was the wild-type protein; the amount of correctly incorporated protein (as determined by trypsin resistance) decreased in the order wild-type PhoE > F330Y > F330V > F330S > F330N > ΔF330. In the last case, hardly any correctly assembled mutant protein could be detected. In vitro assembly studies revealed that this mutant protein was hardly or not affected in the formation of the folded monomer or in outer membrane-induced trimerization, although the stability of the folded monomer was somewhat reduced (Struyvé et al., submitted [a]). However, insertion into isolated outer membranes appeared to be severely defective.

Concluding Remarks

Outer membrane proteins insert into the membrane, probably in a folded configuration (Fig. 1). Consequently, mutations that disrupt the correct folding will interfere with outer membrane insertion. Several point mutations in PhoE interfered with correct folding, as could be demonstrated by using an in vitro assembly system. Interestingly, double mutations that introduced hydrophilic residues at the hydrophobic side of the β-sheet resulted in folding defects. This observation suggests that folding occurs in a hydrophobic environment, e.g., at the surface of a membrane, rather than free in the periplasm.

Although the complete correctly folded protein is required for outer membrane insertion, this does not exclude the possibility that specific sorting or targeting signals exist in these proteins. The

C-terminal consensus sequence, with the phenylalanine residue as the most pronounced feature, could represent such a signal. The observation that deletion of the C-terminal Phe did not affect protein folding but the insertion into the membrane is certainly consistent with this hypothesis.

The contributions of Dirk Bosch and Marja Agterberg to the work described in this chapter are gratefully acknowledged.

Parts of this work were supported by the Netherlands Foundation for Chemical Research, with financial aid from the Netherlands Organization for the Advancement of Research.

REFERENCES

Agterberg, M., H. Adriaanse, E. Tijhaar, A. Resink, and J. Tommassen. 1989. Role of the cell surface-exposed regions of outer membrane protein PhoE of *Escherichia coli* K12 in the biogenesis of the protein. *Eur. J. Biochem.* **185:**365–370.

Agterberg, M., R. Benz, and J. Tommassen. 1987. Insertion mutagenesis on a cell-surface-exposed region of outer membrane protein PhoE of *Escherichia coli. Eur. J. Biochem.* **169:**65–71.

Ames, G. F.-L., E. N. Spudich, and H. Nikaido. 1974. Protein composition of the outer membrane of *Salmonella typhimurium:* effect of lipopolysaccharide mutations. *J. Bacteriol.* **117:**406–416.

Benz, R., A. Schmid, and R. E. W. Hancock. 1985. Ion selectivity of gram-negative bacterial porins. *J. Bacteriol.* **162:**722–727.

Bosch, D., J. Leunissen, J. Verbakel, M. de Jong, H. van Erp, and J. Tommassen. 1986. Periplasmic accumulation of truncated forms of outer-membrane PhoE protein of *Escherichia coli* K-12. *J. Mol. Biol.* **189:**449–455.

Bosch, D., M. Scholten, C. Verhagen, and J. Tommassen. 1989. The role of the carboxy-terminal membrane-spanning fragment in the biogenesis of *Escherichia coli* K12 outer membrane protein PhoE. *Mol. Gen. Genet.* **216:**144–148.

Bosch, D., W. Voorhout, and J. Tommassen. 1988. Export and localization of N-terminally truncated derivatives of *Escherichia coli* K-12 outer membrane protein PhoE. *J. Biol. Chem.* **263:**9952–9957.

Cowan, S. W., T. Schirmer, G. Rummel, M. Steiert, R. Ghosh, R. A. Pauptit, J. N. Jansonius, and J. P. Rosenbusch. 1992. Crystal structures explain functional properties of two *E. coli* porins. *Nature* (London) **358:**727–733.

de Cock, H., T. Claassen, and J. Tommassen. *In vitro* assembly and insertion of PhoE protein of *E. coli* K-12 into the outer membrane. Submitted for publication.

de Cock, H., D. Hekstra, and J. Tommassen. 1990a. *In vitro* trimerization of outer membrane protein PhoE. *Biochimie* **72:**177–182.

de Cock, H., R. Hendriks, T. de Vrije, and J. Tommassen. 1990b. Assembly of an *in vitro* synthesized *E. coli* outer membrane porin into its stable trimeric configuration. *J. Biol. Chem.* **265:**4646–4651.

de Cock, H., N. Quaedvlieg, D. Bosch, M. Scholten, and J. Tommassen. 1991. Glycine-144 is required for efficient folding of outer membrane protein PhoE of *E. coli* K-12. *FEBS Lett.* **279:**285–288.

Heller, K. B. 1978. Apparent molecular weights of a heat-modifiable protein from the outer membrane of *Escherichia coli* in gels with different acrylamide concentrations. *J. Bacteriol.* **134:**1181–1183.

Klose, M., H. Schwarz, S. MacIntyre, R. Freudl, M.-L. Eschbach, and U. Henning. 1988. Internal deletions in the gene for an *Escherichia coli* outer membrane protein define an area possibly important for recognition of the outer membrane by this polypeptide. *J. Biol. Chem.* **263:**13291–13296.

Koplow, J., and H. Goldfine. 1974. Alterations in the outer membrane of the cell envelope of heptose-deficient mutants of *Escherichia coli. J. Bacteriol.* **117:**527–543.

Mizuno, T., M.-Y. Chou, and M. Inouye. 1983. A comparative study on the genes for three porins of the *Escherichia coli* outer membrane. DNA sequence of the osmoregulated *ompC* gene. *J. Biol. Chem.* **258:**6932–6940.

Nikaido, H., and J. Reid. 1990. Biogenesis of prokaryotic pores. *Experientia* **46:**174–180.

Nikaido, H., and H. C. P. Wu. 1984. Amino acid sequence homology among the major outer membrane proteins of *Escherichia coli. Proc. Natl. Acad. Sci. USA* **81:**1048–1052.

Overbeeke, N., and B. Lugtenberg. 1980. Expression of outer membrane protein e of *Escherichia coli* K12 by phosphate limitation. *FEBS Lett.* **112:**229–232.

Park, S., G. Liu, T. B. Topping, W. H. Cover, and L. L. Randall. 1988. Modulation of folding pathways of exported proteins by the leader sequence. *Science* **239:**1033–1035.

Schatz, P. J., and J. Beckwith. 1990. Genetic analysis of protein export in *Escherichia coli. Annu. Rev. Genet.* **24:**215–248.

Schweizer, M., I. Hindennach, W. Garten, and U. Henning. 1978. Major proteins of the *Escherichia coli* outer cell envelope membrane. Interaction of protein II* with lipopolysaccharide. *Eur. J. Biochem.* **82:**211–217.

Sen, K., and H. Nikaido. 1990. *In vitro* trimerization of OmpF porin secreted by spheroplasts of *Escherichia coli. Proc. Natl. Acad. Sci. USA* **87:**743–747.

Struyvé, M., M. Heutink, M. Kleerebezem, T. van der Krift, H. de Cock, and J. Tommassen. Role of the carboxy-terminal phenylalanine in the biogenesis of outer membrane protein PhoE of *Escherichia coli* K-12. Submitted for publication [a].

Struyvé, M., M. Moons, and J. Tommassen. 1991. Carboxy-terminal phenylalanine is essential for the correct assembly of a bacterial outer membrane protein. *J. Mol. Biol.* **218:**141–148.

Struyvé, M., N. Nouwen, M. Veldscholten, M. Heutink, H. de Cock, and J. Tommassen. Mutagenesis of the exterior hydrophobic region of the β-barrel of outer membrane porin PhoE. Submitted for publication [b].

Struyvé, M., J. Visser, H. Adriaanse, R. Benz, and J. Tommassen. 1993. Topology of PhoE porin: the "eyelet" region. *Mol. Microbiol.* **7:**131–140.

Surrey, T., and F. Jähnig. 1992. Refolding and oriented insertion of a membrane protein into a lipid bilayer. *Proc. Natl. Acad. Sci. USA* **89:**7457–7461.

Tommassen, J. 1988. Biogenesis and membrane topology of outer membrane proteins in *Escherichia coli. NATO ASI Ser. H* **16:**351–373.

Tommassen, J., and B. Lugtenberg. 1981. Localization of *phoE,* the structural gene for outer membrane protein e in *Escherichia coli* K-12. *J. Bacteriol.* **147:**118–123.

van der Ley, P., and J. Tommassen. 1987. PhoE protein structure and function, p. 159-163. *In* A. Torriani-Gorini, F. G. Rothman, S. Silver, A. Wright, and E. Yagil (ed.), *Phosphate Metabolism and Cellular Regulation in Microorganisms.* American Society for Microbiology, Washington, D.C.

van Gelder, P., and J. Tommassen. Unpublished observations.

Vogel, H., and F. Jähnig. 1986. Models for the structure of outer-membrane proteins of *Escherichia coli* derived from Raman spectroscopy and prediction models. *J. Mol. Biol.* **190:**191–199.

IX. SIGNAL TRANSDUCTION AND PHOSPHOPROTEINS

Chapter 47

Introduction: Signal Transduction and Phosphoproteins

ANN M. STOCK[1] AND HIDEO SHINAGAWA[2]

Center for Advanced Biotechnology and Medicine and Department of Biochemistry, University of Medicine and Dentistry of New Jersey, Piscataway, New Jersey 08854-5638,[1] and The Research Institute for Microbial Diseases, Osaka University, 3-1, Yamadaoka, Suita, Osaka, Japan 565[2]

Reversible phosphorylation of proteins has long been recognized as a common feature of eukaryotic signal transduction pathways. The realization that many bacterial regulatory systems are controlled by protein phosphorylation involving a common signal transduction pathway is, by comparison, relatively recent. At the time of the last Pho Symposium in 1986, the field of "two-component" bacterial regulatory systems was in its infancy. In the chapters on nitrogen assimilation and regulatory components of the phosphate regulon, Merrick et al. (1986) and Shinagawa et al. (1986) summarized the sequence similarities that had recently been reported among several bacterial regulatory systems (Stock et al., 1985; Makino et al., 1986; Nixon et al., 1986). A note added in proof to the chapter on nitrogen regulation referenced the report by Ninfa and Magasanik (1986) of phosphorylation of the nitrogen response regulator, NtrC, providing the first clue to the molecular mechanism of signal transduction in these systems. Many researchers speculated that the sequence similarities might underlie common structures and functions, and it was hoped that the homology might accelerate elucidation of the molecular mechanism responsible for signal transduction within these diverse regulatory systems. The predictions have to a large extent come true, and a relatively detailed understanding of bacterial signal transduction via phosphotransfer pathways has now been achieved.

Over 60 different regulatory systems in more than 40 different bacterial species have been identified on the basis of sequence similarity as having conserved protein components that are involved in phosphotransfer-mediated signal transduction pathways (for recent reviews, see Bourret et al. [1991] and Parkinson and Kofoid [1992]). The systems share two conserved protein components that provide the central processing pathways that link extracellular or intracellular stimuli to specific responses. The first component is a histidine protein kinase that uses ATP to autophosphorylate at a specific histidine residue. The second component, the response regulator protein, catalyzes the transfer of this high-energy phosphoryl group to one of its own aspartate side chains. Phosphorylation of the response regulator protein generally activates an effector activity in an attached or associated domain.

Although the histidine kinase and response regulator domains are conserved in the different regulatory systems, the ways in which they are integrated into the systems, at both molecular and schematic levels, show a high degree of diversity. The molecules and pathways have been optimized to accommodate the specific requirements of each regulatory system. For instance, although the kinases are most often transmembrane proteins with a periplasmic sensing domain and a cytoplasmic kinase domain, some of the kinases are cytoplasmic proteins which interact with other transmembrane sensors or which sense intracellular signals. In several systems, regulation seems to proceed primarily through a phosphatase activity of the kinase, which inactivates the response regulators by hydrolysis of the acyl phosphate bond, whereas in other systems, this phosphatase activity is undetectable and regulation is achieved through activation of the kinase activity.

Response regulator proteins are typically composed of an N-terminal regulatory domain which contains the site of phosphorylation and a C-terminal domain which mediates the effector activity. The diversity of responses that are mediated by the two-component regulatory systems is reflected by the diversity and modularity of the effector domains themselves. The effector domains are often DNA-binding domains that activate or repress transcription of a specific set of genes. Other response regulator proteins have effector domains with enzymatic activity or have no effector domains at all but instead interact in an intermolecular reaction with downstream proteins that provide the effector functions. In a few cases, response regulator domains are located as attached domains at the C termini of histidine protein kinases. Regulatory domains have an autophosphatase activity which controls the lifetime of the active state. Because of different autophosphatase activities, lifetimes of the phosphorylated response regulator proteins range from seconds to hours, correlating with short- or long-term responses, respectively, required by the specific regulatory systems. In some systems, the longevity of the phosphorylated state of the response regulator is influenced by auxiliary proteins. For instance, in the nitrogen-regulatory system, the lifetime of the phosphorylated response regulator, NtrC, is affected by both the histidine kinase/phosphatase NtrB and the auxiliary protein PII.

Some regulatory systems are composed of only the histidine kinase and the response regulator proteins; an example is the osmoregulatory system, which contains the histidine kinase EnvZ and the response regulator OmpR. Many systems, however, involve more than the two conserved components. There are numerous examples of auxiliary proteins that interact with the conserved components or that act upstream or downstream of the phosphotransfer pathway, as in the cases of the phosphate regulon and chemotaxis systems. There are systems which have multiple kinases that feed into a single response regulator and, conversely, others which have a single kinase that communicates with multiple response regulators, as in the systems related to the initiation of sporulation in *Bacillus subtilis*. Thus, a notable feature of the two-component bacterial signal transduction systems is the adaptable nature of the central phosphotransfer pathway and the modularity of the conserved protein domains.

The large number of histidine kinases and response regulators, together with observations that components from different systems were capable of cross-phosphorylation reactions in vitro, invited speculation that cross talk might provide the basis for global regulatory networks in vivo. The maintenance of regulatory responses in several systems in the absence of the cognate kinases provided further support for this hypothesis. It is also conceivable that the cross-phosphorylation merely reflects the evolutionary relatedness among the two conserved protein components and that cross talk has little physiological significance in most cases. The recent finding that the intracellular metabolite acetyl phosphate can serve as the phospho-donor for response regulator proteins has provided an alternative explanation for kinase-independent regulation (Lukat et al., 1992; Wanner, 1992; McCleary et al., 1993).

Is the involvement of histidine and aspartate phosphorylation in signal transduction pathways in prokaryotes analogous to serine, threonine, and tyrosine phosphorylation in eukaryotes? The answer to this question is clearly no; the systems are distinguishable in several ways. Chemically and energetically, aspartyl phosphate differs from serine and threonine phosphoesters. It has been argued from a thermodynamic viewpoint that the high-energy acyl phosphate bond of phosphoaspartate would be capable of inducing a large conformational change in proteins, not possible with phosphoserine or phosphothreonine. Unlike eukaryotic proteins, which are often multiply phosphorylated, the phosphorylation of bacterial response regulators occurs at a unique site. In eukaryotic systems, protein kinases catalyze the transfer of phosphoryl groups from ATP to target proteins, often exhibiting promiscuous behavior and a general lack of specificity. In the bacterial systems, it is the response regulator protein that catalyzes its own phosphorylation whereas the histidine kinase catalyzes autophosphorylation at a specific histidine residue, thereby providing a pool of high-energy phosphoryl groups for transfer. The lifetime of the phosphoaspartate in the response regulators is much shorter than that of phosphoserines, phosphothreonines, and phosphotyrosines in eukaryotic proteins. The lifetime is limited both by the chemical instability of the acyl phosphate and by the catalytic hydrolysis by the autophosphatase activity of the response regulators and the phosphatase activity of the histidine kinases.

In the 8 years since the Pho '86 Symposium, the phosphotransfer mechanism that forms the central pathway for signal transduction in numerous bacterial regulatory systems has been clearly established. Still, many questions remain to be answered. What controls the interactions between the kinases and the response regulators? What determines the relative kinase versus phosphatase activities of a histidine protein kinase? In systems that involve more than one kinase or response regulator, what controls the flow of phosphate through different branches of the pathway? How is phosphorylation of regulatory domains linked to activation of different effector domains? How are homo- and heteromultimeric protein complexes influenced by phosphorylation? How much cross talk between kinases and regulators of different systems occurs in vivo? Is activation of response regulators by small-molecule phospho-donors physiologically significant? Do the two-component regulatory systems that are so ubiquitous in prokaryotes play a prominent role in eukaryotic signal transduction? Answers to these questions and establishment of the fundamental similarities as well as elucidation of the differences that distinguish these many systems will undoubtedly provide the basis of investigations for years to come.

REFERENCES

Bourret, R. B., K. A. Borkovich, and M. I. Simon. 1991. Signal transduction pathways involving protein phosphorylation in prokaryotes. *Annu. Rev. Biochem.* **60:**401–441.

Lukat, G. S., W. R. McCleary, A. M. Stock, and J. B. Stock. 1992. Phosphorylation of bacterial response regulator proteins by low molecular weight phospho-donors. *Proc. Natl. Acad. Sci. USA* **89:**718–722.

Makino, K., H. Shinagawa, M. Amemura, and A. Nakata. 1986. Nucleotide sequence of the *phoB* gene, the positive regulatory gene for the phosphate regulon of *Escherichia coli* K-12. *J. Mol. Biol.* **300:**900–909.

McCleary, W. R., J. B. Stock, and A. J. Nifa. 1993. Is acetyl phosphate a global signal in *Escherichia coli? J. Bacteriol.* **175:**2793–2798.

Merrick, M. J., S. Austin, M. Buck, R. Dixon, M. Drummond, A. Holtel, and S. MacFarlane. 1986. Regulation of nitrogen assimilation in enteric bacteria, p. 277–283. *In* A. Torriani-Gorini, F. G. Rothman, S. Silver, A. Wright, and E. Yagil (ed.), *Phosphate Metabolism and Cellular Regulation in Microorganisms.* American Society for Microbiology, Washington, D.C.

Ninfa, A. J., and B. Magasanik. 1986. Covalent modification of the *glnG* product, NR_I, by the *glnL* product, NR_{II}, regulates the transcription of the *glnALG* operon in *Escherichia coli. Proc. Natl. Acad. Sci. USA* **83:**5909–5913.

Nixon, B. T., C. W. Ronson, and F. M. Ausubel. 1986. Two component regulatory systems responsive to environmental stimuli share strongly conserved domains with the nitrogen assimilation regulatory genes *ntrB* and *ntrC. Proc. Natl. Acad. Sci. USA* **83:**7850–7854.

Parkinson, J. S., and E. C. Kofoid. 1992. Communication modules in bacterial signaling proteins. *Annu. Rev. Genet.* **26:**71–112.

Shinagawa, H., K. Makino, M. Amemura, and A. Nakata. 1986. Structure and function of the regulatory genes for the phosphate regulon in *Escherichia coli,* p. 20–25. *In* A. Torriani-Gorini, F. G. Rothman, S. Silver, A. Wright, and E. Yagil (ed.), *Phosphate Metabolism and Cellular Regulation in Microorganisms.* American Society for Microbiology, Washington, D.C.

Stock, A., D. E. Koshland, Jr., and J. Stock. 1985. Homologies between the *Salmonella typhimurium* CheY protein and proteins involved in the regulation of chemotaxis, membrane protein synthesis, and sporulation. *Proc. Natl. Acad. Sci. USA* **82:**7989–7993.

Wanner, B. L. 1992. Is cross regulation by phosphorylation of two-component response regulator proteins important in bacteria? *J. Bacteriol.* **174:**2053–2058.

Chapter 48

Signal Transduction in the Phosphate Regulon of *Escherichia coli*: Dual Functions of PhoR as a Protein Kinase and a Protein Phosphatase

HIDEO SHINAGAWA, KOZO MAKINO, MASAMI YAMADA, MITSUKO AMEMURA, TAKEYA SATO, AND ATSUO NAKATA

Research Institute for Microbial Diseases, Osaka University, Suita, Osaka 565, Japan

A large number of the genes (>30) in the phosphate regulon (Pho regulon), which are involved in the uptake and utilization of phosphate and phosphorus compounds, are transcriptionally activated coordinately by phosphorylated PhoB protein when the cell is starved for phosphate (see chapter 2). PhoR regulates the activity of PhoB by promoting phosphorylation and dephosphorylation of PhoB, responding to limiting and excess phosphate conditions, respectively. The environmental concentrations of P_i are monitored by a periplasmic phosphate-binding protein encoded by *pstS*, and the signal for the excess phosphate is transmitted across the cytoplasmic membrane by the products of *pstC*, *pstA*, *pstB*, and *phoU* to PhoR by unknown mechanism (see chapter 4). Thus, PhoB and PhoR proteins are a member of a large family called two-component regulatory systems, which are made up of histidine protein kinases and response regulators involved in various signal transduction pathways (see chapter 47).

Various Modes of Cross Talk in the Signal Transduction Related to the Pho Regulon

It is of great interest to know how so many homologous signal transduction systems in the cell are controlled without resulting in noises due to crossing of the different signals and whether there is any cross talk among the different systems at various levels (Fig. 1). One of the pioneering studies about the cross talk was done by Wanner and Latterell (1980), who discovered a gene originally named *phoM* (later renamed *creC*), which can substitute the positive regulatory function of *phoR*. CreC is a histidine protein kinase like PhoR and can transfer a phosphoryl group to PhoB. CreC has its cognate partner, a response regulator, CreB. Although CreC can phosphorylate both PhoB and CreB, PhoR cannot phosphorylate CreB. Therefore, cross talk is only one direction in this case (Amemura et al., 1990).

In our attempts to clone the *creC* gene, we obtained several chromosomal fragments that induced alkaline phosphatase production in a *phoR creC* double-mutant strain when they were carried on a multicopy plasmid. Two of them were analyzed; one was shown to contain the *ackA* gene encoding acetate kinase (Lee et al., 1990), and the other contained a DNA fragment encoding the carboxyl-terminal 130 amino acids of the β-subunit of glutamate synthetase (unpublished results). Since in a *phoR creC* strain the induction of alkaline phosphatase was caused by mutations that accumulate acetyl phosphate (Wanner and Wilmes-Riesenberg, 1992), and acetyl phosphate acted as substrate for the autophosphorylation of NtrC(NR1), a response regulator of the Ntr regulon (Feng et al., 1992), we examined the possibility that acetyl phosphate was responsible for the cross talk to the Pho regulon in the *phoR creC* mutant. We have demonstrated that acetyl phosphate serves as the substrate for the autophosphorylation of PhoB, which is activated as the transcriptional activator in vitro, as does phospho-PhoR, although the former is less efficient as a phospho-donor than is the latter (unpublished data). These results strongly suggest that the multicopy *ackA* gene elevates the level of acetyl phosphate in the cell, which activates PhoB by phosphorylation in the *phoR creC* strain. In the case of the cross talk by the other cloned DNA, we speculate that the truncated β-subunit of glutamate synthetase interacts abortively with the α-subunit of the enzyme to form inactive dimers, which inhibit the activity of the enzyme in the cell and cause an imbalance in the carbon flow through glycolysis and the tricarboxylic acid cycle, leading to accumulation of acetyl phosphate (McCleary et al., 1993). Since the cross talk due to CreC and acetyl phosphate was observed only in the special mutants and since the functions of PhoR are dominant in vivo in the wild-type strains, the cross talk may not be physiologically significant. The noise due to this

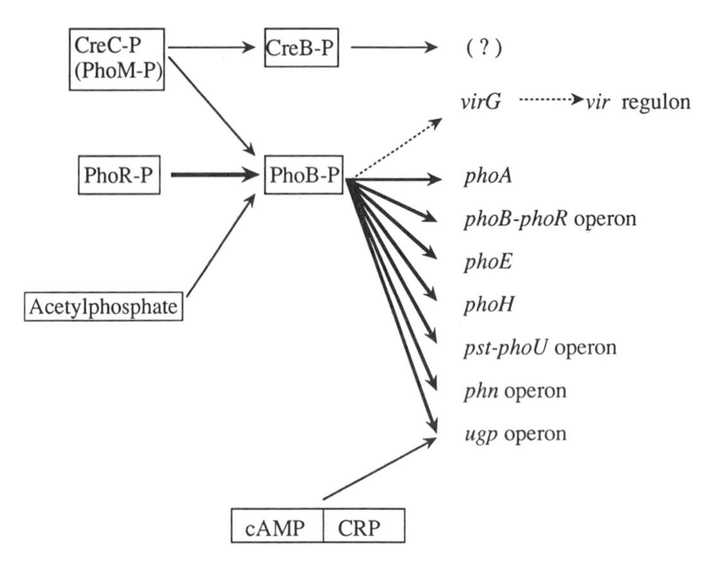

FIG. 1. Cross talk in the signal transduction related to the *pho* regulon. The physiological (or major) pathways of the signal transduction of the *pho* regulon are shown by bold arrows. In the wild-type strain, PhoR-P *trans*-phosphorylates PhoB and PhoB-P in turn activates transcription of the genes in the *pho* regulon. In *A. tumefaciens,* PhoB-P activates the transcription of the *virG* gene under phosphate limitation and the elevated level of VirG activates the transcription of the genes in the *vir* regulon. The *ugp* operon contains two kinds of promoters, one activated by PhoB-P and the other activated by the CRP-cAMP complex.

cross talk should be erased by more powerful straight talks by PhoR.

The *vir* regulon of *Agrobacterium tumefaciens,* which is involved in the transfer of the bacterial DNA into plant cells and is induced by phenolic compounds released from wounded plant cells, is also induced by phosphate starvation (Winans et al., 1988). This cross-regulation is mediated by transcriptional activation of the response regulator gene, *virG*, by PhoB (Aoyama et al., 1991). VirG thus induced activates the transcription of the genes in the *vir* regulon. Since this cross-regulation is observed in the wild-type strain, it should reflect the ability of the normal physiology of the cell to adapt to the conditions of phosphate limitation.

The *ugp* operon is regulated physiologically by phosphate starvation and carbon starvation and genetically by the *pho* system and the cyclic AMP (cAMP)-cAMP receptor protein (CRP) system (Kasahara et al., 1991). It contains the PhoB-binding sites (*pho* box) and a CRP-binding site in the promoter region and has two transcription initiation sites corresponding to the two promoters. PhoB functions as an activator for the *pho* promoter and as a repressor for the CRP-cAMP-regulator promoter, and CRP plays the opposite roles for these promoters. These results suggest that the products of the *ugp* operon are useful for the adaptation of the cell to the conditions not only of phosphate limitation but also of carbon source limitation. Many other examples of cross-regulation at the transcriptional level are already known, and we will find more promoters whose expression is regulated by two or more promoters to respond to different physiological signals as we develop a greater understanding of the *E. coli* genome and physiology.

Membrane Topology of PhoR

The membrane topology of PhoR has been studied by constructing *phoR-phoA* fusions (Scholten and Tommassen, 1993) and by utilizing Tn*phoA* insertions (our unpublished data). The combined data suggest that PhoR possesses two transmembrane regions, one in the region of residues 10 to 50 and the other around residue 70 (Fig. 2). This is consistent with the previous findings that the native PhoR is anchored to the cytoplasmic membrane and that the truncated PhoR lacking the amino-terminal 83 residues is localized in the cytoplasm (Yamada et al., 1990). The protease sensitivity assay of PhoR suggests that it contains only a small area, if any, which is exposed to the periplasmic space (Scholten and Tommassen, 1993). The region flanked by the two transmembrane regions appears to be located in the cytoplasm, since the two *phoR'-'phoA* fusions (at positions 54 and 66) showed very low activities of alkaline phosphatase. These results are consistent with the model that PhoR interacts with PhoU in mediating signal transduction to PhoB and is not involved in directly sensing the external phosphate

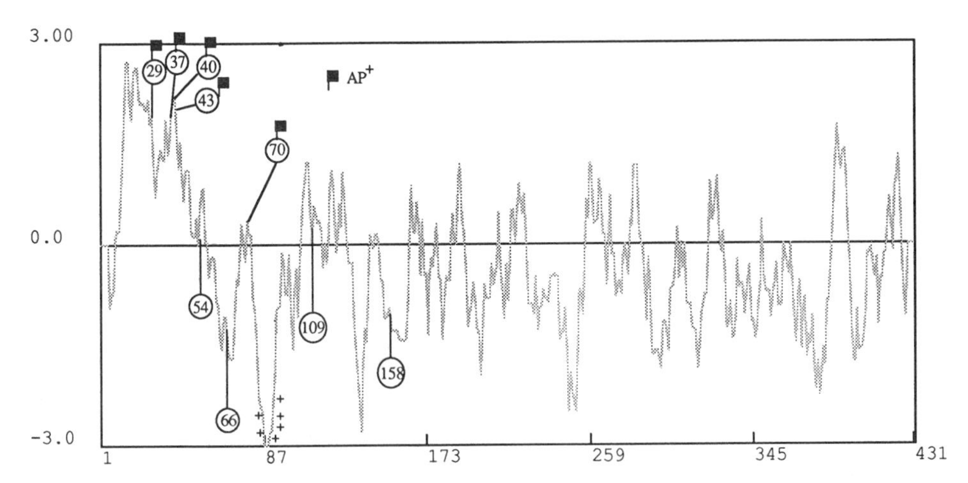

FIG. 2. Hydropathy profile and membrane topology of PhoR as studied by the PhoR-PhoA fusions. The circled numbers indicate the Tn*phoA* insertion sites in *phoR* or the *phoR* fusion sites in *phoR-phoA* fusion genes as expressed in amino acids from the N terminus of PhoR. The insertions with the flags indicate the alkaline phosphatase-positive insertions (AP+) or fusions, which indicates that the sites are exposed to or located in the membrane close to the side of the periplasm (Manoil and Beckwith, 1986). The data except for 43 and 70 (unpublished) were taken from Scholten and Tommassen (1993). +, positively charged residues.

signal in the periplasm. The transmembrane regions may be involved in positioning PhoR for efficient physical interaction with PhoU, which may be associated with the Pst proteins in the membrane and regulate the kinase-phosphatase conversion by reversibly interacting with PhoR. Although the second transmembrane region is not apparent from the hydrophilicity profile, PhoR probably also possesses two transmembrane regions, as do many other histidine protein kinases.

Mutational Analysis of PhoR Functions

To elucidate the mechanisms of the functional interconversion between the protein kinase and phosphatase and the functional domains for the two activities of PhoR, which is composed of 431 amino acids, various types of *phoR* mutants have been isolated and analyzed by us and by others. They are summarized in Fig. 3.

PhoR1084 and PhoR1159 proteins, which lack the 83 and 158 N-terminal amino acids, respectively, behaved as weakly constitutive activators of PhoB in vivo, whereas PhoR1263, lacking the 262 N-terminal amino acids including a putative autophosphorylation site, His-213, was deficient in both functions (Fig. 3B) (Yamada et al., 1990). Most of the classical null-type *phoR* mutants were nonsense mutants (Yamada et al., 1989) and therefore should encode the truncated PhoR proteins, as shown in Fig. 3C. All the internally truncated PhoR proteins with region from 110 to 174 deleted were constitutive kinases (Fig. 3D) (Scholten and Tommassen, 1993).

Three kinds of *phoR* mutants were isolated by treating the cloned *phoR* gene with hydroxylamine (Fig. 3A). Mutants locked in the activator form (constitutive kinase) were isolated as PhoA (alkaline phosphatase) constitutive in a *phoR creC* strain and had the following alterations in PhoR; R204W, T217M, P218S, T220N, and E371K. Mutants locked in the inactivator form (constitutive phosphatase) were isolated as $PhoA^-$ clones in a *phoR* $creC^+$ strain and had an H213Y alteration. Three independent mutants had the same mutation. Since this histidine is best conserved among the histidine kinases, and since PhoR has an autophosphorylation site at the N-3 position of His (Makino et al., 1992), we think that His-213 is the site of autophosphorylation in PhoR. The *phoR* (H213Y) mutant plasmid completely suppressed *phoA* expression in the wild-type strain and in the *phoR creC* strain even under phosphate limitation. This result shows that the phosphatase activity is separable from the kinase activity and that His-213 is completely dispensable for the phosphatase activity.

When we truncated the N-terminal coding region (83 amino acids) from these PhoR mutants, all of them except the G388E mutant retained essentially the same initial properties. The *phoR* (G388E) mutant, which was a null mutant in its full size, became a dominant inactivator upon its truncation of the N-terminal 82 amino acids. Like the H213Y protein, the purified truncated G388E protein has no autophosphorylation activity but possesses a high phosphatase activity. This result suggests that the C-terminal region of PhoR around G388 functionally interacts with the very

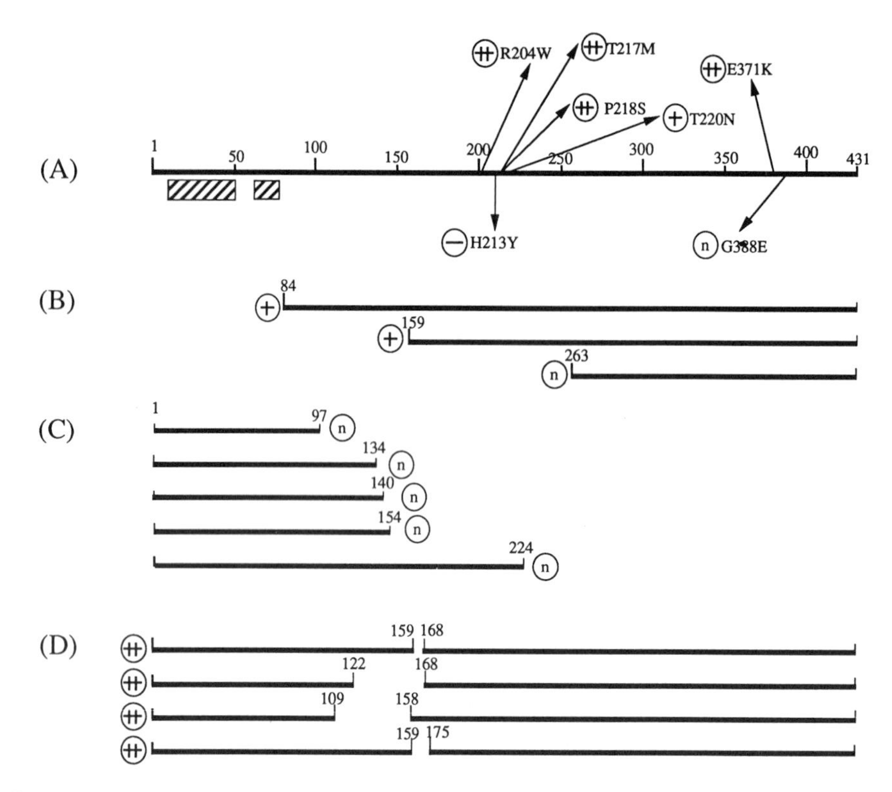

FIG. 3. Mutational analysis of the PhoR functions. (A) Point mutations. The shadowed boxes indicate the transmembrane regions of PhoR. (B) *phoR* mutants encoding N-terminally truncated PhoR proteins (Yamada et al., 1990). (C) Nonsense *phoR* mutants whose coding regions are terminated by the nonsense codons at the indicated positions (Yamada et al., 1989). (D) *phoR* mutants which encode the PhoR proteins with internal deletions (Scholten and Tommassen, 1993). Symbols: ++, mutants locked in the activator (kinase) form, high level; +, mutants locked in the activator (kinase) form, medium level; n, null mutants; −, mutants locked in the inactivator (phosphatase) form.

N-terminal region for the regulation of the phosphatase activity. Since the two types of mutations were generally mapped in close proximity and intermixed locally, the two activities should not constitute separate domains and the functional elements should be intimately interacting with each other for the reversible functional changes. The central and C-terminal regions, where the clusters of point mutations were obtained, are most highly conserved among the histidine kinase family and appear to constitute two functionally important domains.

In general, it appears that the PhoR deletions from the N terminus have lower constitutive activities than do the mutants with internal deletions (Yamada et al., 1990; Scholten and Tommassen, 1993). The N-terminally truncated PhoR mutants also retain some phosphatase activity. Therefore, the story may not be as simple as that the N-terminal one-third suppresses the activator function (kinase) of the rest of the PhoR molecule. There may exist many delicate interactions involving several functional motifs in PhoR to regulate the kinase and phosphatase activities.

This work was supported by Grants-in-Aid for Scientific Research from the Ministry of Education, Science, and Culture of Japan.

REFERENCES

Amemura, M., K. Makino, H. Shinagawa, and A. Nakata. 1990. Cross talk to the phosphate regulon of *Escherichia coli* by PhoM protein: PhoM is a histidine protein kinase and catalizes phosphorylation of PhoB and PhoM-open reading frame 2. *J. Bacteriol.* **172:**6300–6307.

Aoyama, T., M. Takanami, K. Makino, and A. Oka. 1991. Cross-talk between the virulence and phosphate regulons of *Agrobacterium tumefaciens* caused by unusual interaction of transcriptional activator with a regulatory DNA element. *Mol. Gen. Genet.* **227:**385–390.

Feng, J., M. R. Atkinson, W. McCleary, J. B. Stock, B. L. Wanner, and A. J. Ninfa. 1992. Role of phosphorylated metabolic intermediates in the regulation of glutamine synthetase synthesis in *Escherichia coli. J. Bacteriol.* **174:**6061–6070.

Kasahara, M., K. Makino, M. Amemura, A. Nakata, and H. Shinagawa. 1991. Dual regulation of the *ugp* operon by phosphate and carbon starvation at two interspaced promoters. *J. Bacteriol.* **173:**549–558.

Lee, T.-Y., K. Makino, H. Shinagawa, and A. Nakata. 1990. Overproduction of acetate kinase activates the phosphate regulon in the absence of the *phoR* and *phoM* functions in *Escherichia coli. J. Bacteriol.* **172:**2245–2249.

Makino, K., M. Amemura, S. Kim, H. Shinagawa, and A. Nakata. 1992. Signal transduction of the phosphate regulon in *Escherichia coli* mediated by phosphorylation, p. 191–200. *In* S. Papa, A. Azzi, and J. M. Tager (ed.), *Adenine Nucleotides in Cellular Energy Transfer and Signal Transduction.* Birkäuser Verlag, Basel.

Manoil, C., and J. Beckwith. 1986. A genetic approach to analyzing membrane protein topology. *Science* **233:**1403–1408.

McCleary, W. R., J. B. Stock, and A. J. Ninfa. 1993. Is acetyl phosphate a global signal in *Escherichia coli? J. Bacteriol.* **175:**2793–2798.

Scholten, M., and J. Tommassen. 1993. Topology of the PhoR protein of *Escherichia coli* and functional analysis of internal deletion mutants. *Mol. Microbiol.* **8:**269–275.

Wanner, B. L., and P. Latterell. 1980. Mutants affected in alkaline phosphatase expression: evidence for multiple positive regulators of the phosphate regulon in *Escherichia coli. Genetics* **96:**353–366.

Wanner, B. L., and M. R. Wilmes-Riesenberg. 1992. Involvement of phosphotransacetylase, acetate kinase, and acetyl phosphate synthesis in control of the phosphate regulon in *Escherichia coli. J. Bacteriol.* **174:**2124–2130.

Winans, S. C., R. A. Kerstetter, and E. W. Nester. 1988. Transcriptional regulation of the *virA* and *virG* genes of *Agrobacterium tumefaciens. J. Bacteriol.* **170:**4047–4054.

Yamada, M., K. Makino, M. Amemura, H. Shinagawa, and A. Nakata. 1989. Regulation of the phosphate regulon of *Escherichia coli:* analysis of mutant *phoB* and *phoR* genes causing different phenotypes. *J. Bacteriol.* **171:**5601-5606.

Yamada, M., K. Makino, H. Shinagawa, and A. Nakata. 1990. Regulation of the phosphate regulon of *Escherichia coli:* properties of *phoR* deletion mutants and subcellular localization of PhoR protein. *Mol. Gen. Genet.* **220:**366-372.

Chapter 49

Role of Protein Phosphorylation in the Regulation of Aerobic Metabolism by the Arc System in *Escherichia coli*

E. C. C. LIN[1] AND SHIRO IUCHI[2]

Department of Microbiology and Molecular Genetics[1] and Department of Cell Biology,[2] Harvard Medical School, Boston, Massachusetts 02115

Escherichia coli, like many of its cousins, prefers respiratory modes of metabolism (which generate proton motive force with the help of exogenous electron acceptors) to the less energy-efficient fermentative metabolism (which generates ATP at substrate-level phosphorylation with help of dismutation reactions). Although it is still not known how the cell budgets carbohydrates for energy generation and for structural material, an overall picture is beginning to emerge of how the cell switches its metabolism from the fermentative mode to a respiratory mode and how the cell chooses from among available exogenous electron acceptors the one that has the highest oxidative power. In this hierarchy, O_2 (midpoint redox potential, +0.82 V) is dominant over other known acceptors such as nitrate, trimethylamine *N*-oxide, dimethylsulfoxide, and fumarate (Stewart, 1988; Gunsalus, 1992; Iuchi and Lin, 1993). The two-component Arc system plays a critical role in the preferential utilization of O_2.

Discovery of the *arcA* and *arcB* Regulatory Genes and Target Operons of the Regulatory Proteins

It has long been known that the tricarboxylic acid cycle, operating in the same fashion as in mitochondria, functions amply in *E. coli* only during aerobic growth (Hirsch et al., 1963; Amarasingham and Davis, 1965). In particular, it was shown that the cellular activity level of succinate dehydrogenase (encoded by the *sdh* operon) is higher in aerobically grown than anaerobically grown cells. Subsequent development of molecular genetic techniques encouraged us to determine whether a global control exists for adaptation to aerobiosis. To this end we constructed an sdh^+ ϕ(*sdh-lac*) merodiploid strain which has both *sdh* and its fusion gene linked to the β-galactosidase structural gene on the chromosome; we used this strain as the starting strain for a mutant search. Use of the merodiploid is important for several reasons. First, the ϕ(*sdh-lac*) fusion will allow the selection of mutants for enhanced anaerobic growth on lactose. Second, maintenance of the sdh^+ allele avoids perturbation of the network of aerobic metabolism. Third, a change in the behavior of the *sdh* operon will reveal a *trans*-acting regulatory mutation. To isolate mutants, we spread several hundred cells on MacConkey-lactose agar and incubated them anaerobically for 5 days. Red papillae (indicating emergence of mutants with increased expression of the hybrid operon) growing from colonies were purified on the same selective agar. The mutants were then scored for increased activity of succinate dehydrogenase when grown under anaerobic conditions. Two classes of *trans*-acting pleiotropic mutations were identified: *arcA* (aerobic respiration control), located at min 0 (previously sequenced as the *dye* gene [Drury and Buxton, 1985]), and *arcB*, located at min 69.5 (Iuchi and Lin, 1988; Iuchi et al., 1989). Mutations in either gene resulted in elevated anaerobic activity levels of many other enzymes involved in aerobic metabolism, including numerous flavodehydrogenases, members of the tricarboxylic acid cycle and the glyoxylate shunt, a member(s) of the pathways for fatty acid degradation, and cytochrome *o* (Fig. 1).

Protein Components of the Arc System

The Arc system belongs to a signal transduction family comprising a sensor protein and a cognate regulatory protein (Ronson et al., 1987; Stock et al., 1989; Parkinson and Kofoid, 1992). ArcB, like most known sensors, is associated with the cytoplasmic membrane, whereas the cognate ArcA, like most known regulators, is cytosolic. The critical functional regions of the two polypeptides are schematized in Fig. 2. Sensor membrane proteins usually have, in the N-terminal region, two transmembrane segments separated by a periplasmic domain which acts as the signal receptor. The periplasmic segment in ArcB is estimated to be only 7 amino acids long and therefore is not likely to serve as a signal receptor. Furthermore,

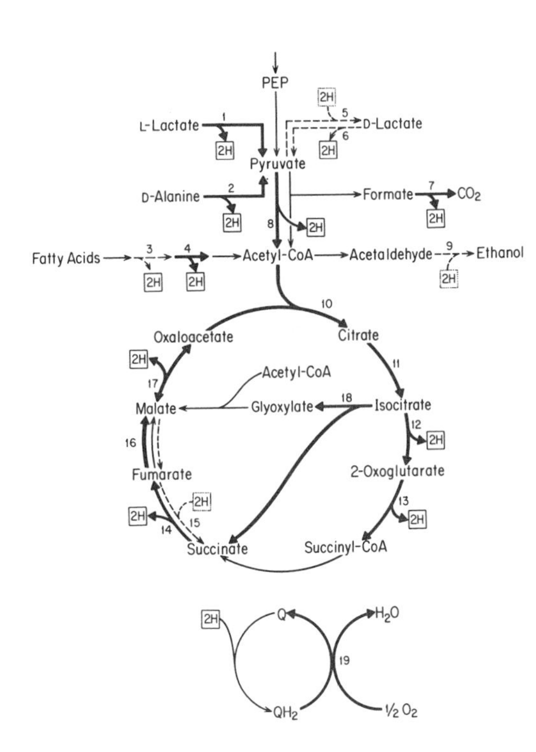

FIG. 1. Diagram of redox pathways. The numbers designate reactions catalyzed by the enzymes: 1, L-lactate dehydrogenase (flavoprotein); 2, D-amino acid dehydrogenase (flavoprotein); 3, acyl coenzyme A (CoA) dehydrogenase (flavoprotein); 4, 3-hydroxyacetyl-coenzyme A dehydrogenase (NAD^+ linked); 5, D-lactate:NAD^+ oxidoreductase; 6, D-lactate dehydrogenase (flavoprotein); 7, formate dehydrogenase (the FDH_O enzyme); 8, pyruvate dehydrogenase; 9, ethanol:NAD^+ oxidoreductase; 10, citrate synthase; 11, aconitase; 12, isocitrate dehydrogenase; 13, 2-oxoglutarate dehydrogenase; 14, succinate dehydrogenase; 15, fumarate reductase; 16, fumarase A (Bell et al., 1989); 17, malate dehydrogenase; 18, isocitrate lyase; 19, ubiquinol-1 oxidase. Reactions catalyzed by enzymes under *arc* control are represented by thick arrows. Reactions catalyzed by enzymes which may or may not be under *arc* control are represented by thin arrows. Reactions catalyzed by enzymes not under *arc* control are represented by dashed arrows. 2H boxed by solid lines represents reducing equivalents yielded by the reaction, and 2H boxed by dotted lines represents reducing equivalents consumed by the reaction (Iuchi et al., 1989; Iuchi and Lin, 1988). PEP, phosphoenolpyruvate.

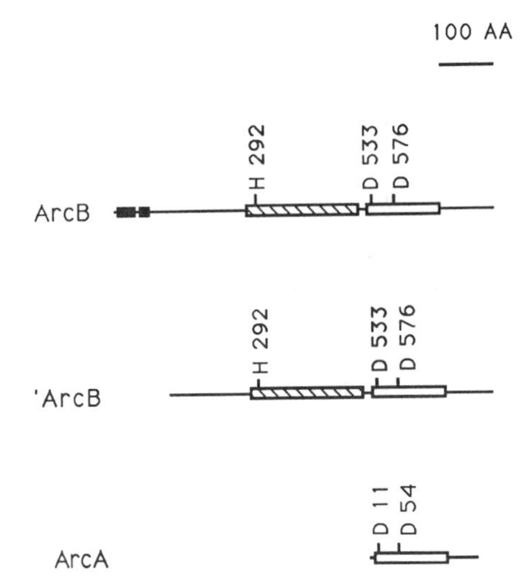

FIG. 2. Schematic representations of the sensor ArcB and the regulator ArcA. ArcB contains 778 amino acid (AA) residues (Iuchi et al., 1990b), in contrast to ArcA, which contains only 238 residues. 'ArcB indicates a truncated ArcB protein lacking its putative transmembrane segments. The gene *arcA* was formerly referred to as *dye* by Drury and Buxton (1985). Symbols: solid boxes, hydrophobic regions (putative membrane-spanning segments); hatched boxes, transmitter modules containing the canonical His residue capable of autophosphorylation; open boxes, receiver modules containing the conserved Asp as a phosphoryl group acceptor.

ArcB has in the C-terminal region an additional region that is highly homologous to the receiver module of ArcA (open box in Fig. 2). In contrast to ArcA, however, there is no helix-turn-helix motif in the C-terminal region of ArcB. The auxiliary ArcA-like region of ArcB is therefore not likely to be involved in DNA binding (Iuchi et al., 1990b).

Probing In Vivo Function of ArcB by Site-Directed Mutagenesis

In a typical two-component regulatory system, upon stimulation the sensor protein undergoes autophosphorylation at a conserved histidyl residue in the transmitter module. The protein then catalyzes the transfer of this phosphoryl group to a conserved aspartyl residue in the receiver module of the regulator. Thereupon the regulator protein becomes functionally active. Site-directed mutagenesis showed that the conserved His-292 of ArcB, as well as the conserved Asp-533 and Asp-576 (part of the canonical acidic pocket of the regulator), is essential for signal transduction (Iuchi and Lin, 1992a). In contrast, the only two Cys residues (at positions 180 and 240), which we thought might be involved in a sensing mechanism by –SH/S–S redox cycling, are not essential.

It is noteworthy that mutant cells synthesizing the truncated protein ArcB′ (whose translation stops at position 516 or 517, thereby deleting the region which is homologous to ArcA), still retains some regulatory function, although the range of control is highly reduced. It thus seems possible that the C-terminal portion of ArcB has a modulating effect on the rate of intermolecular transfer of the phosphoryl group from His-292 to the regulator ArcA (Asp-54 in the acidic pocket).

In Vitro Phosphorylation Experiments

Genetic truncation of ArcB from the N-terminal region (forming protein 'ArcB [Fig. 2]) that removed the two putative transmembrane segments freed the protein from the membrane (without causing loss of autophosphorylating and *trans*-phosphorylating activities. This fortunately allowed us to purify the soluble protein for in vitro experiments. Autophosphorylation of purified 'ArcB was almost instantaneous in the presence of ATP labeled with ^{32}P in the γ but not in the α position. When isolated [^{32}P]'ArcB was incubated at pH 12.5, the rate of dephosphorylation was biphasic (Fig. 3), indicating that the protein was autophosphorylated at two sites, most probably the conserved His and a conserved Asp. When limiting amounts of 'ArcB were incubated with excess ArcA and radioactively labeled ATP, the rate of acquisition of radioactivity by ArcA was proportional to the concentration of 'ArcB in the reaction mixture (Fig. 4). When the labeled [^{32}P]ArcA was incubated at pH 12.5, it showed the rapid monophasic loss of radioactivity expected of [^{32}P]Asp (Iuchi and Lin, 1992b).

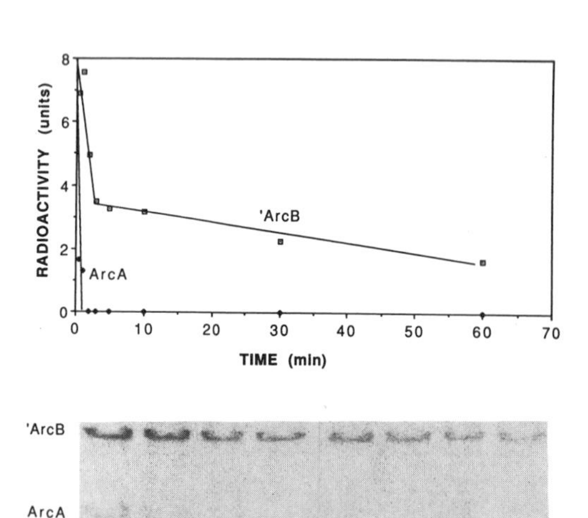

FIG. 3. Rate of alkaline hydrolysis of [^{32}P]'ArcB and [^{32}P]ArcA. The mixture was incubated at pH 12.5 for various periods at 43°C. (Top) The radioactivity retained in the proteins separated by sodium dodecyl sulfate-polyacrylamide gel electrophoresis (SDS-PAGE) (pH 8.8) was determined with a Phosphorimager. Units are expressed as intensity of the Phosphorimager signal in 1 min. (Bottom) The same data visualized by the Phosphorimager (Iuchi and Lin, 1992b).

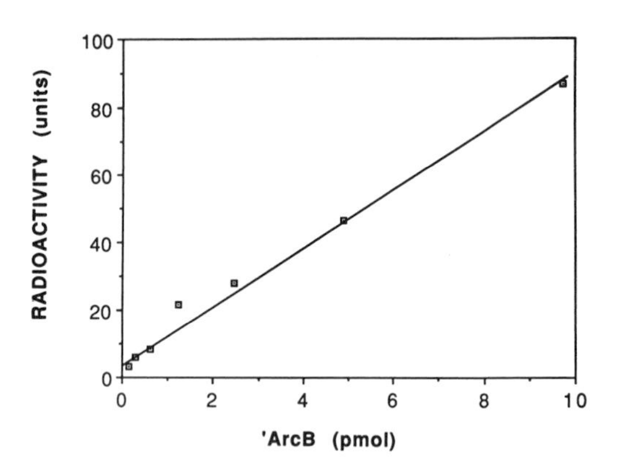

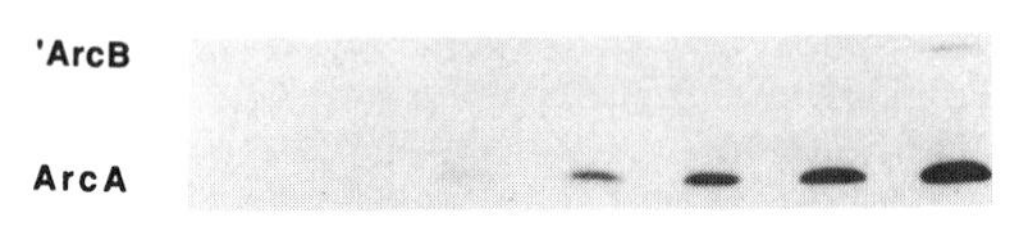

FIG. 4. Phosphorylation rate of ArcA as a function of 'ArcB. Reaction mixtures containing the same amount of ArcA (21 pmol) but different amounts of 'ArcB (0.15, 0.3, 0.61, 1.22, 2.44, 4.9, or 9.8 pmol) were each incubated for 10 min. The radioactivity acquired by ArcA was determined by a Phosphorimager SDS-PAGE process (top) or rendered visible by exposure to X-ray film (bottom).

Signals for ArcB

To test whether the Arc system responds to the concentration of O_2 itself in the growth medium, we compared the wild-type strain with a mutant

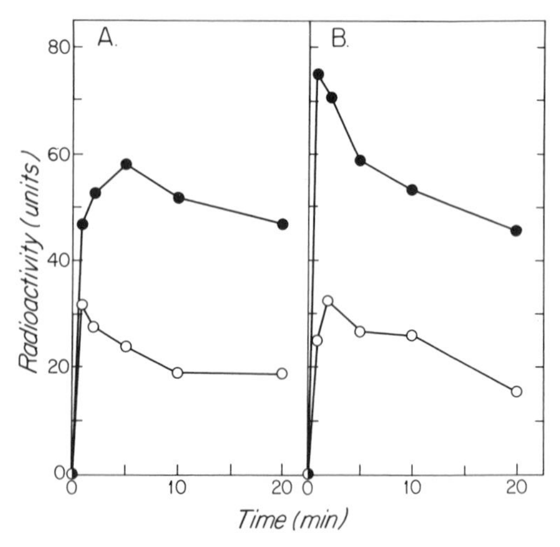

FIG. 5. Effect of D-lactate during the course of autophosphorylation of the sensor proteins. Purified 'ArcB (A) and everted vesicles containing wild-type ArcB protein (B) were phosphorylated in the presence and absence of 10 mM D-lactate.

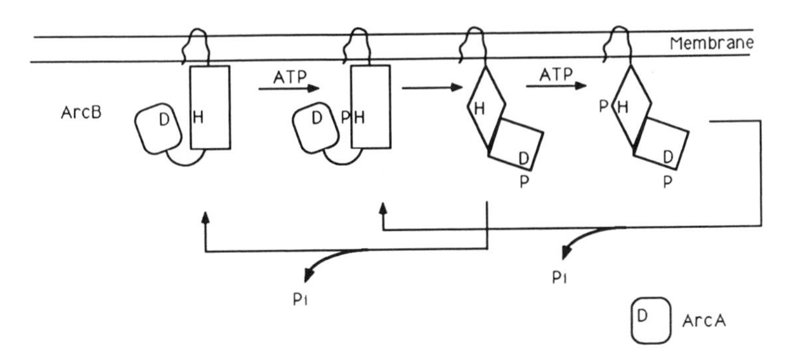

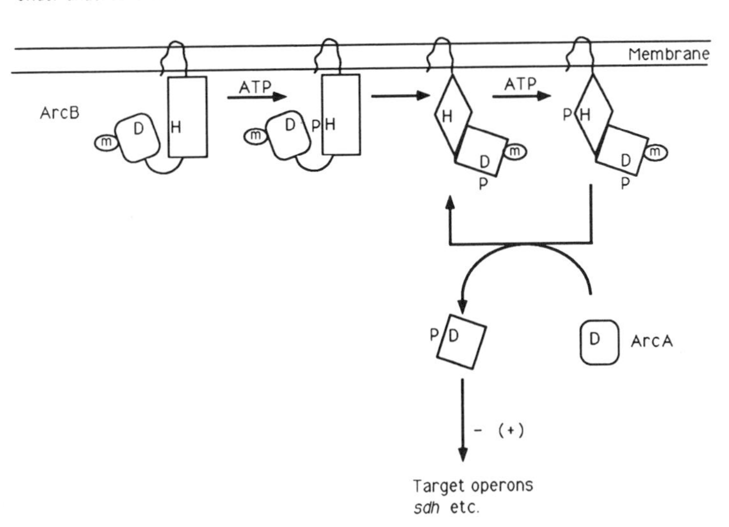

FIG. 6. Model for the signaling process by ArcB under aerobic and anaerobic conditions. H, D, P, and m indicate the conserved His-292 and the conserved Asp-576, amino acid residues, the transferred phosphoryl group, and cellular metabolites such as D-lactate and NADH, respectively. Arrows indicate reaction steps. + and − indicate a positive and negative regulatory effect, respectively. ArcB may form a dimer or oligomer, and the phosphoryl group may be transferred between the subunits (Iuchi, 1993).

doubly deleted in *cyo* and *cyd*. The *cyo* operon encodes the cytochrome *o* complex, and the *cyd* operon encodes the cytochrome *d* complex. Both enzyme complexes mediate terminal transfer of electrons from the transport chain to O_2, but the *d* complex has a higher affinity for O_2 than the *o* complex does (Anraku and Gennis, 1987). In a Δ*cyo* Δ*cyd* mutant, reporter target operons of the Arc system were expressed at approximately the same levels regardless of whether growth was aerobic or anaerobic. It therefore appears that ArcB responds not to molecular oxygen itself but to the redox conditions in the cell (Iuchi et al., 1990a).

An in vitro approach was then tried to characterize the signal(s). We noticed that after autophosphorylation of [^{32}P]′ArcB reached a peak, the protein gradually lost the labeled phosphate. A group of compounds, including lactose, glucose, galactose, glycerol, succinate, D-lactate, pyruvate, NADH, NAD, NADPH, NADP, flavin mononucleotide, and flavin adenine dinucleotide, were separately tested for ability to influence the course of net phosphorylation (Iuchi, 1993). D-Lactate, pyruvate, and NADH enhanced the autophosphorylation level, whereas the other compounds showed no effect. There was a 50% enhancement of phosphorylation by D-lactate at a concentration as low as 0.1 mM, and the stimulation reached a plateau at 1 mM. A control experiment showed that D-lactate had no effect on the low rates of autophosphorylation of several mutant ArcB proteins, one altogether deprived of the receiver module on the C-terminal side, one with Asp-533 changed to Ala, and one with Asp-576 changed to Ala. When D-lactate was tested on the wild-type ArcB in the membrane of everted vesicles, a pattern of stimulation similar to that of purified 'ArcB was observed (Fig. 5).

Working Model for Signal Transduction by the Arc System

A model consistent with all available data on the Arc system is illustrated in Fig. 6. The key observations supporting the model are as follows: (i) 'ArcB without the receiver module could still phosphorylate ArcA, albeit at a greatly diminished rate, and (ii) substitution of the critical Asp residues in the ArcB receiver module resulted in almost complete shutdown of ArcA phosphorylation.

According to the working model, when typical operons under the control of ArcA are highly expressed under aerobic conditions, most of the sensor ArcB is in the quiescent "closed" form with the receiver module (homolog of ArcA) docked on the transmitter module. Occasionally the His-292 in the transmitter module undergoes autophosphorylation at the expense of ATP. This phosphoryl group is rapidly transferred to the conserved Asp-576 in the receiver module, whereupon ArcB assumes an "open" conformation. The phosphoryl group at Asp-576 may be rapidly hydrolyzed, causing the dephosphorylated protein to resume the closed form. In this case, the reaction cycle would have to start again. On the other hand, the His-292 might undergo a second round of autophosphorylation before the Asp-576 loses its phosphoryl group. In this case, either of the following two events might ensue: (i) the Asp-576 might be dephosphorylated, and the freed Asp might be immediately rephosphorylated at the expense of His-292; or (ii) the phosphoryl group at His-292 might be donated intermolecularly to an ArcA (this scenario is less likely). In sum, under aerobic conditions, His-292 autophosphorylation probably operates as futile cycles, probably at low rates (Iuchi, 1993).

Under anaerobic conditions, signal metabolites (e.g., D-lactate or NADH) accumulate and bind to the receiver module of ArcB. When the sensor is in the altered form, ArcB(m), a different course of reactions is more likely to take place. Although autophosphorylation at His-292 is still followed by intramolecular transfer of the phosphoryl group to the Asp, the half-life of the Asp-P is lengthened by the binding of "m" to the receiver module (Iuchi, 1993). As long as the Asp of the receiver is occupied by the phosphoryl group, the ArcB protein remains in the open form. A second round of His-292 autophosphorylation can then occur, and now, because of the lengthened half-life of the Asp-P, intermolecular transfer of the phosphoryl group to ArcA becomes more likely. The product Arc-P then becomes a repressor or an activator of target gene expression.

Work in this laboratory was supported by Public Health Service grants RO1-GM40993 and RO1-GM11983.

REFERENCES

Amarasingham, C. R., and B. J. Davis. 1965. Regulation of α-ketoglutarate dehydrogenase formation in *Escherichia coli. J. Biol. Chem.* **240:**3664–3668.

Anraku, Y., and R. B. Gennis. 1987. The aerobic respiratory chain of *Escherichia coli. Trends Biochem. Sci.* **12:**262–266.

Bell, P. J., S. C. Andrews, M. N. Sivak, and J. R. Guest. 1989. Nucleotide sequence of the FNR-regulated fumarase gene (*fumB*) of *Escherichia coli* K-12. *J. Bacteriol.* **171:**3494–3503.

Drury, L. S., and R. S. Buxton. 1985. DNA sequence analysis of the *dye* gene of *Escherichia coli* reveals amino acid homology between the Dye and OmpR proteins. *J. Biol. Chem.* **260:**4236–4242.

Gunsalus, R. P. 1992. Control of electron flow in *Escherichia coli:* coordinated transcription of respiratory pathway genes. *J. Bacteriol.* **174:**7069–7074.

Hirsch, C. A., M. Rasminsky, B. D. Davis, and E. C. C. Lin. 1963. A fumarate reductase in *Escherichia coli* distinct from succinate dehydrogenase. *J. Biol. Chem.* **238:**3770–3774.

Iuchi, S. 1993. Phosphorylation/dephosphorylation of the receiver module at the conserved aspartate residue controls transphosphorylation activity of histidine kinase in sensor protein ArcB of *Escherichia coli. J. Biol. Chem.* **263:**23972–23980.

Iuchi, S., D. C. Cameron, and E. C. C. Lin. 1989. A second global regulator gene (*arcB*) mediating repression of enzymes in aerobic pathways of *Escherichia coli. J. Bacteriol.* **171:**868–873.

Iuchi, S., V. Chepuri, H.-A. Fu, R. B. Gennis, and E. C. C. Lin. 1990a. Requirement for terminal cytochromes in generation of the aerobic signal for the arc regulatory system in *Escherichia coli:* study utilizing deletions and *lac* fusions of *cyo* and *cyd. J. Bacteriol.* **172:**6020–6025.

Iuchi, S., and E. C. C. Lin. 1988. *arcA* (*dye*), a global regulatory gene in *Escherichia coli* mediating repression of enzymes in aerobic pathways. *Proc. Natl. Acad. Sci. USA* **85:**1888–1892.

Iuchi, S., and E. C. C. Lin. 1992a. Mutational analysis of signal transduction by ArcB: a membrane sensor protein for anaerobic expression of operons involved in the central aerobic pathways in *Escherichia coli. J. Bacteriol.* **174:**3972–3980.

Iuchi, S., and E. C. C. Lin. 1992b. Purification and phosphorylation of the Arc regulatory components of *Escherichia coli. J. Bacteriol.* **174:**5617–5623.

Iuchi, S., and E. C. C. Lin. 1993. Adaptation of *Escherichia coli* to redox environments by gene expression. *Mol. Microbiol.* **9:**9–15.

Iuchi, S., Z. Matsuda, T. Fujiwara, and E. C. C. Lin.

1990b. The *arcB* gene of *Escherichia coli* encodes a sensor-regulator protein for anaerobic repression of the *arc* modulon. *Mol. Microbiol.* **4:**715–727.

Parkinson, J. S., and E. C. Kofoid. 1992. Communication modules in bacterial signaling proteins. *Annu. Rev. Genet.* **26:**71–112.

Ronson, C. W., B. T. Nixon, and F. M. Ausubel. 1987. Conserved domains in bacterial regulatory proteins that respond to environmental stimuli. *Cell* **49:**579–581.

Stewart, V. 1988. Nitrate respiration in relation to facultative metabolism in enterobacteria. *Microbiol. Rev.* **52:**190–232.

Stock, J., A. J. Ninfa, and A. M. Stock. 1989. Protein phosphorylation and regulation of adaptive responses in bacteria. *Microbiol. Rev.* **53:**450–490.

Chapter 50

Role of Histidine Protein Kinases and Response Regulators in Cell Division and Polar Morphogenesis in *Caulobacter crescentus*

AUSTIN NEWTON, GREGORY B. HECHT, TODD LANE, AND NORIKO OHTA

Department of Molecular Biology, Lewis Thomas Laboratory, Princeton University, Princeton, New Jersey 08544

Differentiation in *Caulobacter crescentus* results from the repeated asymmetric cell division of a nonmotile stalked cell to produce the same stalked cell plus a new, motile swarmer at the end of each cell division cycle. The swarmer cell must differentiate into a stalked cell before it can initiate chromosome replication and divide (Fig. 1A). Thus, one sequence of developmental events, including flagellum biosynthesis, bacteriophage ϕCbK receptor site formation, and pilus assembly, leads to formation of the swarmer cell, whereas a second sequence of events, including loss of motility, retraction of the pili, and stalk formation, leads to differentiation of the swarmer cell into a stalked cell (Fig. 1A) (Newton and Ohta, 1990; Shapiro, 1993).

This order of developmental events in *C. crescentus* is invariant and does not depend on cell-cell contact or environmental cues like those required in most other microbial developmental systems, e.g., myxobacteria, *Bacillus subtilis,* and *Anabaena* species. Instead, genetic evidence indicates that the cell division cycle acts as a clock to time many of the developmental events. We review evidence for this conclusion here. We also present recent molecular and biochemical results indicating that genes interconnecting the cell division and developmental pathways in *C. crescentus* encode histidine protein kinases and a response regulator protein that are members of an extensive signal transduction pathway. As discussed below, the involvement of these proteins in essential cell division functions distinguish them from members of the so-called two-component systems described previously (Parkinson and Kofoid, 1992; Stock et al., 1989b) and elsewhere in this volume.

Polar Morphogenesis Depends on Cell Division

The first indication that cell division was required for *Caulobacter* development was the observation that the β-lactam penicillin G, which at low concentrations blocks division to produce long, filamentous cells, also blocked motility and stalk formation (Terrana and Newton, 1976). We later noted similar developmental phenotypes in mutants that were temperature sensitive for an early cell division step (Fig. 1B) (Huguenel and Newton, 1982). An approach to the identification of genes connecting cell division and development was suggested by the phenotype of strains defective in the pleiotropic gene *pleC*. Conditional *pleC* mutants divide normally at the nonpermissive temperature but arrest at the same stage of polar morphogenesis as do division-blocked strains; i.e., they synthesize inactive flagella and are therefore nonmotile, and they do not form stalks (Fig. 1C) (Sommer and Newton, 1989).

Suppressors of Developmental Mutations Map to Cell Division Genes

The possibility that the *pleC* gene plays a key role in the cell cycle regulation of polar morphogenesis was supported by the isolation of cold-sensitive suppressors of a temperature-sensitive *pleC* mutation. These new mutant alleles suppressed the Mot$^-$ and bacteriophage ϕCbK-resistant phenotypes of the *pleC* mutation, and they were also defective in division at the low temperature; the suppressors were mapped to the three new cell division genes, *divJ*, *divK*, and *divL* (Sommer and Newton, 1991). The *divK* allele behaves as a classical bypass suppressor: it suppresses the motility phenotype of all *pleC* mutations examined and confers a severe division defect in the presence or absence of the *pleC* mutation. In contrast, mutations in *divJ* and *divL* suppress only a subset of *pleC* mutations, and the cell division defects of these alleles are much more severe in the original *pleC* mutant than in a *pleC*$^+$ background. The wild-type alleles of *divJ*, *divK*, and *divL*, which are dominant to the suppressor mutations for their respective cell division phenotypes, were cloned by genetic complementation from a pLAFR library of *C. crescentus* DNA (Ohta et al., 1992).

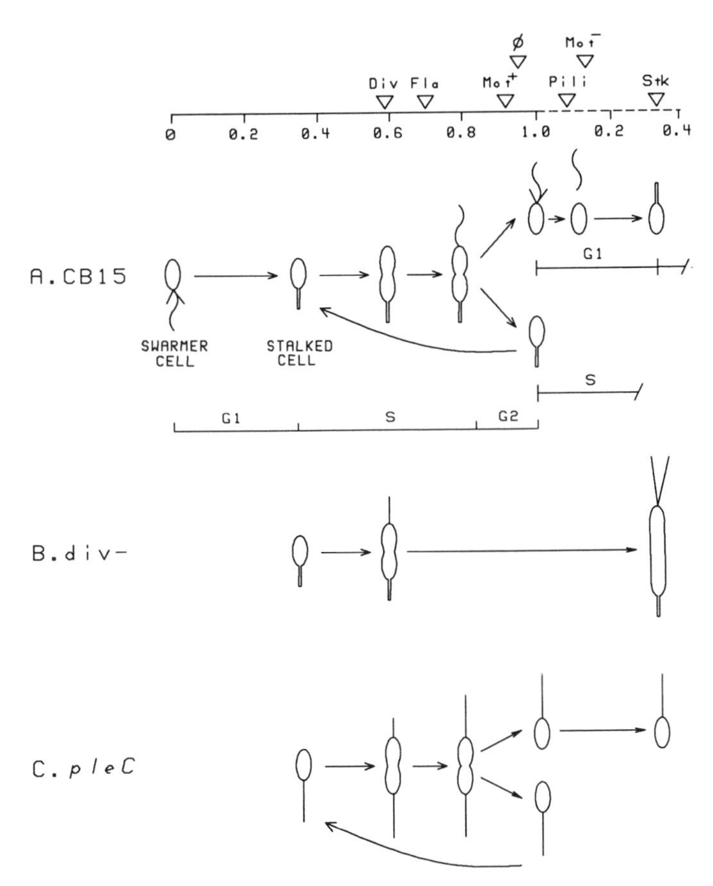

FIG. 1. *C. crescentus* cell cycle. (A) The sequence of developmental events in the wild-type strain CB15 includes flagellum formation (Fla) and activation of flagellum rotation (Mot+) during the DNA synthetic period (S), bacteriophage ϕCbK receptor formation (ϕ), and pilus formation (Pili), loss of motility (Mot−), and stalk formation (Stk) during the subsequent G_1 period of the daughter swarmer cell. (B) Cells blocked early in cell division (div−) assemble an inactive flagellum (designated by straight lines) and do not form a stalk at the cell pole. (C) *pleC* mutants also assemble inactive flagella, they are Mot− and bacteriophage ϕCbK resistant, and they fail to form a stalk, but they divide normally.

DivJ and PleC Are Histidine Protein Kinase Homologs

The pattern of *pleC* suppression by mutations in *divJ* discussed above and by Sommer and Newton (1991) initially suggested that the two genes could encode related functions involved in polar morphogenesis and cell division. Our results on the *divJ* gene (Ohta et al., 1992) and those of Wang et al. (1993) on the *pleC* gene have now shown that both genes encode histidine protein kinases and share 48% amino acid identity over a region of approximately 240 residues containing their kinase domains.

The predicted DivJ protein contains 596 amino acids with a molecular mass of 62 kDa. The C-terminal region of the protein (Ohta et al., 1992) has extensive homology with the histidine protein kinase domains of several proteins involved in bacterial signal transduction (reviewed by Parkinson and Kofoid [1992] and Stock et al. [1989b]). The conserved regions of DivJ include a histidine at position 337 (I); an asparagine at residue 456 (II); and the sequences Asp-Xaa-Gly-Xaa-Gly and Gly-Xaa-Gly (III), located 20 to 50 residues downstream of Asn-456. The N-terminal 150 residues of the protein are strongly hydrophobic; they could contain six membrane-spanning regions and are unlikely to include a periplasmic loop (Fig. 2) (Ohta et al., 1992).

The translated *pleC* gene product also contains a C-terminal histidine protein kinase domain and potential membrane-spanning sequences in the N-terminal region (Wang et al., 1993). PleC, like the DivJ protein discussed above, contains the three conserved regions (I to III) with invariant

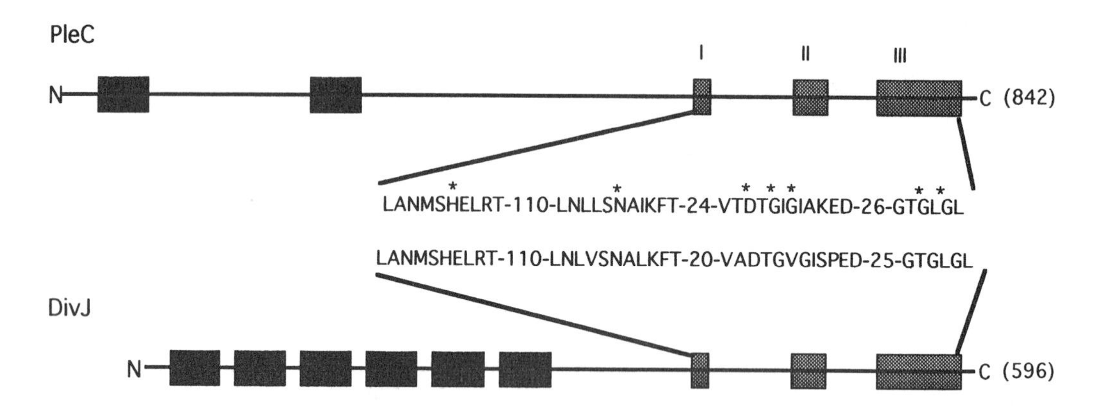

FIG. 2. Schematic sequence organization and alignment between putative transmembrane domains (solid boxes) and histidine protein kinase domains of the PleC and DivJ proteins. Residues marked with * are either invariant or highly conserved among the histidine protein kinases.

residues characteristic of this protein family (see Fig. 2). The majority of known histidine kinases are membrane-associated receptor proteins that usually have two membrane-spanning regions in the N terminus with a periplasmic loop; PleC is predicted to have such transmembrane regions at positions 29 to 50 and 283 to 303 with a periplasmic domain of approximately 230 residues.

The *divK* Gene Encodes an N-Terminal Response Regulator Domain

The predicted DivK protein contains 125 amino acids with 25 to 35% homology to the phosphoacceptor or regulatory domains of the bacterial response regulator proteins (Hecht and Newton, unpublished). In particular, the Glu-9, Asp-10, Asp-53, and Lys-105 residues of DivK correspond to the four typically invariant residues found in response regulators. In addition, DivK contains the hydrophobic sequence motifs conserved in most of these proteins, which are believed necessary for proper folding (Parkinson and Kofoid, 1992; Stock et al., 1989a, 1989b). Histidine protein kinases initiate signal transduction pathways by autophosphorylation of the conserved histidine residue, followed by a phosphotransfer reaction to one of the conserved aspartate residues (corresponding to Asp-53 of DivK) of a cognate response regulator. Although response regulators often function to control transcription of cognate target genes via their C-terminal domains in response to phosphorylation, DivK is unusual in this regard because, like the SpoOF protein of *Bacillus subtilis* and the CheY protein of *Escherichia coli*, it consists entirely of a single N-terminal domain. Thus, this response regulator could act as a phosphorelay protein like SpoOF (Burbulys et al., 1991) or in a manner similar to CheY, in which the response regulator acts to alter the activity of a target protein without carrying out a phosphotransfer to that target (Bourett et al., 1990; Parkinson et al., 1983; Smith et al., 1988; Yamaguchi et al., 1986).

Autophosphorylation and Phosphotransfer Activities of the PleC, DivJ, and DivK Proteins

We have initiated the biochemical characterization of the PleC, DivJ, and DivK proteins based on the sequence analysis discussed above. As a general strategy for the purification of cell division gene products, we first fused the *pleC*, *divJ*, and *divK* genes to the T7 promoter. We then overexpressed the fusion proteins in an *E. coli* strain harboring the T7 polymerase gene under the control of the *lac* promoter and purified the proteins to near homogeneity by column chromatography.

Because DivK was predicted to be a soluble protein from the sequence analysis (Hecht and Newton, unpublished), a DNA fragment containing the entire *divK* gene was fused to the initiation codon of the T7 gene 10 in the plasmid T7-7 (Tabor and Richardson, 1985). The resulting construct produced the intact wild-type DivK protein. A polypeptide with a molecular weight of ca. 15,000 was purified from the supernatant of the sonicated cell extract of *E. coli* grown in the presence of isopropyl-β-D-thiogalactopyranoside (IPTG).

Amino acid sequence analysis predicted both PleC (Wang et al., 1993) and DivJ (Ohta et al., 1992) to be proteins with N-terminal transmembrane domains (Fig. 3). Consequently, we constructed fusions of the C-terminal domains of PleC (PleC*) and DivJ (DivJ*) containing all of

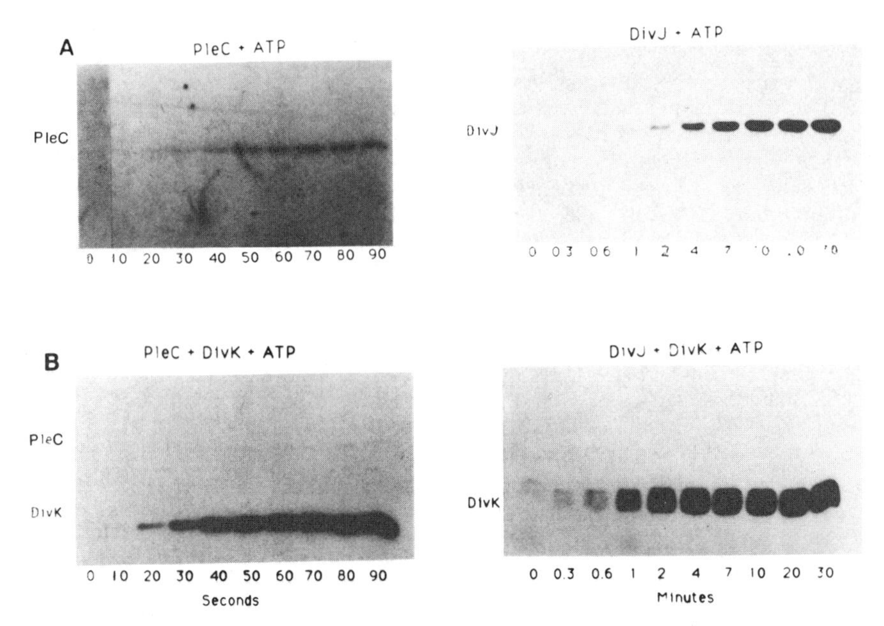

FIG. 3. Autophosphorylation of PleC* and DivJ* (A) and phosphotransfer to DivK (B). All reactions were carried out at room temperature in buffer containing 50 mM Tris·HCl (pH 7.5), 50 mM KCl, 5 mM $MgCl_2$, and 25 μM ATP. Aliquots (20 μl) of each reaction mixture were taken at the intervals indicated, and radiolabeled products were analyzed by sodium dodecyl sulfate-polyacrylamide gel electrophoresis and autoradiography. Each aliquot contained 0.4 pmol of [^{32}P]ATP and either 8 pmol of PleC* or 50 pmol of DivJ* with DivK added in a 10-fold molar excess (B) or without DivK (A). All autoradiographs were exposed for 30 min except for the one for PleC* autophosphorylation in panel A, which was exposed for 3.5 h.

the conserved sequences of typical histidine protein kinases (Parkinson and Kofoid, 1992; Stock et al., 1989b) in frame to a pRSET vector (Kroll et al., 1993) which encodes a polyhistidine nickel-binding domain followed by polylinker cloning sites downstream of a T7 promoter. When induced in the presence of IPTG, the chimeric PleC* protein was overexpressed and formed inclusion bodies. The protein was resolubilized from inclusion bodies in guanidine hydrochloride, purified on a nickel chelate column, and then renatured as described by McCleary and Zusman (1990). The chimeric DivJ* protein was purified in native form by the same procedure.

The purified C-terminal domains, PleC* and DivJ*, were shown to be autophosphorylated in the presence of ATP and Mg^{2+} (Fig. 2A). A LacZ-PleC fusion protein has been shown previously to be autophosphorylated (Wang et al., 1993).

Our data also show that both PleC* and DivJ* are active in a rapid phosphotransfer to the DivK protein, as shown in Fig. 3B. Although both PleC* and DivJ* were autophosphorylated, these activities differed significantly in the presence of DivK. The initial rate of DivK phosphorylation in the reaction with PleC* was 25- to 100-fold higher than the rate of PleC* autophosphorylation in the absence of DivK (Fig. 3: data not shown). This result suggests a synergy between PleC and the DivK protein. In contrast, addition of DivK to the reaction mixture with DivJ* increased the initial rate of phosphorylation only two- to threefold. Preliminary results also indicate that phospho-DivK is less stable in the presence of DivJ* than in the presence of PleC* (Lane et al., unpublished). At present, we can only speculate on whether the apparent phosphatase activities of DivJ* or PleC* reflect the activities of the full-length proteins.

Organization of PleC, DivJ, and DivK in Signal Transduction Pathways

Mutant alleles of *divK* suppress the motility defect of *pleC* mutations and confer a cell division defect. Several lines of evidence suggest that these two functions may be distinct and are executed at different times in the cell cycle. Genetic complementation has shown that all *divK* mutations examined to date are dominant for their ability to suppress the motility defect associated with *pleC* but that the same mutations are recessive for their cold-sensitive cell division phenotype (Hecht and Newton, unpublished). These observations suggest that the *divK* mutations isolated as suppressors of *pleC* simultaneously represent both gain of function with respect to motility and loss of function with respect to cell division. As diagrammed in Fig. 4, our current interpretation of these findings and those discussed below is that wild-type DivK is required for activation of flagellar rotation late in the cell cycle and for the regulation of cell division early in the cell cycle.

Results of both genetic and biochemical experiments indicate that DivK is the cognate response regulator for the PleC histidine protein kinase described by Wang et al. (1993) and that these two proteins are members of a signal transduction pathway controlling motility. First, bypass suppressors of *pleC* mutations map to the *divK* gene, as expected if the two genes were a cognate pair (Sommer and Newton, 1991; Hecht and Newton, unpublished). Second, phospho-PleC* acts as an efficient donor in phosphotransfer to the DivK protein; PleC* autophosphorylation is also strongly stimulated in the presence of DivK (Fig. 3B) (Lane et al., unpublished). Further support for a functional relationship between the PleC and DivK proteins is the timing of synthesis and function of the genes. As reported previously, the temperature-sensitive period for *pleC* activation of motility extends from the mid-S phase (DNA synthetic period) to the end of the cell cycle (Sommer and Newton, 1989), and recent studies of *divK* expression have shown that its transcription coincides with this same period of

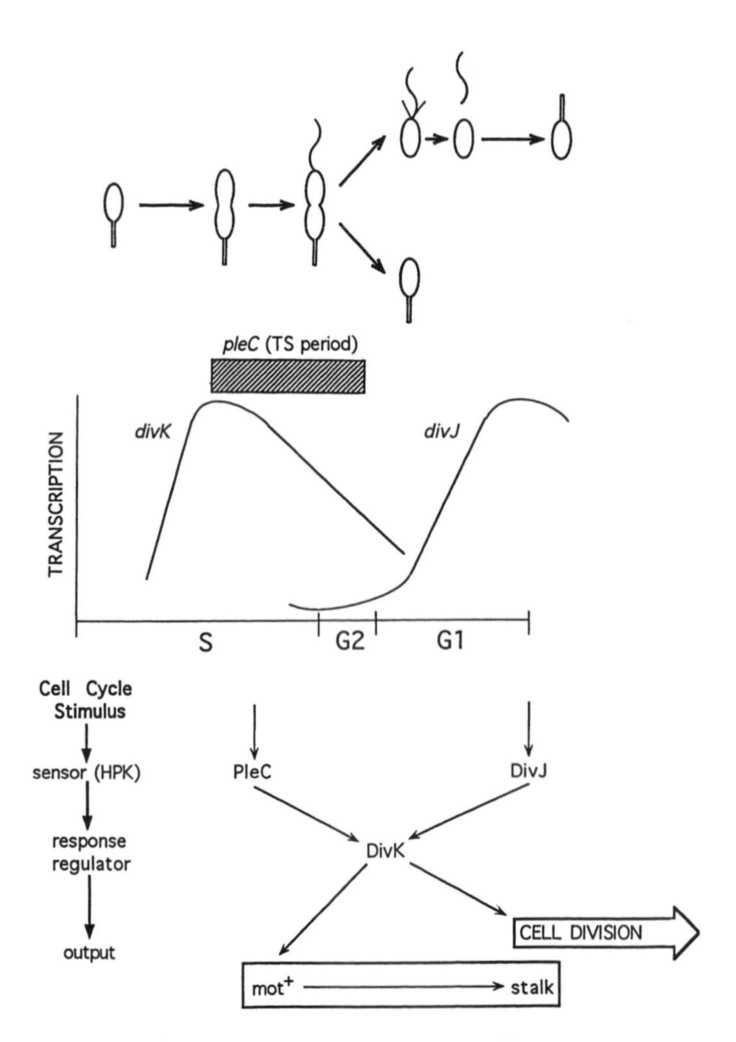

FIG. 4. Model for the organization of signal transduction pathways regulating polar morphogenesis and division in *C. crescentus*. Curves represent the times of *divJ* (Ohta et al., 1992) and *divK* (Hecht and Newton, unpublished) transcription, and the hatched box corresponds to the temperature-sensitive (TS) period for *pleC* function in turning on motility (Sommer and Newton, 1989). See the text for details. HPK, histidine protein kinase.

the cell cycle (Fig. 4) (Hecht and Newton, unpublished).

Our evidence that the DivJ histidine protein kinase and DivK response regulator function early in the cell cycle is less direct. Although we have not determined when DivK is required in the cell cycle for division, strains containing a cold-sensitive *divK* allele form long unpinched filaments at the nonpermissive temperature (Hecht et al., unpublished). This phenotype is consistent with a block early in cell division (Osley and Newton, 1977, 1980). We speculate that DivJ also functions early in the cell cycle, because conditional *divJ* mutations confer an unpinched filamentous phenotype when the strains are examined in a *pleC* background (Ohta et al., 1992; Sommer and Newton, 1991). Consistent with this idea is the observation that *divJ* is transcribed early in the cell cycle, beginning in swarmer cells, with expression peaking in the mid- to late G_1 period (Fig. 4) (Ohta et al., 1992). Thus, the requirements of DivJ and DivK for division appear to be restricted to an early stage of the cell cycle. We do not have direct genetic evidence, however, that the two proteins are a cognate pair, despite the observed phosphotransfer from phospho-DivJ* to DivK in vitro (Fig. 3).

Our working model in Fig. 4 proposes that PleC, DivJ, and DivK are members of signal transduction pathways that regulate polar morphogenesis and motility on the one hand and cell division on the other. Some of the outstanding questions about the role of these proteins are as follows. Does DivK in fact act at two different times in the cell cycle, and, if so, does its differential activity depend on alternative states or levels of phosphorylation? Are the kinase and/or phosphatase activities of the PleC and DivJ proteins restricted to different times in the cell cycle, as suggested in Fig. 4? Is DivK or another protein(s) the cognate response regulator of DivJ? What are the cell cycle signals that modulate the sensor proteins PleC and DivJ? Which proteins or genes are the targets of DivK regulation? It should be possible to address these questions by using many of the genetic and biochemical approaches employed originally to identify these signal transduction proteins.

This research was supported in part by American Cancer Society grant MV-386 and Public Health Service grant GM-22299 from the National Institutes of Health. Todd Lane was supported by postdoctoral fellowship 1 F32 AI08488-03 from the National Institutes of Health, and Gregory B. Hecht was supported by National Institutes of Health training grant T32 GM07388.

REFERENCES

Bourett, R. B., J. F. Hess, and M. I. Simon. 1990. Conserved aspartate residues and phosphorylation in signal transduction by the chemotaxis protein CheY. *Proc. Natl. Acad. Sci. USA* **87:**41–45.

Burbulys, D., K. A. Trach, and J. A. Hoch. 1991. Initiation of sporulation in *B. subtilis* is controlled by a multi-component phosphorelay. *Cell* **64:**545–552.

Hecht, G. B., T. Lane, and A. Newton. Unpublished data.

Hecht, G. B., and A. Newton. Unpublished data.

Huguenel, E. D., and A. Newton. 1982. Localization of surface structures during procaryotic differentiation: role of cell division in *Caulobacter crescentus*. *Differentiation* **21:**71–78.

Kroll, D. J., H. Abdel-Malek Abdel-Hafiz, T. Marcell, S. Simpson, C. Y. Chen, A. Gutierrez-Hartmann, J. W. Lustbader, and J. P. Hoeffler. 1993. A multifunctional prokaryotic protein expression system: overproduction, affinity purification, and selective detection. *DNA Cell Biol.* **12:**441–453.

Lane, T., N. Ohta, and A. Newton. Unpublished data.

McCleary, W. R., and D. R. Zusman. 1990. Purification and characterization of the *Myxococcus xanthus* FrzE protein shows that it has autophosphorylation activity. *J. Bacteriol.* **172:**6661–6668.

Newton, A., and N. Ohta. 1990. Regulation of the cell division cycle and differentiation in bacteria. *Annu. Rev. Microbiol.* **44:**689–719.

Ohta, N., T. Lane, E. G. Ninfa, J. M. Sommer, and A. Newton. 1992. A histidine protein kinase homologue required for regulation of bacterial cell division and differentiation. *Proc. Natl. Acad. Sci. USA* **89:**10297–10301.

Osley, M. A., and A. Newton. 1977. Mutational analysis of developmental control in *Caulobacter crescentus*. *Proc. Natl. Acad. Sci. USA* **74:**124–128.

Osley, M. A., and A. Newton. 1980. Temporal control of the cell cycle in *Caulobacter crescentus:* roles of DNA chain elongation and completion. *J. Mol. Biol.* **138:**109–128.

Parkinson, A. J., S. R. Parker, P. B. Talbert, and S. E. Houts. 1983. Interactions between chemotaxis genes and flagellar genes in *Escherichia coli*. *J. Bacteriol.* **155:**265–274.

Parkinson, J. S., and E. C. Kofoid. 1992. Communication modules in bacterial signaling proteins. *Annu. Rev. Genet.* **26:**71–112.

Shapiro, L. 1993. Protein localization and asymmetry in the bacterial cell. *Cell* **73:**841–855.

Smith, J. M., E. H. Roswell, J. Shiopi, and B. L. Taylor. 1988. Identification of a site of ATP requirement for signal processing in bacterial chemotaxis. *J. Bacteriol.* **170:**2689–2704.

Sommer, J. M., and A. Newton. 1989. Turning off flagellum rotation requires the pleiotropic gene *pleD*: *pleA*, *pleC*, and *pleD* define two morphogenic pathways in *Caulobacter crescentus*. *J. Bacteriol.* **171:**392–401.

Sommer, J. M., and A. Newton. 1991. Pseudoreversion analysis indicates a direct role of cell division genes in polar morphogenesis and differentiation in *Caulobacter crescentus*. *Genetics* **129:**623–630.

Stock, A. M., J. M. Mottonen, J. B. Stock, and C. E. Schutt. 1989a. Three-dimensional structure of CheY, the response regulator of bacterial chemotaxis. *Nature* (London) **337:**745–749.

Stock, J. B., A. J. Ninfa, and A. M. Stock. 1989b. Protein phosphorylation and regulation of adaptive responses in bacteria. *Microbiol. Rev.* **53:**450–490.

Tabor, S., and C. Richardson. 1985. A bacteriophage T7 RNA polymerase/promoter system for controlled exclusive expression of specific genes. *Proc. Natl. Acad. Sci. USA* **82:**1074–1078.

Terrana, B., and A. Newton. 1976. Requirement of cell division for stalk formation in *Caulobacter crescentus*. *J. Bacteriol.* **128:**456–462.

Wang, S. P., P. L. Sharma, P. V. Schoenlein, and B. Ely. 1993. A histidine protein kinase is involved in polar organelle development in *Caulobacter crescentus*. *Proc. Natl. Acad. Sci. USA* **90:**630–634.

Yamaguchi, S., S.-I. Aizawa, M. Kihara, M. Isomura, C. J. Jones, and R. M. Macnab. 1986. Genetic evidence for a switching and energy-transducing complex in the flagella motor of *Salmonella typhimurium*. *J. Bacteriol.* **168:**1172–1179.

Chapter 51

Regulation of Bacterial Nitrogen Assimilation by the Two-Component System NR_I (NtrC) and NR_{II} (NtrB)

EMMANUEL S. KAMBEROV, MARIETTE R. ATKINSON, ELIZABETH G. NINFA, JUNLI FENG, AND ALEXANDER J. NINFA

Department of Biological Chemistry, University of Michigan Medical School, 4310 Medical Science Building 1, 1301 E. Catherine, Ann Arbor, Michigan 48109-0606

In this chapter we discuss recent work on the mechanisms by which the two-component system consisting of the NR_I (NtrC) and NR_{II} (NtrB) proteins control the transcription of nitrogen-regulated genes in *Escherichia coli.* We first present a brief description of the signal transduction switch controlling nitrogen-regulated promoters in *E. coli,* as derived from work from several laboratories. We then review recent work from this laboratory that deals with particular aspects of this regulatory system: (i) the biochemical mechanism for nitrogen regulation independent of the NR_{II} protein, (ii) mutational analysis of the positive and negative regulatory functions of NR_{ii}, and (iii) the mechanism of the autophosphorylation reaction by which NR_{II} is phosphorylated at the expense of ATP. Finally, we discuss several unresolved issues regarding nitrogen regulation in *E. coli.*

A Bicyclic Cascade of Protein Modifications Is Used To Control the Transcription of Nitrogen-Regulated Genes in *E. coli*

The general features of the signal transduction switch that controls the transcription of *glnA*, encoding glutamine synthetase, and the covalent modification and regulation of the glutamine synthetase enzyme have been known for some time (Fig. 1) (for review, see Magasanik [1988], Rhee et al. [1985], and Stock et. al. [1989]). The intracellular nitrogen status of the cell is measured by the product of the *glnD* gene, which is an enzyme with uridylyltransferase (UT) and uridylyl-removing (UR) activities (reviewed by Rhee et al. [1985]). This protein measures the intracellular concentrations of glutamine and 2-ketoglutarate. The only known substrate of the UT and UR activities is the P_{II} protein, the product of *glnB*. Under conditions of nitrogen excess, the intracellular ratio of glutamine to 2-ketoglutarate is high; this results in stimulation of the UR activity and inhibition of the UT activity, and thus the P_{II} protein is present mostly in the unmodified (P_{II}) form. Conversely, under conditions of nitrogen limitation the intracellular ratio of glutamine to 2-ketoglutarate is low; this results in stimulation of the UT activity and inhibition of the UR activity such that the P_{II} protein is present mostly in the modified (P_{II}-UMP) form. The metabolic transformation of the P_{II} protein in response to the nitrogen status of the cell provides the intracellular signal for both the transcriptional regulation of the *glnA* gene and other nitrogen-regulated genes and the regulation of glutamine synthetase activity by covalent modification (reviewed by Magasanik [1988]).

The P_{II} protein regulates the transcription of nitrogen-regulated promoters indirectly by its effect on the activities of the NR_{II} protein, the product of *glnL* (*ntrB*). NR_{II} is a bifunctional positive and negative regulator of *glnA* transcription (Bueno et al., 1985; Ninfa and Magasanik, 1986; MacNeil et al., 1982). The role of NR_{II} is to control the activity of the NR_I protein, the product of *glnG* (*ntrC*), which is the transcription factor directly responsible for the regulation of nitrogen-regulated promoters such as the *glnA* promoter. NR_{II} and NR_I are homologous to many other bacterial regulatory proteins that act in a pairwise fashion to regulate various bacterial adaptive responses, referred to as the two-component systems (reviewed by Ninfa [1991], Stock et al. [1989], and Parkinson and Kofoid [1992]).

NR_{II} drives the transcription of the *glnA* gene by bringing about the phosphorylation of NR_I (Ninfa and Magasanik, 1986). This reaction proceeds as follows. NR_{II} binds ATP and is phosphorylated on a histidine residue (His-139) (Weiss and Magasanik, 1988; Ninfa and Bennett, 1991). The phosphoryl group is then transferred from the His-139 residue of NR_{II} to the Asp-54 residue of NR_I in a reaction that does not require ATP (Weiss and Magasanik, 1988; Sanders et al., 1992). This phosphotransfer is probably catalyzed by NR_I (Feng et al., 1992). The phosphorylation of NR_I

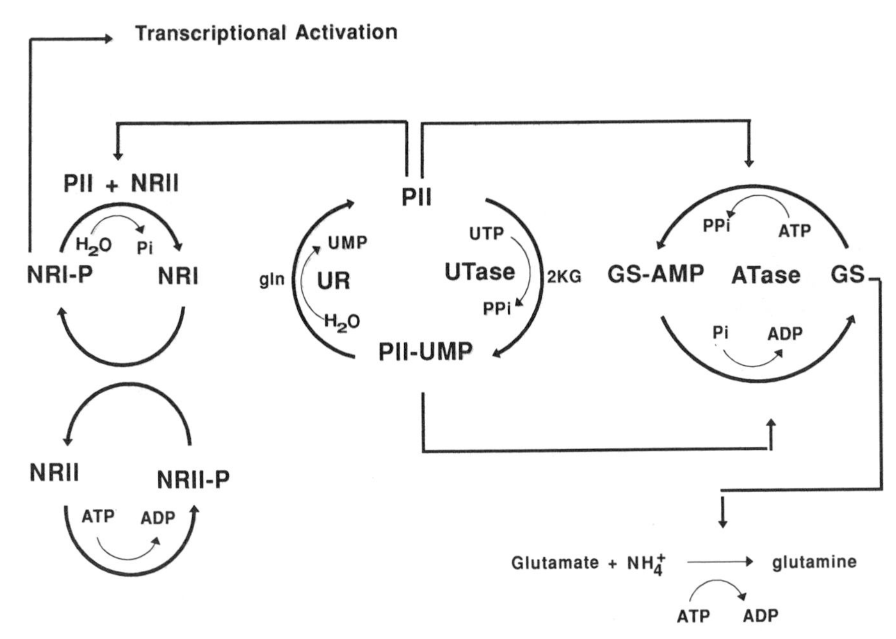

FIG. 1. Signal transduction system that regulates glutamine synthetase (GS) activity and the transcription of *glnA* and the Ntr regulon in *E. coli*. Abbreviations: ATase, adenylyltransferase (*glnE* product); GS, glutamine synthetase (*glnA* product); PII, *glnB* product; NRII, nitrogen regulator II (*glnL* [*ntrB*] product); NRI, nitrogen regulator I transcription factor (*glnG* [*ntrC*] product). Reprinted from Atkinson and Ninfa (1992) with permission.

results in its conversion to the form able to activate transcription of *glnA* and other nitrogen-regulated promoters (Ninfa and Magasanik, 1986). Little is known about the mechanism by which $NR_I \sim P$ activates transcription. However, it is evident that phosphorylation of NR_I stimulates an intrinsic ATPase activity of NR_I that is required for the activation of transcription (Weiss et al., 1991). The ATPase activity of $NR_I \sim P$ is required for the formation of the open transcription complex at the nitrogen-regulated *glnA* promoter (Popham et al., 1989). Several lines of evidence suggest that the ATPase activity of $NR_I \sim P$ and the transcriptional activation of *glnA* by $NR_I \sim P$ require the formation of a tetramer from the typically dimeric NR_I (for example, see Weiss et al. [1992] and Feng et al. [1992]). Thus, a hypothesis for the mechanism of regulation is that phosphorylation of NR_I results in tetramerization of the dimeric NR_I.

The phosphorylated form of NR_I is unstable, with a half-life at neutral pH of approximately 4 min (Keener and Kustu, 1988). This instability is apparently due to an "autophosphatase" activity of $NR_I \sim P$, since denatured $NR_I \sim P$ is stable for several hours. NR_{II} further destabilizes $NR_I \sim P$, and this reaction is greatly stimulated by P_{II} (Ninfa and Magasanik, 1986; Keener and Kustu, 1988; Kamberov et al., submitted). The combination of NR_{II} and P_{II} can be thought of as an "$NR_I \sim P$ phosphatase," although the mechanism of this reaction is unknown.

Dephosphorylation of $NR_I \sim P$ by the combination of P_{II} and NR_{II} results in the loss of its ability to activate transcription (Ninfa and Magasanik, 1986). Thus, a hypothesis for the regulation of *glnA* transcription is that NR_{II} brings about the formation of the activator ($NR_I \sim P$) under conditions of nitrogen limitation (at which P_{II} is mostly in the modified form) and that the combination of NR_{II} and P_{II} brings about the destruction of the activator under nitrogen-excess conditions (at which unmodified P_{II} is present). One tenet of this hypothesis is that the modified form of P_{II} (P_{II}-UMP) is innocuous with regard to NR_{II} (Bueno et al., 1985). Thus, the activation of transcription of *glnA* occurs when P_{II} is converted to the innocuous P_{II}-UMP, permitting the accumulation of the activator, $NR_I \sim P$.

Nitrogen Regulation in the Absence of the NR_{II} Protein

The model described above is derived from numerous studies involving genetic analysis and biochemical characterization of the activities of the regulatory proteins. In experiments with purified components, NR_I has virtually no capacity to activate transcription unless it is phosphorylated. However, in cells lacking NR_{II}, the transcription of *glnA* is regulated by the availability of nitrogen, although this is a slow and imprecise regulation (Bueno et al., 1985; Reitzer and Magasanik, 1985; Atkinson and Ninfa, 1992; Feng et al., 1992). The NR_{II}-independent regulation of *glnA* is due to the direct transfer of phosphoryl groups from acetyl phosphate to NR_I in an autophosphorylation reaction catalyzed by NR_I (Feng et al., 1992). This avenue of research was investigated because previous results with the related Pho regulon and chemotaxis system of bacteria suggested a role for acetyl phosphate in activating homologs of NR_I (Lukat et al., 1992; Wanner and Wilmes-Riesenberg, 1992). Experiments with purified components indicated that acetyl phosphate and certain other small phosphorylated compounds such as carbamyl phosphate and phosphoramidate could drive the activation of transcription by NR_I in the absence of NR_{II} (Feng et al., 1992). These compounds also activate the ATPase activity of NR_I in the absence of NR_{II}. The autophosphorylation of NR_I with acetyl-[^{32}P]phosphate serving as the phosphoryl group donor was directly demonstrated. Furthermore, we demonstrated in experiments with intact cells that the NR_{II}-independent activation of *glnA* required the accumulation of acetyl phosphate (Feng et al., 1992). Interestingly, in cells that contain NR_{II}, mutations affecting the synthesis of acetyl phosphate had no observable effect on the transcription of *glnA*.

Although hard data are lacking, a logical scenario by which the intracellular concentration of acetyl phosphate is altered in response to nitrogen limitation can be imagined. For example, the level of carbon that enters the tricarboxylic acid cycle may be higher under conditions of nitrogen excess as a result of the drawing off of 2-ketoglutarate in the reactions catalyzed by glutamate dehydrogenase and glutamate synthase. Several lines of evidence suggest that the rate of acetyl phosphate synthesis is limited primarily by the availability of the substrate, acetyl coenzyme A CoA (reviewed by McCleary et al. [1993]). One observation resulting from our studies and those of others (Wanner and Wilmes-Riesenberg, 1992) is that cells grown on pyruvate as the sole carbon source apparently have a very high intracellular concentration of acetyl phosphate. In cells lacking NR_{II} such growth results in elevated expression of *glnA* regardless of the availability of nitrogen, yet in cells containing NR_{II} the normal regulation of *glnA* expression is seen. These results again highlight the fact that NR_{II} is both a positive and negative regulator of *glnA* expression.

Structure-Function Analysis of NR_{II}

To identify mutations in *glnL* (which encodes NR_{II}) that affect the negative regulatory function, we selected and characterized spontaneous mutations in *glnL* that suppress a leaky mutation in the *glnD* gene encoding the UT/UR enzyme. The starting cells (*glnD99*::Tn*10*) were unable to efficiently uridylylate P_{II} in response to nitrogen limitation; consequently, they could not fully induce the expression of *glnA* and the Ntr regulon in response to nitrogen limitation. We selected suppressing mutations in *glnL* that restore elevated expression of *glnA* and the Ntr regulon (Atkinson and Ninfa, 1992). The location of these mutations in the *glnL* gene are shown in Fig. 2D. The mutations were found to map either in the N-terminal domain of NR_{II} or in a large cluster flanking the site of autophosphorylation (His-139, Fig. 2D). None of the mutations were found in the highly conserved residues shared by other kinases of two-component systems (Fig. 2A). Among the suppressing mutations, certain mutations seemed to completely eliminate the ability of NR_{II} to interact with P_{II} whereas other mutations appeared, from in vivo experiments, to result in altered NR_{II} proteins that are less sensitive to P_{II} but still able to interact with it when its concentration is very high. Mutations of the latter class actually restored nitrogen regulation in the presence of the *glnD99*::Tn*10* mutation, and this regulation was observed to depend on the presence of both P_{II} and the *glnD99*::Tn*10* mutation (Atkinson and Ninfa, 1992). Essentially, these mutations served to adjust the P_{II}-NR_{II} interaction to the altered levels of P_{II} modification found as a result of the *glnD99*::Tn*10* mutation. Whether these mutations affect the ability of NR_{II} to bind to P_{II} or alter the probability that P_{II}, once bound, can bring about the conversion of NR_{II} to the form that stimulates the dephosphorylation of $NR_I \sim P$ remains to be determined.

The kinases of the two-component systems share a conserved domain, usually located at the C-terminal end of the protein, and also possess unrelated domains. Within the conserved domain, three regions of very strong homology are seen (Fig. 2A). Region 1 contains the conserved histidine residue that is the site of autophosphorylation, region 2 contains a conserved asparagine residue, and region 3 consists of three glycine-rich segments separated by spacers of variable length. To directly ascertain the role of these conserved residues, we systematically altered the eight most highly conserved sites (Fig. 2B) and assessed the positive and negative regulatory activities of the altered NR_{II} proteins in intact cells. None of these highly conserved residues was required for the negative regulatory function of NR_{II}, but the positive regulatory function of NR_{II} was affected to various extents by the alterations (Atkinson and Ninfa, 1993). Alterations at the site of autophosphorylation, His-139, and at the adjacent residue, Glu-140, could result in altered proteins that are entirely lacking the positive regulatory function and are transdominant negative regulators of

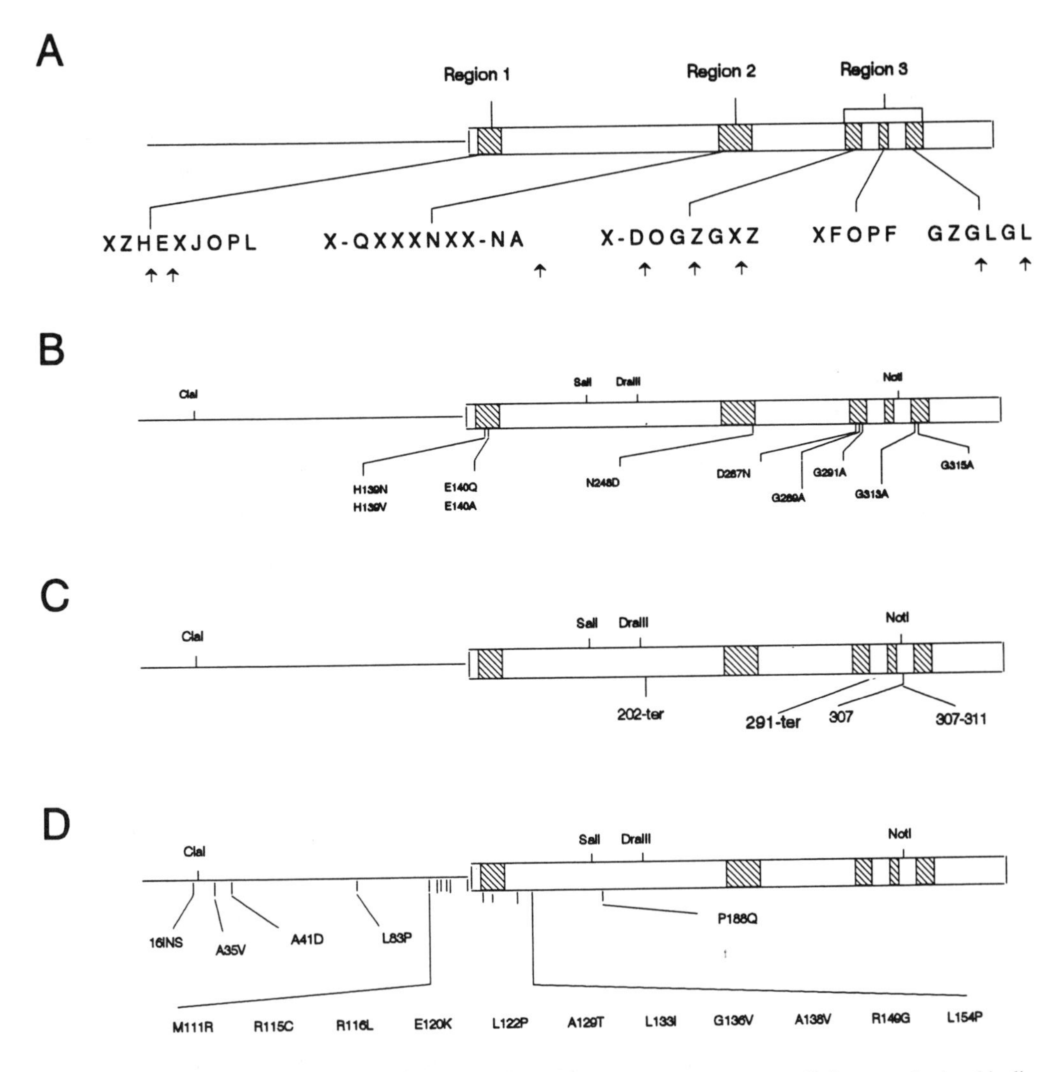

FIG. 2. Mutations in the *glnL* gene of *E. coli* affecting signal transduction. In all four panels the thin line represents the nonconserved N-terminal domain of NR_{II} and the thick line represents the conserved C-terminal kinase/phosphatase domain. (A) Location and consensus sequences of the three conserved regions shared by the related kinases of the two-component systems. The standard single-letter code is used with the following exceptions: X refers to positions where at least 50% have a nonpolar amino acid (I, L, M, V); Z refers to positions where at least 50% of the kinase family have a polar residue (A, G, P, S, T); J refers to positions where at least 50% of the kinase family have a basic amino acid (H, K, R); and O refers to positions where at least 50% of the kinase family have an acidic or amidic amino acid (D, N, E, Q). Positions with less than 50% conservation are indicated by dashes. Arrows indicate the sites where we constructed mutations. Adapted from Parkinson and Kofoid (1992) with permission. (B) Point mutations constructed by site-specific mutagenesis. (C) Mutations resulting from *Bal* 31 mutagenesis of *glnL*. (D) Mutations isolated as suppressors of *glnD99*::Tn*10*. For further discussion of panels B to D, see the text and Atkinson and Ninfa (1992, 1993).

glnA in cells that contain a wild-type copy of the *glnL* gene. Furthermore, a mutation in region 3 (G315A) also resulted in an altered protein that completely lacked the positive regulatory function. Other mutations in region 3 were affected more subtly in the positive regulatory function and retained the wild-type negative regulatory function. These results indicate that the positive and negative regulatory functions of NR_{II} can be genetically separated, as observed previously (MacNeil et al., 1982), and that the conserved residues shared by the other members of the family of proteins play a role only in the positive regulatory function, that is, the kinase activity.

We characterized biochemically several altered NR_{II} proteins (Ninfa et al., 1993, Kamberov et al., submitted). The H139N protein, in which the site of autophosphorylation has been altered, cannot be autophosphorylated or phosphorylate NR_I but brings about the dephosphorylation of $NR_I \sim P$ in the absence of P_{II}. This activity is stimulated approximately 10-fold by P_{II} (Kamberov et al., submitted). The fact that the dephosphorylation of $NR_I \sim P$ is catalyzed to some extent by the H139N protein even in the absence of P_{II} indicates that P_{II} is not directly involved in the catalysis but, rather, plays a strictly regulatory role. Thus, it seems that NR_{II} can spontaneously adopt the conformation necessary for the stimulation of the dephosphorylation of $NR_I \sim P$ and that the role of P_{II} might be to stabilize this conformation of NR_{II}, as suggested previously (Bueno et al., 1985).

A protein with an alteration in region 3, G313A, was also purified and characterized (Ninfa et al., 1993). In intact cells this protein has greatly diminished positive regulatory function but intact negative regulatory function (Atkinson and Ninfa, 1993). The purified protein is capable of autophosphorylation, but the rate of this reaction is very low (Ninfa et al., 1993). This result is due to diminished ATP binding, since the G313A protein was defective relative to the wild-type and H139N proteins in the ability to form a photoadduct with ATP in cross-linking experiments. Thus, the conserved region 3 of this family of proteins has a role in the binding of ATP.

The role of the P_{II} protein in nitrogen regulation was further investigated. We observed that, as in previous work (Ninfa and Magasanik, 1986), P_{II} stimulates the dephosphorylation of $NR_I \sim P$ in a reaction that requires NR_{II}. Furthermore, P_{II} does not stimulate the dephosphorylation of $NR_I \sim P$ when a mutant form of NR_{II} insensitve to P_{II}, NR_{II}2302, is used in place of wild-type NR_{II} (Kamberov et al., submitted). These experiments confirmed the earlier observation that P_{II} is not itself an $NR_I \sim P$ phosphatase but, rather, acts in concert with NR_{II} to bring about the dephosphorylation of $NR_I \sim P$. We examined the effect of P_{II} on the autophosphorylation of NR_{II} and observed that this reaction was not appreciably affected by P_{II}, nor did P_{II} appreciably affect the transfer of phosphoryl groups from $NR_I \sim P$ to NR_I (Kamberov et al., submitted). The presence of P_{II} in reaction mixtures containing NR_{II} and ATP seems to affect only the stability of $NR_I \sim P$ (Kamberov et al., submitted).

Mechanism of the Autophosphorylation Reaction

Work with the CheA kinase revealed that certain pairs of CheA mutants could complement one another in *recA* cells to restore chemotaxis (Smith and Parkinson, 1980). The observation of intramolecular complementation suggested that different subunits of CheA could interact to restore signaling. Since it is known that signaling in this case is due to the autophosphorylation of CheA, this indicates that the autophosphorylation reaction can occur when two mutant subunits, each defective in this activity, are simultaneously present in intact cells. We performed the analogous experiment with mutants with various *glnL* mutations and observed that certain *glnL* mutations could be complemented by other *glnL* mutations in *recA* cells (Atkinson and Ninfa, 1993). In particular, we were able to observe intramolecular complementation between mutations affecting the conserved region 3 of NR_{II} (involved in nucleotide binding) and a mutation (H139V) affecting the site of autophosphorylation.

Three reports indicate that the autophosphorylation of kinases of the two-component systems can occur by the *trans*-phosphorylation of subunits. First, it was observed that a mutant, EnvZ, altered in the site of autophosphorylation could be complemented in vivo by the presence of mutant subunits deleted for region 3 and presumably unable to bind ATP (Yang and Inouye, 1991). In that work, it was directly demonstrated that the truncated subunit lacking region 3 became phosphorylated. Subsequently, two groups demonstrated that certain pairs of CheA mutants, each defective in autophosphorylation, could complement one another for the autophosphorylation activity in vitro (Swanson et al., 1993; Wolfe and Stewart, 1993). In those experiments, it was observed that mutations affecting region 3 could be complemented by mutations that alter the site of autophosphorylation; that is, an increase in the phosphorylation of the region 3 mutants was observed on simple addition of a region 1 mutant protein lacking the site of autophosphorylation. Thus, in aggregate these experiments indicated that ATP bound to a kinase subunit could phosphorylate the active-site histidine residue on a separate subunit. However, these studies did not reveal whether this *trans*-phosphorylation occurs by an intramolecular (within a dimer) or intermolecular (between separate dimers) mechanism or whether *trans*-phosphorylation was an essential feature of the autophosphorylation reaction or an interesting curiosity.

As noted above, the dimeric NR_{II} protein becomes autophosphorylated on His-139 upon incubation with ATP. We demonstrated that this autophosphorylation reaction occurs necessarily by the *trans*-phosphorylation of the NR_{II} monomers within the dimer; that is, a monomer binds ATP and phosphorylates the other monomer within the dimer (Ninfa et al., 1993). Several features of the NR_{II} protein were exploited in these studies. For example, in contrast to CheA, the dissociation constant for the monomer-dimer transition for

NR_{II} is apparently very low (Ninfa, 1992). Unlike the situation observed with CheA (Swanson et al., 1993; Wolfe and Stewart, 1993), when mutant NR_{II} proteins that are defective in autophosphorylation were purified and then combined in vitro, no complementation of the autophosphorylation defect was seen (Ninfa et al., 1993). We identified gentle conditions (2.8 M urea) that result in the reversible dissociation of the NR_{II} dimer. We observed that we could form mixed dimers containing different mutant NR_{II} subunits by mixing different mutant proteins, treating the mixture with urea, and subsequently dialyzing out the urea to permit renaturation (Ninfa et al., 1993). In particular, the G313A mutation (affecting the binding of ATP) could be complemented by the H139N mutation (altering the site of autophosphorylation) in this way. We interpret this finding as indicating that intramolecular complementation between two mutant subunits, each defective in autophosphorylation, can occur only when heterodimers are formed containing the two different mutant subunits (see below).

To further study the *trans*-phosphorylation phenomenon, we fused the maltose-binding protein (MBP) to the N-terminal end of wild-type NR_{II} and purified the fusion protein (MBP-NR_{II}). When MBP-NR_{II} was put through the denaturation-renaturation protocol with normal-sized NR_{II} mutant proteins, hybrid dimers that were a unique molecular weight and could be readily visualized on nondenaturing gels were formed and were purified by excision from gels. Denaturation of the samples followed by separation on denaturing gels then clearly indicated the presence of both types of subunits within the mixed dimers. Experiments in which the normal-sized subunit of such mixed dimers contained particular mutations were most revealing. When the normal-sized subunit lacked the proper site of autophosphorylation, the hybrid dimer was phosphorylated only on the MBP-NR_{II} subunit. Conversely, when the normal-sized subunit contained the G313A mutation, greatly reducing the binding of ATP, the hybrid dimer was autophosphorylated mostly on the mutant subunit. That is, the effect of the ATP-binding-site mutation was observed within the hybrid dimer but effect was not observed on the subunit that bore this mutation but, rather, on the other (wild-type) subunit within the dimer. Such a result strongly implies that the autophosphorylation reaction occurs by *trans*-phosphorylation of the monomers within the dimer. Additional experiments indicated that intermolecular phosphorylation (between different dimeric molecules) does not occur. For example, complementation of the autophosphorylation defect of the region 3 G313A mutation is observed only for that fraction of the protein that is in hybrid dimers after the urea/dialysis protocol. The homodimeric fraction of the renatured protein from the same reaction mixtures had properties identical to those of the untreated G313A mutant; that is, it was not autophosphorylated. In aggregate, the data indicate that the autophosphorylation reaction necessarily proceeds by *trans*-phosphorylation of subunits within the dimer (Ninfa et al., 1993). Such a finding is consistent with our observation of intramolecular complementation by different mutant *glnL* alleles in *recA* cells (Atkinson and Ninfa, 1993).

Unresolved Issues

Perhaps the most important issue remaining is the nature of the phosphatase activity that results in the dephosphorylation of $NR_I \sim P$ in the presence of NR_{II} and P_{II}. Our results with the H139N protein suggest that P_{II} plays only a regulatory role in this activity. One possibility is that the phosphatase activity represents a stimulation of the normal "autophosphatase" activity of $NR_I \sim P$. Thus, it might prove useful to isolate and examine mutations that affect the autophosphatase activity, to see if they similarly affect the NR_{II}-stimulated dephosphorylation of $NR_I \sim P$.

Curiously, in cells lacking P_{II}, the UT/UR protein must still be present to obtain the elevated expression of certain nitrogen-regulated genes in response to nitrogen starvation (Bueno et al., 1985; Atkinson and Ninfa, unpublished). Conceivably, *E. coli* may contain another P_{II}-like protein that is modified by the UT/UR protein. If so, this protein and its role in nitrogen regulation should be further characterized. Furthermore, mutants lacking the UT/UR protein have a leaky phenotype (Bueno et al., 1985; Son and Rhee, 1987; Atkinson and Ninfa, 1992). Therefore, the possibility cannot be excluded that *E. coli* also contains another UT/UR protein.

The P_{II} protein is remarkable in that it interacts with at least three other proteins: NR_{II}, the UT/UR enzyme, and the adenylyltransferase enzyme (the *glnE* product) responsible for the reversible modification of glutamine synthetase in response to nitrogen availability (reviewed by Rhee et al. [1985]). It will be of interest to determine the mechanism by which these three proteins interact with P_{II} and whether the same mechanism is used in all three cases.

Work in our laboratory has been supported by grants from the NIH (GM47460) and NSF (DMB9004048).

We thank Boris Magasanik for many helpful discussions.

REFERENCES

Atkinson, M. R., and A. J. Ninfa. 1992. Characterization of *Escherichia coli glnL* mutations affecting nitrogen regulation. *J. Bacteriol.* **174:**4538–4548.

Atkinson, M. R., and A. J. Ninfa. 1993. Mutational analysis of the bacterial signal-transducing protein kinase/phosphatase nitrogen regulator II (NR_{II} or NTRB). *J. Bacteriol.* **175:**7016–7023.

Atkinson, M. R., and A. J. Ninfa. Unpublished data.

Bueno, R., G. Pahel, and B. Magasanik. 1985. Role of *glnB* and *glnD* gene products in the regulation of the *glnALG* operon of *Escherichia coli. J. Bacteriol.* **164:**816–822.

Feng, J., M. R. Atkinson, W. R. McCleary, J. B. Stock, B. L. Wanner, and A. J. Ninfa. 1992. Role of phosphorylated metabolic intermediates in the regulation of glutamine synthetase synthesis in *Escherichia coli. J. Bacteriol.* **174:** 6061–6070.

Kamberov, E. S., M. R. Atkinson, and A. J. Ninfa. Reconstitution of the signal-transducing switch regulating bacterial nitrogen assimilation: improved method for the purification of the P_{II} protein and further characterization of the role of P_{II} in nitrogen regulation. Submitted for publication.

Keener, J., and S. Kustu. 1988. Protein kinase and phosphoprotein phosphatase of nitrogen regulatory proteins NTRB and NTRC of enteric bacteria: roles of conserved amino terminal domain of NTRC. *Proc. Natl. Acad. Sci. USA* **85:**4976–4980.

Lukat, G. S., W. R. McCleary, A. M. Stock, and J. B. Stock. 1992. Phosphorylation of bacterial response regulator proteins by low molecular weight phosphodonors. *Proc. Natl. Acad. Sci. USA* **89:**718–722.

MacNeil, T., G. P. Roberts, D. MacNeil, and B. Tyler. 1982. The products of *glnL* and *glnG* are bifunctional regulatory proteins. *Mol. Gen. Genet.* **188:**325–333.

Magasanik, B. 1988. Reversible phosphorylation of an enhancer-binding protein regulates the transcription of nitrogen regulated genes. *Trends Biochem. Sci.* **13:**475–479.

McCleary, W. R., J. B. Stock, and A. J. Ninfa. 1993. Is acetyl phosphate a global signal in *Escherichia coli*? *J. Bacteriol.* **175:**2793–2798.

Ninfa, A. J. 1991. Protein phosphorylation and the regulation of cellular processes by the homologous two-component regulatory systems of bacteria. *Genet. Eng.* **13:**39–72.

Ninfa, A. J., and R. L. Bennett. 1991. Identification of the site of autophosphorylation of the bacterial protein kinase/phosphatase NR_{II}. *J. Biol. Chem.* **266:**6888–6893.

Ninfa, A. J., and B. Magasanik. 1986. Covalent modification of the *glnG* product, NR_I, by the *glnL* product, NR_{II}, regulates the transcription of the *glnALG* operon in *Escherichia coli. Proc. Natl. Acad. Sci. USA* **83:**5909–5913.

Ninfa, E. G. 1992. Doctoral dissertation. Princeton University, Princeton, N.J.

Ninfa, E. G., M. R. Atkinson, E. S. Kamberov, and A. J. Ninfa. 1993. Mechanism of autophosphorylation of *Escherichia coli* nitrogen regulator II (NR_{II} or NtrB): *trans*-phosphorylation between subunits. *J. Bacteriol.* **175:**7024–7032.

Parkinson, J. S., and E. C. Kofoid. 1992. Communication modules in bacterial signaling proteins. *Annu. Rev. Genet.* **26:**71–112.

Popham, D. L., D. Szeto, J. Keener, and S. Kustu. 1989. Function of a bacterial activator protein that binds to transcriptional enhancers. *Science* **243:**629–635.

Reitzer, L. J., and B. Magasanik. 1985. Expression of *glnA* in *Escherichia coli* is regulated at tandem promoters. *Proc. Natl. Acad. Sci. USA* **82:**1979–1983.

Rhee, S.-G., S. C. Park, and J. H. Koo. 1985. The role of adenylyltransferase and uridylyltransferase in the regulation of glutamine synthetase in *Escherichia coli. Curr. Top. Cell. Regul.* **27:**221–232.

Sanders, D. A., B. L. Gillece-Castro, A. L. Burlingame, and D. E. Koshland, Jr. 1992. Phosphorylation site of NtrC, a protein phosphatase whose covalent intermediate activates transcription. *J. Bacteriol.* **174:**5117–5122.

Smith, R. A., and J. S. Parkinson. 1980. Overlapping genes at the CheA locus of Escherichia coli. *Proc. Natl. Acad. Sci. USA* **77:**5370–5374.

Son, H. S., and S. G. Rhee. 1987. Cascade control of glutamine synthetase. Purification and properties of PII protein and nucleotide sequence of its structural gene. *J. Biol. Chem.* **262:**8690–8695.

Stock, J. B., A. J. Ninfa, and A. M. Stock. 1989. Protein phosphorylation and the regulation of adaptive responses in bacteria. *Microbiol. Rev.* **53:**450–490.

Swanson, R. V., R. B. Bourret, and M. I. Simon. 1993. Intermolecular complementation of the kinase activity of CheA. *Mol. Microbiol.* **8:**435–441.

Wanner, B. L., and M. R. Wilmes-Riesenberg. 1992. Involvement of phosphotransacetylase, acetate kinase, and acetyl phosphate synthesis in control of the phosphate regulon in *Escherichia coli. J. Bacteriol.* **174:**2124–2130.

Weiss, D. S., J. Batut, K. E. Klose, J. Keener, and S. Kustu. 1991. The phosphorylated form of the enhancer-binding protein NtrC has an ATPase activity that is essential for activation of transcription. *Cell* **67:**155–167.

Weiss, V., F. Claverie-Martin, and B. Magasanik. 1992. Phosphorylation of nitrogen regulator I of *Escherichia coli* induces strong cooperative binding to DNA essential for the activation of transcription. *Proc. Natl. Acad. Sci. USA* **89:**5088–5092.

Weiss, V., and B. Magasanik. 1988. Phosphorylation of nitrogen regulator I (NR_I) of *Escherichia coli. Proc. Natl. Acad. Sci. USA* **85:**8919–8923.

Wolfe, A. J., and R. C. Stewart. 1993. The short form of the CheA protein restores kinase activity and chemotactic ability to kinase deficient mutants. *Proc. Natl. Acad. Sci. USA* **90:**1518–1522.

Yang, Y., and M. Inouye. 1991. Intermolecular complementation between two defective mutant signal-transducing receptors of *Escherichia coli. Proc. Natl. Acad. Sci. USA* **88:**11057–11061.

Chapter 52

Structural Basis for the Mechanism of Phosphoryl Transfer in Bacterial Chemotaxis

ANN H. WEST, ERIK MARTINEZ-HACKERT, AND ANN M. STOCK

Center for Advanced Biotechnology and Medicine, University of Medicine and Dentistry of New Jersey, 679 Hoes Lane, Piscataway, New Jersey 08854-5638

One of the most common means of communication between biological macromolecules involves the transfer of a phosphoryl group. In bacteria, this is evident from the existence of over 60 different signal transduction pathways that regulate a variety of responses to changing environmental conditions (Bourret et al., 1991; Parkinson and Kofoid, 1992; Stock et al., 1990). These so-called two-component systems utilize a common mechanism of phosphoryl transfer between two conserved protein components: a histidine protein kinase and a response regulator protein. The study of the bacterial chemotaxis pathway has served as a model for gaining an understanding of the biochemical and structural basis of how phosphorylation and dephosphorylation reactions control cellular responses. The response regulator, CheY, is currently the only member of the two-component systems for which a high-resolution protein structure is known. It is expected that this information might be generally applicable to many of the other two-component regulatory systems and perhaps to other enzymatic reactions that similarly involve phosphoryl group transfer.

Bacterial Chemotaxis

In *Escherichia coli* and *Salmonella typhimurium,* phosphoryl transfer between the histidine protein kinase, CheA, and the response regulator protein, CheY, forms the core of the signal-transducing pathway that controls cell motility (for reviews, see Bourret et al. [1991], Stock et al. [1991], Macnab [1987], and Stewart and Dahlquist [1987]). The directed movement, toward attractants and away from repellents in the extracellular environment, is achieved by influencing the direction of rotation of the flagellar filaments. Cells normally swim smoothly in one direction by counterclockwise rotation of the flagellar motor but display brief periods of tumbling which occur upon transient clockwise rotation of the flagellar motor. This allows for random reorientation and then resumption of smooth swimming.

The chemotaxis signaling pathway is modulated by changes in the concentration of chemoeffectors in the medium. The system consists of a family of transmembrane receptors with different ligand specificities, six cytoplasmic signal-transducing proteins, and the flagellar switch proteins at the base of the flagellar rotary motor apparatus (Fig. 1). Central to the pathway is the phosphotransfer reaction between the kinase, CheA, and the regulator, CheY. The other proteins serve auxiliary roles in regulating the activities of CheA and CheY. The homologous receptors (Tar, Tsr, Trg, and Tap) that detect and respond to certain amino acids, metals, sugars, and dipeptides, exist as homodimers within the cytoplasmic membrane. The ligand-binding domain faces the periplasmic space, whereas the signaling domain is exposed to the cytoplasm. Specific glutamate residues within the cytoplasmic domains of the receptors are subject to covalent modification by a methyltransferase protein, CheR. The autophosphorylating activity of the histidine kinase CheA is modulated by both the level of ligand occupancy and the methylation state of the receptors. When stimulated, CheA uses ATP to autophosphorylate a specific histidine residue located near its N terminus. This phosphoryl group is subsequently transferred to a specific aspartic acid residue of the response regulator protein, CheY. In its phosphorylated state, CheY is able to induce clockwise rotation of the flagellar motor and tumbling cell behavior, most probably via an interaction with the flagellar switch components, FliG, FliM, and FliN. In vivo, phospho-CheY is extremely short-lived as a result of an inherent autophosphatase activity and the phosphatase-accelerating effects of another protein component, CheZ. Alternatively, phosphoryl transfer within the pathway can occur between CheA and CheB. Phosphorylation of CheB activates an attached C-terminal effector domain with receptor methylesterase activity. The reversible methylation and demethylation of the receptors constitute the biochemical basis for cellular adaptation by serving to desensitize the receptors. Recent evidence has indicated that

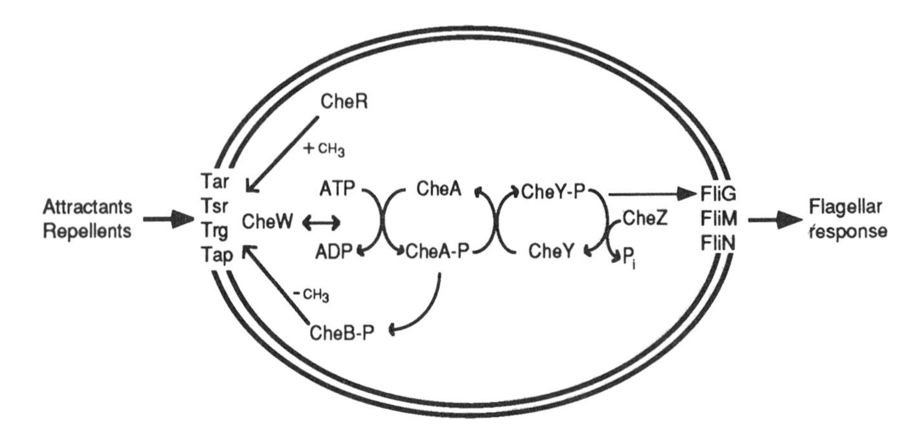

FIG. 1. Signaling pathway in bacterial chemotaxis.

the conserved histidine protein kinase CheA is part of a stable ternary complex with CheW and the membrane receptor, with CheW functioning to couple CheA to the receptor (Gegner et al., 1992). In vitro evidence has also indicated that CheY associates with CheA within the receptor-CheA-CheW complex (Swanson et al., 1993). Phosphorylation of CheY results in its release from the complex, which would enable a diffusible signal to be carried to the flagellar motor (Schuster et al., 1993).

Structure and Function of the Response Regulator Protein CheY

Both CheY and CheB are members of the large family of response regulator proteins that share a common regulatory domain and function as phosphorylation-activated molecular switches to control the activities of attached C-terminal domains or of downstream targets (Bourret et al., 1991; Parkinson and Kofoid, 1992; Stock et al., 1985; Volz, 1993). CheY is composed solely of the regulatory domain, which is more generally found as an N-terminal domain attached to an effector domain in other response regulator proteins. The three-dimensional structures of both *S. typhimurium* and *E. coli* CheY have been determined (Stock et al., 1989; Volz and Matsumura, 1991). CheY is an α/β-domain protein composed of a central core of five parallel β-strands surrounded by five α-helices (Fig. 2). Like many other α/β proteins, the active site is located in a crevice formed at the C-terminal ends of β-strands that join to α-helices lying on opposite sides of the β-sheet ($\beta1$ and $\beta3$ in this case). Residues that are invariant within the response regulator family correspond to Asp-57 (the site of phosphorylation), Asp-13, and Lys-109 in CheY. The structure of CheY revealed that these conserved residues, along with another highly conserved carboxylate-containing residue, Asp-12, are located on separate loops between $\beta1$ and αA, $\beta3$ and αC, and $\beta5$ and αE but are close together in three-dimensional space. These residues together constitute an acidic active-site pocket, consistent with results of mutational analyses that had indicated that these residues were critical for function (Bourret et al., 1990; Lukat et al., 1991).

Several experimental observations have contributed to the important discovery that the regulatory domain of the response regulator proteins catalyzes the phosphotransfer and phosphate hydro-

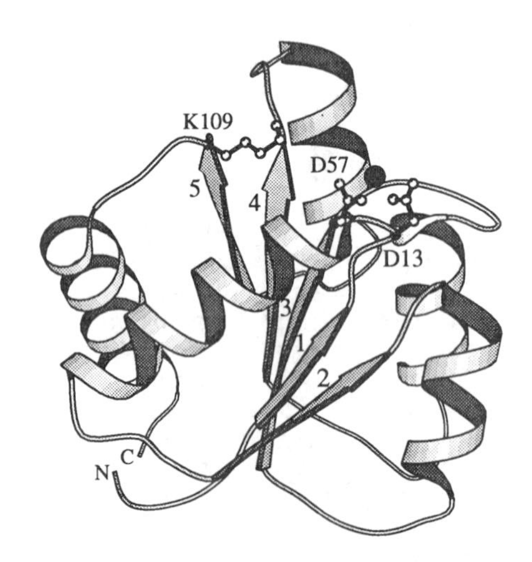

FIG. 2. Tertiary structure of CheY. The parallel β-sheet (numbered strands) is flanked on both sides by α-helices. The diagram includes side chains of the active-site residues, Lys-109, Asp-57, and Asp-13, which are conserved throughout the family of bacterial response regulator proteins. The solid circle represents the catalytic metal ion.

lysis reactions. First, a small subdomain of CheA devoid of the kinase activity but containing the phosphorylated histidine residue could serve as a phosphoryl donor to CheY in vitro (Hess et al., 1988). More recently, it was demonstrated that CheY, in the absence of any other phosphoprotein donor, could be phosphorylated by using small-molecule high-energy phospho-donors, such as acetyl phosphate, carbamoyl phosphate, or phosphoramidate (Fig. 3) (Lukat et al., 1992). CheB could also be phosphorylated with phosphoramidate, and this phosphorylation event resulted in an activation of its associated enzymatic activity. Other examples of response regulators that are capable of catalyzing their own phosphorylation from small-molecule phospho-donors, include NR_I, the regulator of the nitrogen assimilation pathway (Feng et al., 1992), and AlgR, the regulator of alginate production in *Pseudomonas aeruginosa* (Deretic et al., 1992). These observations indicate that autocatalysis of phosphoryl transfer is a general feature of the response regulator domains.

Although the phosphorylated histidine kinase is probably the primary phosphoryl donor under normal conditions in vivo, accumulating evidence indicates that small-molecule donors, acetyl phosphate in particular, may also contribute to the phosphorylation of response regulator proteins (McCleary et al., 1993; Wanner, 1992). In the osmoregulation, nitrogen assimilation, and phosphate assimilation systems, strains with mutations that eliminate a particular histidine kinase activity still retain residual activity of the associated response regulator (Backman et al., 1983; Forst et al., 1990). This could be explained by cross talk coming from other two-component kinases. Alternatively, the kinase-independent activation of the response regulators could be correlated with conditions that result in a metabolic increase in the intracellular levels of acetyl phosphate (Feng et al., 1992; Lee et al., 1990; Wanner and Wilmes-Riesenberg, 1992).

Structure of the Active Site of CheY and Mechanism of Phosphorylation

Most enzymatic and nonenzymatic phosphoryl transfer reactions require divalent cations. Biochemical studies have firmly established this requirement for the phosphorylation and dephosphorylation of CheY by using both CheA and small-molecule phosphate donors (Lukat et al., 1990, 1992). In addition, sensitive spectroscopic assays and site-specific mutagenesis of active-site residues have indicated that the active-site pocket is the locus for metal binding (Kar et al., 1992; Lukat et al., 1990, 1992). With the knowledge that these regulatory domains themselves carry out the phosphotransfer reaction, we were interested from a structural perspective in the mechanism of the metal-catalyzed phosphoryl transfer reaction. We have determined the crystal structure to 1.8-Å resolution of *S. typhimurium* CheY with Mg^{2+} bound at the active site (Stock et al., 1993). Although the overall structure of the molecule remains the same upon addition of metal, small but significant rearrangements occur in the active site. The Mg^{2+} ion replaces an ordered water molecule present in the nonmetal structure and coordinates to three protein ligands and three water molecules in an octahedral geometry (Fig. 4). There is a noticeable shift in the position of the

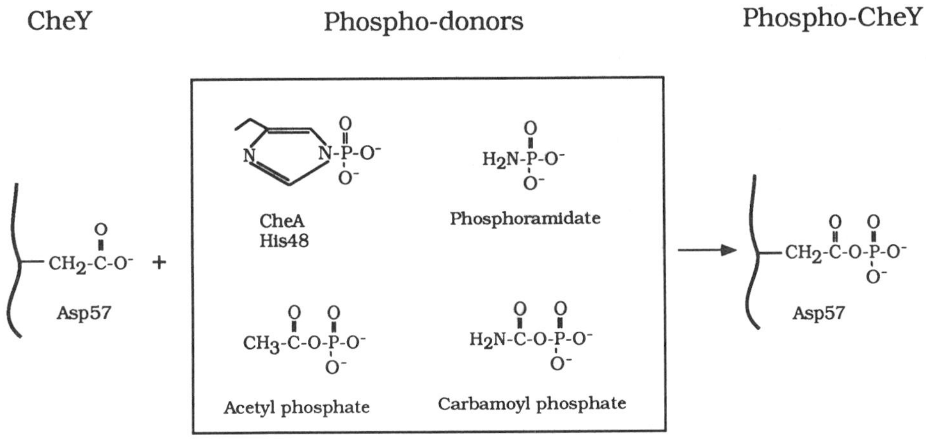

CheY | Phospho-donors | Phospho-CheY

Asp57 + [CheA His48; Phosphoramidate; Acetyl phosphate; Carbamoyl phosphate] → Asp57

FIG. 3. Phosphorylation of CheY. Asp-57 is the site of phosphorylation in CheY (Sanders et al., 1989). The chemical structures of known phosphoryl donors are shown: the phosphorylated histidine residue, His-48, of the kinase CheA (Hess et al., 1988) and three small molecule phosphoryl donors (Lukat et al., 1992).

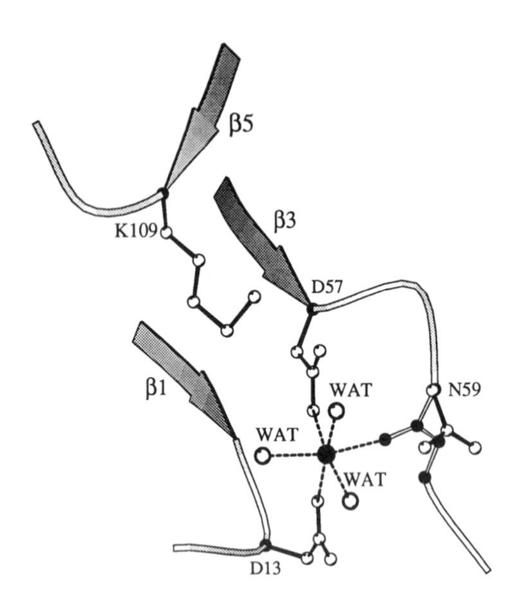

FIG. 4. Octahedral Mg^{2+} coordination within the active site of CheY. The refined atomic model of Mg^{2+}-bound CheY (Stock et al., 1993) shows only the loops corresponding to the C-terminal ends of β-strands 1, 3, and 5. The dashed lines indicate the octahedral coordination to the Mg^{2+} ion (shown as a solid circle). Side chain carboxylate oxygens of Asp-57 and Asp-13, a backbone carbonyl oxygen of Asn-59, and three water molecules (open circles labeled WAT) provide the six ligands to the metal. The side chain atoms of the highly conserved lysine residue, Lys-109, are also shown.

side chain of Asp-57 in order for one of the carboxylate oxygens to directly coordinate to the metal. This movement results in the loss of an ionic interaction between Asp-57 and the ε-amino group of Lys-109 and has the desired effect of leaving one carboxylate oxygen available to serve as the attacking nucleophile in the phosphoryl transfer reaction. A carboxylate oxygen of Asp-13 and the backbone carbonyl oxygen from Asn-59 are the other two protein ligands to the metal. A carboxylate oxygen of Asp-12 forms part of the second coordination sphere by making an indirect contact through a water molecule.

Because of the reduced complexity, studies of phosphoryl transfer reactions between small molecule analogs have provided much of the information regarding the functions that metal ions can provide during catalysis in enzymatic reactions (Cooperman, 1976; Knowles, 1980). Besides functioning in a charge-shielding capacity for the incoming negatively charged substrate or negatively charged leaving group, metal ions can catalyze reactions by activating the nucleophile and by providing a template for the transition state. It should be noted that although many of these functions could potentially be provided by protein side chains, in CheY the only positively charged residue in the vicinity is Lys-109 and substitution of this residue with an arginine or neutral glutamine does not substantially reduce the rate of phosphorylation (Lukat et al., 1991).

Most phosphoryl transfer reactions are thought to proceed via the formation of a pentavalent transition state, with the attacking nucleophilic oxygen and the leaving group occupying axial positions and the remaining oxygens forming a trigonal planar arrangement with the phosphorus atom (Knowles, 1980). In a kinetics study of the nonenzymatic transfer of a phosphoryl group between carboxylate nucleophiles and phosphorylated pyridine analogs, Herschlag and Jencks (1990) concluded that the catalytic role of the Mg^{2+} ion was in stabilizing the transition state. Their results were consistent with a mechanism involving a transition state in which the Mg^{2+} ion bridges both the oxygen of the attacking carboxylate and the equatorial oxygens of the transition-state phosphorus (Fig. 5). This chemical reaction mechanism mimics the phosphoryl transfer between a phosphohistidine and an aspartic acid side chain of a response regulator protein. There are several attractive features of this transition state model that can be discussed in the context of the Mg^{2+}-bound structure of CheY. Positioning of a pentavalent phosphorus within the active site of CheY, consistent with the proposed transition-state model, can be accomplished by simply replacing two of the water molecules that are direct ligands to the metal with the equatorial oxygens of the pentavalent phosphorus (Fig. 6). A simple rotation of the carboxylate side chain of Asp-57 would place the attacking oxygen in the proper axial position for an in-line nucleophilic attack while allowing for a bulky leaving group, such as an imidazole ring of a histidine, to depart from a sterically unobstructed region at the other axial position. This transition-state model also offers a rationale for the specificity of phosphorylation at Asp-57. Positioning the pentavalent phosphorus near Asp-13, for example, would result in an unfavorable steric collision with nearby side chain

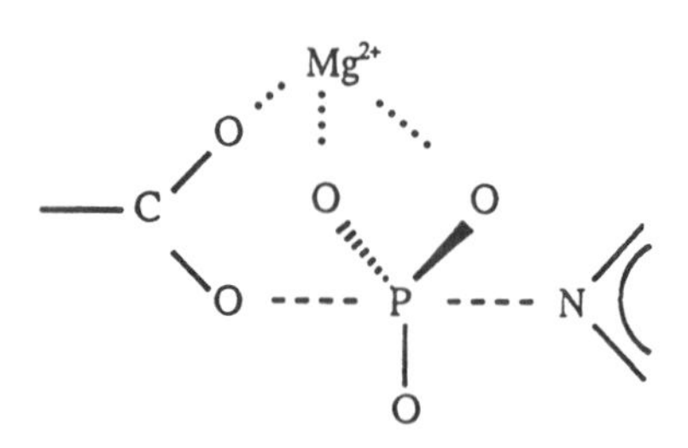

FIG. 5. Pentavalent phosphorus transition state proposed for Mg^{2+}-catalyzed phosphoryl transfer to a carboxylate (Herschlag and Jencks, 1990).

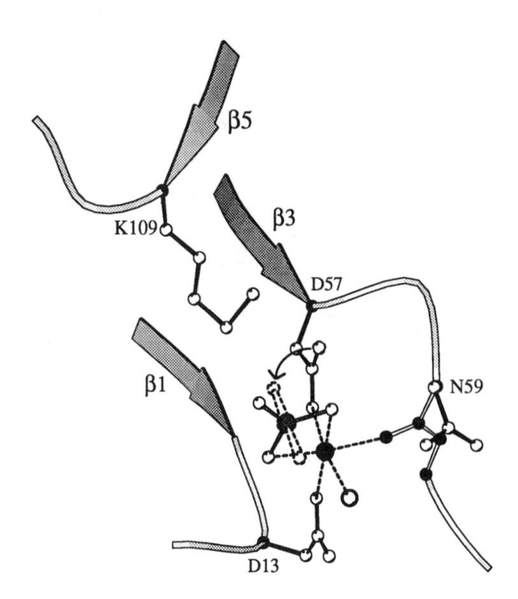

FIG. 6. Postulated transition state for phosphorylation of CheY at Asp-57. A pentavalent phosphorus is shown positioned in the CheY-Mg^{2+} model of the active site in a manner consistent with the proposed transition state shown in Fig. 5. The Mg^{2+} ion simultaneously coordinates to the attacking carboxylate nucleophile (from Asp-57) and to equatorial oxygens of the bipyramidal transition-state phosphorus. The arrow indicates the rotation of the side chain of Asp-57 that would be necessary to position the nucleophilic oxygen for an in-line attack at the axial position of the phosphorus atom (stippled circle). The other axial position would be occupied by the leaving group (indicated by dashed lines).

atoms, and nucleophilic attack could not be facilitated by a mere rotation of the carboxylate side chain of Asp-13 but, rather, would involve movement of backbone atoms.

Roles of the Conserved Active-Site Residues

In summary, metal ions, because of their high charge density and precise coordination geometry, have the effect of providing an organizing center within an active site. In many cases, catalysis is the result of bringing the proper substrates into alignment and overcoming the thermodynamic barriers to the transition state. The Mg^{2+}-bound structure of CheY has provided a basis for a proposed mechanism of phosphorylation and has more clearly defined the roles of the conserved carboxylate residues (aspartates 12, 13, and 57 in CheY) (Stock et al., 1993). Their function in the phosphoryl transfer reaction is to correctly position the essential Mg^{2+} ion which provides the catalytic function.

Although the structure of CheY with Mg^{2+} bound at the active site provides insight into the mechanism of phosphoryl transfer, it provides no direct information about steps subsequent to phosphorylation. Biochemical and structural studies have suggested no significant role for the invariant lysine residue (Lys-109 in CheY) in the phosphoryl transfer mechanism (Stock et al., 1989, 1993; Lukat et al., 1991; Volz and Matsumura, 1991). However, physiological analyses of Lys-109 mutants indicate that Lys-109 is essential for activation of CheY (Lukat et al., 1991). It would be expected that phosphorylation would induce significant rearrangements within the active site, most probably involving side chain interactions and metal recoordination around the phosphate. It is tempting, but presumptuous in the absence of structural information, to speculate about these interactions. Unfortunately, direct structural analysis of the phosphorylated form of CheY has been hindered by its short lifetime. Elucidation of the structure of a phosphorylated response regulator domain will inevitably define additional roles for the conserved active-site residues in the activation and dephosphorylation mechanisms.

This work was supported by grants from the National Institutes of Health (GM47958) and the Lucille P. Markey Charitable Trust to A.M.S. and by a postdoctoral research fellowship to A.H.W. from The Jane Coffin Childs Memorial Fund for Medical Research.

REFERENCES

Backman, K. C., Y.-M. Chen, S. Ueno-Nishio, and B. Magasanik. 1983. The product of *glnL* is not essential for regulation of bacterial nitrogen assimilation. *J. Bacteriol.* **154:**516–519.

Bourret, R. B., K. A. Borkovich, and M. I. Simon. 1991. Signal transduction pathways involving protein phosphorylation in prokaryotes. *Annu. Rev. Biochem.* **60:**401–441.

Bourret, R. B., J. F. Hess, and M. I. Simon. 1990. Conserved aspartate residues and phosphorylation in signal transduction by the chemotaxis protein CheY. *Proc. Natl. Acad. Sci. USA* **87:**41–45.

Cooperman, B. S. 1976. The role of divalent metal ions in phosphoryl and nucleotidyl transfer. *Metal Ions Biol. Syst.* **5:**79–125.

Deretic, V., J. H. J. Leveau, C. D. Mohr, and N. S. Hibler. 1992. *In vitro* phosphorylation of Alar, a regulator of mucoidy in *Pseudomonas aeruginosa,* by a histidine protein kinase and effects of small phospho-donor molecules. *Mol. Microbiol.* **6:**2761–2767.

Feng, J., M. R. Atkinson, W. McCleary, J. B. Stock, B. L. Wanner, and A. J. Ninfa. 1992. Role of phosphorylated metabolic intermediates in the regulation of glutamine synthetase synthesis in *Escherichia coli. J. Bacteriol.* **174:**6061–6070.

Forst, S., J. Delgado, A. Rampersaud, and M. Inouye. 1990. In vivo phosphorylation of OmpR, the transcription activator of the *ompF* and *ompC* genes in *Escherichia coli. J. Bacteriol.* **172:**3473–3477.

Gegner, J. A., D. R. Graham, A. F. Roth, and F. W. Dahlquist. 1992. Assembly of an MCP receptor, CheW, and kinase CheA complex in the bacterial chemotaxis signal transduction pathway. *Cell* **70:**975–982.

Herschlag, D., and W. P. Jencks. 1990. The effects of Mg^{2+}, hydrogen bonding, and steric factors on rate and equilibrium constants for phosphoryl transfer between carboxylate ions and pyridines. *J. Am. Chem. Soc.* **112:**1942–1950.

Hess, J. F., R. B. Bourret, and M. I. Simon. 1988. Histidine phosphorylation and phosphoryl group transfer in bacterial chemotaxis. *Nature* (London) **336:**139–143.

Kar, L., P. Matsumura, and M. E. Johnson. 1992. Bivalent-metal binding to CheY protein. Effect on protein conformation. *Biochem. J.* **287:**521–531.

Knowles, J. R. 1980. Enzyme-catalyzed phosphoryl transfer reactions. *Annu. Rev. Biochem.* **49:**877–919.

Lee, T.-Y., K. Makino, H. Shinagawa, and A. Nakata. 1990. Overproduction of acetate kinase activates the phosphate regulon in the absence of the *phoR* and *phoM* functions in *Escherichia coli. J. Bacteriol.* **172:**2245–2249.

Lukat, G. S., B. H. Lee, J. M. Mottonen, A. M. Stock, and J. B. Stock. 1991. Roles of the highly conserved aspartate and lysine residues in the response regulator of bacterial chemotaxis. *J. Biol. Chem.* **266:**8348–8354.

Lukat, G. S., W. R. McCleary, A. M. Stock, and J. B. Stock. 1992. Phosphorylation of bacterial response regulator proteins by low molecular weight phospho-donors. *Proc. Natl. Acad. Sci. USA* **89:**718–722.

Lukat, G. S., A. M. Stock, and J. B. Stock. 1990. Divalent metal ion binding to the CheY protein and its significance to phosphotransfer in bacterial chemotaxis. *Biochemistry* **29:**5436–5442.

Macnab, R. M. 1987. Motility and chemotaxis, p. 732–759. *In* F. C. Neidhardt, J. L. Ingraham, K. B. Low, B. Magasanik, M. Schaechter, and H. E. Umbarger (ed.), *Escherichia coli and Salmonella typhimurium: Cellular and Molecular Biology.* American Society for Microbiology, Washington, D.C.

McCleary, W. R., J. B. Stock, and A. J. Ninfa. 1993. Is acetyl phosphate a global signal in *Escherichia coli*? *J. Bacteriol.* **175:**2793–2798.

Parkinson, J. S., and E. C. Kofoid. 1992. Communication modules in bacterial signaling proteins. *Annu. Rev. Genet.* **26:**71–112.

Sanders, D. A., B. L. Gillece-Castro, A. M. Stock, A. L. Burlingame, and D. E. Koshland, Jr. 1989. Identification of the site of phosphorylation of the chemotaxis response regulator protein, CheY. *J. Biol. Chem.* **264:**21770–21778.

Schuster, S. C., R. V. Swanson, L. A. Alex, R. B. Bourret, and M. I. Simon. 1993. Assembly and function of a quaternary signal transduction complex monitored by surface plasmon resonance. *Nature* (London) **365:**343–347.

Stewart, R. C., and F. W. Dahlquist. 1987. Molecular components of bacterial chemotaxis. *Chem. Rev.* **87:**997–1025.

Stock, A., D. E. Koshland, Jr., and J. Stock. 1985. Homologies between the *Salmonella typhimurium* CheY protein and proteins involved in the regulation of chemotaxis, membrane protein synthesis, and sporulation. *Proc. Natl. Acad. Sci. USA* **82:**7989–7993.

Stock, A. M., E. Martinez-Hackert, B. F. Rasmussen, A. H. West, J. B. Stock, D. Ringe, and G. A. Petsko. 1993. Structure of the Mg^{2+}-bound form of CheY and the mechanism of phosphoryl transfer in bacterial chemotaxis. *Biochemistry* **32:**13375–13380.

Stock, A. M., J. M. Mottonen, J. B. Stock, and C. E. Schutt. 1989. Three-dimensional structure of CheY, the response regulator of bacterial chemotaxis. *Nature* (London) **337:**745–749.

Stock, J. B., G. S. Lukat, and A. M. Stock. 1991. Bacterial chemotaxis and the molecular logic of intracellular signal transduction networks. *Annu. Rev. Biophys. Biophys. Chem.* **20:**109–136.

Stock, J. B., A. M. Stock, and J. M. Mottonen. 1990. Signal transduction in bacteria. *Nature* (London) **344:**395–400.

Swanson, R. V., S. C. Schuster, and M. I. Simon. 1993. Expression of CheA fragments which define domains encoding kinase, phosphotransferase, and CheY binding activities. *Biochemistry* **32:**7623–7629.

Volz, K. 1993. Structural conservation in the CheY superfamily. *Biochemistry* **32:**11741–11753.

Volz, K., and P. Matsumura. 1991. Crystal structure of *Escherichia coli* CheY refined at 1.7-Å resolution. *J. Biol. Chem.* **266:**15511–15519.

Wanner, B. L. 1992. Is cross regulation by phosphorylation of two-component response regulator proteins important in bacteria? *J. Bacteriol.* **174:**2053–2058.

Wanner, B. L., and M. R. Wilmes-Riesenberg. 1992. Involvement of phosphotransacetylase, acetate kinase, and acetyl phosphate synthesis in control of the phosphate regulon in *Escherichia coli. J. Bacteriol.* **174:**2124–2130.

Chapter 53

Phosphate Taxis and Its Regulation in *Pseudomonas aeruginosa*

JUNICHI KATO, YUKIHIRO SAKAI, TOSHIYUKI NIKATA, AGUS MASDUKI, AND HISAO OHTAKE

Department of Fermentation Technology, Hiroshima University, Higashi-Hiroshima, Hiroshima 724, Japan

Phosphorus compounds are essential constituents in organisms. In nature, P_i is often found to be a growth-limiting factor for organisms. To deal with P_i limitation, bacteria have evolved complex regulatory systems to assimilate P_i very efficiently (Wanner, 1987).

Pseudomonas aeruginosa PAO1 is attracted to P_i (Fig. 1). This chemotactic response to P_i was detected by a rapid computer-assisted capillary technique (Nikata et al., 1992). The specificity of chemoreceptors for P_i was high. No other tested phosphorus compounds elicited similar responses to those for P_i (Kato et al., 1992). Although *P. aeruginosa* is known to be attracted to most of the 20 common amino acids (Moulton and Montie, 1978), competition experiments showed that the chemoreceptors for P_i appeared to be different from those for amino acids.

The chemotactic response was induced by P_i limitation. No positive response to P_i was observed by *P. aeruginosa* PAO1 cells grown in P_i-sufficient medium (Fig. 2). During growth under P_i limitation, the strength of P_i taxis, assessed by the number of bacteria accumulating near the mouth of the capillary (Nikata et al., 1992), increased in parallel with alkaline phosphatase (AP) activity. P_i-starved cells were also attracted to arsenate. However, P_i competitively inhibited the response for arsenate, indicating that both P_i and arsenate were detected by the same chemoreceptor (Fig. 3).

AP-constitutive *P. aeruginosa* mutants were selected by the procedure of Poole and Hancock (1984). APC1, an AP-constitutive mutant,

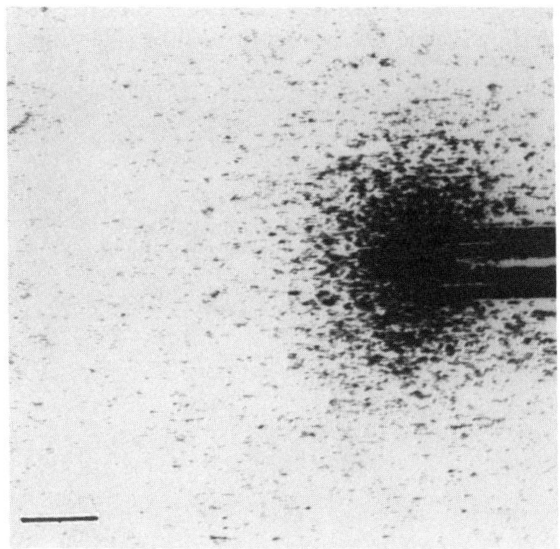

FIG. 1. Attraction of *P. aeruginosa* to phosphate. The capillary tube, with a diameter of 10 μm, contained 10 mM phosphate. Bar, 10 μm.

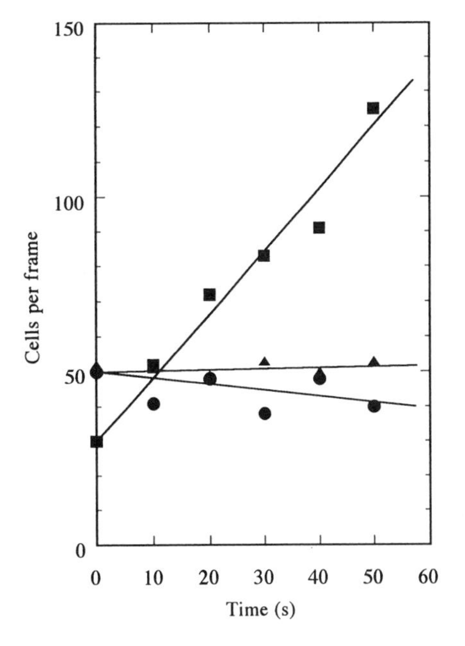

FIG. 2. Chemotactic responses of *P. aeruginosa* wild-type PAO1 (●), AP-constitutive APC1 (■), and APC1(pPT06.5) (▲) toward 10 mM P_i. Cells were grown in T_{10} minimal medium containing 10 mM P_i at 37°C with shaking (Kato et al., 1992). The chemotactic response was assessed by the rapid capillary technique (Nikata et al., 1992). For cell counting, one videotape frame was analyzed for each given time point.

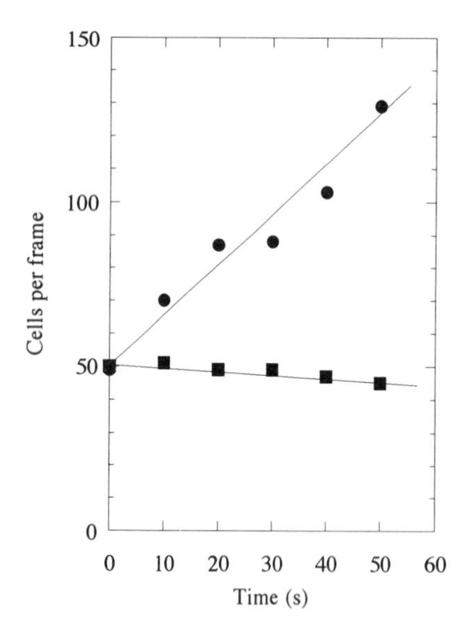

FIG. 3. Chemotactic response of *P. aeruginosa* wild-type PAO1 toward arsenate. Cells grown overnight in T_{10} medium were inoculated into T_0 medium containing no P_i and incubated at 37°C with shaking. The capillary containing 1 mM arsenate was inserted into the assay chamber filled with chemotaxis buffer containing either 1 mM (■) or no (●) P_i.

showed a chemotactic response to P_i, regardless of whether the cells were starved for P_i (Fig. 2). Like *Escherichia coli, P. aeruginosa* possesses the regulatory proteins PhoB and PhoR, which are responsible for inducing AP synthesis under P_i limitation (Anba et al., 1990). To investigate the regulatory mechanism of P_i taxis, *phoB* and *phoR* deletions in the *P. aeruginosa* chromosome were created. A 2.7-kb *Stu*I-*Sal*I chromosomal fragment, containing the entire *phoB* gene and the 5′ portion of the *phoR* gene, was cloned into plasmid vector pUC119. A kanamycin resistance cassette was inserted in either the *Sal*I site of the *phoB* gene or the *Bam*HI site of the *phoR* gene. The resulting recombinant plasmids were introduced into *P. aeruginosa* PAO1 cells by electroporation, and Km^r transformants were selected. The *phoB* and *phoR* deletions were confirmed by Southern hybridization to be properly constructed. As expected, neither the *phoB* nor *phoR* mutant induced AP synthesis under conditions of P_i limitation. Nevertheless, both mutants were fully induced by P_i limitation for P_i taxis (Kato and Ohtake, unpublished).

A DNA fragment, which complements the mutation in APC1, was cloned from the *P. aeruginosa* genome by using the broad-host-range cosmid vector pCP19. A recombinant plasmid, pPT06, containing a DNA insert of approximately 20 kb, was reintroduced into *P. aerugninosa* PAO1, and the cells were screened for the ability to repress AP synthesis in APC1 under conditions of P_i excess. The computer-assisted capillary assay revealed that APC1(pPT06) cells were attracted to P_i only when they were starved for P_i. The DNA insert was digested with various restriction enzymes, and fragments were subcloned in cosmid vector pCP19 (Fig. 4). Results from complementation tests in APC1 showed that the 3.2-kb *Eco*RI- *Bgl*II fragment in pPT06.5 restored the ability of APC1 to repress both AP synthesis and P_i taxis under conditions of P_i excess (Fig. 2).

The nucleotide sequence of this 3.2-kb region was determined. Computer analysis revealed two potential open reading frames in this region. The polypeptides encoded by these open reading frames had 73 and 24% identities with the products of *E. coli pstB* and *phoU* genes, respectively. In *E. coli,* the PhoU protein has been identified as

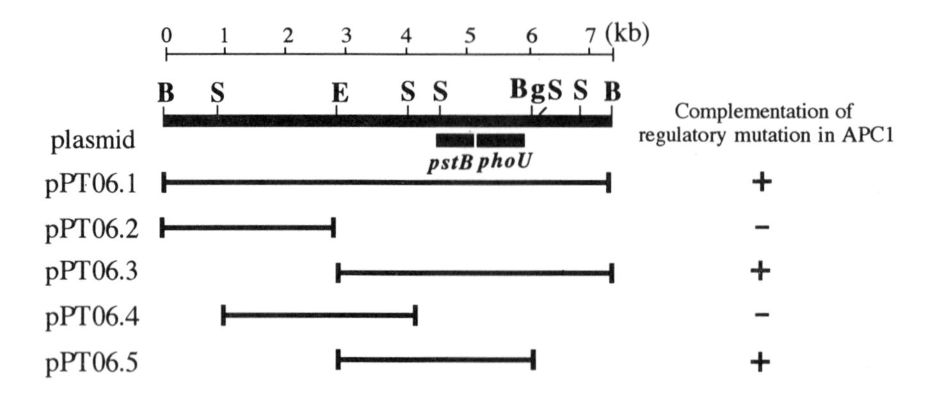

FIG. 4. Restriction map of the *P. aeruginosa* chromosomal DNA cloned into pCP19 and the ability of recombinant plasmids to complement the mutation in APC1. Restriction sites: B, *Bam*HI; Bg, *Bgl*II; E, *Eco*RI; S, *Sal*I. The locations of *pstB*- and *phoU*-like genes are shown below the restriction map.

a negative regulator of AP synthesis (Muda et al., 1992). It is of interest that the *pstB*- and *phoU*-like genes may be involved in the negative regulation of both AP synthesis and P_i taxis in *P. aeruginosa*. Unlike the *pho* regulon in *E. coli*, P_i taxis was not positively regulated by PhoB and PhoR. It is unclear whether a positive regulator exists for P_i taxis in *P. aeruginosa*. We also found P_i taxis in *Enterobacter cloacae* IFO3320 but not in *E. coli* strains.

REFERENCES

Anba, J., M. Bidaud, M. L. Vasil, and A. Lazdunski. 1990. Nucleotide sequence of the *Pseudomonas aeruginosa phoB* gene, the regulatory gene for the phosphate regulon. *J. Bacteriol.* **172:**4685–4689.

Kato, J., A. Ito, T. Nikata, and H. Ohtake. 1992. Phosphate taxis in *Pseudomonas aeruginosa*. *J. Bacteriol.* **174:**5149–5151.

Kato, J., and H. Ohtake. Unpublished data.

Moulton, T. T., and T. C. Montie. 1978. Chemotaxis by *Pseudomonas aeruginosa*. *J. Bacteriol.* **137:**274–280.

Muda, M., N. N. Rao, and A. Torriani. 1992. Role of PhoU in phosphate transport and alkaline phosphatase regulation. *J. Bacteriol.* **174:**8057–8064.

Nikata, T., K. Sumida, J. Kato, and H. Ohtake. 1992. Rapid method for analyzing bacterial behavioral responses to chemical stimuli. *Appl. Environ. Microbiol.* **58:**2250–2254.

Poole, K., and R. E. W. Hancock. 1984. Phosphate transport in *Pseudomonas aeruginosa:* involvement of a periplasmic phosphate-binding protein. *Eur. J. Biochem.* **144:**607–612.

Wanner, B. L. 1987. Phosphate regulation of gene expression in *Escherichia coli*, p. 1326–1333. *In* F. C. Neidhardt, J. L. Ingraham, K. B. Low, B. Magasanik, M. Schaechter, and H. E. Umbarger (ed.), *Escherichia coli and Salmonella typhimurium: Cellular and Molecular Biology*, vol. 2. American Society for Microbiology, Washington, D.C.

X. STRUCTURE-FUNCTION RELATIONSHIPS

Chapter 54

Introduction: Structure and Function of *Escherichia coli* Alkaline Phosphatase

EVAN R. KANTROWITZ

Department of Chemistry, Merkert Chemistry Center, Boston College, Chestnut Hill, Massachusetts 02167

Since the last phosphate meeting in 1986 the resolution of the crystal structure of *Escherichia coli* alkaline phosphatase has been extended and refinement of the structure has allowed the use of site-specific mutagenesis to probe the mechanism of the enzyme and to investigate the role of the metals in structural stabilization and catalysis, as well as the relationship between the more active mammalian enzymes and the *E. coli* enzyme.

The following nomenclature system is used in this chapter. The notation used to name a mutant version of alkaline phosphatase is, for example, K328H. The original amino acid is indicated by the one-letter amino acid code to the left of the location in the sequence, and the new amino acid is indicated by the one-letter code to the right of the location. Hence K328H/D153H is the double-mutant alkaline phosphatase in which Asp-153 and Lys-328 are both replaced by histidine residues. E—P is the covalent complex containing the phosphoserine at position 102, and E·P is the noncovalent complex between the enzyme and P_i.

Eunice Kim (Vertex Pharmaceuticals), who has worked for a number of years with Harold Wyckoff at Yale University, presented a paper on the refinement and the analysis of the X-ray structure of the enzyme at the most recent phosphate conference in Woods Hole, Mass. Recent improvements in methods of data collection for large proteins provided the means to increase the resolution of the structure of the *E. coli* enzyme from 2.8 to 2.0 Å (Kim and Wyckoff, 1989, 1991). The refinement of the X-ray structure of the phosphate-bound enzyme led to clear density for all the residues except a few (1 to 4 and 404 to 408 and some of the side chain atoms). Also, the active site of the enzyme has been clarified. The knowledge of the structure of the enzymes determined in the presence of phosphate, in the absence of phosphate, and with cadmium replacing the normal metals has also provided insights into the catalytic mechanism of the enzyme: this provided the subject of talks by Wyckoff and Coleman.

Figure 1 shows the overall structure of the alkaline phosphatase dimer based on the 2.0-Å data determined in the presence of phosphate (Kim and Wyckoff, 1991). The overall molecule is approximately 100 by 50 by 50 Å with the two active sites approximately 30 Å apart. The enzyme has an α/β structure with 10 strands of sheet making up the central β-sheet. This β-sheet is connected by 15 α-helical regions. The top portion has been retraced, and it turns out that there is another domain which is composed of three strands of sheet and a short helix. The extensive interface between the two monomers is composed of both hydrophobic and hydrophilic residues. Furthermore, the interface has approximately 50 buried water molecules and 40 more associated around the edge. Thus, not all of the water need be extruded during assembly, and evolutionary drift is facilitated since the monomer surfaces need not be precisely complementary.

In addition to her work on the refinement of the structure of *E. coli* alkaline phosphatase, Kim reported on her work comparing the primary structures of a variety of bacterial and mammalian alkaline phosphatases (Kim and Wyckoff, 1989). Although there is only approximately 25 to 30% identity between the bacterial and mammalian sequences, the residues in the active site, including all the metal-binding ligands, are the same except for that at position 155, which is Thr in bacteria and Ser in mammals. Most of the gross differences between the bacterial and mammalian enzymes involve insertions and/or deletions of loop

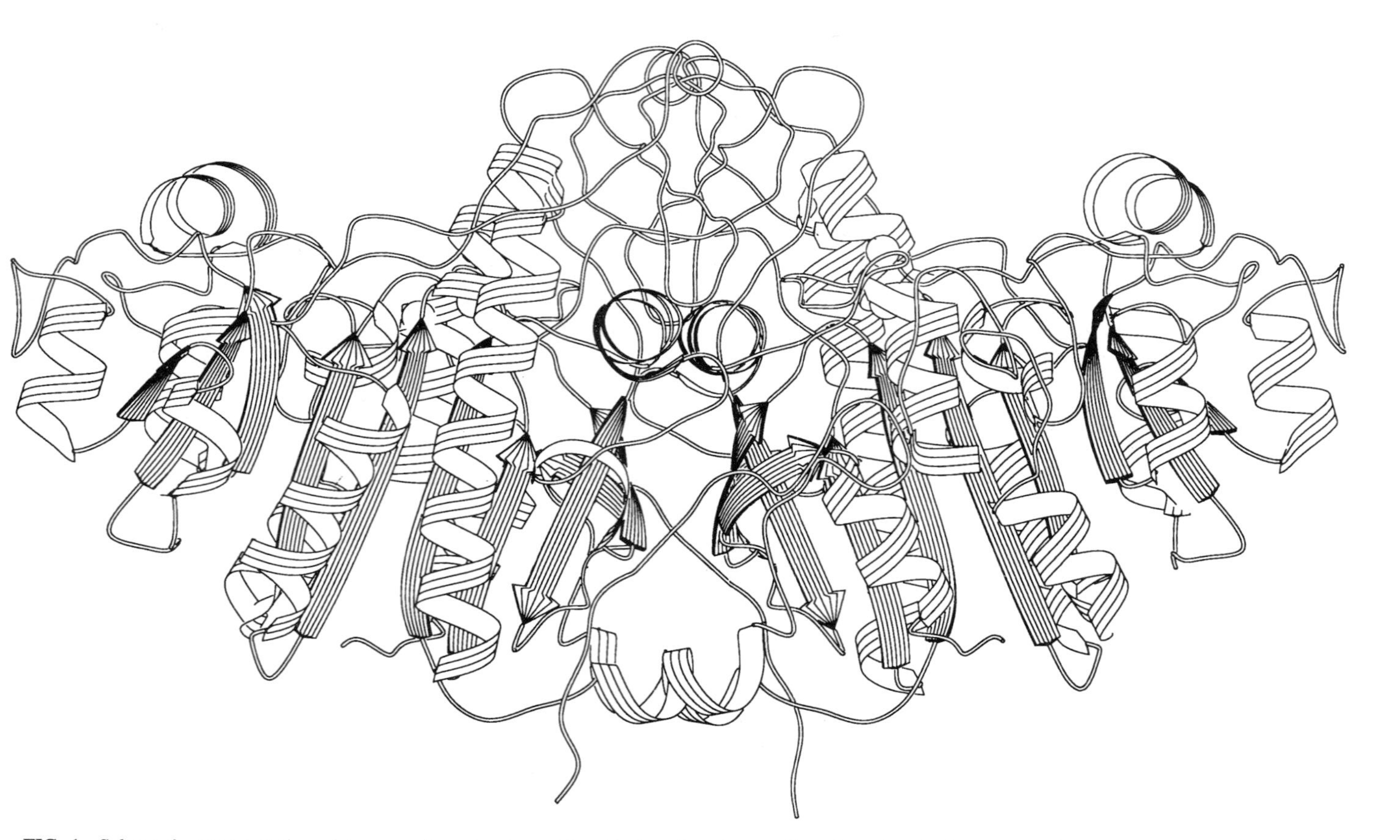

FIG. 1. Schematic representation of the secondary structure of the alkaline phosphatase dimer based on the X-ray crystallographic data of Kim and Wyckoff (1991). The β-sheet is shown as a ribbon with an arrow denoting the direction; the α-helices are shown idealized. This figure was drawn with the program SETOR (Evans, 1993).

regions. Close to the active site there are two noticeable differences in the sequences between the bacterial and mammalian enzymes. In *E. coli,* positions 153 and 328 are Asp and Lys, respectively, but in the mammalian enzyme, His is present at both of these positions. When the mammalian sequence is modeled onto the *E. coli* structure, the active region is almost identical, except at positions 153 and 328.

Wyckoff started his talk by reminding us that the first crystals of alkaline phosphatase were grown by Coleman in 1969, using some of the 7 g of enzyme that had been prepared for nuclear magnetic resonance spectroscopy (NMR) studies. Although two crystal forms were found with space groups $P3_121$ and I222, it was the I222 form that was ultimately solved at high resolution. After a brief introduction to the structure, Wyckoff's talk concentrated on three topics, the differences between the bacterial and mammalian enzymes, the role of the metals, and the catalytic mechanism. As Kim had mentioned, the insertions and deletions observed in the comparison between the bacterial and mammalian enzymes are, for the most part, found in loop regions. Wyckoff pointed out that the insertions in the mammalian sequences not only occur at loops but also are entirely on a single exon. Furthermore, the mammalian exon boundaries do not fall in the middle of these loops but rather at the ends, or beyond, within the more conserved regions. In one case, however, a whole exon is "inserted" into an *E. coli* loop to change a 30-residue minidomain into a 90-residue domain. The evolutionary significance of these findings is perhaps substantial but is still obscure. The differences between gram-positive bacteria, gram-negative bacteria, yeasts, and mammals undoubtedly involve both drift and positive fixation. At present we do not know the specific functions of the various isozymes and cannot speculate on the functional significance of specific variations. It is clear, however, that retention of phosphate may be as important as hydrolysis in a phosphate deprivation response in *E. coli,* whereas rapid turnover and release may be the biologically favored mode of operation in other cases. Clearly, *E. coli* had the opportunity to experience and accept point mutations in the active-site region that increase the turnover rate 10- to 20-fold as seen in mammals and discussed below.

Figure 2 shows the three metal-binding sites in *E. coli* alkaline phosphatase. With respect to the pH dependence of the equilibrium binding of Mg, Wyckoff pointed out that the affinity increases with pH in the range from 7 to 9. The necessary conclusion is that a proton is released during complex formation. This could be ionization of a hydroxyl group of the protein or a bound water molecule, deionization of a histidine abnormally protonated in the free enzyme, or ionization of an abnormally protonated carboxyl group. In light of the observed structure of the fully metalated enzyme, the most probable cause is ionization of Thr-155 when Mg is added. Further analysis of the binding equilibria would be needed to determine if a full proton release is involved at each pH. It is possible, for example, that in the free apo enzyme Thr-155 is fully protonated over this entire range of pH and that it is only partially ionized at some pHs within this range. If the plot of log (K) versus pH is linear with a slope of 1, a full ionization is involved and all that one can deduce about the pK of the complex is that it is below the range of observation. If the slope is less than 1 within all or part of the range, the pK of the complex (or the components) is within the range. One method to confirm that the ionization of Thr-155 is responsible for the pH dependence of Mg binding would be to use site-specific mutagenesis to replace Thr-155 by Ser and determine whether this substitution alters the pH dependence of Mg binding.

Wyckoff also pointed out that the equilibrium between E—P and E•P is also pH dependent. As Coleman has determined by NMR, at a pH of approximately 5 there is an equal mixture or E—P and E•P and at pH 7 there is very little E—P. The fact that this equilibrium is pH dependent must mean that there is a net proton involvement in the reaction, which should be written into the equation to fully describe the system. As in the case of the Mg binding equilibria a plot of log (K) versus pH would reveal whether a full proton is involved over the entire range. If so, nothing can be said about the pK of a specific group other than that no significant pK is within the range of observation. The apparent pK of 5.0 or 5.5 would be the pK of the reaction. Since E—P is favored at low pH, this requires that the charge on E—P be less negative than that on E•P. In light of the structures of the complexes, how can this be? The simplest explanation is that the charge on P_i in the complex is -3 and, of course, the charge on E—P is -2. This proposition and the implications are discussed further below.

Wyckoff continued the discussion of the phosphate complex pointing out that P_i is bound much more tightly than sulfate. For example, the crystals used for the structure determination contain approximately 3 M sulfate (as ammonium sulfate), yet sulfate is not observed bound in the active site even in the absence of phosphate. Phosphate is bound in the active site not only by two zinc atoms but also by Arg-166 with two salt links, the amide NH of Ser-102, two water molecules linked to Mg and Lys-328 and Asp-153, and at least one more well-positioned water molecule (Fig. 3). At alkaline pH, the release of phosphate from the noncovalent complex is part of the rate-determining process. At pH 7.0 the slow hydro-

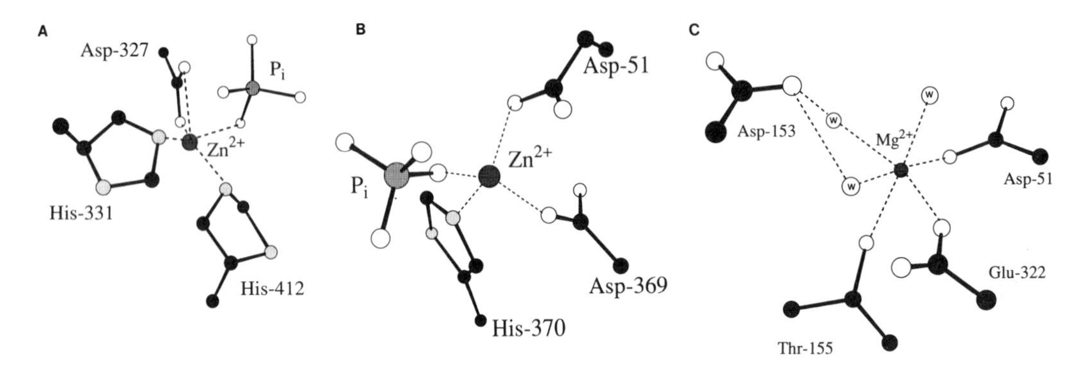

FIG. 2. Metal-binding sites in *E. coli* alkaline phosphatase (Kim and Wyckoff, 1991). (A) the Zn_1-binding site. The Zn^{2+} is coordinated to both carboxyl oxygens of Asp-327, imidazole nitrogens of His-331 and His-412, and one of the phosphate oxygens. The hydroxyl (not shown) that hydrolyzes the phosphoserine is also coordinated to this zinc atom. (B) The Zn_2-binding site. The Zn^{2+} is coordinated to a phosphate oxygen, a carboxyl oxygen of Asp-369, an imidazole nitrogen of His-370, and one of the carboxyl oxygens of Asp-51. The other carboxyl oxgyen of Asp-51 is coordinated to the Mg^{2+}. This Zn^{2+} is also involved in deprotonation of Ser-102 by direct interaction with the hydroxyl oxygen. (C) The magnesium-binding site. The Mg^{2+} is octahedrally coordinated to a carboxyl oxygen of Asp-51 and Glu-322, to the hydroxyl of Thr-155, and to three water molecules. Asp-153 interacts via hydrogen bonds with two of the water molecules coordinated to the magnesium. Water molecules are indicated by an open circle containing the letter w. In mammalian alkaline phosphatases, Thr-155 is replaced by serine and Asp-153 is replaced by His.

lysis of E—P is dominant while at pH 9.0 the slow release of P_i is dominant. In between, there is a change in the rate-limiting step with both rate constants involved. The slow off rate and the slow phosphorylation of Ser-102 by P_i in the reverse reaction are both consistent with and partially explained by the hypothesis that bound P_i is the trianion over the entire range from pH 5.0 to 10.0 and higher. This hypothesis is also supported by the ^{31}P NMR chemical shift data of Gettins and Coleman (1983), which showed very little change over all of the range observable from pH 5 to 10.

Gettins et al. (1985) showed by NMR inversion transfer that the off rate of P_i from *E. coli* alkaline phosphatase is approximately the same at pH 6.5 and 8.8 and propose that it is sensibly invariant in between, although the data presented would allow for a dip by a factor of 2 in between. At pHs above 8.5 the complex with P_i is weakened by increasing the hydroxyl or chloride ion concentration. The pH effect requires a net increase in negative charge upon release of P_i. This is counter to the trianion hypothesis, since the pK of free P_i is 12.5. If P_i is protonated during release, then more than one negative charge must have developed. Wyckoff proposed that one event could be the deprotonation of Ser-102 as P_i leaves. This requires that Ser-102 be protonated in the E·P complex much of the time. This would help in the current dilemma and also help explain the very low rate of the E·P → E—P reaction. This reaction, according to the current scheme, would require protonation of the appropriate O of P_i so that OH^- can depart and the deprotonation of Ser-102 so that it can react during formation of the SerOP bond. E·P would be a nonproductive complex most of the time in either the forward or reverse reaction. The second event required to obtain a net increase in negative charge during P_i departure would be the ionization of a water molecule binding to the Zn sites vacated by the P_i. This would be consistent with the enhancement of the departure rate by chloride ion shown by Coleman's laboratory (see below).

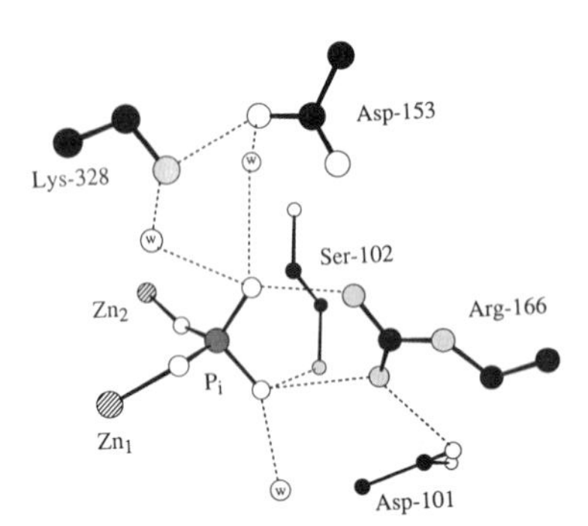

FIG. 3. Phosphate-binding site of *E. coli* alkaline phosphatase based on the 2.0-Å resolution X-ray structure determined in the presence of phosphate at pH 7.5 (Kim and Wyckoff, 1991). The phosphate is coordinated to both Zn_1 and Zn_2 and also directly interacts with Arg-166. Lys-328 and Asp-153 each form a water-mediated interaction with a phosphate oxygen. Water molecules are indicated by an open circle containing the letter w.

There is one further implication of much of what has been said above, namely, that the equilibrium ratio of [E—P] to [E·P] is much lower than the steady-state ratio in the pH range from 7.0 upward. If the rate constant for departure of P_i is constant over the pH range at which the overall steady-state rate increases by a factor of 10, the fraction of the enzyme in the E·P complex must increase by a factor of 10 also. Since we are considering V_{max} and not postulating a free protein isomerization step, we must conclude that the sum of [E·S] + [E—P] decreases from 91 to 9% over this pH range. It is generally accepted that [E·S] is not a substantial fraction of [E_{total}] on the basis of many lines of evidence including burst kinetics with phosphate free enzyme at pH 8.0 (Bloch and Schlesinger, 1973). In the steady state, therefore, [E—P] must be approximately equal to [E_{total}](k_{cat} at pH 10 − k_{cat} at the experimental pH). At the pH where k_{cat} is $k_{max}/2$, [E—P] = [E·P]. At this pH the equilibrium ratio of [E—P] to [E·P] is near zero by observation. If no pK values of either of these compounds changed in this region and at equilibrium [E—P] = [E·P] at pH 5.0, at pH 7.5 the calculated ratio would be $10^{-2.5} \approx 1/300$.

Wyckoff also discussed the multiplicity of interactions of P_i with the enzyme. When Arg-166 is mutated to Ala, the affinity for P_i is greatly reduced, as is the hydrolase activity. The transferase activity is affected much less strongly. This implies that there is enough redundancy in the active-site interactions with substrates and intermediates to allow catalysis with a good leaving group in the first step and "attacking" group in the second without any assistance from Arg-166. With a poor attacking group in the second step, namely, water, assistance from Arg-166 is effective and is almost required.

Wyckoff also reported the structures of two mutant versions of alkaline phosphatase that were determined by Kim. The first had Arg-166 replaced by Ala, and the second had Asp-101 replaced by Ala. The R166A enzyme exhibits a 30-fold reduction in hydrolysis activity (Chaidaroglou et al., 1988) and weaker binding of phosphate. This reduction in phosphate affinity would be predicted from the structure since Arg-166 directly stabilizes the bound phosphate. In the presence of a phosphate acceptor, the difference in the activity between the mutant and wild type is much smaller, suggesting that this mutation has a larger influence on hydrolysis than on transphosphorylation. The structure of the R166A enzyme was virtually identical to that of wild type except at the site of the substitution. The D101A enzyme exhibits enhanced activity (Chaidaroglou and Kantrowitz, 1989), and its overall structure is the same as that of the wild-type enzyme, except that the position of Arg-166 is altered. In the wild-type structure, Asp-101 stabilizes the position of Arg-166 via a salt link. Obviously this link cannot occur in the D101A enzyme, and the loss of this interaction allows a disordering of the side chain of Arg-166. This structure is in agreement with the structure of another mutant of alkaline phosphatase in which Asp-101 was replaced by Ser (Chen et al., 1992). In the structure of the D101S enzyme, Arg-166 is also disordered; this enzyme has an increased catalytic rate and weaker binding affinity for phosphate (Mandecki et al., 1991).

Coleman also discussed the detailed catalytic mechanism of the enzyme, based on both NMR and structural studies (see below). He stressed the point that Cd NMR has been very useful in studying the catalytic mechanism of alkaline phosphatase because the chemical shift of the Cd at each of the three metal-binding sites is different (Coleman et al., 1983). In addition to Cd, studies with Mn have provided insights into the catalytic mechanism, because the paramagnetism of Mn has provided a means to observe the waters coordinated to Zn_1 (Schulz et al., 1989). Since a water or hydroxide coordinated to Zn_1 is critical in the proposed mechanism, it was important to confirm its presence through direct observation. In the absence of phosphate, two water molecules are coordinated to Zn_1, whereas on addition of phosphate, only one is coordinated. Similar results were also obtained with Cl NMR. In the absence of phosphate two Cl^- ions are bound to Zn_1, but in the presence of phosphate only one is bound.

The reaction catalyzed by the enzyme with Cd at all three metal sites has greatly reduced kinetic parameters; in effect, this is like having the reaction go in slow motion. As Coleman pointed out, this kinetic behavior is perfect for NMR experiments, since it allows the catalytic reaction of alkaline phosphatase to be observed on the NMR time scale. In fact, using the Cd-substituted enzyme, over a period of days one can monitor the formation of the phosphoenzyme by ^{31}P NMR.

Figure 4 shows the proposed mechanism of alkaline phosphatase based on the solution studies of Coleman and the structural studies of Wyckoff. As seen in this figure (upper left), the substrate is bound with the ester oxygen coordinated to Zn_1 and another phosphate oxygen coordinated to Zn_2. The phosphate is stabilized by Arg-166 (R166), and one molecule of water is shown coordinated to Zn_1. This arrangement positions Ser-102 (S102) in the proper orientation for apical attack on the phosphorus, while Zn_1 is positioned correctly to stabilize the developing negative charge on the leaving RO^-. In the E—P complex (Fig. 4, upper right), a second water molecule can coordinate to Zn_1. The pK_a of this water should be between 8 and 9, on the basis of results obtained with model compounds (Kimura et al., 1991) and the observed pH dependence of the enzymatic activity. Thus, a hydroxide nucleophile is available for at-

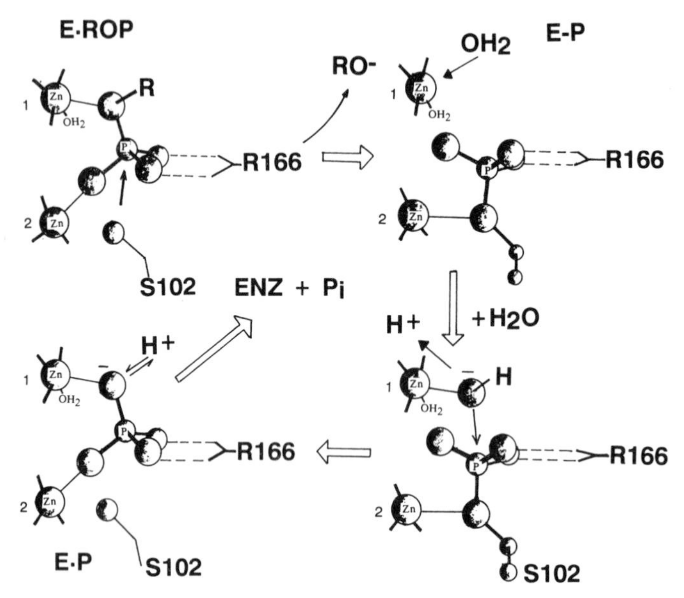

FIG. 4. Major intermediates in the proposed mechanism of action of alkaline phosphatase. The dianion of a phosphate monoester, $ROPO_4^{2-}$, forms E·ROP, in which the ester oxygen coordinates Zn_1, a second oxygen coordinates Zn_2 to form a phosphate bridge between the Zn_1 and Zn_2, and the other two phosphate oxygens form hydrogen bonds (---) with the guandium group of R166. Ser-102 is the nucleophile in the first half of the reaction and would occupy the position opposite to the leaving group, RO^-, in a five-coordinate intermediate. Upon formation of E—P, the phosphate (as the dianion of a phosphoseryl residue) moves slightly into the cavity of the active center but maintains a coordinate bond between Zn_2 and the ester oxygen as well as the two hydrogen bonds with R166. A water molecule coordinates Zn_1 in the position occupied originally by the ester oxygen of the substrate. At alkaline pH the water dissociates a proton to become Zn-$^-$OH, in position to be the nucleophile for the hydrolysis of the phosphoseryl ester. E·P is formed as the phosphate moves away from the serine and one of the phosphate oxygens again coordinates Zn_1 to reestablish a phosphate bridge. E·P is pictured as a phosphate trianion for the reasons discussed in the text. Dissociation of phosphate from E·P is the slowest, $\sim 35\ s^{-1}$, and therefore the rate-limiting step.

tack on the phosphorus (Fig. 4, lower right), again in the apical position, to displace Ser-102 from the opposite side. In this case, Zn_2 is coordinated to the hydroxyl of Ser-102 and thus can stabilize the developing negative charge. Following two inversions, the reaction as proposed goes with overall retention of configuration at the phosphorus, as was observed experimentally (Jones et al., 1978). As seen in Fig. 4 (lower left), the phosphate is bound in the E·P as the trianion.

Additional evidence for the phosphate bound to E·P as the trianion comes from studies of the kinetics of the reverse reaction. Phosphorylation of the enzyme by phosphate is dramatically slower than phosphorylation by similar compounds, such as methylphosphate, that cannot exist in trianionic form. Thus, protonation of the trianionic phosphate is required for release of phosphate from the E·P complex.

Albert Matlin (Oberlin College) presented a paper on his studies involving site-specific mutagenesis to probe the active site of *E. coli* alkaline phosphatase. One of the interesting questions yet to be answered is why the mammalian alkaline phosphatases are more active than the bacterial enzymes. As was previously mentioned by Kim (see above), the active sites of the bacterial and mammalian enzymes are conserved except for Lys-328 and Asp-153 in the *E. coli* sequence, both of which are histidines in mammalian enzymes. As seen in Fig. 3, Asp-153 is not a direct ligand of the Mg at the M3 site but rather is an indirect ligand coordinated via two water molecules. To probe the role of these interactions, the D153N and D153A mutants of *E. coli* alkaline phosphatase were generated by site-specific mutagenesis (Matlin et al., 1992). As seen in Table 1, the k_{cat} of the D153A enzyme is elevated 13.5-fold compared with that of the wild-type enzyme. The more modest replacement of D153 by Asn results in an enzyme with slightly decreased activity compared with the wild type. The D153A enzyme also exhibits a 7.7-fold increase in K_m; therefore, the catalytic efficiency, as measured by the k_{cat}/K_m ratio, is actually better for the D153A enzyme than for the wild-type enzyme. As predicted from the X-ray structure of the wild-type enzyme (Kim and Wyckoff, 1989, 1991), these

TABLE 1. Kinetic parameters of the wild-type and the D153N and D153A mutant enzymes at pH 8[a]

Enzyme	k_{cat} (s^{-1})	K_m (μM)	$10^6 k_{cat}/K_m$ ($M^{-1} s^{-1}$)
Wild type	59 ± 4	20.5 ± 4	2.9
D153N	30.0 ± 1.0	20.4 ± 5.0	1.5
D153A	799 ± 50	153 ± 15	5.2

[a]Assays were performed at 25°C in 1 M Tris buffer (pH 8) with 10 mM $MgCl_2$ and 50 μM $ZnSO_4$ and with *p*-nitrophenyl phosphate as the substrate.

mutant enzymes require the addition of Mg for full activity.

The binding of phosphate as measured by the K_i is greatly reduced in the D153A enzyme, and this decreased affinity of the enzyme for phosphate may provide a partial explanation for the enhanced activity of these mutant enzymes. However, ^{31}P NMR analysis of the enzyme-catalyzed reaction in 1 M Tris (*p*-nitrophenyl phosphate as the substrate at pH 8.0) indicates that the transphosphorylation/hydrolysis ratio increases fourfold in the D153A enzyme. Therefore, the increased k_{cat} of the D153A enzyme must be due in part to an increase in the rate of the transphosphorylation pathway. The pH profile of activity for both the D153A and D153N enzymes is also altered, with the pH for optimal activity shifted toward higher pH.

Matlin concluded by observing that these experiments have indicated that the aspartic acid residue at position 153 in the *E. coli* enzyme is partially responsible for the lower activity of the bacterial alkaline phosphatases compared to the mammalian enzymes. He suggested two possible explanations for this alteration in activity. First, by removal of Asp-153 the interaction of the active site with the phosphate via a water molecule coordinated to the Mg could be lost (Fig. 3). This would weaken the binding of phosphate, and since phosphate release is rate determining at this pH the activity of the enzyme would be enhanced. Second, if phosphate is bound as the trianion, it must take up a proton before it can leave. The removal of the side chain of residue 153 could alter the pK_a of the water that is coordinated to the Mg and interacts with the phosphate. If the water could more easily donate a proton, the phosphate might be able to leave more readily. The second explanation could also account for the observed changes in the pH-versus-activity profile of the D153A enzyme compared with the wild-type enzyme.

I would like to pick up where Matlin left off. One question we have addressed in my laboratory at Boston College is why the mammalian alkaline phosphatases are 10- to 20-fold more active than the bacterial enzymes. In addition, why are the pH-versus-activity profile of the mammalian enzymes shifted to higher pH? Our strategy was to use site-specific mutagenesis to replace Asp-153 and Lys-328 in the *E. coli* enzyme with histidine residues. As mentioned by Kim (see above), these are the only differences between the *E. coli* and mammalian enzymes in the vicinity of the active site.

The K328H enzyme exhibits both enhanced activity and a shift in the pH activity profile. We have studied this enzyme in detail (Xu and Kantrowitz, 1991) and found that the binding of the substrate is altered only slightly, whereas the binding of phosphate is dramatically weakened. As seen in Fig. 3, Lys-328 in the *E. coli* structure, interacts with the phosphate via a water molecule. Thus, it is possible that for initial binding of the substrate to the enzyme this water-mediated interaction is not important but that once phosphate is bound, the water-mediated interaction with Lys-328 gains significance for the binding of phosphate in the E·P complex. In fact, we may now speculate that this interaction is one of many that are necessary to stabilize the phosphate as the trianion.

Although the K328H enzyme is more active than the wild-type enzyme, it is not as active as the mammalian enzyme. We have also built a mutation at position 153, but in our experiment we introduced a histidine residue. In addition, we constructed the double mutant with histidine residues at both positions 153 and 328. A summary of the kinetic parameters for these mutants is given in Table 2. As was the case for the K328H enzyme, the pH activity profile of both of these enzymes is shifted toward higher pH. The double mutant with histidines at both positions is the most active, exhibiting a k_{cat} within a factor of 2 of the activity of the mammalian enzymes. In addition, the binding of phosphate is dramatically weakened in the double mutant but is actually stronger in the D153H enzyme (Janeway et al., 1993).

Both for the D153H mutant and the D153H/K328H double mutant, additional Mg must be added to obtain the maximal activity. In fact, the Mg activation is time dependent but is independent of the Mg concentration, suggesting that there must be a fast bimolecular step involving the formation of a Mg·E complex followed by a slower unimolecular step resulting in the formation of the active enzyme.

To determine in more detail the reasons for the observed behavior of the D153H enzyme, we crystallized it and determined its X-ray structure

TABLE 2. Kinetic parameters of the wild-type and mutant enymes at pH 10.0[a]

Enzyme	k_{cat}[b] (s^{-1})	K_m (μM)	$10^6\ k_{cat}/K_m$ ($M^{-1}\ s^{-1}$)	Mg^{2+} concn (mM)
Wild type	82 ± 7	106 ± 7	0.77	1
Wild type	86 ± 4	86 ± 5	1.0	10
D153H/K328H	17 ± 1	407 ± 14	0.04	<0.1
D153H/K328H	262 ± 10	405 ± 20	0.65	1[c]
D153H/K328H	508 ± 44	608 ± 65	0.84	10[c]
D153H	48 ± 7	477 ± 52	0.10	1[c]
D153H	286 ± 31	448 ± 76	0.64	10[c]

[a]Assays were performed at 25°C in 0.1 M (cyclohexylamino) propane-sulfonic acid buffer (pH 10.0) (ionic strength adjusted to 0.5 M with NaCl) with *p*-nitrophenyl phosphate as the substrate.

[b]The k_{cat} values are calculated from the V_{max} by use of a dimer molecular weight of 94,000 (Bradshaw et al., 1981). The k_{cat} per active site would be half of the value indicated.

[c]The enzymes were preincubated for 2 h at 24°C in 10 mM Mg^{2+} in 0.1 M (cyclohexylamino) propanesulfonic acid buffer (pH 10.0) before the assays were performed.

to 2.4 Å. The structure has now been refined to an R factor of 0.197 with root mean square deviation of angles of 3.2° and root mean square deviation of bonds of 0.015 Å (Murphy et al., 1993). As seen in Fig. 5, the metal at the M3 site is no longer Mg but Zn. The octahedral Mg site has been converted into a tetrahedral Zn site with one of the imidazole nitrogens of the new histidine at 153 acting as a ligand to the Zn. This structure represents the inactive form of the enzyme, since large amounts of Mg must be added to activate the D153H enzyme. The activation process may involve the slow replacement of Zn by Mg after the initial rapid binding of Mg. Another explanation would invoke a slow conformational change of the enzyme after Mg has replaced Zn. However, there are very few overall structural differences between the wild-type and D153H structures except in the active-site area.

The X-ray structure of the D153H enzyme has also provided an explanation for the enhanced binding of phosphate. As demonstrated in the structure (Murphy et al., 1993) and previously suggested by molecular dynamics (Janeway et al., 1993), when the salt link between Lys-328 and Asp-153 is broken in the D153H enzyme, Lys-

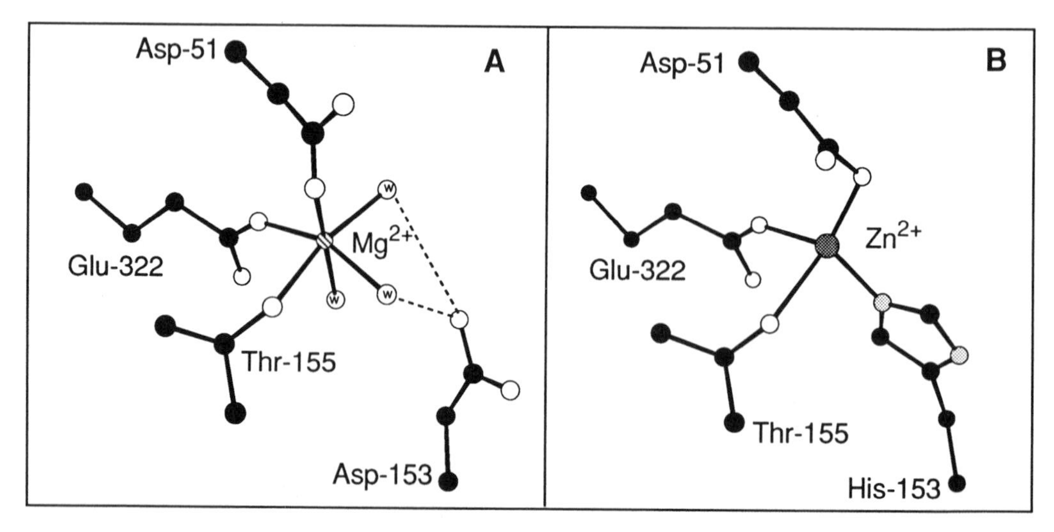

FIG. 5. (A) Structure of wild-type *E. coli* alkaline phosphatase in the region of the magnesium-binding site based on the structure of Kim and Wyckoff (1991). The ligands to the octahedral magnesium include the carboxylate oxygens of Asp-51 and Glu-322, the hydroxyl oxygen of Thr-155, and the oxygens of three water molecules (circled letter w). Asp-153 interacts indirectly with the magnesium via two of these water molecules. (B) Structure of the D153H alkaline phosphatase in the same region of the structure as shown in panel A. The ligands to the tetrahedral zinc include the carboxylate oxygens of Asp-51 and Glu-322, the hydroxyl oxygen of Thr-155, and a nitrogen of the imidazole ring of His-153.

328 can swing down and directly interact with the phosphate (Murphy et al., 1993). This would add stabilization to the interactions with the phosphate and explain the enhanced binding to the enzyme.

In summary, the replacement of Asp-153 and Lys-328 by histidine residues creates a mutant enzyme that is almost as active as the mammalian enzyme and exhibits properties characteristic of the mammalian enzymes, including a shift in the pH of optimal activity. Furthermore, at least some of the mammalian enzymes (Brunel and Cathala, 1973; Cathala et al., 1975) are activated in a time-dependent fashion by Mg in a manner completely analogous to that observed with the D153H and D153H/K328H enzymes. At low Mg concentrations, these mammalian alkaline phosphatases may actually exist with three Zn atoms in each active site. Activation would result from the replacement of one of these Zn atoms by Mg, just as in the D153H and D153H/K328H enzymes.

Catherine Brennan (Abbott Laboratories) discussed the use of the three-dimensional structure of the enzyme along with sequence comparisons between different species to develop a hybrid alkaline phosphatase system; the activity of this hybrid is modulated by the binding of antibody specific to an epitope inserted into a loop region of the structure. Two regions of the structure were chosen for possible insertions, the loops at 91 to 93 and 407 to 408 (Fig. 6). These positions were selected because they were close to the active site and were predicted to be sites that can probably tolerate insertions, on the basis of sequence comparisons between the *E. coli* and other bacterial and mammalian alkaline phosphatases. One of these hybrids, API1, has a 13-amino-acid sequence, an epitope from the human immunodeficiency virus type 1 gp120 protein, inserted between residues 407 and 408. API1 maintains the full activity of the wild-type alkaline phosphatase, but in the presence of the anti-gp120 antibody the enzyme activity is inhibited by 40 to 50%. Thus, the hybrid enzyme can be used to detect the presence of antibody in solution. Brennan also showed some hypothetical models of the alkaline phosphatase anti gp120 antibody complex. The one likely reason for the loss of activity upon complex formation is a conformational change of the protein near the active site.

Brennan also summarized the structural work being done at Abbott by Cele Abad-Zapatero on the D153G enzyme. Crystallographic data have been collected from crystals of the mutant with the metals present in the active site. These crystals loose the active-site metals very easily, without apparent damage. Data from experiments with the demetallized crystals and from crystals soaked only in Zn^{2+} and only in Mg^{2+} have also been collected. Efforts are under way to characterize

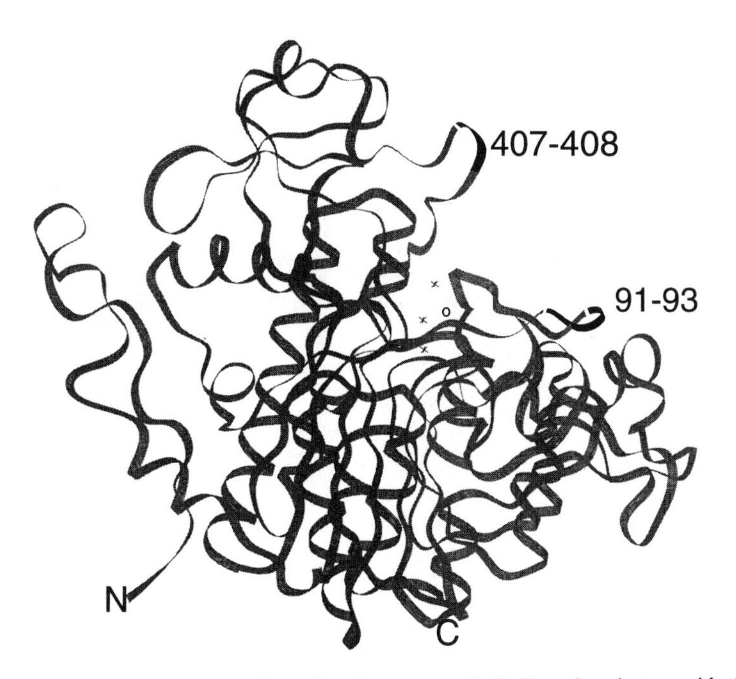

FIG. 6. Ribbon diagram of the three-dimensional structure of alkaline phosphatase with the insertion sites labeled. The positions of the phosphate and metal atoms in the active site are marked. Symbols N and C indicate the amino and carboxy termini. Only one subunit is shown. The interface with the other subunit is in front. This figure was generated by using InsightII (Biosym) on a Silicon Graphics computer.

the structural changes taking place on the mutant enzyme under these conditions.

The availability of structural data for multiple forms of *E. coli* alkaline phosphatase has been critical for our understanding of this protein. However, the catalytic mechanism of *E. coli* alkaline phosphatase could not have been deduced by crystallography alone. It has been necessary to combine the crystallographic data with other techniques such as NMR, kinetic analysis, and site-specific mutagenesis to provide functional data that can be directly related to the structural data. This combined approach has provided us with a very detailed description of this most important enzyme.

The work in my laboratory was supported by grant GM42833 from the National Institute of General Medical Sciences and in part by Pittsburgh Supercomputing Center grant number 1 P41RR06009 from the NIH National Center for Research Resources.

I thank each of the participants, particularly H. Wyckoff, for carefully reviewing and editing their portion of the manuscript.

REFERENCES

Bloch, W., and M. J. Schlesinger. 1973. The phosphate content of *Escherichia coli* alkaline phosphatase and its effect on stopped flow kinetic studies. *J. Biol. Chem.* **248:**5794–5805.

Bradshaw, R. A., F. Cancedda, L. H. Ericsson, P. A. Newman, S. P. Piccoli, K. Schlesinger, and K. A. Walsh. 1981. Amino acid sequence of *Escherichia coli* alkaline phosphatase. *Proc. Natl. Acad. Sci. USA* **78:**3473–3477.

Brunel, C., and G. Cathala. 1973. Activation and inhibition processes of alkaline phosphatase from bovine brain by metal ions (Mg^{2+} and Zn^{2+}). *Biochim. Biophys. Acta* **309:**104–115.

Cathala, G., C. Brunel, D. Chappelet-Tordo, and M. Lazdunski. 1975. Bovine kidney alkaline phosphatase. Catalytic properties, subunit interactions in the catalytic process, and mechanism of Mg^{2+} stimulation. *J. Biol. Chem.* **250:**6046–6053.

Chaidaroglou, A., J. D. Brezinski, S. A. Middleton, and E. R. Kantrowitz. 1988. Function of arginine in the active site of *Escherichia coli* alkaline phosphatase. *Biochemistry* **27:**8338–8343.

Chaidaroglou, A., and E. R. Kantrowitz. 1989. Alteration of aspartate 101 in the active site of *Escherichia coli* alkaline phosphatase enhances the catalytic activity. *Protein Eng.* **3:**127–132.

Chen, L., D. Neidhart, W. M. Kohlbrenner, W. Mandecki, S. Bell, J. Sowadski, and C. Abad-Zapatero. 1992. 3-D Structure of a mutant (Asp101→Ser) of *E. coli* alkaline phosphatase with higher activity. *Protein Eng.* **5:**605–610.

Coleman, J. E., K. I. Nakamura, and J. F. Chlebowski. 1983. ^{65}Zn(II), ^{115m}Cd(II), ^{60}Co(II), and Mg(II) binding to alkaline phosphatase of *Escherichia coli:* structure and functional effect. *J. Biol. Chem.* **258:**386–395.

Evans, S. V. 1993. SETOR: Hardware-lighted three-dimensional solid model representations of macromolecules. *J. Mol. Graphics* **11:**134–138.

Gettins, P., and J. E. Coleman. 1983. ^{31}P nuclear magnetic resonance of phosphoenzyme intermediate of alkaline phosphatase. *J. Biol. Chem.* **258:**408–416.

Gettins, P., M. Metzler and J. E. Coleman. 1985. Alkaline phosphatase. ^{31}P NMR probes of the mechanism. *J. Biol. Chem.* **260:**2875–2833.

Janeway, C. M. L., X. Xu, J. E. Murphy, A. Chaidaroglou, and E. R. Kantrowitz. 1993. Magnesium in the active site of *Escherichia coli* alkaline phosphatase is important for both structural stabilization and catalysis. *Biochemistry* **32:**1601–1609.

Jones, S. R., L. A. Kindman, and J. R. Knowles. 1978. Stereochemistry of phosphoryl group transfer using a chiral [^{16}O, ^{17}O, ^{18}O] stereochemical course of alkaline phosphatase. *Nature* (London) **275:**564–565.

Kim, E. E., and H. W. Wyckoff. 1989. Structure of alkaline phosphatase. *Clin. Chim. Acta* **186:**175–188.

Kim, E. E., and H. W. Wyckoff. 1991. Reaction mechanism of alkaline phosphatase based on crystal structures. *J. Mol. Biol.* **218:**449–464.

Kimura, E., Y. Kurogi, M. Shionoya, and M. Shiro. 1991. Synthesis, properties, and complexation of a new imidazole-pendant macrocyclic 12-membered triamine ligand. *Inorg. Chem.* **30:**4524–4530.

Mandecki, W., M. A. Shallcross, J. Sowadski, and S. Tomazic-Allen. 1991. Mutagenesis of conserved residues within the active site of *E. coli* alkaline phosphatase yields enzymes with increased K_{cat}. *Protein Eng.* **4:**801–804.

Matlin, A. R., D. A. Kendall, K. S. Carano, J. A. Banzon, S. B. Klecka, and N. M. Solomon. 1992. Enhanced catalysis by active-site mutagenesis at Asp-153 in *E. coli* alkaline phosphatase. *Biochemistry* **31:**8196–8200.

Murphy, J. E., X. Xu, and E. R. Kantrowitz. 1993. Conversion of a magnesium binding site into a zinc site by a single amino acid substitution in *Escherichia coli* alkaline phosphatase. *J. Biol. Chem.* **268:**21497–21500.

Schulz, C., I. Bertini, M. S. Viezzoli, R. D. Brown, S. H. Koenig, and J. E. Coleman. 1989. Manganese(II) as a probe of the active center of alkaline phosphatase. *Inorg. Chem.* **28:**1490–1496.

Xu, X., and E. R. Kantrowitz. 1991. A water-mediated salt link in the catalytic site of *Escherichia coli* alkaline phosphatase may influence activity. *Biochemistry* **30:**7789–7796.

Chapter 55

Phosphoporin: a Catalog of Open Questions

JURG P. ROSENBUSCH

Biozentrum, University of Basel, CH-4056 Basel, Switzerland

Phosphoporin, the product of the *phoE* gene, shares many characteristics with other porins from *Escherichia coli,* with respect to both its structure and its function (Cowan et al., 1992). The molecule is a trimer, each of the three identical subunits forming a β-barrel which consists of 16 antiparallel β-strands. The N and C termini of the polypeptide (340 residues long) are close enough to form a salt bridge, so that the barrel is quasicircular. The part of the barrel surface which is exposed to the hydrophobic core consists of small aliphatic residues, with the hydrogen-bonding potential of its backbone fully saturated by interstrand bonds. At the lipid-water interphase on either side of the hydrophobic zone, two bands of aromatic residues encircle the trimer. They consist of tyrosines and phenylalanines, with the hydroxyl groups of the former pointing toward the aqueous phase. This array probably plays a significant role in anchoring the porins in the membrane. The three parts with water-exposed surfaces are rather small and may be distinguished as follows: the part lining the channel which connects the two major aqueous compartments on both sides of the membrane; the part facing the periplasmic space (~8 Å high), which consists of eight sharp turns (two to four residues long) and contains several anionic charges; and on the outside, a dome-like structure (~20 Å in height) is exposed. It consists of six longer loops (9 to 16 residues) which are tightly packed and on which several anionic and a few cationic charges are found. The small channel entrance is located at the apex of this dome, where solutes may enter the vestibule through the mouth of the pore (Fig. 1). With the exception of one short α-helical segment (1.5 turns), these loops are aperiodic. Another loop (L2) bends away from the axis of the pore and latches into the neighboring subunit. The longest loop (L3; 35 residues, containing ca. two turns of α-helix) bends into the channel, where, approximately in the middle of the membrane, it presses toward the barrel wall, forming a bulge in the channel lumen which functions as a constriction site or selectivity gate in the transmembrane pathway and constitutes a brace which stabilizes the protein, as it interacts very tightly, both by electrostatic and hydrophobic interactions, with the barrel wall. Its structural role is clear from the effect of deletion mutations in the loop region of the OmpF porin: deletions between 6 and 15 residues increasingly affect the heat stability of the protein (Benson et al., 1988). In the constriction site of the channel, a strong electrostatic field exists: two carboxylate groups (D-113 and E-117) are situated on the loop (L3) opposite to the barrel wall (near the threefold molecular symmetry axis) where a cationic cluster (R-42, R-82, R-132, K-16, and K-18) exists in phosphoporin. The diameter of the constriction in the latter is 6 by 10 Å, slightly smaller than the corresponding dimension (7 by 11 Å) in OmpF porin. This difference is accounted for by a single lysyl residue (K-131; numbering according to the OmpF porin) which protrudes into the constricting area (Fig. 2).

Subunit interactions within the trimeric molecules are both hydrophobic and polar in character: near the threefold molecular axis and at the monomer interfaces, hydrophobic residues of neighboring subunits interdigitate, forming a protein mass in the center of the trimer so solid that it excludes water and detergent molecules (except for a few well-defined detergent monomers near the boundaries). Moreover, the L2 loop (see above) latches into the gap in the β-barrel of the neighboring subunit, a gap which arises from the bending of the constriction loop (L3) into the channel. The carboxyl group (E-71) at the tip of the latch in L2 forms a salt bridge with the Arg-cluster (R-42, R-82, and R-132) in the accepting monomer, which in itself is stabilized by two carboxylate groups (Fig. 3). Although the porin trimer consists of three thin-walled β-barrels, it is a very sturdy protein and, despite its overall polarity, not only is compatible with the environment with a low dielectric constant in the membrane core but also resides very stably in this environment. The coexistence of the topologically well-segregated interactions appears to allow the protein to withstand not only an extreme pH range (1.8 to 12.4) but also chaotropic agents (up to 5 M guanidinium chloride). Porins are not inactivated by treatments with most solvent systems, including organic solvents such as benzene and toluene, and they maintain their native oligomeric structure in harsh detergents such as sodium dodecyl sulfate. Native trimers are completely protease resistant (Rosenbusch, 1974). The structure of phosphoporin is highly conserved relative to that of the OmpF porin, and the overall architecture (at a resolution of 6 to 8 Å) is similar to that of maltoporin (Jap et al., 1991; Pauptit et al., 1991) and a porin from *Rhodobacter capsulatus* (Weiss et al., 1991). Details of the structures of *E. coli* porins have been published (Cowan et al., 1992).

If a solute molecule is followed along its pathway across the outer membrane barrier, it first passes the channel mouth, which, according to the X-ray structure, is small (Fig. 1), although it is not possible to determine the effect of its dynamic properties and hence the significance of its contribution to selectivity. In the crystal structure, the loops appear very rigidly packed, whereas mapping the surface topology by atomic force microscopy seems to indicate that conformational changes may occur (Schabert et al., in press). Solutes probably diffuse freely through the vestibule (~15 by 18 Å) and, after crossing the constriction, reach the wide, bell-shaped part of the lumen (diameter, ~15 by 22 Å) which communicates freely with the periplasmic space. Functionally, small polar solutes (<600 Da) pass porin channels essentially unhindered (Schindler and Rosenbusch, 1978), whereas hydrophobic solutes are slowed significantly (Nikaido, 1992). The channel conductance of phosphoporin is 0.6 nS (Dargent et al., 1986), intermediate between the OmpF and OmpC conductance values (0.8 and 0.5 nS, respectively [Buehler et al., 1991]). The channels close above characteristic threshold potentials, and although the functional significance of this phenomenon remains to be determined, the values of the critical voltages (100, 90, and 160 mV for OmpF, PhoE, and OmpC, respectively) are well defined for any particular porin. Phosphoporin channels are slightly selective for anions, whereas OmpF and OmpC channels exhibit a preference for cations (Benz et al., 1988). Larger phosphorylated compounds such as ATP inhibit the ion conductance significantly. Their respective affinities have been evaluated as inhibition constants by conductance measurements in a planar lipid bilayer system and are in the millimolar range (Dargent et al., 1986). Polyphosphates do not inhibit the ion conductance unless magnesium ions are present, and the selectivity is not restricted to phosphorylated compounds but is observed for other anions as well (Bauer et al., 1988). This observation is interesting, although the suggestion does not seem too far-fetched that phosphate, as the specific (negative) effector in the regulation of the expression of phosphoporin and the other gene products of the pho regulon, may have played a significant role in evolutionary terms. The effects of divalent cations will be discussed below. Lysyl residues have been invoked to explain ion selectivity, on the basis of both genetic evidence (Overbeeke et al., 1983) and the results of chemical modifications (Page and Rosenbusch, 1986). A specific substitution of the lysyl residue, which the X-ray structure revealed to protrude into the constriction site, by a carboxylate group (K131E) indeed inverts the selectivity of phosphoporin to a preference for cations (Bauer et al., 1989). Structural and functional results thus seem to afford an attractive basis for an understanding of the ion selectivity and may be considered a paradigm for molecular pathology.

Mutants in the OmpF porin that affect permeation have been isolated by the application of strong selective pressure: bacteria have been grown on carbon sources (maltose oligomers) larger than the exclusion limit of porin channels (Benson et al., 1988). The several mutations found are dispersed over the first one-third of the linear sequence of the protein, but in the three-dimensional structure they all map in the constriction site. In the point mutations found (Fig. 2), charged residues are substituted by neutral ones (R42S, R82C, R132P, D113G), and deletions, covering 6 to 15 residues, are all located in the constricting loop L3 (e.g., Δ108–114 and Δ118–133). This experimental approach thus immediately yields a survey of many of the critical residues involved in defining the constriction site, indicating the kind of interactions involved. Site-specific mutations, chosen on the basis of genetic or biochemical evidence, have a reduced frequency of affecting vital properties of the protein (e.g., in the active site), but instead probe areas where the actual functional consequences differ from those expected. Thus, reversing a charge by replacing a Lys by a Glu residue (K18E) in the constriction has a moderate effect on ion selectivity, similar to that of a mutation in loop L1 (K34E). Replacing a Lys group on the base of loop 2 (K69E) has a much stronger effect, whereas the exchange of two Arg residues or of a Gly in loops L4, L5, and L7 does not affect selectivity at all (Bauer et al., 1989). Although it is, of course, possible to rationalize these results, they demonstrate that in vitro site-specific mutagenesis is very useful in delineating the functional topology of areas which would elude approaches such as the application of strong selection procedures and which the high-resolution structure may not necessarily be expected to provide.

Although the current resolution of phosphoporin has recently improved considerably (to 2.0 Å [Nunn and Schirmer, personal communication]), many questions remain to be solved or, rather, can now be approached meaningfully. In the following, some of these questions will be discussed briefly.

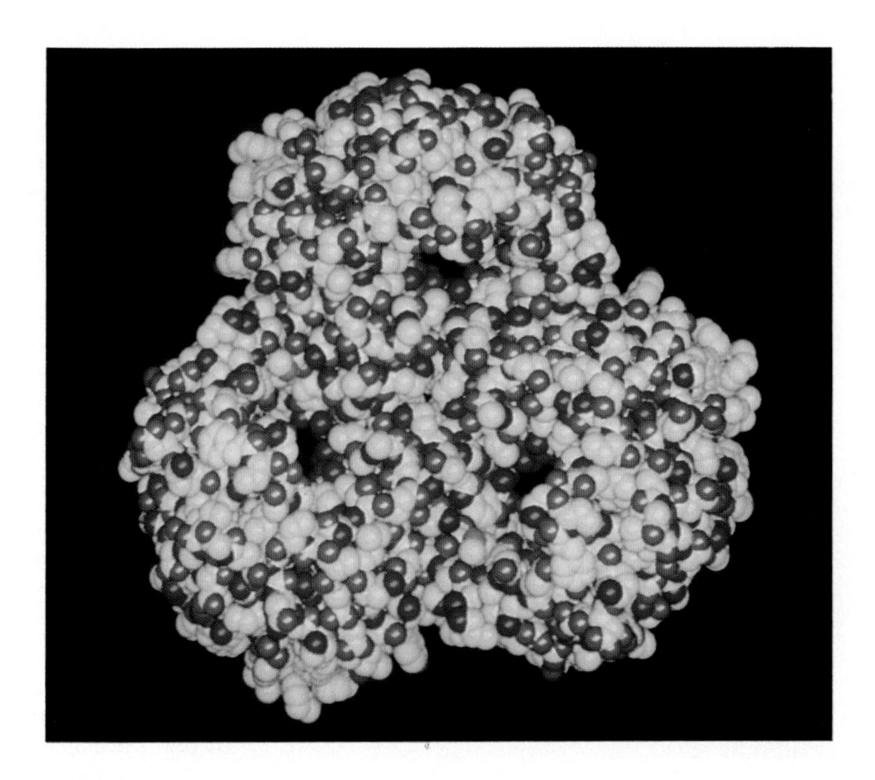

FIG. 1. Space-filling model of the trimer of the OmpF-porin (matrix porin), viewed from the outside of the cell. The threefold symmetry axis (in the center of the cloverleaf) runs perpendicular to the plane of the membrane (the plane of the paper). The three channel entrances (one per monomer) appear as small apertures in a massive protein. The impression of the sturdiness of the protein reflects its unusual stability adequately but gives no clue of its dynamic properties, which have not yet been determined. A ring of aromatic residues (Tyr, Phe) runs along the protein-lipid interphase, i.e., around the entire trimer. Color code: yellow, carbon; white, carbon in aromatic rings; red, oxygen; blue, nitrogen.

(i) What is the significance of a binding site in a channel protein? Teleologically speaking, tight binding of a solute appears undesirable, because it would preclude high rates of transport such as are observed not only in porins but also in specific ion channels (Hille, 1984). Phosphoporin appears to be a good example to illustrate several aspects. First, it has been argued that molecules such as polyphosphates (average size, 5 or 15 phosphates [types P_5 and P_{15}, respectively]) do not bind in the channel unless magnesium ions are present (Bauer et al., 1988). This observation illustrates our present inability to distinguish between two topologically different sites where such an effect may occur. The Stokes radii of polyphosphates are larger in the absence than in the presence of divalent cations, so that in view of the small channel entrance (Fig. 1), the substrate may be unable to enter the pore at all. When complexed with divalent cations, on the other hand, it may pass through the entrance, reach the selectivity gate, and exert its inhibitory effect there. In the presence of Mg^{2+}, inhibition constants, defined by reducing channel conductance by one-half, have been determined to be 1.7×10^{-4} and 3.4×10^{-4} M for P_5 and P_{15}, respectively. Second, since Lys-131 protrudes into the channel "trajectory" and since a change of its charge (K131E) causes an inversion of the ion selectivity (Bauer et al., 1989), this residue may be viewed as binding site for (poly)anions. However, the cationic cluster, consisting of three Arg and two Lys residues, is located at the same height in the channel (see above). Even if their in situ pK values are highly abnormal, it appears questionable to assume that the single lysyl residue (K-131) represents a classical binding site. A different viewpoint may be derived from the model calculations of Karshikoff et al. (in press). If, on the basis of all charges in the channel of OmpF porin, the electrostatic field is calculated along the axis in the pore, it appears that this field turns screwlike around this axis. This may suggest that in OmpF porin, small polar solutes, such as glucose, are whirled through the

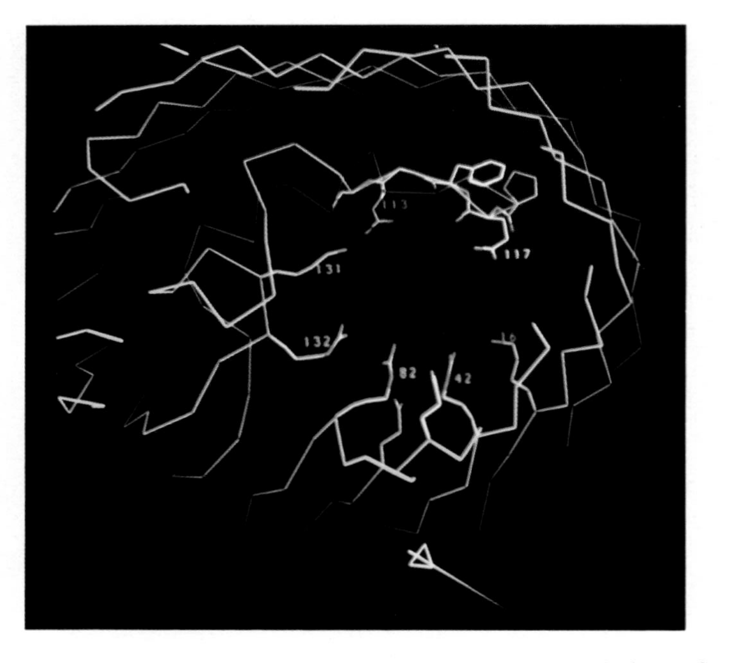

FIG. 2. Transverse section through a monomer. The constriction site is seen from the vestibulum of the channel. The threefold symmetry axis (bottom right) is shown as a triangle, with a white line in the direction of the axis. The backbone of the barrel (yellow) is slightly elliptic. The constriction loop narrows the diameter of the channel, with two carboxylate residues (D-113 and E-117) and two rings (F-115 and P-116) near its tip (P-$_{116}$EFGG-$_{120}$). From near the barrel wall emerge the basic residues: K-131, which protrudes into the channel, and R-132, which contributes to the cationic cluster formed together with R-82, R-42, and K-16. Code: white, special residues; blue, nitrogen; red, oxygen.

FIG. 3. Sagittal section through the channel in a monomer. The threefold symmetry axis (on the left) runs perpendicular to the plane of the membrane and, in conjunction with Fig. 2, allows the arrangement of the channel in the membrane to be verified. Loop L3 is seen on the right, and K-131 is in the center. On the left, the cationic cluster (K-16, R-42, R-82, and R-132) is shown sandwiched between two carboxyl groups, one below (E-62) and one above (E-64). This arrangement stabilizes the cluster.

channel as if in a vortex and that this phenomenon prevents collisions of the solutes with the wall, possibly avoiding exchange of water molecules between the hydration shells of the channel lining and the solutes. This would explain why the permeability coefficient of glucose across porin is indistinguishable from its diffusion coefficient in bulk solvent (Schindler and Rosenbusch, 1978). Whether such a mechanism may actually significantly enhance the rate of solute translocation cannot be determined presently.

(ii) Rate enhancements have been postulated to occur in specific channel proteins where solute transport would be accelerated at low substrate concentrations, whereas the rates decrease at saturating concentrations (Nikaido, 1992). Inhibition constants of maltose derivatives in maltoporin (LamB porin), determined by the methods described above for phosphates, reveal analogous results: maltotriose and maltoheptaose bind more tightly than the disaccharide maltose (K_i, 7.6×10^{-4} and 2.4×10^{-4} M versus 10^{-2} M, respectively [Dargent et al., 1987]). Inhibitor binding may thus be envisaged to occur as to a string of subsites, analogous to substrate (or inhibitor) binding in lysozyme (Phillips, 1967). Each of these subsites in isolation would have very weak affinity for the sugars. The binding of maltose oligosaccharides to maltoporin would thus block a presumed one-dimensional diffusion of the disaccharide along the string of subsites. This concept may be confirmed, or refuted, once the three-dimensional structure is available.

(iii) So far, it has been assumed that the effect of counterions bound to the ionic residues is negligible. Whether this assumption is correct is completely unknown. Because of the poor scattering power of the ions used in crystallization and their short resident times at any particular site, determining this may not be trivial. However, it is clearly critical for many aspects of the understanding of solute transport, and it is presently under further study.

(iv) Dehydration and rehydration, mentioned above, may be a critical phenomenon, particularly in channels with small diameters (sodium and potassium channels in axonal membranes are examples [Hille, 1984]). How valid porin may be as a model for these highly specific channels, particularly with respect to the influence of crystal radii, the dimensions of the hydrated solutes, and the free energy change of hydration, must await the structure determination of these specific proteins at high resolution. Porin-type channels are probably poor as immediate models, but the mechanisms mentioned may allow the phenomena in channels of low specificity to be understood conceptually.

(v) One of the most intriguing questions concerns channel dynamics, that is, the opening and closing of pores. Hypotheses about the advantage to cells to have a regulatable compartment on the outside of the plasma membrane are not difficult to devise, but, because of the dimensions of the periplasmic space (with the consequent inaccessibility to electrodes in patch-clamp experiments), they are not easy to test experimentally. However, if the structures of different conformations could be studied, for instance by atomic force microscopy, this would seems to hold great promise (Schabert et al., in press). Of course, the more established methods, involving different crystal forms, evolutionarily related proteins, or site-specific mutations, are likely to contribute to resolving these problems in the shorter term. In the long term, approaches such as molecular dynamics and solid-state nuclear magnetic resonance techniques may provide experimental answers.

The generous contribution of Fig. 1 to 3 by T. Schirmer and S. W. Cowan, Biozentrum, Basel, Switzerland is gratefully acknowledged.

The studies were supported by grants from the Swiss National Science Foundation.

REFERENCES

Bauer, K., M. Struyvé, D. Bosch, R. Benz, and J. Tommassen. 1989. One single lysine residue is responsible for the special interaction between polyphosphate and the outer membrane porin PhoE of *Escherichia coli*. *J. Biol. Chem.* **264:**16393–16398.

Bauer, K., P. van der Ley, R. Benz, and J. Tommassen. 1988. The *pho*-controlled outer membrane protein PhoE does not contain specific binding sites for phosphate or polyphosphates. *J. Biol. Chem.* **263:**13046–13053.

Benson, S. A., J. L. L. Occi, and B. A. Sampson. 1988. Mutations that alter the pore function of the OmpF porin of *Escherichia coli* K12. *J. Mol. Biol.* **203:**961–970.

Benz, R., A. Schmid, and R. E. W. Hancock. 1988. Ion selectivity of gram-negative bacterial porins. *J. Bacteriol.* **162:**722–727.

Buehler, L. K., S. Kusumoto, H. Zhang, and J. P. Rosenbusch. 1991. Plasticity of *E. coli* porin channels: dependence of their conductance on strain and lipid environment. *J. Biol. Chem.* **266:**24446–24450.

Cowan, S. W., T. Schirmer, G. Rummel, M. Steiert, R. Ghosh, R. A. Pauptit, J. N. Jansonius and J. P. Rosenbusch. 1992. Crystal structures explain functional properties of two *E. coli* porins. *Nature* (London) **358:**727–733.

Dargent, B., W. Hofmann, F. Pattus, and J. P. Rosenbusch. 1986. The selectivity filter of voltage-dependent channels formed by phosphoporin (PhoE protein) from *E. coli*. *EMBO J.* **5:**773–778.

Dargent, B., J. P. Rosenbusch, and F. Pattus. 1987. Selectivity for maltose and maltodextrins of maltoporin, a pore-forming protein of *E. coli* outer membrane. *FEBS Lett.* **220:**136–142.

Hille, B. 1984. *Ionic Channels of Excitable Membranes,* 2nd ed. Sinauer Associates, Sunderland, Mass.

Jap, B. K., P. J. Walian, and K. Gehring. 1991. Structural architecture of an outer membrane channel as defined by electron crystallography. *Nature* (London) **350:**167–170.

Karshikoff, A., V. Spassov, S. W. Cowan, R. Ladenstein, and T. Schirmer. Electrostatic properties of two porin channels from *E. coli*. *J. Mol. Biol.*, in press.

Nikaido, H. 1992. Porins and specific channels of bacterial outer membranes. *Mol. Microbiol.* **6:**435–442.

Nunn, R., and T. Schirmer. Personal communication.

Overbeeke, N., H. Bergmans, F. Mansfield, and B. Lugtenberg. 1983. Complete nucleotide sequence of phoE, the structural gene for the phosphate limitation inducible outer membrane pore protein of *Escherichia coli* K-12. *J. Mol. Biol.* **163:**513–522.

Page, M. G. P., and J. P. Rosenbusch. 1986. Topographic labelling of pore-forming proteins from the outer membrane of *Escherichia coli. Biochem. J.* **235:**651–661.

Pauptit, R. A., T. Schirmer, J. N. Jansonius, J. P. Rosenbusch, M. A. Parker, A. D. Tucker, D. Tsernoglou, M. S. Weiss, and G. E. Schulz. 1991. A common channel-forming motif in evolutionary distant proteins. *J. Struct. Biol.* **107:**136–145.

Phillips, D. C. 1967. The hen egg-white lysozyme molecule. *Proc. Natl. Acad. Sci. USA* **57:**484–495.

Rosenbusch, J. P. 1974. Characterization of the major envelope protein from *Escherichia coli*. Regular arrangement on the peptidoglycan and unusual dodecyl sulfate binding. *J. Biol. Chem.* **249:**8019–8029.

Schabert, F., J. H. Hoh, S. Karrasch, A. Hefti, and A. Engel. Scanning force microscopy of *E. coli* porin in buffer solution. Proc. Int. Conf. Scanning Tunnel Microscopy (STM '93). *J. Vac. Sci. Technol. B,* in press.

Schindler, H., and J. P. Rosenbusch. 1978. Matrix protein from *Escherichia coli* outer membranes form voltage-controlled channels in lipid bilayers. *Proc. Natl. Acad. Sci. USA* **75:**3751–3755.

Weiss, M. S., U. Abele, J. Weckesser, W. Welte, E. Schiltz, and G. E. Schulz. 1991. Molecular architecture and electrostatic properties of a bacterial porin. *Science* **254:**1627–1630.

Chapter 56

Zn(II)-Mediated Protein Interactions in *Escherichia coli* Signal Transduction: Cation-Promoted Association of the Phosphotransferase System Regulatory Protein III^{Glc} with Target Protein Glycerol Kinase

DONALD W. PETTIGREW,[1] MICHAEL FEESE,[2] NORMAN D. MEADOW,[3] S. JAMES REMINGTON,[2] AND SAUL ROSEMAN[3]

Department of Biochemistry and Biophysics, Texas A&M University, College Station, Texas 77843[1]; Institute of Molecular Biology and Departments of Physics and Chemistry, University of Oregon, Eugene, Oregon 97403[2]; and Department of Biology and McCollum-Pratt Institute, The Johns Hopkins University, Baltimore, Maryland 21218[3]

Regulatory Properties of the PTS

In enteric bacteria, such as *Escherichia coli* and *Salmonella typhimurium,* the phosphoenolpyruvate (PEP):glucose phosphotransferase system (PTS) catalyzes the vectorial phosphorylation of several sugars (Meadow et al., 1990; Postma et al., 1993; Roseman and Meadow, 1990; Saier, 1989). This primary metabolic function for glucose is shown schematically in Fig. 1. The vectorial phosphorylation simultaneously transports the glucose into the cell and phosphorylates it to yield glucose 6-phosphate, thus priming it for subsequent metabolism via glycolysis. The phosphoryl group is transferred by the proteins of the PTS from PEP to glucose; this pathway involves sequential phosphorylation-dephosphorylation cycles for the PTS proteins I, HPr, III^{Glc}, and, perhaps, II^{Glc}. (The notation III^{Glc} is used for the glucose-specific PTS protein that carries the phosphoryl group from HPr to II^{Glc}. Other nomenclature used for this protein includes enzyme III^{Glc}, factor III^{Glc}, and IIA^{Glc}. These systems of nomenclature are discussed in chapter 26.)

In addition to this primary metabolic role, the PTS regulates the uptake of non-PTS sugars and the expression of genes involved in their metabolism. For example, *E. coli* cells are unable to metabolize glycerol where grown on plates that also contain glucose (Zwaig and Lin, 1966), and cells that are grown on glucose express low levels of the enzymes that are involved in glycerol metabolism (Lin, 1976). This phenomenon has been observed for several other non-PTS sugars including lactose, maltose, and melibiose (Meadow et al., 1990; Saier and Roseman, 1976). This regulation is also depicted schematically in Fig. 1 for metabolism of glycerol. It involves the PTS regulatory protein III^{Glc} in its dephosphorylated state. When glucose is being transported, the concentration of the dephosphorylated III^{Glc} is relatively higher than that of the phosphorylated form, $III^{Glc}{\sim}P$. This affects glycerol metabolism and expression of the genes of the *glp* regulon at two related points. First, it maintains catabolite repression because $III^{Glc}{\sim}P$ is required for the activation of adenyl cyclase; thus, the level of cyclic AMP (cAMP) remains low. Second, III^{Glc} inhibits GlpK, blocking synthesis of G3P, which also has two effects. Phosphorylation of glycerol is the required first step in glycerol metabolism, and G3P is the inducer for expression of the genes of the *glp* regulon.

This model is termed inducer exclusion because it prevents entry of the inducer into the cell (Postma et al., 1993; Roseman and Meadow, 1990; Saier, 1989). This model for PTS regulation of glycerol metabolism differs from those for lactose, maltose, and melibiose only with respect to the target for III^{Glc} binding; in these cases, III^{Glc} inhibits the permease or transporter system to block inducer uptake. In all cases, inhibition of inducer uptake by III^{Glc} requires the sugar (DeBoer et al., 1986; Saier, 1989).

The relation of this regulatory function of the PTS to expression of the genes of the *glp* regulon is shown schematically in Fig. 1. The absence of glucose and concomitant vectorial phosphorylation changes the relative amounts of III^{Glc} and $III^{Glc}{\sim}P$. This results in activation of adenyl cyclase to relieve catabolite repression and relief of inhibition of GlpK, which allows synthesis of the inducer G3P. GlpF, the glycerol facilitator, appears to be required for maximal activation of GlpK in vivo (Voegele et al., 1993). The analogous situation exists for other non-PTS sugars; the absence of glucose relieves inhibition of inducer uptake.

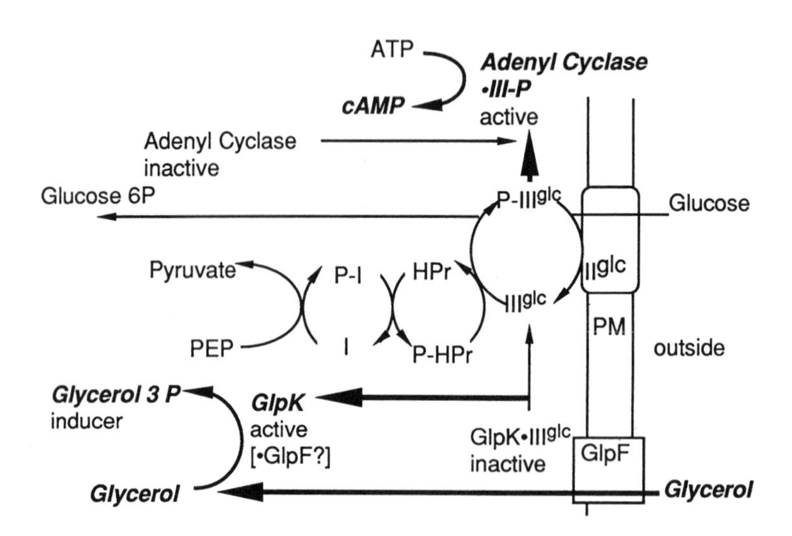

FIG. 1. Relation between vectorial phosphorylation and signal transduction functions of the PTS. When glucose is available, it is vectorially phosphorylated by the PTS to yield glucose 6-phosphate (Glucose 6P). If glycerol is also available, the activity of glycerol kinase is inhibited by the dephosphorylated state of the PTS regulatory protein, III^{Glc}, thus preventing accumulation of the inducer, glycerol 3-phosphate (Glycerol 3P). If glucose becomes unavailable, the relative concentrations of P-III^{Glc} and III^{Glc} change such that the pathways indicated by the bold arrows obtain and the molecules and complexes shown in boldface italics accumulate, resulting in induction of the expression of the elements of the *glp* regulon. Activation of GlpK may involve GlpF, as indicated by the question mark (Voegele et al., 1993). Abbreviations: GlpF, glycerol facilitator; GlpK, glycerol kinase; PM, plasma membrane; PTS components, I, HPr, II^{Glc}, III^{Glc}, where the phosphorylated state is indicated by P.

The regulatory effects of the PTS on uptake of non-PTS sugars and expression of the genes involved in their metabolism represent a signal transduction system which relays information about the availability of carbon sources to the gene expression systems of the bacterium. Thus, studies of the molecular, structural, and biochemical properties of this regulatory system are of interest. One area of particular interest is the mechanism(s) by which a small protein such as III^{Glc} (molecular weight, 18,100) recognizes so many different target proteins, including HPr, II^{Glc}, LacY, GlpK, MalK, and MelB (Meadow et al., 1990).

III^{Glc} Structure

III^{Glc} is the central regulatory element in PTS regulation of non-PTS sugar uptake and gene expression. The structures of *E. coli* III^{Glc} and the equivalent *Bacillus subtilis* protein have been determined by using both X-ray diffraction and NMR methods (Fairbrother et al., 1991; Liao et al., 1991; Pelton et al., 1991a, 1991b; Worthylake et al., 1991). For the *E. coli* protein, the polypeptide backbone is formed into a thin antiparallel β-sandwich. The structure of the molecule appears to be resistant to conformational changes. Nuclear magnetic resonance studies show that the conformation of the *E. coli* protein is not affected by phosphorylation of the active-site histidine residue (Pelton et al., 1992), and modeling studies show that the phosphoryl group can be built into the structure of the protein from *B. subtilis* without altering the protein structure (Herzberg, 1992; Liao et al., 1991).

Interactions between III^{Glc} and Its Target Proteins

III^{Glc} interactions with its target proteins can be considered in two categories, depending on the metabolic function of the resulting complex (Meadow et al., 1990).

Covalent interactions with PTS proteins. The first category considers the interactions between III^{Glc} and its neighbors in the PTS, HPr~P and II^{Glc}. These interactions have been characterized as covalent because they involve the covalent transfer of the phosphoryl group from HPr~P to II^{Glc} by III^{Glc}. The interaction between HPr~P and III^{Glc} from *B. subtilis* has been modeled by docking the separately determined structures of the dephospho forms of the two proteins and using energy minimization to optimize the fit between them (Herzberg, 1992). Relations between this modelled structure and the crystal structure of the

GlpK-III^{Glc} complex are considered in the Discussion.

Regulatory interactions with non-PTS proteins: the GlpK-III^{Glc} crystal structure. One of the implications of the inducer exclusion model is formation of a complex between dephospho-III^{Glc} and its regulatory target proteins. The existence of such a complex with GlpK was indicated by in vitro studies showing that purified III^{Glc} inhibits GlpK catalysis (Novotny et al., 1985; Postma et al., 1984). Complex formation was recently verified by crystallographic studies of the cocrystal formed between GlpK and dephospho-III^{Glc}.

The crystal structure of the ternary complex of GlpK–glycerol–dephospho-III^{Glc} with and without bound ADP was described recently (Hurley et al., 1993). This structure shows that III^{Glc} binds to the GlpK tetramer with a stoichiometry of one molecule of III^{Glc} per GlpK subunit, and the III^{Glc} molecules that are bound to neighboring GlpK subunits do not interact with one another. A ribbon diagram showing III^{Glc} bound to one GlpK subunit is presented in Fig. 2. The structure of III^{Glc} in the complex is the same as that of uncomplexed III^{Glc} (Worthylake et al., 1991). Thus, formation of the complex does not affect the conformation of III^{Glc}. The GlpK-binding site for III^{Glc} is one face of a 3_{10} helix that is located more than 30 Å from the GlpK active site; thus, the inhibition of GlpK by III^{Glc} must involve conformational changes in GlpK that are transmitted for a considerable distance through the protein.

The III^{Glc} binding site on GlpK is formed by amino acid residues R-402 and 472 to 481, with the primary structure . . . P-472–GIETTERNY-481. . . . Hydrophobic interactions appear to dominate the formation of the complex; the side chains of P-472, I-474, and Y-481; the aliphatic portions of E-478 and R-479; and backbone atoms of GlpK form hydrophobic interactions with the III^{Glc} side chains V-40, F-41, I-45, V-46, F-71, F-88, and V-96. Few polar interactions are formed in the complex; salt bridges are formed between R-402 and R-479 of GlpK and E-43 and D-38 of III^{Glc}, whereas a long (3.35-Å) hydrogen bond is formed between N-480 of GlpK and D-94 of III^{Glc}. Thus, the interaction involves a total of 7 GlpK and 10 III^{Glc} amino acid residues. This seems to be remarkably few amino acid residues given the large number of regulatory target proteins that III^{Glc} recognizes.

We describe here results of recent crystallographic and biochemical studies showing that the assembly of the GlpK-III^{Glc} complex forms an intermolecular Zn(II)-binding site. These results indicate that a heretofore unrecognized factor may be involved in the formation of complexes of III^{Glc} with its regulatory target proteins and suggest a mechanism for recognition of the targets.

The GlpK-III^{Glc} Complex Contains an Intermolecular Zn(II)-Binding Site

In the course of crystallographic studies of the GlpK-III^{Glc} complex, crystals were soaked with $MnCl_2$ to determine the cation-binding site(s). Upon analysis of the difference electron density map, a large (11σ) positive peak was located at the interface between GlpK and III^{Glc}. The size of the density difference feature suggested that a contaminating metal ion that is much more electron dense than Mn(II) was bound between the two proteins. The apparent geometry of the site was consistent with Zn(II)-binding sites in metalloproteases.

This putative intermolecular Zn(II)-binding site was further investigated in a subsequent experiment in which a solution containing 5 mM Zn(II) as the only divalent cation was added to the cocrystals of the protein complex. In this case, the difference electron density showed a 22σ positive difference peak at the intermolecular site, confirming that Zn(II) is bound at the interface between the two proteins. In addition, a 7σ positive-electron-density difference peak was observed at a different location that also binds Mg(II) and Mn(II), showing that Zn(II) binds at the active site.

The structure of the intermolecular Zn(II) site is shown in Fig. 3. The site is formed on association of the two proteins, and each protein contributes ligands that occupy the coordination positions of the Zn(II) ion. Two of the coordinating ligands are the active-site histidine residues of III^{Glc}, His-75 and His-90. One of the ligands is glutamate 478 of GlpK, and the fourth coordination position is occupied by a water molecule. The water molecule is held in place by hydrogen bonds to backbone amides of III^{Glc} and thus is likely to be a hydroxide ion. The coordination geometry of the Zn(II) ion is a distorted tetrahedron and is identical to that seen in thermolysin (Holmes and Matthews, 1982). The three protein atoms that coordinate the Zn(II) in the GlpK-III^{Glc} cocrystal and in thermolysin superimpose to a root mean square error of 0.19 Å, which is well within the estimated coordinate errors of the models.

Zn(II) Enhances III^{Glc} Inhibition of GlpK

The structural studies described above show that assembly of the GlpK-III^{Glc} complex forms an intermolecular Zn(II)-binding site which is occupied by Zn(II). Evidence that the Zn(II) site has functional significance is provided by the effect of Zn(II) on III^{Glc} inhibition of GlpK, which is shown in Fig. 4. As the concentration of III^{Glc} increased, GlpK activity was inhibited. In the absence of added Zn(II), the extent of inhibition and III^{Glc} concentration dependence of the inhibition

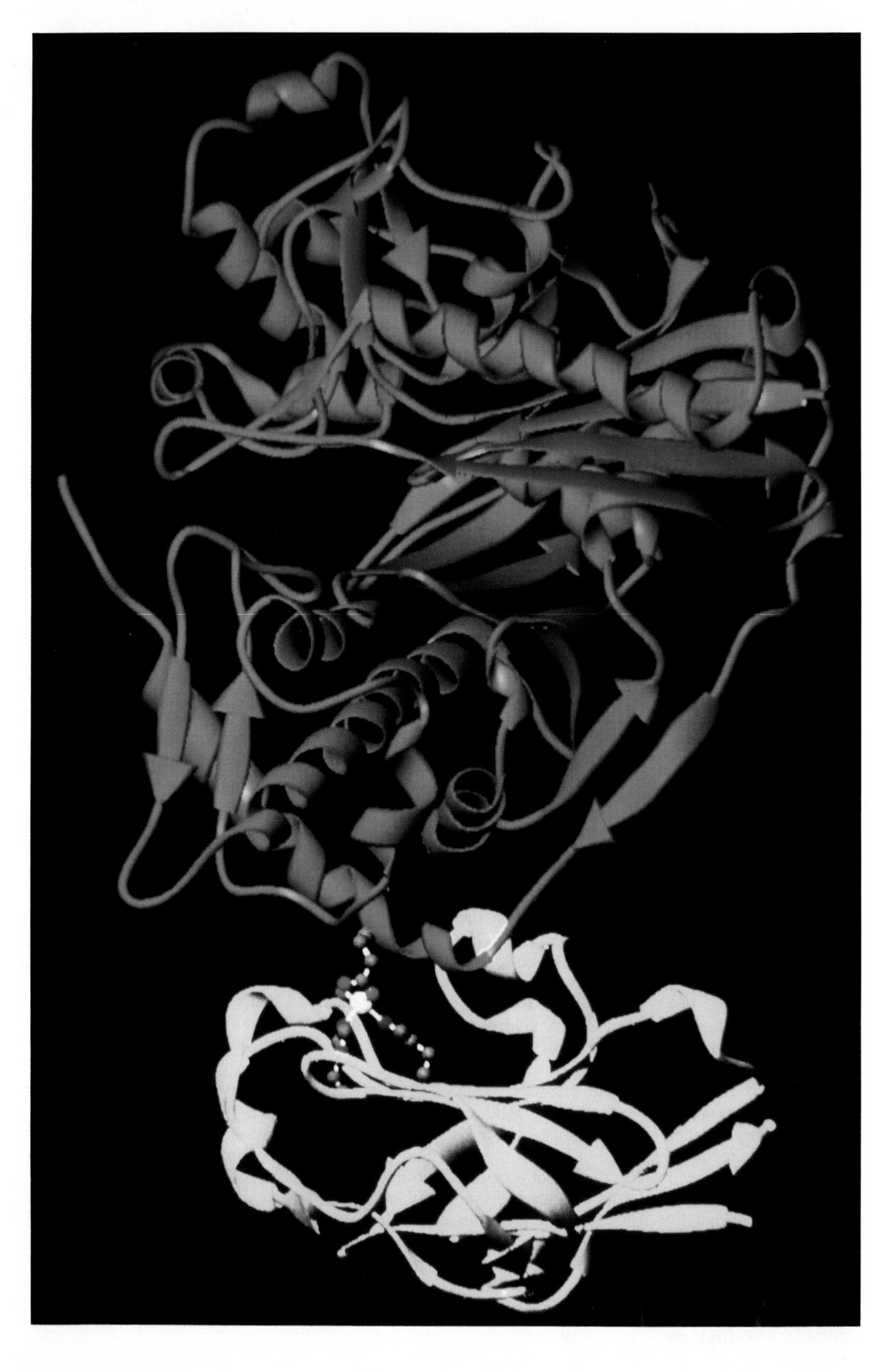

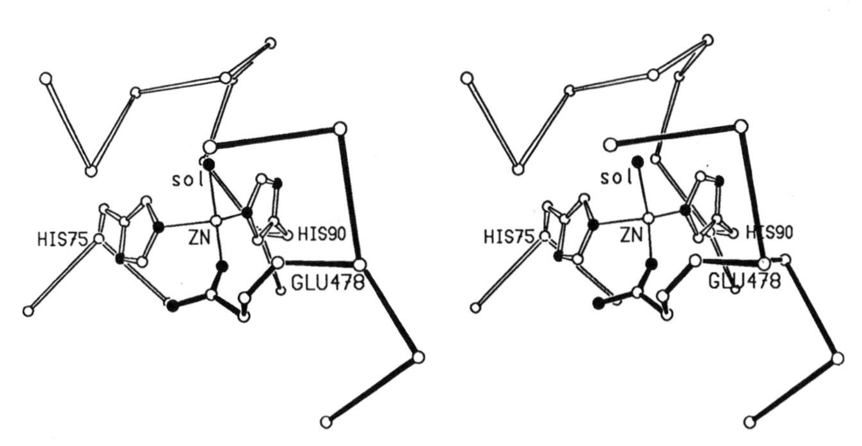

FIG. 3. Stereo view of the structure of the intermolecular Zn(II)-binding site. The intermolecular Zn(II) ion is shown as the open circle labeled ZN. The side chains that are contributed by III^{Glc}, His-75 and His-90, are shown by the open bonds, and the side chain contributed by GlpK, Glu-478, is shown by the solid bonds. The nitrogen or oxygen atoms of the ligand-attaching side chains are shown as solid circles. The solid circle labeled sol is the water molecule (or hydroxide ion) that occupies the fourth coordination position of the Zn(II). The structure was determined as described previously (Feese et al., in press).

agreed well with earlier reports (Novotny et al., 1985; Postma et al., 1984). Addition of Zn(II) enhanced the inhibition by III^{Glc}. In the absence of III^{Glc}, Zn(II) shows modest inhibition of GlpK activity, which reflects the fact that ZnATP is a substrate for GlpK, with a smaller K_m and k_{cat} than those for MgATP (Pettigrew, unpublished).

A preliminary steady-state kinetics study of the effect of Zn(II) on III^{Glc} inhibition of GlpK has been conducted, and some of the results of those studies are summarized in Table 1. As the Zn(II) concentration is increased, the K_i for III^{Glc} inhibition of GlpK decreases. This is consistent with enhancement of complex formation by the intermolecular binding of Zn(II), as expected from the structural studies. The apparent K_d for Zn(II) for complex formation under the conditions of the steady-state kinetics assay is about 10 μM. Results of preliminary studies show that III^{Glc} inhibition of GlpK is also enhanced by Co(II), Cd(II), and Ni(II); however, much higher concentrations of these divalent cations are required, indicating that the affinity of the intermolecular site is much lower for these cations than for Zn(II).

Mutation of a Zn(II)-Coordinating Ligand of III^{Glc} Abolishes Zn(II) Enhancement of III^{Glc} Inhibition of GlpK

A mutation of one of the III^{Glc} Zn(II)-coordinating ligands, active-site histidine 75, to glutamine (H75Q) was constructed and characterized (Presper et al., 1989). The effect of this mutation on Zn(II) enhancement of III^{Glc} inhibition of GlpK is also shown in Fig. 4. In the absence of Zn(II), the extent and III^{Glc} concentration dependence of the inhibition are the same for mutant H75Q as for wild-type III^{Glc}. Thus, the mutation does not appear to affect the hydrophobic and polar interactions between the two proteins. However, 100 μM Zn(II) does not affect the inhibition by H75Q. Thus, the mutation abolishes the enhancement by Zn(II) at this concentration, which is consistent with the role of the histidine residue shown by the structure of the complex. The effect of the cation appears to be separate from, independent of, and in addition to the direct interactions between the amino acid residues in the complex.

FIG. 2. Ribbon diagram of the structure of the GlpK-III^{Glc} complex. The figure shows one subunit of the GlpK tetramer in blue with the bound III^{Glc} in yellow. α-Helices are shown as coils, and β-conformation is shown as arrows. The active site of GlpK is the large central cleft. The side chains of each protein that form the intermolecular Zn(II)-binding site, His-75 and His-90 from III^{Glc} and Glu-478 from GlpK, are shown. The Zn(II) ion is the white sphere located at the interface between the proteins, and the solvent molecule which occupies the fourth coordination position of the Zn(II) is shown as a red sphere.

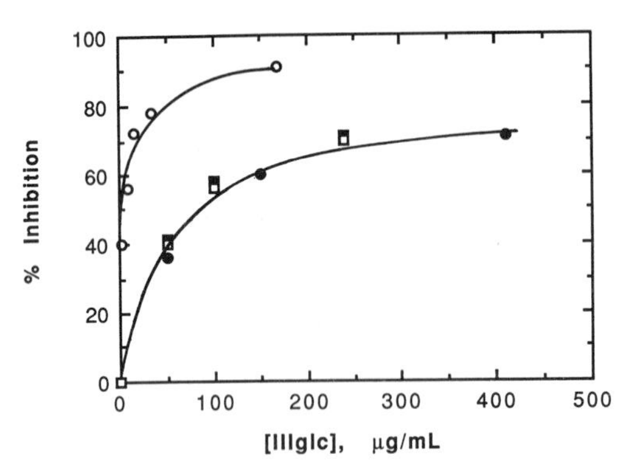

FIG. 4. Effect of Zn(II) on III^Glc inhibition of GlpK. GlpK specific activity at 1 μg/ml was determined by using an ADP-coupled spectrophotometric assay (Pettigrew et al., 1990) at pH 7.0 and 25°C with 2 mM glycerol, 2.5 mM ATP, 5 mM $MgCl_2$, and additions of $ZnCl_2$ and III^Glc as shown; results are expressed as percent inhibition relative to no added III^Glc. Symbols: ●, wild-type III^Glc, 0 mM Zn(II); ○, wild-type III^Glc, 100 μM Zn(II); ■, H75Q III^Glc, 0 mM Zn(II); □, H75Q III^Glc, 100 μM Zn(II).

TABLE 1. Effect of zinc on III^Glc inhibition of glycerol kinase[a]

Zn(II) concn (mM)	K_i[b]	
	μg/ml	μM
0	300	16
0.1	5	0.28

[a]The activity of glycerol kinase was determined by using an ADP-coupled spectrophotometric assay at pH 7 and 25°C, as previously described (Pettigrew et al., 1990).
[b]The concentration of III^glc giving 50% inhibition of glycerol kinase. The values were determined from steady-state kinetics studies of the inhibition with respect to ATP (Pettigrew, unpublished).

Discussion

The structural and biochemical results that have been described above show that an intermolecular Zn(II)-binding site is formed upon assembly of the GlpK-III^Glc complex, the site is occupied by a Zn(II) ion, and occupancy of the site increases the association constant for complex formation. We suggest that this mechanism will prove to be common in interactions between regulatory proteins and their target proteins and that it be termed cation-promoted association (CPA).

Other examples of CPA have been described recently. Binding of gelsolin to actin forms an intermolecular binding site for Ca(II), and both proteins contribute metal-coordinating ligands (McLaughlin et al., 1993). Assembly of the hexamer of the human cell cycle-regulatory protein CksHs2 involves intermolecular cation- and anion-binding sites, and formation of its complex with the catalytic subunit of the cyclin-dependent protein kinase may involve intermolecular anion-binding sites (Parge et al., 1993). Finally, the Zn(II)-binding site in the crystal of α-lactalbumin is made up of residues from two different protein molecules. It has been suggested that the Zn(II) may mediate the interaction of α-lactalbumin with galactosyltransferase to form lactose synthase, although there is no clear evidence for this role at present (Ren et al., 1993).

CPA has interesting implications. It allows the formation of a very tight, very specific complex by using few amino acid residues. It allows low concentrations of cations to play roles in regulation and may not require modulation of the cation concentration. The cation-binding site is disrupted by dissociation of the complex; thus, concentrations of cations which are sufficient for enhancing the formation of the complex will not interfere with other functions, e.g., phosphorylation of III^Glc in its phosphotransferase function.

Because the target protein contributes only one of the Zn(II)-coordinating ligands, CPA may allow considerable latitude in target recognition sequences. This implication is of particular interest for binding of III^Glc to its target proteins. Dephospho-III^Glc binds to several target proteins, including GlpK, MalK, LacY, and MelB, in its noncovalent regulatory function. Putative III^Glc-binding sites on MalK, LacY, and MelB have been deduced from mutational studies. These regions are summarized in Table 2. These results show two points of interest.

First, different conclusions have been reached by different laboratories about which region of MalK and MelB is the III^Glc-binding site. In one case, the MalK III^Glc-binding site was postulated to begin at amino acid 275, and, in the other case, amino acid 121 is proposed to be the start of the site. The region of GlpK that starts with amino acid 370 was described as showing weak similarity to the first postulated III^Glc-binding site on MalK. However, the crystal structure of the complex shows that this is not the III^Glc-binding site on GlpK. Second, the last line in Table 2 shows the actual III^Glc binding site as determined from the structure of the GlpK-III^Glc complex. This region of GlpK shows no sequence similarity to either of the postulated regions of MalK or MelB; furthermore, it shows no detectable sequence similarity with any region of any of these III^Glc target proteins or any other III^Glc target protein.

This apparently paradoxical result seems to suggest that either the mechanism by which III^Glc recognizes its many regulatory targets differs for each target; e.g., different regions of III^Glc bind to different target proteins, or the binding site on III^Glc can recognize primary structures that do not appear to be similar. The interactions between

TABLE 2. III^{Glc}-binding sites on target proteins

Target protein	Recognition sequence	Basis	Reference(s)
MalK	$_{275}$VQ_VGANMSL_GIRP__EHLL	Mutants	Dean et al., 1990; Kuehnau et al., 1991
LacY	$_{198}$ANAVGANHSAFSLKLALE_LF	Mutants	Dean et al., 1990; Kuehnau et al., 1991
MelB	______$_{211}$GS_HL__TLK_AIVALI	Sequence similarity	Dean et al., 1990
GlpK	______$_{370}$GVNANH_IIRATLESIA	Sequence similarity	Dean et al., 1990
MalK	$_{121}$LQLAHLLDR_KPKA	Mutants	Kuroda et al., 1992
MelB	$_{439}$IQI_HLLDKYR_KT	Sequence similarity	Kuroda et al., 1992
GlpK	$_{472}$PGIETTERNY	X-ray structure	Hurley et al., 1993

GlpK and III^{Glc} in the cocrystal of the complex are consistent with the second alternative. The target protein contributes a substantial hydrophobic interaction surface involving only a few amino acid side chains, two or three polar interactions that are clustered in a small area of the interface, and a single coordinating ligand for the Zn(II) ion. These requirements could be met in many ways such that a canonical primary structure may not exist.

This hypothesis is also supported by comparison of the structure of the GlpK-III^{Glc} complex with that modeled for the HPr ~ P ~ III^{Glc} complex of *B. subtilis* (Herzberg, 1992). The III^{Glc} from *B. subtilis* will substitute for that in *E. coli* (reviewed by Liao et al. [1991]), so the comparison of the structures of the complexes is reasonable and reveals similarities in the interactions of III^{Glc} with its target proteins. Both complexes involve hydrophobic interactions with the patch surrounding the III^{Glc} active site; thus, the same region of III^{Glc} forms the contact in the complex. An Arg residue in each of the targets interacts with the same region of III^{Glc}: R-17 of HPr ~ P interacts with D-38 and D-95 of III^{Glc}, while R-479 of GlpK interacts with D-38. The interacting structure on each target involves a helix; it is the 3_{10} helix described above for GlpK and an α-helix for HPr ~ P. However, except for the Arg residue, there is no similarity between the primary structures of the III^{Glc} binding sites. Thus, it appears that III^{Glc} does recognize different primary structures at the same binding site. It should also be noted that the modeled structure shows interactions between III^{Glc} and HPr ~ P involving amino acids of HPr ~ P that are about 20 residues further toward the carboxy terminus; thus, recognition may not be limited to the single helical region in all targets.

The apparent absence of a canonical primary structure for recognition of regulatory target proteins is also observed for calmodulin (Meador et al., 1992). However, in that case, the interaction involves a large number of amino acid residues from each protein (Ikura et al., 1992; Meador et al., 1992). Thus, it seems that CPA may enhance highly specific recognition of different primary structures consisting of few amino acid side chains. This may be important for regulation under physiological conditions. Inhibition of GlpK by III^{Glc} clearly occurs in the absence of Zn(II), as shown in Fig. 4 and in earlier investigations (Novotny et al., 1985; Postma et al., 1984). However, Fig. 4 shows that at the concentrations of III^{Glc} that are estimated to exist in vivo (from 0.08 mg/ml in exponential-phase cells grown on glycerol to 0.25 mg/ml when the cells are grown on glucose), only about 40 to 70% inhibition is obtained in vitro in the absence of Zn(II). This contrasts with the complete inhibition seen in vivo when α-methyl glucoside is added to cultures growing on glycerol (Saier and Roseman, 1976) or by the inability of colonies to metabolize glycerol in the presence of glucose (Zwaig and Lin, 1966). This difference between in vivo and in vitro behavior is reduced greatly by Zn(II); Fig. 4 shows that about 90% inhibition is obtained at 0.08 mg of III^{Glc} per ml.

This work was supported in part by grants GM 49992 (D.W.P.), GM 42618 (S.J.R.), and GM 38759 (S.R.) from the National Institutes of Health and by the Texas Agricultural Experiment Station (D.W.P.).

REFERENCES

Dean, D., J. Reizer, H. Nikaido, and M. H. Saier, Jr. 1990. Regulation of the maltose transport system of *Escherichia coli* by the glucose-specific enzyme III of the phosphoenolpyruvate-sugar phosphotransferase system. *J. Biol. Chem.* **265:**21005–21010.

DeBoer, M., C. Broekhuizen, and P. Postma. 1986. Regulation of glycerol kinase by enzyme III^{Glc} of the phosphoenolpyruvate:carbohydrate phosphotransferase system. *J. Bacteriol.* **167:**393–395.

Fairbrother, W., J. Cavanagh, H. Dyson, A. Palmer III, S. Sutrina, J. Reizer, M. H. Saier, Jr., and P. Wright. 1991. Polypeptide backbone resonance assignments and secondary structure of *Bacillus subtilis* enzyme III^{Glc} determined by two-dimensional and three-dimensional heteronuclear NMR spectroscopy. *Biochemistry* **30:**6896–6907.

Feese, M., D. W. Pettigrew, N. D. Meadow, S. Roseman, and S. J. Remington. Cation promoted association (CPA) of a regulatory and target protein is controlled by protein phosphorylation. *Proc. Natl. Acad. Sci. USA*, in press.

Herzberg, O. 1992. An atomic model for protein-protein phosphoryl group transfer. *J. Biol. Chem.* **267:**24819–24823.

Holmes, M. A., and B. Matthews. 1982. Structure of thermolysin refined at 1.6Å resolution. *J. Mol. Biol.* **160:**623.

Hurley, J., H. Faber, D. Worthylake, N. D. Meadow, S. Roseman, D. W. Pettigrew, and S. J. Remington. 1993. Structure of the regulatory complex of *Escherichia coli* III^{Glc} with glycerol kinase. *Science* **259:**673–677.

Ikura, M., G. Clore, A. Gronenborn, G. Zhu, C. Klee, and A. Bax. 1992. Solution structure of a calmodulin-target peptide complex by multidimensional NMR. *Science* **256:**632–638.

Kuehnau, S., M. Reyes, A. Sievertsen, H. Shuman, and W. Boos. 1991. The activities of the *Escherichia coli* MalK protein in maltose transport, regulation, and inducer exclusion can be separated by mutations. *J. Bacteriol.* **173:**2180–2186.

Kuroda, M., S. de Waard, K. Mizushima, M. Tsuda, P. Postma, and T. Tsuchiya. 1992. Residence of the melibiose carrier to inhibition by the phosphotransferase system due to substitutions of amino acid residues in the carrier of *Salmonella typhimurium. J. Biol. Chem.* **267:**18336–18341.

Liao, D., G. Kapadia, P. Reddy, M. H. Saier, Jr., J. Reizer, and O. Herzberg. 1991. Structure of the IIA domain of the glucose permease of *Bacillus subtilis* at 2.2 Å resolution. *Biochemistry* **30:**9583–9594.

Lin, E. C. C. 1976. Glycerol dissimilation and its regulation in bacteria. *Annu. Rev. Biochem.* **30:**535–578.

McLaughlin, P., J. Gooch, H. Mannherz, and A. Weeds. 1993. Structure of gelsolin 1-actin complex and the mechanism of filament severing. *Nature* (London) **364:**685–692.

Meador, W., A. Means, and F. Quiocho. 1992. Target enzyme recognition by calmodulin: 2.4 Å structure of a calmodulin-peptide complex. *Science* **257:**1251–1255.

Meadow, N. D., D. Fox, and S. Roseman. 1990. The bacterial phosphoenolpyruvate:glycose phosphotransferase system. *Annu. Rev. Biochem.* **59:**497–542.

Novotny, M., W. Frederickson, E. Waygood, and M. H. Saier, Jr. 1985. Allosteric regulation of glycerol kinase by enzyme III^{Glc} of the phosphotransferase system in *Escherichia coli* and *Salmonella typhimurium. J. Bacteriol.* **162:**810–816.

Parge, H., A. Arvai, D. Murtari, S. Reed, and J. Tainer. 1993. Human CksHs2 atomic structure: a role for its hexameric assembly in cell cycle control. *Science* **262:**387–394.

Pelton, J., D. Torchia, N. D. Meadow, and S. Roseman. 1992. Structural comparison of phosphorylated and unphosphorylated forms of III^{Glc}, a signal-transducing protein form *Escherichia coli,* using three-dimensional NMR techniques. *Biochemistry* **31:**5215–5224.

Pelton, J., D. Torchia, N. D. Meadow, C. Wong, and S. Roseman. 1991a. ^{1}H, ^{15}N, and ^{13}C NMR signal assignments of III^{Glc}, a signal-transducing protein of *Escherichia coli,* using three-dimensional triple-resonance techniques. *Biochemistry* **30:**10043–10057.

Pelton, J., D. Torchia, N. D. Meadow, C. Wong, and S. Roseman. 1991b. Secondary structure of the phosphocarrier protein III^{Glc}, a signal transducing protein from *Escherichia coli,* determined by heteronuclear three-dimensional NMR spectroscopy. *Proc. Natl. Acad. Sci. USA* **88:**3479–3488.

Pettigrew, D. W. Unpublished data.

Pettigrew, D. W., G. Yu, and L. Youguo. 1990. Nucleotide regulation of Escherichia coli glycerol kinase: initial-velocity and substrate binding studies. *Biochemistry* **29:**8620–8627.

Postma, P., W. Epstein, A. Schuitema, and S. Nelson. 1984. Interaction between III^{Glc} of the phosphoenolpyruvate:sugar phosphotransferase system and glycerol kinase of *Salmonella typhimurium. J. Bacteriol.* **158:**351–353.

Postma, P. W., J. W. Lengeler, and G. R. Jacobson. 1993. Phosphoenolpyruvate:carbohydrate phosphotransferase systems of bacteria. *Microbiol. Rev.* **57:**543–594.

Presper, K., C. Wong, L. Liu, N. D. Meadow, and S. Roseman. 1989. Site-directed mutagenesis of the phosphocarrier protein, III^{Glc}, a major signal-transducing protein in *Escherichia coli. Proc. Natl. Acad. Sci. USA* **86:**4052–4055.

Ren, J., D. Stuart, and K. Acharya. 1993. α-Lactalbumin possesses a distinct zinc binding site. *J. Biol. Chem.* **268:**19292–19298.

Roseman, S., and N. D. Meadow. 1990. Signal transduction and the bacterial phosphotransferase system. *J. Biol. Chem.* **265:**2993–2996.

Saier, M. H., Jr. 1989. Protein phosphorylation and allosteric control of inducer exclusion and catabolite repression by the bacterial phosphoenolpyruvate:sugar phosphotransferase system. *Microbiol. Rev.* **53:**109–120.

Saier, M. H., Jr., and S. Roseman. 1976. Sugar transport. Inducer exclusion and regulation of the melibiose, maltose, glycerol, and lactose transport systems by the phosphoenolpyruvate:sugar phosphotransferase system. *J. Biol. Chem.* **251:**6606–6615.

Voegele, R., G. Sweet, and W. Boos. 1993. Glycerol kinase of *Escherichia coli* is activated by interaction with the glycerol facilitator. *J. Bacteriol.* **175:**1087–1094.

Worthylake, D., N. D. Meadow, S. Roseman, D. Liao, O. Herzberg, and S. J. Remington. 1991. Three-dimensional structure of the *Escherichia coli* phosphocarrier protein III^{Glc}. *Proc. Natl. Acad. Sci. USA* **88:**10382–10386.

Zwaig, N., and E. C. C. Lin. 1966. Feedback inhibition of glycerol kinase, a catabolic enzyme in *Escherichia coli. Science* **153:**755–757.

AUTHOR INDEX

SUBJECT INDEX